Metallothionein IV

Edited by
C.D. Klaassen

Springer Basel AG

Editor

Curtis D. Klaassen
University of Kansas
Medical Center
3901 Rainbow
Kansas City, KS 66160-7417
USA

Library of Congress Cataloging-in-Publication Data
Metallothionein IV / edited by C.D. Klaassen.
    p.   cm.
    Includes bibliographical references and index.
    ISBN 978-3-0348-9799-0   ISBN 978-3-0348-8847-9 (eBook)
    DOI 10.1007/978-3-0348-8847-9
(hardcover : alk. paper)
    1. Metallothionein. I. Klaassen, Curtis D.
    QP552.M47M473 1999
    572'.68--dc21

Deutsche Bibliothek Cataloging-in-Publication Data
Metallothionein IV / ed. by C.D. Klaassen. - Basel ; Boston ; Berlin :
Birkhäuser, 1999
    ISBN 978-3-0348-9799-0

 ISBN 978-3-0348-9799-0
9 8 7 6 5 4 3 2 1

Table of contents

Metallothionein nomenclature and structure

Chemical and analytical aspects of metallothionein

Role of metallothionein in reproduction, development and cell proliferation

Metallothionein in brain disease

Role of metallothionein in copper metabolism, diabetes and autoimmune disease

Role of metallothionein in oxidative stress

Metallothionein and carcinogenesis

Metallothioneins as biomarkers

List of Contributors

Asim B. Abdel-Mageed, Department of Pharmacology, Tulane University School of Medicine, New Orleans, LA 70112, USA

Josef Abel, Division of Toxicology, at the Heinrich-Heine-University, 40225 Düsseldorf, Germany

David L. Adelson, Division of Animal Production, CSIRO, Prospect, NSW 2149, Australia

Krishna C. Agrawal, Department of Pharmacology, Tulane University School of Medicine, New Orleans, LA 70112, USA
Tel: +1 504 584-2628, Fax: +1 504 588-5283, E-mail: agrawal@mailhost.tcs.tulane.edu

Adriano Aguzzi, Neuropathologie USZ, Schmelzbergstrasse 12, CH-8091 Zürich, Switzerland

Hiroyuki Aikawa, Department of Environmental Health, School of Medicine, Tokai University, Bohseidai, Isehara, Kanagawa 259-11, Japan
Tel: +81 0463-93-1121 Ext. 2612, Fax: +81 0463-93-8796

Glen K. Andrews, Department Biochemistry and Molecular Biology, University of Kansas Medical Center, Kansas City, KS 66160-7417, USA

Miquel D. Antoine, Department of Chemistry and Biochemistry, University of Maryland Baltimore County, 1000 Hilltop Circle, Baltimore, MD 21250, USA

Margarita D. Apostolova, Department of Pathology, University of Western Ontario, London, Ontario N6A 5C1 Canada

Ana Rosa Linde Arias, Department of Environmental Medicine and Informatics, Graduate School of Environmental Earth Science, Hokkaido University, Sapporo 060, Japan

Nobuyuki Arima, Department of Pathology and Cell Biology, University of Occupational and Environmental Health, School of Medicine, 1-1 Iseigaoka, Yahatanishi-ku, Kitakyushu, 807 Japan

Ian M. Armitage' University of Minnesota, Department of Biochemistry, 4-225 Millard Hall, 435 Delaware St. S.E., Minneapolis, MN 55455, USA
Tel: +1-612 624-5977, Fax: +1-612 625-2163, E-mail: Ian@dimer.biochem.umn.edu

René Saint-Arnaud, Genetics Unit, Shriners Hospital for Crippled Children, Montréal, Québec, Canada H3G 1A6

S. Asami, Department of Occupational Health and Toxicology, School of Allied Health Sciences, Kitasato University, 1-15-1 Kitasato, Sagamihara, Kanagawa 228, Japan

Michael Aschner, Department of Physiology and Pharmacology, Bowman Gray School of Medicine, Medical Center Blvd., Winston-Salem, NC 27157-1083, USA
Tel: +1 910 716 8530, Fax:+1 910 716 8501, E-mail: maschner@bgsm.edu

Sílvia Atrian, Departament de Genètica, Facultat de Biologia, Universitat de Barcelona, Av. Diagonal 645, E-08071 Barcelona, Spain

Dasha A. Avramova, Department of Ecology of the Institute of Nuclear Power Engineering, Studgorodok-1, Obninsk, 249020, Kaluga region, Russia

Samuel C. Balderman, Erie County Medical Center, Department of Surgery, State University of New York at Buffalo, Buffalo, NY 14215, USA

John H. Beattie, Trace Element and Gene Expression Group, Rowett Research Institute, Greenburn Road, Bucksburn, Aberdeen AB21 9SB, Scotland, UK
Tel: +44-1224-716631, Fax: +44-1224-716629, E-mail:J.Beattie@rri.sari.ac.uk

M.J. Bebianno, U.C.T.R.A.-University of Algarve, Campus de Gambelas, 8 000 Faro, Portugal
Fax: +351-89-818 353, E-mail: mbebian@mozart.ulg.si.pt

Laura L. Bennett, Department of Chemistry, Dartmouth College, Hanover, NH 03755, USA

Burkhard Berger, Institut für Zoologie u. Limnologie der Universität Innsbruck, Technikerstraße 25, A-6020 Innsbruck, Austria

Maryka H. Bhattacharyya, Center for Mechanistic Biology and Biotechnology, Argonne National Laboratory, 9700 South Cass Avenue, Argonne, IL 60439-4833, USA
Tel: +1 630 252-3923, E-mail: bhatt@anlcmb.bim.anl.gov

Pierre-Alain Binz, Laboratoire Central de Chimie Clinique, Hôpital Cantonal Universitaire, rue Micheli-du-Crest 24, CH-1211 Geneve 14, Switzerland
Fax: 41-22-372 73 99, E-mail: pabinz@bioc.unizh.ch

Douglas Bittel, Department Biochemistry and Molecular Biology[2], University of Kansas Medical Center, Kansas City, KS 66160-7417, USA

Horst Bluethmann, Department of Biology, Pharmaceutical Research Gene Technology, F. Hoffmann-La Roche Ltd., Basel, Switzerland

Carmen A. Blum, Center for Mechanistic Biology and Biotechnology, Argonne National Laboratory, Argonne, IL 60439-4833, USA

Samuel Blumenthal, Department of Chemistry, University of Wisconsin-Milwaukee, Milwaukee, WI 53201, USA

Horst Blüthmann, Hoffmann La-Roche PRTB, CH-4002 Basel, Switzerland

Roger Bofill, Departament de Química, Universitat Autònoma de Barcelona, E-08193 Bellaterra, Barcelona, Spain

Ralf Bogumil, Biochemisches Institut der Universität Zürich, Winterthurerstrasse 190, CH-8057 Zürich, Switzerland

Guy Bordin, European Commission, Joint Research Centre, Institute for Reference Materials and Measurements, Retieseweg, B-2440 Geel, Belgium
Tel: +32-14 571 201, Fax +32 14 584 273, E-mail: bordin@irmm.jrc.be

Meira Bosnic, Department of Veterinary Anatomy and Pathology, University of Sydney, NSW 2006, Australia

Antony G. Breen, Departments of Medicine and Therapeutics and Biomedical Sciences, Polwarth Building, Foresterhill, Aberdeen AB25 2ZD, UK

Ian Bremner, Trace Element and Gene Expression Group, Rowett Research Institute, Greenburn Road, Bucksburn, Aberdeen AB21 9SB, Scotland

Arend Bruinink, Institute of Toxicology, ETH and University of Zurich, Switzerland

Lu Cai, Department of Pathology, University of Western Ontario, London, Ontario N6A 5C1 Canada

Yongjiu Cai, Department of Veterinary PathoBiology, University of Minnesota, 295 AS/VM Building, 1988 Fitch Avenue, Saint Paul, Minnesota 55108, USA
Tel: +1 612 625-0204, Fax: +1 612 625-1714, E-mail: caixx008@tc.umn.edu

Graham R. Cam, Division of Animal Production, CSIRO, Prospect, NSW 2149, Australia

Iain L. Campbell, Department of Neuropharmacology, The Scripps Research Institute, La Jolla, CA 92037, USA

Antonio Capasso, CNR Institute of Protein Biochemistry and Enzymology, via Marconi 10, I-80125 Naples, Italy

Clemente Capasso, CNR Institute of Protein Biochemistry and Enzymology, via Marconi 10, I-80125 Naples, Italy

Mercè Capdevila, Departament de Química, Universitat Autònoma de Barcelona, E-08193 Bellaterra, Barcelona, Spain

Vincenzo Carginale, CNR Institute of Protein Biochemistry and Enzymology, via Marconi 10, I-80125 Naples, Italy

Javier Carrasco, Departamento de Biología Celular y Fisiología, Unidad de Fisiología Animal, Facultad de Ciencias, Universidad Autónoma de Barcelona, Bellaterra, Barcelona, Spain 08193

C.-C. Chang, Department of Life Sciences, National Tsing Hua University, Hsin-Chu 300 Taiwan, ROC

John M. Charnock, Department of Chemistry, University of Manchester, England

Je-Hsin Chen, Institute of Radiation Biology, National Tsing Hua University, Hsinchu, Taiwan, Republic of China

Shi Cheng, Department of Biophysics, Beijing Medical University, Beijing, 100083, P.R. of China
Tel: +86 10 6209 1414 (O), +86 10 6209 1840 (H), Fax: +86 10 6209 5681

Yiwu Chen, National Laboratory of Protein Engineering, College of Life Sciences, Peking University, Beijing 100871, P.R. China

M. George Cherian, Department of Pathology, University of Western Ontario, London, Ontario N6A 5C1 Canada
Tel: +1 519 661-2030, Fax: +1 519 661-3370, E-mail: mcherian@julian.uwo.ca

Zhong Chongxia, Department of Biophysics, Beijing Medical University, Beijing, 100083, P.R. of China

K.H. Andy Choo, Murdoch Institute for Research into Birth Defects, Royal Children's Hospital, Flemington Road, Parkville, 3052, Australia

W. Chu, University of Kansas Medical Center, Kansas City, KS, USA

Steven D. Cohen, University of Connecticut, Storrs, CT 06269, USA

Olga M. Collins, Department of Oncology, University of Western Ontario, London Regional Cancer Centre, 790 Commissioners Road East, London, Ontario, Canada N6A 4L6

Neus Cols, Departament de Genètica, Facultat de Biologia, Universitat de Barcelona, Av. Diagonal 645, E-08071 Barcelona, Spain

D.R Conklin, Department of Physiology and Pharmacology, Bowman Gray School of Medicine, Medical Center Blvd., Winston-Salem, NC 27157-1083, USA

Timothy P. Coogan, National Cancer Institute at the National Institute of Environmental Health Sciences111 Alexander Drive, PO Box 12233, Research Triangle Park, NC 27709, USA

Elizabeth H. Cox, Department of Chemistry, Dartmouth College, Hanover, NH 03755, USA

P. Coyle, Institute of Medical and Veterinary Science, Adelaide, SA, Australia

Robert S. Cross, Department of Pharmacology, Toxicology, and Therapeutics, University of Kansas Medical Center, 3901 Rainbow Boulevard, Kansas City, KS 66160-7417 USA

K.C. Crowthers, Department of Molecular and Cell Biology, University of Connecticut, Storrs, CT 06269, USA

M. Daggett, University of Kansas Medical Center, Kansas City, KS, USA

Reinhard Dallinger, Institut für Zoologie u. Limnologie der Universität Innsbruck, Technikerstraße 25, A-6020 Innsbruck, Austria
Tel: ++43-512-507-6182, Fax: ++43-512-507 - 3960, E-mail: reinhard.dallinger@uibk.ac.at

T. Dalton, University of Kansas Medical Center, Kansas City, KS, USA

Isabelle Delisle, Centre de recherche en cancérologie, CHUQ, Pavillon Hôtel-Dieu de Québec, Québec, Québec, Canada G1R 2J6

Janice M. DeMoor, Department of Oncology, University of Western Ontario, London Regional Cancer Centre, 790 Commissioners Road East, London, Ontario, Canada N6A 4L6

Eugene DeRose, Department of Chemistry and the UWM-NIEHS Marine and Freshwater Biomedical Core Center, University of Wisconsin-Milwaukee, Milwaukee, WI, 53201-0413, USA

Bhalchandra A. Diwan, SAIC Frederick, NCI-FCRDC, Frederick, MD, USA

Claudia Dohle, Diabetes Forschungsinstitut, Auf'm Hennekamp 65, D-40225 Düsseldorf, Germany

Munira Dughish, Department of Chemistry, University of Wisconsin-Milwaukee, Milwaukee, WI 53201, USA

Jackie S. Duncan, Molecular Physiology Group, Rowett Research Institute, Greenburn Road, Bucksburn, Aberdeen AB21 9SB, Scotland

Manuchair Ebadi, Department of Pharmacology, University of Nebraska College of Medicine, 600 S. 42nd Street, Omaha, NE 68198-6260
Tel (Lab): 402-559-8245, Tel (Office): 402-559-5140, Fax: 402-559-7495,
E-mail: mebadi@mail.unmc.edu

John Ejnik, Department of Chemistry and the UWM-NIEHS Marine and Freshwater Biomedical Core Center, University of Wisconsin-Milwaukee, Milwaukee, WI, 53201-0413, USA

Mitchell R. Emerson, Department of Pharmacology, Toxicology, and Therapeutics, University of Kansas Medical Center, 3901 Rainbow Boulevard, Kansas City, KS 66160-7417 USA

Tadasu Emoto, Department of Environmental Medicine and Informatics, Graduate School of Environmental Earth Science, Hokkaido University, Kita 10, Nishi 5, Kita-ku, Sapporo 060, Japan

Fumio Endo, Department of Pediatrics, Kumamoto University Shool of Medicine, Kumamoto 860, Japan

Jay C. Erickson, Howard Hughes Medical Institute, University of Washington, Seattle, WA 98195, USA

Daniele Fabris, Department of Chemistry and Biochemistry, University of Maryland Baltimore County, 1000 Hilltop Circle, Baltimore, MD 21250, USA

Peter Faller, Biochemisches Institut der Universität Zürich, Winterthurerstrasse 190, CH-8057 Zürich, Switzerland

Xingwang Fang, Department of Technical Physics, Peking University, Beijing 100871, P.R. China

Raffaella Faraonio, Centre de recherche en cancérologie, CHUQ, Pavillon Hôtel-Dieu de Québec, Québec, Québec, Canada G1R 2J6

Catherine Fenselau, Department of Chemistry and Biochemistry, University of Maryland
Baltimore County, 1000 Hilltop Circle, Baltimore, MD 21250, USA
Tel: +1 410-455-3236, Fax: +1 410-455-1091, E-mail: fenselau@umbc.edu

David Fowle, Department of Chemistry, The University of Western Ontario, London, ON,
N6A 5B7 CANADA

I. Carmen Fuentealba, Department of Pathology and Microbiology, Atlantic Veterinary
College, University of Prince Edward Island, Charlottetown, Prince Edward Island, C1A
4P3, Canada
Tel: +1 902-566-0868, Fax: +1 902-566-0851, E-mail: cfuentealba@upei.ca

C. David Garner, Department of Chemistry, University of Manchester, England

Scott H. Garrett, Department of Pathology, West Virginia University, P.O. Box 9203,
Morgantown, West Virginia 26506-9203, USA
Tel: +1 304-293-7078, Fax: +1 304-293-6249, E-mail: sgarrett@wvu.edu

L. Gedamu, Department of Biological Sciences, The University of Calgary, Calgary, Alberta,
Canada, T2N 1N4
Tel: 403-220-5556, Fax: 403-289-9311, E-mail: lgedamu@acs.ucalgary.ca

Oleg Georgiev, Institut für Molekularbiologie, Abteilung II der Universität Zürich,
Winterthurstrasse 190, CH-8057 Zürich, Switzerland

M. Gerpe, CONICET and Marine Sciences Department, FCEyN, University of Mar del Plata,
Funes 3350, (7600) Mar del Plata, Argentina

Mercedes Giralt, Departamento de Biología Celular y Fisiología, Unidad de Fisiología
Animal, Facultad de Ciencias, Universidad Autónoma de Barcelona, Bellaterra, Barcelona,
Spain 08193

Helga Gleichmann, Diabetes Forschungsinstitut, Auf'm Hennekamp 65, D-40225
Düsseldorf, Germany
Tel: +49-211-33821, Fax: +49-211-603, E-mail: gleich@dfi.uni-duesseldorf.de

Ekaterina I. Goncharova, Nelson Institute of Environmental Medicine and Kaplan Cancer
Center, New York University Medical Center, 550 First Avenue, New York, NY 10016, USA

Berta González, Unidad de Histología, Facultad de Medicina, Universidad Autónoma de
Barcelona, Bellaterra, Barcelona, Spain 08193

Pilar González-Duarte, Departament de Química, Universitat Autònoma de Barcelona,

E-08193 Bellaterra, Barcelona, Spain
Tel: +34-3-5811363, Fax: +34-3-5813101, E-mail: IQIPILAR@CC.UAB.ES

Roser González-Duarte, Departament de Genètica, Facultat de Biologia, Universitat de
Barcelona, Av. Diagonal 645, E-08071 Barcelona, Spain

Robert A. Goyer, National Institute of Environmental Health Sciences, Research Triangle
Park, NC, USA

Anna-Rae Green, Department of Chemistry, The University of Western Ontario, London,
ON, N6A 5B7 CANADA

Ziqi Gui, Department of Chemistry, The University of Western Ontario, London, ON, N6A
5B7 CANADA

Çagatay Günes, Institut für Molekularbiologie, Abteilung II der Universität Zürich,
Winterthurstrasse 190, CH-8057 Zürich, Switzerland

Sultan S. Habeebu, Department of Pharmacology, Toxicology & Therapeutics University of
Kansas Medical Center Kansas City, KS 66160-7417, USA

George Haleblian, Department of Chemistry, Dartmouth College, Hanover, NH 03755, USA

Tetsuo Hamada, Department of Pathology and Cell Biology, University of Occupational and
Environmental Health, School of Medicine, 1-1 Iseigaoka, Yahatanishi-ku, Kitakyushu, 807
Japan

Ruyi Hao, Department of Neurology, University of Tennessee College of Medicine,
Memphis, TN, USA

John L. Harwood, School of Molecular and Medical Biosciences, University of Wales
Cardiff, P.O. Box 911, Cardiff CF1 3US, UK

Daniel W. Hasler, Biochemisches Institut der Universität Zürich, Winterthurerstrasse 190,
CH-8057 Zürich, Switzerland

Akira Hata, Department of Prevented Medicine, School of Medicine, Hokkaido University,
Sapporo 060, Japan

Yetrib Hathout, Department of Chemistry and Biochemistry, University of Maryland
Baltimore County, 1000 Hilltop Circle, Baltimore, MD 21250, USA

C. Haux, University of Göteborg, Zoological Institute, Department of Zoophysiology,

Medicinaregatan 18, S-41390 Göteborg, Sweden

Gabrielle M. Hawksworth, Dept of Medicine and Therapeutics, Polwarth Building, Foresterhill, Aberdeen AB25 2ZD
Tel: +44 1224663123, ext 52487, Fax: +44 1224699884, E-mail g.m.hawksworth@abdn.ac.uk

Tao He, Department of Chemistry and Biochemistry, University of Maryland Baltimore County, 1000 Hilltop Circle, Baltimore, MD 21250, USA

Ulrich Heinzmann, Institute of Pathology, GSF-National Research Center for Environment and Health, D-85764 Neuherberg, Germany

Joaquín Hernández, Departamento de Biología Celular y Fisiología, Unidad de Fisiología Animal, Facultad de Ciencias, Universidad Autónoma de Barcelona, Bellaterra, Barcelona, Spain 08193

Rainer Heuchel, Institut für Molekularbiologie, Abteilung II der Universität Zürich, Winterthurstrasse 190, CH-8057 Zürich, Switzerland

Juan Hidalgo, Departamento de Biología Celular y Fisiología, Unidad de Fisiología Animal, Facultad de Ciencias, Universidad Autónoma de Barcelona, Bellaterra, Barcelona, Spain 08193
Tel: +34 3 581 20 37, Fax: +34 3 581 23 90, E-mail: Hidalgo@cc.uab.es

Brian T. Hill, Department of Chemistry, Dartmouth College, Hanover, NH 03755, USA

Midori Hiramatsu, Institute of Life Support Technology, Yamagata Technopolis Fdn., Yamagata, Japan

Toshitsugu Hirauchi, Department of Environmental Medicine and Informatics, Graduate School of Environmental Earth Science, Hokkaido University, Sapporo 060, Japan

Harumi Hisada, Department of Radiobiochemistry, School of Pharmaceutical Sciences, University of Shizuoka, 52-1 Yada, Shizuoka 422, Japan

Shuk-mei Ho, Department of Biology, Tufts University, Medford, MA 02155, USA
Tel: +1-617-627-3540, Fax: +1-617-627-3805, E-mail: sho@emerald. tufts. edu

C. Hogstrand, T.H. Morgan School of Biological Sciences, 101 Morgan Building, University of Kentucky, Lexington, Kentucky 40506-0225, USA

Adele F. Holloway, Human Immunology, Hanson Centre, Frome Rd, Adelaide, Australia,

5000

N. Hu, University of Kansas Medical Center, Kansas City, KS, USA

Qing Huai, Department of Technical Physics, Peking University, Beijing 100871, P.R. China

Guojian Huang, Institute of Clinical Medical Sciences, China-Japan Friendship Hospital, Beijing 100029, P.R. China

Meilin Huang, Department of Chemistry, University of Wisconsin-Milwaukee, Milwaukee, WI 53201, USA

P.C. Huang, Department of Biochemistry, Johns Hopkins University, Baltimore, MD 21205, USA

Peter E. Hunziker, Universität Zürich, Biochemisches Institut, Winterthurerstr.190, CH-8057 Zürich, Switzerland
Tel: +41-1-365 55 20, Fax: +41-1-365 68 05, E-mail: phunzi@bioc.unizh.ch

Tupur Husain, Molecular and Cellular Biology Program, Tulane University School of Medicine, New Orleans, LA 70112, USA

Norio Itoh, Faculty of Pharmaceutical Sciences, Osaka University, Yamada-oka, Suita, Osaka 565, Japan

Patric L. Iversen, Department of Pharmacology, University of Nebraska College of Medicine, 600 South 42nd Street, Omaha, NE 68198, USA

William Jenner, Glaxo Wellcome R&D, Ware, Herts, SG12 0DP, UK

Laran T. Jensen, University of Utah Health Sciences Center, Salt Lake City UT 84132, USA

Shaohua Jin, Department of Pharmacology, Toxicology, and Therapeutics, University of Kansas Medical Center, 3901 Rainbow Boulevard, Kansas City, KS 66160-7417 USA

Jeffrey A. Johnson, Department of Pharmacology, Toxicology and Therapeutics, University of Kansas Medical Center, G017B Breidenthal, 3901 Rainbow Blvd., Kansas City, KS 66160-7417
Tel: +1-913 588-7517, Fax: +1-913 588-7501,E-mail: jjohnso2@kumc.edu

Jeremias H.R. Kägi, Biochemisches Institut der Universität Zürich, Winterthurerstrasse 190, CH-8057 Zürich, Switzerland
Tel: +41-1-635-5550, Fax: +41-1-635-6805, E-mail: jhrkagi@bioc.unizh.ch

Naoki Kameda, Department of Environmental Health, Jichi Medical School, Tochigi-Ken 329-04, Japan

Satomi Kameo, Department of Environmental Medicine and Informatics, Graduate School of Environmental Earth Science, Hokkaido University, Kita-10, Nishi-5, Kita-ku, Sapporo, 060, Japan
Tel: +81 (0)11-706-2244, Fax: +81 (0)11-717-0629

Masako Kanekiyo, Faculty of Pharmaceutical Sciences, Osaka University, Yamada-oka, Suita, Osaka 565, Japan

Y. James Kang, Department of Medicine, University of Louisville School of Medicine, 530 S. Jackson St., Louisville, KY 40202, USA
Tel: +1 502 852-6991, Fax: +1 502 852-6904, E-mail: yjkang01@homer.louisville.edu

Keiko Kasutani, Faculty of Pharmaceutical Sciences, Osaka University, Yamada-oka, Suita, Osaka 565, Japan

Peter Kille, School of Molecular and Medical Biosciences, University of Wales Cardiff, P.O. Box 911, Cardiff CF1 3US, UK
Tel. +44-(0)1222-874507, Fax. +44-(0)1222-874116, E-mail: Kille@cardiff.ac.uk

Masami Kimura, Department of Molecular Biology, Keio University School of Medicine, Tokyo 160, Japan

Shinzo Kimura, Department of Environmental Medicine and Informatics, Graduate School of Environmental Earth Science, Hokkaido University, Sapporo 060, Japan

Tomoki Kimura, Faculty of Pharmaceutical Sciences, Osaka University, Yamada-oka, Suita, Osaka 565, Japan

Curtis D. Klaassen, Department of Pharmacology, Toxicology & Therapeutics, University of Kansas Medical Center, Kansas City, Kansas 66160-7417
Tel: (913) 588-7714, Fax: (913) 588-7501, E-mail: cklaasse@kumc.edu

Stefan Klauser, Biochemisches Institut der Universität Zürich, Winterthurerstrasse 190, CH-8057 Zürich, Switzerland

Dominik Klein, Institute of Toxicology and Environmental Hygiene, Technical University 80636 Munich, Germany

P. Kling, Department of Cellular and Developmental Biology, Umea University, S-901 87 Umea, Sweden

Zorana Kljaković, Institute of Oceanography and Fisheries, HR-21000 Split, P.O. Box 500, Croatia

Kinji Kobayashi, Safety Research Laboratory, Tanabe Seiyaku Co. Ltd., Yodogawa-ku, Osaka 532, Japan

K.A. Koch, The University of Michigan Medical School, Ann Arbor, MI, USA

Naoko Koizumi, Department of Public Health, Hyogo College of Medicine, Nishinomiya, Hyogo, 663 Japan

Shinji Koizumi, Division of Hazard Assessment, National Institute of Industrial Health, 6-21-1, Nagao, Tama-ku, Kawasaki 214, Japan
Tel: +81-044-865-6111, Fax: +81-044-865-6116, E-mail: snkzm@mxa.meshnet.or.jp

Seiji Kojima, Department of Veterinary Pathology, Tottori University, Tottori 680, Japan

Yutaka Kojima, Department of Environmental Medicine and Informatics, Graduate School of Environmental Earth Science, Hokkaido University, Kita-10, Nishi-5, Kita-ku, Sapporo, 060, Japan

James Koropatnick, Department of Oncology, University of Western Ontario, London Regional Cancer Centre, 790 Commissioners Road East, London, Ontario, Canada N6A 4L6
Tel: +1-519 685-8654, Fax: +1-519 685-8646, E-mail: jkoropat@julian.uwo.ca

Eric P. Kowack, Department of Chemistry, Dartmouth College, Hanover, NH 03755, USA

Sonja Kozar, Ruđer Bošković Institute, Center for Marine Research Zagreb, HR-10001 Zagreb, P.O. Box 1016, Croatia

Susan Krezoski, Department of Chemistry, University of Wisconsin-Milwaukee, Milwaukee, WI 53201, USA

Sara Krull, Department of Chemistry, University of Wisconsin-Milwaukee, Milwaukee, WI 53201, USA

Hiroki Kubota, National Institute of Health Sciences, Kamiyoga 1-18-1, Setagaya, Tokyo 158, Japan

M.V. Ramana Kumari, Institute of Life Support Technology, Yamagata Technopolis Fdn., Yamagata, Japan

Masaaki Kurasaki, Department of Environmental Medicine and Informatics, Graduate School

of Environmental Earth Science, Hokkaido University, Kita-10, Nishi-5, Kita-ku, Sapporo, 060
Tel: +81-11-706-2243, Fax: +81-11-717-0629

Željko Kwokal, Ruđer Bošković Institute, Center for Marine Research Zagreb, HR-10001 Zagreb, P.O. Box 1016, Croatia

Frank Laib, Department of Chemistry and the UWM-NIEHS Marine and Freshwater Biomedical Core Center, University of Wisconsin-Milwaukee, Milwaukee, WI, 53201-0413, USA

Donald L. Lamm, Department of Urology, West Virginia University, P.O. Box 9203, Morgantown, West Virginia 26506-9203, USA

W.J. Langston, Plymouth Marine Laboratory – Citadel Hill, Plymouth PL1, 2PB, UK

Olivier Larochelle, Centre de recherche en cancérologie, CHUQ, Pavillon Hôtel-Dieu de Québec, Québec, Québec, Canada G1R 2J6

Alice A. Larson, Department of Veterinary PathoBiology, University of Minnesota, Saint Paul, Minnesota 55108, USA

John C. Lau, Department of Pathology, University of Western Ontario, London, Ontario N6A 5C1 Canada

Mary M. Lauro, Department of Biology, Boston College, Chestnut Hill, Massachusetts 02167, USA

John S. Lazo, Department of Pharmacology, E1340 Biomedical Science Tower, University of Pittsburgh School of Medicine, Pittsburgh, PA 15261, USA
Tel: +1 412 648-9319, Fax: +1 412 648-2229, E-mail: lazo@pop.pitt.edu

J.C. Lee, Department of Molecular and Cell Biology, University of Connecticut, Storrs, CT 06269, USA

Kai-Fai Lee, Department of Biology, Tufts University, Medford, MA 02155, USA

Rain Lehtme, Institute of Molecular and Cell Biology, University of Tartu, Vanemuise 46, Tartu EE24+, Estonia

Àngels Leiva, Departament de Química, Universitat Autònoma de Barcelona, E-08193 Bellaterra, Barcelona, Spain

Laura S. Levy, Molecular and Cellular Biology Program, Tulane University School of Medicine, New Orleans, LA 70112, USA

Donna Lewand, Department of Chemistry, University of Wisconsin-Milwaukee, Milwaukee, WI 53201, USA

Q. Li, University of Kansas Medical Center, Kansas City, KS, USA

Sai Li, Erie County Medical Center, Department of Surgery, State University of New York at Buffalo, Buffalo, NY 14215, USA

Josef Lichtmannegger, GSF-National Research Center for Environment and Health, Institute of Toxicology, Ingolstädter Landstr. 1, D-85764 Neuherberg, Germany

Hou Lin, Department of Biophysics, Beijing Medical University, Beijing, 100083, P.R. of China

Lih-Yuan Lin, Institute of Radiation Biology, National Tsing Hua University, Hsinchu, Taiwan, Republic of China

Jie Liu, Department of Pharmacology, Toxicology and Therapeutics, University of Kansas Medical Center, Kansas City, KS 66160-7417, USA
Tel: +1 913 588-7529, Fax: +1 913 588-7501, E-mail: jliu@kumc.edu

Xiaoyan Liu, Department of Chemistry, Dartmouth College, Hanover, NH 03755, USA

Yaping Liu, Department of Pharmacology, Toxicology & Therapeutics University of Kansas Medical Center Kansas City, KS 66160-7417, USA

Tracy Lloyd, Erie County Medical Center, Department of Surgery, State University of New York at Buffalo, Buffalo, NY 14215, USA

Angela Lucas, University of Connecticut, Storrs, CT 06269, USA

Michael A. Lynes, Department of Molecular and Cell Biology, U-125, 75 North Eagleville Rd., University of Connecticut, Storrs, CT 06269-3125
Tel: 860-486-4350, Fax: 860-486-4331, E-mail: Lynes@uconnvm.uconn.edu

Pavel Mader, Department of Chemistry, Faculty of Agronomy, Czech University of Agriculture, 165 21 Prague 6-Suchdol, Czech Republic

Tamio Maitani, National Institute of Health Sciences, Kamiyoga 1-18-1, Setagaya, Tokyo 158, Japan

Tel: +81-3-3700-9409, Fax: +81-3-3707-6950, E-mail: maitani@nihs.go.jp

Silvia Marino, Neuropathologie USZ, Schmelzbergstrasse 12, CH-8091 Zürich, Switzerland

John E. Mata, Department of Pharmacology, University of Nebraska College of Medicine, 600 South 42nd Street, Omaha, NE 68198, USA

Ichiro Matsuda, Department of Pediatrics, Kumamoto University Shool of Medicine, Kumamoto 860, Japan

G.D. Mayer, T.H. Morgan School of Biological Sciences, 101 Morgan Building, University of Kentucky, Lexington, Kentucky 40506-0225, USA

R. McCabe, The Jackson Laboratory, 600 Main St., Bar Harbor, ME 04609,

Jane S. Merkel, Department of Chemistry, Dartmouth College, Hanover, NH 03755, USA

Michael V. Miceli, Department of Ophthalmology, Tulane University School of Medicine, New Orleans, LA 70115, USA

Anna E. Michalska, Murdoch Institute for Research into Birth Defects, Royal Children's Hospital, Flemington Road, Parkville, 3052, Australia

Adrian T. Miles, Departments of Medicine and Therapeutics and Biomedical Sciences, Polwarth Building, Foresterhill, Aberdeen AB25 2ZD, UK

Alan M. Miller, Molecular and Cellular Biology Program, Tulane University School of Medicine, New Orleans, LA 70112, USA

Janine M. Miller, Biochemistry, University of Tasmania, GPO Box 252-58, Hobart, Australia 7001

Kyong-Son Min, Department of Nutrition, Kobe Gakuin University, Japan
Tel: +81-78-974-1551, Fax: +81-78-974-5689, E-mail min@nutr.kobegakuin.ac.jp

Seiki Minamide, Department of Laboratory Sciences, Gunma University School of Health Sciences, 3-39-15, Showa machi, Maebashi, Gunma, 371 Japan

Takeshi Minami, Lab. of Cell Biology, Dept. of Anatomy, Nara Medical University, 840 Shijo-Cho, Kashihara, Nara 634, Japan
Tel: +81-7442-2-3051 (Ext. 2229), Fax: +81-7442-4-8432,
E-mail: minamita@nmu-gw.cc.naramed-u.ac.jp

Jeffrey D. Moehlenkamp, Department of Pharmacology, Toxicology and Therapeutics, University of Kansas Medical Center, G017B Breidenthal, 3901 Rainbow Blvd., Kansas City, KS 66160-7417

Pierre Moffatt, Centre de recherche en cancérologie, CHUQ, Pavillon Hôtel-Dieu de Québec, Québec, Québec, Canada G1R 2J6

Amalia Molinero, Departamento de Biología Celular y Fisiología, Unidad de Fisiología Animal, Facultad de Ciencias, Universidad Autónoma de Barcelona, Bellaterra, Barcelona, Spain 08193

John Morgan, PABIO, U.W.C., Cardiff CF1 3US, Wales, UK

Takehito Morita, Department of Veterinary Pathology, Tottori University, Tottori 680, Japan

Ceri A. Morris, School of Molecular and Medical Biosciences, University of Wales Cardiff, P.O. Box 911, Cardiff CF1 3US, UK

Josef Müller-Höcker, Institute of Pathology, University of Munich, D-80337 Munich, Germany

Julia E. Mullins, Department of Pathology and Microbiology, Atlantic Veterinary College, University of Prince Edward Island, Charlottetown, Prince Edward Island, C1A 4P3, Canada

Karl-Heinz Müller, Institut für Molekularbiologie, Abteilung II der Universität Zürich, Winterthurstrasse 190, CH-8057 Zürich, Switzerland

Amalia Muñoz, Department of Chemistry and the UWM-NIEHS Marine and Freshwater Biomedical Core Center, University of Wisconsin-Milwaukee, Milwaukee, WI, 53201-0413, USA

Norio Muto, Faculty of Pharmaceutical Sciences, Osaka University, Yamada-oka, Suita, Osaka 565, Japan

Fumio Naganuma, Department of Laboratory Sciences, Gunma University School of Health Sciences, 3-39-15, Showa machi, Maebashi, Gunma, 371 Japan

Koji Nagano, Department of Pediatrics, Kumamoto University Shool of Medicine, Kumamoto 860, Japan

Takeaki Nagamine, National Cancer Institute at the National Institute of Environmental Health Sciences111 Alexander Drive, PO Box 12233, Research Triangle Park, NC 27709, USA

Tomomi Nagashima, Research Section of Pathology, Institute of Immunological Science, Hokkaido University, Sapporo 060, Japan

Katsuyuki Nakajima, Department of Laboratory Sciences, Gunma University School of Health Sciences, 3-39-15, Showa machi, Maebashi, Gunma, 371 Japan

Kyoumi Nakazato, Division of Pathology, Department of Laboratory Sciences, Gunma University School of Health Sciences, 3-39-15 Maebashi, Gunma 371, Japan

Ryuji Nakano, Department of Pathology and Cell Biology, University of Occupational and Environmental Health, School of Medicine, 1-1 Iseigaoka, Yahatanishi-ku, Kitakyushu, 807 Japan

Yukiko Nakahara, Department of Nutrition, Kobe Gakuin University, Kobe, 651-21, Japan

April M. Newman, Trace Element and Gene Expression Group, Rowett Research Institute, Greenburn Road, Bucksburn, Aberdeen AB21 9SB, Scotland

David A. Newsome, M.D., Touro Infirmary, 1401 Foucher St., New Orleans, LA 70115 Fax: +1 504-897-7637, Tel: +1 504-897-8179, E-mail: seru@mailhost.tcs.tulane.edu

Beate Nicolaus, School of Molecular and Medical Biosciences, University of Wales Cardiff, P.O. Box 911, Cardiff CF1 3US, UK

Daotai Nie, University of South Carolina, Dept. of Chemistry and Biochemistry, 730 S. Main St., Columbia SC 29208, USA

Tadashi Niioka, Department of Environmental Medicine and Informatics, Graduate School of Environmental Earth Science, Hokkaido University, Sapporo 060, Japan Tel: +81-11-706-2242, Fax: +81-11-717-0629, E-mail: niioka@ees.hokudai.ac.jp

Nikša Odžak, Institute of Oceanography and Fisheries, HR-21000 Split, P.O. Box 500, Croatia

Ruriko Ninomiya, Department of Public Health, Hyogo College of Medicine, Nishinomiya, Hyogo, 663 Japan

Hisao Nishimura, Department of Human Sciences, Aichi Mizuho University, Toyota-shi, Aichi 470-03, Japan Tel: +81-565-43-0134, Fax: +81-565-46-5220, E-mail: kgf00676@niftyserve.or.jp

Kayo Nishida, Department of Nutrition, Kobe Gakuin University, Kobe, 651-21, Japan

Noriko Nishimura, Environmental Health Sciences Division, National Institute for Environmental Studies, 16-2, Onogawa, Tsukuba, Ibaraki 305, Japan

Hiroko Nomiyama, Department of Environmental Health, Jichi Medical School, Tochigi-Ken 329-04, Japan

Kazuo Nomiyama, Department of Environmental Health, Jichi Medical School, Tochigi-Ken 329-04, Japan
Tel: +81-285-44-2111, ext. 3139, Fax: Int-81-285-44-8465, E-mail: nomiyama@jichi.ac.jp

Fumitomo Odawara, Department of Environmental Medicine and Informatics, Graduate School of Environmental Earth Science, Hokkaido University, Sapporo 060, Japan

Yasumitsu Ogra, Laboratory of Toxicology and Environmental Health, Faculty of Pharmaceutical Sciences, Chiba University, 1-33, Yayoi-cho, Inage-ku, Chiba 263, Japan

Patricia Ohly, Diabetes Forschungsinstitut, Auf'm Hennekamp 65, D-40225 Düsseldorf, Germany

Hisayoshi Ohta, Department of Occupational Health and Toxicology, School of Allied Health Sciences, Kitasato University, 1-15-1 Kitasato, Sagamihara, Kanagawa 228, Japan
Tel. and Fax: 0427-78-8070, E-Mail: hohta@medcc.kitasato-u.ac.jp

Masashi Okabe, Department of Environmental Medicine and Informatics, Graduate School of Environmental Earth Science, Hokkaido University, Kita 10, Nishi 5, Kita-ku, Sapporo 060, Japan

Shoji Okada, Department of Radiobiochemistry, School of Pharmaceutical Sciences, University of Shizuoka, 52-1 Yada, Shizuoka 422, Japan

Masashi Okamoto, Department of Laboratory Sciences, Gunma University School of Health Sciences, 3-39-15, Showa machi, Maebashi, Gunma, 371 Japan

Yuko Okazaki, Faculty of Pharmaceutical Sciences, Kinki University, 3-4-1 Kowakae, Higashi-Osaka, Osaka 577, Japan

P.-E. Olsson, Department of Cellular and Developmental Biology, UmeåUniversity, S-901 87 Umeå, Sweden
Tel: + 46 90 7869545, Fax: +46 90 7866691, E-mail: per-erik.olsson@biology.umu.se

Satomi Onosaka, Department of Nutrition, Kobe Gakuin University, Kobe, 651-21, Japan

Noriko Otaki, Department of Occupational Disease, National Institute of Industrial Health,

Kawasaki 214, Japan

Fuminori Otsuka, Department of Environmental Toxicology, Faculty of Pharmaceutical Sciences, Teikyo University, Sagamiko, Kanagawa 199-01, Japan

Gülin Öz, Department of Biochemistry, 4-225 Millard Hall, 435 Delaware St. S.E., Minneapolis, MN 55455, USA

Toshihiko Ozawa, National Institute of Radiological Sciences, Chiba 263, Japan

Vìra Pacáková, Department of Analytical Chemistry, Charles University, Faculty of Science, 128 40 Prague 2, Czech Republic

Marina Paić, Ruđer Bošković Institute, Center for Marine Research Zagreb, HR-10001 Zagreb, P.O. Box 1016, Croatia

Oscar Palacios, Departament de Química, Universitat Autònoma de Barcelona, E-08193 Bellaterra, Barcelona, Spain

Amy E. Palmer, Department of Chemistry, Dartmouth College, Hanover, NH 03755, USA

Richard D. Palmiter, Howard Hughes Medical Institute and Department of Biochemistry, University of Washington, Box 357370, Seattle WA 98195, USA
E-mail: palmiter@u.washington.edu

Peep Palumaa, Institute of Molecular and Cell Biology, University of Tartu, Vanemuise 46, Tartu EE2400, Estonia
Tel./Fax: +372-7-465 838, E-mail: peep.palumaa@ut.ee

W. Paramchuk, Department of Biological Sciences, The University of Calgary, Calgary, Alberta, Canada, T2N 1N4

Elio Parisi, CNR Institute of Protein Biochemistry and Enzymology, Via Marconi 10, I-80125 Naples, Italy, Tel: +39 81 7257323, Fax: +39 81 2396525, E-mail: parisi@iigbna.iigb.na.cnr.it

Gianluca Passaretti, CNR Institute of Protein Biochemistry and Enzymology, via Marconi 10, I-80125 Naples, Italy

Jasenka Pavičić, Ruđer Bošković Institute, Center for Marine Research Zagreb, HR-10001 Zagreb, P.O. Box 1016, Croatia

Thomas L. Pazdernik, Department of Pharmacology, Toxicology, and Therapeutics,

University of Kansas Medical Center, 3901 Rainbow Boulevard, Kansas City, KS 66160-7417, USA
Tel: +1 913 588-7504, Fax: +1 913 588-5677, E-mail: tpazdern@kumc.edu

M.M.O. Peña, The University of Michigan Medical School, Ann Arbor, MI, USA

David H. Petering, Department of Chemistry, University of Wisconsin-Milwaukee, P.O. Box 413, Milwaukee, WI 53201-0413
Tel: +1 414-229-585, Fax: +1 414-229-5530, E-mail: Petering@csd.uwm.edu

Ronald F. Pfeiffer, Department of Neurology, University of Tennessee College of Medicine, Memphis, TN, USA

J.C. Philcox, Institute of Medical and Veterinary Science, Adelaide, SA, Australia

Bruce R. Pitt, Department of Pharmacology, University of Pittsburgh School of Medicine, Pittsburgh, Pennsylvania 15261, USA

Donald J. Plocke, Department of Biology, Boston College, Chestnut Hill, Massachusetts 02167, USA
Tel: +1-617-552-3559, Fax: +1-617-552-2011, E-mail: donald.plocke@bc.edu

Matthew C. Posewitz, Department of Chemistry, Dartmouth College, Hanover, NH 03755, USA

Dean L. Pountney, Department of Biochemistry, 4-225 Millard Hall, 435 Delaware St. S.E., Minneapolis, MN 55455, USA

P. Anthony Presta, Department of Chemistry, The University of Western Ontario, London, ON, N6A 5B7 CANADA

Guido di Prisco, CNR Institute of Protein Biochemistry and Enzymology, via Marconi 10, I-80125 Naples, Italy

He Qihua, Department of Biophysics, Beijing Medical University, Beijing, 100083, P.R. of China

Fernando Cordeiro Raposo, European Commission, Joint Research Centre, Institute for Reference Materials and Measurements, Retieseweg, B-2440 Geel, Belgium

Biserka Raspor, Ruđer Bošković Institute, Center for Marine Research Zagreb, HR-10001 Zagreb, P.O. Box 1016, Croatia
Tel: +385-1-46 80 216, Fax: +385-1-46 80 242, E-mail RASPOR@OLIMP.IRB.HR

V.E. Reeve, Department of Veterinary Anatomy and Pathology, B14, University of Sydney, NSW 2006, Australia
Tel: +61 2 9351 2084, Fax: +61 2 9351 7348, E-mail: v.reeve@vetp.usyd.edu.au

Lifen Ren, Department of Chemistry, University of Wisconsin-Milwaukee, Milwaukee, WI 53201, USA

C.A. Richardson, Department of Molecular and Cell Biology, University of Connecticut, Storrs, CT 06269, USA

Vicente Rodilla, Departments of Medicine and Therapeutics and Biomedical Sciences, Polwarth Building, Foresterhill, Aberdeen AB25 2ZD, UK

Adela Rodriguez, European Commission, Joint Research Centre, Institute for Reference Materials and Measurements, Retieseweg, B-2440 Geel, Belgium
Tel: +32-14-571 200, Fax: +32-14-584 273

A.M. Rofe, Institute of Medical and Veterinary Science, Adelaide, SA, Australia

Valentina A. Romantsova, Medical Radiological Research Center of the Russian Academy of Medical Sciences,Obninsk, 249020, Kaluga Region, Russia

Núria Romero-Isart, Departament de Química, Universitat Autònoma de Barcelona, E-08193 Bellaterra, Barcelona, Spain

Toby G. Rossman, Nelson Institute of Environmental Medicine and Kaplan Cancer Center, New York University Medical Center, 550 First Avenue, New York, NY 10016, USA
Tel: +1 914 351-2380, Fax: +1 914 351-3489, E-mail: rossman@charlotte.med.nyu.edu

Gennady M. Rott, Medical Radiological Research Center of the Russian Academy of Medical Sciences,Obninsk, 249020, Kaluga Region, Russia

Joann E. Roy, Department of Chemistry, Dartmouth College, Hanover, NH 03755, USA

Binggen Ru, National Laboratory of Protein Engineering, College of Life Sciences, Peking University, Beijing 100871, P.R. China
Tel: +86-010-62751842, Fax: +86 10 62751850, E-mail: rubing@public.east.cn.net

Ming Rui, Faculty of Pharmaceutical Sciences, Chiba University, Chiba 263, Japan

Selma Sadovic, Department of Biomedical Sciences, College of Pharmacy, University of Rhode Island, Kingston, RI 02881, USA

Shigeru Saito, Department of Environmental Medicine and Informatics, Graduate School of Environmental Earth Science, Hokkaido University, Sapporo 060, Japan

Takeshi Saito, Department of Hygiene, School of Medicine, Hokkaido University, Sapporo 060, Japan

Fred E. Samson, Ralph L. Smith Research Center, University of Kansas Medical Center, Kansas City, KS 66160-7417 USA

S.L-A. Samson, Department of Biological Sciences, The University of Calgary, Calgary, Alberta, Canada, T2N 1N4

Takakazu Sasaguri, Department of Pathology and Cell Biology, University of Occupational and Environmental Health, School of Medicine, 1-1 Iseigaoka, Yahatanishi-ku, Kitakyushu, 807 Japan

Yasuyuki Sasaguri, Department of Pathology and Cell Biology, University of Occupational and Environmental Health, School of Medicine, 1-1, Iseigaoka, Yahatanishi-ku, Kitakyushu, Japan
Tel: +81 093-691-7426, Fax: +81 093-603-8518, E-mail: yasu-s@med.uoeh-u.ac.jp

Hiroshi Satoh, Department of Environmental Health Sciences, Tohoku University Graduate School of Medicine, Sendai 980-77, Japan

Kyoko Sato, National Institute of Health Sciences, Kamiyoga 1-18-1, Setagaya, Tokyo 158, Japan

Masahiko Satoh, Environmental Health Sciences Division, National Institute for Environmental Studies, 16-2 Onogawa, Tsukuba, Ibaraki 305, Japan
Tel: +81-298-50-2448, Fax: +81-298-50-2588, E-mail: masahiko@nies.go.jp

Shin Sato, Department of Environmental Medicine and Informatics, Graduate School of Environmental Earth Science, Hokkaido University, Kita 10, Nishi 5, Kita-ku, Sapporo 060, Japan

Glenn R. Sauer, University of South Carolina, Dept. of Chemistry and Biochemistry, 730 S. Main St., Columbia SC 29208, USA
Tel: +1 803 777-0977, Fax: +1 803 777-9521, E-mail: gsauer@psc.sc.edu

Walter Schaffner, Institut für Molekularbiologie II der Universität Zürich, Winterthurstrasse 190, CH-8057 Zürich, , Switzerland

S. Schieman, Department of Biological Sciences, The University of Calgary, Calgary,

Alberta, Canada, T2N 1N4

I.B. Schweitzer, The Jackson Laboratory, 600 Main St., Bar Harbor, ME 04609,

Rosaria Scudiero, CNR Institute of Protein Biochemistry and Enzymology, via Marconi 10, I-80125 Naples, Italy

Carl Séguin, Centre de recherche en cancérologie, Pavillon Hôtel-Dieu de Québec, 11 côte du Palais, Québec (Québec) Canada G1R 2J6
Tel: +1-418 691-5554. Fax: +1-418 691-5439, E-mail: Carl.Seguin@crhdq.ulaval.ca

Y. Seki, Department of Occupational Health and Toxicology, School of Allied Health Sciences, Kitasato University, 1-15-1 Kitasato, Sagamihara, Kanagawa 228, Japan

Donald A. Sens, Department of Pathology, West Virginia University, P.O. Box 9203, Morgantown, West Virginia 26506-9203, USA

Mary Ann Sens, Department of Pathology, West Virginia University, P.O. Box 9203, Morgantown, West Virginia 26506-9203, USA

Ivana Šestáková, J. Heyrovsky Institute of Physical Chemistry, Academy of Sciences of the Czech Republic,182 23 Prague 8, Czech Republic
E-mail: SESTAKOV@ jh-inst.cas.cz

Zahir A. Shaikh, Department of Biomedical Sciences, College of Pharmacy, University of Rhode Island, Kingston, RI 02881, USA
Tel: +1 401-874-5036, Fax: +1 401-874-2181, E-mail: ZShaikh@URIACC.URI.EDU

C. Frank Shaw III, Department of Chemistry and the UWM-NIEHS Marine and Freshwater Biomedical Core Center, University of Wisconsin-Milwaukee, Milwaukee, WI, 53201-0413, USA
Tel: +1-414-229-5037, Fax: +1-414-229-5530, E-mail: cfsiii@csd.uwm.ed

Stacey E. Shehin-Johnson, University of Connecticut, Storrs, CT 06269, USA

Akinori Shimada, Department of Veterinary Pathology, Tottori University, Tottori 680, Japan
Tel: +81-857-31-5423, Fax: +81-857-31-5424, E-mail: aki@agr.tottori-u.ac.jp

Hideaki Shimada, National Cancer Institute at the National Institute of Environmental Health Sciences111 Alexander Drive, PO Box 12233, Research Triangle Park, NC 27709, USA

Shohei Shimajiri, Department of Pathology and Cell Biology, University of Occupational and Environmental Health, School of Medicine, 1-1 Iseigaoka, Yahatanishi-ku, Kitakyushu, 807

Japan

Noriyuki Shiraishi, National Cancer Institute at the National Institute of Environmental Health Sciences111 Alexander Drive, PO Box 12233, Research Triangle Park, NC 27709, USA

Ge Shujun, Department of Biophysics, Beijing Medical University, Beijing, 100083, P.R. of China

L.D. Shultz, The Jackson Laboratory, 600 Main St., Bar Harbor, ME 04609,

Cuthbert O. Simpkins, Erie County Medical Center, Department of Surgery, 462 Grider St., Buffalo, NY 14215, USA
Tel: +1 716-8983846, Fax: +1 716-8985029 or 716-6311147, E-mail: cuthbert@acsu.buffalo.edu

Seema Somji, Department of Pathology, West Virginia University, P.O. Box 9203, Morgantown, West Virginia 26506-9203, USA

Fiona A. Stennard, Wellcome/CRC Institute, Tennis Court Rd, Cambridge CB 1QR, UK

Martin J. Stillman, Department of Chemistry, The University of Western Ontario, London, ON, N6A 5B7 CANADA
Tel: +1-519 661-3821, Fax: +1-519 661-3022, E-mail: martin.stillman@uwo.ca

M. Sturkenboom, Institute of Medical and Veterinary Science, Adelaide, SA, Australia

Stephen Stürzenbaum, School of Molecular and Medical Biosciences, University of Wales Cardiff, P.O. Box 911, Cardiff CF1 3US, UK

Yawara Sumi, Department of Chemistry, St. Marianna University School of Medicine, Kawasaki 216, Japan

Karl H. Summer, GSF-National Research Center for Environment and Health, Institute of Toxicology, Ingolstädter Landstr. 1, D-85764 Neuherberg, Germany
Tel: +49-89-3187-2664, Fax: +49-89-3187-3449, E-mail: summer@gsf.de

Junko S. Suzuki, Environmental Health Sciences Division, National Institute for Environmental Studies, 16-2, Onogawa, Tsukuba, Ibaraki 305, Japan

Kaoru Suzuki, Division of Hazard Assessment, National Institute of Industrial Health, 6-21-1, Nagao, Tama-ku, Kawasaki 214, Japan

Kazuo T. Suzuki, Faculty of Pharmaceutical Sciences, Chiba University, Inage, Chiba 263, Japan
Tel/Fax: +81-43-290-2891, E-mail: ktsuzuki@p.chiba-u.ac.jp

Keiji Suzuki, Division of Pathology, Department of Laboratory Sciences, Gunma University School of Health Sciences, 3-39-15 Maebashi, Gunma 371, Japan
Tel: +81-027-220-8936, Fax: +81-027-220-8999, E-mail: ksuzuki@news.sb.gunma-u.ac.jp

Mika Suzuki-Kurasaki, Department of Environmental Medicine and Informatics, Graduate School of Environmental Earth Science, Hokkaido University, Sapporo 060, Japan

Tadashi Suzuki, Department of Laboratory Sciences, Gunma University School of Health Sciences, 3-39-15, Showa machi, Maebashi, Gunma, 371, Japan
Tel: +81-27-220-8951, Fax: +81-27-220-8999, E-mail address: tasuzuki @news.sb.gunma-u.ac.jp

Boris I. Synsynys, Department of Ecology of the Institute of Nuclear Power Engineering, Studgorodok-1, Obninsk, 249020, Kaluga region, Russia

Atsushi Takeda, Department of Radiobiochemistry, School of Pharmaceutical Sciences, University of Shizuoka, 52-1 Yada, Shizuoka 422, Japan
Tel: +81-54-264-5700, Fax: +81-54-264-5705, E-mail: takedaa@ys7.u-shizuoka-ken.ac.jp

H. Tanaka, Department of Occupational Health and Toxicology, School of Allied Health Sciences, Kitasato University, 1-15-1 Kitasato, Sagamihara, Kanagawa 228, Japan

Hidenori Tanaka, Faculty of Pharmaceutical Sciences, Kinki University, 3-4-1 Kowakae, Higashi-Osaka, Osaka 577, Japan

Keiichi Tanaka, Faculty of Pharmaceutical Sciences, Osaka University, Yamada-oka, Suita, Osaka 565, Japan

Weifeng Tang, Department of Biomedical Sciences, College of Pharmacy, University of Rhode Island, Kingston, RI 02881, USA

Akihide Tanimoto, Department of Pathology and Cell Biology, University of Occupational and Environmental Health, School of Medicine, 1-1 Iseigaoka, Yahatanishi-ku, Kitakyushu, 807 Japan

David J. Tate, Jr., Sensory and Electrophysiology Research Unit, Touro Infirmary, New Orleans, Louisiana 70115, USA

Arata Teranishi, Arata Teranishi, Department of Public Health, Hyogo College of Medicine,

Nishinomiya, Hyogo, 663 Japan
Tel: +81-798-45-6566, Fax: +81-798-45-6567, E-mail: atera@hyo-med.ac.jp

D.J. Thiele, The University of Michigan Medical School, Ann Arbor, MI, USA

E.D. Thompson, T.H. Morgan School of Biological Sciences, 101 Morgan Building,
University of Kentucky, Lexington, Kentucky 40506-0225, USA

John H. Todd, Department of Pathology, West Virginia University, P.O. Box 9203,
Morgantown, West Virginia 26506-9203, USA

Setsuko Tohno, Laboratory of Cell Biology, Department of Anatomy, Nara Medical
University, 840 Shijo-Cho, Kashihara, Nara 634, Japan

Yoshiyuki Tohno, Laboratory of Cell Biology, Department of Anatomy, Nara Medical
University, 840 Shijo-Cho, Kashihara, Nara 634, Japan

Chiharu Tohyama, Environmental Health Sciences Division, National Institute for
Environmental Studies, 16-2, Onogawa, Tsukuba, Ibaraki 305, Japan
Tel: +81-298-50-2336; Fax: +81-298-56-4678, E-mail: ctohyama@nies.go.jp

Chiharu Tohyama, Environmental Health Sciences Division, NIES, Tsukuba, 305, Japan

Paul Trayhurn, Molecular Physiology Group, Rowett Research Institute, Greenburn Road,
Bucksburn, Aberdeen AB21 9SB, Scotland

Jun-ichi Ueda, National Institute of Radiological Sciences, Chiba 263, Japan

Takashi Uemura, Department of Veterinary Pathology, Tottori University, Tottori 680, Japan

Irena Ujević, Institute of Oceanography and Fisheries, HR-21000 Split, P.O. Box 500,
Croatia

Takashi Umemura, Department of Veterinary Pathology, Tottori University, Tottori 680,
Japan

Ken-ichi Urakami, Terumo Corporation, Research and Development Center, Terumo
Corporation, Kanagawa 259-01, Japan

Milan Vašák, Biochemisches Institut der Universität Zürich, Winterthurerstrasse 190, CH-
8057 Zürich, Switzerland
Tel.: +41-1-635-5552, Fax: +41-1-635-6805

Rubén A. Velázquez, Department of Veterinary PathoBiology, University of Minnesota, Saint Paul, Minnesota 55108, USA

Sudha Venkatesh, Department of Chemistry, University of Wisconsin-Milwaukee, Milwaukee, WI 53201, USA

Vesa Virtanen, European Commission, Joint Research Centre, Institute for Reference Materials and Measurements, Retieseweg, B-2440 Geel, Belgium
Tel. +32-14-571207, Fax: +32-14-584273, E-mail: virtanen@davinci.irmm.jrc.be

Hana Vodièková, Department of Chemistry, Faculty of Agronomy, Czech University of Agriculture, 165 21 Prague 6-Suchdol, Czech Republic

Michael P. Waalkes, National Cancer Institute at the National Institute of Environmental Health Sciences111 Alexander Drive, PO Box 12233, Research Triangle Park, NC 27709, USA

Waihei Chu, Department of Pharmacology, Toxicology and Therapeutics, University of Kansas Medical Center, G017B Breidenthal, 3901 Rainbow Blvd., Kansas City, KS 66160-7417

P. Walsh, RSMAS/MBF University of Miami, 4600 Rickenbacker Causeway, Miami, Florida 33149, USA

Ju-Feng Wang, Departments of Medicine, and Pharmacology and Toxicology, University of Louisville School of Medicine, 530 South Jackson St., Louisville, KY, 40202, USA

Wenqing Wang, Department of Technical Physics, Peking University, Beijing 100871, P.R. China
Fax: +861062751873, Tel: +861062752457, E-Mail: WANGWQ@SUN.IHEP.AC.CN

Zhiyong Wang, Diabetes Forschungsinstitut, Auf'm Hennekamp 65, D-40225 Düsseldorf, Germany

Ian Watt, School of Molecular and Medical Biosciences, University of Wales Cardiff, Cardiff, CF1 3US, Wales, UK

Adrian K. West, Biochemistry, University of Tasmania, GPO Box 252-58, Hobart, Australia 7001
Tel: +61-3-62262595, Fax: +61-3-62262703, E-mail: awest@utas.edu.au

Karen E. Wetterhahn, Department of Chemistry, Dartmouth College, Hanover, NH 03755, USA

Dean E. Wilcox, Department of Chemistry, Dartmouth College, Hanover, NH 03755, USA
Tel: +1 603-646-2874, Fax: +1 603-646-3946, E-mail: dean.wilcox@dartmouth.edu

Annemarie Willi, Universität Zürich, Biochemisches Institut, Winterthurerstr.190, CH-8057
Zürich, Switzerland

Allison K. Wilson, Center for Mechanistic Biology and Biotechnology, Argonne National
Laboratory, Argonne, IL 60439-4833, USA

Dennis R. Winge, University of Utah Health Sciences Center, Salt Lake City UT 84132,
USA

Elizabeth S. Woo, Department of Pharmacology, University of Pittsburgh School of
Medicine, Pittsburgh, Pennsylvania 15261, USA

Anne M. Wood, Trace Element and Gene Expression Group, Rowett Research Institute,
Greenburn Road, Bucksburn, Aberdeen AB21 9SB, Scotland

Hui-Yun Wu, Departments of Medicine, and Pharmacology and Toxicology, University of
Louisville School of Medicine, 530 South Jackson St., Louisville, KY, 40202, USA

Jilan Wu, Department of Technical Physics, Peking University, Beijing 100871, P.R. China

Licia N. Y. Wu, University of South Carolina, Dept. of Chemistry and Biochemistry, 730 S.
Main St., Columbia SC 29208, USA

Roy E. Wuthier, University of South Carolina, Dept. of Chemistry and Biochemistry, 730 S.
Main St., Columbia SC 29208, USA

Yi Xiong, National Laboratory of Protein Engineering, College of Life Sciences, Peking
University, Beijing 100871, P.R. China

Takashi Yamada, National Institute of Health Sciences, Kamiyoga 1-18-1, Setagaya, Tokyo
158, Japan

Rie Yamaguchi, Department of Environmental Medicine and Informatics, Graduate School of
Environmental Earth Science, Hokkaido University, Sapporo 060, Japan

Yoshiaki Yamano, Department of Metabolic Biochemistry, Tottori University, Tottori 680,
Japan

Futoshi Yamasaki, Department of Environmental Medicine and Informatics, Graduate School
of Environmental Earth Science, Hokkaido University, Sapporo 060, Japan

H. Yoshikawa, Department of Occupational Health and Toxicology, School of Allied Health Sciences, Kitasato University, 1-15-1 Kitasato, Sagamihara, Kanagawa 228, Japan

Minoru Yoshida, Department of Chemistry, St. Marianna University School of Medicine, 2-16-1 Sugao, Miyamae-ku, Kawasaki 216, Japan
Fax: +81-44-977-8133. E-mail: m2yosida@marianna-u.ac.jp

T. Yoshida, Department of Environmental Health, School of Medicine, Tokai University, Bohseidai, Isehara, Kanagawa 259-11, Japan

J. Youn, Department of Molecular and Cell Biology, University of Connecticut, Storrs, CT 06269, USA

Chih-Wen Yu, Institute of Radiation Biology, National Tsing Hua University, Hsinchu, Taiwan, Republic of China

Joseph Zaia, Department of Chemistry and Biochemistry, University of Maryland Baltimore County, 1000 Hilltop Circle, Baltimore, MD 21250, USA

Shu Zhou, Institute of Clinical Medical Sciences, China-Japan Friendship Hospital, Beijing 100029, P.R. China

Yanjiao Zhou, National Laboratory of Protein Engineering, College of Life Sciences, Peking University, Beijing 100871, P.R. China

THE FOURTH INTERNATIONAL METALLOTHIONEIN MEETING

President:
Curtis D. Klaassen, University of Kansas Medical Center, Kansas City, KS

Secretary/Treasurer:
Jie Liu, University of Kansas Medical Center, Kansas City, KS

Organizing Committee:
George M. Cherian, University of Western Ontario, London, Ontario, Canada
Bruce A. Fowler, University of Maryland, Baltimore, MD
Peter Goering, USFDA, Center for Devices and Radiological Health, MD
James D. Koropatnick, London Regional Cancer Institute, Ontario, Canada
F. William Sunderman, Jr, University of Connecticut, Farmington, CT
Kazuo T. Suzuki, University of Chiba, Inage, Chiba, Japan
Michael P. Waalkes, National Cancer Institute at NIEHS, Research Triangle Park, NC

Scientific Committee:
Glen Andrews, Kansas City, KS
Detmar Beyersmann, Bremen, Germany
Maryka H. Bhattacharyya, Argonne, IL
Guy Bordin, Joint Research Center IRMM, Retieseweg, Geel, Belgium
Ian Bremner, Aberdeen, United Kingdom
Manuchair Ebadi, Omaha, NE
Ernest Foulkes, Cincinnati, OH
Lashitew Gedamu, Alberta, Canada
J.P. Groten, Zeist, The Netherlands
Beth A. Hart, Burlington, Vermont
Juan Hidalgo, Barcelona, Spain
P.C. Huang, Baltimore, MD
Peter E. Hunziker, Zurich, Switzerland
Nobumasa Imura, Tokyo, Japan
Jeremias H.R. Kägi, Zurich, Switzerland
Y. James Kang, Louisville, KY
Masami Kimura, Tokyo, Japan
Yutaka Kojima, Sapporo, Japan
John S. Lazo, Pittsburg, PA
Akira Naganuma, Tokyo, Japan
Kazuo Nomiyama, Tochigi, Japan
Gunnar F. Nordberg, Umea, Sweden
Satomi Onosaka, Kobe, Japan
Richard D. Palmiter, Seattle, WA
David H. Petering, Milwaukee, WI
Allan M. Rofe, Adelaide, Australia
Toby G. Rossman, New York, NY
BingGen Ru, Beijing, China
Masao Sato, Fukushima, Japan
Carl Segiun, Quebec, Canada
Zahir A. Shaikh, Kingston, RI
Naoki Sugawara, Sapporo, Japan
Karl H. Summer, Neuherberg, Germany
Keiichi Tanaka, Osaka, Japan
Dennis J. Thiele, Ann Arbor, MI
Chiharu Tohyama, Ibaraki, Japan
Bert L. Vallee, Boston, MA
Milan Vašák, Zurich, Switzerland
Dennis R. Winge, Salt Lake City, UT

MT-97 Meeting Sponsors

The 4th International Metallothionein Meeting (MT-97) thanks the following organizations for their generous sponsorship of activities at the MT-97 in Kansas City, Missouri. The sponsors include:

Monsanto
National Institute of Environmental Health Science
National Cancer Institute
Bayer
Coca-Cola
Society of Toxicology
Searle
DowElanco
Nickel Producers Environmental Research Association
University of Kansas Medical Center
Westin Crown Center Hotel

Endorsed by the Society of Toxicology (USA) and the International Union of Toxicology (IUTOX)

Metallothionein nomenclature and structure

Nomenclature of metallothionein: Proposal for a revision

Yutaka Kojima, Pierre-Alain Binz and Jeremias H.R. Kägi

Biochemisches Institut der Universität Zürich, Winterthurerstrasse 190, CH-8057 Zürich, Switzerland

Summary. The present rules concerning the nomenclature of metallothionein were formulated at the Second International Meeting of Metallothionein in 1985. Since that time, a large number of additional amino acid sequences of metallothionein have become known yielding ample phylogenetic information. On the basis of this development, a new revision of the nomenclature is proposed.

The original recommendations concerning the nomenclature of metallothionein were made by the plenum of the First International Meeting on Metallothionein and Other Low Molecular Weight Metal-binding Proteins in 1978 [1]. Subsequently, a revised version was adopted by the Committee on the Nomenclature of Metallothionein appointed at the General Discussion Session of the Second International Meeting on Metallothionein and Other Low Molecular Weight Metal-binding Proteins in 1985 [2, 3]. Since that time, more than 170 amino acid sequences of metallothionein from over 50 species have become known and have been subjected to phylogenetic analysis [4]. On the basis of this new knowledge, the need emerges for another revision of the nomenclature of metallothionein. A proposal for modified recommendations is as follows:

Definition

The term metallothionein was used initially to designate the cadmium-, zinc- and copper-containing sulfur-rich protein from equine renal cortex [5].

This protein was characterized as follows [3, 6–8]:

- low molecular weight
- high metal content
- characteristic amino acid composition (high cysteine content, no aromatic amino acids nor histidine)
- unique amino acid sequence (characteristic distribution of cysteinyl residues such as cys-x-cys)
- spectroscopic features characteristic of metal thiolates (mercaptides)
- metal thiolate cluster

Definition of metallothionein

Polypeptides resembling equine renal metallothionein in several of their features can be designated as „metallothionein".

Nomenclature

Metallothionein

- The term metallothionein should be used for all polypeptides fitting the above definition. This designation should be employed in publication titles, running titles, and as a key word for computer filing.
- The metal-free form (apoprotein) may be designated either as apometallothionein or thionein.
- More specific terms, such as cadmium-metallothionein (or cadmium-thionein), zinc-metallothionein (or zinc-thionein) etc., are appropriate for metallothioneins that contain only one metal. The metal prefixed to the word thionein should not be used to designate the metal employed for metallothionein induction.
- The molar metal content can be specified by a subscript, i.e. Cd_7-metallothionein or Zn_7-metallothionein. When the metallothionein contains more than one metal, for example, 5.3 mol cadmium and 1.7 mol zinc per mol, terms such as (cadmium, zinc)-metallothionein, (cadmium, zinc)-thionein, $(Cd_{5.3}, Zn_{1.7})$-metallothionein and $(Cd_{5.3}, Zn_{1.7})$-thionein are recommended.
  The sequence in which the metals are listed should indicate the order of their relative amounts.
- In contexts where specification of the metal composition is not available or is of no special interest, the term metallothionein should be used.

Multiple forms of metallothionein

The nomenclature adopted is based on the recommendation of the Nomenclature of Multiple Forms of Enzymes, IUPAC-IUB Commission on Biochemical Nomenclature (CBN) [9].
- The terms multiple forms of metallothionein or isometallothioneins should be used as broad terms covering all metallothioneins occurring naturally in a single species. These terms should apply only to forms of metallothionein arising from genetically determined differences in primary structure and not to forms that differ only in metal composition or are derived by posttranslational modification of the same primary sequence.
- Different isometallothioneins should be defined by the term metallothionein followed by an arabic numeral. Where subdivisions occur, the major groups should be designated by numbers and the individual forms whithin the subdivision by letters (metallothionein-1A, metallothionein-1B, metallothionein-1C; metallothionein-2A, metallothionein-2B, etc.).

Metallothionein genes

The nomenclature, in general, follows that of multiple forms of metallothionein.
* For cases in which the isolation and sequencing of an isometallothionein-encoding gene precedes the identification, isolation and sequencing of the protein, the gene should be designated as metallothionein and specified, if feasible, both by a number indicating the pertaining isometallothionein subdivision as inferred from the nucleic acid sequence, and by a letter identifying the supposititious isoform encoded by the gene.
* Isometallothionein-encoding, but not expressed, sequences lacking introns or essential control sequences flanking the metallothionein-encoding regions will be designated as ψmetallothionein followed, if feasible, by an arabic numeral and a letter denoting the correspondence of the decoded amino acid sequence to an already known isometallothionein.

Classification

Metallothioneins can be divided into families.

Family (division) and subdivision (subfamily, subgroup) should preferably be based on criteria relating to phylogenetics and justifiable differences in protein and/or gene structure. These families and subdivisions were registered in the SWISS-PROT Protein Sequence database [4, 10].

Note

The subdivisional designations (subfamily, subgroup) should be used exclusively for metallothioneins checked by the above primary structure criteria and should not be employed for differentiating unidentified species arising during purification procedures.

References

1. Nordberg M, Kojima Y (eds) (1979) Metallothionein and other low molecular weight metal-binding proteins. *In*: JRH Kägi, M Nordberg (eds): *Metallothionein (Experientia, Suppl. 34).* Birkhäuser Verlag, Basel Boston Stuttgart, 41–124.
2. Fowler BA, Hildebrand CE, Kojima Y, Webb M (1987) Nomenclature of metallothionein. *In*: Y Kojima, JHR Kägi (eds): *Metallothionein II (Experientia, Suppl. 52).* Birkhäuser Verlag, Basel Boston Stuttgart, 19–22.
3. Kojima Y (1991) Definitions and nomenclature of metallothioneins. *In*: JF Riordan, BL Vallee (eds): *Methods in Enzymology, Volume 205.* Academic Press, San Diego New York Boston London Sydney Tokyo Toronto, 8–10.
4. Binz P-A, Kägi JHR (1997) *this volume.*
5. Kägi JHR, Vallee BL (1960) Metallothionein: a cadmium- and zinc-containing protein from equine renal cortex. *J Biol Chem* 235: 3460–3465.
6. Kägi JHR, Himmelhoch SR, Whanger PD, Bethune JL, Vallee BL (1974) Equine hepatic and renal metallothioneins. Purification, molecular weight, amino acid composition, and metal content. *J Biol Chem* 249: 3537–3542.
7. Kojima Y, Berger C, Vallee BL, Kägi JHR (1976) Amino-acid sequence of equine renal metallothionein-1B. *Proc. Natl. Acad Sci USA* 73: 3413–3417.

Y. Kojima et al.

8. Vašák M, Kägi JHR (1983) Spectroscopic properties of metallothionein. *In*: H Sigel (ed.): *Metal ions in Biological Systems*. Marcel Dekker, Inc., New York Basel, 213–273.
9. IUPAC-IUB Commission on Biochemical Nomenclature (CBN) (1977) Nomenclature of multiple forms of enzymes. Recommendations. *J.Biol Chem* 252: 5939–5941.
10. SWISS PROT Protein Sequence database, File metallo.txt.

Metallothionein IV
C. Klaassen (ed.)
© 1999 Birkhäuser Verlag Basel/Switzerland

Metallothionein: Molecular evolution and classification

Pierre-Alain Binz [1] and Jeremias H.R. Kägi[2]

[1]*Laboratoire Central de Chimie Clinique, Hôpital Cantonal Universitaire, rue Micheli-du-Crest 24, CH-1211 Geneve 14, Switzerland*
[2]*Biochemisches Institut der Universität Zürich, Winterthurerstrasse 190, CH-8057 Zürich, Switzerland*

Introduction

The nomenclature system for metallothionein was first adopted in 1978 [1] and extended in 1985 by introducing a subdivision of MTs into three classes [2]. According to this convention class I includes all proteinaceous MTs with locations of Cys closely related to those in the mammalian forms. Class II comprises proteinaceous MTs which lack this property. Class III subsumes metallopolyisopeptides containing gammaglutamyl-cysteinyl units resembling in their features proteinaceous MTs.

As the number of MT structures has grown and now approaches 200 sequences this subdivision has become inadequate. The length of the sequences, the amino acid compositions, the numbers and repartitions of the Cys are now known to vary appreciably. Moreover the gene sequences now available display a number of additional elements potentially related to functional properties of the MTs which are not included in the present classification criteria. Thus, to differentiate the MTs better we are proposing a new classification system based on sequence similarities and phylogenetic relationships. This system divides the MT superfamily into families of evolutionary related sequences which are further subdivided into subfamilies, subgroups and isolated isoforms and alleles. SWISS-PROT [3] provides in a specific document (metallo.txt) the description and definition of the proposed classification and an index of MT sequences.

As explained below the analysis started with (1) the screening of the protein and nucleotide databases for MT entries followed by the alignment of the empirically similar sequences into distinct groups. Then (2) phylogenetic algorithms and other comparison methods were used to critically establish the structural and evolutionary relationships among the protein and nucleotide MT sequences. Based on such information, families and subdivisions of families were (3) discerned with a number of common characteristics and (4) definitions were established for each level of subdivision. (5) As a consequence, modifications of the existing classification are proposed. (6) The feasibility of grouping MTs into clans on the basis of phenotypical features is also discussed. Finally (7) guidelines are given for the classification of new sequences.

Creation of the sets of evolutionary related sequences

Information on partial and total sequences of MT was collected by screening sequence databases (SWISS-PROT, PIR, OWL, GenBank, EMBL) and publications. After translation of the nucleotide sequences, removal of redundancy and checking for accuracy by consulting the original publications all sequences were deposited in SWISS-PROT. Multiple alignments were calculated first for sets of highly similar sequences. Then by using profile analysis and sequence pattern searches with PROSITE and by applying appropriate alignment procedures all MTs were segregated into a number of non-overlapping families of structurally related sequences ranging in size between 1 and more than 80 members.

Assessment of the phylogenetic relationships

Phylogenetic relationships of the protein and of the coding sequences were deduced by applying maximum parsimony, neighbor-joining, Fitch and Margoliash and maximum likelihood methods as provided in the PHYLIP package [4]. The bootstrap method was used for assessing the quality of the branching system. Subdivisions were deduced from the occurrence of distinct monophyletic groups in the phylogenetic trees (Fig. 1). In particular cases the phylogenetic analyses were also performed on alignments of untranslated portions of the corresponding genes, such as the 5'untranslated region of the mRNA (5'UTR). Pairwise comparison of gene sequences using dotplots was employed for the detection of relationships between intron sequences. The numbers, the positions and the directions of potential MREs in the 5'untranscribed portion of the genes (5'UT) of the vertebrate sequences were also informative.

Interpretation of the results: MT families, subfamilies and subgroups

The comprehensive analysis of the sequence data provides unambiguous models for the evolutionary development of MTs and allows thus for a natural grouping of the MTs into a number of families. Such families are characterized by the following features.
- a unique sequence pattern
- a taxonomic range
- a set of common sequence characteristics obtained from both amino acid and polynucleotide sequences
- a multiple sequence alignment (for families containing at least two members)
- a phylogenetic tree (for cases where the number of sequences exceeds 4 members)

Some of the MT-families allow a further subdivision into a number of phylogenetically separable subfamilies with different structural features or other genetic properties (i.e. similarity of 5'UTR, of introns sequences, or of number, position and direction of regulatory elements like MREs, and so one).

In some particular cases, based on phylogenetic separation, the subfamilies can be further subdivided into subgroups.

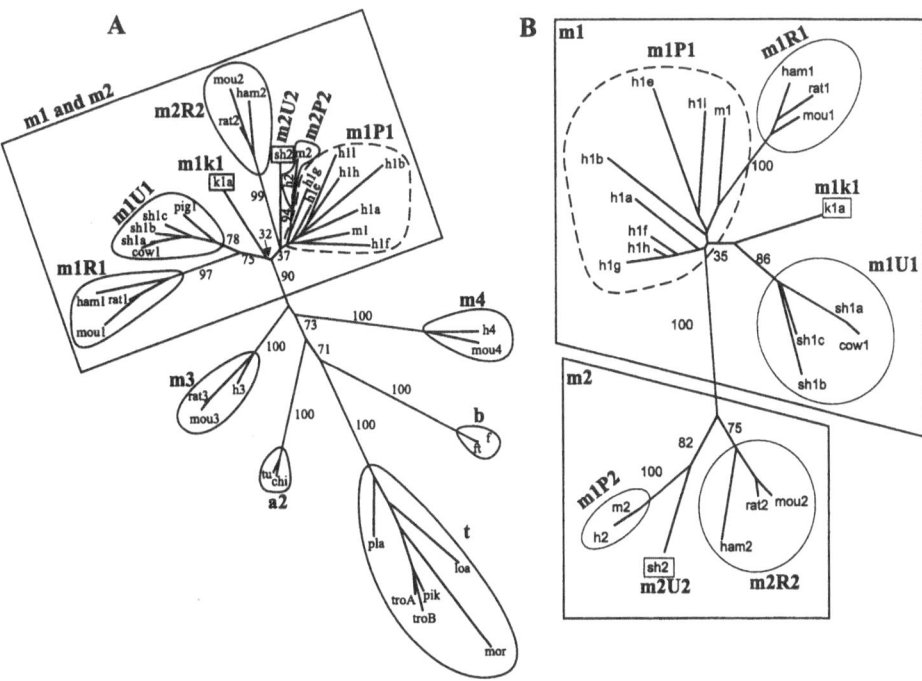

Figure 1. Phylogenetic trees of vertebrate MTs calculated from coding exon sequences (A) and from 5'UTR sequences (B). The unrooted trees serve to illustrate that analysis of different gene regions can yield complementary information on the molecular evolution of MT. (A) documents the subdivision of coding exon sequences of vertebrate MTs into encircled subfamilies and subgroups labelled as in Table 1 and Figure 2. The large rectangle encloses the crowded portion of the tree containing the sequences of the mammalian subfamilies m1 and m2. While the branching clearly separates several taxonomic subgroups of m1 and m2 the analysis did not yield the complete separation of the two paraloguous subfamilies. By contrast, in (B), the tree calculated from the 5'UTR of mammalian m1 and m2 MT-genes clearly demonstrates this subdivision in the mammalian phylogeny. Both trees were calculated by the method of Fitch and Margoliash implemented in the PHYLIP package. The numbers indicate bootstrap values serving as a measure of the statistical significance of the branching system. Because of the remaining ambiguities in the branching the m1P1 (primate MT-1) subgroup is encircled with a stippled line. In A the branch lengths in the rectangle were doubled for clearer reading. Sequences are represented with an abbreviation of the species name followed by the label of the corresponding isoform. The abbreviations used are: h, human; m, green monkey; sh, sheep; cow, bovine; pig, pig; rat, rat; mou, mouse; ham, hamster; tu, turkey; chi, chicken; f and ft, frog; pla, plaice; loa, stone loach; pik, pike; tro, rainbow trout; mor, atlantic cod

Definition of the metallothionein superfamily and its subdivisions

As before, the **metallothionein superfamily** is defined phenomenologically as comprising all polypeptides which resemble equine renal metallothionein in several of their features [1, 2]. Such general features are low molecular weight, high metal content, characteristic amino acid composition (high Cys content, low content of aromatic amino acid residues), unique amino acid sequence with characteristic distribution of Cys, i.e. Cys-X-Cys and spectroscopic manifestations characteristic of metal thiolate clusters.

Table 1. Set of presently defined MT families and subfamilies

Family 1:	vertebrate MTs	Family 7:	ciliata MTs
m1:	mammalian MT-1	ci:	ciliata MT
m2:	mammalian MT-2		
m3:	mammalian MT-3	Family 8:	fungi-I MTs
m4:	mammalian MT-4	f1:	fungi-I MT
m:	n.d. mammalian MT		
a1:	avian MT-1	Family 9:	fungi-II MTs
a2:	avian MT-2	f2:	fungi-II MT
a:	n.d. avian MT		
b:	batracian MT	Family 10:	fungi-III MTs
t:	teleost MT	f3:	fungi-III MT

Family 2:	mollusc MTs	Family 11:	fungi-IV MTs
mo1:	mussel MT-1	f4:	fungi-IV MT
mo2:	mussel MT-2		
mog:	gastropod MT	Family 12:	fungi-V MTs
mo:	n.d. mollusc MT	f5:	fungi-V MT

Family 3:	crustacean MTs	Family 13:	fungi-VI MTs
c1:	crustacean MT-1	f6:	fungi-VI MT
c2:	crustacean MT-2		
c:	n.d. crustacean MT	Family 14:	prokaryota MTs
		pr:	prokaryota MT

Family 4:	echinodermata MTs	Family 15:	planta MTs
e1:	echinodermata MT type 1	p1:	plant MT type 1
e2:	echinodermata MT type 2	p2:	plant MT type 2
		p2v:	plant MT type 2 variant, described
Family 5:	diptera MTs		as a clan of p2
d1:	diptera MT type 1	p3:	plant MT type 3
d2:	diptera MT type 2	p21:	plant MT type 2x1
		pec:	plant EC MT-like protein
Family 6:	nematoda MTs		
n1:	nematoda MT type 1	Family 99:	phytochelatins and other non-
n2:	nematoda MT type 2		proteinaceous MTs

The n.d. labelled subfamilies refer to sets of sequences for which because of insufficient phylogenetic or structural information an assignment to an already specified subfamily is not feasible. An index of the sequences and their classification is provided by the documentation file metallo.txt in SWISS-PROT. The detailed description of the classification criteria is available as a file published on the World Wide Web and linked from the file metallo.txt in SWISS-PROT. New sequences and propositions for their classification can be sent to SWISS-PROT. Proposed modifications of the classification criteria for new sequences can be sent to mtpage@bioc.unizh.ch.

A **MT family** subsumes MTs which share a particular set of sequence-specific characters. Members of a family can belong to only one family and are thought to be evolutionary related. The inclusion of a MT in a family presupposes that its amino acid sequence is alignable with that of all members. A common and exclusive sequence pattern, a profile and a phylogenetic tree can therefore be connected with each family. Each family is identified by its number and its taxonomic range. An example is Family 1: vertebrate MTs (see Tab. 1).

A **MT subfamily** contains MTs which in addition to the family characters share a set of more stringent phylogenetic features. These extra criteria are usually specific monophyletic relationships among the sequences of proteins and/or of nucleotide segments in the genes (5' or 3' untranscribed portion of the genes, 5' or 3' untranslated portions of the nucleotide sequences, exons, introns). If relevant other differentiating criteria can also be included, such as presence, conservation or repetition of sequence patterns. A subfamily is usually abbreviated with a letter character followed, if necessary, with an arabic number. An example is m1: mammalian MT-1 (see Tab. 1 and Fig. 2).

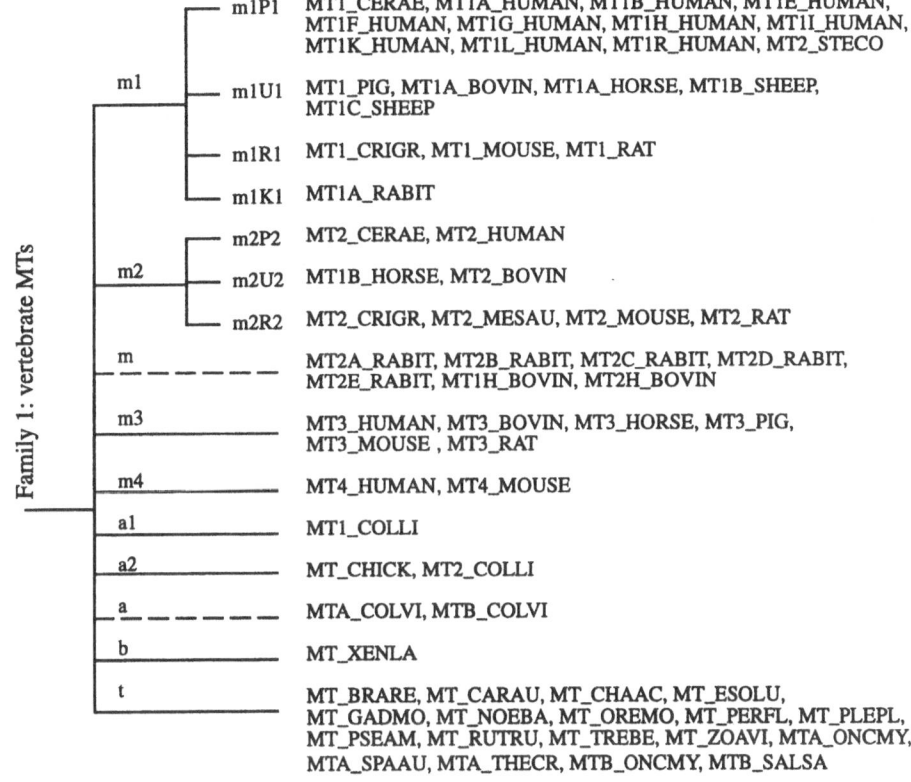

Family Subfamily Subgroup   Isoforms, SWISS-PROT entries

		m1P1	MT1_CERAE, MT1A_HUMAN, MT1B_HUMAN, MT1E_HUMAN, MT1F_HUMAN, MT1G_HUMAN, MT1H_HUMAN, MT1I_HUMAN, MT1K_HUMAN, MT1L_HUMAN, MT1R_HUMAN, MT2_STECO
	m1	m1U1	MT1_PIG, MT1A_BOVIN, MT1A_HORSE, MT1B_SHEEP, MT1C_SHEEP
		m1R1	MT1_CRIGR, MT1_MOUSE, MT1_RAT
		m1K1	MT1A_RABIT
		m2P2	MT2_CERAE, MT2_HUMAN
	m2	m2U2	MT1B_HORSE, MT2_BOVIN
		m2R2	MT2_CRIGR, MT2_MESAU, MT2_MOUSE, MT2_RAT
	m		MT2A_RABIT, MT2B_RABIT, MT2C_RABIT, MT2D_RABIT, MT2E_RABIT, MT1H_BOVIN, MT2H_BOVIN
	m3		MT3_HUMAN, MT3_BOVIN, MT3_HORSE, MT3_PIG, MT3_MOUSE , MT3_RAT
	m4		MT4_HUMAN, MT4_MOUSE
	a1		MT1_COLLI
	a2		MT_CHICK, MT2_COLLI
	a		MTA_COLVI, MTB_COLVI
	b		MT_XENLA
	t		MT_BRARE, MT_CARAU, MT_CHAAC, MT_ESOLU, MT_GADMO, MT_NOEBA, MT_OREMO, MT_PERFL, MT_PLEPL, MT_PSEAM, MT_RUTRU, MT_TREBE, MT_ZOAVI, MTA_ONCMY, MTA_SPAAU, MTA_THECR, MTB_ONCMY, MTB_SALSA

Family 1: vertebrate MTs

Figure 2. Proposed classification of the Family1: vertebrate MTs. The subfamilies are abbreviated as in Table 1 and as described in the SWISS-PROT database. The subgroups are abbreviated as follows: m1P1: primate and dolphin MT1; m1U1: ungulate MT-1; m1R1: rodent MT-1; m1k1: lagomorph MT-1; m2P2: primate MT-2; m2U2: ungulate MT-2; m1R2: rodent MT-2. The dotted lines indicate the subfamilies containing sequences not as yet unambiguously classified. The sequences are therefore temporarily assigned to these incompletely specified subfamilies. The sequences are represented by the corresponding SWISS-PROT identification codes.

A **MT subgroup** represents, as a result of statistically validated phylogenetic analyses, a branch of MT sequences of a subfamily which is clearly distinguishable in a tree by its monophyletic character. An example is m2U2: ungulate MT-2, subgroup of the m2 subfamily (see Fig. 2).

Isoforms or allelic forms are specifiable as members of subgroups, subfamilies and families. They are named using the nomenclature system defined in [5], i.e. human MT-1E.

In addition, in cases where it is justifiable, one can define **clans**. A clan is a set of partial or total amino acid or polynucleotide sequences, subgroups, subfamilies, families or combinations of them which share characters not defined by the above classification criteria. They can be related to common spatial structure, thermodynamic properties, metal binding properties, functionally related characters or other relevant features. A clan is defined by the property common to its members. The abbreviation should reflect this property.

Proposals for modifications of the classification of MTs

The current division of the MTs into three classes should be replaced by the new classification into families. The class I and class II sequences thus far known are distinguishable into 15 families (see Table 1). The presently as class III defined metallopolyisopeptides can be classified as a special MT-family, i.e. Family 99: phytochelatins and other non-proteinaceous MTs.

The criteria for differentiating the vertebrate MT-1 and MT-2 subforms should also be modified (see Fig. 1 and Fig. 2). The differentiation was based initially on the order of their elution from ion exchangers, and subsequently on the absence or the presence of an acidic amino acid at position 10 or 11 of the polypeptide chain causing the charge difference [6]. At the time this definition was adopted, no spatial structures were determined and the existence of the additional vertebrate subforms MT-3 and MT-4 was as yet unknown. The phylogenetic analyses of the vertebrate sequences have now based this division on firmer grounds and established the mammalian MT-1, MT-2, MT-3 and MT-4 as MT subfamilies which separated prior to the radiation of the mammalian order [7]. The subdivision into MT-1 and MT-2 is clearly demonstrated by the specific differences in the introns and the 5'UTR and 5'UT regions of the mammalian gene sequences. Interestingly, owing to the lack of sufficient informative sites, the coding portions of the genes and the protein sequences alone do not permit an unambiguous discrimination between the two subfamilies. The sole phenotypical indication of difference between the members of these two subfamilies is the presence and absence of a net negative charge located at the tip of the N-terminal domain of MT-2 and of MT-1, respectively. The phylogenetic separation of the mammalian MT-1 and MT-2 subfamilies is unrelated to the MT-1 and MT-2 subfamilies of other vertebrate species.

The detailed classification criteria can be obtained as a file published on the World Wide Web and linked from the file metallo.txt in SWISS-PROT.

Classification of MT-clans

While the principal classification system of the MTs must be based on phylogenetic analysis, the option of defining clans as complementary classification collectives provides a much needed means to catalogue and identify the MTs on the basis of their diverse phenotypical features. Thus, MTs sharing certain compositional or structural similarities or related chemical or biological properties can be assembled to clans. When defining a new clan, the relevant common features of its members, its allocated number, and name and the list of its members should be added to the file metallo.txt in SWISS-PROT.

Classification of new sequences

For classifying a new sequence the following procedure may be used: If the sequence is a nucleotide sequence, submit it to EMBL or GenBank preferably. Submit its translation to SWISS-PROT. If the sequence is determined as a protein primary structure, submit it to SWISS-PROT. Search for the possibility of placing the sequence into an existing family. Search the list of sequence patterns for a pattern matching with your sequence. For an additional and complementary control, perform a Fasta search, a Blast search or a profile analysis. If matching is unsuccessful, define a new family for this sequence. Otherwise align the sequence with the other members of its family. Check the criteria of the description of the family and try to adjust them if needed. Perform a phylogenetic analysis in order to position the new sequence in the tree. If the family contains subdivisions, check if the sequence matches with one of them. Otherwise adjust the criteria of the best matching subdivision or, if needed, specify a new one. For an update of the classification submit the sequence and the suggested subdivision to SWISS-PROT together with your proposals for a revision of the specifications of the subdivision to mtpage@bioc.unizh.ch or to SWISS-PROT.

Acknowledgements
We are grateful to Amos Bairoch for his help in the design of the classification and description file in SWISS-PROT and to Yutaka Kojima for his critical review.

References

[1] Kägi JHR, Kojima Y (1979) Nomenclature of metallothionein: a proposal. *Exp Suppl* 34: 141–142.
[2] Fowler BA, Hildebrand CE, Kojima Y, Webb M (1987) Nomenclature of metallothionein. *Exp Suppl* 52: 19–22.
[3] Bairoch A, Apweiler R (1996) The SWISS-PROT protein sequence data bank and its new supplement TrEMBL. *Nucl Acid Res* 24: 21–25.
[4] Felsenstein J (1988) Phylogenies from molecular sequences: inference and reliability Annu Rev Gent 22: 521–565 b Felsenstein J (1993) PHYLIP (Phylogeny Inference Package) version 3.5c. Distributed by the author. Department of Genetics, University of Washington, Seattle.
[5] Kojima Y, Binz P-A, Kägi JHR (1997) Nomenclature of metallothionein: Proposal for a revision. *This volume*.
[6] Kägi JHR, Kojima Y (1987) Chemistry and Biochemistry of metallothionein. *Exp Suppl* 52: 25–61.
[7] Binz P-A (1996) Metallothioneins: studies on molecular evolution and on the spectroscopic properties of its metal thiolate clusters *PhD thesis*, University of Zürich, Switzerland.

Metallothionein IV
C. Klaassen (ed.)
© 1999 Birkhäuser Verlag Basel/Switzerland

Structural and biological studies on native bovine Cu,Zn-metallothionein-3

Milan Vašák[1], Ralf Bogumil[1], Peter Faller[1], Daniel W. Hasler[1], Pierre-Alain Binz[1], Stefan Klauser[1], Arend Bruinink[2], John M. Charnock[3] and C. David Garner[3]

[1]*Institute of Biochemistry, University of Zurich, Switzerland*
[2]*Institute of Toxicology, ETH and University of Zurich, Switzerland,*
[3]*Department of Chemistry, University of Manchester, England*

Introduction

Following the discovery of the neuronal growth inhibitory factor, classified as metallothionein-3 (MT-3), by Uchida et. al. 1991 [1] we have focussed our research on the structure-function analysis of this protein. MT-3 is a central nervous system (CNS) specific metallothionein-like protein, which has been linked to Alzheimer's disease (AD) [1]. The amino acid sequence of human MT-3 (68 amino acids) exhibits 70% sequence identity to mammalian metallothioneins (MT-1 and MT-2) (61 amino acids), including the preserved array of 20 Cys residues. In MT-1 and MT-2 these 20 Cys residues are involved in the formation of two metal-thiolate clusters, i.e. $Me^{II}_3(Cys)_9$ and $Me^{II}_4(Cys)_{11}$ [2]. Relative to mammalian MT-1 and MT-2 sequences, MT-3 contains two conserved inserts, a single Thr in the N-terminal and a Glu-rich hexapeptide in the C-terminal region (see Fig. 1). Despite their high sequence identity only MT-3 exhibits a growth inhibitory activity in neuronal cell culture studies, indicating that the – so far unknown – structure of MT-3 must differ from that of the other MTs. Furthermore, it has been demonstrated that the N-terminal metal binding domain MT-3(1–32) is sufficient for the growth inhibitory activity [3, 6]. In contrast to MTs, which usually contain 7 Zn(II) ions, native MT-3 isolated from bovine brain possesses an unusual metal ion composition, i.e. 4–5 Cu(I) and 2–2.5 Zn(II) per polypeptide chain, suggesting the presence of different cluster structures in this protein [4].

Using a variety of spectroscopic techniques we have studied both native MT-3 isolated from bovine brain as well as a synthetic peptide corresponding to the biologically active N-terminal domain of human MT-3 (residues 1–32). These studies provided information regarding the metal binding properties, cluster formation and cluster structures in these MT-3 derivatives allowing inferences to be drawn as to unique structural features present in MT-3, likely related to its growth inhibitory activity. In parallel to the structure oriented studies, we have been developing a new assay system to assess the biological activity of MT-3. This article provides a short overview of the current state of both the spectroscopic/structural and the biological studies on MT-3. Details of the experimental work can be obtained from the original references.

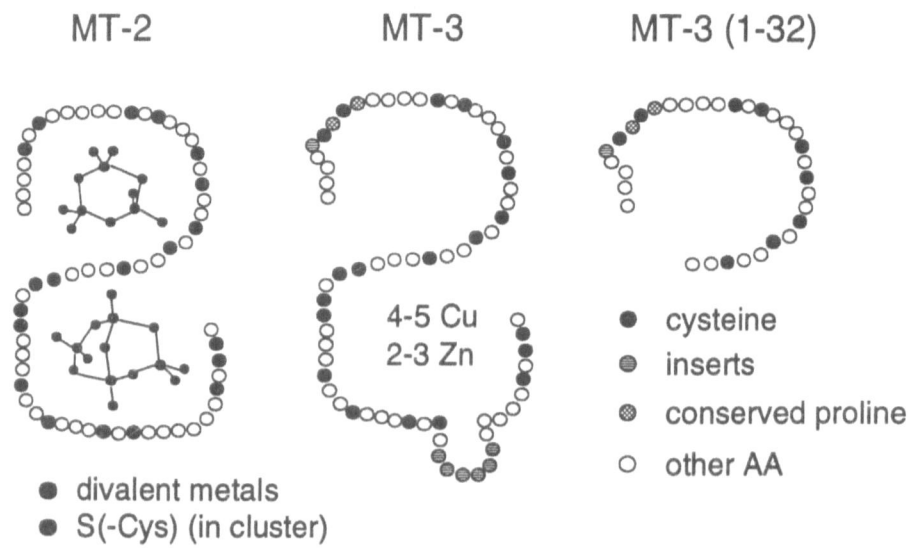

Figure 1. Schematic comparison of MT-2, MT-3 and the N-terminal domain of MT-3 (1–32).

Biological studies

So far very little is known about the origin of the biological activity of MT-3 and about possible binding partners of MT-3 in the brain. A major problem is that there is no well established test system, which would allow a good quantification of biological activity important for comparative studies [1]. In the original neural assay, the survival of low-density cell cultures of neonatal rat cortical neurons was assessed in the presence of AD brain extract and increasing Cu,Zn-MT-3 concentrations. In the absence of brain extract, only a few cells survived over the testing period of 3–5 days. Although using this assay a biological activity of recombinant Zn_7-MT-3 was also found, no comparative studies with native Cu,Zn-MT-3 exist. Therefore, we have been developing a different assay system using serum-free flat sedimented cultures of embryonic chick cerebral cells in which the AD brain extract is substituted by marmoset hippocampal extract [5, unpublished observations]. Results of our first studies, in which native Cu,Zn-MT-3 isolated from bovine brain was used, are summarized in Figure 2. As revealed by the upper panel, the native Cu,Zn-MT-3 exhibits a concentration dependent inhibitory effect in cerebral cell cultures (followed by mitochondrial MTT-activity), whereas no effect was observed in control cells (neural retina cell cultures, results not shown), indicating that the MT-3 effect is tissue specific [unpublished observations]. The lower panel shows that the number of viable cells in the cell culture (measured by neutral red uptake) and the total cell mass was unaffected, clearly establishing that the native Cu,Zn-MT-3 is nontoxic to neural cells. The latter result contrast the previous report in which Cu(I) or Cu(I) and Zn(II) containing MT-3, obtained by a metal reconstitution of the recombinant material, were found to be cytotoxic [6].

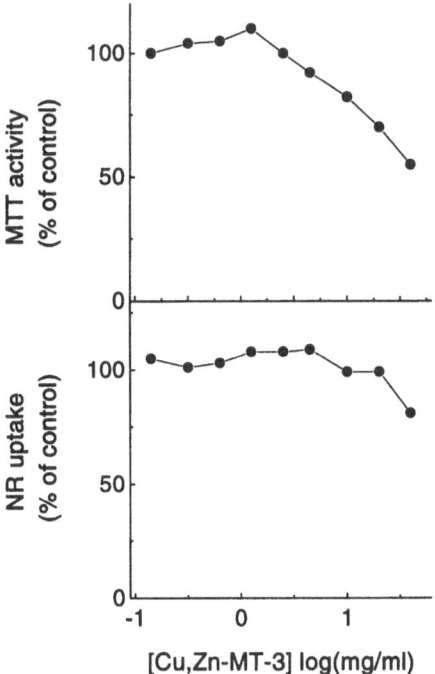

Figure 2. Effect of native Cu,Zn-MT-3 on chick cerebral cell cultures in the presence of marmoset hippocampal extract. For details see text

Further studies will include a comparison of the biological activity of native Cu,Zn-MT-3 with those of different metal derivatives of recombinant MT-3 and of the synthetic N-terminal domain (MT-3, 1–32).

Structural studies

Several spectroscopic techniques like electronic absorption, circular dichroism (CD), magnetic circular dichroism (MCD), and in the case of Cu(I) complexes also luminescence spectroscopy can be used to study the formation and organization of the metal-thiolate clusters in MTs [2]. Furthermore, a more detailed structural information about the metal centres in MT-3 can be obtained from extended X-ray absorption fine structure (EXAFS) studies. Analysis of EXAFS spectra can provide an insight into the coordination geometry of each metal type, which is derived from the measured metal-ligand distances (first shell backscattering). The analysis of the second shell backscattering affords further information about the metal-metal distances in cluster structures [7]. The biologically relevant N-terminal domain of MT-3 (1–32) was synthesized and used as a model system.

Using spectroscopic titration experiments, the metal-thiolate clusters formed both with monovalent (Cu(I)) and divalent metal ions (Zn(II), Cd(II)) in the N-terminal domain of human MT-3 (1–32) were investigated [8]. The spectral features of these metal derivatives were compared with those obtained with the native Cu,Zn-MT-3 [4]. In the case of the N-terminal domain, the titration with Cu(I) ions results in the formation of two well defined clusters involving all 9 cysteine ligands of GIF(1–32), i.e. Cu_4S_9- and Cu_6S_9-clusters. The respective electronic absorption and CD-spectra are shown in Figure 3 and the Cu(I) luminescence spectra in Figure 4. For a comparison, the corresponding spectra of native Cu_{4-5},Zn_{3-2}-MT-3 are also included. The absorption and CD-bands in the high energy region (below 280 nm) originate predominantly from CysS-Cu(I) ligand-to-metal charge-transfer (LMCT) bands, whereas the spectral features in the low energy region (above 300 nm), showing weak molar absorptivities but strong CD bands, originate from formally spin-forbidden 3d-4s metal cluster-centered transitions brought about by Cu(I)-Cu(I) interactions in polynuclear Cu(I) complexes. Thus, the occur-

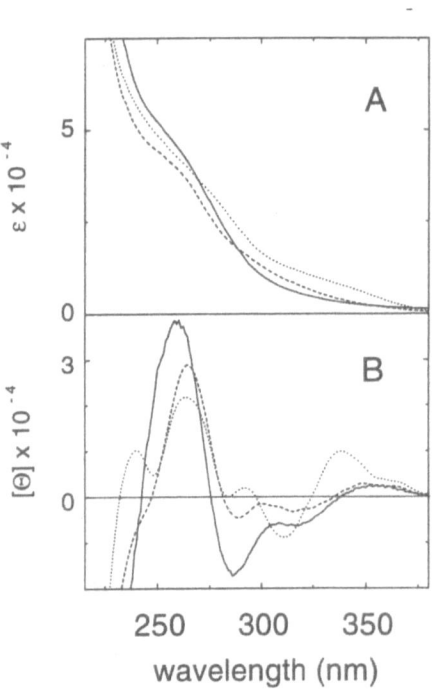
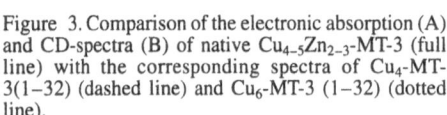
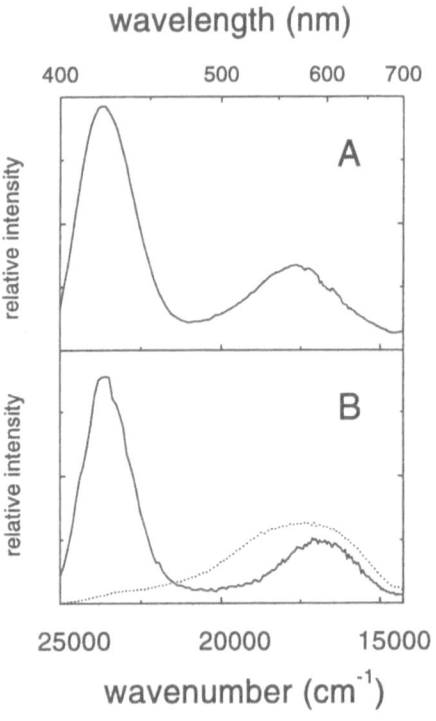

Figure 3. Comparison of the electronic absorption (A) and CD-spectra (B) of native $Cu_{4-5}Zn_{2-3}$-MT-3 (full line) with the corresponding spectra of Cu_4-MT-3(1–32) (dashed line) and Cu_6-MT-3 (1–32) (dotted line).

Figure 4. Comparison of the low temperature (77 K) luminescence emission spectra of native $Cu_{4-5}Zn_{2-3}$-MT-3 (A) with the corresponding spectra of Cu_4-MT-3(1–32) (B, full line) and Cu_6-MT-3 (1–32) (B, dotted). Excitation was at 300 nm.

rence of low-energy features (Fig. 3) indicates the existence of a Cu(I)-cluster in all three protein structures studied.

Cu(I) containing MTs can also be studied by luminescence emission spectroscopy, since upon excitation in the UV (at 300 nm) they emit one or more largely red-shifted luminescence bands mainly attributable to the Cu(I)-thiolate chromophore. The shape and position of these bands, are influenced by the Cu(I) coordination geometry. In Figure 4 a close correspondence between the spectra of the Cu_4-MT-3 (1–32) peptide and the native Cu_{4-5},Zn_{2-3}-MT-3 is seen, whereas the Cu_6-MT-3 (1–32) peptide exibits different spectral features. In the corresponding electronic absorption and CD spectra (Fig. 3), a close similarity between native MT-3 and the Cu_4-MT-3 (1–32) peptide also exists. Moreover, the Cu_4S_9-cluster in Cu_4-MT-3 (1–32) shows a remarkable stability to air oxidation, a property also found with native Cu,Zn-MT-3 but not with the Cu_6S_9-cluster. Overall, the comparison of the spectral features allows to propose the presence of a Cu_4S_9-cluster in native Cu_{4-5},Zn_{2-3}-MT-3.

These conclusions are also supported by the EXAFS studies of these MT-3 derivatives. In Table 1 the results of the EXAFS investigations performed at the Daresbury Synchrotron Radiation Laboratory (U.K.) are summarized.

Table 1. Summary of the EXAFS results at the Cu- and Zn-K-edge

		Cu K-edge			Zn K-edge		
Sample	Scatterer	N^a (Å)	Distance		Scatterer	N^a (Å)	Distance
Cu,Zn-MT-3	S	2.5	2.26		S	3.8	2.34
	Cu	0.4	2.67				
Cu_4-peptide	S	2.8	2.24				
	Cu	1.3	2.66				
	Cu	1.7	4				
Zn_3-peptide					S	3.9	2.33
and					S	1.1	2.86
Zn_7-MT-3					Zn	0.3	3.27

^{a}N = Number of backscatterers (±1)

Native Cu,Zn-MT-3, Zn(II) substituted MT-3 as well as the N-terminal peptide containing either Cu(I) or Zn(II) were analyzed [unpublished observations]. The metal K-edge EXAFS reveals that both Cu(I) and Zn(II) are exclusively ligated by thiolates with a primarily trigonal Cu(I) and a tetrahedral Zn(II) cordination. The presence of a Cu⋯Cu distance of ca. 2.7 Å in native Cu,Zn-MT-3 and the Cu_4-MT-3 (1–32) peptide provides clear evidence for a cluster structure. The Cu⋯Cu distance is in the same range as that observed in Cu-K-edge EXAFS of other Cu-MTs [9]. The absence of a backscattering contribution of ca. 2.7 Å at the Zn K-edge in native Cu,Zn-MT-3 (Tab. 1), strongly suggest that no mixed Cu,Zn-cluster is formed, rather

Zn ⋯ Zn   3.24 Å

Figure 5. Proposed structure for the Zn$_3$-cluster in Zn-MT-3.

that the Cu(I) and Zn(II) ions are organized in separate clusters. An unexpected and interesting result is the Zn-backscattering interaction at 3.28 Å found in the Zn-substituted MT-3 form and in the N-terminal peptide loaded with 3 Zn (Tab. 1). Such a short Zn⋯Zn distance is not consistent with the Zn$_3$-cluster usually found in other mammalian MTs, where a Zn⋯Zn distance of about 3.8 Å exits. In recent Zn-K-edge EXAFS studies of the transcriptional activator PPR1, which contains a binuclear (Zn$_2$(Cys)$_6$ cluster, a Zn⋯Zn distance of 3.16 Å has been determined [10]. The presence of two Cys bridges between the metal ions is responsible for the short Zn⋯Zn distance. In Figure 5 a tentative structure for the Zn$_3$-cluster in Zn$_7$-MT-3 is shown, which is consistent with the EXAFS results. Note, that an additional Zn-sulfur backscattering interaction at ca. 2.9 Å has also been detected (Tab. 1). In this model a binuclear cluster as found in PPR1 is linked to a third tetrahedral Zn-site. It is likely that two conserved prolyl residues in MT-3 are responsible for a different polypeptide fold and hence cluster structure compared to that determined for MT-1 or MT-2. Clearly, more studies are necessary to confirm the presence of this novel type of cluster and its relationship to the specific biological activity of MT-3.

Acknowledgement
This work was supported by Swiss National Science Foundation Grant 31-49460.96, the EMDO-Stiftung and by the Stipendienfonds der Basler Chemischen Industrie.

References

1. Uchida Y, Takio K, Titani Ihara Y, Tomonaga M (1991) The growth inhibitory factor that is deficient in the Alzheimer's disease is a 68 amino acid metallothionein-like protein. *Neuron 7*: 337–347.
2. Vašák M, Kägi JHR (1994) Metallothioneins. *In*: RB King (ed.): *Encyclopedia of Inorganic Chemistry*, John Wiley and Sons Ltd., Vol 4: 2229–2241.
3. Uchida Y, Ihara Y (1995) The N-terminal portion of the growth inhibitory factor is sufficient for biological activity. *J Biol Chem 270*: 3365–3369.
4. Bogumil R, Faller PA, Pountney DL, Vašák M (1996) Evidence for Cu(I) clusters and Zn(II) clusters in neuronal growth inhibitory factor isolated from bovine brain. *Eur J Biochem 238*: 698–705.
5. Bruinink A, Reiser P (1991) Ontogeny of MAP2 and GFAP antigens in primary cultures of embryonic chick brain. Effect of substratum, oxygen tension, serum and Ara-C. *Int J Dev Neurosci 9*: 269–279.
6. Sewell AK, Jensen LT, Erikson JC, Palmiter RD, Winge DR (1995) Bioactivity of metallothionein-3 correlates with its novel β-domain sequence rather than metal binding properties. *Biochemistry 34*: 4740–4747.
7. Garner CD (1991) X-ray absorption spectroscopy and the structure of transition metal centers in proteins. *Adv Inorg Chem 36*: 303–339.
8. Faller P, Vašák M. (1997) Distinct metal-thiolate clusters in the N-terminal domain of neuronal growth inhibitory factor (GIF). *Biochemistry* 36: 13341–13348.
9. Winge DR, Dameron CT, George GN (1994) The metallothionein structural motif in gene expression. *Adv Inorg Biochem 10*: 1–48.
10. Ball LJ, Diakun GP, Gadhavi PL, Young NA, Armstrong EM, Garner CD, Laue ED (1995) Zinc co-ordination in the DNA binding domain of the yeast transcription activator PPR1. *FEBS Lett 358*: 278–282.

Metallothionein IV
C. Klaassen (ed.)
© 1999 Birkhäuser Verlag Basel/Switzerland

Circular dichroism, emission, and exafs studies of Ag(I), Cd(II), Cu(I), and Hg(II) binding to metallothioneins and modeling the metal binding site

Martin J. Stillman, Anna Rae Green, Ziqi Gui, David Fowle and P. Anthony Presta

Department of Chemistry, The University of Western Ontario, London, ON, N6A 5B7 CANADA

Summary. Spectroscopic evidence supports the formation of metal-thiolate cluster structures involving 7, 12, and 18 metals in the 20-cysteine mammalian metallothioneins. The three-dimensional tertiary structure of the protein is established completely by the presence of the cross-linking S_{cys}-M-S_{cys} bonds that form a network inside a cage formed by the connecting peptide chain. Circular dichroism (CD) and luminescence spectra provide rich detail of a complicated metal binding chemistry when metals are added directly to the metal free or zinc-containing protein. CD spectral data unambiguously identify key metal to protein stoichiometric ratios that result in well-defined structures. Emission spectra in the 450–750 nm region have been reported for metallothioneins containing Ag(I), Au(I), Cu(I), and Pt(II), at both room temperature and cryogenic temperatures. The luminescence of Cu-MT can also be detected directly from mammalian and yeast cells. Analysis of the temporal changes in the emission spectrum of Cu_n-MT (n = 1–20) in the 600 nm region provides direct evidence for mobility of Cu(I) atoms within the metal binding sites in mammalian metallothionein. Energy-minimized structures for Zn_7-MT, Cd_7-MT, Cu_{12}-MT, and Hg_7-MT have been calculated using molecular mechanics (MM2) techniques. Deep crevices that expose the metal-thiolate clusters are seen in each structure.

Introduction

The primary sequence for the mammalian protein is dominated by the presence of 20 cysteines out of a 61 (MT 1) or 62 (MT 2A) peptide chain [1]. *In vivo* it is generally the metals in groups 11 and 12 (Cu(I), Zn(II), Cd(II), and Hg(II)) that bind, however, *in vitro*, in addition, metals as diverse as Ag(I), Au(I), Bi(III), Co(II), Fe(II), Pb(II), Pt(II), and Tc(IV), bind with stoichiometries of 7, 12, or even 18 [2–6]. The precise determination of the stoichiometric ratio between the bound metals and the number of accessible cysteinyl sulfurs (S_{cys}) is critical in our understanding of the chemistry, biochemistry, and physiological chemistry of this exceptional protein. The absence of aromatic amino acids in the primary sequence allows access to the sulfur to metal charge transfer transitions between 220 and 400 nm for UV-absorption, circular dichroism (CD), magnetic circular dichroism (MCD) and emission spectroscopies [2–6].

The three-dimensional structure of *metal-free* metallothionein is essentially that of a random chain, however, spectroscopic measurements from a variety of techniques, including [113]Cd NMR, ^{1}H NMR, and x-ray crystallography, show that metal binding involves a metal-thiolate cluster binding site with both bridging and terminal thiolate groups [7, 8]. Formation of S_{cys}-M-S_{cys} bonds as part of digonal (2-), trigonal (3-) or tetrahedral (4-) co-ordination, cross-links the peptide chain resulting in formation of a three-dimensional metal binding site. *This means that the tertiary structure of the metallated protein is dominated by metal-thiolate clusters*, bound in a single domain for the yeast or fungus proteins and in two domains for mammalian and crustacean proteins [5–8].

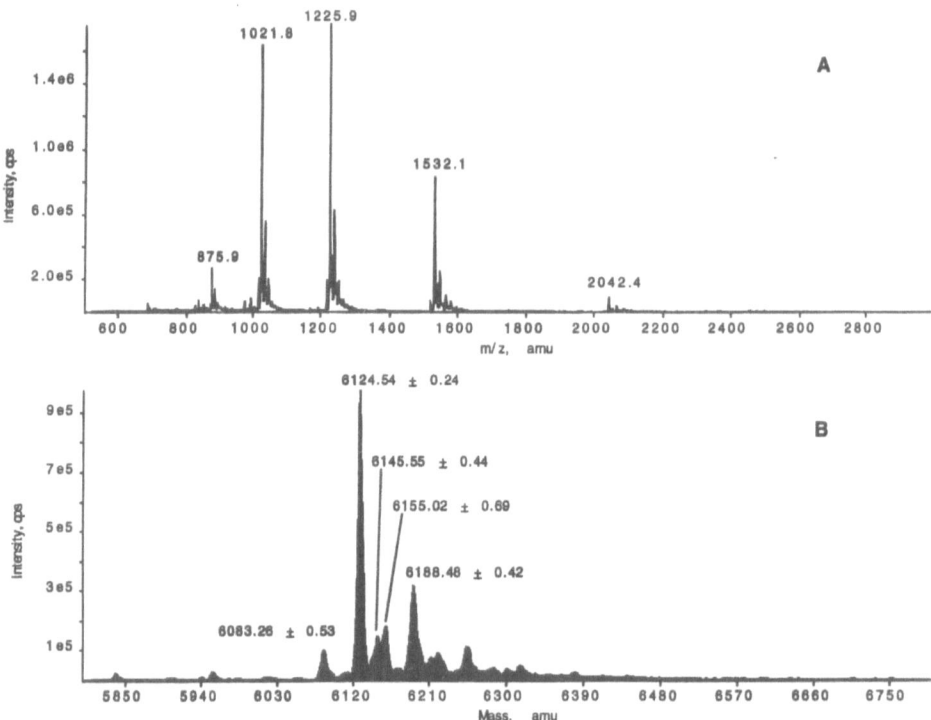

Figure 1. Electrospray mass spectrum of metal-free, rabbit liver MT 2A. (A) The recorded spectrum, the proto-nated species distribute across the 3+ (2042), 4+ (1532), 5+ (1226), 6+ (1022), and 7+ (876) ions; (B) the calcu-lated parent ions. Reproduced with permission [10].

Isolation and purity

Metallothionein was induced in rabbits and isolated as previously described [9–13]. Zn_7-MT 1 and Zn_7-MT 2 were further purified using DEAE anion exchange cartridges (MemSep LC-100, Millipore) [10]. Electrospray mass spectra of the NH- and N-acetylated terminal proteins were obtained using lyophilized protein samples dissolved in dilute formic acid. Figure 1 shows the mass spectral data obtained for rabbit liver MT 2A [10]. Each isoform of rabbit liver MT com-prises 1 of 6 co-eluting subisoforms [1]. Essentially complete isolation of Zn_7-MT 2A is achieved on the MemSep anion exchange cartridge with only minor components identified as MT 2B and 2C. ES-MS data show that the fraction identified as MT 1 comprises substantial amounts of isoforms 2D and 2E while Zn_7-MT 2A is the sole protein present with a mass of 6124.7) ± 0.7 Da [10]. In current studies, the fraction of MT 2A prepared using this technique has increased to close to 100% (P.A. Presta, D. Beerna, M.J. Stillman, unpublished data).

Spectroscopic properties: probing the metal-binding reaction

The spectroscopic signatures of metal binding to the metal-free and metalated proteins with Ag(I), Au(I), Bi(III), Cd(II), Cu(I), Co(II), Fe(II), and Hg(II) have been described in a number of papers [2–4]. In general, two strategies are used when studying the metal binding reactions of metallothioneins *in vitro*:

(i) Metals bind directly to the metal-free or apo-protein.

(ii) Metals with greater binding constants than Zn(II) are added to displace the Zn(II) in Zn_7-MT, because the binding constants follow the general order Hg(II) > Ag(I) > Cu(I) > Cd(II) > Zn(II) [2–4, 9, 11–13]. Because Zn(II) is tetrahedrally coordinated by the thiolate groups, considerable reorganization must occur in the binding site to accommodate metals like Cu(I) and Ag(I), metals that generally exhibit trigonal or digonal coordination geometries.

CD, MCD, emission, and XAFS spectroscopies are metal and S_{cys} dependent [3, 4] so that the spectral data report on the metal-sulfur coordination directly from the metal binding site. The emission from cellular systems and animal organs is sensitive to the *in vivo* concentration of Cu(I) [18, 19]. The emission spectrum has also provided detailed information about the metal binding process because non-luminescent Cu(I) becomes luminescent when bound to S_{cys} [3, 4]. Temporal measurements of the Cu-S emission intensity recorded in the 600 nm region can be used to follow the changes in the environment of Cu(I) atoms as the cysteines rearrange to accommodate the incoming Cu(I) [14, 15]. XAFS data provide average bond lengths and coordination geometries following absorption by either the metal or the coordinating sulfur [16, 17]. Spectroscopic data from a number of techniques have been recently summarised [3].

Metal binding: stoichiometric ratios

Metal binding to metallothionein is complicated: spectroscopic signal maxima are observed at metal to protein molar ratios of 7, for Hg(II), Cd(II), Co(II), and Fe(II), 12, for Cu(I) and Ag(I), at 17 for Ag(I) and 18 for Hg(II) [2–4].

This complexity in metal binding is illustrated in the optical data recorded during titrations of both metal-free MT and Zn_7-MT. Metal binding to apo-MT follows different pathways than metal binding to Zn_7-MT [2–4]. The CD spectrum in particular is sensitive to the overall wrapping of the cysteines around the metals. As S_{cys}-M-S_{cys}-M- cross-linking occurs with increased metal loading, re-orientation of each cysteine within the binding site results in large changes in the chirality, e.g. refs. [9, 11–13]. Saturation of CD band intensities is observed when three-dimensional structures form as metal is added. Dramatic changes take place when Hg(II) is added to metal-free rabbit liver metallothionein 2A, Figure 2. A metal-dependent CD signal intensifies at the 7 Hg(II) to MT point. The collapse of this signal as the Hg:MT ratio rises indicates that the structure associated with 7 Hg(II) also collapses to be replaced by a structure without significant three-dimensional order. If the titration is carried out below pH 7, then a completely new dependence of the CD intensity on the Hg:MT ratio is observed [11]. In addition to bands at Hg:MT = 7, a strong signal grows in at the 18 Hg:MT point.

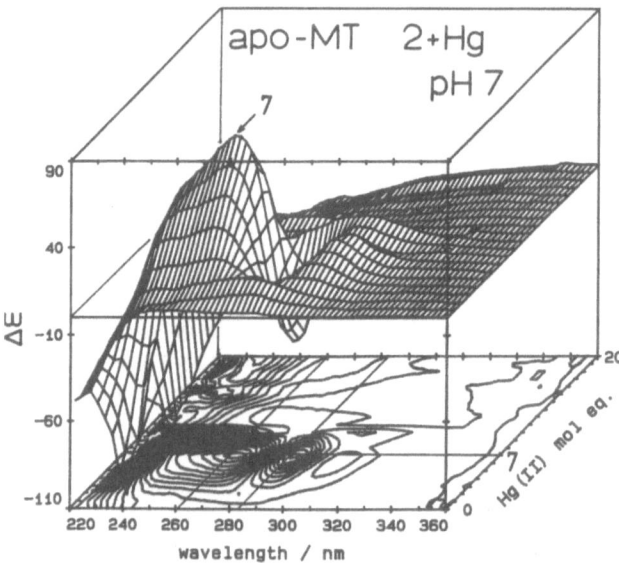

Figure 2. CD spectra recorded for apo-MT as a function of Hg(II) loading carried out at pH7 (the z-axis). Reproduced with permission [11].

Peptide folding is important in determining the thermodynamic minima for different metal to protein stoichiometries. CD spectra show that when Cu(I) is titrated into Zn_7-MT at temperatures between 3 °C and 52 °C, Figure 3 [9], the prior presence of the 7 Zn(II) results in the peptide chain wrapping in a manner that precludes formation of some copper-related structures unless the orientation of the chain can change. At 3 °C, the CD spectrum shows a maximum signal intensity at Cu_{12}-MT; further Cu(I) results in formation of a species at the 15 Cu(I) point. At 52 °C, a new species dominates the titration: Cu_9,Zn_2-MT forms with a CD intensity that is 50% greater than the intensity of the Cu_{12}-MT, subsequently Cu_{15}-MT forms. Clearly, the thermal energy available at 52 °C allows for rearrangements of the chain. These data indicate the importance of characterizing the species being prepared for spectroscopic analysis, for example, in the case of NMR or EXAFS studies.

The emission spectrum measured for Cu-MT as Cu(I) is added to Zn_7-MT also provides detailed information about the Cu(I) to MT stoichiometric ratios that lead to the formation of well defined structures. Figure 4 shows the emission of the copper-thiolate clusters in Cu-MT as Cu(I) is added to Zn_7-MT. Clearly, the maximum at the 12 Cu(I):MT point identifies the formation of the Cu_{12}-MT species. However, the dependence of the intensities on the number of copper atoms added is far from linear [15].

Combining the CD and emission spectral data provides greater detail of the possible structures formed. While it might be expected that Ag(I) would follow the structural pattern of Cu(I), the CD and emission data shown in Figure 5 indicate a preference for the formation of a species at the 17 Ag:MT point rather than at 12 Ag:MT [12, 17]. The emission spectra do show that

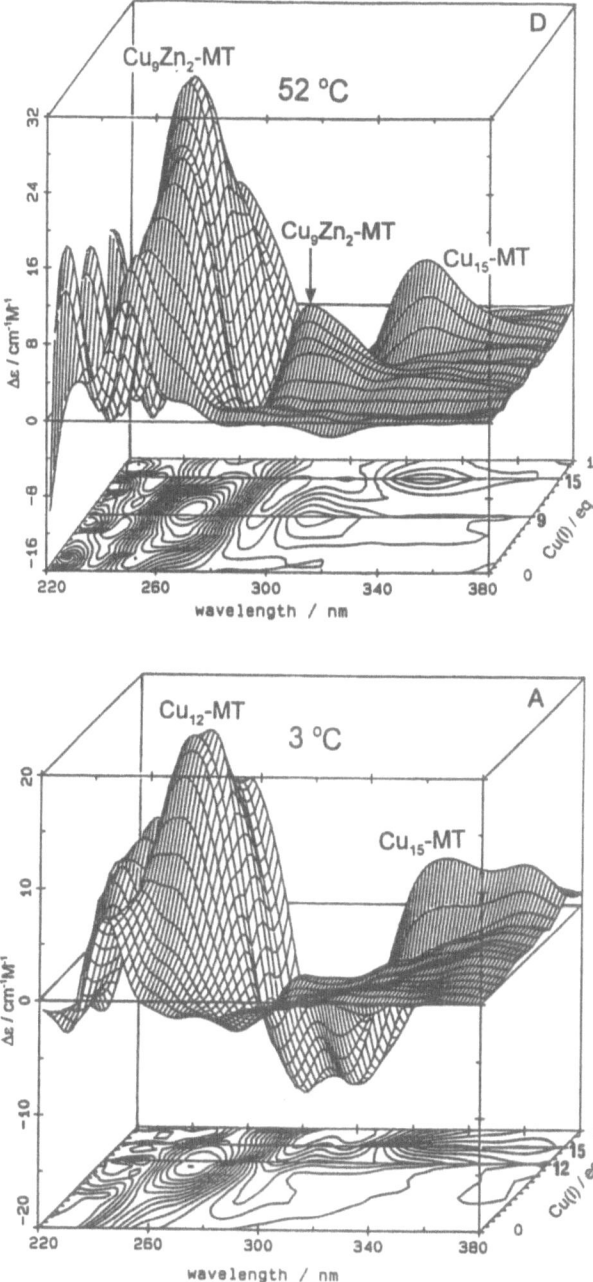

Figure 3. CD spectra recorded as a function of Cu(I) loading (the z-axis) at 3 °C and 52 °C. Reproduced with permission [9].

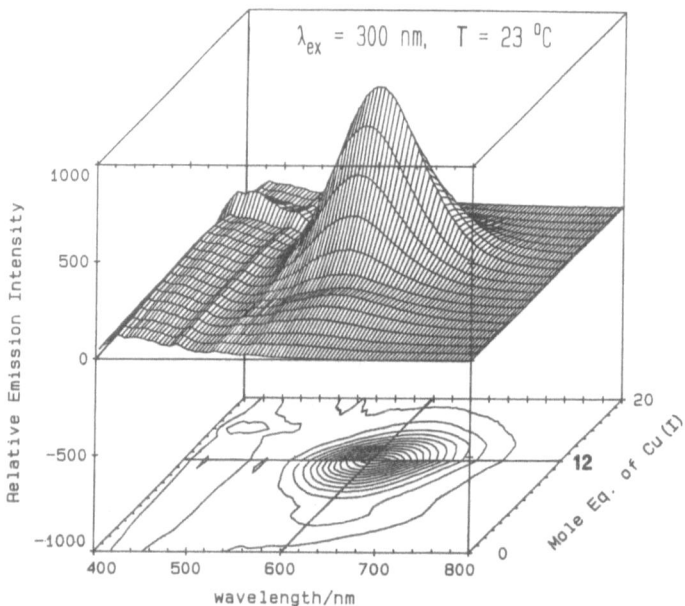

Figure 4. Emission spectra recorded as a function of Cu(I) loading (the z-axis) at 23 °C. Reproduced with permission [15].

there is a change in the structure between the 12 and 17 points that could indicate that at the 12 Ag:MT point a structure forms that is unique for Ag_{12}-MT, but is similar enough in its coordination geometry and silver-sulfur connectivities that there is no major change in the spectral data between 12 and 17 Ag(I). The XAFS data supports a digonal co-ordination for both species [17].

Structural implications of the spectroscopic data

The optical data described above provide very exact values for the metal to protein stoichiometric ratios. The corresponding XAFS data from both the coordinating thiolate sulfur and the bound metal provide structurally-significant information unavailable from other techniques [16, 17]. The bond lengths are averages, which for the multi-metal clusters in metallothionein is a problem, and unless large differences exist in the metal-thiolate bond lengths, a mixture of different coordinations may be difficult to determine. Co-ordination numbers obtained from XAFS data also may have wide confidence limits. Other spectroscopic techniques must be used to determine the co-ordination – purity of the clusters examined and inorganic models must be used to correlate calculated bond lengths with co-ordination geometry. Recent XAFS data have been summarized [2, 3].

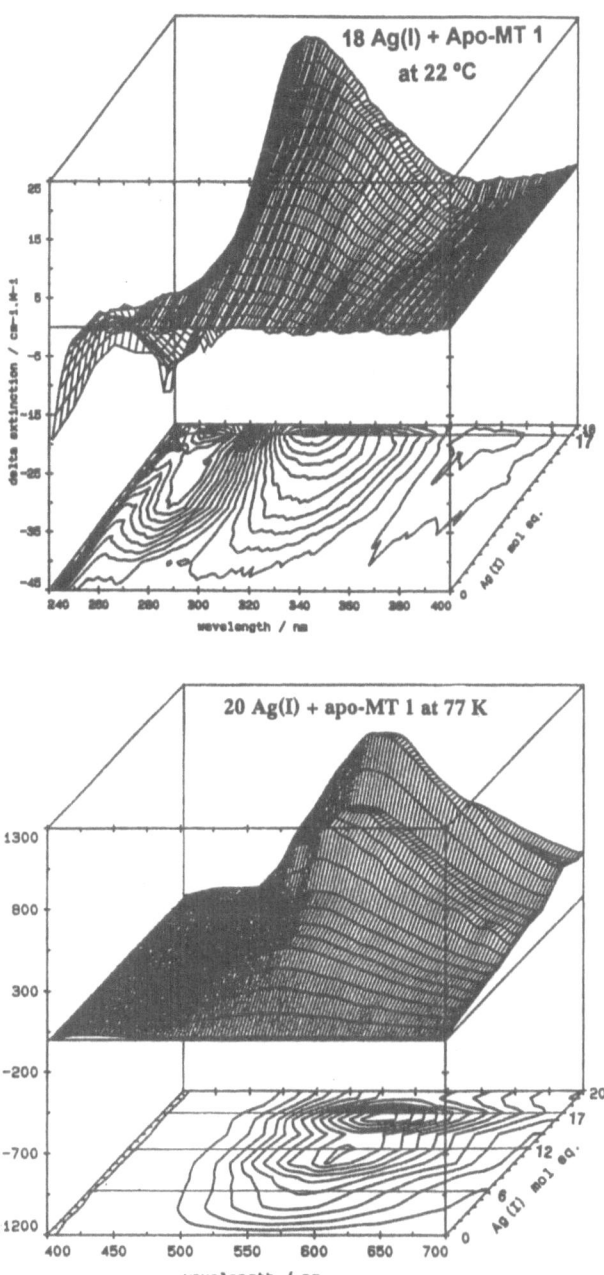

Figure 5. CD (at 22 °C) and emission (at 77 K) spectra recorded as a function of Ag(I) loading (the z-axis). Reproduced with permission [17].

The dynamics of metal binding

The temperature dependence of the emission spectrum intensity and wavelengths and the circular dichroism patterns [9, 13–15] indicate that the kinetics of metal binding is complicated. While the thermodynamic product is found *in vivo* it may be the kinetic product that is studied in many cases *in vitro*. CD data indicate that at room temperature Cd(II) displaces Zn(II) from Zn_7-metallothionein in a random manner, with Cd(II) atoms binding evenly across the two domains of the protein. When the temperature is raised, the metallothionein-bound Cd(II) and Zn(II) rearrange into the domain-specific species with a filled Cd_4-α domain. The emission

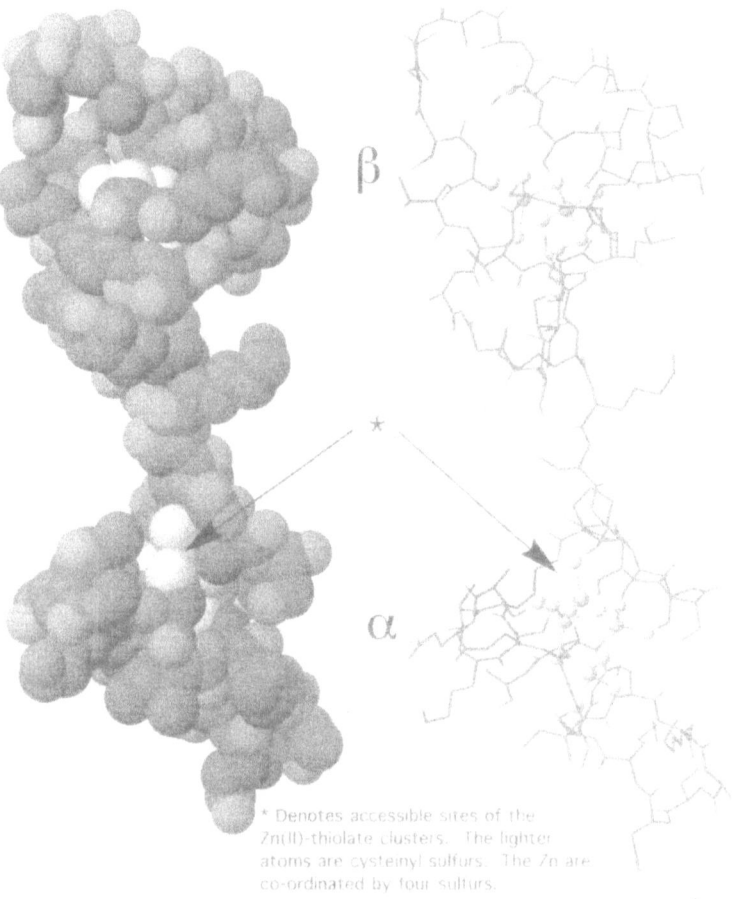

* Denotes accessible sites of the
Zn(II)-thiolate clusters. The lighter
atoms are cysteinyl sulfurs. The Zn are
co-ordinated by four sulfurs.

Figure 6. Calculated structure of Zn_7-MT. Reproduced with permission [20].

spectral intensity characteristic of the copper-thiolate cluster in metallothionein follows a complicated relationship with temperature and the concentration of Cu(I) added. In experiments that monitor the change in intensity of the emission at 600 nm at room temperature following addition of single atoms of Cu(I) to Zn_7-MT 2A [14], it was found that the intensity diminishes over 20–40 min for the first 7 Cu(I), and increases for Cu(I) up to 12. Our interpretation is that the Cu(I) binds to both domains initially resulting in a rapid increase in emission intensity. Over the next 20 min, Cu(I) migrates from the α domain to the β domain. The β domain is less emissive and so the intensity falls until the domain is filled at the 6 Cu(I) point after which the α domain fills.

Modeling the structures

To date, Stout has reported the only crystal structure, which was determined for the mixed metal rat liver Cd_5Zn_2-MT [6, 8]. Significantly, the three-dimensional, space-filling model drawn for

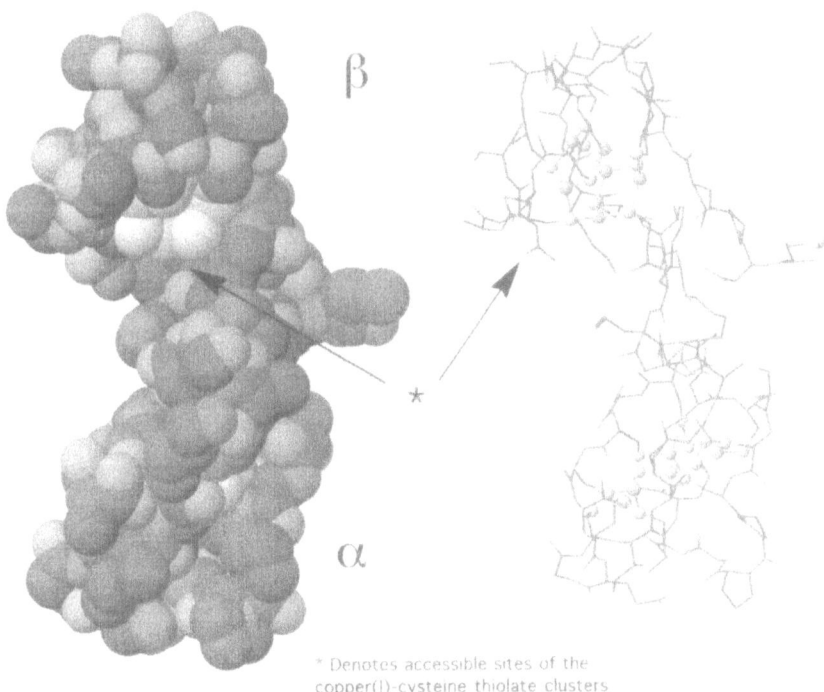

* Denotes accessible sites of the
copper(I)-cysteine thiolate clusters

Rabbit Liver Cu_{12}-MT 2a (View - methionine right)

Figure 7. Calculated structure of Cu_{12}-MT 2A. Reproduced with permission [21].

Cd$_5$Zn2-MT clearly showed deep crevices when the peptide chain was wrapped round the two metal-thiolate cores. Molecular modelling provides a means of (i) testing possible structures that can accommodate the optical, NMR, and XAFS data, (ii) investigating structural changes necessary for during metal-exchange reactions, and (iii) determining locations around the binding site region that can be used during competitive ligation reactions, for example, the reaction of dimercaptopropanol. We have reported structures calculated for Zn$_7$-MT, Cd$_7$-MT, and Hg$_7$-MT [20] and for Cu$_{12}$-MT 2A in which each Cu(I) is trigonally co-ordinated [21]. We are cur-

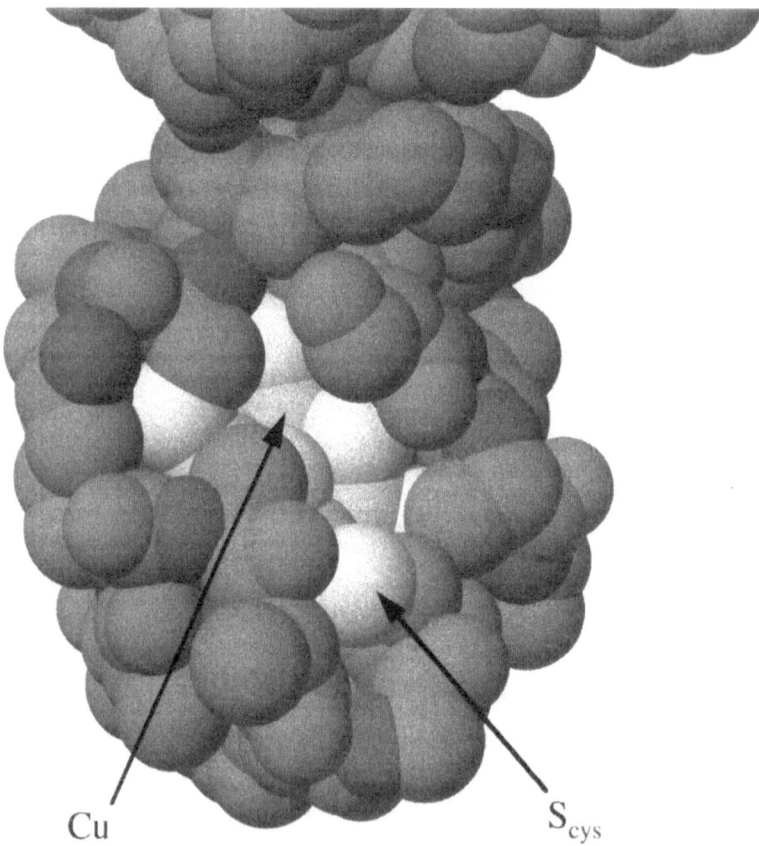

Cu   S$_{cys}$

Alpha domain in Cu$_{12}$-MT 2 showing the exposed face of the Cu$_6$S$_{11}$ cluster. The light S$_{cys}$ atoms are seen in front of the co-ordinated Cu(I) atoms.

Figure 8. Calculated structure of Cu$_{12}$-MT 2A showing the crevice in the α domain.

rently investigating improvements in the force fields to be applied to metals bound in metal-lothionein. Key features of the models have been the presence of crevices in both domains. Figure 6 shows the space-filling and line representation of Zn_7-MT, Figure 7 shows the same views for Cu_{12}-MT 2A, Figure 8 shows the α domain of Cu_{12}-MT 2A with the exposed cop-per-thiolate clusters and Figure 9 shows the orientation of the 6 Cu(I) within the binding site.

The space-filling representation of Hg_7-MT shows that tetrahedral coordination strains the peptide chain resulting in regions where the clusters are completely exposed to the solvent [20].

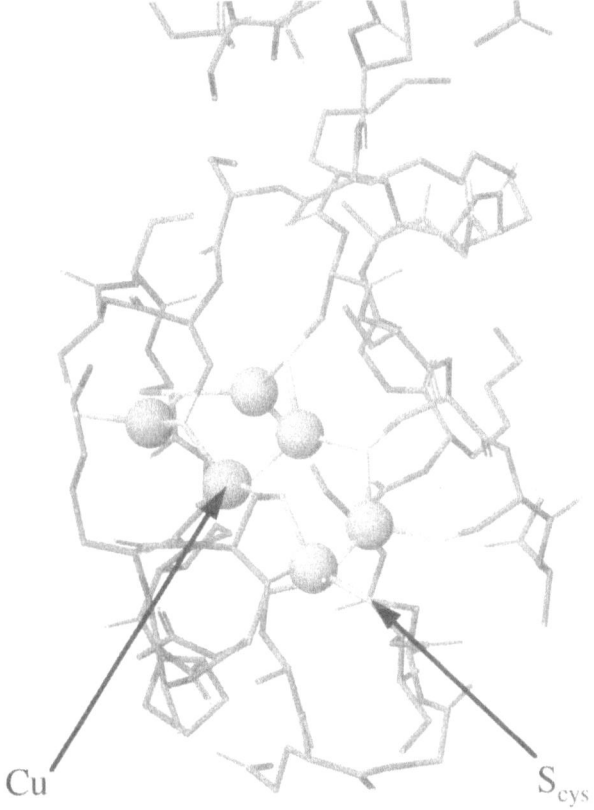

Cu S_{cys}

Alpha domain in Cu_{12}-MT 2 showing the structure of the exposed face of the Cu_6S_{11} cluster. The Cu(I) atoms are shown as spheres, the S_{cys} atoms are shown as light bridging and terminal bonds

Figure 9. Calculated structure of Cu_{12}-MT 2A showing the orientation of the 6 Cu(I) atoms within the α domain.

This result appears to provide support for the analysis of the XAFS data in which a two-long, two-short bond geometry fitted the data [16]. Stout's structures [8], that calculated for Cd_7-MT 2A [20], and the structure calculated based on NMR data for Cd_7-MT 2A [8] provide comparison of structures from the solid state, gas phase, and solution phase, respectively [20].

The model structures also provide further detail of the environment just outside the binding site that is formed by the peptide chain as it wraps round the metals. Spectroscopic titrations do not indicate how the metal exchange reaction operates at the molecular level. In particular, how does the incoming metal approach the existing metal-thiolate cluster during metal exchange reactions? The space-filling representation of our models provide a view of the outside of the binding sites showing quite clearly crevices that expose the metal-thiolate. Emission intensity measurements suggest that the clusters are porous towards solvent but to varying degrees depending on the metal loading. This suggests that the crevices change in size as the metals are exchanged. The structures shown in Figures 8 and 9 suggest that with the first step of the reaction the incoming metal approaches the crevice and forms bonds with the exposed thiolates from the outside.

Acknowledgements
We gratefully acknowledge financial support of this research from NSERC of Canada and the Academic Development Fund at U.W.O. MJS is a member of the Centre for Chemical Physics and the Photochemistry Unit at U.W.O.

References

1. Kägi JHR (1993) Evolution, structure and chemical reactivity of class I metallothioneins: An overview. *In*: KT Suzuki, N Imura and M Kimura (eds): *Metallothionein III*, Birkhäuser Verlag, Basel, 29–55.
2. Stillman MJ (1997) Spectroscopic Properties of Ag(I), Cd(II), Cu(I), Hg(II) and Zn(II) Metallothioneins. *In*: ND Hadjiliadis (ed.): *Cytotoxic, Mutagenic, and Carcinogenic Potential of Heavy Metals Related to Human Environment*. Kluwer, Dordecht, 139–194.
3. Stillman MJ (1995) Metallothioneins. *Coord Chem Rev* 144: 461–511.
4. Stillman MJ (1992) Optical spectroscopy of metallothioneins. *In*: MJ Stillman, CF Shaw, KT Suzuki (eds): *Metallothioneins. Synthesis, structure and properties of metallothioneins, phytochelatins and metal-thiolate complexes* V.C.H. Publishers, New York, pp. 55–127.
5. Riordan JF, Vallee BL (eds) (1991) *Metallobiochemistry Part B. Metallothionein and Related Molecules. Meth Enzymol* , vol. 205 Academic Press, New York.
6. Stillman MJ, Shaw CF, Suzuki KT (eds) (1992) *Metallothioneins. Synthesis, structure and properties of metallothioneins, phytochelatins and metal-thiolate complexes*. V.C.H. Publishers, New York.
7. Otvos JD, Armitage IM (1980) Structure of the metal clusters in rabbit liver metallothionein MT, [113]Cd NMR. *Proc Natl Acad Sci U.S.A.* 77: 7094–7098.
8. Braun W, Vašák M, Robbins AH, Stout CD, Wagner G, Kägi JHR, Wuthrich K (1992) Comparison of the NMR solution structure and the X-ray crystal structure of rat metallothionein-2. *Proc Natl Acad Sci USA* 89: 10124–10128.
9. Presta A, Green AR, Zelazowski AJ, Stillman MJ (1995) Copper binding to rabbit liver metallothionein. Formation of a continuum of copper(I)-thiolate stoichiometric species. *Eur J Biochem* 227: 226–240.
10. Le Blanc, Y.J.C., Presta, A., Veinot, J., Gibson, D., Siu, K.W.M., Stillman, M.J. (1997) Identification of the Isoforms and Subisoforms of Rabbit Liver Metallothionein using Electrospray Mass Spectrometry. *Protein Pept Lett* 4: 313–320.
11. Lu W, Zelazowski AJ, Stillman MJ (1993) Mercury binding to metallothioneins: Formation of Hg_{18}-MT. *Inorg Chem* 32: 919–926.
12. Zelazowski AJ, Gasyna Z, Stillman MJ (1989) Ag binding to rabbit liver MT. CD and emission study of Ag-thiolate cluster formation with apoMT and the alpha and beta fragments. *J Biol Chem* 264: 17091–17099.
13. Stillman MJ, Zelazowski AJ (1988) Domain specificity in metal binding to metallothionein. A CD and MCD study of Cd and Zn binding at temperature extremes. *J Biol Chem* 263: 6128–6133.

14. Green AR, Stillman MJ (1996) Mobility of copper in binding sites in rabbit liver metallothionein 2. *Inorg Chem* 35: 2799–2807.
15. Green AR, Presta A, Gasyna Z, Stillman MJ (1994) Luminescent probe of copper-thiolate cluster formation within mammalian metallothionein. *Inorg Chem* 33: 4159–4168.
16. Jiang DT, Heald SM, Sham TK, Stillman MJ (1994) Structures of the cadmium, mercury, and zinc thiolate clusters in metallothionein: XAFS study of Zn_7-MT, Cd_7 -MT, Hg_7 -MT and Hg_{18} -MT formed from rabbit liver metallothionein. *J Amer Chem Soc* 116: 11004–11013.
17. Gui Z, Green AR, Kasrai M, Yang B, Feng X, Bancroft GM, Stillman MJ (1996) Sulfur K-Edge EXAFS Studies of Cadmium-, Zinc-, Copper-, and Silver-Rabbit Liver Metallothioneins. *Inorg Chem* 35: 6520–6529.
18. Stillman MJ, Gasyna Z, Zelazowski AJ (1989) A luminescence probe for MT in liver tissue: emission intensity measured directly from Cu(I)-MT induced in rat liver. *FEBS Lett* 257: 283–286.
19. Presta PA, Stillman MJ (1996) Incorporation of copper into the yeast Saccharomyces cerevisiae. Identification of Cu(I)-MT in intact yeast cells. *J Inorg Biochem* 66: 231–240.
20. Fowle DA, Stillman MJ (1997) Comparison of the Structure and Stability of the Metal-Thiolate Binding Site in Zn(II)-, Cd(II), and Hg(II)-Metallothioneins Using Molecular Modeling Techniques. *J Biomol Struct Dyn* 14: 393–406. 75.
21. Presta A, Fowle DA, Stillman MJ (1997) Structural Model of Rabbit Liver Copper Metallothionein. *J Chem Soc Dalton Trans* 977–984.

Metallothionein IV
C. Klaassen (ed.)
© 1999 Birkhäuser Verlag Basel/Switzerland

Metallothionein structure update

Gülin Öz, Dean L. Pountney, Ian M. Armitage

University of Minnesota, Department of Biochemistry, 4-225 Millard Hall, 435 Delaware St. S.E., Minneapolis, MN 55455, USA

Metallothioneins (MTs) are low molecular weight (2–7 kDa), cysteine-rich proteins which bind with high affinity to metal ions, such as Zn(II), Cd(II), Cu(I) and Ag(I), in unusual metal-thiolate clusters [1–3]. MT biosynthesis is induced by a wide range of factors, including metal ions, oxidative stress, glucocorticoid hormones and cytokines, leading to their implication in many diverse cellular processes. However, the primary function of MTs under normal physiological conditions is in the transport and storage of the essential metals: zinc and copper.

To date, complete three-dimensional structures have been elucidated for four distinct metallothioneins: yeast (*Saccharomyces cerevisiae*) MT [4, 5], blue crab (*Callinectes sapidus*) MT-1 [6, 7], sea urchin (*Strongylocentrotus purpuratus*) MTA [8] and mammalian (rabbit, rat and human) MT-2 [9–12]; less detailed structural information is also available for fungal (*Neurospora crassa*) MT [13] and mammalian MT-3 [14–16]. Amino acid sequences of each of these proteins are given in Table 1. Solution structures were obtained using multidimensional/multinuclear NMR methods, with the mammalian structure being confirmed by X-ray crystallography. All of the current structures show exclusively cysteine coordination of the bound mono- or divalent metal ions in polynuclear metal-thiolate clusters.

Zn/Cd MT structures

Mammalian and invertebrate MTs

The structures of the crab, sea urchin and mammalian MTs are of the cadmium-containing proteins and in each case the specific cysteine-cadmium connectivities were obtained using ^{1}H-^{113}Cd heteronuclear multiple quantum coherence (HMQC) experiments. Conventional ^{1}H NMR experiments (COSY, TOCSY, NOESY) were used to obtain the sequential proton assignments and to provide distance geometry constraints which combined with the specific metal-cysteine connectivities were used in the structure calculations. The cadmium-derivatives were chosen due to the lack of a suitable NMR-active zinc isotope, however, the highly similar chemistry of these two metals results in essentially identical structures for the zinc- and cadmium-containing proteins [17]. In each structure, all of the metal ions are bound with pseudotetrahedral, tetrathiolate coordination in polynuclear metal clusters involving both bridging and terminal cysteine ligands. In the 61 amino acid polypeptide sequence of the mammalian MTs, the 20 cysteines bind 7 divalent metal ions in two discrete M_3Cys_9 and M_4Cys_{11} clusters in the N-terminal (β) and C-terminal (α) domains, respectively. In the 58 amino acid sequence of the crab

Table 1. Amino acid sequences of several structurally characterized MTs

mouse MT-1	MDP_NCSCSTGG_SC__T__CTSSCACKN_CKCTSCKKSCCSCCPVGCSKCAQGCVCKGA_____ADKCTCCA
rat MT-2	MDP_NCSCATDG_SC__S__CAGSCKCKQ_CKCTSCKKSCCSCCPVGCAKCSQGCICKEA_____SDKCSCCA
human MT-3	MDPETCPCPSGG_SC__T__CADSCKCEG_CKCTSCKKSCCSCCPAECEKCAKDCVCKGGEAAEAEAEKCSCCQ
N. crassa	GDCGCSGAS_SC__N__CGSGCSCSN_CGSK
S. cerevisiae	QNEGHECQCQC_G___SCKNNEQCQKSCSCPTGCNSDDKCPCGNKSEETKKSCCSGK
crab MT-1	PGPCCNDKCVCQEGG__CKAGCQCTS_CRC_SPC_____QKCTSGCKCATKEECSKT__CT_____KPCSCCPK
sea urchin MTA	PDVKCVCCTEGKECACFGQDCCVTGECCKDGTCCGICTNAACKCANGCKCGSGCSCT_____EGNCAC

MT, the 18 cysteines bind 6 metal ions again in two distinct α and β domains with each containing M_3Cys_9 clusters. In spite of the evolutionary distance between vertebrates and invertebrates, comparison of the polypeptide folds of the crab and mammalian proteins (see Fig. 1) reveals in addition to some significant differences, several perhaps functionally important similarities between the two structures. Whilst the polypeptide in both MT β domains wraps around the metal cluster in a right-handed spiral, in the crab MT there are much narrower and shallower clefts and the metals are more completely engulfed, thus, less solvent accessible than in the mammalian structure. The crab MT α-domain is more compact primarily as a result of the presence of only 3 metal ions and, unlike the mammalian α-domain which exhibits a clearly left-handed twist, has no obvious handedness of the polypeptide fold. In this domain, the crab protein exhibits a short α-helix from residues 42–48 in contrast to the mammalian structure which has no regular secondary structure elements. There are no clefts which extend to the metal cluster in the α-domain of the crab structure and in line with the broader 113-Cd resonances of the β cluster, we predict that, as in the mammalian protein, the crab β-domain would exhibit more facile metal exchange than the α domain. This common feature of rapid metal exchange of MT β domains may be important to a role in mediating interprotein metal transfer reactions. Interestingly, the recently determined structure of sea urchin MT also shows two separate domains containing 3- and 4-metal clusters but in a reverse orientation with respect to the mammalian proteins [8].

MT-3 (GIF)

Due to a possible link to the etiology of Alzheimer's disease, considerable interest is currently focused on the brain-specific mammalian MT isoform MT-3. Since the protein's discovery by Uchida et al. in 1991 [18], limited structural information concerning this protein has become available. Circular dichroism and luminescence spectroscopic data have shown the ability of the protein to form Cd(II)- and Cu(I)-thiolate clusters similar to the mammalian MT-1 and MT-2 isoforms [14, 15]. The 113-Cd NMR spectrum of the 113-Cd substituted protein (Fig. 2) [16] shows seven resonances at analogous positions compared to MT-1 and MT-2, confirming the presence of similar Cd_3- and Cd_4-clusters. Notably, the resonances observed from the metals in the β domain (II, III and IV) are significantly broader than in MT-1 or MT-2, indicating more rapid metal exchange. The resonances from metals in the α domain (I, V, VI and VII) are somewhat shifted with respect to MT-1 or MT-2, especially CdVII, possibly as a result of the acidic hexapeptide insert (residues 55–60) in this domain. Studies to elucidate the full 3D solution structure of this protein are in progress.

Cu/Ag MT structures

Yeast MT

The structure of the yeast MT was determined for the native copper-containing form (CuMT) and also for the silver-substituted derivative (AgMT) [4, 5]. As with the crab and mammalian

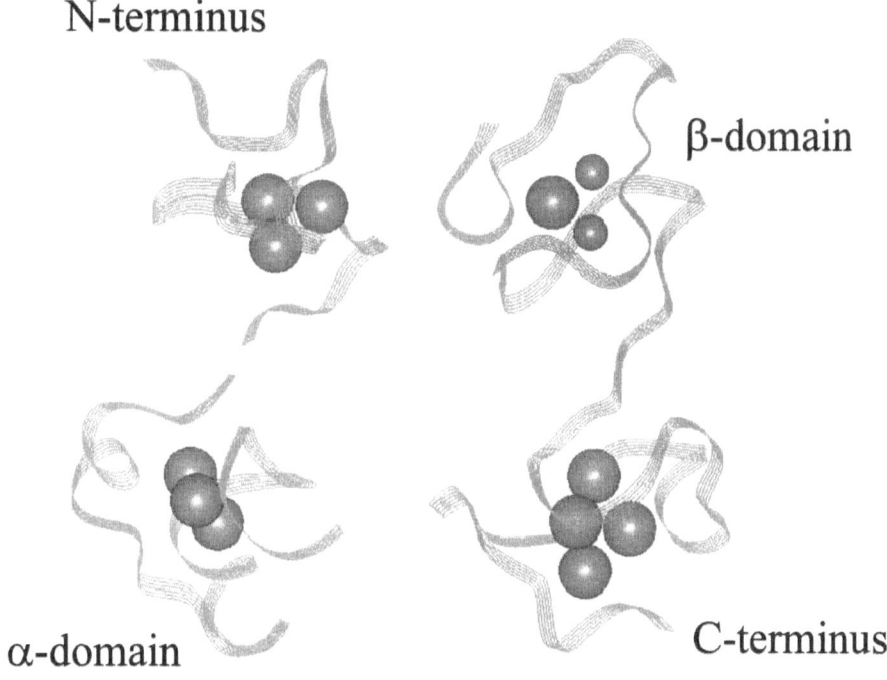

Figure 1. Structural models of Cd/Zn MTs: (right) rat Cd_5Zn_2-MT-2 based on the crystallographic data [24]; (left) crab Cd_6-MT NMR structure of α-(bottom) and β-(top) domains [7]. Cd(II) ions are shown as large spheres and Zn(II) ions as small spheres.

proteins, full definition of the protein structure required the specific metal-ligand bonding to be established. Comparison of chemical shift, coupling constant and proton exchange data showed both metalloforms to be isostructural, enabling the metal-cysteine connectivities obtained from the NMR-active [109]Ag nucleus and [1]H-[109]Ag HMQC experiments on AgMT to be used in the calculation of both structures. The 53 amino acid protein binds 7 Cu(I) or Ag(I) ions and coordinates these metal ions in a single cluster structure involving 10 out of the 12 available cysteines and both digonal and trigonal coordination geometries (Fig. 3 A). The C-terminus of the protein (beyond residue 41) which contains the two free cysteines was found to occur in two major conformations, thus permitting a unique protein fold to be calculated only for the first 40 amino acids. The polypeptide fold (Fig. 3 B) shows two large, parallel loops, separated by a deep cleft occupied by the metal cluster. No regular secondary structural elements were observed except for several type I turns. Only minor differences were observed between the silver and copper metalloforms, with the slightly larger silver cluster resulting in a less than 6.5% increase in overall protein volume. It is interesting to speculate on the function of the two

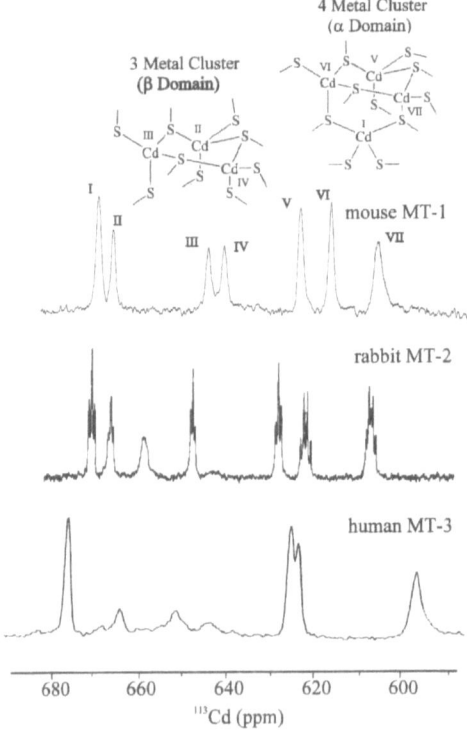

Figure 2. 113-Cd NMR spectra of mammalian Cd_7-MTs: (top) recombinant mouse MT-1 at 133.2 MHz (G. Öz, unpublished data); (middle) rabbit MT-2 at 55.5 MHz [25]; (bottom) recombinant human MT-3 at 110 MHz [16]. Schematic structures of the α- and β-clusters are illustrated.

C-terminal cysteines which are not involved in binding metals. The mobility of the C-terminal region may allow these residues to play a dynamic role in loading metal ions into the protein or perhaps they function to bind additional or different metal ions under certain cellular conditions. Further studies aimed at defining each of the major conformations of the full-length protein may be able to shed some light on this question.

Fungal MT

The fungal MT from *Neurospora crassa* at 25 amino acids is the smallest known MT gene product. When the fungus is exposed to high copper levels, the protein contains 6 Cu(I) ions bound to 7 cysteine residues. The positions of the cysteines are identical to the first 7 cysteines of mammalian MTs (β domain) (Tab. 1), suggesting that there may be structural similarities to mammalian copper MTs (see below). The small size of this protein results in negligible ^{1}H-^{1}H NOEs at 500 MHz ($\omega\tau_c \sim 1$) requiring the use of 2D rotating frame experiments and transverse NOE data for sequential proton assignments [13]. Only five non-sequential transverse NOEs were observed, insufficient to permit a tertiary structural model to be calculated, but consistent with the presence of one large polypeptide loop and several tight turns. Inspection of the copper-cluster of yeast MT (Fig. 3 A) and comparison of the amino acid sequences of the *S. cerevisiae* and *N. crassa* MTs provokes some speculation as to the structure of the fungal protein. Deletion of the first cysteine (Cys7) and removal of CuI from the yeast cluster would allow Cys20 to bridge directly between CuVII and CuVI with minimal disturbance of the overall structure. This is in good agreement with the alignment of the *N. crassa* amino acid sequence to the N-terminus of the yeast sequence. The absence of the final two metal-binding cysteines of the yeast protein (Cys36 and Cys38) from the shorter *N. crassa* amino acid sequence would result in the conversion of CuIII, CuIV, CuV and CuVI from trigonal to digonal coordination. This speculative model of the *N. crassa* protein would preserve the order of cysteine-metal connectivities along the polypeptide chain relative to the yeast structure and would result in one large polypeptide loop with overlapping ends encompassing the Cu_6-cluster.

Cu binding to mammalian and invertebrate MT

Mammalian MTs bind copper as Cu(I) with exceptionally high affinity, displacing zinc or cadmium and preferentially occupying the N-terminal domain in mixed metalloforms. Although under normal conditions zinc is the predominant metal ion bound to MTs in mammals, stoichiometric amounts of copper can be found in neonates, in copper-overloaded individuals, and in MT-3 [19]. Monometallic copper derivatives containing 8–12 Cu(I) have been characterized by optical methods, but are rather unstable [20]. Recently, stable rabbit MT-2 copper/zinc, copper/cadmium, silver/zinc and silver/cadmium mixed-metal hybrids have been characterized which bind 6 Cu(I)/Ag(I) and 4 Zn(II)/Cd(II) ions in the N- and C-termini, respectively [21].

Invertebrates such as crab, lobster and snail, which use hemocyanin as their oxygen carrier, express organ-specific copper-containing MT isoforms which bind up to 6 Cu(I) ions per pro-

A

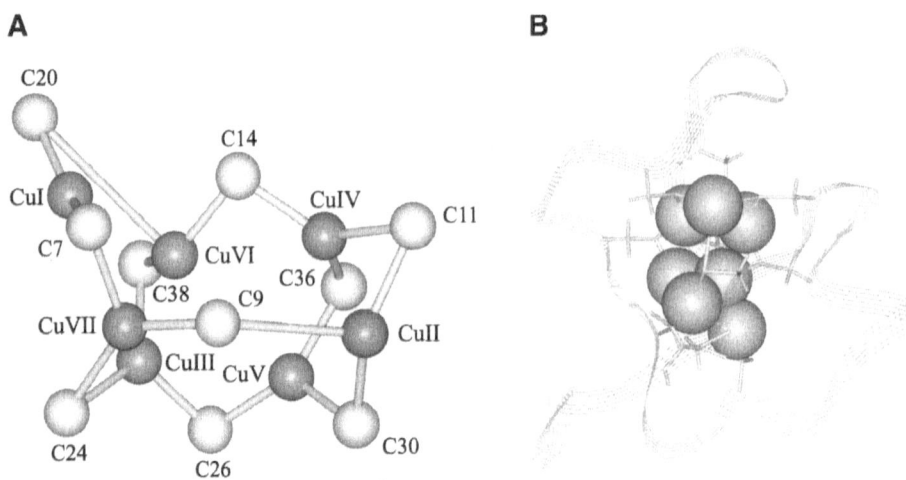

B

Figure 3. Yeast CuMT structure [5]: (A) Copper-thiolate cluster model, showing the cysteine sulfur (light) and Cu(I) (dark) atoms. The C20-CuVI and C9-CuII connectivities were deduced from the calculated structure; (B) Ribbon diagram of the first 40 amino acids, illustrating the cysteine side-chains. The bound Cu(I) ions are shown as spheres.

tein domain [22, 23]. It has been proposed that invertebrate copper-MTs may be important in supplying copper to apo-hemocyanin and in the tight regulation of free copper in copper-rich organisms, however, little is known about the structures of these proteins (M. Brouwer, personal communication).

The functional significance of the observed structural variations between different metallothionein isoforms and between species remains opaque.

Acknowledgements
The authors are indebted to the numerous past researchers in the Armitage laboratory who have contributed to this field of research. Special acknowledgement is also extended to past and ongoing collaborators, Drs. Cynthia W. Peterson, Dennis R. Winge, Marius Brouwer and James D. Otvos, who have enabled many of the studies reported. Partial support for these studies came from the National Institutes of Health grant (DK18778) to IMA.

References

1. Otvos JD and Armitage IM (1979) [113]Cd NMR of metallothionein: diirect evidence for the existence of polynuclear metal binding sites. *J Amer Chem Soc* 101: 7734–7736.
2. Kägi JH, Schäffer A (1988) Biochemistry of metallothionein. *Biochemistry* 27: 8509–8515.
3. Pountney DL, Kägi JHR, Vašák M (1995) Metallothioneins. *In:* G Berthon (ed.): *Handbook of Metal-Ligand Interactions in Biological Fluids; Bioinorganic Chemistry, Vol. 1*. Marcel Dekker Inc., New York, pp. 431–450.
4. Narula SS, Winge DR, Armitage IM (1993) Copper- and silver-substituted yeast metallothioneins: sequential H NMR assignments reflecting conformational heterogeneity at the C terminus. *Biochemistry* 32: 6773–6787.
5. Peterson CW, Narula SS, Armitage IM (1996) 3D solution structure of copper and silver-substituted yeast

metallothioneins. *FEBS Lett* 379: 85–93.

6. Narula SS, Armitage IM, Brouwer M, Y, Enghild JJ (1993) Establishment of two distinct protein domains in blue crab *Callinectes sapidus* metallothionein-I through Heteronuclear ([1]H-[113]Cd) and Homonuclear (^{1}H-^{1}H) correlation NMR experiments. *Magn Reson Chem* 31: S96-S103.

7. Narula SS, Brouwer M, Hua Y, Armitage IM (1995) Three-dimensional solution structure of *Callinectes sapidus* metallothionein-1 determined by homonuclear and heteronuclear magnetic resonance spectroscopy. *Biochemistry* 34: 620–631.

8. Kägi JHR, Riek R, Prêcheur B, Wang Y, Mackay E, Güntert P, Wider G and Wüthrich K (1997) NMR solution structure of metallothionein MTA from the sea urchin *Strongylocentrotus pupuratus*, this volume.

9. Braun W, Wagner G, Wörgötter E, Vašák M, Kägi JH, Wüthrich K (1986) Polypeptide fold in the two metal clusters of metallothionein-2 by nuclear magnetic resonance in solution. *J Mol Biol* 187: 125–129.

10. Schultze P, Wörgötter E, Braun W, Wagner G, Vašák M, Kägi JH, Wüthrich K (1988) Conformation of [Cd₇]-metallothionein-2 from rat liver in aqueous solution determined by nuclear magnetic resonance spectroscopy. *J Mol Biol* 203: 251–268.

11. Messerle BA, Schäffer A, Vašák M, Kägi JH, Wüthrich K (1990) Three-dimensional structure of human [[113]Cd7]metallothionein-2 in solution determined by nuclear magnetic resonance spectroscopy. *J Mol Biol* 214: 765–779.

12. Braun W, Vašák M, Robbins AH, Stout CD, Wagner G, Kägi JH, Wüthrich K (1992) Comparison of the NMR solution structure and the x-ray crystal structure of rat metallothionein-2. *Proc Natl Acad Sci USA* 89: 10124–10128.

13. Malikayil JA, Lerch K, Armitage IM (1989) Proton NMR studies of metallothionein from *Neurospora crassa*: sequence-specific assignments by NOE measurements in the rotating frame. *Biochemistry* 28: 2991–2995.

14. Pountney DL, Fundel SM, Faller P, Birchler NE, Hunziker P, Vašák M (1994) Isolation, primary structures and metal binding properties of neuronal growth inhibitory factor (GIF) from bovine and equine brain. *FEBS Lett* 345: 193–197.

15. Bogumil R, Faller P, Pountney DL, Vašák M (1996) Evidence for Cu(I) clusters and Zn(II) clusters in neuronal growth-inhibitory factor isolated from bovine brain. *Eur J Biochem* 238: 698–705.

16. Shen G (1995) Chapter 5, Properties of recombinant neuronal growth inhibitor factor (GIF). PhD thesis, North Carolina State University.

17. Messerle BA, Schäffer A, Vašák M, Kägi JH, Wüthrich K (1992) Comparison of the solution conformations of human [Zn7]-metallothionein-2 and [Cd₇]-metallothionein-2 using nuclear magnetic resonance spectroscopy. *J Mol Biol* 225(2): 433–443.

18. Uchida Y, Takio K, Titani K, Ihara Y, Tomonaga M (1991) The growth inhibitory factor that is deficient in the Alzheimer's disease brain is a 68 amino acid metallothionein-like protein. *Neuron* 7: 337–347.

19. Koch KA, Pena MMO, Thiele DJ (1997) Copper binding motifs in catalysis, transport, detoxification and signalling. *Chem Biol* 4: 549–560.

20. Pountney DL, Schauwecker I, Zarn J, Vašák M (1994) Formation of mammalian Cu8-metallothionein *in vitro*: evidence for the existence of two Cu(I)4-thiolate clusters. *Biochemistry* 33: 9699–9705.

21. Li H, Otvos JD (1996) [111]Cd NMR studies of the domain specificity of Ag⁺ and Cu⁺ binding to metallothionein. *Biochemistry* 35: 13929–13936.

22. Brouwer M, Schlenk D, Ringwood AH, Brouwer-Hoexum T (1992) Metal-specific induction of metallothionein isoforms in the blue crab *Callinectes sapidus* in response to single- and mixed-metal exposure. *Arch Biochem Biophys* 294: 461–468.

23. Dallinger R, Berger B, Hunziker P, Kägi JH (1997) Metallothionein in snail Cd and Cu metabolism. *Nature* 388: 237–238.

24. Robbins AH, McRee DE, Williamson M, Collett SA, Xuong NH, Furey WF, Wang BC, Stout CD (1991) Refined crystal structure of Cd, Zn metallothionein at 2.0 Å resolution. *J Mol Biol* 221: 1269–1293.

25. Otvos JD, Engeseth HR, Nettesheim DG, Hilt CR (1987) Interprotein metal exchange reactions of metallothionein. *EXS* 52: 171–178.

Metallothionein IV
C. Klaassen (ed.)
© 1999 Birkhäuser Verlag Basel/Switzerland

Simulation of Zn/Cd binding in mammalian metallothionein domains

C.C. Chang[1] and P.C. Huang[2]

[1]*Department of Life Sciences, National Tsing Hua University, Hsin-Chu 300 Taiwan, ROC*
[2]*Department of Biochemistry, Johns Hopkins University, Baltimore, MD 21205, USA*

Summary. Metallothionein naturally binds seven gram atoms of divalent ions such as Zn and Cd through cysteines forming two metals binding clusters, each constitutes a domain. Four of the metals [M1, M5, M6, M7] are found in alpha, the C-terminal domain, and three [M2, M3, M4] in beta, the N-terminal domain. The level of avidity is site specific. By semiempirical MNDO calculations, we find the relative binding stability for Cd to be M4 > M2 > M3 in the beta-cluster and M5 > M7 > M1,M6 in the alpha-cluster. This is reflected by energy differences computed with a series of simulated structures derived from either X-ray crystallography or NMR coordinates. Thus, replacement of Zn by Cd can be expected to follow the order: M4 → M2 → M3 in the beta-domain and M5 → M7 → M1 or M6 in the alpha-domain. Bridging cysteines are stronger than terminal cysteines in their relative average binding strength. Among the terminal cysteines, the strongest binding strength is found in Cys 21 to Cd[M4] and in Cys 26 to Zn[M3], both in the beta-domain of metallothionein. Although the binding strength of cysteine in the alpha domain has not yet been determined, we are able to predict from this study differential contribution of individual cysteines to metal binding. Such a prediction can be further tested by experiments with site-directed mutagenesis.

Introduction

The omnipresence of transition II B metals in nature challenges the biological systems on earth to evolve specific mechanism(s) with which to use or disuse these elements. Thus zinc (Zn) and copper are adopted as essential, while cadmium (Cd)and mercury rejected as toxic. Silver and gold, present in relatively rate quantities, are valued as precious and traded as such by humans, although their medicinal effects have also been recognized. It is envisioned that the essential, toxic and therapeutic effects of these metals are exerted through specific complexing of these metals to peptides/proteins in the cell. Indeed, in human genome alone over one thousand transition metal binding proteins are coded as shown by direct or deduced sequence analyses. All of these proteins contain cysteine (Cys, C) or histidine (His) or both, which is responsible for the binding of metals. There is evidence to show that binding to metals stabilizes the folded conformation of protein domains and facilitates intermolecular interactions (Berg and Shi, 1996).

At least three basic forms of metal-binding are now known: as fingers, twists, or, as exemplified by transcription activating factors, GAL-4 regulatory protein and metallothionein (MT), respectively (Vallee and Auld, 1993). While sequence specificity for zinc binding has been demonstrated with zinc-finger proteins through combinatorial approaches (Kim and Berg, 1996), preference for metal species at specific sites has also been shown in MT by semi-empirical MNDO (Modified Neglect of Diatomic Overlap) computation (Chang and Huang, 1996) and by isomorphous replacement and MM2 (molecular mechanics) techniques (Fowle and

Stillman, 1997). We extend the above findings to examine the relative binding energy of individual Cys, thus the stability, to Zn versus Cd among the two domains of MT.

The two domains of MT correspond to two metal binding clusters. Although the isoforms of mammalian MT, MT I to IV, differ in size, all of them maintain 20 Cys positioned with a high degree of conservativeness. Since MT is devoid of His, metal binding relies on the coordination with the thio group of Cys. In its native form, MT binds seven divalent metal ions via it 20 Cys. Eleven of the twenty Cys are present in the carboxyl half, the alpha-domain; the remaining 9 in the amino-half, the beta-domain. X-ray (Robbins et al., 1991) and NMR (Schultze et al., 1988) analyses show that each cluster consists of tetrahedral tetrathiolate structures. Three of the nine Cys in the beta-domain are bridging Cys, binding 2 metal ions, while the other 6 Cys are terminal, each linking with a single metal ion. The seven metals so coordinated occupy, respectively, positions M1 to M7, as resolved by 111,113Cd NMR. Thus, four of the metals [M1, M5, M6, M7] are found in the alpha-domain, and three [M2, M3, M4] in the beta-domain (Otvos, and Armitage, 1980; Boulanger et al., 1983) (Fig. 1)

The binding strength for individual metal to this protein at each site is unknown. However, it is known that the overall metal affinity with MT varies with the metal species (Kägi and Schaffer, 1988). When purified, Zn-7 MT can be completely replaced by Cd, forming a Cd-7 MT. However, such a replacement is often incomplete since MT isolated from cells exposed to Cd shows a mixed metal content in the form of Zn-2/Cd-5 (Nielson and Winge, 1983; Good et al., 1991; Pountney and Vašák, 1992; Cismowski and Huang, 1991). X-ray crystallographic data indicate that both of these two Zn ions reside in the beta-domain (Robbins et al., 1991). The nature of this site preference has been an enigma.

We showed in a previous study that the preference for binding sites by Zn and Cd in one, but not both, of the two metal binding domains of MT could be demonstrated on the basis of energetics (Chang and Huang, 1996). The preference may be caused by the size of the metal ions and/or the thiolated binding affinity. In the present study we compute the binding energies between individual Cys and Zn/Cd, two naturally bond divalent metal ions, testing the hypothesis that Cys vary in their contribution to metal binding due to a difference in binding energy, depending on the site its occupies and the metal species it binds.

In order to obtain the binding energies between cysteine residues located at specific sites and Zn/Cd, we concentrated our effort on the beta domain in this study as this is the domain in which mixed metal was noted. Since high-level *ab initio* calculation of an entire metal-bound metalloprotein was not feasible, we reduced the beta-domain to beta-metal binding cluster and computed the binding energies of Cys-Metal by MOPAC93/MNDO-PM3 (Dewar and Thiel, 1977; Setwart, 1989, 1991, 1993). The computation was based on a method has been applied to examine designed mimics of suitable regions in the ligand binding domains of a variety of proteins, with the dynamics of the whole structure reconstructed (Warshel and Levitt, 1976; Field et al., 1990; Kollman, 1993; Hwang and Pan, 1995; Chang and Huang, 1996). Thus molecular modeling program CHARMm (Brooks et al., 1983) was used in replacing all the missed atoms in the crystallographic derived structure, and the energies computed upon insertion of either Zn or Cd or a mixture of both. The metal(s) were then dissociated by molecularly teasing the residue away from the metal along the orientation of the mercaptide bond. A series of conformers at defined intervals of dissociation was generated and their binding ener-

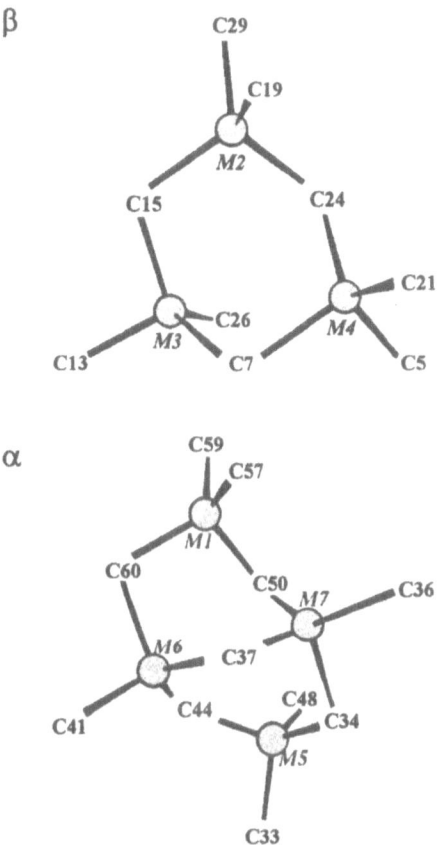

Figure 1. The calculation model for beta and alpha-domains of MT. The number of each Cys indicates of the amino acid residue of the protein.

gies obtained with the potential well theory. Pairwise conformations consisting of the same metallic atom types were compared, and the relative energy difference obtained.(Fig. 2)

Both homo-metal clusters, i.e. a cluster with either three Zn or three Cd and hetero-metal clusters, i.e. mixed metals in the cluster, were examined.

In homo-metal clusters, the relative average binding energy was the strongest for C21 to Cd[M4] and C15 and C26 to Zn[M3], unless the Cys involved are of the bridging type. The bridging Cys are much stronger than terminal ones, except C7 in Cd3 conformer. Among the terminal Cys, the relative binding strength between Zn and C26/C29 are stronger than the other four. The relative binding strength between all terminal Cys and Cd are at the similar level, except C19. C21 is higher than the other terminal Cys. We inferred that C19 may not contribute much to the selective metal binding in the beta-cluster of MT.

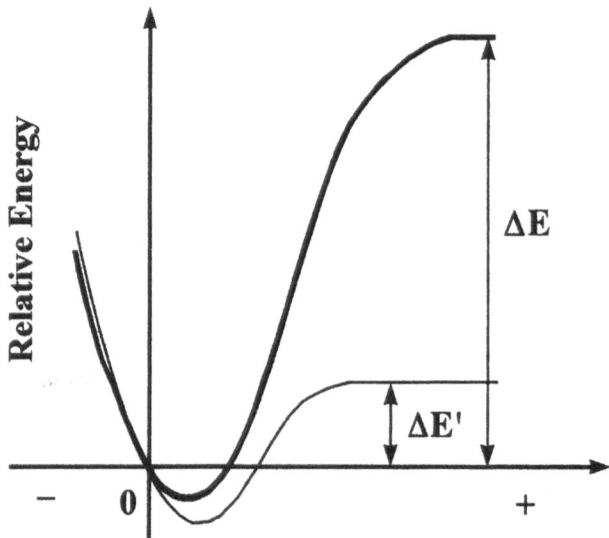

Distance from Initial Position

Figure 2. Theoretical curve of energy change for teasing two atoms. The zero point means the native (initial) structure of the conformer. Base on this point the binding strength of atom A (bold line) is stronger than B (thin line).

In mixed metal clusters (Zn2,Cd1 and Zn1,Cd2), the overall binding energy is strongest for the conformer Zn[M3] than the other two Cd2,Zn1 conformers. On the other hand, a conformer with Cd located at M4 is stronger in binding energy than the other two Zn2,Cd1 conformers for the beta-cluster.

By comparing the overall binding energy we may reach the following conclusions:

1. The bridging Cys play a more important role in structural stability than terminal Cys for their relatively higher binding energy.
2. Not all terminal Cys are the same; C26 is most critical to Zn binding preference, C21 to Cd binding preference, and C19 may not contribute much to metal binding in the beta-domain, and
3. In beta-domain, M3 is a Zn preference site and M4 is a Cd preference site.

These results suggest that binding site preference for Zn/Cd is determined by binding strength between specific Cys and metal ion species.

The above observations are in general consistent with earlier studies with site-directed mutagenesis (Cismowski and Huang, 1991; Cismowski et al., 1991; Chernaik and Huang, 1991; Johnson and Huang, in preparation). While present study with site preference of metals in MT lends credence to the validity of simulated energy computation, we hope to extend the semi-empirical method to other rational design of structure/function studies in which the identification of specific ligand binding residues requires systematic substitution, rendering it feasible carry out mutatgenesis in silico. In this sense, the use of molecular simulation in predicting ligand-protein interaction would have wide implications.

Acknowledgment
We thank JK Hwang for his valuable comments, SF Chen and WF Liao for assistance in programming, and National Center of High-Performance Computing for the use of the facilities. This study was supported by grants NSC 86-2113-M-007-027 and NSC 86-2311-B-007-029 from the National Science Council, ROC.

References

Berg JM and Shi Y (1996) The galvanization of biology: a growing appreciation for the roles of zinc., *Science* 271: 1081–5.
Boulanger Y, Goodman CM, Forte CP, Fesik SW and Armitage IM (1983) Model. for mammalian metallothionein structure. *Proc Natl Acad Sci USA* 80: 1501–1505.
Brooks BR, Bruccoleri RE, Olafson BD, States DJ, Swaminathan S and Karplus M (1983) CHARMm: a program for macromolecular energy, minimization, and dynamics calculations. *J Comput Chem* 4: 187–217.
Chang CC and Huang PC (1996) Semi-empirical simulation of Zn/Cd binding site preference in the metal binding domains of mammalian metallothionein. *Protein Eng* 9: 1165–1172.
Chernaik ML and Huang PC (1991) Differential effect of cysteine-to-serine substitutions in metallothionein on cadmium resistance. *Proc Natl Acad Sci USA* 88: 3024–3028.
Cismowski MJ and Huang PC (1991) Effect of cysteine replacements at positions 13 and 50 on metallothionein structure. *Biochemistry* 30: 6626–6632.
Cismowski MJ, Narula SS, Armitage IM, Chernaik ML and Huang PC (1991). Mutation of invariant cysteines of mammalian metallothionein alters metal binding capacity, cadmium resistance, and 113Cd NMR spectrum. *J Biol Chem* 266: 24390–24397.
Dewar MJS and Thiel W (1977) Ground States of Molecules. 38. The MNDO method. Approximations and parameters. *J.Am.Chem.Soc.* 99: 4899–4917.
Field MJ, Bash PA and Karplus MA (1990) A combined quantum mechanical and molecular mechanical potential for molecular dynamics simulations. *J Comput Chem* 11: 700–733.
Fowle DA and Stillman MJ (1997) Comparison of the structures of the metal-thiolate binding site in Zn(II)-, Cd(II)-, and Hg(II)-metallothioneins using molecular modeling techniques. *J Biomol Struct Dyn* 14: 393–406.
Good M, Hollenstein R and Vašák M (1991) Metal selectivity of clusters in rabbit liver metallothionein. *Eur J Biochem* 197: 655–659.
Hwang JK and Pan JJ (1995) Simulation of chemical reactions in solution by a combination of classical and quantum mechanical approach. *Chem Phys Lett* 243: 171–175.
Kägi JHR and Schaffer A (1988) Biochemistry of metallothionein. *Biochemistry* 27: 8509–8515.
Kim CA and Berg JM (1996) A 2.2 A resolution crystal structure of a designed zinc finger protein bound to DNA. *Nature Struct Biol* 3: 940–5.
Kollman P (1993) Free energy calculations: Applications to chemical and biochemical phenomena. *Chem Rev* 93: 2395–2417.
Nielson KB and Winge DR (1983) Preferential binding of copper to the B domain of metallothionein. *J Biol Chem* 258: 13063–13069.
Otvos JD and Armitage IM (1980) Structure of the metal clusters in rabbit liver metallothionein. *Proc Natl Acad Sci* 77: 7094–7098.
Pountney DL and Vašák M (1992) Spectroscopic studies on metal distribution in Co(II)/Zn(II) mixed-metal clusters in rabbit liver metallothionein 2. *Eur J Biochem* 209: 335–341.
Robbins AH, McRee DE, Williamson M, Collett SA, Xuong NH, Furey WF, Wang BC and Stout CD (1991) Refined crystal structure of Cd, Zn metallothionein at 2.0 * resolution. *J Mol Biol* 221: 1269–1293.
Schultze P, Worgotter E, Braun W, Wanger G, Vašák M, Kägi JHR and Wuthrich K (1988) Conformation of [Cd7]-metallothionein-2 from rat liver in aqueous solution determined by nuclear magnetic resonance spectroscopy. *J Mol Biol* 203: 251–268.
Stewart JJP (1989) Optimization of parameters for semiempirical methods I. Method. *J Comp Chem* 10: 210–220.
Stewart JJP (1991) Optimization of parameters for semiempirical methods. III Extension of PM3 to Be, Mg, Zn, Ga, Ge, As, Se, Cd, In, Sn, Sb, Te, Hg,Tl, Pb, and Bi. *J Comp Chem* 12: 320–341.
Stewart JJP (1993) MOPAC 9300 Manual Fujisu Limited Tokyo Japan.
Vallee BL and Auld DS (1993) New perspective on zinc biochmistry- cocatalytic site in multi-zinc enzyme. *Biochemistry* 32: 6493–6500.
Warshel A and Levitt M (1976) Theoretical studies of enzymic reactions: dieletric, electrostatic and steric stabilization of carbonium ion in the reaction of lysozyme. *J Mol Biol* 103: 227–249.

Metallothionein IV
C. Klaassen (ed.)
© 1999 Birkhäuser Verlag Basel/Switzerland

The growth inhibitory activity of metallothionein-3 correlates with its novel β domain sequence rather than metal binding properties

Laran T. Jensen[1], Jay C. Erickson[2], Richard D. Palmiter[2] and Dennis R. Winge[1]

[1]*University of Utah Health Sciences Center, Salt Lake City UT 84132, USA*
[2]*Howard Hughes Medical Institute, University of Washington, Seattle, WA 98195, USA*

A neuronal survival assay consisting of cultured neonatal rat cortical neurons supplemented with Alzheimer brain extracts was used to identify a protein designated growth inhibitory factor (GIF) as a candidate growth inhibitory component in normal brain tissue [1]. Subsequent characterization of GIF revealed a 70% sequence homology with mammalian metallothionein (MT) including the 20 metal-binding cysteine residues in conserved Cys-X-Cys and Cys-X-X-Cys motifs. The GIF gene also has a similar intron/exon structure to that of mammalian MT-1 and MT-2 leading to its classification as MT-3 [2]. Relative to mammalian MTs, MT-3 contains two inserts, a single threonine residue in the N-terminal β-domain and a glutamine rich hexapeptide in the C-terminal α-domain.

In contrast to other mammalian MTs which are predominantly expressed in liver and kidney, MT-3 is found almost exclusively in the brain [1–3]. Furthermore, MT-3 mRNA levels were unaffected by the administration of heavy metal ions and other inducers of MT biosynthesis [2]. In the neuronal survival assay MT-1 and MT-2 do not exhibit the neuronal growth inhibitory activity seen with MT-3 [1, 4]. Thus, MT-3 is the only known MT to exhibit an activity that may be distinct from metal ion buffering. To determine the basis of the growth inhibitory activity of MT-3 we tested whether the MT-3 activity resulted from either unusual metal-binding properties or a distinct amino acid sequence compared with MT-1 or MT-2.

Properties of purified MT-3

Human and mouse MT-3 proteins were produced in *E. coli* by expression of the cDNA from a T7-based expression vector. MT-3 proteins were purified and shown to be homogenous by the appearance of a single symmetrical elution peak on C_{18} reverse-phase HPLC, the expected amino acid composition, and a single component of the correct mass by electrospray mass spectroscopy [5]. A typical yield from 6 liters of bacterial culture was 30 to 40 mg of MT-3. Human and mouse MT-3 proteins differ in 9 out of 68 possible positions. These sequence variations are distributed throughout the protein. The inactive rat MT-1 differs from human MT-3 in 19 positions excluding the insertion residues.

Purified human and mouse MT-3 were tested in the neuron survival assay as Zn(II), Cd(II), or Zn(II)/Cu(I) complexes. Tissue from the same AD brain was used for all survival assays. Both human and mouse Zn(II) MT-3 complexes antagonized the ability of Alzheimer's disease brain extract to promote increased survival, although the activity of human MT-3 was some-what more pronounced [4, 5]. Neither rat Zn(II) MT-1 nor Zn(II) MT-2 exhibited any activity. In the absence of added AD brain extract, all Zn(II) MTs stimulated survival of neurons, demonstrating that the inhibitory activity of MT-3 is not due to a direct toxic effect on neurons [5]. MT-3 purified from brain has been reported to contain 4 Cu and 3 Zn ions [6]. To determine the metal requirements for the growth inhibitory activity of MT-3, both MT-1 and MT-3 were reconstituted with 4 Cu(I) and 3 Zn(II) ions. While both Zn- and Cu,Zn-MT-3 appeared active in the bioassay, Zn_7-MT-3 was more active. Also, Cu,ZnMT-1 inhibited survival of neurons, but this apparent activity is likely due to Cu-induced toxicity to neurons.

Human and mouse MT-3 bound 7 mol equivalents of either Zn(II) or Cd(II) [5]. Thus, MT-3 binds the same number of divalent metal ions as the well-characterized MT-1 and MT-2. Metallothionein binds Cu(I) and Ag(I) ions with higher stoichiometry than divalent metal ions such as Cd(II) and Zn(II) [7]. To determine whether MT-3 forms analogous Cu(I) complexes, human and mouse apoMT-3 were titrated with increasing quantities of Cu(I) presented as Cu(I)-acetonitrile and the emission of Cu(I)-MT3 complexes was monitored [8]. Maximal emission was observed with 12 Cu(I) ions for both human and mouse MT-3. This is the same maximal stoichiometry as for MT-1 and MT-2 [9]. Thus, the maximal metal binding stoichiometry of both human and mouse MT-3 are identical to those of MT-1 and MT-2.

Metal ions bound to MT-1 and MT-2 are known to be kinetically reactive [10]. To determine whether the unique activity of MT-3 correlated with altered kinetic reactivity of MT-bound Zn(II) ions, the reactivity of MT metal complexes with EDTA was evaluated. The kinetics of the reaction were complex, suggesting at least two phases. An initial rate of Zn(II) transfer to EDTA occurs within the time of mixing and accounts for less than 10% of MT-bound Zn(II). Monitoring the second, slower phase of the reaction revealed similar rates of Zn(II) loss from human and mouse MT-3 and rat MT-1 [5]. Likewise, the kinetics of EDTA-mediated Cd(II) loss from human and mouse MT-3 and rat MT-1 were similar. The kinetic reactivity of the metal ions bound to MT has been suggested to be a result of solvent-accessible cysteine thiolate groups [11]. The thiol reactivity of Zn(II) and Cd(II) complexes of MT-1 and MT-3 with Ellman's reagent was similar for human and mouse MT-3 and rat MT-1 [5].

Competition of the metal bound cysteine thiolates with proton ions is another means of determining the stability of the metal cluster in MTs [12]. The ligand to metal charge transfer bands in MT-3 and MT-1 reconstituted with 7 Cd(II) ions were measured for pH ranging from 2 to 8. The loss of charge transfer transitions in the ultraviolet was monitored at 255 nm. Both MT-3 and MT-1 reacted similarly to a decrease in pH. The loss of half of the bound Cd(II) occurred at approximately pH 3.8 for both MT-3 and MT-1.

Domain peptides of MT-3

The similarity of MT-3 with MT-1 and MT-2 suggested that the structure of MT-3 may resemble MT-2 in having two distinct domains [11]. Consequently, the bioactivity of MT-3 may be conferred by either one domain or the combination of both domains. Domain peptides from human MT-3 were isolated to determine whether the activity maps to either one of the two candidate domains. T7-based expression vectors were constructed with DNA encoding the N-terminal β domain and the C-terminal α domain peptides. Only the purification of the α peptide was successful. There was no accumulation of the β domain peptide from cultures induced with either cadmium or copper salts added to the bacterial medium. Cd(II) complexes of αMT-3 domain peptides were purified from *E. coli* but in low yield. Since isolation of the β-domain peptide by bacterial expression was unsuccessful, the 32-residue polypeptide was synthesized by solid-phase peptide synthesis using standard Fmoc chemistry. Amino acid analysis of αMT-3 and βMT-3 gave the expected composition. Electrospray mass spectroscopy yielded masses of both the αMT-3 (3811.0 Da) and βMT-3 (3263.3 Da) which agreed well with the predicted masses of 3811.5 Da for αMT-3 and 3263.8 Da for βMT-3.

In titration studies apo domain peptides were found to bind 3 Zn(II) ions in the β domain and 4 Zn(II) ions in the α domain. Each of the two domain peptides exhibited maximal luminescence at 6 mol equivalents Cu(I) in titration studies with Cu(I)-acetonitrile. This is also the predominant Cu coordination complex for MT-1 and MT-2 [9]. The observed binding of Cd(II), Zn(II), and Cu(I) at stoichiometries expected for the domains of MT-2 supports the prediction that MT-3 exist as a two domain structure.

Zn(II) complexes of the α and β domains of MT-3 were tested for their ability to inhibit the increased survival of cortical neurons when cultured in the presence of AD brain extract [5]. The β domain Zn(II) complex exhibited a concentration dependent inhibition, whereas the α domain complex had little effect. The β domain of rat MT-2 was devoid of bioactivity. In the absence of added AD extract, domain peptide complexes stimulated neuronal survival. The mapping of activity to the β domain implies that the six-residue insertion in the α domain is without consequence. The growth inhibitory activity of the β domain of MT-3 suggests that structural differences within the β domains of MT-1 and MT-3 are a determinant of bioactivity.

Mutational analysis of MT-3

In addition to the one-residue insertion in the MT-3 β domain, there are five sequence positions at which significant changes occur between MT-1 or MT-2 and MT-3. One striking change is the C_6-P-C-P_9 tetrapeptide sequence in MT-3. The C-P-C-P sequence is novel in the MT family of proteins. To determine whether this sequence contributes to the activity of MT-3, a β domain peptide variant in which the C-P-C-P sequence was changed to C-T-C-T was synthesized. Serines, threonines, and alanines are commonly found in these positions in other MTs. The P7T,P9T β domain peptide was purified to a single component and had amino acid analysis composition and mass by electrospray mass spectroscopy consistent with the expected values. The mutant β domain peptide bound 3 Zn(II) ions. The kinetic reactivities of the mutant Zn(II) β-MT3 complex with EDTA and DTNB were similar to those of the wild-type β domain peptide. The survival of neurons cultured in the present of AD brain extract was unaffected by the P7T,P9T mutant β domain peptide. A second β mutant containing the C-S-C-A sequence was also devoid of inhibitory activity. As expected, the mutant β domain peptides stimulated neuronal survival in the absence of AD extracts.

To verify that the Pro residues at positions 7 and 9 in MT-3 were indeed critical for the growth inhibitory activity, these residues were changed in the full length MT-3. Mutations of Pro at position 7 to Ser and Pro at position 9 to Ala creating P7S,P9A MT-3 did abolish the growth inhibitory activity that is seen in the wild type MT-3. To determine if one or both of the proline residues were critical for activity single mutations at each position were generated. Neither P7S MT-3 nor P9A MT-3 showed any growth inhibitory activity suggesting that both Pro residues are necessary for the bioactivity of MT-3. The amino acid sequence of MT-3 contains a single amino acid insert consisting of a Thr residue at position 5 (relative to the MT-3 sequence) compared to MT-1 and MT-2. A deletion of Thr5 caused a loss of growth inhibitory activity suggesting a role for this residue in the bioactivity of MT-3. Since Thr and Ser residues are often the targets for phosphorylation, a Thr-to-Ala substitution was generated and found to have attenuated bioactivity.

To test directly whether the N-terminal sequence of MT-3 was the critical determinant of bioactivity, amino acid substitutions were generated in MT-I to determine whether growth inhibitory activity could be created in MT-I. Pro residues were inserted at positions 7 and 9. S6P,S8P MT-1 did not contain any growth inhibitory activity. Since the Thr at position 5 in MT-3 appears to also be important for activity, a Thr was inserted in S6P,S8P MT-1 at the same position as it occurs in MT-3 generating S6P,S8P+T5 MT-1. The insertion of the Thr generated a molecule with bioactivity. Preliminary results suggest that S6P,S8P+T5 MT-1 contains greater growth inhibitory activity than wild type MT-3.

Thus, the ability of MT-3 to inhibit the survival of rat cortical neurons *in vitro* map to the β domain. The unique Thr in MT-3 and Pro residues at positions 7 and 9 are critical for bioactivity. Inserting these MT-3 residues in a MT-1 backbone appears to generate bioactivity in the modified MT-1 showing that these residues in MT-3 may be the key determinant of bioactivity. Bioactivity does not appear to be related to the metal-binding properties of MT-3. The mechanism of growth inhibition of neurons by MT-3 remains unresolved. Therefore, no conclusions can be reached on the structural basis for the MT-3 N-terminal sequence in the novel bioactivity.

References

1. Uchida Y, Takio K, Titani K, Ihara Y and Tomonaga M (1991) *Neuron* 7,337–347.
2. Palmiter RD, Findley SD, Whitmore TE and Durnam DM (1992) *Proc Natl Acad Sci USA* 89: 6333–6337.
3. Tsuji S, Kobayashi H, Uchida Y, Ihara Y and Miyatake T (1992) *EMBO J* 11: 4843–4850.
4. Erickson JC, Sewell AK, Jensen LT, Winge DR, Palmiter RD (1994) *Brain Res* 649: 297–304.
5. Sewell AK, Jensen LT, Erickson JC, Palmiter RD and Winge DR (1995) *Biochemistry* 34: 4740–4747.
6. Bogumil R, Faller P, Pountney DL and Vašák M (1996) *Eur J Biochem* 238: 698–705.
7. Nielson KB, Atkin CL and Winge DR (1985) *J Biol Chem* 260: 5342–5350.
8. Stillman MJ and Gasyna Z (1991) *Methods Enzymol* 205: 540–555.
9. Nielsen KB and Winge DR (1985) *J Biol Chem* 260: 8698–8701.
10. Shaw CF, III, Savas MM and Petering DH (1991) *Methods Enzymol* 205: 401–414.
11. Robbins AH, McRee DE, Williamson M, Collett SA, Xuong NH, Furey WF, Wang BC and Stout CD (1991) *J Mol Biol* 221: 1269–1293.
12. Law AY and Stillman MJ (1980) *Biochem Biophys Res Commun* 94: 138–143.

Recombinant synthesis and metal-binding abilities of mouse metallothionein 1 and its α- and β-domains

Sílvia Atrian[1], Roger Bofill[2], Mercè Capdevila[2], Neus Cols[1], Pilar González-Duarte[2], Roser González-Duarte[1], Àngels Leiva[2], Oscar Palacios[2] and Núria Romero-Isart[2]

[1]*Departament de Genètica, Facultat de Biologia, Universitat de Barcelona, Av. Diagonal 645, E-08071 Barcelona, Spain*
[2]*Departament de Química, Universitat Autònoma de Barcelona, E-08193 Bellaterra, Barcelona, Spain*

Summary. Heterologous expression in *E.coli*, coupled with spectroscopic analyses, has enabled the metal binding properties of the recombinant mouse Zn-MT, Zn-αMT and Zn-βMT to be characterized. Cd(II) binding properties of Zn_7-MT and Zn_4-αMT have shown that they behave like the native proteins and those of the Zn_3-βMT fragment have been reported for the first time. In contrast, the behavior of these proteins towards Hg(II), Cu(I) and Ag(I), which is strongly influenced by the pH and the stabilization time, differs from that of the native forms. Furthermore, recombinant Cd-βMT has also been expressed and characterized.

Introduction

To date most of the spectroscopic and chemical studies on the metal-binding properties of MT have been carried out with proteins isolated from mammalian organs. Relatively few studies have been undertaken with synthetically prepared MT and even less with the protein obtained by genetic engineering. The first attempts to express mammalian MT in *E.coli* had limited success, probably due to translational and stability constraints of this singular eukaryotic protein in a prokaryotic environment. In 1990, Kille et al. described the synthesis of a non-mammalian MT (rainbow-trout) in *E.coli* as an independent protein, and the same procedure was further used for human MT [1] and sea urchin MT [2]. Unfortunately, Kille's expression system was based on Cd(II) addition to the culture medium and therefore the MT recovered was Cd-complexed, and thus unsuitable for titration experiments. Recently, Odawara et al. [3] described another expression system in *E.coli*, equally unsuccessful for the Zn-coordinated form. Recombinant DNA methods could also provide individual MT domains, however, the expression systems reported (also Cd-dependent) were unsuccessful for β domain synthesis [4]. We have designed a new expression system and successfully recovered mouse Zn_7-MT and the corresponding α (Zn_4-αMT) and β (Zn_3-βMT) domains. Their metal-binding abilities towards several metal ions have been analyzed.

Recombinant synthesis of Zn-MT forms

Mouse MT 1 cDNA was amplified from plasmid pBX [5]. The expression vector pGEX-4T-1 (Pharmacia) containing the glutathione-S-transferase (GST) coding region was used for fusion protein synthesis. Besides, DNA fragments coding for the mouse αMT 1 (from Lys31 to the C-terminus) and βMT 1 (from Met1 to Lys30) were constructed by one-step mutagenic PCR using the full-length MT cDNA. *E.coli* JM105 cells were transformed and recombinant plasmids were recovered and analyzed. To overexpress MT 1 proteins, the protease-deficient strain *E.coli* BL21 was used. Cultures in LB at 37°C were supplemented with ampicillin, induced with IPTG and finally supplemented with $ZnCl_2$. Expression assays, purification of the fusion protein, and recovery and characterization of the MT portion are described in [6]. Aliquots of the protein-containing FPLC fractions were pooled, aliquoted and kept at $-70\,°C$ for further use.

Cd(II) binding

The spectral changes in absorption (UV-VIS) and circular dichroism (CD) recorded during the titrations of recombinant mouse Zn_7-MT and Zn_4-αMT with Cd(II) showed that they behave analogously to the proteins obtained from mammalian organs [6, 7]. Furthermore, while Cd_4-αMT species are completely resistant to the effects of additional Cd(II), exposure of Cd_7-MT to Cd^{2+} resulted in further binding of these ions to the protein although saturation was not achieved on the addition of up to 22 mol eq of Cd^{2+} to Zn_7-MT. Titration of recombinant mouse Zn_3-βMT with Cd(II) showed formation of Cd_3-βMT previously unreported. The addition of excess Cd(II) led to Cd_4-βMT which, with the extra loading of Cd(II), unraveled to give rise isodichroically to Cd_9-βMT [6, 8] that exhibits a characteristic CD spectrum.

Hg(II) binding

In contrast to Cd(II), titration of the recombinant proteins with Hg(II), Ag(I) and Cu(I) has afforded different results than those found with the native proteins [9]. The Hg(II) binding abilities of MT at pH 7 and pH 3 have been examined using CD and UV-VIS spectroscopies. The constancy of the pH along the titrations and the consideration of the time required for thermodynamic or kinetic control have revealed characteristic features of the Hg(II)-thiolate coordination in MT. Different pathways are followed for the formation of the Hg-MT species depending on the stabilization time allowed.

In the first stages of the titrations at pH 7, Figure 1A, the Zn(II) remains bound to MT to afford the species: Zn_7Hg_2-MT, Zn_xHg_5-MT$(x \leq 7)$ and Zn_yHg_7-MT $(y < 7)$, all with characteristic spectral fingerprints. Further additions beyond 7 mol equiv. of Hg(II) cause the progressive loss of the Zn(II) ions, which gives rise to the Hg_{9-10}-MT, Hg_{11}-MT and Hg_{15-16}-MT species.

In the titrations at pH 3, Figure 1B, Hg_5-MT [$(Hg_4)_\alpha(Hg)_\beta$-MT] is formed irrespective of the stabilization time, t = 0 or 24 h. However, if thermodynamic control is allowed the Hg(II)-binding pathway follows the sequence: Hg_7-MT [$(Hg_1)_\alpha(Hg_6)_\beta$-MT], Hg_{10}-MT [$(Hg_4)_\alpha(Hg_6)_\beta$-MT], Hg_{18}-MT. Conversely, if kinetic control is imposed the sequence is: Hg_7-MT [$(Hg_4)_\alpha(Hg_3)_\beta$-MT], Hg_{11}-MT, Hg_{13}-MT and Hg_{18}-MT. Both Hg_{18}-MT species have the same structure.

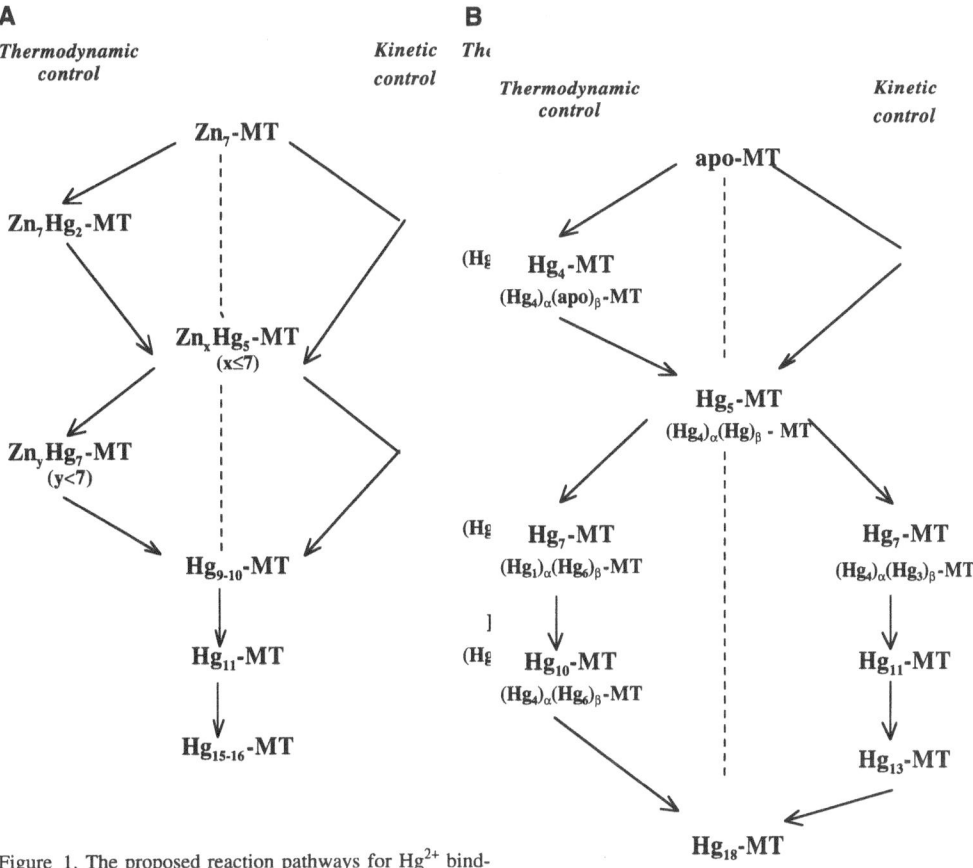

Figure 1. The proposed reaction pathways for Hg^{2+} binding to recombinant mouse: A) Zn_7-MT 1 at pH 7 and B) apo-MT 1 at pH 3. The proposed distribution between domains is based on CD and UV-Vis data. The species at the arrow ends occur at saturation points for spectral signals and thus represent changes from one structure to the next. The reaction pathways adopted for metal binding

One striking example of the influence of the pH value and stabilization time is afforded by the Hg_7-MT species: in the titration of Zn_7-MT with Hg(II) at pH 7 the Zn_xHg_7-MT (x < 7) species is obtained. In the titrations at pH 3, when thermodynamic control is allowed the Hg_7-MT species presents the $(Hg_1)_\alpha(Hg_6)_\beta$-MT distribution between domains, while if kinetic control is imposed the Hg_7-MT species is also formed but with a different metal-distribution, $(Hg_4)_\alpha(Hg_3)_\beta$-MT. These variables also play a significant role in the Hg-MT species formed along the titrations of the Zn_4-αMT and Zn_3-βMT fragments with Hg(II).

Cu(I) and Ag(I) binding

Numerous studies have indicated that Ag(I) and Cu(I) bind similarly if not identically to MT [10]. In contrast, work now in progress with recombinant Zn_7-MT and corresponding fragments clearly suggests a different behavior for both metal ions. Also, it has been observed that the Zn(II) ions remain bound to the protein in the first stages of all the titrations at pH 7 to give rise to species of general formula Zn_xM_y-MT (M = Ag, Cu) as described for Hg(II). In all these cases electrospray mass spectrometry (ESI-MS) provides additional information on the nature of the species formed along the titration. Preliminary results on the titration of apo-βMT with Cu(I) at pH 3 and Zn_3-βMT with Ag(I) at pH 7 are shown in Figure 2.

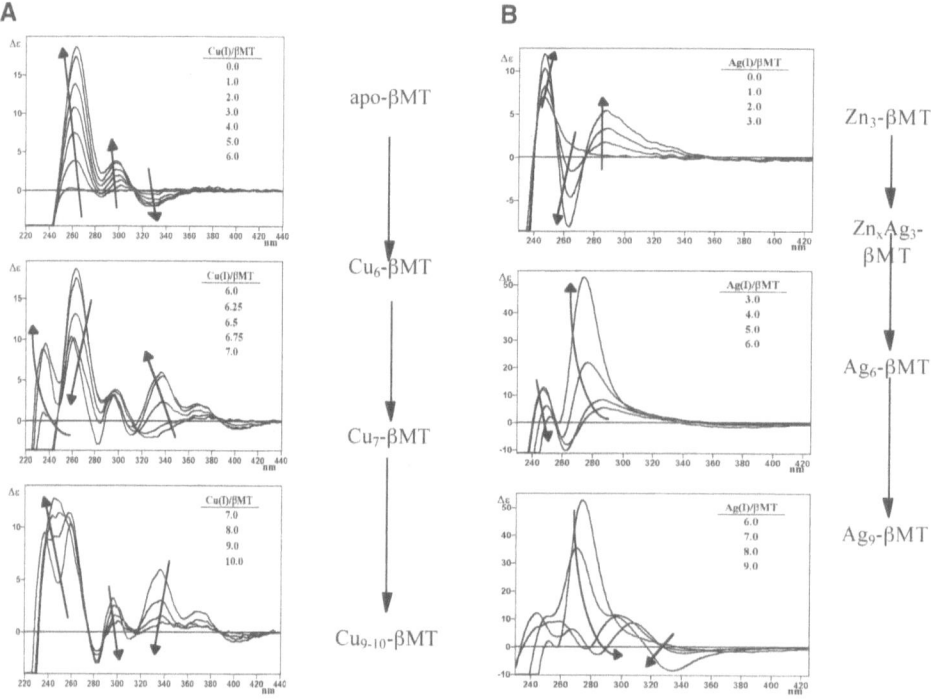

Figure 2. CD spectra recorded during titration of recombinant mouse: (A) apo-βMT 1 at pH 3 with Cu^+ and (B) Zn_3-βMT 1 at pH 7 with Ag^+. The M(I) (M = Cu, Ag) to βMT ratios are given within each frame. The proposed reaction pathways are also indicated.

Cd-βMT biosynthesis

All the expression systems assayed for MT domain synthesis [iv] had in common the presence of Cd in the culture medium, and thus implied the synthesis of Cd-coordinated forms. In contrast, our Zn-based expression system successfully rendered good yields of Zn-βMT, indicating that a poor Cd-binding capacity of the β-domain *in vivo* could probably be the ultimate reason of the reported problems. To further analyze this hypothesis, we also assayed the recombinant production of Cd-coordinated βMT. *E.coli* cells (JM105 and BL21) transformed with the same constructs used to express MT forms in the presence of Zn were grown in Cd supplemented media. No GST/βMT fusion protein was detected in the homogenates of bacterial JM105 cells (protease-wild-type genotype). However, when BL21 cells (protease-deficient genotype) were used, the presence of GST/Cd-βMT was clearly observed which indicates that biosynthesis of the Cd-βMT complex can be also successfully attained, provided that full proteolysis is avoided. The Cd-βMT has been purified and characterized in order to compare its structure to that of the Cd$_3$-βMT complex obtained by Zn/Cd replacement of the Zn$_3$-βMT form. The recombinant β-domain recovered from cadmium-loaded protease-free cells is very different from the Cd$_3$-βMT cluster obtained by titration of the recombinant Zn$_3$-βMT cluster. Protein yields are low, Cd-binding is partially hampered and oxidation goes up to 63%.

Overall, problems concerning β-domain expression arise from a synergic effect of three factors: 1) the nature of the β-domain – as the biosynthesis of the entire MT and of the α-domain occurs in the presence of Cd-, 2) the presence of cadmium in the cell – as the biosynthesis of the β-domain occurs in the absence of Cd, either Zn-supplemented or even metal-free media- and 3) the presence of proteases in the cell – as the biosynthesis of the β-domain is only performed in Cd-supplemented media using an *E.coli* protease-free strain (BL21)-. Based on these facts, our hypothesis is that cadmium-binding to the MT β-domain is deficient and slow, and therefore a partially-complexed peptide results from recombinant expression. Furthermore, cadmium presence in the cell could modify some proteases, which, for some reason, would be activated and cause the absolute degradation not only of the βMT portion, but of the total-length fusion protein. In 1986, Bond et al. [11] reported a 240–970% activation of a Cd-substituted carboxypeptidase A in front of thioamide compounds, situation which resembles that of βMT: an only partially metal-loaded peptide, with free thiol-groups inside the cells, and a high cadmium concentration that might be involved in protease modification.

Our results show that recombinant synthesis of βMT in *E.coli* is possible even in the presence of cadmium, and provide a feasible explanation to the repeatedly reported phenomenon of the unsuccessful β-domain synthesis. Besides, they agree with the data available in the literature about the behavior *in vitro* of synthetic or native βMT in front of Cd [12, 13]. The inherent inability of the β-domain to coordinate Cd(II) supports the hypothesis that Cd(II)-binding was not one of the primitive functions of this metal-chelating protein whose ancestral building unit is supposed to be the β fragment.

Site-directed mutant forms of mouse MT

Protein engineering is one of the most powerful tools to study function/structure relationships and to slightly modify a protein to better adapt it to some biotechnological purpose. Thus, we have constructed in our laboratory several mouse MT site-directed genes and biosynthesized them in the expression system described above. In different experiments, Cys in positions 5, 13, 33, 36, 41 and 57 have been independently substituted by His according to molecular modeling analyses [14] which predicted that a metal cluster architecture similar to that found for the native domains could be achieved. We are currently analyzing their coordination abilities to unravel further details of the MT metal-binding mechanisms and to produce mutant forms less susceptible of oxidation in order to be used in heavy metal detoxification of water.

Acknowledgments
We are grateful to Dr. D. Palmiter (University of Washington, USA) who kindly provided a mouse MT 1-cDNA clone. This work was supported by the Catalan Government (Departament de Medi Ambient, Generalitat de Catalunya) and the Spanish *Comisión Interministerial de Ciencia y Tecnología* (CICYT # BIO92-0591-CO2). We also thank the Serveis Científico-Tècnics de la Universitat de Barcelona. MC and NR-I are indebted to the Generalitat de Catalunya for a post-doctoral and a pre-doctoral CIRIT scholarship, respectively.

References

1. Kille P, Lees WE, Darke BM, Winge DR, Dameron CT, Stephens P, Kay J (1992) Sequestration of cadmium and copper by recombinant rainbow trout and human metallothioneins and by chimeric (mermaid and fishman) proteins with interchanged domains. *J Biol Chem* 267: 8042–8049.

2. Wang YJ, Mackay EA, Kurasaki M, Kägi JHR (1994) Purification and characterization of recombinant sea urchin metallothionein expressed in *Escherichia coli*. *Eur J Biochem* 225: 449–457.

3. Odawara F, Kurasaki M, Suzuki-Kurasaki M, Oikawa S, Emoto T, Yamasaki F, Linde Arias AR, Kojima Y (1995) Expression of human metallothionein-2 in *Escherichia coli*: cadmium tolerance of transformed cells. *J Biochem* 118: 1131–1137.

4. a) Sewell AK, Jensen LT, Erickson JC, Palmiter RD, Winge DR (1995) Bioactivity of metallothionein-3 correlated with its novel β domain sequence rather that metal binding properties. *Biochemistry* 34: 4740–4747.
   b) Kurasaki M, Emoto T, Linde-Arias AR, Okabe M, Yamasaki F, Oikawa S, Kojima Y (1996) Independent self-assembly of cadmium-binding α-fragment of metallothionein in *Escherichia coli* without participation of b-fragment. *Protein Engng* 9: 1173–1180.

5. Glanville N, Durnam DM, Palmiter RD (1981) Structure of mouse metallothionein-I gene and its mRNA. *Nature* 292: 267–269.

6. Capdevila M, Cols N, Romero-Isart N, González-Duarte R, Atrian S, González-Duarte P (1997) Recombinant synthesis of mouse Zn_3-β and Zn_4-α metallothionein 1 domains and characterization of their cadmium(II) binding capacity. *Cell Mol Life Sci* (formerly *Experientia*) 53: 681–688.

7. Cols N, Romero-Isart N, Capdevila M, Oliva B, González-Duarte P, González-Duarte R, Atrian S (1997) Binding of excess cadmium(II) to Cd_7-metallothionein from recombinant mouse Zn_7-metallothionein 1. UV-VIS absorption and circular dichroism studies and theoretical location approach by surface accessibility analysis. *J Inorg Biochem* 68: 157–166.

8. Capdevila M, Romero N, Cols N, Atrian S, Stillman MJ, González-Duarte R, González-Duarte P (1996) Cadmium(II) binding to recombinant mouse Zn_3-βMT metallothionein 1. *Anal Quim Int Ed* 92: 199–201.

9. a) Lu W, Stillman MJ (1993) Mercury thiolate cluster in metallothionein. Analysis of circular dichroism spectra of complexes formed between α-metallothionein, apometallothionein, zinc metallothionein and cadmium metallothionein and Hg^{2+}. *J Amer Chem Soc* 115: 3291–3299.
   b) Lu W, Zelazowski A, Stillman MJ (1993) Mercury binding to metallothioneins: Formation of the Hg_{18}-MT species. *Inorg Chem* 32: 919–926
   c) Zelazowski A, Stillman MJ (1992) Silver binding to rabbit liver zinc metallothionein and zinc a and b fragments. Formation of silver metallothionein with Ag(I): Protein ratios of 6, 12 and 18 observed using circular dichroism spectroscopy. *Inorg Chem* 31: 3363–3370.
   d) Li YJ, Weser U (1992) Circular dichroism, luminiscence, and electronic absorption of copper binding sites. *Inorg Chem* 31: 5526–5533.

10. Li H, Otvos JD (1996) HPLC characterization of Ag^+ and Cu^+ metal exchange reactions with Zn- and Cd-metallothioneins. *Biochemistry* 35: 13937–13945.

11. Bond MD, Holmquist B, Vallee BL (1986) Thioamide substrate probes of metal-substrate interactions in carboxypeptidase A catalysis. *J Inorg Biochem* 28: 97–105.

12. a) Okada Y, Ohta M, Yagyu M, Min KS, Onosaka S, Tanaka K (1985) Synthesis of a nanocosapeptide (β-fragment) corresponding to the N-terminal sequence 1–29 of human liver metallothionein II and its heavy metal-binding properties. *FEBS Lett* 183: 375–378.
    b) Okada Y, Ohta N, Iguchi S, Tsuda Y, Sasaki H, Kitagawa T, Yagyu M, Min KS, Onosaka S, Tanaka K (1986) Amino acids and peptides XIII. Synthesis of a nonacosapeptide corresponding to the N-terminal sequence 1–29 (β-fragment) of human liver metallothionein II (hMTII) and its heavy metal-binding properties. *Chem Pharm Bull* 34: 986–998.

13. a) Kull FJ, Reed HF, Elgree TE, Ciardelli TL, Wilcox DE (1990) Solid-phase peptide synthesis of the α and β domains of liver metallothionein 2 and the metallothionein of *Neurospora crassa*. *J Amer Chem Soc* 112:

2291–2298.

b) Stillman MJ, Cai W, Zelazowski AJ (1987) Cadmium binding to metallothioneins. Domain specificity in reactions of α and β fragments, apometallothionein, and zinc metallothionein with Cd^{2+}. *J Biol Chem* 262: 4538–4548.

14. Romero N, Capdevila M, González-Duarte P, Oliva B (1996) Computational analysis of cysteine substitutions modeled on the α and β domains of Cd_5,Zn_2-metallothionein 2. *J Mol Model* 2: 1–5.

Biochemical characterization of monomeric and dimeric cadmium-substituted metallothionein-3

Peep Palumaa and Rain Lehtme

Institute of Molecular and Cell Biology, University of Tartu, Vanemuise 46, Tartu EE2400, Estonia

Summary. Human brain-specific metallothionein (MT-3) has been expressed in *E. coli* in the presence of cadmium salt and purification of protein combining ethanol precipitation, gel filtration and ionic exchange chromatography has been elaborated. Results obtained demonstrate the occurrence of protein in distinct metastable monomeric and dimeric forms which could be successfully separated by ionic exchange chromatography. Monomeric and dimeric MT-3 has been characterized by UV, CD spectroscopy as well as by sulphydryl content. UV spectrum of monomeric form is similar to the spectrum of common Cd_7MT-2 but exposes slightly lower extinction coefficent. In CD spectrum of monomeric MT-3 the lowest energy transition is blue-shifted (maximum at 254 nm) as compared to Cd_7MT-2 (maximum at 259 nm) and CD band at 225 nm is negative. Dimeric MT-3 exposes in UV spectrum a shoulder at 250 nm and a new transition with maximum at 285 nm. CD spectrum of dimeric MT-3 exposes positive band at 259 nm and negative band at 230 nm and broad shoulder in region up to 300 nm. CD band at 225 nm is negative. Dimeric MT-3 dissociates into monomeric form by spontaneous incubation as well as by the action of sulphydryl reagents (2-mercaptoethanol and dithiothreitol) pointing towards existence of intermolecular disulphide bridges in dimeric MT-3.

Introduction

Metallothionein-3 (MT-3) was discovered as a protein with levels drastically decreased in the brain of patients with Alzheimer's disease [1]. It has been demonstrated that MT-3 exhibits growth inhibitory activity towards rat cultured neurons and initially the protein has been named as growth inhibitory factor (GIF) [1–4].

MT-3 protein is characterized only on the level of primary structure which exhibits ca. 65% similarity with primary structures of other mammalian MTs [5]. MT-3 contains a pattern of conserved cysteines and lysines identical to other MTs, but it differs from other members of MT family in two aspects. First, MT-3 protein has two insertions relative to all other known mammalian MTs: a single amino acid insert after the fourth amino acid and a block of six amino acids near the carboxyl terminus [1, 5–6]. Secondly, significant differences occur in N-terminal nanopeptide sequence Met(1)-Asp-Pro-Glu-Thr-Cys-Pro-Cys-Pro which is specific for MT-3. It has been demonstrated that this proline-rich sequence is related with growth inhibitory activity of MT-3 [7] where Pro7 and Pro9 play a crucial role [8].

Isolated MT-3 from brain contains usually 2–5 Zn ions and 1–5 Cu ions in species-specific ratios [1, 5, 9]. Mixed metalloforms of MT-3 are stable in air and they have been characterized by UV, CD and luminescence spectroscopy [5, 9]. Solely Zn- or Cd-containing metalloforms of MT-3 have been expressed in bacterial expression systems [2, 8, 10] but peculiar reversible aggregation of the protein has hindered the isolation and biochemical characterization of homogeneous metalloforms [8].

In the present work we have developed ionic exchange chromatography-based procedure for isolation of homogenous monomeric and dimeric forms of cadmium MT-3. Isolated MT-3 forms have been characterized by UV and CD spectroscopy.

Purification of different MT-3 forms

Human MT-3 cDNA, was kindly donated by Richard D. Palmiter (University of Washington), cloned into pET-11c expression vector, transfected into *E. coli* strain BL-21 and protein was expressed in LB medium by using standard techniques [11].

Protein has been purified by using two step precipitation with ethanol [12] and crude MT-3 has been purified by gel filtration on Sephadex G-75 column. Although gel filtration of crude MT-3 on Sephadex G-75 column exposes one main cadmium-containing protein peak, according to analytical gel filtration on Superdex 75 this fraction is composed from two protein forms corresponding to monomeric and dimeric MT-3. For further separation the cadmium-containing peak from Sephadex G-75 column was pooled and applied to a Q-Sepharose (Pharmacia) ion exchange column. Proteins were eluted with the linear NaCl gradient from 100 to 500 mM. Two peaks eluting at 190 mM and 310 mM NaCl have been obtained (Fig. 1). Size exclusion chromatography and electroforetical experiments without reducing agents demonstrates that the first peak corresponds to the MT-3 monomer, and the second peak to MT-3 dimer. Stokes radius of monomeric MT-3 seems larger than that of Cd_7MT-2.

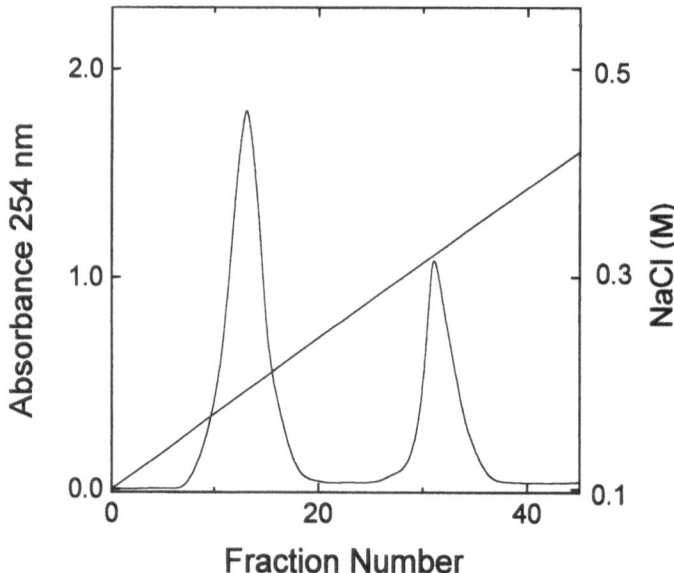

Figure 1. Anion exchange chromatography of MT-3 containing sample on Q-Sepharose column.

Isolated molecular forms of MT-3 are stabile in timescale of days. However, rechromatography of MT-3 dimer sample on Superdex 75 column exposes slow time dependent appearance of MT-3 monomer peak and concomitant decrease of the peak of MT-3 dimer. Phenomenon observed speaks for spontaneous dissociation of MT-3 dimer to monomeric form which leads to 30% dissociation of dimer during seven days. Dissociation of MT-3 dimer to monomer has also been initiated by sulphydryl reagents 2-mercaptoethanol and dithiothreitol which accelerate the dissociation process. Complete dissociation of MT-3 dimer to monomers has not reached in conditions applied and final equilibrium position corresponding to 40–50% of dimer has been reached by the influence of sulphydryl reagents. Isolated monomeric MT-3 is more stabile compared with dimeric form but exposes concentration dependent dimerization leading to 10% dimerization during ten days (MT-3 concentration 10 μM). Ionic exchange chromatography of spontaneously dissociated sample of dimeric MT-3 or dimerized sample of MT-3 monomer exposes again two peaks characteristic for MT-3 monomer and dimer.

Spectroscopic characterisation of different MT-3 forms

UV spectrum of monomeric MT-3 is similar to that of Cd_7MT-2 spectrum with characteristic absorption shoulder in 250 nm region (Fig. 2). However, molar extinction coefficient is somewhat lower as compared to Cd_7MT-2 which points towards existence of lower number of Cd-Cys transitions in protein structure [13]. According to SH measurements MT-3 monomer

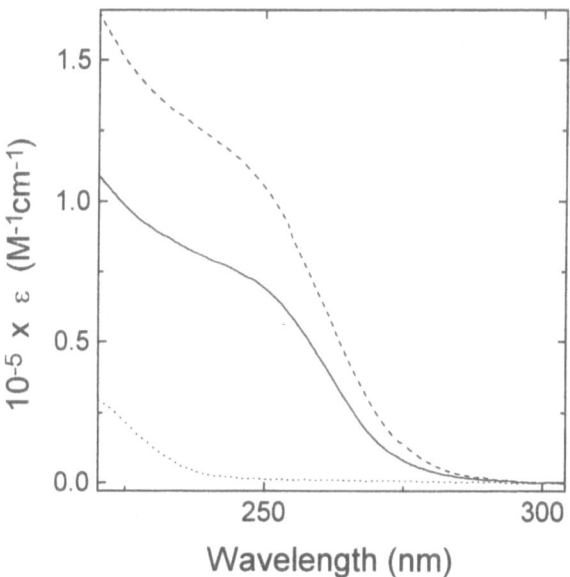

Figure 2. UV spectra of monomeric MT-3. Conditions: 20 mM TRIS-HCl, 100 mM NaCl buffer, pH 8.4 – solid line. 20 mM TRIS-HCl, 100 mM NaCl, pH 2.0 – dotted line. Cd_7MT-2 in 20 mM TRIS-HCl, 100 mM NaCl buffer, pH 8.4 – dashed line.

contains one disulphide bond and consequently the number of Cd-Cys transitions should be lower as in case of Cd_7MT-2. We suggest that in metal clusters of MT-3 also carboxylates can participate in metal coordination.

In CD spectrum of monomeric MT-3 the lowest energy transition is blue-shifted (maximum at 254 nm) as compared to Cd_7MT-2 (maximum at 259 nm) and CD band at 225 nm is negative (Fig. 3). Observed blue shift of lowest energy transition is in agreement with putative participation of carboxylates in metal coordination.

Dimeric MT-3 exposes in UV spectrum a shoulder at 250 nm region, characteristic for MTs and a new transition with maximum at 285 nm not described earlier in case of CdMTs (Fig. 4). As lowering of the pH to 2.0 leads to disappearance of this feature transition should originate from metal-ligand transitions. The lowest energy charge transfer transitions in CdMT-2 are centered around 250 nm [13, 14] and in Cd_7MT-2 spectrum there is practically no absorbtion around 280 nm. Consequently bridging or terminal thiolates could not explain the origin of the transition observed. Consequently the origin of latter band could be explained with metal-disulphide charge transfer transitions. Features listed demonstrate existence of new Cd cluster structure in dimeric MT-3, which deserves detailed structural studies.

CD spectra of dimeric MT-3 exposes the lowest energy transition at 259 nm, similar to Cd_7MT-2 and broad shoulder up to 300 nm (Fig. 5). CD band at 225 nm is negative.

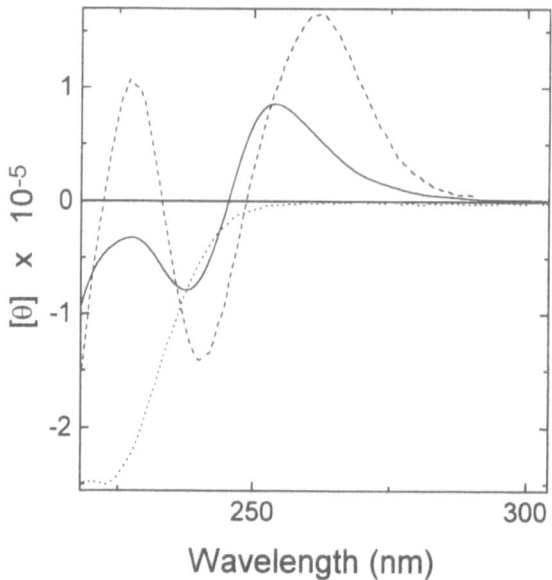

Figure 3. CD spectra of monomeric MT-3. Conditions: 20 mM TRIS-HCl, 100 mM NaCl buffer, pH 8.4 – solid line. 20 mM TRIS-HCl, 100 mM NaCl, pH 2.0 – dotted line. Cd_7MT-2 in 20 mM TRIS-HCl, 100 mM NaCl buffer, pH 8.4 – dashed line

UV and CD spectra of monomeric as well as dimeric MT-3 are sensitive to pH. Acidification of the sample to pH 2 leads to total disappearance of absorption above 240 nm. Latter behaviour is typical to MTs and suggests that metal-ligand charge transfer bands are responsible for absorbtion as well as CD transitions above 240 nm.

Conclusions

Present data demonstrate, that CdMT-3 could be isolated in monomeric and dimeric molecular forms.

Monomeric MT-3 elutes from Q-Sepharose column at 190 mM of NaCl, exposes slightly larger Stokes radius in size exclusion chromatography as compared with common Cd$_7$MT-2 (apparent molecular weight on Superdex 75 column 17 kDa and 11 kDa respectively) and contains one disulphide bond. In this form there are fewer Cd-Cys transitions in protein structure and participation of carboxylates in metal coordination is suggested.

MT-3 dimer elutes from Q-Sepharose column at 310 mM of NaCl, exposes apparent molecular weight 28.8 kDa. MT-3 dimer is bound with disulphide bridges and dissociates spontaneously and by the action of sulphydryl reagents to monomeric MT-3 form. Dissotiation is always incomplete, speaking for existence of equilibria between monomeric and dimeric forms.

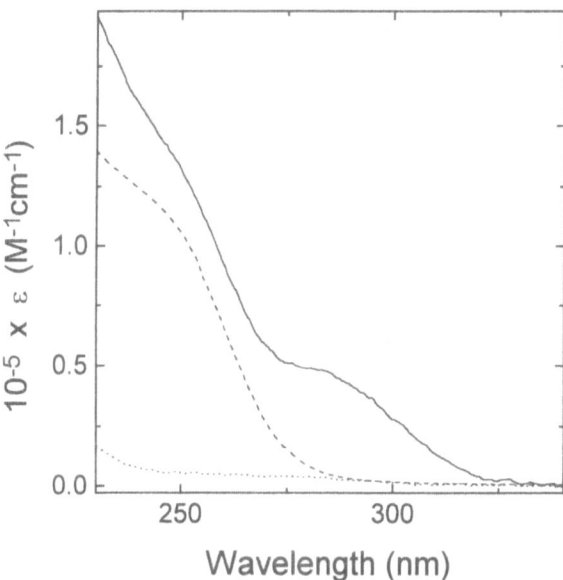

Figure 4. UV spectra of dimeric MT-3. Conditions: 20 mM TRIS-HCl, 100 mM NaCl buffer, pH 8.4 – solid line. 20 mM TRIS-HCl, 100 mM NaCl, pH 2.0 – dotted line. Cd$_7$MT-2 in 20 mM TRIS-HCl, 100 mM NaCl buffer, pH 8.4 – dashed line.

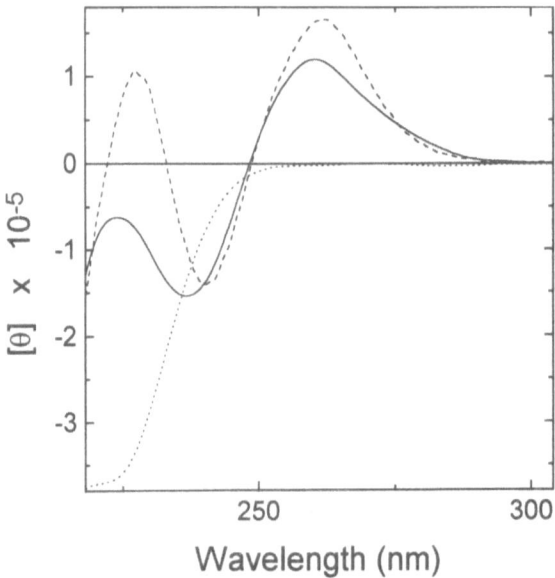

Figure 5. CD spectra of dimeric MT-3. Conditions: 20 mM TRIS-HCl, 100 mM NaCl buffer, pH 8.4 – solid line. 20 mM TRIS-HCl, 100 mM NaCl, pH 2.0 – dotted line. Cd$_7$MT-2 in 20 mM TRIS-HCl, 100 mM NaCl buffer, pH 8.4 – dashed line.

Peculiar feature observed in spectroscopy of MT-3 dimer is an intense metal-dependent absorption band in UV region with maximum at 285 nm which has earlier not been observed in case of CdMTs. Origin of latter transition is unclear and could be clearified by detailed structural studies.

Acknowledgements
The authors gratefully acknowledge prof. R.D. Palmiter (University of Washington) for human MT-3 cDNA and prof. A. Gräslund (University of Stockholm) for kind help in performing CD measurements. This work was supported by Grant No 707 from Estonian Science Foundation.

References

1. Uchida Y, Takio K, Titani K, Ihara Y and Tomonaga M (1991) *Neuron* 7: 337–347.
2. Tsuji S, Kobayashi H, Uchida Y, Ihara Y and Miyatake T (1992) *EMBO J* 11: 4843–4850.
3. Kobayashi H, Uchida Y, Ihara Y, Nakajima K, Kohsaka S, Miyatake T and Tsuji S (1993) *Brain Res Mol Brain Res* 19: 188–194.
4. Uchida Y (1994) *Biol Signals* 3: 211–215.
5. Pountney DL, Fundel SM, Faller P, Birchler NE, Hunziker P and Vašák M (1994) *FEBS Lett* 345: 193–197.
6. Palmiter RD, Findley SD, Whitmore TE and Durnam DM (1992) *Proc Natl Acad Sci USA* 89: 6333–6337.
7. Uchida Y and Ihara Y (1995) *J Biol Chem* 270: 3365–3369.
8. Sewell AK, Jensen LT, Erickson JC, Palmiter RD and Winge DR (1995) *Biochemistry* 34: 4740–4747.
9. Bogumil R, Faller P, Pountney DL and Vašák M (1996) *Eur J Biochem* 238: 698–705.
10. Erickson JC, Sewell AK, Jensen LT, Winge DR and Palmiter RD (1994) *Brain Res* 649: 297–304.

11. Sambrook J, Fritsch EF and Maniatis T (1989) *Molecular Cloning. A Laboratory Manual*. Cold Spring Harbor Laboratory Press.
12. Vašák M (1991) *Methods Enzymol* 205: 41–44.
13. Henehan CJ, Pountney DL, Zerbe O and Vašák M (1993) *Protein Sci* 2: 1756–1764.
14. Willner H, Vašák M and Kägi JH (1987) *Biochemistry* 26: 6287–6292.

Chemical and analytical aspects of metallothionein

Metallothionein IV
C. Klaassen (ed.)
© 1999 Birkhäuser Verlag Basel/Switzerland

Mammalian metallothionein sub-isoform separation by RP-HPLC with on-line UV and electrochemical detection

Guy Bordin, Fernando Cordeiro Raposo and Adela Rodriguez

European Commission, Joint Research Centre, Institute for Reference Materials and Measurements, Retieseweg, B-2440 Geel, Belgium

Summary. The present article reviews the most significant results obtained over the last five years on the use of HPLC to study mammalian MT polymorphism.

Firstly, it has been shown that reverse phase HPLC with electrochemical (EC, coulometric) detection allowed us to follow the chemical evolution of a trithiolic hexapeptide intrinsic to the MT structure. The EC response is proportional to the number of free thiols, which permits the identification of the reduced and oxidised forms of the peptide which are well separated.

This EC detection was then applied to MT characterisation. Four MTs from rabbit liver and horse kidney were subjected to RP-HPLC (TFA/acetonitile) with on-line UV and EC detection. They were found to exhibit a different polymorphism. Actually, two types of peaks were observed, those equally detected by UV and by EC being attributed to original thiol containing sub-isoforms and those less hydrophobic, detected by UV but hardly at all by EC, being attributed to oxidised forms containing disulphides instead of thiols, hence not detectable in our EC mode. The elutions were then carried out at various temperatures between 25 and 60 °C. This appears to have two main effects: small retention time decrease of all peaks with increasing temperature and large effect on peak "detectability": when the temperature rises, a drastic alteration of the oxidised peaks is observed. As a result, between 20 and 40 °C all peaks are detected and resolved, while at 60 °C only those peaks assumed to be the original sub-isoforms, hence heat-stable, are present.

Previous to that, separation of MT by size exclusion chromatography had shown that they were all partially dimerised (5 to 15% of molar rates of dimers). Although small, these levels of non-monomer MT are not negligible.

Introduction

Metallothioneins (MT) very often exist as various isoforms arising from genetic evolution [1]. In mammals, two main isoMTs are commonly encountered, namely MT-1 and MT-2, differing by a single negative charge at neutral pH [2]. Furthermore, in many species, the presence of sub-isoforms has been shown, leading to the existence of a significant micropolymorphism.

However, the biological roles of the various isoforms and sub-isoforms remain largely unknown [3]. Improvement of this knowledge will obviously rely on having reliable separation methods with adequate detection systems. Historically HPLC with a reverse phase column was initially used to study MT sub-polymorphism [4–7] and was later followed by capillary electrophoresis [8, 9].

Nowadays both techniques are used intensively for MT investigations [e.g. 10–14]. For a few years, our team has been widely involved in this type of research, using both RP-HPLC [15–17] and CZE [18]. This paper will only review our most significant results obtained by chromatography.

To our knowledge, electrochemistry has never been used as a means of detection in an MT separation procedure. We have therefore configured an on-line combination of ultra-violet (UV) and electrochemical (EC) modes of detection. While UV is a universal detection mode, EC can be much more selective since only functional groups which undergo an electrochemical reaction will be detected.

In a first step, we studied the chromatographic behaviour of the trithiolic hexapeptide Lys–Cys–Thr–Cys–Cys–Ala [56–61] MT-1 intrinsic to the mouse liver MT structure, employing electrochemical detection [19]. We showed the advantage of using a coulometric mode in order to detect thiol groups at a porous graphite electrode according to the reaction $2\ R\text{-}SH \rightarrow R\text{-}SS\text{-}R + 2\ H^+ + 2\ e^-$.

In a second step, we applied the same EC system, along with UV detection, to identify and quantify the sub-isoforms of four MTs [16]. Elutions were carried out with a gradient of TFA/acetonitrile, meaning that actually apo-sub-isoforms were detected.

The effect of column temperature on the retention of the various sub-isoforms of the MTs has also been studied [17]. The role of this parameter on the MT isoform separation had never been investigated *per se*. It has been observed that the entire chromatograms are affected by the temperature variations.

Besides, the presence of MT dimers has been shown using size exclusion chromatography [15].

Experimental

Instrumentation

Elutions were carried out on a Kontron chromatograph. The detectors were a UV capillary detector Model 433 (Kontron) and a Coulochem Model 5100 A electrochemical detector (ESA Inc., Bedford, USA) used with a Model 5020 guard cell and a Model 5011 analytical cell containing two coulometrically efficient porous graphite working electrodes.

Samples

The peptide Lys–Cys–Thr–Cys–Cys–Ala [56–61] MT-1, acetate salt, synthetised from mouse liver and containing mercaptoethanol as a stabiliser (L4512 Lot 39F5810), metallothioneins of rabbit liver Cd, Zn MT (RL, Lot 23H9550) and its purified isoforms MT-1 (RL-1, Lot 94H9504) and MT-2 (RL-2, Lot 90H9605) and metallothionein of horse kidney Cd, Zn MT (HK, Lot 82H9590) were purchased from Sigma (St. Louis, MO, USA). All details have been published elsewhere [15–17, 19].

Results and discussion

Study of redox properties and transformation of the Lys–Cys–Thr–Cys–Cys–Ala [56–61] MT–1 peptide

The basic chromatogram of the hexapeptide consists of two peaks (Fig. 1a). 100% of fully reduced peptide would elute as a single peak. We made the hypothesis that the first peak corresponds to a partially oxidised form with an intramolecular disulphide bond between the two neighbouring cysteines. Several general considerations support this hypothesis [20], and different experiments have been carried out to check it.

Evolution of the peptide chromatogram pattern with time
There is a drastic change in the proportion of the two peaks, before arriving at the complete disappearance of any signal (Fig. 1). Two phases can be observed: a sharp decrease of the main peak (A) accompanied by a moderate increase of peak (B), and the total decrease of (B). The pattern is typical of a consecutive reaction (A) → (B) → (C) where (B) is only an intermedi-

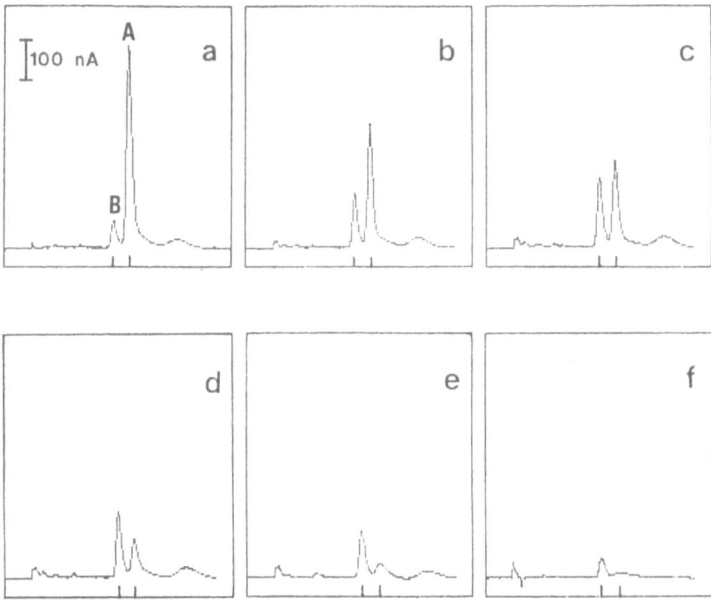

Figure 1. Evolution with time of the hexapeptide Lys-Cys-Thr-Cys-Cys-Ala chromatogram when the sample ($c = 6 \cdot 10^{-5}$ M) is kept at room temperature. [Mobile phase: NaH_2PO_4 0.1 M/8.5% Methanol, pH = 4.8, 1 ml min^{-1}, EC detection at E = 0.6 V, column: Bio-SilC$_{18}$HL 90-5S, 5 μm particle size, 150x 4.6 mm I.D (Bio-Rad)]. (a) t_0 (initial time); (b) t_0 + 30 h; (c) t_0 + 48 h; (d) t_0 + 72 h; (e) t_0 + 144 h; (f) t_0 + 168 h. (Reproduced from [19]

ate and (C) is a compound undetectable by our detection mode, meaning the absence of free
SH groups. A possible (B) → (C) reaction is therefore a dimer formation. Under inert atmosphere, the (A) → (B) reaction does not occur.

Peptide behaviour in basic pH
At pH = 9.5, the reduced peptide is quickly oxidised: disappearance of peak (A) and increase
of peak (B) with different magnitudes. After two hours, (B) has also disappeared and the
hexapeptide probably exists exclusively as dimeric disulphide (C).

Addition of a reducing agent to a peptide solution
To a peptide solution containing species (B) only, an excess of 2-mercaptoethanol is added.
Asuming our hypothesis to be correct we should recover the fully reduced peptide (A). This is
what occurs as shown in Figure 2.

When 2-mercaptoethanol is added to a solution containing (C) exclusively, the same result
is obtained.

All these chemical evolutions have been shown to be perfectly quantitative, the coulometric response being proportional to the number of free thiols in the case of this small molecule
[19].

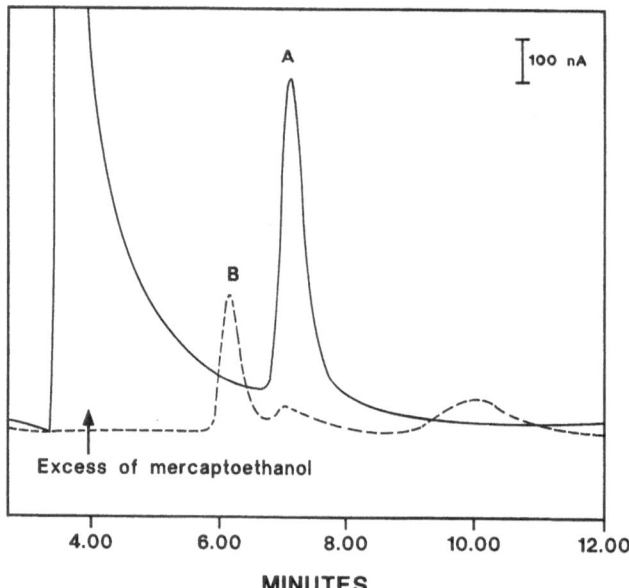

Figure 2. Evolution of the hexapeptide chromatogram (c = 6·10^{-5} M) after addition of: mercaptoethanol (0.1%
v/v). – – – – – peptide alone: only the intramolecular disulphide (B) exists (pH = 9.5); ———— after addition of
the reducing agent: recovery of the reduced peptide (A). (Reproduced from [19] with permission.)

The overall reaction scheme can be summarised as follows (C-SH: cysteine; X: other aminoacids):

$$
\begin{array}{ccc}
\text{(A)} & \xrightarrow{[O_x]} & \text{(B)} \\
\text{X–C–X–C–C–X} & \underset{[Red]}{\rightleftharpoons} & \text{X–C–X–C–C–X} \\
\quad | \quad | \; | & & \quad | \quad | \; | \\
\text{SH} \quad \text{SH SH} & & \text{SH} \quad \text{S—S}
\end{array}
$$

$$[Red] \nwarrow \qquad \nearrow [O_x]$$

$$
\begin{array}{c}
\text{X–C–X–C–C–X} \\
| \qquad | \; | \\
\text{S} \qquad \text{S—S} \\
| \\
\text{S} \qquad \text{S—S} \\
| \qquad | \; | \\
\text{X–C–X–C–C–X} \\
\text{(C)}
\end{array}
$$

Size-exclusion chromatography (SEC) of the MTs

Submitted to an elution on a SEC column, the four MTs show the same chromatographic pattern (UV detection at 250 nm): a first small peak eluted at a retention time corresponding to a molecule having a molecular weight doubling that of the second and main peak (Fig. 3). The proportion of the first peak does not vary with the concentration.

The addition of 2-mercaptoethanol to the MT solution thereafter re-eluted shows the disappearance of the first peak coupled with an increase of the major peak (recovery: $102 \pm 4\%$, $n = 13$). These experiments indicate that the first species eluted in SEC is a disulphide containing dimeric form of the MT.

From the peak areas, it was then possible to calculate the molar proportion of dimer for each MT, given in Table 1: it is variable, depending on the MT sample.

Table 1. Molar proportions of the dimeric form for each MT

MT	% of dimer	n
RL-1	7.5 ± 0.5	8
RL-2	15.3 ± 0.8	12
RL	13.1 ± 0.4	13
HK	5.1 ± 0.6	8

(Reproduced from [15] with permission.)

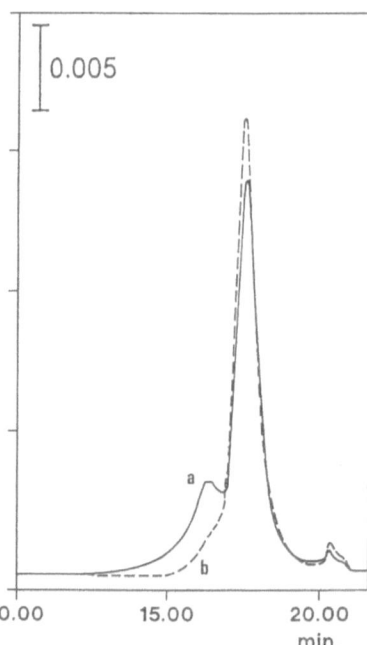

Figure 3. Size exclusion chromatogram of RL metallothionein (c = 200 mg.l⁻¹). UV detection (λ = 250 nm; 0.005 AUFS). [Mobile phase: Na_2HPO_4 0.15 M, $Na_2H_2PO_4$ 0.04 M, NaCl 0.15 M, pH = 7.15, 0.5 ml.min⁻¹, column: Bio-SiL SEC 125 (300 × 7.8 mm)]. **a**: original pattern; **b**: after addition of 2-mercaptoethanol (0.5% v/v). (Reproduced from [15] with permission.)

Finally, the composition of the MT-RL in its two isoforms can be estimated. RL being a "mixture" of RL-1 and RL-2 and knowing the various dimer ratios, one can calculate the respective proportions of the two isoforms (x% RL-1, y% RL-2).

This leads to the identification of an MT RL as being composed of 26% (±5%) RL-1 and 74% (± 12%) RL-2 [15].

MT sub-isoform separation by reverse-phase HPLC

All MTs were prepared at neutral and acid pH and eluted with a linear gradient of TFA (pH = 3) and acetonitrile. On line UV (λ = 230 nm) -electrochemical (EC) detection was used.

MT samples prepared at neutral pH
The four MTs exhibit a different polymorphism (Fig. 4). Due to the electrochemical properties of SH group, the EC chromatograms should help to qualify the nature of some peaks, that is, to answer the question: does a given peak correspond to an original sub-isoMT or not?

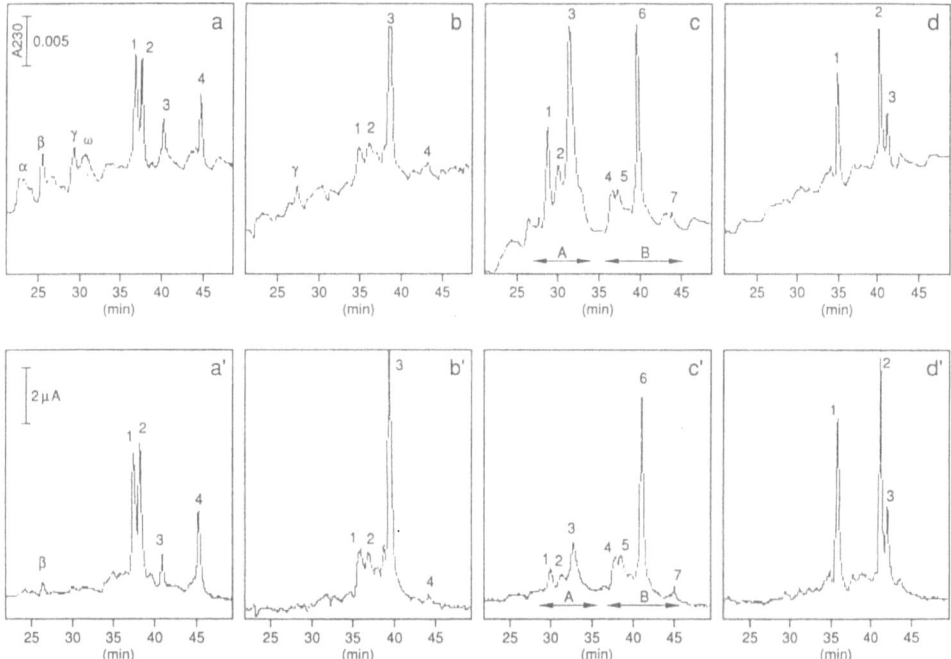

Figure 4. Reverse phase chromatograms of four metallothioneins prepared at neutral pH. (**a, b, c, d**) UV detection ($\lambda = 230$ nm; 0.005 AUFS). (**a', b', c', d'**) EC detection (E = 0.6 V). (**a-a'**) RL-1; (**b-b'**) RL-2; (**c-c'**) RL; (**d-d'**) HK (c = 500 mg.l^{-1}). Elution: from 90% A (0.1% TFA) – 10% B (0.1% TFA in acetonitrile) to 70% A–30% B in 70 min. Column: Hi-Pore RP 318 (250 × 4.6 mm), 300 Å pore size, 5 μm particle size (Bio-Rad). (Reproduced from [16] by courtesy of Marcel Dekker Inc)

On the whole, UV and EC chromatograms have comparable morphologies in the cases of RL-2 and of HK while they display some different aspects for RL-1 and RL.

- RL-2 comprises of one dominant peak (3), surrounded by several minor peaks (1, 2, 4). One small peak, γ, which is eluted earlier, is only detected by UV.
- HK mainly exhibits three peaks on both chromatograms.
- RL-1 is more complex than the previous proteins. Four peaks (1–4) are detected by both EC and UV modes. Furthermore, the UV chromatogram exhibits some early smaller peaks (α, β, γ, ω) which are not or only slightly detected by EC.
- Both chromatograms of RL show two groups of peaks, groups A (peaks 1, 2 and 3) and B (peaks 4–7). Relative to B peaks, group A peaks clearly give much less intense EC than UV responses.

MT samples prepared at acid pH
Compared to neutral pH no changes are noticed for RL-2 and HK, while for RL-1 and RL significant evolutions are observed.

- for RL-1, the UV picture is characterised by a high increase of the α–ω peaks relative to the "main" peaks 1–4. In EC, the α–ω peaks remain rather negligible.
- for RL, both EC and UV chromatograms, registered just after preparation, are rather similar to the neutral pH situation with quite stable group B/group A peak ratios. However, after a short time, one can observe a decrease of both EC and UV B peaks and an increase of A peaks in UV.

Discussion on the nature of the various peaks

The MTs show a variable number of peaks with variable response depending on the detection mode. Certain peaks only are equally detected by both EC and UV with similar peak proportions and are eluted by the same range of organic phase proportions. We can tentatively attribute these peaks to putative original sub-isoforms of the MTs.

From this point of view, HK and RL-2 represent relatively simple cases, HK having three well resolved sub-isoMTs and RL-2 only one dominant form, whatever the pH of the sample.

For the two other MTs, only the more hydrophobic species are detected by both EC and UV modes with similar peak signal ratios. The comparison of the retention times of the peaks of RL-1, RL-2 and RL show that group B peaks of RL, peaks 4, 5 and 7 on one hand and peak 6 on the other hand correspond to sub-isoforms of RL-1 and of RL-2 respectively. On the contrary, group A peaks can hardly result directly from a simple contribution of either RL-1 or RL-2.

It can be observed that RL's peak 3A appears isolated, with no corresponding peaks on either RL-1 or RL-2 pictures.

Stable at neutral pH, an acidic medium enhances the formation of the species α–ω of RL-1 and of group A peaks of RL. The formation of intra- or intermolecular disulphide containing compounds could perhaps explain these results: thiols which are chemically oxidised in disulphides will not be able to give an EC response anymore.

Attribution of peaks in rabbit liver MT

Two solutions comprising of 20% RL-1 and 80% RL-2 were prepared at neutral and at acid pH respectively.

The chromatogram of RL-1 + RL-2 corresponds very well to the superposition of those of RL-1 and RL-2: peak 3 coming from RL-2 is dominant and α, β and ω from RL-1 are very small, while γ, found in both isoMTs is significant in the UV mode.

When the chromatogram of the mixture is compared to that of an original RL solution (Fig. 5), a perfect correspondence of peaks 1, 2, 3 and 4 of RL-1 + RL-2 with group B peaks of RL is observed for both detection modes. γ corresponds qualitatively to 1A and ω to 2A, while 3A remains with no "partner" peak in the mixture. Mixing the two isoforms does not generate peak 3A which is largely found in the original RL.

Effect of temperature on rabbit liver MT sub-isoform separation by RP-HPLC

Increasing the temperature of the column (25–60 °C) leads to the observation of two main effects: a logical, but moderate, decrease in retention times of all peaks (6–7 min) and a progressive alteration of the "non-original" sub-isoforms (α-ω in RL-1, γ in RL-2 and A peaks in RL).

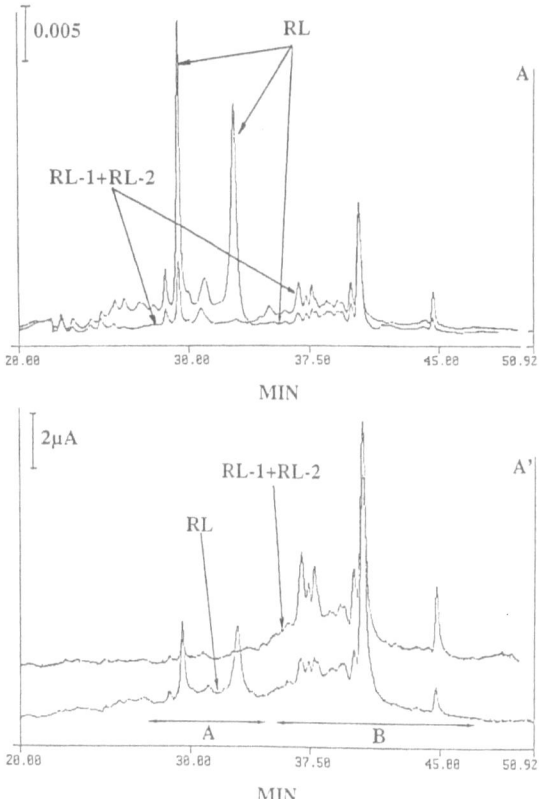

Figure 5. Comparison of the chromatogram of a RL solution with that of a mixture of 20% RL-1 + 80% RL-2 at 35 °C, both prepared at neutral pH. (**A**) UV detection; (**A'**) EC detection.

The case of RL-1 is illustrated here (Fig. 6), all other detailed results being published elsewhere [17].

- Peaks 1, 2 3 and 4: when the temperature is increased, moderate augmentations of the peak intensities can be observed up to 50–55 °C before they decrease slightly.
- Peaks α, β, γ and ω: they reach a maximum of intensity between 30 and 40 °C, before decreasing quickly. Their EC signals are always very small.

On the whole, at an average temperature of 35 °C, conditions are attained where all original and chemically oxidised sub-isoforms are separated and detected. At 60 °C, only those peaks assumed to be original sub-isoforms, hence heat-stable, are present.

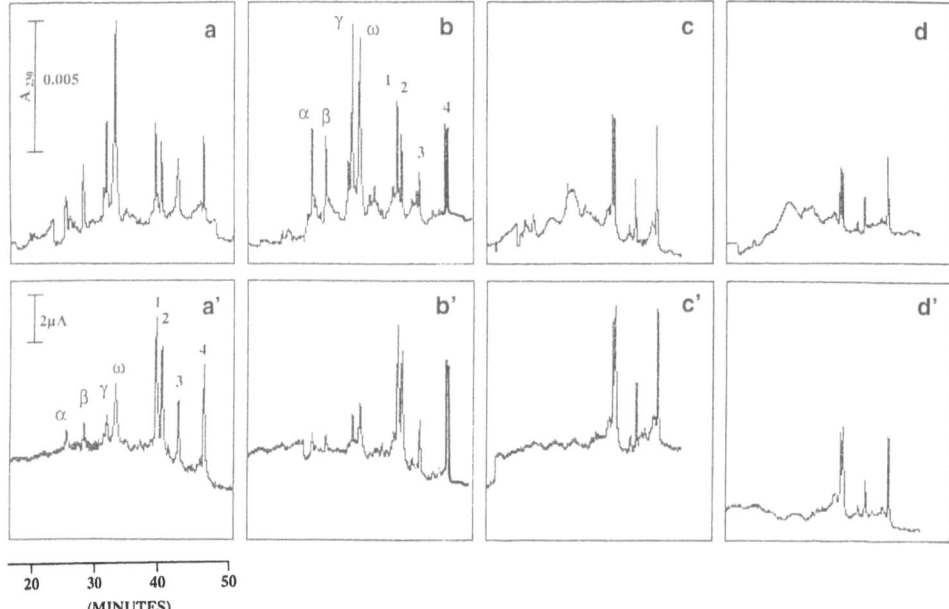

Figure 6. Reverse phase chromatograms of the RL-1 metallothionein at various temperatures for samples pre-pared at acidic pH. (**a, b, c, d**) UV detection; (**a', b', c', d'**) EC detection. (**a-a'**) 25 °C; (**b-b'**) 35 °C; (**c-c'**) 55 °C; (**d-d'**) 60 °C.

Conclusion

- The proportion of the dimeric form of rabbit liver and horse kidney MT is not negligible (5–15% in molar proportions), confirming the few results reporting the existence of MT dimers [21–22];
- The usefulness of using a coulometric detector for thiolic molecules has been shown with the hexapeptide. The electrochemical response allows one to identify quantitatively the reduced and oxidised forms of the peptide which are well separated on a RP column.
- The MTs from rabbit liver (RL, RL-1 and RL-2) and from horse kidney (HK) were found to exhibit a different polymorphism. Two types of peaks were observed, those equally detect-ed by UV and EC and those eluted earlier, detected by UV but not or only very little by EC. The first peaks should then correspond to the original sub-isoforms of the MT, bound to met-als at neutral pH and protonated in acidic pH. The other peaks, not electrochemically active in our system, do not contain thiol groups anymore and might therefore correspond to oxi-dised forms containing disulphide bonds.
  - HK exhibits three peaks of the first type. These results agree with or improve upon those of other teams [6, 10].
  - RL-2 has one single highly dominant peak in both modes, preceeded by a small but still significant peak observed in the UV mode only. These findings are in good agreement with most of others [4, 5].

- In RL-1, three main peaks give both UV and EC signals, while four less hydrophobic species are detected by UV only. The original polymorphism of RL-1 is then made up of three sub-isoforms, which is in excellent agreement with the results of others [4]. Some authors had resolved two isoforms only at acidic pH [9].
- RL also contains two groups of peaks: the four B peaks giving UV and EC responses, being attributed to either RL-1 (3 peaks) or to RL-2 (1 peak) and the three A peaks containing few remaining thiol groups. One of the oxidised A peaks of RL does not have any corresponding species in either RL-1 or RL-2. Therefore, the reactivity of the RL metallothionein is not simply the "sum" of the respective reactivities of RL-1 and RL-2.

For all rabbit liver MTs, preparing the samples in acidic medium enhances the formation of the oxidised molecules.

- The variation of temperature on MT sub-isoform separation has a significant effect on the type of species detected: when it rises, an alteration of the oxidised peaks is observed, especially above 50 °C. As a result, between 20 and 40 °C all peaks are perfectly detected and resolved, while at 60 °C only putative original thiol containing sub-isoforms are present.
- On the whole, the on-line UV-EC detection system constitutes a very good tool for MT sub-isoform identification. Used alone, the UV mode could in some cases overestimate the number of putative original sub-isoMTs. On the contrary, the coulometric detection used alone would provide the right number of thiol containing forms but would also only give partial information on the other species. Only the combination of the two detection modes, the universal UV and the more selective EC, can give access to the double information.

References

1. Suzuki KT, Imura N, Kimura M (1993) *Metallothionein III*. Birkhäuser Verlag, Basel, and references cited therein.
2. Nordberg M, Kojima Y (1979) Metallothionein and other low molecular weight metal-binding proteins. *In*: JHR Kägi, M Nordberg (eds): *Metallothionein*. Birkhäuser Verlag, Basel.
3. Cherian MG, Chan HM (1993) Biological functions of metallothionein – A review. *In*: KT Suzuki, N Imura, M Kimura (eds): *Metallothionein III*. Birkhäuser Verlag, Basel.
4. Klauser S, Kägi JHR, Wilson KJ (1983) Characterisation of isoprotein patterns in tissue extracts and isolated samples of metallothionein by reverse-phase high-pressure liquid chromatography. *Biochem J* 209: 71–80.
5. Richards MP, Steele NC (1987) Isolation and quantification of metallothionein isoforms using reversed-phase high-performance liquid chromatography. *J Chromatogr* 402: 243–256.
6. Van Beek H, Baars AJ (1988) Isolation and quantification of cadmium-, zinc- and copper-metallothioneins by high-performance liquid chromatography – atomic absorption spectrometry. *J Chromatogr* 442: 345–352.
7. Richards MP (1989) Characterisation of the metal composition of metallothionein isoforms using reversed-phase high-performance liquid chromatography with atomic absorption spectrophotometric detection. *J Chromatogr* 482: 87–97.
8. Beattie JH, Richards MP, Self R (1993) Separation of metallothionein isoforms by capillary zone electrophoresis. *J Chromatogr* 632: 127–135.
9. Richards MP, Beattie JH (1993) Characterisation of metallothionein isoforms – Comparison of capillary zone electrophoresis with reversed-phase liquid chromatography. *J Chromatogr* 648: 459–468.
10. Richards MP (1991) Purification and quantification of metallothioneins by reversed-phase high-performance liquid chromatography. *Methods Enzymol* 205: 217–238.
11. Wan M, Kägi JHR, Hunziker PE (1993) Resolution and quantification of four metallothionein isoforms from rabbit liver kidney cells. *Protein Express Purif* 4: 38–44.
12. Liu G, Wang W, Shan X (1994) Factor influencing the separation of metallothioneins by capillary zone electrophoresis. *J Chromatogr* 653: 41–46.
13. Beattie JH, Richards MP (1994) Separation of metallothionein isoforms by micellar electrokinetic capillary chromatography. *J Chromatogr* 664: 129–134.

14. Minami T, Matsubara H, O-higashi M, Otaki N, Kimura M, Kubo K, Okabe N, Okazaki Y (1996) Identification of metallothionein isoforms with capillary zone electrophoresis using a polyacrylamide-coated tube. *J Chromatogr* 685: 353–359.
15. Bordin G, Cordeiro Raposo F, Rodriguez AR (1994) Contribution à l'étude du polymorphisme de quelques métallothionéines par chromatographie liquide de haute performance. *Can J Chem* 72: 1238–1245.
16. Bordin G, Cordeiro Raposo F, Rodriguez AR (1996) Characterisation of metallothionein isoforms by reverse phase high performance liquid chromatography with on-line UV and electrochemical detection. *J Liq Chromatogr Relat Technol* 19: 3085–3104.
17. Bordin G, Cordeiro Raposo F, Rodriguez AR (1998) Effect of temperature variation on metallothionein sub-isoform separation by reverse phase high performance liquid chromatography. *J Liq Chromatogr Relat Technol* 21: 2039–2060.
18. Virtanen V, Bordin G, Rodriguez AR (1996) Separation of metallothionein isoforms with capillary zone electrophoresis using an uncoated capillary column. Effects of pH, temperature, voltage, buffer concentration and buffer composition. *J Chromatogr* 734: 391–410.
19. Bordin G, Cordeiro Raposo F, Rodriguez AR (1994) Chromatographic characterisation of a hexapeptide containing three thiol groups, intrinsic to the metallothionein structure. *Chromatographia* 39: 146–154.
20. Lee KK, Black JA, Hodges RS (1991) Separation of intrachain disulfide bridged peptides from their reduced forms by reverse phase chromatography. *In*: CT Mant, RS Hodges (eds*): High Performance Liquid Chromatography of peptides and proteins*. CRC Press, 389–398.
21. Suzuki KT, Yamamura M (1980) Isolation and characterisation of metallothionein dimers. *Biochem Pharmacol* 29: 689–692.
22. Klein D, Sato S, Summer SH (1994) Quantification of oxidised metallothionein by a cadmium saturation method. *Anal Biochem* 221: 405–409.

Metallothionein IV
C. Klaassen (ed.)
© 1999 Birkhäuser Verlag Basel/Switzerland

Characterisation of mammalian Cd, Zn metallothioneins using differential pulse polarography

Adela Rosa Rodriguez

European Commission, Joint Research Centre, Institute for Reference Materials and Measurements, Retieseweg, B-2440 Geel, Belgium

Summary. Differential pulse polarography allows one to characterise Cd, Zn MT. In fact, polarograms exhibit different peaks at characteristic potentials corresponding to electrochemical reactions for established systems: the oxidation of the mercury electrode in the presence of the MT's thiol groups and the reduction of metal cations either free or bound to the MT. The polarographic response can therefore be used to distinguish the chemical forms of compounds (speciation) and for monitoring changes of different species as a function of several parameters. Five commercially available (Sigma Co.) MTs of animal origin i.e. rabbit liver and horse kidney were used, as well as eight of human origin i.e. foetal liver and adult kidney.
On the one hand the dissociation and the formation of MT from their metal depleted form was investigated in order to assess the reversibility of the complexation equilibrium MT ↔ M + T. This equilibrium has been upset by a mere change in the pH, going from basic to acid solution and *vice versa*. On the other hand, the influence of gradual addition of cadmium and/or zinc was studied in order to obtain a better understanding of the binding properties and ion exchange of these molecules. In some cases two different electrochemical responses due to the reduction of complexed Cd(II) as well as to the Zn(II) complexes were distinguished. The predominance of these two species for both cations seems to depend on the total concentration, on the ratio between zinc and cadmium concentration and on the solution pH.

Introduction

Metallothioneins (MTs) are considered to be key molecules involved in some major biological processes in relation to the metabolism of essential and toxic metal ions [1]. The characteristics and physico-chemical properties of metallothioneins have formed the subject of several recent monographies [2–6] and reviews [7–14]. Among the analytical methods used for the MT characterisation, electroanalysis has, until recently, scarcely been applied for this purpose [15–16]. Nevertheless the total quantification of MT has been carried out using the modified Brdicka procedure [17–24], the differential pulse polarography (DPP) technique [25] and the cathodic stripping voltammetry (CSV) at the mercury electrode in the presence of copper [26]. Other studies on electrochemical characterisation of MT have been performed at the mercury and carbon electrodes by cyclic voltammetry (CV) and DPP [27–31].

In fact, electrochemical methods offer an extensive range of applications, such as the characterisation of chemical systems and the investigation of chemical phenomena in solution. In the case of Cd, Zn Metallothioneins the organic part of the molecule is electroactive due to the presence of thiol groups and the reduction of the cations, cadmium and zinc, is electrochemically reversible at the mercury electrode. For the study of these complex molecules, the peptidic fragment Lys-Cys-Thr-Cys-Cys-Ala [56–61] MT I (FT), a molecule intrinsic to the mouse liver MT structure, cluster α, has been taken as a simpler model of MT by our team. The studies of FT alone and in the presence of Cd and/or Zn at different ratios of metal ion: peptide using direct current polarography (DC) and DPP at the dropping mercury electrode (DME) [32–34] as well as square wave voltammetry (SWV) and CV at the hanging mercury drop electrode (HMDE) [35–37] have been widely used for the identification and assignment of polarographic peaks, for the elucidation of electrochemical mechanisms, for the study of diffusion-adsorption processes and for assessing the reversibility of each electrochemical system.

The DPP technique has been used by our team under experimental conditions in which the adsorption phenomena are not prevalent. In this case polarograms of Cd, Zn MT exhibit different peaks at characteristic potentials corresponding to electrochemical reactions for established systems. Therefore this technique has been applied in the systematic investigation of the electrochemical behaviour of mammalian MT and for monitoring changes of different species as a function of different parameters [38–46]. On the one hand, the dissociation and the formation of these MTs from their depleted form was investigated in order to assess the reversibility of the complexation equilibrium. This equilibrium was upset by a mere change of pH, going from basic to acid solutions and *vice versa*. On the other hand the influence of gradual additions of zinc and/or cadmium to the MT was studied in order to obtain a better understanding of the binding properties and ion exchange of metallothioneins. The most significant results obtained in this previous investigation are presented in this paper.

Samples

Several MT from different origin, rabbit liver, horse kidney, human foetal liver and human adult kidney were used. These MTs have different isoforms depending on their amino acid sequence. The total content of cadmium and zinc is very variable, containing either only zinc in the molecule – human foetal liver, rabbit liver – only cadmium – human adult kidney – or the mixture of both – rabbit liver, horse kidney, human adult kidney and human foetal liver.

Metallothioneins from Sigma Co (St. Louis MO. USA): RL MT 1 + 2, RL MT 1 and RL MT 2 from rabbit liver. HK MT from horse kidney. RL Zn MT 2 from rabbit liver.

Metallothioneins from the Department of Biochemistry, Catholic University of Leuven, Belgium: They were isolated and characterised following the procedure described in [47–48].

HFL MT 0, HFL MT 1 and HFL MT 2 from human foetal liver, native form. HAK MT 0, HAK MT 1 and HAK MT 2 from human adult kidney, native form. HFL Cd MT 1 from human foetal liver, loaded with Cd. HAK Cd MT 1 from human adult kidney, loaded with Cd.

The metal concentrations in the metallothionein were determined by electrothermal atomic absorption spectrometry and given in [39–46]. The concentration of thiolate groups was determined in some of the samples by the Brdicka method [49] using RL MT 1 as a standard. The molar concentration of MT was calculated taking 6.700 Da as the molecular weight.

The experimental part is described elsewhere [39–44].

Assignment of polarographic peaks

An example of some of the polarograms obtained is given in Figure 1 corresponding to HAK Cd MT 1. In this Figure all peaks which could be found are shown. The dashed lines correspond to the peak potentials of the reduction of free Cd^{2+} and Zn^{2+} in the same supporting electrolyte. From anodic to cathodic potentials the peaks are denoted MT(Cd), MT(Zn), CdT', CdT, ZnT' and ZnT. Under our experimental conditions – low concentration of MT ($< 10^{-5}$ M) and the use of a dropping mercury electrode – a diffusion controlled mechanism is assumed. Peaks labelled MT(Cd) and MT(Zn) have been attributed to the oxidation of the mercury electrode in the presence of the thiol groups contained in the organic part of the molecule. MT is considered as a chelating agent having acid-base properties.

Peaks labelled CdT' and CdT, on the one hand, and ZnT' and ZnT, on the other hand, are due to the reduction of two different complexes of Cd and Zn, respectively, with the protein.

The oxidation and reduction processes have been formulated in [32, 33, 39]. The peak potential of the reduction of complexed cations, $E'_{0(M)}$, varies in a predictable manner as a function of pH: $E'_{0(M)} = \text{cte} + 0.059$ m pH/n (at $25\,°C$), where m is the number of protons, hence the

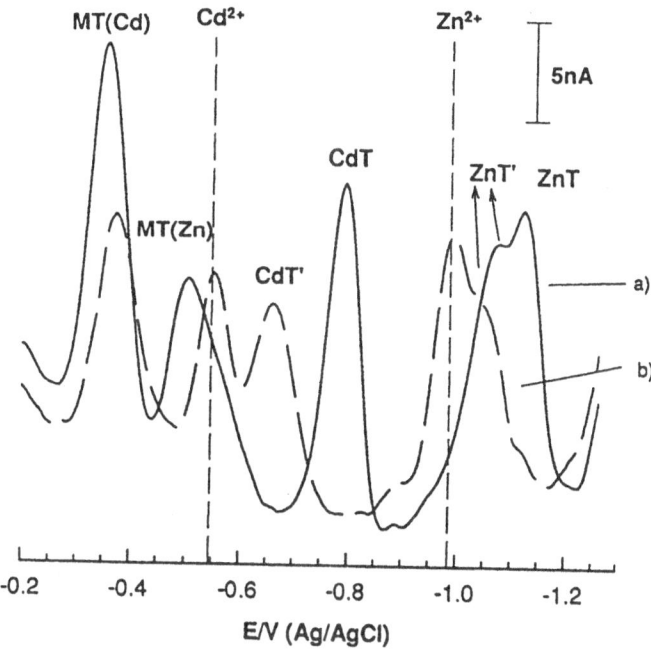

Figure 1 DPP polarograms of HAK Cd MT 1, (a) MT $6.2\cdot10^{-6}$ M, (b) after addition of 1.25 mol Cd/mol MT. Hepes-phosphate 10^{-2} pH = 7.15.

number of SH groups, involved in the reduction mechanism and n the number of electrons transferred.

Conclusions

Differential pulse polarography allows one to characterise Cd, Zn metallothioneins. A solution of MT gives polarographic features exhibiting different peaks corresponding to well established electrochemical reactions. Two electrochemical systems can be distinguished: the first one corresponds to the oxidation of the mercury electrode in the presence of thiol groups contained in the MT, being either bound or not to the metal ions, depending on the pH. The second one is due to the reduction of cations, cadmium and zinc, at the dropping mercury electrode, either free or complexed with the MT. DC polarographic waves confirm the oxidation and reduction mechanism.

Under our experimental conditions, low concentration of compounds and the use of a dropping mercury electrode, a diffusion controlled mechanism is proposed and, in this case, the adsorption phenomena are considered as not prevalent. The polarographic response – peak potential and peak intensity – can therefore be used to distinguish the chemical form of compounds (speciation) and particularly to follow the evolution of different species in solution as a function of different parameters. DPP is an easy analytical method allowing one to clearly distinguish between the CdT and ZnT complexes. Along with that it is possible to obtain information on the organic part of the molecule.

Because the rate at which chemical equilibrium is reached is very high, significant modifications in the polarograms are observed after a short period of time on slightly changing the conditions, such as the solution pH, metal ion concentration, etc.

The polarographic behaviour of all studied MTs is analogous, regardless of the different origin, the different isoforms and on the very variable metallic content. However, small differences exist between mammalian MT from Sigma and human MT. While for CdT and ZnT the polarographic responses are similar in both mammalian animal and human MT, in the case of the peaks in which the organic part of the molecule is involved, the only MT peak found for MT from Sigma is split into two well defined peaks, named MT(Cd) and MT(Zn) in the case of some of the human metallothioneins. The existence of both signals and the predominance of one of them depend on the ratio between the initial content of cadmium and zinc in the metallothionein.

Human metallothioneins are not stable, neither at very basic pH nor at acidic pH. Nevertheless MT from Sigma are stable over a large range of pH. For the latter MT, the equilibrium MT $\leftrightarrow$ M + T is reversible and consequently the protein in acid solution is still able to keep its metal-binding capacity and in neutral and basic solution to complex metal ions through its thiol groups. However, the electrochemical response monitoring the evolution of this equilibrium, from acid to basic solutions is not the same as for this native form. The release of metal ions from the metallothioneins as well as the formation of the metallothioneins from the metal depleted form occur, but when the pH is varied from acid to basic pH, the Cd, Zn MT obtained probably possesses properties and characteristics which are different to its native form. In whatever way, throughout a narrow pH range (~ 7–10) all mammalian, animal and human MT behave as a simple complex of a metal ion with a ligand having acid-basic properties. Peak potentials of CdT and ZnT complexes vary linearly with the pH as can be expected from the electrochemical mechanism previously formulated, in which it was assumed that cadmium and zinc formed only a simple complex with the whole organic molecule. Even if the structure of MT is complicated the electrochemical behaviour of CdT and ZnT as a function of the solution pH is simple.

Assuming the electrochemical mechanism formulated for the reduction of cations complexed by the thiol groups, the value of the slopes $Ep = f(pH)$ allows us to determine m, the number of protons involved in the electrochemical reactions and hence the number of SH groups in which the protons are replaced by a metal ion. m lies between 2.3 and 2.6 for all studied MTs for CdT complexes, values comparable to those derived from the accepted structure of Cd_5Zn_2MT, 2.75 (-S/M) for the α domain. For the ZnT complex in the case of Sigma MT, the ratio found, 2.9, is very close to that of the β domain which is equal to 3. However this ratio for human MT lying between 2.1 and 2.4 is lower than that derived from the MT structure. In the case of these MTs the simultaneous presence of ZnT' and ZnT complexes could provoke a change in the structure and hence in the co-ordination of zinc thiolate bonds.

The human MTs exhibit low stability when the pH of the solution reaches acid values. These molecules do not preserve the metal binding capacity and hence the equilibrium of formation of human MTs from their depleted form does not take place. Results obtained using electrochemical techniques suggest that changes in the molecules occur but do not allow one to formulate hypotheses on changes in the structural features, but only to notice this fact.

For all studied MTs, two responses corresponding to the reduction of Cd(II) have been found and attributed to two different complexes of Cd-metallothioneins. An equilibrium exists between both species CdT $\leftrightarrow$ CdT'. It is possible to provoke the change of one complex into the other varying the following parameters: change of pH, addition of zinc and excess of cadmium. Furthermore the exchange is very rapid and the equilibrium is quickly reached. In the case of some human MTs two different responses due to the reduction of Zn(II) complexes have also been distinguished. The influence of the pH is similar for both kinds of complexes: CdT and ZnT are more stable at basic pH, which means CdT' $\rightarrow$ CdT and ZnT' $\rightarrow$ ZnT at pH > 8.5. The shift of potential for zinc complexes is less clearly seen because ZnT' and ZnT peaks are poorly resolved. Likewise the appearance or disappearance of one of the two Cd and Zn species can be provoked by a mere change of pH. It is pointed out that the range of pH over which these transformations occur is very narrow and corresponds to the complete deprotonation of SH groups.

When cadmium and/or zinc are added, all MTs are able to incorporate the metal ions into their structure. This implies a reorganisation of the molecule in which cadmium and zinc complexes are involved. The electrochemical method does not allow one to know if the incorporation of added cations occurs in the same way, with the same kind of S-H co-ordination as for the cadmium and zinc initially bound to the thionein, although the electrochemical responses, before and after the additions, can be identified corresponding to CdT and/or CdT' peaks and ZnT and/or ZnT' peaks. The combination of results obtained by other analytical techniques such as two-dimensional [113]Cd or [111]Cd-^{1}H NMR methods, allowing the determination of cadmium involved in the binding as well as the evaluation of the length of S-M bonds under given experimental conditions (where either CdT or CdT' is detected) could be very useful for a better understanding of phenomena described in our electroanalytical study.

When cadmium and/or zinc are added, it seems difficult to establish clear trends for the general characteristics of mammalian MTs taking the complexity and the heterogeneity of these molecules into account. However, the transformation of the CdT complex into the CdT' species in the presence of added zinc is a common phenomena for all studied MTs. For human MTs, the addition of cadmium provokes the disappearance of the MT(Zn) peak when this peak is initially detected.

If the detection of free Cd^{2+} is considered as the criterium of the maximum concentration of this cation which can be bound by the MT, the total concentration (Cd_{in} + Cd_{Cd}) is similar (4.5–5.0 mol Cd/mol MT) for RL MTs, HFL Cd MT 1, HAK Cd MT 1 and HAK MT 1 and independent of the order of the addition of cadmium and zinc metal ions.

Data presented in this paper indicate that electrochemical methods are a very useful tool for metallothionein characterisation, particularly with respect to the occurrence of two CdT and ZnT complexes, exhibiting different stability and having different properties, not mentioned in works on MT using analytical methods other than electroanalysis.

Acknowledgements
This investigation was made possible with the significant contribution of the following people: A. Muñoz, J. Chivot, J. Mendieta, C. Ruiz, O. Nieto, G. Hellemans, M. Dabrio, C. Harlyk and G. Bordin.

References

1. Florence TM (1989) Trace elements speciation in biological systems. *In*: GE Batley (ed.): *Trace Element Speciation: Analytical Methods and Problems*. CRC Press. Boca Raton FL, 331–332.
2. Kägi JHR, Nordberg M (eds) (1979) *Metallothionein*. Birkhäuser Verlag, Basel.
3. Kägi JHR, Kojima Y (eds) (1987) *Metallothionein II*. Birkhäuser Verlag, Basel.
4. Stillman MJ, Shaw III CF, Suzuki KT (eds) (1992) *Metallothioneins, Synthesis, Structure and Properties of Metallothioneins, Phytochelatins and Metal-Thiolate Complexes*. VCH, New York.
5. Riordan JF, Vallee BL (eds) (1991) *Meth Enzymol , Vol. 205. Metallobiochemistry, Part B, Metallothioneins and Related Molecules*. Academic Press, London.
6. Suzuki KT, Imura N, Kimura M (Eds) (1993) *Metallothionein III, Biological Roles and Medical Implications*. Birkhäuser Verlag, Basel.
7. Kägi JHR, Kojima Y, Kissling MM, Lerch K (1980) Metallothionein: an exceptional metal thiolate protein. *Excepta Medica* 223: 223–237.
8. Kägi JHR, Vašák M, Lerch K, Gilg DEO, Hunziker P, Bernhard WR, Good M (1984) Structure of mammalian metallothioneins. *Environ Health Perspect* 54: 93–103.
9. Karin M (1985) Metallothioneins: Proteins in search of function. *Cell* 41: 9–10.
10. Hamer DH (1986) Metallothionein. *Ann Rev Biochem* 55: 913–951.
11. Otvos JD, Petering DH, Shaw CF (1989) Structure-reactivity relationships of metallothionein, a unique metal-binding protein. *Comment Inorg Chem* 9: 1–35.
12. Kay J, Cryer A, Darke BM, Kille P, Less WE, Norey CG, Stark JM (1991) Naturally occurring and recombinant matallothioneins: Structure, immunoreactivity and metal-binding functions. *Int J Biochem* 23: 1–5.
13. Kille P, Hemmings A, Lunney EA (1994) Memories of metallothionein. *Biochim Biophys Acta* 1205: 151–161.
14. Stillman MJ (1995) Metallothioneins. *Coord Chem Rev* 144: 461–511.
15. Armstrong FA (1990) Probing Metalloproteins by voltammetry. *In*: MJ Clark et al. (eds): *Structure and Bonding, Vol. 72*. Springer, Berlin, 137–230.
16. Sequaris JM (1992) Analytical Voltammetry of Biological Molecules. *In*: MR Smyth, JG Vos (eds): *Wilson and Wilson's Comprehensive Analytical Chemistry, Vol. 27 Analytical Voltammetry*. Elsevier, Amsterdam,

115–157.

17. Olafson RW, Sim RG (1979) An electrochemical approach to quantification and characterization of metallothioneins. *Anal Biochem* 100: 343–351.
18. Olafson RW (1981) Differential pulse polarographic determination of marine metallothionein induction kinetics. *J Biol Chem* 256: 1263–1268.
19. Onasaka S, Cherian MG (1982) Comparison of metallothionein determination by polarographic and cadmiun saturation methods. *Toxicol Appl Pharmacol.* 63: 270–274.
20. Cosson P, Thompson JAJ (1983) Utilisation de l'électrode PAR 303 pour la détection et le dosage des métallothionéines dans les tissus d'invertébrés marins. *Analusis* 11: 33–35.
21. Thompson JAJ, Cosson P (1984) An improved electrochemical method for the quantification of metallothioneins in marine organisms. *Marine Environ Res* 11: 137–152.
22. Bebiano MJ, Langston WJ (1989) Quantification of metallothioneins in marine invertebrates using differential pulse polarography. *Portugal Electrochim Acta* 7: 59–64.
23. Magemann R, Dick JG, Klaverkamp JF (1994) Metallothionein E stimates in marine mammal and fish tissues by three methods: ^{203}Hg displacement, polarography and metal-summation. *Anal Chem* 54: 147–160.
24. Raspor B, Pavicic J (1996) Electrochemical methods for quantification and charactesitation of metallothioneins induced in *Mytilus Galloprovincialis*. *Fresenius J Anal Chem* 354: 529–534.
25. Muñoz A, Rodríguez AR (1995) Quantitative determination of cadmiun, zinc thioneins using differential pulse polarography. *Analyst* 120: 529–532.
26. Scarano G, Morelli E (1996) Cathodic stripping voltammetric determination of metallothioneins. *Electroanalysis* 8: 396–398.
27. Olafson RW (1988) Electrochemical characterisation of metallothionein, metal-mercaptide complexes: application of cyclic voltammetry to investigation of metalloproteins. *Bioelectrochem Bioenerg* 19: 111–125.
28. Chou CL, Guy RD, Uhte JF (1991) Isolation and characterisation of metal-binding proteins (metallothioneins) from lobster digestive gland *(Homarus Americanus)*. *Sci Total Envir* 105: 41–59.
29. Šestáková I, Milholová D, Vodicková H, Mader P (1995) Electrochemical behaviour of metallothioneins at mercury and carbon electrodes. *Electroanal* 7: 237–246.
30. Fedurco M, Šestáková I (1996) Adsorption of Cd, Zn-metallothioneins on covered Hg electrodes and its voltammetric determination. *Bioelectrochem Bioenerg* 40: 223–232.
31. Studincková M, Turánek J, Zábrsová H, Krejcí M, Kysel M (1997) Rat liver metallothioneins are metal dethiolene clusters. *J Electroanal Chem* 421: 25–32.
32. Mendieta J, Chivot J, Muñoz A, Rodríguez AR (1995) Electrochemical behaviour of metallothioneins and related molecules. Part I: Lys-Cys-Thr-Cys-Cys-Ala thionein fragment [56–61] MT I. *Electroanal* 7: 663–669.
33. Muñoz A, Rodríguez AR (1995) Electrochemical behaviour of metallothioneins and related molecules. Part II: Lys-Cys-Thr-Cys-Cys-Ala thionein fragment [56–61] MT I in the presence of cadmiun and/or zinc. *Electroanal* 7: 670–673.
34. Mendieta J, Rodríguez AR (1996) Electrochemical study of the binding properties of metallothionein I related peptide with cadmium or/and zinc. *Electroanal* 8: 473–479.
35. Nieto O, Rodríguez AR (1996) Complexation properties of the metallothionein fragment Lys-Cys-Thr-Cys-Cys-Ala [56–61] MT I with zinc using square wave voltammetry. *Bioelectrochem Bioenerg* 40: 215–222.
36. Harlyk C, Bordin G, Nieto O, Rodríguez AR (1997) Cyclic voltammetry study of the peptide Lys-Cys-Thr-Cys-Cys-Ala [56–61] MT I in the presence of cadmium. *Electroanal* 9: 608–613.
37. Harlyk C, Bordin G, Nieto O, Rodríguez AR (1998) Electrochemical behaviour of a metallothionein related peptide in the presence of cadmium using cyclic voltammetry. *J Electroanal Chem* 446: 139–150.
38. Muñoz A (September 1992) PhD Dissertation Complutense Univ. Madrid.
39. Muñoz A, Rodríguez AR (1995) Electrochemical behaviour of metallothioneins and related molecules. Part III: Metallothioneins. *Electroanal* 7: 674–680.
40. Ruiz C (October 1996) PhD Dissertation Alcalá de Henares Univ. Madrid.
41. Ruiz C, Mendieta J, Rodríguez AR (1995) The electrochemical behaviour of Cd, Zn thioneins depending on the solution pH using differential pulse polarography. *Anal Chim Acta* 305: 285–294.
42. Ruiz C, Rodríguez AR (1996) Study of the additions of cadmium and zinc to the Cd,Zn metallothioneins using differential pulse polarography. Influence of the pH. *Anal Chim Acta* 325: 43–51.
43. Ruiz C, Rodríguez AR (1997) Characterisation of human metallothioneins form foetal liver and adult kidney using differential pulse polarography. *Anal Chim Acta* 350: 305–317.
44. Nieto O, Bordin G, Rodríguez AR, Hellemans G, De Ley M (1997) Characterisation of human foetal liver Zn-metallothioneins using differential pulse polarography. *Talanta* 46: 315–324.
45. Nieto O, Bordin G, Rodríguez AR, Hellemans G, De Ley M, Characterisation of human adult kidney Cd,Zn-metallothioneins using differential pulse polarography. *Bioelectrochem Bioenerg*; *in press*.
46. Dabrio M, PhD Thesis, in preparation.
47. Van Houdt K, Nicasi I, Van Mechelen E, Veulemans B, De Ley M (1992) Monoclonal antibodies against metallothioneins from the human liver. *Immunol Lett* 32: 21–26.
48. Hellemans G (June 1997) PhD Dissertation, Katholieke Univ. Leuven.
49. Palecek E, Pecham Z (1971) Estimation of nanogram quantities of proteins by pulse-polarographic technique. *Anal Biochem* 42: 59–51.

Metallothionein IV
C. Klaassen (ed.)
© 1999 Birkhäuser Verlag Basel/Switzerland

Comparison of CZE and MECC separations of MT isoforms

Vesa Virtanen and Guy Bordin

European Commission, Joint Research Centre, Institute for Reference Materials and Measurements, Retieseweg, B-2440 Geel, Belgium

Introduction

In order to characterize the biological functions of individual isoforms and especially sub-isoforms of metallothioneins (MT), analytical methods that provide a high degree of resolution are needed. Capillary zone electrophoresis (CZE) and micellar electrokinetic capillary chromatography (MECC) have been successfully adapted to the separation of MT isoforms by several scientists. Earlier MECC using borate buffer with sodium dodecylsulphate (SDS) micelles at pH 8.4 was adapted to MT isoform separation [1–2]. Even though the results obtained were good there was still a lack of separation of individual sub-isoforms. Method development for metallothionein isoform analysis by capillary electrophoresis can be divided into two groups. Firstly, methods which give quantitative and qualitative information for each isoform and secondly methods which produce results for each individual sub-isoform. This work studies the applicability of tris-tricine buffer for isoform separation. It also looks at the effects of additions of another anionic surfactant, sodium cholate (SC) and organic solvent (MeOH), to the tris-tricine running buffer in order to gain separation of the sub-isoforms.

Theory of MECC

MECC, first reported by Terabe et al. [3], is usually categorized as a part of electrophoresis, but the separation principle is more similar to reversed phase liquid chromatography than to electrophoresis. Separation, in the case of capillary electrophoresis (CE) is based on the difference of electrophoretic mobilities of analytes, whereas in the case of MECC it is mainly based on difference in the distribution constants of analytes into the micellar pseudo-phase. Compounds that have the same charges and similar structures often migrate at more or less the same speed in capillary electrophoresis. By adding micelles into the running solution, the differences in distribution constants of these compounds to the micellar phase leads, in many cases, to successful separations. At concentrations above the so-called critical micelle concentration, CMC, (8–9 mM for SDS and 14 mM for SC), spherical aggregates of individual surfactants are formed. This spontaneous aggregation is caused by increasing hydrophobic interactions of the surfactants at higher concentrations. The average number of molecules per micelle is termed the aggregation number (AN). The size of the micelles formed by SC and SDS differ quite a lot since the AN is 3 for SC and 62 for SDS [4].

Experimental

MECC was performed on a P/ACE System 5000 capillary electrophoresis instrument (Beckman, Fullerton, CA, USA) equipped with a P/ACE Station software. Uncoated capillaries of 50 μm I.D. and 375 μm O.D. containing polyimide cladding were obtained from Beckman (Fullerton, CA, USA). Overall capillary length was 57 cm with on-line detection at 50 cm. Pressure injection mode was used (0.5 psi). Separated components were detected at 214 nm. All reagents used were of research grade. All buffers were prepared with ultrapure water obtained from a Milli-Q Plus 185 water purification system (Millipore S.A., Molsheim, France) and degassed in an ultrasonic bath. The buffers were filtered using a 0.2 μm filter (Gelman Sciences, Ann Arbor, MI, USA). MT samples were obtained from Sigma (St. Louis, MO, USA).

TRIS – tricine buffer system

The electropherograms of rabbit liver MT-1 and rabbit liver MT-2 obtained by using tris – tricine buffer are shown in Figure 1. As it can be seen 6 peaks can be observed for rabbit liver MT-1(solid line). When comparing the peaks obtained with those obtained from rabbit liver MT-2 sample (dotted line) it can be seen that peaks numbered 1,2 and 3 are corresponding only to rabbit liver MT-1 but peak 4 is found as a small peak in MT-2 which indicates that in MT-2 sample some residues of MT-1 are found. Peaks 5 and 6 found in MT-1 isoform sample are more or less residues of isoform MT-2, which are retained even after fractioning and cleaning procedures [5]. Considering the fact that rabbit liver MT-1 is supposed to contain 6 sub-isoforms, it is clear that tris-tricine system is capable of giving information about the total value

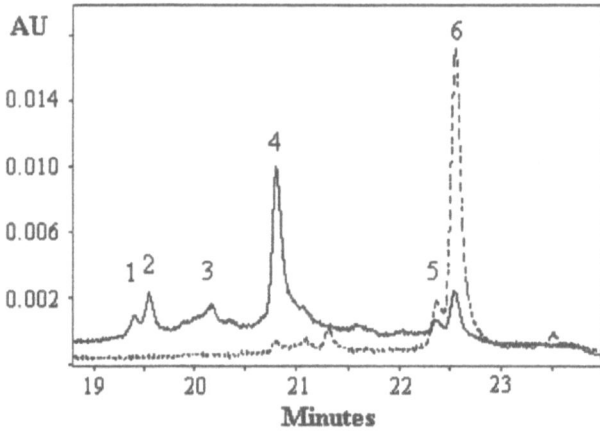

Figure 1. Electropherograms of rabbit liver MT-1 (solid line) and rabbit liver MT-2 (dotted line) obtained by using tris – tricine buffer 150 mM–150 mM, pH 7.75. Temperature 15 °C and running voltage 10 kV.

of MT-1 but not of each individual sub-isoform. Addition of MeOH to the buffer system enhanced the separation of peak groups 1,2 and 5,6 but peak 4 stayed nearly similar (data not shown, unpublished observations).

MECC with sodium cholate micelles

The choice of surfactant in MECC defines the physico-chemical nature of the micelles and is analogous to the choice of stationary phase in HPLC. Bile salts, to which sodium cholate belongs, is a class of surfactants which have not attained much attention in CE so far. Addition of sodium cholate, SC, into the buffer solution causes changes in the electrolyte system used. As long as the concentration stays below the critical micelle concentration, CMC, sodium cholate does not aggregate and does not form micelles. This means also that the so-called micellar pseudo-phase is not formed. When concentration exceeds CMC limit, micelles occur and micellar pseudo-phase is formed. As the concentration of SC increases the separation of MT-1 sub-isoforms seems to improve. This indicates that MT-1 sub-isoforms have differences in their dissociation constants into the micellar pseudo-phase formed by sodium cholate. The main peak corresponding to MT-1 in Figure 1, labelled 4, starts to split into several smaller peaks. Figure 2 shows the separation of rabbit liver MT-1 (solid line) and rabbit liver MT-2 (dotted line) when sodium cholate concentration in the buffer is 150 mM. Several peaks are observed for MT-1 isoform in the area where there was one big peak (labelled 4) in Figure 1. MT-2 shows enhanced separation of peaks which were co-migrating when aqueous tris – tricine buffer was used. These give possibilities to get quantitative and qualitative data from each individual peak. Using diode-array detector a UV scan can be performed from each peak. This UV scanning gives

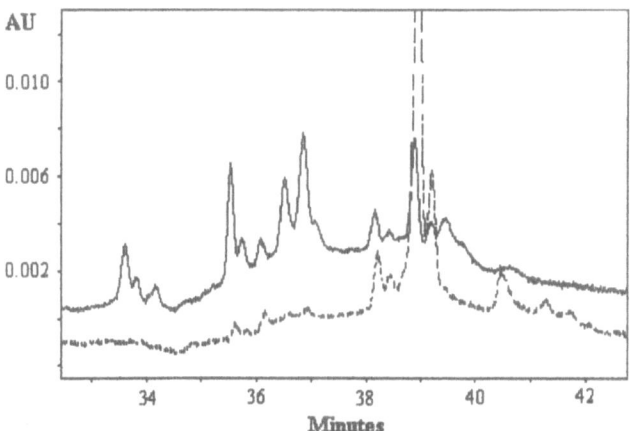

Figure 2. Electropherograms of rabbit liver MT-1 (solid line) and rabbit liver MT-2 (dotted line) obtained by using MECC with SC. Conditions: Tris-tricine – SC 150 mM–150 mM - 150 mM, pH 8.00, temperature 18 °C and running voltage 10 kV.

information about the bound metals (Zn and/or Cd) as the cation-thiol bonds cause shoulders in the UV spectrum of the MT isoforms (Zn-thionein at 225–230 nm and Cd-thionein at 255 nm).

The addition of organic solvent, in this case MeOH, increases the viscosity of the buffer and alters the polarity of the buffer. This causes changes in the double-layer of the capillary affecting directly zeta potential and the electroosmotic flow, EOF [6–8]. The zeta potential decreases with an increasing content of organic solvent since the inflection points corresponding to pK' of the silanol groups shift towards higher pH upon addition of the solvent [6]. EOF decreases with increasing MeOH content which in itself enhances separation of analytes migrating with quite similar speed. At introduction of MeOH into the MECC buffer solution the separation is improved at some parts of the electropherogram as seen in Figure 3. At higher organic solvent concentrations the separation performance seems to reduce slightly indicating possible deterioration of the micelles.

Conclusions

Capillary zone electrophoretic separation of MT isoforms by using tris-tricine buffer offers good separation of isoforms but not so good for sub-isoforms. Compared to capillary electrophoresis, micellar electrokinetic capillary chromatography (MECC) gives an opportunity to enhance the separation of compounds with, the same charge and similar or very close, structures. Using tris-tricine-sodium cholate buffer an enhanced separation of metallothionein isoforms and sub-isoforms has been obtained offering good future prospects for obtaining advanced information. Addition of organic solvent, MeOH, to the MECC buffer solution enhanced the separation partially.

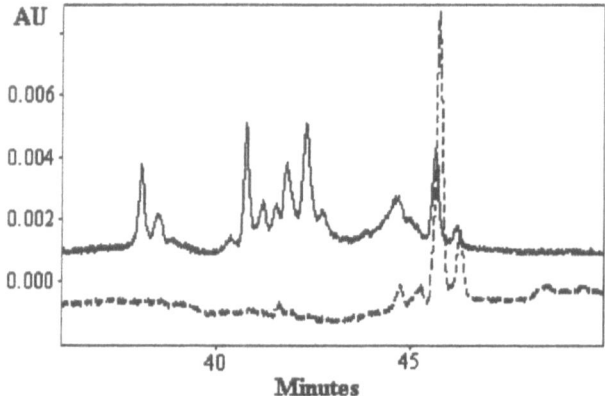

Figure 3. Electropherograms of rabbit liver MT-1 (solid line) and rabbit liver MT-2 (dotted line). Conditions: Tris -tricine – SC – MeOH 150 mM - 150 mM - 150 mM - 15%(v/v), pH 8.00, temperature 18 °C and running voltage 15 kV.

References

1. Beattie JH and Richards MP (1994) *J Chromatogr A* 664: 129.
2. Beattie JH and Richards MP (1995) *J Chromatogr A* 700: 95.
3. Terabe S, Otsuka K, Ichikawa K, Tsuchiya A and Ando T (1984) *Anal Chem* 56: 111.
4. Kuhn R and Hoffstetter-Kuhn S (1993) *In: Capillary Electrophoresis: Principles and Practice*, Spinger-Verlag, Heidelberg, Germany, pp. 191–194.
5. Virtanen V, Bordin G and Rodriguez AR (1996) *J Chromatogr A* 734: 391.
6. Schwer C and Kenndler E (1991) *Anal Chem* 63: 1801.
7. Fujiwara J and Honda S (1987) *Anal Chem* 59: 487.
8. Virtanen V, Bordin G and Rodriguez AR (1996) *In*: P Sandra and G Devos (eds): *18th International Symposium on Capillary Chromatography*, Vol.III, Huethig Publishing, Heidelberg, pp. 1976–1986.

Synthetic peptides and domains of metallothionein for structure function studies: rabbit liver peptide MT-IIA 49–61 and lobster MT β_N β_C and $\beta_{C\rightarrow N}$ domains

Amalia Muñoz, Frank Laib, Eugene DeRose, David H. Petering, John Ejnik and C. Frank Shaw III

Department of Chemistry and the UWM-NIEHS Marine and Freshwater Biomedical Core Center, University of Wisconsin-Milwaukee, Milwaukee, WI, 53201-0413, USA

Summary. Mammalian and crustacean metallothioneins are complex proteins which form two independent domains with 3 or 4 metal ions (Cd^{+2} or Zn^{+2}) in the native domains. Synthesis of the individual domains and even smaller fragments allows investigation of specific structure-function questions that can not be addressed easily with site-directed mutagenesis of the entire two-domain structure. Residues 49–61 of mammalian MTs, which include the only sequential four cysteines bound to the same metal ion (Cd^I), are a putative nucleation site. We have synthesized rabbit liver MT-II$_{49-61}$ (Ac-Ile-Cys-Lys-Gly-Ala-Ser-Asp-Lys-Cys-Ser-Cys-Cys-Ala-COOH) and examined its binding to Cd^{+2} and other metal ions. Physiochemical studies (^{113}Cd NMR, UV-vis, CD, ES-MS, LSIMS and chemical reactions) of Cd-Pep$_{49-61}$ demonstrate that it exhibits properties consistent with a role in Cd-induced MT folding. To examine the possible influence of (1) the position of the cysteine residues and (2) the steric and electrostatic effects of neighboring amino acids on the folding and stability of MT clusters, the stable lobster β_C and β_N domains (each having 9 cysteines and binding 3 M^{+2} ions) and a modified domain $\beta_{C\rightarrow N}$ (in which the cysteines of the C-terminal domain are relocated so they are spaced as in the N-terminal domain) have been synthesized and characterized. The synthetic native domains (Cd_3-β_C and Cd_3-β_N) show metal binding properties and reactivity similar to the holo-protein, but the modified Cd_3-$\beta_{C\rightarrow N}$ domain is unusually reactive and in the presence of Chelex aggregates to a $Cd_5(\beta_{C\rightarrow N})_2$ dimer. These differences in structure and reactivity demonstrate that the requirements for formation of a stable beta cluster are more stringent than simply the sequential positions of the cysteines along the peptide chain and include specific interactions with neighboring amino acids.

Introduction

Metallothionein (MT) is a small protein with an unusual composition characterized by the lack of aromatic amino acids and a very high content of cysteine residues (30%), which confer a high capacity for metal binding [1–3]. MT is induced by *in vivo* exposure to heavy metals such as Cd^{+2}, Au(I), Pt(II) or Hg^{+2} and a great variety of hormones, other chemical stimuli and various forms of radiation, *inter alia* [2–4]. The primary biological roles of MT are not clear yet, but it is involved in the detoxification of heavy metals and the regulation of zinc and copper metabolism including donation of zinc to newly synthesized apoenzymes, while other evidence supports putative roles in the expression of the genetic message, radical scavenging, the stress response and in the pharmacology of metallodrugs and alkylating agents [2–4].

Mammalian MTs consist of 61 amino acids, of which 20 are cysteines, and bind 7 Cd^{+2} or Zn^{+2} ions distributed in two discrete Cd-clusters located at the α- and β-domains, respectively [5–7]. The thiolate side chains of the protein act as bridging and terminal ligands to form a $Cd_3(S_{t,b})_9$ cluster (at the N-terminus) and a $Cd_4(S_{t,b})_{11}$ cluster (at the C-terminus), with each Cd^{+2} tetrahedrally coordinated to 4 cysteine residues [5–7]. Apo-MT lacks tertiary structure and folds into its two domain structure only in the presence of 7 Cd^{+2} or Zn^{+2} ions.

Conventional solution methods for chemical synthesis of polypeptides were used in prior studies[10, 11] to prepare several metallothionein fragments of different lengths (5 to 30 amino acids). The resulting polypeptides showed lower metal-binding capacity than the native protein. Subsequently, solid phase methods were successfully applied to the preparation of synthetic α and β domains of mammalian MTs [12, 13]. Here we describe the synthesis of MT peptide 49–61 and two native and one modified lobster domains for structure-function studies.

Under certain conditions the formation of independent $M(S_t)_4$ sites precedes the formation of the MT clusters of Cd^{+2}, Zn^{+2} or Co^{+2} [14–17]. From the X-ray structure [6, 7], Robbins et al. proposed that Cys50, Cys57, Cys59 and Cys60, the only sequential set of four bound to the same metal ion (Cd^I), are where the first metal ion binds to MT to initiate protein folding. Since the proteolytic isolation of the peptide 49–61 from native MT is impractical, that portion of rabbit liver MT-IIA was prepared synthetically. The structure and reactivity of the Cd-peptide 49–61 complex are described in Part I of this abstract.

Crustacean MTs are useful models for studying MT because they exhibit important variations in sequence and structure [8, 9, 18, 19]. They contain only 18 cysteines and bind 6 M^{2+} ions, from which they generate two 3-metal domains, designated β_n and β_c. Surprisingly, these domains fold differently than one another or the mammalian β domain [8]. Thus, in nature there are three stable beta domain structures. This raises the possibility that the criteria for forming a beta-domain with Cd_3S_9 stoichiometry are minimal and that many sequences will generate such a structure driven solely by the stability of Cd thiolate interactions. In Part II, we describe the structure-reactivity relationships of the synthetic native lobster β_c and β_n domains and a modified $\beta_{c \to n}$ domain.

Structure and reactivity of Cd-peptide$_{49-61}$

Synthesis

The peptide 49–61 of rabbit liver MT-IIA was synthesized by solid-phase, FMOC methods. Removal of the Acm-groups from the full peptide with Hg(II) acetate and removal of Hg(II) with 2-mercaptoethanol generated the apo-peptide. After chromatographic purification,

Pep$_{49-61}$: Ac-Ile-Cys^{50}-Lys-Gly-Ala-Ser-Asp-Lys-Cys^{57}-Ser-Cys^{59}-Cys^{60}-Ala-COOH the amino acid composition (all residues within 10%) was in good agreement with the theoretical values. The UV-vis spectrum and Hg-analysis by GF-AAS indicate that all the Hg(II) was removed. The ES-MS spectrum shows a single species at m/z = 663(+2), with a +2 charge envelope due to the presence of two protonatable groups, Lys-51 and Lys-56 with an additional peaks at m/z = 674(+2) and 685(+2) from Na^+/H^+ exchange.

Characterization of the Cd-peptide 49–61

Reaction of the apo-peptide with $CdCl_2$ under reducing conditions followed by chromato-graphic removal of any free-Cd^{+2} generates a species with stoichiometry Cd_x-pep$_y$ (x/y- 1.2–.5) and SH/Cd^{+2}.2.6–3.4. Treatment of Cd_x-pep$_y$ with Chelex removes the adventitiously bonded Cd^{+2} and yields the 1:1 adduct. The retention times of the 1:1 Cd-peptide and apo-peptide on an analytical C_8 reverse-phase HPLC column (t_R = 6.27 and 7.3 min respectively) indicate structural differences due to metal chelation and the absence of potential contaminants. The Cd-peptide was chemically characterized by analyzing for Cd^{+2} (by AAS, 138.8 ± 4.1 μM Cd), cys-teine content (by the DTNB reaction, 543.6 ± 16.2 μM) and free amino groups (by Ruhemann's purple method, 266.5 ± 8 μM). Since the peptide includes two lysines and four cysteines, the ratios SH/Cd^{+2} = 3.92, NH$_2$/Cd^{+2} = 1.92 and SH/NH$_2$ = 2.04 are in good agreement with the theoretical values of 4, 2 & 2, respectively.

The ^{111}Cd-1D NMR spectrum of the ^{111}Cd-peptide (Fig. 1) consists of a single resonance (δ_{Cd} = 716 ppm). The chemical shift indicates that the ^{111}Cd^{+2} in the metal site is terminally coordinated to four side-chain thiolates of the cysteine residues, whereas shifts for the holo-protein occur at higher field, because in MT at least two Cys-thiolates are bridging in the coor-dination sphere of each Cd^{+2} ion [5–7, 20]. The experimental molecular weight 1220 ± 40 (determined by analytical gel-exclusion chromatography with GSSG, B-12, lobster Cd_3-β_n, and insulin as standards; theory 1441) establishes that the complex is a monomer with all four cys-teines donated by a single peptide chain.

During the pH-titration of the Cd-peptide, monitored spectrophotometrically at the metal-to-ligand charge transfer band, 245 nm, cadmium starts to dissociate from the thiolate clusters below pH 3.6 and is completely dissociated at pH 2.8. Back-titration with NaOH regenerates the Cd-peptide and demonstrates that the Cd^{+2} binding/dissociation process is reversible and pH dependent. The curve shows a single inflection point and the pH of half-dissociation for the Cd in the Cd-peptide is 3.3, similar to that of the holoprotein (pH = 3.0) [21].

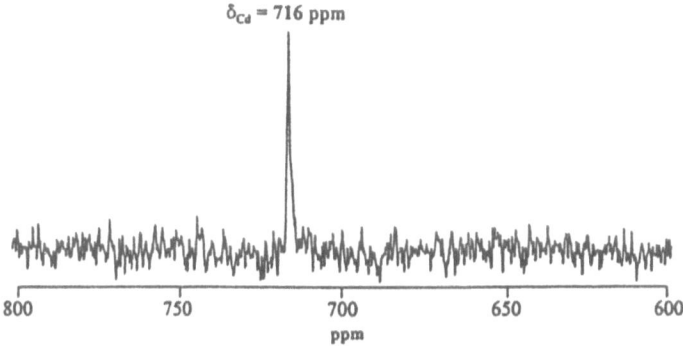

Figure 1. ^{111}Cd-1D NMR spectrum measured at 106.1019 MHz vs. 0.1 M ^{111}Cd(ClO$_4$)$_2$ of 1.8 mM ^{111}Cd-pep-tide solution in 5 mM Tris-HCl and pH 7.4. The spectrum was taken using a GE 500 MHz NMR at 25°C with 69632 acquisitions.

Titration of the apo-peptide with cadmium (as $CdCl_2$), monitored at 245 nm, shows a rapid rise in A_{254} between 0 and 1 equiv of Cd^{2+} and a smaller rise between 1 and 2 equiv, which indicates that the peptide can react *in situ* with at least 2 equivalents of Cd^{+2}. Chelex treatment of the final solution regenerates the 1:1 Cd-peptide complex. The extinction coefficient at 245 nm ($\varepsilon = 12{,}610 \ M^{-1}cm^{-1}$) in the UV-vis spectrum is similar to the per cadmium coefficient for mammalian Cd_7MT species. The CD spectrum ($\lambda_{max} = 245$ nm) confirms that a chiral structure is formed about the cadmium ion. Molecular modeling with SYBYL using the Gasteiger-Hückel approach for charges, shows that the clockwise chirality corresponding to that in the holoprotein is of lower energy than the counterclockwise chirality for coordination of the tetradentate chelate about the tetrahedral Cd^{+2}.

The ES-MS of the Cd-peptide (net charge +1 at pH ~ 3.5, which is close to the half-dissociation pH, 3.3), shows peaks due to four different species. Major peaks at m/z = 666(+2) and m/z = 1331 correspond to the apo-peptide with +2 and +1 charges, respectively. Peaks at m/z = 1440, 1549 and 1658 arise from Cd-peptide, Cd_2-peptide and Cd_3-peptide, respectively, formed with displacement of protons at each addition. The low pH favors a rapid metal exchange analogous to the labile intra-cluster interchange observed in the mammalian MT β-clusters [22]. The resulting equilibria, n = 2 and 3, generate the apo-peptide and Cd_2-peptide and Cd_3-peptide species: pH 3.5

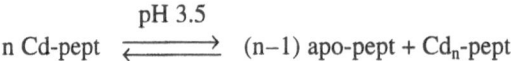

$$\text{n Cd-pept} \ \underset{\longleftarrow}{\overset{\text{pH 3.5}}{\longrightarrow}} \ \text{(n–1) apo-pept + Cd}_n\text{-pept}$$

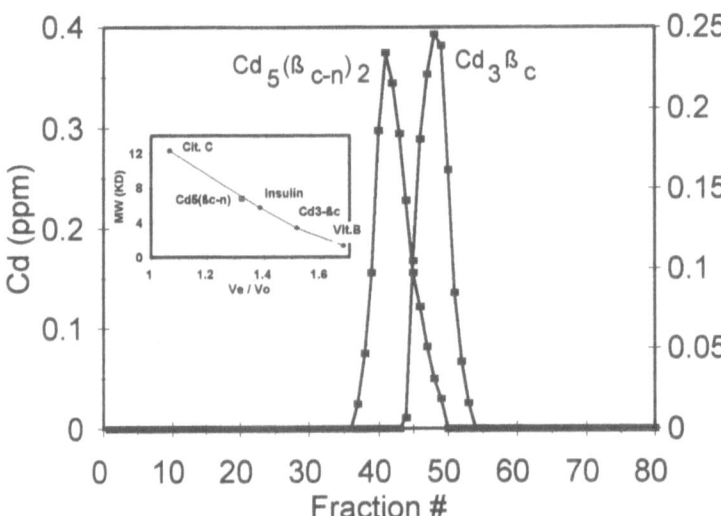

Figure 2. Elution profile for the gel-filtration chromatographic separation of $Cd_3\beta_c$ and $Cd_3(\beta_{c \to n})_2$ carried out over a Sephadex G-25 column (111X1 cm) pre-equilibrated with 5 mM Tris-HCl (pH 7.5). Inset: Calibration curve. Peptide mass (KD) vs. relative elution volumes.

The ^{111}Cd NMR chemical shift of 716 ppm for Cd-peptide indicates that the coordination of the first Cd^{+2} involves only cysteine residues. The binding of the second Cd^{2+} during titration of apo-peptide or upon acidification for ES-MS sampling also involves the thiolates since A_{254} increases, but the third Cd^{2+} does not increase the number of Cd-S interactions. At pH 3.5, the C-terminal carboxylic group, aspartic acid and two lysines can be deprotonated upon coordination of the second and third Cd^{+2}.

Kinetic reactivity

Because MT is a dynamic and unusually reactive metalloprotein, it is interesting to compare the reactivity of the Cd-peptide complex to that of the larger holo-protein. Reactions with EDTA [23, 24], DTNB [25, 26] and apo-carbonic anhydrase [24, 27] were chosen for this purpose.

DTNB reaction
Reactions between Cd-peptide and increasing concentrations of DTNB (pseudo-first order excess) were carried out at 25 °C and monitored at 412 nm, the absorbance of

$$\text{x5DTNB + Cd-peptide} \longrightarrow \text{apo-peptide}_{ox} + 9TNB + 3Cd^{+2}$$

the reaction product TNB (2-nitro-5-thiobenzoate). Stopped-flow spectrometry was required to resolve a step too fast for conventional spectroscopic measurements. Analysis of the absorbance via first-order plots indicates that the reaction is biphasic. The plot of the k_{obs} values vs [DTNB] indicates that the ultrafast step is DTNB-concentration independent, while the second phase has DTNB-concentration-dependent and -independent steps:

$$\text{rate} = (k_{uf} + k_{1,f} + k_{2,f} [DTNB]) [peptide].$$

The values calculated for each component of the rate constant were compared with those of the Cd_4-α and Cd_7MT (Tab. 1). The two rate constants designated fast are similar to those obtained for the Cd_4-α domain of MT [26] and the fast phase of the holoprotein [25, 26], but the ultrafast step is unprecedented for the DTNB reaction with any native MT or isolated MT-domain.

Table 1. Rate contants for Cd-peptide 49–61 reaction with DTNB[a]

Protein	k_{uf} /s^{-1}	$k_{1,f}$ H 10^3/s^{-1}	$k_{2,f}$/M^{-1}s^{-1}	$k_{1,s}$ H 10^3/s^{-1}	$k_{2,s}$/M^{-1}s^{-1}
Cd-pep$_{49-61}$	0.15	2.59	0.88	-	-
$Cd_4\alpha$[b]	-	0.64	1.12	-	-
Cd_7MT[c]	-	1.26	1.75	0.42	0.12

[a][Cd-peptide]$_o$ = 12μM, [DTNB]$_o$ = 0.3–2 mM in 5 mM Tris·HCl/0.1 M KCl at pH 7.4 and 25.0 ± 0.1 °C. [b]ref.26. [c]ref.25.

EDTA reaction

The reaction between Cd-peptide and EDTA was monitored by following the removal of Cd^{+2} from the peptide at 254 nm. The reaction goes to completion in about 30 min and is therefore much faster than in the case of Cd_7MT which requires 4 h under similar conditions [23].

Carbonic Anhydrase reactions

Previous studies indicate possible roles of MT in donating zinc to metalloenzymes [27–29] and in extracting Cd^{+2} incorrectly inserted into apo-Zn^{+2}-enzymes [28, 29]. In this work, competition of the peptide 49–61 and carbonic anhydrase (CA) for Cd^{+2} was examined in both directions as a model of the latter function. The reactions were monitored by following the formation or loss of Cd-peptide binding at 250 nm. When Cd-CA was reacted with equimolar quantities of apo-peptide 49–61, an equilibrium situation was established rapidly.

$$\text{Cd-CA + apo-peptide} \rightleftharpoons \text{apo-CA + Cd-peptide}$$

However, for the reaction between apo-CA and equimolar quantities of Cd-peptide 49–61, the slow transfer of Cd^{+2} from the peptide complex to CA takes ~ 96 h to reach equilibrium. In either direction, the equilibrium was reached with 60–80% of Cd bound to the apo-peptide. Thus, the affinities of the two species (apo-CA and apo-peptide) for Cd^{2+} are similar, but the formation of Cd-peptide is kinetically and thermodynamically favored.

Reactions with other metal ions

Like MT [1, 30, 31], peptide 49–61 binds to other heavy metals. The stoichiometries, UV maxima and molecular weights obtained by ES-MS for Hg-pept, Hg_2-pept; Au_2-pept, Zn-pept, and $(Et_3PAu)_2$-pept are consistent with the properties of the related MT complexes of these metals.

Discussion

Synthetic apo-peptide 49–61 forms a stable Cd-peptide derivative with a 1:1 stoichiometry. The Cd-thiolate interactions were confirmed by UV-vis, CD, ES-MS, and ^{111}Cd-NMR spectroscopies. Additional Cd^{+2} ions bind weakly, but are easily displaced by Chelex:

$$\text{apo-peptide + Cd}^{+2} \rightleftharpoons \text{Cd-peptide} \overset{\text{xs Cd}^{+2}}{\underset{\text{Chelex}}{\rightleftharpoons}} \text{Cd}_n\text{-peptide (n = 2, 3)}$$

Cooperative binding of Cd^{+2}, Zn^{+2}, Cu^+ and Ag^+ in metal thiolate clusters generated within each of the MT-domains [14, 15, 32] indicates that the formation of the clusters is thermodynamically favored over independent metal-thiolate sites. The pH of half dissociation of ~ 3.3 for the

Cd-peptide is similar to that of Cd$_7$MT (pH $\approx$ 3.0), which indicates that the clusters in MT show only a slightly higher thermodynamic stability than Cd-peptide 49–61.

Cd^{2+} binding to MT at neutral pH generates the typical Cd(S$_{t,b}$)$_4$ sites of the MT-clusters (which involve terminal and bridging thiolates), while at basic pH values exchange broadening is attributed to independent terminally-coordinated tetrathiolate sites, Cd(S$_t$)$_4$ in equilibrium with the clusters [17]. Independent binding of sub-stoichiometric quantities of Cd^{2+} and Co^{2+} has been demonstrated spectroscopically [15, 33]. Cysteines 50, 57, 59 and 60 in mammalian MTs are the only sequential set of four bound to the same metal ion, Cd^I [6, 7, 34]. From the X-Ray structure, Robbins et al. [6, 7] suggested the following mechanism for folding of the α-domain cluster, M$_4$S$_{11}$: First, the Cd^I binds to these four cysteines (50, 57, 59, 60) while Cd^{VII} can bind trigonally to Cys34, Cys36 and Cys37 without alteration of the intervening random coil. Then, formation of the α/3$_{10}$ helix by residues 41–47 will lead to binding of Cd^{VI} and Cd^V with the formation of bridging cysteine thiolates and finally Cd^I and Cd^{VII} can associate with the Cd^{VI}-Cd^V nucleus to form the complete cluster. This mechanism requires that Cd^{+2} should bind to cysteines 50, 57, 59 and 60 without involvement of residues 1 to 48. Thus, the folding of peptide 49–61 into a tetradentate, terminally coordinated 1:1 peptide complex with the same chirality as in the holo-protein, supports a pathway for cluster formation via the initial independent binding of several metal ions.

Although the Cd-binding affinities for both the apo-peptide and apo-MT are similar, the kinetic reactivity of the Cd-peptide 49–61 with EDTA and DTNB is greater than that observed for Cd$_7$MT. Clearly, the Cd-thiolate bonds in the Cd-peptide are more labile than in the case of the holoprotein. This difference can be explained by (1) the small molecular weight of the peptide, about one-fifth that of the apo-MT, which can account for a steric effect, (2) the tetradentate mononuclear coordination in the peptide compared to the clusters of MT and (3) the greater accessibility of the peptide thiolates and Cd(II) to the solvent. In particular, all four cysteines (50, 57, 59 and 60) are terminally coordinated to Cd^{2+} in Cd-peptide, but in the holoprotein or α-domain Cys50 and Cys60 are bridging and have much smaller solvent contact surfaces. The enhanced reactivity of the Cd-peptide complex reported here indicates that the sequestration of Cd^{2+} by MT has an important kinetic component.

In conclusion, Cd-peptide 49–61 forms a stable complex, in the absence of the remaining protein structure, with Cd^{+2} tetracoordinated by four cysteines corresponding to Cys50, Cys57, Cys59 and Cys60 with the same stereochemistry as in rabbit liver MT-II. While the thermodynamic stability is relatively unperturbed, the kinetic reactivity of the Cd(S$_t$)$_4$ structure is substantially enhanced when it is independent of the complete domain (Cd$_4\alpha$ or Cd$_7$MT). Thus, the structure and chemical properties are consistent with the putative function of MT residues 49–61 as a nucleation site.

Structure function studies of native and modified lobster MT-I domains

Solid Phase methods using Fmoc-protected amino acids were used to synthesize the following β-domains related to lobster MT-I sequence determined by amino acid sequencing [35] and 2-D NMR [8]:

β_c Ac-P-C-E$^\ominus$K$^\oplus$C-T-S-G-C-K$^\oplus$C-P-S-K$^\oplus$D$^\ominus$E$^\ominus$C-A-K$^\oplus$T-C-S-K$^\oplus$P-C-S-C-C-P-T-coo$^\ominus$

$\beta_{c\to n}$ Ac-P-K$^\oplus$E$^\ominus$C-C-T-S-G-C-K$^\oplus$C-P-S-K$^\oplus$D$^\ominus$C-E$^\ominus$A-K$^\oplus$C-T-C-K$^\oplus$P-C-S-C-S-P-T-coo$^\ominus$

β_n Ac-P-G-P-C-C-K$^\oplus$D$^\ominus$K$^\oplus$C-E$^\ominus$C-A-E$^\ominus$G-G-C-K$^\oplus$T-G-C-K$^\oplus$C-T-S-C-R$^\oplus$C-A-coo$^\ominus$

β_c and β_n are the synthetic native lobster domains, while $\beta_{c\to n}$ is a modified domain with the amino acid sequence of β_c, but with the cysteine residues in the positions corresponding to the β_n. They were chemically synthesized using Merrifield solid-phase techniques, cleaved and then purified by reverse phase-HPLC. Sequential treatment with $Hg(CH_3COO)_2$ and β-mercaptoethanol removed the Acm-groups of the cysteine residues and resulted in the corresponding apo-β domains. ES-MS carried out at each stage of peptide deblocking and the amino acid analysis of the resulting apo-peptides show a good agreement between the experimental and theoretical values. Cadmium reconstitution of all three apo-domains yields $Cd_3\beta$ derivatives which are stable during gel-exclusion chromatography. But treatment with Chelex-100 [7] to remove adventitiously bound Cd^{+2} shows a marked difference between native and modified sequences (Tab. 2). Analytical chromatography of the chelex-treated domains showed that $Cd_3\beta_c$ and $Cd_3\beta_n$ elute at ~ 3400 daltons, while $Cd_{2.6}\beta_{c\to n}$ elutes at ~ 6800 daltons, confirming that the modified sequence forms a dimer, $Cd_5(\beta_{c\to n})_2$. Thus, only the $Cd_3\beta_{c\to n}$ aggregates to a dimer, $Cd_5(\beta_{c\to n})_2$ with loss of one Cd^{+2} per dimer, while the native sequences retain their Cd^{2+} and remain monomeric. This may account for other differences in their spectroscopic and kinetic behavior described below.

UV-spectra for the $Cd_3\beta_c$ and $Cd_3\beta_n$ (native domains) and $Cd_5(\beta_{c\to n})_2$ and $Cd_3\beta_{c\to n}$ (modified sequence) show absoption bands at ~ 250 nm characteristic of metal to ligand (Cd-S) charge transfers. These UV bands and CD spectral features arising from the chiral folding of the peptide about the metal ions indicate the formation of Cd-thiolate clusters similar to those observed in the holo-protein (lobster Cd_6MT). The differences in the λ_{max} values observed in the UV and CD (Fig. 3) spectra of the Cd-$\beta_{c\to n}$ derivatives also suggest structural differences in the cluster structures derived from the native and modified sequences.

Characterization of the $Cd_3\beta_c$ derivative by ES-MS at pH 3.5, which is close to the pH of half titration, exhibited evidence of metal scrambling to generate metal-rich and metal-poor species. Under the conditions of this experiment the following equilibrium takes place:

$$Cd_3\beta_c\ (+2)\ 4 \quad \underset{\xrightarrow{\hspace{1.5cm}}}{\overset{\text{pH } 3.5}{\rightleftharpoons}} \quad Cd_4\beta_c\ (+4) + Cd_2\beta_c\ (0)$$

Table 2. Stoichiometries and properties of the synthetic domains

After size exclusion chromatography	After Chelex treatment	Molecular weight	UV $\lambda_{Cd\text{-}S}$ (nm)	CD $\lambda_{Cd\text{-}S}$ (nm)
$Cd_3\beta_c$	$Cd_3\beta_c$	~3400	247	258
$Cd_3\beta_{c\to n}$		-	248	248
	$Cd_5(\beta_{c\to n})_2$	~6800	254	247
$Cd_3\beta_n$	$Cd_3\beta_n$	~3400	247	262

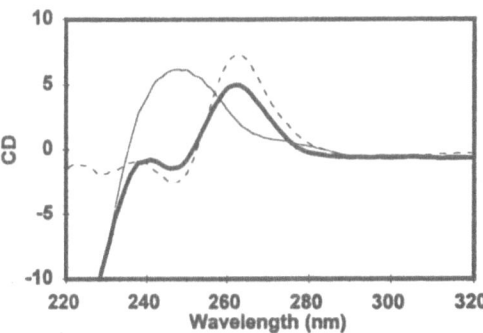

Figure 3. CD spectra of the lobster MT-1 (—), $Cd_3\beta_c$ (----) and $Cd_3(\beta_{c \to n})_2$ (—) in 5 mM TrisAHCl, 0.1M KCl at pH 7.5 measured in a Jasco spectropolarimeter 70.

The m/z values for $Cd_3\beta_c$ and $Cd_4\beta_c$ are in good agreement with the expected theoretical masses.

The ^{113}Cd-1D NMR chemical shifts of the ^{113}Cd$_3\beta$ domains indicate that the Cd^{+2} is tetra-coordinated to four cysteines with some cysteines bridging between two Cd^{+2} ions. The three resonances exhibited by the synthetic native lobster domains, ^{113}Cd$_3\beta_c$ and ^{113}Cd$_3\beta_n$, are consistent with $Cd_3(S_{t,b})_9$ clusters as found in the β_c- and β_n-domains of lobster Cd_6MT [8]. Thus, ^{113}Cd-NMR spectra and ^{1}H-NMR spectra (which show a good dispersion at the amide region) confirm that the synthetic native domains have formed stable clusters similar to those of the holoprotein.

However, NMR techniques show significant differences between the modified sequence ^{113}Cd$_5$ $(\beta_{c \to n})_2$ and the native domains of the holoprotein. The ^{113}Cd$_5$ $(\beta_{c \to n})_2$ exhibits a very broad signal in the ^{113}Cd-NMR (Tab. 3) and in the amide region of the ^{1}H-NMR spectrum.

Table 3. Chemical shifts (δ_{Cd}) for the holoprotein(Cd_6MT) and the synthetic $Cd_3\beta$ domains

	^{111}Cd$_6$MT	^{113}Cd$_3\beta_c$	^{113}Cd$_3\beta_n$	^{113}Cd$_3\beta_{c \to n}$
	663.2	664.4	-	
δ_{Cd} β_c (ppm)	645.2	642.7	-	
	621.1	616.2	-	~640
				$\Delta V_{1/2}$ = 1330 Hz
	649.1	-	659.2[a]	
δ_{Cd} β_n (ppm)	645.2	-	650.1[a]	
	631.4	-	646.6[a]	

[a] These resonances for the N-acetylated β_n cluster shift over a period of 6 months before stabilizing at 649.7, 646.7 and 641.6 ppm. The source of this unique chemistry is under investigation.

These NMR characteristics suggest rapid exchange of Cd^{2+} ions between sites in the monomers and multiple peptide conformations in solution.

The reactions of the Cd-β species with DTNB are biphasic for each of the species studied. The overall rate expression for the reaction of the $Cd_3\beta_x$ domains, as for Cd_6MT, is found to be

$$Rate = (k_{1f} + k_{2f}[DTNB] + k_{1s} + k_{2s}[DTNB])[Cd_3\beta]$$

while, in the case of $Cd_5(\beta_{cvn})_2$, there is an ultrafast step which occurs for this domain alone:

$$Rate = (k_{uf} + k_{1s} + k_{2s}[DTNB][Cd_5\beta_2]).$$

In pH-titrations, carried out to study the stability of the Cd-β clusters, the behavior of $Cd_3\beta_c$ and $Cd_3\beta_n$ is comparable to that observed for the holoprotein [18]: the clusters start to disappear at pH < 4. However, the pH-titration of the $Cd_3\beta_{c \to n}$ or $Cd_5(\beta_{c \to n})_2$ takes place *via* two separate stages, the first one is the displacement of Cd^{+2} by the protonation of the Cd-S cluster as the pH decreases from 7 to 2. The second one is due to the deprotonation of two free cysteines as the pH values increase from 7.5 to 9.5, which is consistent with the ultrafast kinetic step in its reaction with DTNB and the labile exchange of Cd^{2+} observed in the NMR spectrum.

Discussion

Stable $Cd_3\beta_c$ and $Cd_3\beta_n$ clusters were generated from the synthetic native lobster MT-I domain peptides. Spectroscopic characterization of these Cd derivatives demonstrates (1) an absorption band at 250 nm due to MLCT, (2) a CD-signal centered at ~ 260 nm due to interactions between asymmetrically coupled pairs of Cd-S units within the binding site and (3) ^{113}Cd-1D NMR chemical shifts between 610 and 670 ppm reflecting the formation above pH 4, of Cd-thiolate clusters analogous to those found in lobster MT. Lobster and mammalian MTs both react in a biphasic fashion with DTNB [18, 25, 26]. The rate constants calculated for the native sequences are comparable to those observed for the holoprotein. Each domain in lobster MT exhibits biphasic reactions, while each domain of mammalian MT reacts in a single phase.

The modified $Cd_3\beta_{c \to n}$ domain loses cadmium to Chelex and forms a dimer, $Cd_5(\beta_{c \to n})_2$. This Cd-derivative is also formed via thiolate interactions, but its differential spectroscopic features (UV- and CD-λ_{max}) and a very broad ^{113}Cd-NMR signal imply structural changes when it forms. There are also marked differences between synthetic native and modified domains in their reactivity toward DTNB and pH changes. The ultrafast reaction with DTNB and the second step in the pH-titration (~ pH 9) are attributed to the presence of free thiols. In turn, they can catalyze Cd^{+2} exchange between the two peptides in the dimer, causing the unusually broad Cd-NMR signal.

Therefore, we have demonstrated that not all the possible arrangements of cysteine generate stable structures. Both the folding of the peptides to create the Cd-thiolate bonds and the reactivity of the domains reflect subtle interactions between the 9 cysteines and the intervening amino acids.

Acknowledgments
This research was supported by the US NIH grants ES 04026, ES 04184 and DK51308. We thank Dr. Linda Olson for her assistance with the protein synthesizer.

References

1. Kojima Y, Berger C, Vallee BL and Kägi JHR (1976) *Proc Natl Acad Sci USA* 73: 3413–3417.
2. Shaw CF III, Stillman MJ, Suzuki KT (1991) *In*: MJ Stillman, CF Shaw and KT Suzuki (eds): *Metallothioneins: Synthesis, Structure and Properties of Metallothioneins, Phytochelatins and Metal-Thiolate Complexes*. VCH publishers: New York, pp. 1–13.
3. Cherian MG, Chan HM (1993) *In*: KT Suzuki, N Imura, M Kimura (eds): Metallothionein *III. A Biological Functions of Metallothionein*. Verlag Basel: Birkhäuser, Switzerland, pp. 87–109.
4. Hamer D (1986) *Ann Rev Biochem* 55: 913–951.
5. Otvos JD, Armitage IM (1980) *Proc Natl Acad Sci USA* 77: 7094–7098.
6. Robbins AH, McRee DE, Williamson M, Collett SA, Xuong NH, Furey WF, Wang BC, Stout CD (1991) *J Mol Biol* 221: 1269–1293.
7. Robbins AH, Stout CD (1991) *In*: MJ Stillman, CF Shaw and KT Suzuki (eds): *Metallothioneins: Synthesis, Structure and Properties of Metallothioneins, Phytochelatins and Metal-Thiolate Complexes*. VCH publishers: New York, pp. 31–54.
8. Zhu Z, DeRose EF, Mullen GP, Petering DH, Shaw CF III (1994) *Biochemistry* 33: 8858–8865.
9. Narula SS, Brouwer M, Hua Y, Armitage IM (1995) *Biochemistry* 34: 620–631.
10. Matsumoto S, Nakayama S, Nishiyama Y, Okada Y, Min K-S, Onosaka S, Tanaka K (1992) *Chem Pharm Bull* 40: 2694–2700. Matsumoto S, Nishiyama, Okada Y, Min K-S, Onosaka S, Tanaka, K. (1992) *Chem Pharm Bull* 40: 2701–2706.
11. Onaka Y, Tanaka K, Sawada J-I, Kikuchi Y (1991) *In*: MJ Stillman, CF Shaw and KT Suzuki (eds): *Metallothioneins: Synthesis, Structure and Properties of Metallothioneins, Phytochelatins and Metal-Thiolate Complexes*. VCH publishers: New York, pp. 195–255.
12. Li, Y.-J.; Weser, U. (1992) *Inorg Chem* 31: 5526–5533.
13. Kull FJ, Reed MF, Elgren TE, Ciardelli TL, Wilcox DE (1990) *J Amer Chem Soc USA* 112: 2291–2298.
14. Willner H, Bernhard WR, Kägi JHR (1991) *In*: MJ Stillman, CF Shaw and KT Suzuki (eds): *Metallothioneins: Synthesis, Structure and Properties of Metallothioneins, Phytochelatins and Metal-Thiolate Complexes*. VCH publishers: New York, pp. 128–143.
15. Willer H, Vašák M, Kägi JHR (1987) *Biochemistry* 26: 6287–6292.
16. Bernhard WR, Vašák M, Kägi JHR (1985) *Biochemistry* 25: 1975–1980.
17. Good M, Hollenstein R, Sadler PJ, Vašák M (1988) *Biochemistry* 27: 7163–7166.
18. Zhu Z, Goodrich M, Isab AA, Shaw CF III (1992) *Inorg Chem* 31: 1662–1667.
19. Zhu Z, Petering DH, Shaw CF III (1995) *Inorg Chem* 34: 4477–4483.
20. Summers MF (1988) *Coord Chem Rev* 86: 43–134.
21. Vallee BL (1979) *Metallothionein: Proceedings of the First International Meeting on Metallothionein and other Low Molecular Weight Metal-binding Proteins*. Birkhäuser Verlag: Basel-Boston-Stuttgart, pp. 64–70.
22. Otvos JD, Liu X, Li H, Sleen G, Basti M (1993) *In*: KT Suzuki, N Imura, M Kimura (eds): *Metallothionein III. Dynamic aspects of Metallothionein Structure*. Verlag Basel: Birkhäuser, Switzerland, pp. 57–74.
23. Gan T, Muñoz A, Shaw CF III, Petering DH (1995) *J Biol Chem* 270: 5339–5345.
24. Li T-Y, Kraker AJ, Shaw CF III, Petering DH (1980) *Proc. Natl. Acad. Sci., USA*. 77: 6337–6338.
25. Li T-Y, Minkel DT, Shaw CF III, Petering DH (1981) *Biochem J* 193: 441–446.
26. Savas MM, Petering DH, Shaw CF III (1991) *Inorg Chem* 30: 581–583.
27. Udom A, Brady FO (1980) *Biochem J* 187: 329–335.
28. Huang Z-X, Hu H-Y, Gu W-Q, Sun J-Q (1993) *Chin J Chem* 11: 246–250.; Huang Z-X, Hu H-Y, Gu W-Q, Wu G (1994) *J Inorg Biochem* 54: 147–155.
29. Ejnik J (1996) *PhD. Thesis*, University of Wisconsin-Milwaukee,.
30. Shaw CF III, Savas MM (1991) *In*: MJ Stillman, CF Shaw and KT Suzuki (eds): *Metallothioneins: Synthesis, Structure and Properties of Metallothioneins, Phytochelatins and Metal-Thiolate Complexes*. VCH publishers: New York, pp. 144–163.
31. Cai W, Stillman MJ (1988) *J Amer Chem Soc* 110: 7872–7873.
32. Nielson KB, Winge DR (1984) *J Biol Chem* 259: 4941–4946.
33. Bertini I, Luchinat C, Messori L, Vašák M (1989) *J Amer Chem Soc* 111: 7296–7303.
34. Messerle BA, Schäffer A, Vašák M, Kägi JHR, Wüthrich K (1990) *J Mol Biol* 214: 765–779.
35. Brouwer M, Winge DR, Gray WR (1989) *J Inorg Biochem* 35: 289–303.
36. Bühler RHO, Kägi JHR (1979) *Experientia Suppl* 34: 211–220.

Metallothionein IV
C. Klaassen (ed.)
© 1999 Birkhäuser Verlag Basel/Switzerland

Voltammetric methods in isolation and identification of plant metallothioneins from alga *Chlorella*

Ivana Šestáková[1], Pavel Mader[2], Hana Vodičková[2] and Vìra Pacáková[3]

[1]*J. Heyrovsky Institute of Physical Chemistry, Academy of Sciences of the Czech Republic, 182 23 Prague 8, Czech Republic*
[2]*Department of Chemistry, Faculty of Agronomy, Czech University of Agriculture, 165 21 Prague 6-Suchdol, Czech Republic*
[3]*Department of Analytical Chemistry, Charles University, Faculty of Science, 128 40 Prague 2, Czech Republic*

In response to the environmental pollution by metals such as Cd, Cu, Pb, Ni, and As, special thiol-rich peptides are synthesized in plants and certain fungi. Their general structure is (γ-Glu-Cys)$_n$-R, where R is an amino acid residue (Gly, Ala, Ser, Glu). These peptides are known as plant metallothioneins, class III metallothioneins or γ-EC peptides [1]. Cd and Cu are the metals most frequently bound to plant metallothioneins from different sources and are well characterized by UV spectroscopy due to special absorption of the Cd-S and Cu-S bonds. Another possibility is offered by voltammetric methods which can distinguish between free metal ion and metal ion bound in complexes of different stability. For Cd complexes of γ-EC peptides structural models have been proposed [2] in which each cadmium ion is coordinated by four sulfur atoms. No voltammetric studies have been performed with γ-EC peptides so far. With animal metallothioneins, where tetracoordinated cadmium is present in molecules consisting of 60–62 amino acids, several polarographic and voltammetric studies have been performed. In our first study [3], we worked with rabbit liver metallothionein-2, prepared from rabbits exposed to cadmium salt, and with Cd-MT and apo-MT prepared from rabbit. We compared the voltammetric behavior on several types of carbon electrodes with that on the hanging mercury drop electrode (HMDE); active role of mercury electrode, confirmed also in other papers, can be best demonstrated with cyclic voltammetry (Fig. 1). An extreme reactivity of thiols with mercury is well known and detailed studies on reactions of cystine [4] and cysteine [5] have been performed. Also, it has been demonstrated, that under certain conditions, compact triphenyphosphine oxide films formed at the Hg/solution interface, can block thiol or disulfide adsorption on Hg electrodes [4]. In voltammetric study of Cd, Zn-metallothionein on such oxide film covered mercury electrode, no peaks compared to those on bare electrodes were observed and very sensitive determination of Cd containing MT, using adsorptive stripping voltammetry, could be worked out [6].

Based on our experience with reproducible cultivation and processing of green alga, experiments with cultivation of this algae in Cd enriched media were performed with the final goal of isolating the Cd-plant metallothionein from such a source of biological material. Preparative separation was performed using size exclusion HPLC on the column HEMA BIO 40 with UV

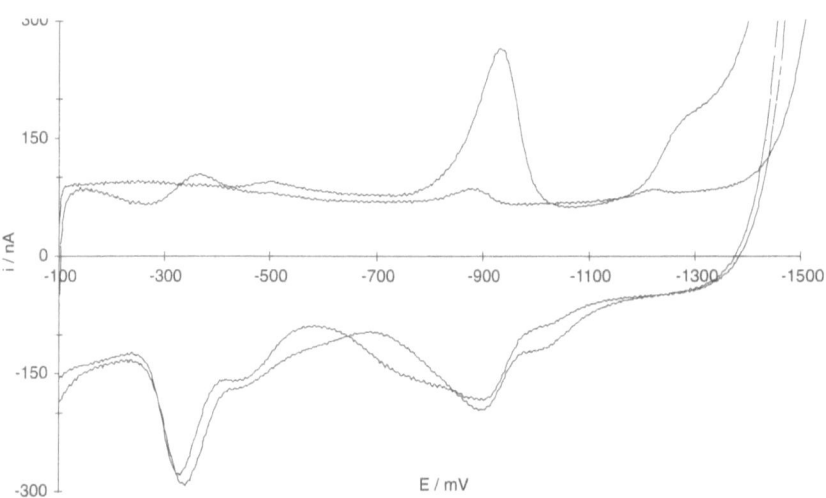

Figure 1. Cd,Zn-Metallothionein on hanging mercury electrode. 1.2×10^{-5}M Cd,Zn-MT, borate buffer pH 9.1, CV starting from –100 mV, scan rate 500 mV/s. Thick curves- 1st cycle, thin curves -3rd cycle

detection at 254 nm. The presence of absorption maximum in the region 250–260 nm has been

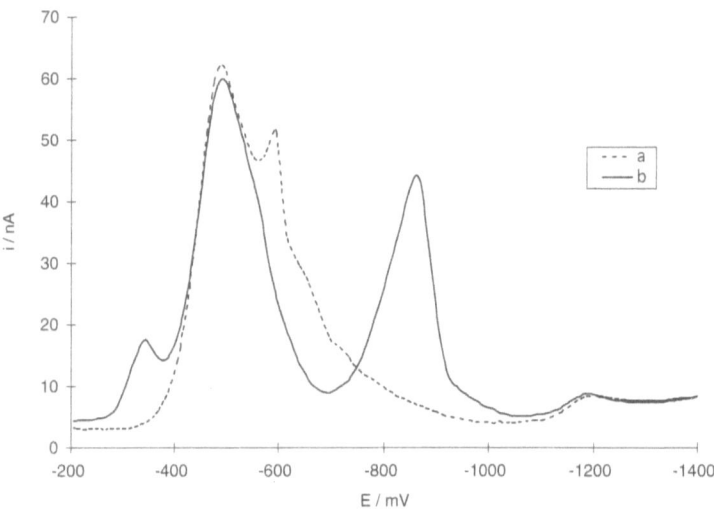

Figure 2. Cd-(γ-EC)$_2$G on hanging mercury electrode. 2.0×10^{-6} M peptide-curve a, 1.0×10^{-6} M Cd^{2+} added-curve b, borate buffer pH 8.5, electrolysis 120 s at –1500 mV, then DPV from 0 V

confirmed for each peak.

Voltammetric studies were performed with HPLC fractions of algal extracts and with synthetically prepared (γ-Glu-Cys)$_2$-Gly and (γ-Glu-Cys)$_3$-Gly on HMDE using computer-controlled PC-ETP analyzer. Complexation of synthetic peptides with cadmium ions, followed voltammetrically (cadmium ions were added to the buffer solution containing peptide, or peptide solutions were added to cadmium ions present in buffer solutions), exhibited similar features to those described for Cd complex in animal metallothionein [7] and in one study of complexation of MT fragment [8]. For (γ-Glu-Cys)$_2$-Gly and Cd^{2+} in concentration ratio 2:1, where coordination of cadmium ion with four S atoms occurs, the reduction peak of Cd complex is observed in the same potential region as in the case of animal metallothioneins (Fig. 2). This reduction peak can be used for speciation of Cd plant metallothionein by voltammetric methods and conditions for such determination in algal extracts were found (Fig. 3). The amounts of Cd-peptides found voltammetrically correlated well with the amounts of Cd in fractions determined by AAS. Also, our results are in agreement with reduction of the Cd-peptide complex prepared from alga *Pheodactylum tricornutum* and roots of graminae *Agrostis capillaris* [9] by HPLC.

We observed more complex behavior for Cu and Zn complexes of the synthetic peptides studied (γ-Glu-Cys)$_2$-Gly and (γ-Glu-Cys)$_3$-Gly.

For Zn complexes, three different reduction peaks were observed, only two of them having a parallel in animal Zn-containing metallothioneins. Very negative Zn reduction peak (with cor-

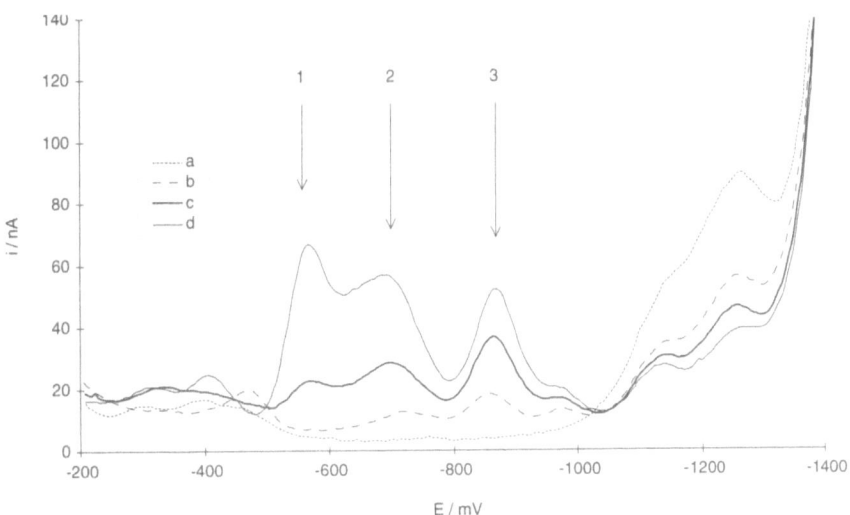

Figure 3. HPLC fraction from alga Chlorella (with highest content of Cd by AAS), after acidification and separation on Metalfix column. Voltammetric curve after pH adjustment to 8.5- curve a, addition of Cd^{2+} [10^{-6} M] into analyzed solution: curve b-0.5, c-1.0, d-2.0. Electrolysis 240 s at –1500 mV, then DPV from –150 mV, scan rate 20 mV/s. Peaks denoted as: 1-uncomplexed Cd, 2-Cd-cysteine or glutathione, 3-Cd-peptide

responding anodic peak at −1350 mV observed by cyclic voltammetry) has been found for high peptide and zinc ion concentrations.

Cu complexes of (γ-EC)₂G and (γ-EC)₃G peptides are, contrary to what was expected, influenced by addition of higher concentrations of zinc ions. Evidence for complexes in which both Zn and Cu are simultaneously present was found by voltammetric investigation of HPLC fractions from the alga *Chlorella*.

Complexation reactions of amino acids with cupric ions are used in electrochemical detection in HPLC [10], and copper electrodes have been introduced [11]. For sulfur containing amino acids, the method of cathodic stripping voltammetry is performed with HMDE in the presence of added Cu^{2+} ions [12]. This method has been applied also to animal MT [13] or phytochelatin isolated from *Pheodactylum tricornutum* [14].However serious interference we encountered with HPLC fractions from *Chlorella*. For successful determination, replacement of Tris buffer (which complexes Cu^{2+} ions) by non-complexing buffer, omission of interfering dithiothreitol, and removal of zinc ions were necessary.

For voltammetric studies of fractions in borate buffer with ascorbic acid and after removal of Zn ions by Metalfix, the method of cathodic stripping voltammetry can be used and presence of Cu-peptides verified (reduction peaks at −780 mV, −850 mV and −910 mV).Values of the height of these peaks suggest that there are many more peptides with SH-groups in fraction F3 than in fraction F2. The Cu-content in these two fractions as determined by AAS, has exact-

Table 1. Amounts of Cd and Cu in the fractions F1–F3 obtained after separation of the concentrate by SE-HPLC in *borate buffer*

Sample (E) – injection volume 100 μl, 194 ng Cd and 124 ng Cu

Control sample (C) – injection volume 100 μl, 15 ng Cd and 91 ng Cu

Sample (E)	fraction	F1	F2	F3
	volume (ml)	0.86	1.34	1.3
	Cd conc. (ng/ml)	37.2	94.7	10.6
	Cd content (ng)	32	127	14
	volume (ml)	0.86	1.34	1.3
	Cu conc. (ng/ml)	14.2	75.1	8.9
	Cu content (ng)	12.2	100.6	11.6
Control sample (C)	fraction	F1	F2	F3
	volume (ml)	0.86	1.34	1.3
	Cd conc. (ng/ml)	18.8	10.5	3.7
	Cd content (ng)	16.2	14.1	4.8
	volume (ml)	0.86	1.34	1.3
	Cu conc. (ng/ml)	13.8	73.2	6.8
	Cu content (ng)	11.9	98	8.8

ly the opposite trend (Tab. 1). Representation of the SH/SS amino acids in fractions F2 and F3 after acid hydrolysis of these two fractions, corresponds to the voltammetric results.

Acknowledgements
Financial support for this work, by grants from Grant Agency of the Czech Republic 204/97/KO 84 and 306 95 0268 (IS and VP) and by grant VaV/340/2/1996 from Ministry of the Environment of the Czech Republic (PM and HV), is greatly acknowledged. For AAS determinations, we thank Dr. Miholová, Dr. Kolihová and Dr. Száková of the Czech Agriculture University in Prague.

References

1. Rauser WE (1995) Phytochelatins and Related Peptides. *Plant Physiol* 109: 1141–1149.
2. Hayashi Y, Winge DR (1992) (γ - EC)$_n$ G Peptides. *In*: MJ Stillman, CF Shaw III, KT Suzuki (eds): *Metallothioneins*. VCH, New York, pp. 271–283.
3. Šestáková I, Miholová D, Vodičková H, Mader P (1995) Electrochemical Behavior of Metallothioneins at Mercury and Carbon Electrodes. *Electroanal* 7: 237–246.
4. Heyrovský M, Mader P, Veselá V, Fedurco M (1994) The reactions of cystine at mercury electrodes. *J Electroanal Chem* 369: 53–70.
5. Heyrovský M, Mader P, Vavřička S, Veselá V, Fedurco M (1997) The anodic reactions at mercury electrodes due to cysteine. *J Electroanal Chem* 430: 103–117.
6. Fedurco M, Šestáková I (1996) Adsorption of Cd, Zn- metallothionein on covered Hg electrodes and its voltammetric determination. *Bioelectrochem Bioenerg* 40: 223–232.
7. Munoz A, Rodriguez AR (1995) Electrochemical Behavior of Metallothioneins and Related Molecules. Part III. Metallothionein. *Electroanal* 7: 674–680.
8. Munoz A, Rodriguez AR (1995) Electrochemical behavior of metallothioneins and related molecules. PART II: Lys-Cys-Thr-Cys-Cys-Ala thionein fragment [56–61] MT I in the presence of cadmium and/or zinc. *Electroanal* 7: 670–673.
9. Nyberg S, Zhou LZ (1995) Polarography as a tool in peptide and protein analysis- studies on metal-chelating substances induced by cadmium in the algae *Pheodactylum Tricornutum* and the graminae *Agrostis Capillaris*. *Ecotoxicol Environ Safety* 32: 147–153.
10. Štulík K, Pacáková V (1987) *Electroanalytical measurements in flowing liquids*, Ellis Horwood Ltd, Chichester, England.
11. Štulík K, Pacáková V, Le K, Hennissen B (1988) Amperometric flow detection with a copper working electrode-response mechanism and application to various compounds. *Talanta* 35: 455–460.
12. Van den Berg CMG, Househam BC, Riley JP (1988) Determination of cystine and cysteine in seawater using cathodic stripping voltammetry in the presence of Cu (II). *J Electroanal Chem* 239: 137–148.
13. Scarano G, Morelli E (1996) Cathodic stripping voltammetric determination of metallothioneins. *Electroanal* 8: 396–398.
14. Scarano G, Morelli E (1996) Determination of phytochelatins by cathodic stripping voltammetry in the presence of copper (II). *Anal Chim Acta* 319: 13–19.

Metallothionein IV
C. Klaassen (ed.)
© 1999 Birkhäuser Verlag Basel/Switzerland

Characterization of isometallothioneins in Chang liver cells and the transient accumulation of metallothionein and glutathione during cell proliferation

Annemarie Willi and Peter E. Hunziker

Universität Zürich, Biochemisches Institut, Winterthurerstr.190, CH-8057 Zürich, Switzerland

Characterization of isometallothioneins in Chang liver cells

Human Chang liver cells continuously grown in the presence of 150 to 200 µM zinc synthesize four metallothionein (MT) isoforms differing in their primary structure (Fig. 1). The sequences of three isoforms were identical with the known hepatic forms MT-2, MT-1e and MT-1l and correspond to the N-terminally acetylated gene products of the MT-2A, MT-1E and MT-1X gene, respectively [1]. The fourth isoform, designated as MT-1m, differs from MT-1l by the replacement of the conserved glutamic acid residue in position 23 with glutamine and has not yet been observed in human tissues.

As reported earlier [2] these isoforms eluted from the reversed-phase column in four pairs of resolved fractions. Using mass spectrometry coupled on-line to reversed-phase HPLC these paired fractions were identified as corresponding subfractions of the same isoMT differing in mass by 42 Da (Tab. 1) revealing that the isoMTs in Chang liver cells occur as N-terminally acetylated and non-acetylated subforms. Judged from the mass difference of the metal-containing (holo-MT) and the metal-free (apo-MT) forms both the acetylated and the non-acetylated subforms were shown to bind exactly seven Zn^{2+} ions (Tab. 1).

The same concurrence of a metal-containing N-terminally modified and non-modified form was also reported for a mollusc MT [3]. In MT isolated from gills of oysters continuously grown

MT-2   Ac-MDPNCSCAAGDSCTCAGSCKCKECKCTSCKKSCCSCCPVGCAKCAQGCICKGASDKCSCCA

MT-1e   Ac-MDPNCSCA**TGG**SCTCAGSCKCKECKCTSCKKSCCSCCPVGCAKCAQGC**V**CKGAS**E**KCSCCA

MT-1l   Ac-MDPNCSC**SPVG**SC**A**CAGSCKCKECKCTSCKKSCCSCCPVGCAKCAQGCICKG**T**SDKCSCCA

MT-1m   Ac-MDPNCSC**SPVG**SC**A**CAGSCKCK**Q**CKCTSCKKSCCSCCPVGCAKCAQGCICKG**T**SDKCSCCA

Figure 1. Primary structures of isometallothioneins from Chang liver cells. Chang liver cells were continuously grown in medium containing 150–200 µM zinc. MT isoforms were isolated as described [19]. The primary structures were deduced from the sequences of proteolytic peptides determined by collision-induced tandem mass spectrometry (C. Zehnder et al., unpublished data). Amino acid residues differing from the MT-2 sequence are printed in bold letters.

Table 1. Mass determination of isometallothioneins from Chang liver cells

Isoform		Mass of holo-MT	Mass of apo-MT	Mass difference	Zinc ions bound
MT-2	acetylated	6526	6084	442	7
	non-acetylated	6484	6042	442	7
MT-1e	acetylated	6499	6055	444	7
	non-acetylated	6455	6013	443	7
MT-1l	acetylated	6553	6110	443	7
	non-acetylated	6511	6068	443	7
MT-1m	acetylated	6552	6109	443	7
	non-acetylated	6509	6067	442	7

The masses of the isometallothioneins were determined at pH 6.0 (Holo-MT) and at pH 2.0 (Apo-MT) [18]. The number of bound zinc ions was calculated from the mass difference of the holo- and the apo-form. As the binding of Zn^{2+} to apoMT releases two protons/metal ion, the net mass increment/Zn^{2+} was 63.

in the presence of cadmium the ratio of the N-terminally acetylated and the non-acetylated subform changed with the duration of metal exposure. Similarly in Chang liver cells grown at elevated zinc concentrations the proportion of the non-acetylated subform of each isoform decreased from 30% in proliferating cells to 5% in cells attaining density-mediated growth arrest.

The biological significance of the unusual occurrence of an acetylated and a non-acetylated subform of a single protein is unknown. It remains to be seen whether this variation in post-translational modification serves a physiological purpose in modulating the rate of degradation of metal-saturated MT or if the reduced efficiency of the acetylating process in dividing cells is a result of the non-physiological conditions of metal induction employed.

Metallothionein accretion in Chang liver cells during proliferation

The development of a sensitive HPLC method [4] enabled us to analyse basal amounts of MT in crude cell extracts (Fig. 2). Of the four isoMTs synthesized in cultured Chang liver cells (Fig. 1) grown in the presence of 200 µM Zn^{2+} [2] only the two most abundant isoforms, MT-2 and MT-1e, are detected in unstimulated cells.

With the sensitivity provided by this method it has become possible to monitor and to compare the contents of MT isoforms in resting and in proliferating cells [5]. As shown in Figure 3 MT-2 increased from the basal content of 23 pmol/10^7 cells in freshly plated cells to an approximately four times higher content of 94 pmol/10^7 cells after 48 h of growth. Beyond this time MT-2 content declined sharply and approached the initial content at growth arrest (96 h of growth) while the total amount of cellular protein remained fairly constant over the same culture period. The increase in cellular MT-2 content up to a maximum at 48 h is accounted for

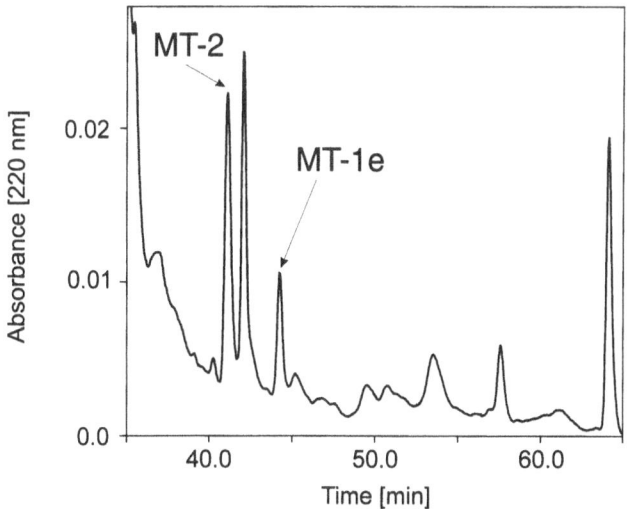

Figure 2. HPLC separation of MTs obtained from proliferating Chang liver cells. A sample derived from 5.5×10^6 cells was prepared as described [4] and injected on a Spheri-5 RP-18 column (2.1 (i.d.) $\times$ 220 mm). MTs were eluted at a flow rate of 400 µl/min with a gradient formed between buffer A (25 mM Tris/HCl, pH 7.5) and buffer B (same as A, containing 60% (v/v) acetonitrile). The gradient used was as follows: 0% buffer B for 15 min; 0 to 23% buffer B during 45 min.

by an enhanced rate of net biosynthesis during proliferation (Fig. 4) and coincides with the enhanced rate of DNA synthesis (Fig. 4, insert). The about 10-fold increase in MT-2 synthesis during the first two days of proliferation together with the preceding transient increase in the concentration of MT mRNA [5] is indicative of an activation of transcription in dividing cells.

The relative abundance of the MTs in resting cells and their accretion during proliferation is in keeping with the earlier findings of a strongly increased MT content in regenerating liver after partial hepatectomy [6, 7] and the reported emergence of MT or MT mRNA in specialized tissues undergoing growth and differentiation [8, 9]. Together with recent data obtained on synchronized human colonic cancer cells [10] the transient accretion of MT suggests that the proliferation associated expression of this protein may provide an adjustable store for zinc and serves a role in arranging for its availability in the massive biosynthetic processes preceding and accompanying cell proliferation.

Changes in glutathione content during proliferation

The tripeptide glutathione is the most prevalent cellular thiol in mammalian cells. Its intracellular concentration ranges from 0.5 mM to 10 mM [11] with the bulk amounts found in the cytosol. Usually over 99% of the total amount of glutathione is present in its reduced form, and thus, it represents the quantitatively most relevant reducing agent in the cell functioning as the

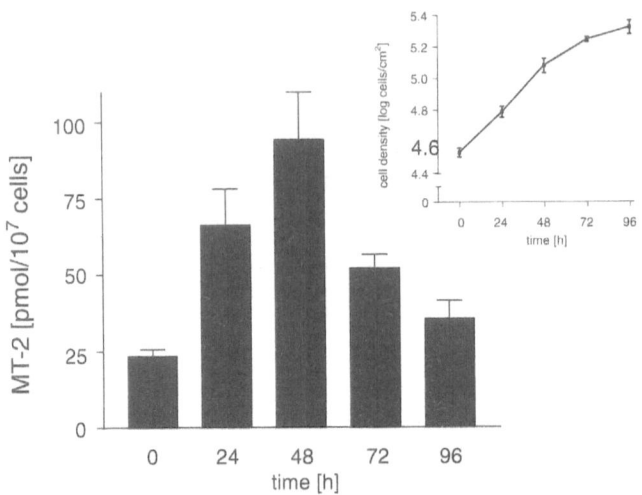

Figure 3. Net cytosolic MT-2 contents at successive growth stages and growth curve (insert) of Chang liver cell culture. Maximum rates of proliferation were achieved by seeding cells from a confluent culture at a density of 34000 ± 4700 cell/cm^2. Under these conditions cellular growth slowed down after 48 h and by 96 h a state of density-mediated growth arrest was reached (insert). For simplicity, MT measurements were restricted to MT-2 which was three to four times as abundant as MT-1e (Fig. 1) at all growth states. Samples were prepared as described [5]. The amount of MT-2 was assessed by integration of the MT-2 HPLC peak (Fig. 1.) with purified human MT-2 as external standard. The values shown are means ± S.E.M. of at least four independent experiments.

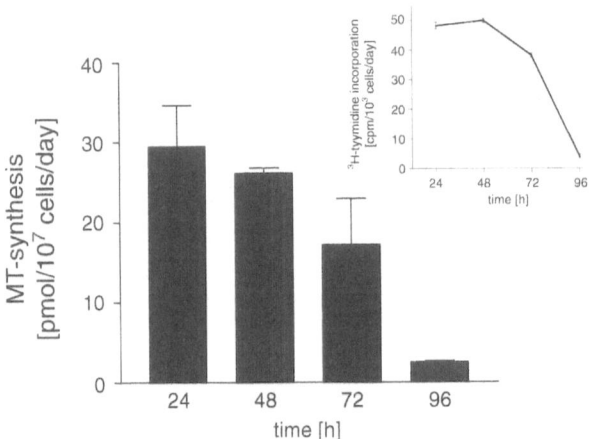

Figure 4. Rates of MT-2 synthesis and [^{3}H]-thymidine incorporation (insert) at successive growth states of a Chang liver cell culture. Samples were prepared as described [5]. The net rate of MT-2 synthesis was calculated from the amount of incorporated [^{35}S]-cysteine assuming 20 mol of cysteine/mol of MT. The rate of [^{3}H]-thymidine incorporation (insert) was determined after a 24 h pulse. The data shown are means ± S.E.M. of at least two independently performed experiments.

major determinant of the intracellular redox potential. Besides its involvement in a variety of metabolic processes, glutathione functions also as a polydentate ligand for a large number of different metal ions (for review see [12], and references therein). A biological significance for its complexation of metal ions other than detoxification is as yet unproven, however. Binding of Cu^{2+} by glutathione was shown to play a role both in the metabolism of Cu^{2+} and glutathione [13], and the reduction of Cu^{2+} to Cu^+ by glutathione is suggested as the necessary step for the binding of copper to metallothionein [14]. In contrast, the complexation of Zn^{2+} by glutathione has not received much attention. The actual structure of zinc-glutathione complexes are still unclear [12]. However, the high concentration of glutathione in cells and its relatively high cumulative binding constant for Zn^{2+} yielding at pH 7 an apparent net binding constant of the order of $10^7 \, M^{-1}$ as determined by potentiometric titration [15] makes it very likely that complexation by glutathione is important in the biological speciation of zinc.

The development of a chromatographic method based on gelfiltration (A. Willi et al., unpublished data) allowed us to separate glutathione and a zinc-glutathione complex from crude cell-homogenates. Both glutathione and its zinc complex were quantified by measuring the incorporation of $[^{35}S]$-cysteine and $[^{65}Zn]$, respectively.

In cultured Chang liver cells the total glutathione content increased 2.5- to 5-fold at the onset of proliferation reaching a maximum of approximately 270 nmol/10^7 cells after 24 h of growth and returned to initial values at density mediated growth arrest (Fig. 5). Similar changes have also been observed in other human cell lines [16] supporting the suggested involvement of glutathione in cellular proliferation [17]. Compared to the total glutathione the concentration of the zinc-glutathione complex ranged around an avarage of 40 pmol/10^7 cells at all growth states

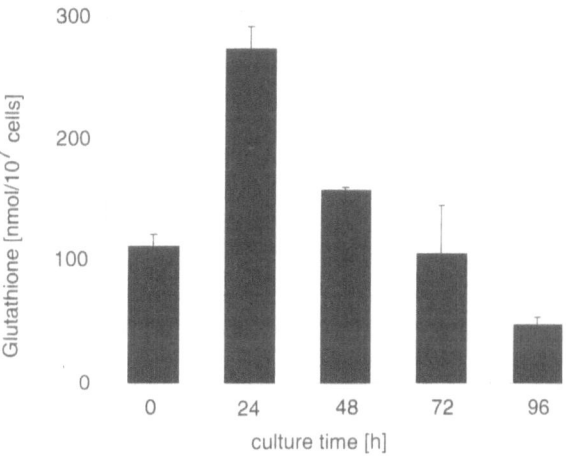

Figure 5. Cellular glutathione content at successive growth states of a Chang liver cell culture. Samples were obtained by ultracentrifugation of cell lysates. Glutathione was separated by gelfiltration at pH 8.5 and quantified by measuring the radioactivity of incorporated $[^{35}S]$-cysteine. The data shown are means ± S.E.M. of at least two independent experiments.

with a tendency towards lower concentrations in proliferating cells. In view of the proposed equilibria between zinc and glutathione [15] these results suggest a significant decrease in the availability of cellular zinc at the onset of cell proliferation.

Changes in zinc partitioning during proliferation

Over the entire observation period of 96 h the total cellular zinc content of about 6 nmol/10^7 cells remained constant, implying a balancing influx of zinc from the culture medium into the cell undergoing proliferation. However, there was a substantial repartitioning of the metal in the different growth states. The proportion of zinc bound to MT increased from 2.5% in resting cells to as much as 10% in the exponential growth phase. Glutathione-bound zinc remained with a proportion of about 0.7% nearly constant, but because of the proliferation associated increase in total glutathione, the concentration of free Zn^{2+} must be lowered appreciably. Based on the known binding equilibria of glutathione-zinc complexes [15] and the actual glutathione concentration free Zn^{2+} concentrations can be calculated to drop from 20 nM in resting cells to about 1 nM in proliferating cells.

These results suggest that the transient accretion of glutathione and zinc scavenging thionein may play important roles in the regulation of the intracellular availability of zinc in different growth states and, by modulating the intracellular chemical potential of zinc, may provide a major driving force for the assimilation of zinc during cell proliferation.

Acknowledgements
This work was supported by Swiss National Foundation Grant 3100-040807.94/1 and the Kanton Zürich. We are grateful to Prof. J.H.R. Kägi for his critical review of the manuscript.

References

1. West AK, Stallings R, Hildebrand CE, Chiu R, Karin M, Richards RI (1990) Human metallothionein genes: structure of the functional locus at 16q13. *Genomics* 8: 513–518.
2. Cavigelli M, Kägi JHR, Hunziker PE (1993) Cell- and inducer-specific accretion of human isometallothioneins. *Biochem J* 292: 551–554.
3. Roesijadi G, Vestling MM, Murphy CM, Klerks PL, Fenselau CC (1991) Structure and time-dependent behavior of acetylated and non- acetylated forms of a molluscan metallothionein. *Biochim Biophys Acta* 1074: 230–236.
4. Studer R, Hunziker PE (1997) Quantification of basal amounts of isometallothioneins in cultured cells by reversed-phase Hplc. *J Liq Chromatogr Relat Technol* 20: 617–625.
5. Studer R, Vogt CP, Cavigelli M, Hunziker PE, Kägi JHR (1997) Metallothionein accretion in human hepatic cells is linked to cellular proliferation. *Biochem J* 328: 63–67.
6. Ohtake H, Hasegawa K, Koga M (1978) Zinc-binding protein in the livers of neonatal, normal and partially hepatectomized rats. *Biochem J* 174: 999–1005.
7. Cain K, Griffiths BL (1984) A comparison of isometallothionein synthesis in rat liver after partial hepatectomy and parenteral zinc injection. *Biochem J* 217: 85–92.
8. Tohyama C, Suzuki JS, Hemelraad J, Nishimura NNH (1993) Induction of metallothionein and its localization in the nucleus of rat hepatocytes after partial hepatectomy. *Hepatology* 18: 1193–1201.
9. Nishimura H, Nishimura N, Tohyama C (1989) Immunohistochemical localization of metallothionein in developing rat tissues. *J Histochem Cytochem* 37: 715–722.
10. Nagel WW, Vallee BL (1995) Cell cycle regulation of metallothionein in human colonic cancer cells. *Proc Natl Acad Sci USA* 92: 579–583.

11. Meister A, Anderson ME (1983) Glutathione. *Annu Rev Biochem* 52: 711–760.
12. Rabenstein DL (1989) *In*: D Dolphin, R Poulsen, O Avramovic (eds): *Glutathione: Chemical, Biochemical and Medical Aspects*. John Wiley and Sons, New York, NY, pp. 147–186.
13. Christie NT, Costa M (1984) *In vitro* assessment of the toxicity of metal compuonds: IV. Disposition of metals in cells: Interactions with membranes, glutathione, metallothionein, and DNA. *Biol Trace Elem Res* 6: 139–158.
14. Ferreira AM, Ciriolo MR, Marcocci L, Rotilio G (1993) Copper(I) transfer into metallothionein mediated by glutathione. *Biochem J* 292: 673–676.
15. Perrin DD, Watt AE (1971) Complex formation of zinc and cadmium with glutathione. *Biochim Biophys Acta* 230: 96–104.
16. Duthie SJ, Collins AR (1997) The influence of cell growth, detoxifying enzymes and DNA repair on hydrogen peroxide-mediated DNA damage (measured using the comet assay) in human cells. *Free Radical Biol Med* 22: 717–724.
17. Porreca E, Di Febbo C, Pandolfi A, D'Orazio A, Martelli N, Mezzetti A, Cuccurullo F, Poggi A (1993) Differences in the glutathione system of cultured aortic smooth muscle cells from young and aged rats. *Atherosclerosis* 100: 141–148.
18. Hess D, Studer R, Hunziker PE (1994) Identification and characterization of metal protein complexes of crude cell extracts by mass-spectrometry. *J Prot Chem* 13: 531.
19. Hunziker PE, Kaur P, Wan M, Kanzig A (1995) Primary structures of seven metallothioneins from rabbit tissue. *Biochem J* 306: 265–270.

Metallothionein IV
C. Klaassen (ed.)
© 1999 Birkhäuser Verlag Basel/Switzerland

Quantification of cytosolic metallothionein contents by a developed assay system using immobilized antibody

Tomomi Nagashima[2], Masashi Okabe[1], Shigeru Saito[1], Masaaki Kurasaki[1], Toshitsugu Hirauchi[1], Shinzo Kimura[1] and Tadashi Niioka[1]

[1]*Department of Environmental Medicine and Informatics, Graduate School of Environmental Earth Science, Hokkaido University, Sapporo 060, Japan*
[2]*Present Address: Research Section of Pathology, Institute of Immunological Science, Hokkaido University, Sapporo 060, Japan*

Introduction

Metallothioneins (MTs) are low molecular weight and metal-binding proteins of a high cysteine content [1]. Although the biological functions of the proteins are not fully understood, it is believed to play an important role in the homeostasis of essential heavy metals such as Zn and Cu, as well as the detoxification of heavy metals such as Cd and Hg [2]. The biosynthesis of MT can be induced by a variety of physical, chemical, sensory and psychological stressors [2–5].

For determining MT contents enzyme-linked immunosorbent assay (ELISA) methods have been evolved using MT antibodies prepared in each laboratory [6–9]. Whereas a monoclonal antibody to MT has been commercially available by Zymed Laboratories Inc., which has been proved useful for histochemical studies [10, 11]. However, if the antibody is used for ELISA to quantify the MT contents in body fluids as well as in cytosol, unknown factors prevent determination of MT contents in those biological materials (data not shown). Although we found that MT contents were able to be determined in MT fraction after gel filtration treatment of these samples by using the antibody, simplified methods for measuring MT contents without this pretreatment are desirable especially for large numbers of samples.

Recently Shields et al [12] reported that the immobilized antibody on blotting membrane had benefits for measuring protein contents. In this article, it was examined whether the MT content in a biological material was able to be determined without pretreatment by a new method using the commercial antibody to MT and immobilizing it on blotting membrane.

Assay system of MT contents

Procedures for the quantification of MT by using immobilized antibody on blotting membrane are as follows.

The dot blotter (Easy-Titer[TM] Unit, Pierce) utilized a sheet of Zeta probe (positively charged nylon) membrane (Bio-Rad) or polyvinylidene fluoride (PVDF) membrane (Bio-Rad) between two plates as an adsorptive surface. The upper plate was a sample application plate with 96 wells in the typical 8 × 12 format. The lower plate was a membrane support plate through which 96 individual cannulae passed into wells of a microtiter plate enclosed with a collection chamber. The two plates binding a membrane and the collection chamber were clamped together. Bare sites of the membrane in the unit were washed with 200 μl of buffer consisting of 40 mM Tris/33 mM HCl and 154 mM NaCl, and then were coated with 1 ×, 10 ×, 20 × or 100 × diluted mouse anti-MT IgG (*clone: E9*, Zymed) with the same buffer. The membrane with the immobilized antibody on it, was blocked with 200 μl of 5 mg/ml bovine serum albumin (BSA) in the buffer and then washed three times with the buffer. Reconstituted Cd-MTs, described later in detail, of 0.1 to 100 μg in 100-μl buffer or in 100-μl cytosol diluted with the same buffer were put into the wells. The membrane was incubated for 1 or 2 h, or overnight. After pulling the solution through the membrane, the membrane was washed three times with the buffer. At the final step, a 300-μl portion of 0.1 M HNO_3 was added and pulled through the membrane. The HNO_3 solution containing released metals was collected by the microtiter plate held in the collection chamber. The Cd content in the solution was measured with a Hitachi atomic absorption spectrophotometer, model 180-80 or a Seiko inducibility coupled plasma (ICP) mass spectrophotometer, model SPQ-6500.

Experimental conditions of the assay system

In the final studies, the experimental conditions used by the assay for quantification of MT contents are listed in Table 1. Several experimental conditions (concentration of antibody, sort of membrane, reaction time of each step, *etc.*) were examined before the final experimental conditions were adopted.

BSA was used for blocking reagent in the assay. It showed a high performance in the current method. BSA was reported also to be used to block nonspecific binding sites in an ELISA system [13].

Reconstituted Cd-MT was used as a standard antigen. It was prepared from rabbit Cd, Zn-MT (Sigma) using $CdCl_2$ as a Cd(II) donor by the method of Miura et al [14]. The Cd, Zn-MT was incubated with 10 mol equivalent of Cd (II) for 20 min. The mol equivalent indicates 1 mol of metal ions per mol of protein. Chelex 100 was added to the mixture to remove the

Table 1. Experimental conditions used by the assay system

Item	Condition
Membrane	PVDF
Concentration of antibody	a dilution of 1: 30 with the buffer
Reaction time of antigen-antibody	over night
Immobilizing time of antibody	30 min
Blocking solution	5-mg BSA/ml

unbound metals. After centrifugation, the Cd content bound to MT was measured with the atomic absorption spectrophotometer.

Standard curve of the assay system for MT in buffer solution

The relationship between concentrations of MT and of released Cd ions was linear ranging from 10- to 1,000-μg MT/ml (1- to 100-μg MT/100 μl) with a correlation coefficient of 0.997 (n = 7) using reconstituted Cd-MT as a standard antigen in the buffer. The reproducibility of the standard curve was confirmed when the assay was carried out in quadruplicate between 0- and 250-μg MT/ml. An ICP mass spectrophotometer was employed to measure Cd content in the range of 0.1- to 10-μg MT/ml. The standard curve was also linear in this range.

High amounts of MT were able to be determined by the method presented here. In LEC rat, which was a model animal of Wilson disease (i.e. genetic disorder in Cu metabolism) the amount of Cu accumulated in the liver was reported to be approximately 150 to 250 μg/g wet weight [15, 16]. The accumulated Cu was believed to be bound to MT in the tissue. In this case, the method presented here was thought to be useful for measuring MT content.

The present assay is not very sensitive compared with the ELISA for rat MT reported by other researchers. The useful range for MT quantification in ELISA has been reported to be 0.1- to 20.0-ng MT [7], and 0.05- to 50-ng MT [6]. However, in control mice the concentration of MT in the liver was observed to be 3.8-μg MT/ml (0.43 μg Cd/ml with 1:4 [w/v] homogenization) [8]. Under stressful conditions, MT contents were observed to be 3.73- (control) to 31.8-μg MT/g wet weight liver of rats [4]. These values were high enough for the detection by the assay presented here.

Standard curve of the assay system for MT in cytosol of saline-injected rat liver

The standard curve was linear (r = 0.993, n = 5) as to Cd-MT concentrations from 0 to 200 μg/ml when the assay was performed using the hepatic cytosol of a saline-injected rat. The obtained standard curve corresponded well with the one derived from Cd-MT resolved in the buffer. Recently, we have found that MT contents in urine and hepatic cytosol were not able to be determined by ELISA using the E9 antibody. In this study, the present assay showed a good relationship between MT content resolved in cytosol and Cd content released from the MT.

Comparison between MT contents determined by the assay system and by gel filtration assay

The content of MT determined by the present method was compared with that by gel filtration assay to verify the MT content determined by the new method for hepatic cytosol of rat administrated intraperitoneally with 2.0-mg Cd/kg body weight as $CdCl_2$ in saline. The MT content was determined to be 182.2 μg/ml by the method using blotting membrane after the Cd-saturated treatment [17]. The value corresponded well with the one (216.5 µg/ml) obtained from the gel filtration assay using a Sephadex G-50 column (1.2 × 40 cm) equilibrated with 10 mM Tris/5 mM HCl. The results indicated that the developed assay system was useful to determine MT contents in cytosol. In addition, differently from the ELISA, this method has a potential for direct analysis of the metal composition of MT examined.

In conclusion, a new method for quantification of MT has been developed using immobilized antibody on blotting membrane. The useful measurement ranged from 1- to 1000-µg MT/ml.

Acknowledgment
The authors wish to acknowledge Dr. Yutaka Kojima (emeritus professor at Hokkaido University) for his valuable discussions and suggestions.

References

1. Kägi JHR, Kojima Y (1979) Nomenclature of metallothionein. *In*: JHR Kägi, M Nordberg (eds): *Metallothionein*. Birkhäuser Verlag, Basel, pp.141–142.
2. Kägi JHR, Kojima Y (1987) Chemistry and biochemistry of metallothionein. *In*: JHR Kägi, Y Kojima (eds): *Metallothionein II*. Birkhäuser Verlag, Basel, pp. 25–61.
3. Oh SH, Deagen JT, Whanger PD, Weswig PH (1978) Biological function of metallothionein. V. Its induction in rats by various stresses. *Amer J Physiol* 234: E282-E285.
4. Niioka T, Kojima Y (1991) Studies on induction of zinc metallothionein by sensory and psychological stresses in rat liver. *In*: CD Klaassen, KT Suzuki (eds): *Metallothionein in Biology and Medicine*. CRC Press, Boca Raton, pp. 265–269.
5. Arizono K, Tanahe A, Ariyoshi T, Moriyama M (1991) Induction of metallothionein by emotional stress. *In*: CD Klaassen, KT Suzuki (eds): *Metallothionein in Biology and Medicine*. CRC Press, Boca Raton, pp. 271–282.
6. Garvey JS (1984) Metallothionein: structure/antigenicity and detection/quantitation in normal physiological fluids. *Environ Health Perspect* 54: 117–127.
7. Thomas DG, Linton HJ, Garvey JS (1986) Fluorometric ELISA for the detection and quantitation of metallothionein. *J Immunol Method* 89: 239–247.
8. Baer KN, Benson WH (1987) Influence of chemical and environmental stressors on acute cadmium toxicity. *J Toxicol Environ Health* 22: 35–44.
9. Chan HM, Cherian MG, Bremner I (1992) Quantification of metallothionein isoforms using an enzyme-linked immunosorbent assay (ELISA) with two specific antisera. *Toxicol Appl Pharmacol* 116: 267–270.
10. Nakayama K, Okabe M, Aoyagi K, Yamanoshita O, Okui T, Ohyama T, Kasai N (1996) Visualization of yellowish-orange luminescence from cuprous metallothioneins in liver of Long-Evans Cinnamon rat. *Biochim Biophys Acta* 1289: 150–158.
11. Okabe M, Nakayama K, Kurasaki M, Yamasaki F, Aoyagi K, Yamanoshita O, Sato S, Okui T, Ohyama T, Kasai N (1996) Direct visualization of copper-metallothionein in LEC rat kidneys: application of autofluorescence signal of copper-thiolate cluster. *J Histochem Cytochem* 44: 865–873.
12. Shields MJ, Siegel JN, Clark CR, Hines KK, Potempa LA, Gewurz H, Anderson B (1991) An appraisal of polystyrene-(ELISA) and nitrocellulose-based (ELIFA) enzyme immunoassay systems using monoclonal antibodies reactive toward antigenically distinct forms of human C-reactive protein. *J Immunol Method* 141: 253–261.
13. Cousins RJ (1991) Measurement of human metallothionein by enzyme-linked immunosorbent assay. *Methods Enzymol* 205: 131–140.
14. Miura T, Yamasaki F, Kurasaki M (1996) Copper-binding metallothionein expressed in *Escherichia coli* grown in medium containing copper and cadmium. *Int. J Biochromatogr* 2: 67–76.
15. Li Y, Togashi Y, Sato S, Emoto T, Kang J, Takeichi N, Kobayashi H, Kojima Y, Une Y, Uchino J (1991) Spontaneous hepatic copper accumulation in Long-Evans Cinnamon rats with hereditary hepatitis. *J Clin Invest* 87: 1858–1861.
16. Yamada T, Suzuki Y, Agui T, Matsumoto K (1992) Elevation of metallothionein gene expression associated with hepatic copper accumulation in Long-Evans Cinnamon mutant rat. *Biochim Biophys Acta* 1131: 188–191.
17. Onosaka S, Cherian MG (1982) Comparison of metallothionein determination by polarographic and cadmium-saturation methods. *Toxicol Appl Pharmacol* 63: 270–274.

Metallothionein IV
C. Klaassen (ed.)
© 1999 Birkhäuser Verlag Basel/Switzerland

Protein engineering of metallothionein to study the metal-binding ability

Masaaki Kurasaki, Futoshi Yamasaki, Rie Yamaguchi, Ana Rosa Linde Arias[*], Masashi Okabe, Tadasu Emoto, Fumitomo Odawara, Mika Suzuki-Kurasaki, Shigeru Saito and Yutaka Kojima

Department of Environmental Medicine and Informatics, Graduate School of Environmental Earth Science, Hokkaido University, Sapporo 060, JAPAN
[*]*Present address: Department of Biological Function, Faculty of Medicine, Ovied University, Oviedo 33006, Spain*

Introduction

It is well known that metals play an important role in post-transcriptional modifications of proteins stabilizing their structure and physiological activity [1]. The mechanism by which biosynthesized proteins interact with metals and originate such cluster structures like metallothionein (MT) remain unclear.

A prominent structural feature of MTs is the arrangement of thiolate-metal coordination into two distinct fragments [2, 3]. The amino-terminal half, β-fragment containing 9 cysteines, and the carboxyl half, α-fragment containing 11 cysteines bind 3 and 4 ions of Cd (II) or Zn (II), respectively (see in Fig. 1A). In previous attempts, the α- and β-fragments showed many similar biochemical properties to native MT [3, 4, 5].

Most mammalian MTs consist of 61 or 62 amino acids containing 20 cysteines at invariant positions and highly conserved basic amino acids [6]. The mammalian forms of MTs have the characteristic 7 Cys-x-Cys, where x is amino acids other than cysteines [7].

Recently, expression systems using *E. coli* and yeast were used to investigate the physiological roles and metal-binding abilities of vertebrate MTs [8, 9, 10], their mutant MTs [11, 12, 13, 14, 15, 16] and chimeric MT [17].

To evaluate the physiological significance of mammalian MT, we have attempted to express human MT-2 and its mutant MTs in *E. coli*. In this study, we constructed *E. coli* expression plasmids of fourteen mutant MTs, and analyzed whether the expressed mutant MTs in *E. coli* configured the Cd-binding form. In addition, the significance of the results is discussed.

Confirmation of human MT expression system in *E. coli*

Human MT-IIA cDNA kindly provided by Prof. Karin was inserted into pKK223-3 at a position downstream of the *tryptophan-lactose (tac)* promoter. After induction with isopropyl β-D-thiogalactoside (IPTG), the soluble sonic extract of the cells was applied to a Sephadex G-75 column. In the MT fraction as shown in Figure 2A, Zn and Cu were hardly detected. After a DEAE-Sephadex A-25 chromatography, the amino acid composition of the purified MT are in good agreement with the theoretical values of human MT-2. In ultraviolet absorption spectrum

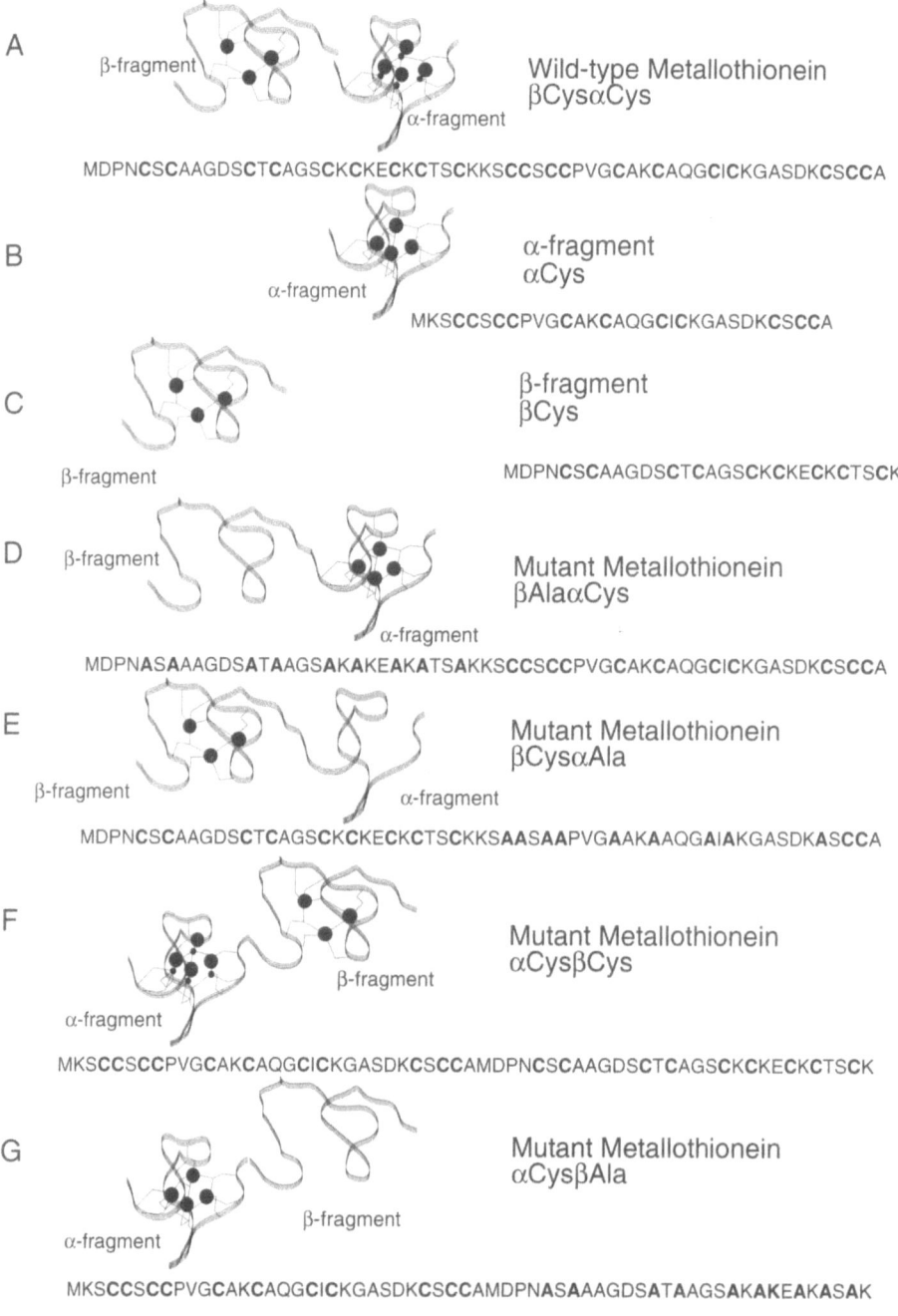

A β-fragment / α-fragment — Wild-type Metallothionein βCysαCys

MDPNCSCAAGDSCTCAGSCKCKECKCTSCKKSCCSCCPVGCAKCAQGCICKGASDKCSCCA

B α-fragment — α-fragment αCys

MKSCCSCCPVGCAKCAQGCICKGASDKCSCCA

C β-fragment — β-fragment βCys

MDPNCSCAAGDSCTCAGSCKCKECKCTSCK

D β-fragment / α-fragment — Mutant Metallothionein βAlaαCys

MDPNASAAAGDSATAAGSAKAKEAKATSAKKSCCSCCPVGCAKCAQGCICKGASDKCSCCA

E β-fragment / α-fragment — Mutant Metallothionein βCysαAla

MDPNCSCAAGDSCTCAGSCKCKECKCTSCKKSAASAAPVGAAKAAQGAIAKGASDKASCCA

F α-fragment / β-fragment — Mutant Metallothionein αCysβCys

MKSCCSCCPVGCAKCAQGCICKGASDKCSCCAMDPNCSCAAGDSCTCAGSCKCKECKCTSCK

G α-fragment / β-fragment — Mutant Metallothionein αCysβAla

MKSCCSCCPVGCAKCAQGCICKGASDKCSCCAMDPNASAAAGDSATAAGSAKAKEAKASAK

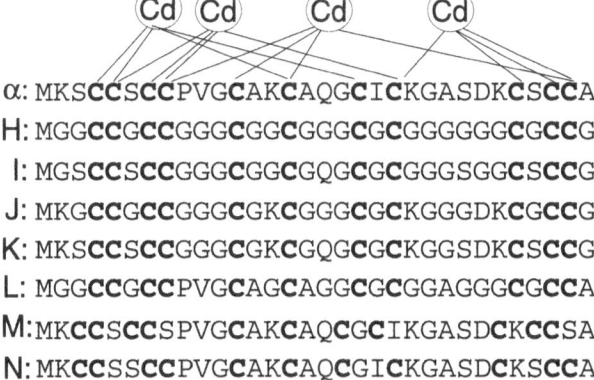

Figure 1. (see also previous page) Schemes of the structure of each wild-type MT and mutant MT and their amino acid sequences. (A): wild-type MT, (B): α-fragment, (C): β-fragment, (D): mutant D, (E): mutant F, (G): mutant G, (H): mutant H, (I): mutant I, (J): mutant J, (K): mutant K, (L): mutant M, and (N): mutant N. Cys is indicated by bold character.

of the protein, high Cd-mercaptide absorption at 250 nm was observed. The partial amino acid sequence (Met-Asp-Pro-Asn-Cys-Ser-Cys-Ala-Ala-Gly-Asp-Ser-Cys-Thr-Cys-Ala-) of the protein matches exactly with that of the native human MT-2. It was calculated to contain approximatly 6.8 mol of Cd per mol of the protein based on its -SH and Cd contents. From 4.3 g wet weight of cells, about 340 µg of the Cd-binding MT-2 was obtained.

Construction of mutant MT expression plasmids

The complementary oligonucleotides used to make the mutant expression plasmids were synthesized with an DNA Synthesizer. The purified oligonucleotides were phosphorylated at their 5'-ends and were annealed with the complementary oligonucleotide. The resulting fragment was ligated with the human MT expression vector [18].

Features of fourteen mutant MTs as shown in Figure 1 were (B) α-fragment only, (C) β-fragment only, (D) all cysteines in the β-fragment were replaced by alanines (βAlaαCys), (E) the first 9 cysteines of the α-fragment were replaced by alanines (βCysαAla), (F) the mutant MT which has the inverse fragment composition (αCysβCys), (G) in mutant F, all cysteines in the β-fragment were replaced by alanines (αCysβAla), (H) all amino acids other than cysteines of the α-fragment, were replaced with glycines, (I) all amino acids other than serines, glutamine and cysteines of the α-fragment, were replaced with glycines, (J) all amino acids other than lysines, aspartic acid and cysteines of the α-fragment, were replaced with glycines, (K) all amino acids other than lysines, aspartic acid, serines, glutamine and cysteines of the α-fragment, were replaced with glycines, (L) all amino acids other than the nonpolar amino acids and cysteines of the α-fragment, were replaced with glycines, (M) three Cys-x-Cys sequences of

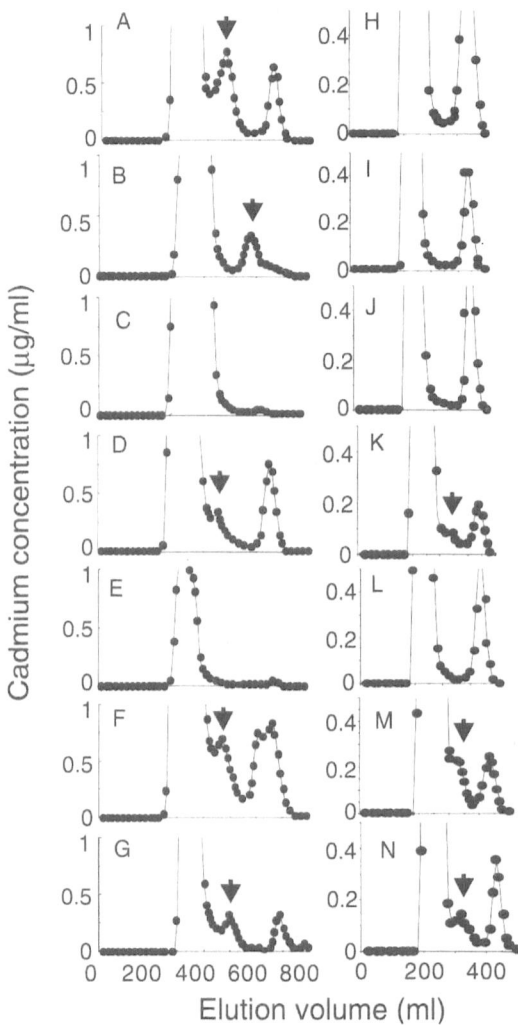

Figure 2. Sephadex G-75 or Shephadex G-50 column chromatography of the soluble fraction from *E. coli* cells transformed with each expression plasmid, (A): wild-type MT, (B): α-fragment, (C): β-fragment, (D): mutant D, (E): mutant F, (G): mutant G, (H): mutant H, (I): mutant I, (J): mutant J, (K): mutant K, (L): mutant M, and (N): mutant N. The elution profiles of Cd (●) are indicated. The expressed wild-type MT and mutant MTs are presented by arrow. The data of mutant A to G and mutant H to N were published in *Protein Eng* 10: 413–416, 1997 and *Cell Molec Life Sci* 53: 459–465, 1997, respectively.

the α-fragment were shifted one residue in the direction of the N-terminal, (N) all y-Cys-x - Cys sequences of the α-fragment were replaced with Cys-y-x-Cys sequences, and (O) the mutant MT with the two conserved serines (at positions 32 and 35) replaced by leucines.

Significance of both fragment of MT in Cd-binding

To examine whether six mutant MTs (B to G) expressed in *E. coli* formed the Cd-binding configuration, gel filtration study was carried out as shown in Figure 2A to G. The mutant D (βAlaαCys), F (αCysβCys) and G (αCysβAla) were observed at 0.42, 0.47 and 0.46 of dissociation coefficient (*K*d) values [3], respectively, in the same manner as the *K*d value (0.49) of the expressed wild-type human MT-2. The α-fragment had a Kd value 0.73, half the size of intact MT. In contrast with that the β-fragment produced by *in vitro* techniques [19] and chemical synthesis [20] was reconstituted as a three Cd-binding form, the β-fragment (C) and mutant E (βCysαAla) as metal-binding form were undetectable in *E. coli*, although the mRNA of the two proteins was detected from the *E. coli*. In addition, the Cd-binding mutant MTs was scarcely produced in form of inclusion bodies.

Each purified Cd-binding protein was identified by their sequence, amino acid analyses, and/or desorption/ionization time-of-flight mass spectrometry. Approximately 200–400 μg of the purified proteins were obtained from about 2.5 g wet weight of *E. coli*. Ultraviolet absorption spectrum of each purified mutant MTs expressed as Cd-binding form had an absorption shoulder at 250 nm. Upon acidification, the spectrum indicated a clear loss of the shoulder. The pH-dependent displacement of Cd ions suggested that the Cd-binding mutant proteins showed Cd-thiolate chelation [21]. Number of bound Cd ions per each expressed protein was also shown in Table 1. The obtained ratios of mutant B, D and G were in good agreement with the theoretical value, 4.0, of α-fragment, and the ratio of mutant F agreed with that of wild-type MT. From the results, these mutant proteins maintained a structure of metal-thiolate cluster found in mammalian MTs [2, 22].

In order to estimate the metal-binding ability of the expressed mutant proteins, initial velocities and the pseudo-first-order rate constants of the mutant proteins reacted with DTNB were determined [12, 23] (Tab. 1). From the results, the liganding strength or the coordination state of the cluster structure of the mutant D, F and G was stronger than that of wild-type MT and similar to that of the α-fragment. It is of interest to note that although the mutant F also consisted of two fragments, and only differed by the order of the fragments (β-α in wild-type MT, α-β in the mutant F), the mutant F exhibited more than 5 times greater Cd-binding ability than the wild-type MT, indicating that Cd-binding of the mutant was tighter than that of wild-type MT. The metal-binding ability of the β-fragment was reported to be weaker than that of the α-fragment [24]. The biological functions of MT may be regulated by the β-fragment which more easily release binding-metals.

From this study and reports by other researchers, we postulate that the maintenance of metal-binding clusters of MTs depends on the α-fragment which then confers proper Cd-binding conformation to the β-fragment (i. e., a trans-domain interaction). Rhee *et al.* [25] postulated that the two fragments of MT were active independently. In contrast, the mutant MTs containing α-fragment lacking cysteines could not bind Cd. In addition, the results of mutant F and G indicate that the Cd-binding to MT in biosynthetic processes needs the presence of intact α-fragment itself but not the sequential order of the two fragments. This is supported by results obtained from *in vitro* experiments where Cd bound to apo-MTs initially in the α-fragment followed by the β-fragment [4, 17].

Table 1. Cd-binding of mutant and wild-type MTs

Mutant MT	Obtained Cd-binding-protein from 1 l culture[a]	Cd-binding	Cd bound per protein[b]	Initial velocity	Pseudo first order rate constant	Cd-binding protein as inclusion body
	µg			nmol/min		
			mean ± SD	mean ± SD	mean ± SD	
Wild-type MT	340	Cd-thiolate	6.9 ± 0.26 (3)	2.58 ± 0.34 (3)	1.25 ± 0.26 (3)	N. D.
A	200	Cd-thiolate	4.2 ± 0.16 (3)	0.22 ± 0.04 (3)	0.34 ± 0.05 (3)	N. D.
B	not detectable	–	–	–	–	not detectable
C	240	Cd-thiolate	4.6 ± 0.31 (3)	0.32 ± 0.06 (3)	0.49 ± 0.08 (3)	N. D.
D	not detectable	–	–	–	–	not detectable
F	340	Cd-thiolate	6.8 ± 0.18 (3)	0.34 ± 0.02 (3)	0.48 ± 0.07 (3)	N. D.
G	370	Cd-thiolate	4.3 ± 0.24 (3)	0.36 ± 0.09 (3)	0.50 ± 0.17 (3)	N. D.
H	not detectable	–	–	–	–	N. D.
I	not detectable	–	–	–	–	N. D.
J	not detectable	–	–	–	–	N. D.
K	50	Cd-thiolate	3.1	2.27	0.54	N. D.
L	not detectable	–	–	–	–	N. D.
M	140	Cd-thiolate	3.7	3.50	0.57	N. D.
N	80	Cd-thiolate	3.3	5.21	0.41	N. D.
O	240	Cd-thiolate	6.4 ± 0.32 (3)	4.15 ± 0.45 (3)	N. D.	not detectable

[a]MT content was calculated on the basis of Cd content in the purified proteins. [b]The ratio was calculated on the basis of Cd and, cysteine or -SH contents. N.D; not determined. In initial velocity and pseudo first order rate constant of mutant K, M and N, the values were adjusted in comparison with values of α-fragment in this table, because the values were changed in the different experiment.
The data of mutant A to G and mutant H to N were published in *Protein Eng* 10: 413–416, 1997 and Cell Molec Life Sci 53: 459–465, 1997, respectively.

Significance of cysteine and other amino acid residues of MT in Cd-binding

In the elution profiles of the gel filtration for the sonic extracts of mutant H to N, the Cd-binding mutant K, M and N were eluted at the elution volume in the same manner as the wild-type MT (Fig. 2H to N). On the other hand, no Cd binding peak appeared to be produced by mutant H, I, J and L.

The expressed mutant K, M and N as Cd-binding form also had an absorption shoulder at 245–255 nm. The molar ratios of fragment M was similar to that of the α-fragment. The initial velocities and the pseudo first order rate constants of the mutants were also determined by the reaction with DTNB. The rate constant of mutant N coincided with that of the α-fragment and it was lower than that of mutant K or M (Tab. 1). The Cd-binding cluster structure of mutant K and M was partially disrupted due to their mutations compared with that of mutant N, indicating the mutant N had a more stable Cd-binding form than the mutant K and M. These results suggest that the extreme alterations of constitutive amino acids in MTs would be more critical than the shifts of invariant positions of cysteines in MTs in the Cd-binding capacity and maintaining configuration, although the invariant positions of 20 cysteines in MTs are thought to be important for the metal-binding abilities of MTs.

The basic amino acids, e.g. lysine, are highly conserved in MTs. Lysine may play a role in restraining the structural expansion of the MT molecule [26]. The mutant K that all nonpolar amino acid residues replaced by glycines, was obtained as a Cd-binding form. It was suggested that the polar amino acid residues were one of the principal factors of the metal-binding in MTs. It has been reported that MT mutant, which had all three lysines in α-fragment replaced by glutamates, binds seven Cd ions more labilely than the α-fragment of wild-type MTs [15]. However, the Cd-binding affinity of the MT mutant was due to only three amino acid substitutions, in contrast mutant K having Cd-binding affinity contained seven amino acid substitutions.

In addition, there was no evident difference between the mutant O with the two conserved serines replaced by leucines (at positions 32 and 35) and wild-type MT from the results of the molar ratio and reaction with DTNB. However the results of pH titration and long term storage indicate that highly conserved serines could contribute to the stability of the cluster structure of MT.

Conclusion

In order to evaluate the metal-binding ability of mammalian MT by site directed mutagenesis, the expression plasmids of fourteen mutant MTs were constructed. We examined whether the mutant MTs were expressed as a Cd-binding form in *E. coli* grown in the medium containing Cd by the addition of IPTG as an inducer. Furthermore, the expressed mutant proteins as Cd-binding form were purified and analyzed for their biochemical, spectroscopic, and metal-binding properties.

From these results, the obtained conclusions are as follows; (i) the α-fragment is indispensable component in metal-binding processes of Cd-MT, (ii) The Cd-binding to MT needs the presence of intact α-fragment itself but not the sequential order of the two fragments, and (iii) The position of cysteine residues are less critical than expected in Cd-binding

Acknowledgment
The authors indebted to Prof. Jeremias H. R. Kägi from Biochemisches Institut der University Zürich for his useful suggestions in this work, and to Prof. Michael Karin for the donation of the plasmid phMTII-3.

References

1. Vallee BL, Auld DS (1990) Zinc coordination, function and structure of zinc enzymes and other proteins. *Biochemistry* 29: 5647–5659.
2. Otvos JD, Armitage IM (1980) Structure of the metal cluster in rabbit liver metallothionein. *Proc Natl Acad Sci USA* 77: 7094–7098.

3. Winge DR, Miklossy KA (1982) Domain nature of metallothionein. *J Biol Chem* 257: 3471–3476.
4. Nielson KB, Winge DR (1983) Order of metal binding in metallothionein. *J Biol Chem* 258: 13063–13069.
5. Zelazowski AJ, Szymanska JA, Law AYC, Stillman MJ (1984) Spectroscopic properties of the α fragment of metallothionein. *J Biol Chem* 259: 12960–12963.
6. Kojima Y, Kägi JHR (1978) Metallothionein. *Trends Biochem Sci* 3: 90–93.
7. Kägi JHR (1993) Evolution, structure and chemical activity of class I metallothioneins: An overview. *In*: KT Suzuki, N Imura, M Kimura (eds): *Metallothionein III*. Birkhäuser Verlag, Basel, pp. 29–55.
8. Murooka Y, Nagaoka T (1987) Expression of cloned monkey metallothionein in *Escherichia coli*. *Appl Environ Microbiol* 53: 204–207.
9. Hou Y, Kim R, Kim S (1988) Expression of the mouse metallothionein-I gene in *Escherichia coli*: increased tolerance to heavy metals. *Biochim Biophys Acta* 951: 230–234.
10. Kille P, Stephens P, Cryer A, Kay J (1990) The expression of a synthetic rainbow trout metallothionein gene in *E. coli. Biochim Biophys Acta* 1048: 178–186.
11. Chernaik ML, Huang PC (1991) Differential effect of cysteine-to-serine substitutions in metallothionein on cadmium resistance. *Proc Natl Acad Sci USA* 88: 3024–3028.
12. Cismowski MJ, Huang PC (1991) Effect of cysteine replacements at positions 13 and 50 in metallothionein structure. *Biochemistry* 30: 6626–6632.
13. Cismowski MJ, Narula SS, Armitage IM, Chernaik ML, Huang PC (1991) Mutation of invariant cysteines of mammalian metallothionein alters metal binding capacity, cadmium resistance, and ^{113}Cd NMR spectrum. *J Biol Chem* 266: 24390–24397.
14. Cody CW, Huang PC (1993) Metallothionein detoxification function is impaired by replacement of both conserved lysines with glutamines in the hinge between the two domains. *Biochemistry* 32: 5127–5131.
15. Cody CW, Huang PC (1994) Replacement of all α-domain lysines with glutamate reduces metallothionein detoxification function. *Biochem Biophys Res Commun* 202: 954–959.
16. Pan PK, Hou FY, Cody CW, Huang PC (1994) Substitution of glutamic acids for the conserved lysines in the α domain affects metal binding in both α and β domains of mammalian metallothionein. *Biochem Biophys Res Commun* 202: 621–628.
17. Kille P, Lees WE, Darke BM, Winge DR, Dameron CT, Stephens P, Kay J (1992) Sequestration of cadmium and copper by recombinant rainbow trout and human metallothioneins and by chimeric (mermaid and fishman) proteins with interchanged domains. *J Biol Chem* 267: 8042–8049.
18. Odawara F, Kurasaki M, Suzuki-Kurasaki M, Oikawa S, Emoto T, Yamasaki F, Linde Arias AR, Kojima Y (1995) Expression of human metallothionein-2 in *Escherichia coli*: Cadmium tolerance of transformed cells. *J Biochem* 118: 1131–1137.
19. Nielson KB, Winge DR (1984) Preferential binding of copper to the β domain of metallothionein. *J Biol Chem* 259: 4941–4946.
20. Okada Y, Ohta N, Yagyu M, Min KS, Onosaka S, Tanaka K (1985) Synthesis of a nonacosapeptide (β-fragment) corresponding to the N-terminal sequence 1–29 of human liver metallothionein II and its heavy metal-binding properties. *FEBS Lett* 183: 375–378.
21. Vašák M, Galdes A, Hill HAO, Kägi JHR, Bremner I, Young BW (1980) Investigation of the structure of metallothioneins by proton nuclear magnetic resonance spectroscopy. *Biochemistry* 19: 416–425.
22. Nielson KB, Winge DR (1985) Independence of the domain of metallothionein in metal binding. *J Biol Chem* 260: 8698–8701.
23. Kurasaki M, Emoto T, Linde-Arias AR, Okabe M, Yamasaki F, Oikawa S, Kojima Y (1996) Independent self assembly of cadmium-binding α-fragment of metallothionein without participation of β-fragment in *Escherichia coli*. *Protein Eng*.9: 1173–1180.
24. Wang Y, Mackay EA, Kurasaki M, Kägi JHR (1994) Purification and characterization of recombinant sea urchin metallothionein expressed in *Escherichia coli*. *Eur J Biochem* 225: 449–457.
25. Rhee IK, Lee KS, Huang PC (1990) Metallothioneins with interdomain hinges expanded by insertion mutagenesis. *Protein Eng*.3: 205–213.
26. Vašák M, McClelland CHE, Hill HAO, Kägi JHR (1985) Role of lysine side chains in metallothionein. *Experientia* 41: 30–40.

Metallothionein IV
C. Klaassen (ed.)
© 1999 Birkhäuser Verlag Basel/Switzerland

Metal binding proteins: Molecular engineering of improvements in metal specificity

Ian Watt and Peter Kille

School of Biosciences, Cardiff University, Cardiff, CF1 3US, Wales, UK

Summary. The results discussed here provide three main conclusions about the nature of the metal binding properties of the two domain types present in native human MT. Firstly, the constructs of the tail domain of MT will bind cadmium with higher affinity than a corresponding head domain construct. This selectivity of metal binding was shown to be independent of multimer size. Secondly, maximum cluster stability was achieved when two domains were present, irrespective of domain type. Increasing the multimer size to a trimeric molecule decreased the stability of the metal clusters, and proteins with four domains did not accumulate upon expression. Finally the loss of cadmium from the tail domain appears to be co-operative in nature, whereas this is not the case for the head domain of MT. Future studies will investigate the stability of the metal clusters of the constructs described here with regard to other competitive metal chelating agents. The stability of mono-valent metal ion clusters within these novel proteins will also be addressed.

Introduction

Heavy metals are essential for life, however large quantities can be toxic to biological systems. Thus the study of the systems used by organisms to control heavy metals is of great interest. Although many heavy metal binding proteins have been discovered and characterised, the basis of selective metal binding remains largely unknown. The focal point of this article is the investigation of the properties of a family of metal binding proteins, the metallothioneins (MTs). Since the initial discovery of this small cadmium binding protein in equine kidney cortex [1], very closely related proteins have been reported in a wide range of organisms, from mammals to fish, plants and micro-organisms. All vertebrate MTs discovered thus far have approximately 60 amino acids, with very few hydrophobic residues and no aromatic residues. Roughly one third of the total amino acids present in MT are cysteine residues, and it is through these residues that metal chelation is achieved. Furthermore, the position of the cysteine residues is conserved in all vertebrate MTs. In the presence of metal ions the polypeptide chain wraps itself around the metal ions to form two distinct clusters (Fig. 1) [2,3]. In the presence of divalent ions, such as cadmium, the N-terminal (head) domain contains three ions bound by 9 cysteine residues. The C-terminal (tail) domain however contains four metal ions bound by 11 cysteine residues [2].

Studies on MT that had been stripped of metal ions (*apo*-thionein) showed that when reconstituted with cadmium the tail domain fills preferentially and co-operatively [4,5,6]. Further studies have also shown that the binding sites in the tail domain are 30 times more stable than those in the head domain [7]. In the presence of mono-valent ions, such as copper(I) or silver, each domain can bind up to six ions [8,9], but the order of filling is reversed with the head domain exhibiting preferential binding [10].

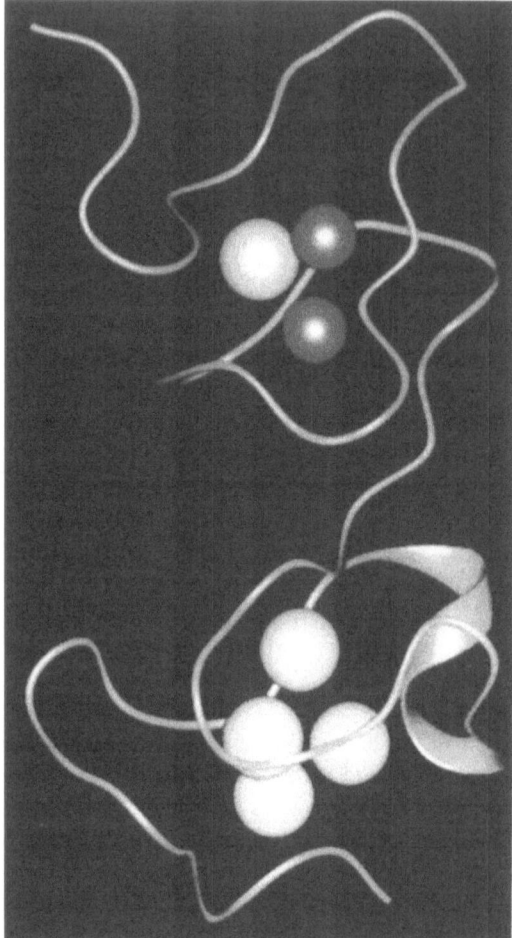

Figure 1. Ribbon diagram of (Zn$_2$Cd$_5$) rat MT-IIA as derived from X-ray crystallography data.

Results and discussion

Studies on the metal binding properties of the native molecule are limited due to the heterologous nature of normal MT. Therefore the aim of this study was to create an environment in which each domain of the heterologous MT molecule could be created in isolation, and the metal binding properties studied. Previously individual domains have been prepared by limited proteolysis or chemical synthesis [8,9,11,12]. These studies have shown that each of the domains function independently and the stoichiometry of metal binding is unchanged between individual domains and intact molecules. Recent studies have shown that the expression of individual domains could confer cadmium resistance in tobacco plants [13], but not on cadmium sensitive yeast strains [14], where it was reported that this could only be achieved by the expression of the a multi-domain MT molecule [14]. In addition, it was also shown that a genetically engineered multimer of tail domains encoded considerable cadmium resistance *in planta* [15]. Hence, the approach in this study was to engineer a family of homo-multimers for each of the domains of MT.

In order to facilitate the study of individual domain types, an expression system for the production of large quantities of recombinant protein had to be identified. The system used is the pET29a expression vector (Novagen), where the protein is produced as a fusion protein with the S-Tag encoded on the vector. This expression system has been successfully used within the group to express and purify a variety of MTs, including rat MT III, human MT III, and a macroalgae MT (Fig. 2). For this study the pET29a vector had been modified to include a *Sfi*I restriction site and an in frame termination codon proximal to the *Sfi*I site, and was renamed pP1.

Genes encoding novel proteins based on naturally occurring MTs were constructed. A family of genes encoding homo-multimers (from one to four domains) were engineered (Fig. 3) from individual domains that had been synthesised by PCR. The primers used in the PCR reaction contained a recognition site for the restriction endonuclease *Sfi*I. Digestion of the PCR products with *Sfi*I produced a DNA product encoding individual domains (both head and tail) with *Sfi*I 'sticky ends'. The formation and cloning of multimers was achieved in one ligation reaction, containing the vector pP1 cut with *Sfi*I and *Sfi*I cut domain. Directional control over the ligation of individual domains to form multimers was achieved by the use of the restriction

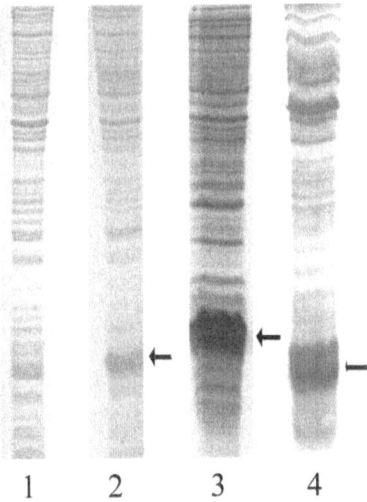

Figure 2. SDS polyacrylamide gel electropheresis of total cell extracts from expression studies on three genes from different sources of MT. The MT in each preparation is marked by an arrow. Lane 1: uninduced sample; Lane 2: Rat MT III; Lane 3: Dimeric construct of two human head domains; Lane 4: *Fucus vesiculosis* MT.

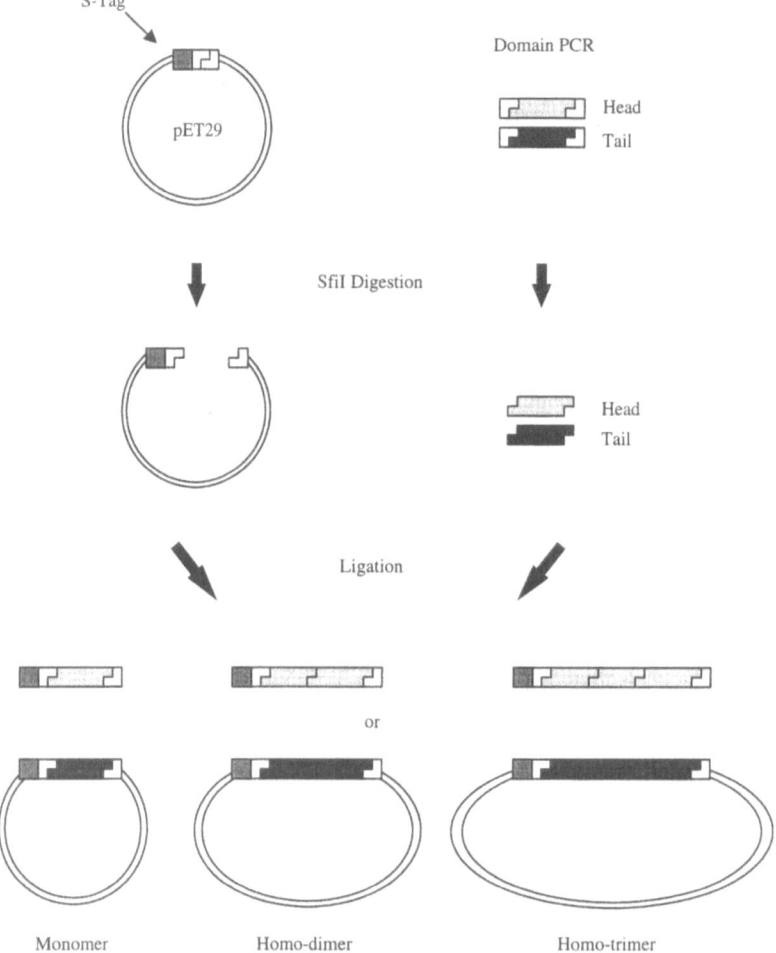

Figure 3. Schematic representation of the approach used for the construction of homo-domain multimers.

enzyme *SfiI* which upon cleavage of its recognition site produces non-complementary antiparallel overhangs. The same process was also used to produce and isolate a clone encoding the native human MT.

Recombinant protein for each of the homo-multimers and the native MT was produced by expression in an *E. coli* host. The clones encoding four domains in tandem repeat did not accumulate protein upon expression, and this was taken to indicate the instability of proteins consisting of greater than three domains. The purification procedure took advantage of the stabil-

Panel A.

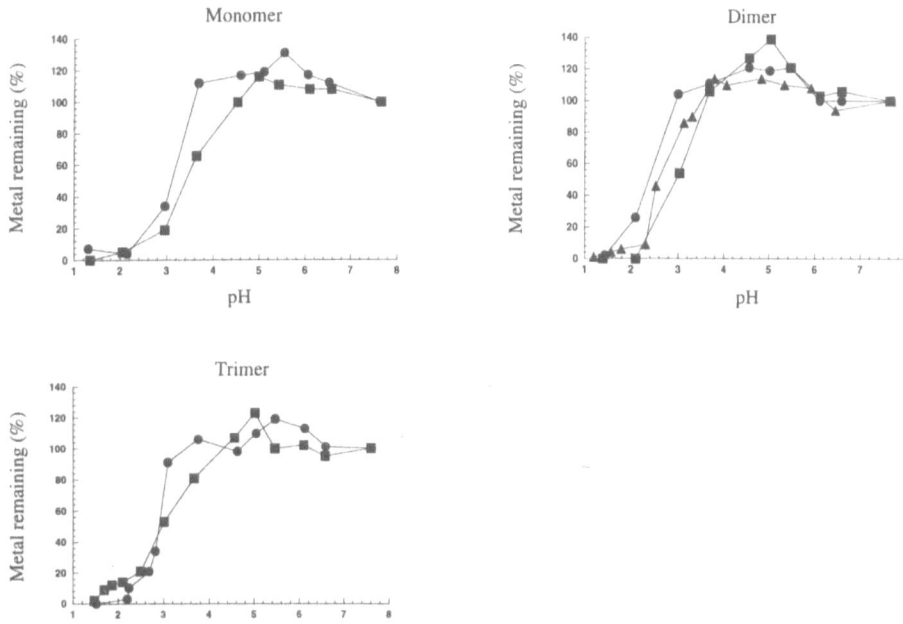

Panel B.

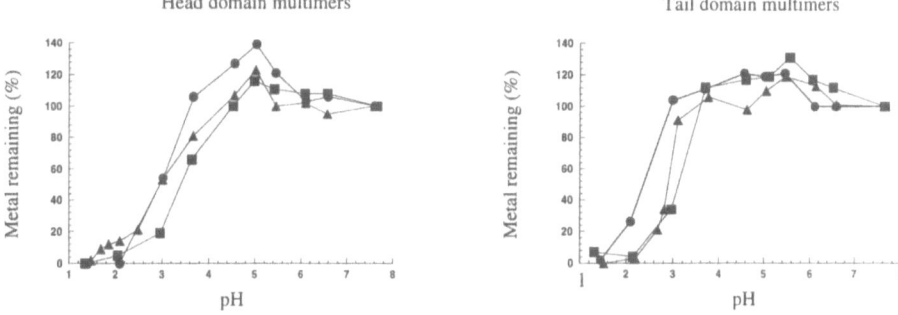

Figure 4. Homo-domain multimers were titrated against an increasing hydrogen ion concentration in order to assess the stability of the metal clusters. (Panel A) Relative stability of constructs of the same size formed from either the head or tail domains of human MT (Head ●, Tail ■, Human MT ▲). Native Human MT is included for comparison in the dimer graph. Metal remaining was calculated as a percentage of the metal present at the highest pH. (Panel B) Relative stability of homo-multimers of varying size (Monomer ●, Dimer ■, Trimer ▲), formed from either the head or tail domain is shown. Metal remaining was calculateed as a percentage of the metal present at the highest pH.

ity of the metal clusters of MT with regard to high temperature [16]. A single heating operation followed by cation exchange chromatography resulted in the isolation of pure protein.

Titration of MT domain multimers against an increasing hydrogen ion concentration caused the metal ions sequestered by the recombinant MTs to be competitively released. The pattern of this release was markedly different for each type of domain multimer studied. When the data was plotted as percentage of metal remaining against pH, all data fell as a roughly sigmoidal curve. However the point at which only 50% of the metal was remaining varied according to multimer size and domain type. The value at which 50% of the cadmium remains associated with the native MT produced here was approximately pH 3.0, which is in accordance with previous studies of native MTs [17]. As shown in Figure 4 the curve for a domain dimer is shifted to the left with respect to the curve for a monomer. This suggests that a homodimer is able to bind cadmium more strongly than a single domain. A trimer of domains however, shows no further increase in cluster stability. This pattern was observed for multimers of both the head and tail domains. Comparison of the cadmium binding characteristics of head and tail constructs revealed that proteins based on the tail domain bound cadmium more strongly when compared to those formed from the head domain. This is also in agreement with previous studies [4]. Closer inspection of the shape of the curves reveals a further difference between head and tail domain multimers. Over the range in which the bulk of the metal ions are lost the curve is steep when titrating a protein formed from a tail domain. However when the same experiment is performed with a head domain protein that section of the curve is noticeably less steep. This suggests that the metal ions in a tail domain cluster are lost in a co-operative fashion, and that as the cluster is broken open the competing agent removes all of the cadmium ions. The principal of co-ordinated loss of cadmium ions from the tail domain is in agreement with the property of cooperative metal binding and cluster formation previously demonstrated for the tail domain in the reconstitution of *apo*-thionein [4,5,6]. Yet in the head domain cluster each of the cadmium binding sites appears to lose its associated metal ion independently of the other sites.

Acknowledgements
This study was financially supported by a special PhD studentship funded by the Biotechnology and Biological Sciences Research Council. The project supervision was facilitated by the NERC continued support for Dr. Kille through the Advanced Fellowship program (GT5/94/ALS).

References

1. Margoshes M and Valleee BL (1957) *J Amer Chem Soc* 79: 4813–4814.
2. Robbins AH, McRee DE, Williamson M, Collet SA, Xuong NH, Furey WF, Wang BC and Stout CD (1991) *J Mol Biol* 221: 1269–1293.
3. Messerle BA, Schaffer A, Vašák M, Kägi JHR and Wuthrich K (1990) *J Mol Biol* 214: 781–786.
4. Nielson KB and Winge DR (1983) *J Biol Chem* 258: 13063–13069.
5. Stillman MJ, Cai W and Zelazowski AJ (1987) *J Biol Chem* 262: 4538–4548.
6. Good M, Hollenstein R, Sadler PJ and Vašák M (1988) *Biochemistry* 27: 7163–7166.
7. Bernhard WR, Vašák M and Kägi JHR (1986) *Biochemistry* 25: 1975–1980.
8. Nielson KB and Winge DR (1985) *J Biol Chem* 260: 8698–8701.
9. Li YJ and Weser U (1992) Inorg. Chem. 31: 5526–5533.
10. Nielson KB, Atkin CL and Winge DR (1985) *J Biol Chem* 260: 5342–5350.

11. Matsumoto S, Nakayama S, Nishiyama Y, Okada Y, Min KS, Onosaka S and Tanaka K (1992) *Chem Pharm Bull* 40: 2694–2700.
12. Matsumoto S, Nishiyama Y, Okada Y, Min KS, Onosaka S and Tanaka K (1992) *Chem Pharm Bull* 40: 2701–2706.
13. Pan A, Tie F, Duau Z, Yang M, Wang Z, Li L, Chen Z and Ru B (1994) *Mol Gen Genet* 242: 666–674.
14. Rhee KI, Lee KS and Huang PC (1990) *Protein Eng* 3: 221–226.
15. Pan A, Tie F, Yang M, Luo J, Wang Z, Ding X, Li L, Chen Z and Ru B (1993) *Protein Eng* 6: 755–762.
16. Gasull T, Pardy K, Hernandez J, Hidalgo J and Kille P (1997) *J Biochem Biophys Meth*; *in press*.
17. Tommey AM, Shi J, Lindsay WP, Urwin PE and Robinson NJ (1991) *FEBS Lett* 292: 48–52.

Metallothionein IV
C. Klaassen (ed.)
© 1999 Birkhäuser Verlag Basel/Switzerland

Molecular splicing of metallothionein – study on domains of metallothionein

Yi Xiong, Yanjiao Zhou, Yiwu Chen and Binggen Ru

National Laboratory of Protein Engineering, College of Life Sciences, Peking University, Beijing 100871, P.R. China

Mammalian metallothioneins (MTs) are a class of low-molecular weight proteins containing 20 cysteine residues out of a total of 61 amino acid residues. MTs fold into two separate domains in the presence of certain metal ions [1,2]. Each domain binds metal ions in a polynuclear metal-thiolate cluster with ligation through thiolates of the 20 cysteines residues [3]. Several NMR studies showed that the seven Cd(II) ions in the protein were positioned into the two clusters: α domain-with four Cd(II) ions and 11 cysteine residues, and β domain-with three Cd(II) ions and 9 cysteine residues [4–6]. The two spherical domains, with a similar diameter of 1.5–2.0 nm, are connected by a hinge region consisting of a conserved Lys-Lys-Ser (KKS) segment in the middle of the polypeptide chain [7].

Many studies on the domains of MT have been reported. The α and β domain obtained by proteolysis of native MT showed many similar biochemical properties to native MT [8–10]. The metal-binding properties of chemically synthesized α and β domain were also examined [11,12]. Recently, expression systems using *E. coli*, yeast and plants were set up to investigate the physiological roles and metal-binding abilities of domains and mutants of MT [13–17]. The linker region was also studied by insertion, site-mutation or chemical modification, and the ability of these mutants in conveying resistance to metals have been measured [18–20]. All the results indicate that the two domains act independently and differ in metal-binding capacity and preference.

In this paper, the biochemical characteristics and metal-binding properties and structures of a series of MT domains and mutants were studied including the following areas: (1) the α and β domains obtained by proteolysis of native rabbit MT and by expression in *E.coli*; (2) the α domain mutants connected by different linker regions and expressed in *E.coli*.

Studies on proteolysed domains of native MT and expressed α-domain from *E.coli*

The preparation of α and β domains from rabbit liver MT

Metallothionein was purified as Cd, Zn-protein from rabbits injected with $CdCl_2$ [21]. Metals were removed from MT by incubating with 0.1N HCl at 37 for 15 min and gel filtration on Sephadex G-25 eluted with 0.01N HCl. The α domain was prepared by reconstitution of apometallothionein with 3 mol eq of Cd^{2+} to form Cd_4-α, digested by subtilisin, and subjected to chromatography on superfine Sephadex G-75 column. All the solutions used must be N_2-bubbled to avoid polymerism of MT. The β domain was prepared by a similar method except that apometallothionein was reconstituted with 6 mol eq of Ag^+ to form Ag_6-β.

The peak with the maximum atomic absorption and UV absorption at 254 nm had a Kd value of 0.7, which agrees with earlier reports [2]. Some modifications, such as decreasing the ratio of metal to apoMT in reconstitution, increasing the amount of subtilisin and cutting down the reaction time of hydrolysis, made the peaks of poly MT and native MT disappear in the elution profile, and the recovery of α and β domain increased.

The molecular weight of 3154 Da agrees with the α-fragment containing two lysines in the hinge region. Analysis of amino acid composition and the content of metal and sulfhydral groups agree with the expected value. All the identifications proved that α and β domains of MT-I and MT-II from rabbit liver were obtained.

The structures of α and β domain

New advanced method – atomic force microscopy was used to determine the appearance of the α and β domain and a single global structure was observed. In the region above 210 nm (Fig. 1), UV absorption and CD intensity are considered to arise from LMCT transitions between the metal and the coordinating thiolate groups. With increasing pH from acidic to neutral, Cd^{2+}

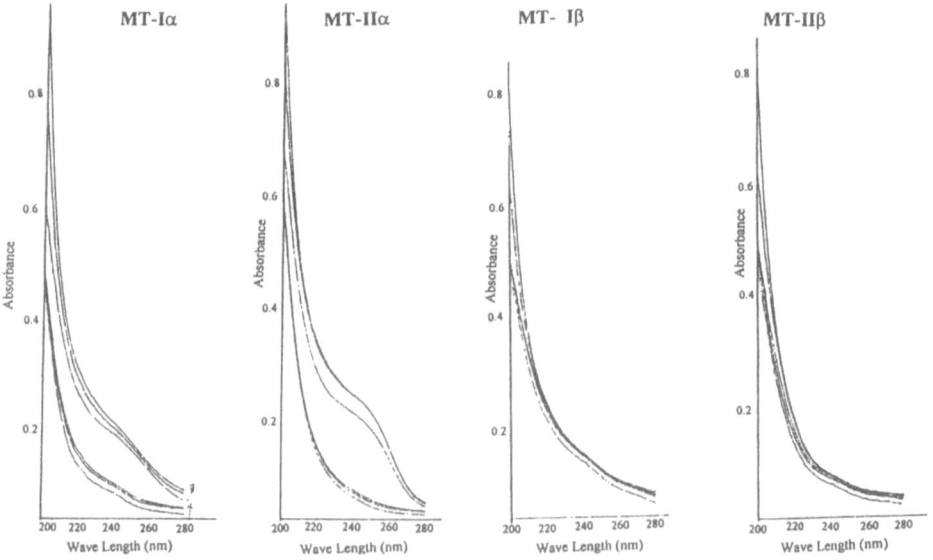

Figure 1. UV absorption spectroscopes of separate α and β domain. Protein concentration is 0.1 mg/ml.

chelated with apoMT or apo-α and then formed the absorption band at 254 nm. But because the LMCT transition between Ag^+ and thiolate groups was not apparent, Ag_6-β domain had no absorption band above 210 nm. UV absorption and CD spectra of separate α and β domains are similar to native MT, indicating the independent structures of the two domains.

The comparison of metal binding affinity between MT and α domain

To reveal the difference of metal-binding strength between MT and α domain, the reaction with DTNB was examined. The pseudo initial velocity of each reaction was calculated according to the reaction profile in Figure 2 and the result is listed below:

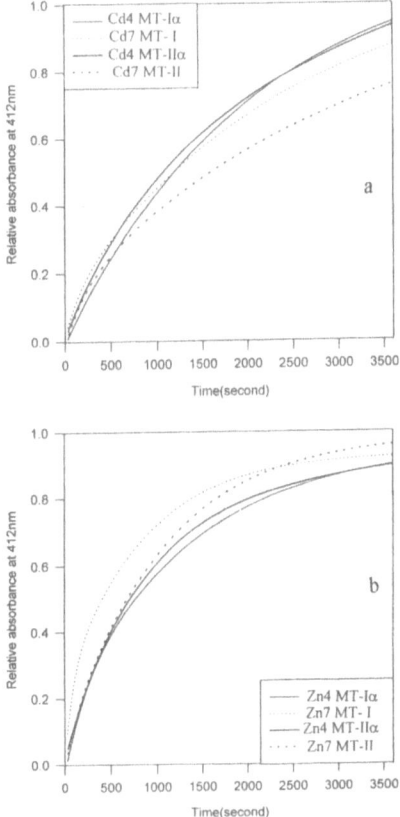

Figure 2. Reaction with DTNB of α-fragment and native MT binding with Cd (a) and Zn (b). 10 nmol of the purified protein was dissolved in 100 μl of degassed 10 mM phosphate buffer, pH7.5, and the reaction was started by adding 25 nmol of DTNB. The absorbance at 412 nm was recorded on a Shimadzu UV-240 spectrophotometer at 10 s intervals for 90 min at room temperature, beginning at 20 s after adding DTNB.

The Cd-binding strength: MT-I < MT-II < MT-IIα < MT-Iα
The Zn-binding strength: MT-I < MT-II < MT-Iα < MT-IIα

Studies on expressed α domain purified from *E.coli*

The α domain cDNA of mouse MT-I was synthesized by PCR and ligated into the fusion expression vector pGEX-4T-1. The fusion protein GST/α was expressed and extracted, and the α domain fragment was cleaved and purified as described previously [17]. The amino acid composition, molecular mass (3462 Da), UV and CD spectra were all studied, and demonstrated that the α domain has the same character and structure as native α domain. The divalent metal-binding stoichiometry was determined to be 3.95:1, also the same as in native α domain.

Studies on expressed mouse MT-I and the three α-α mutants from *E.coli*

Construction and preparation of mMT-I and α-α domain mutants

The mMT-I cDNA was synthesized by PCR using pBX/MT as template and inserted into fusion expression vector pGEX-4T-1 to construct plasmid pGMA8. The β domain of mMT-I was replaced by α domain and an initial codon ATG was added before the first α domain gene, to construct a whole α-α domain mutant. The native linker region (KKS) and two different linker regions with one or two amino acids elongation (SKKS and GKKST) were inserted between the two α domains. As shown in Figure 3a, mouse MT-I gene contains two Alu I recognition sites (ACGT); one is inside the genes of β domain and the other is in the linker region. One copy of the α-α domain mutant gene was cut from the mouse MT-I gene by Alu I (Fig. 3b). Another was amplified by PCR using the same 5' primer and different 3' primers (Fig. 3c) for different linker regions, and then digested with EcoR I. The two α domain genes and the pGEX-4T-1 vector were ligated to produce expression plasmids. The expressed fusion proteins were extracted and the mMT-I and α-α domain mutants were cleaved and purified as described above.

The expressed fusion protein was approximately 51% of the total cellular proteins and the production of purified mMT-I and α-α mutants were 3–4 mg/L culture. Since the expression level of MT in *E.coli* are always limited because of its toxic high content of thiol groups, a fusion expression vector was selected and a much higher yield was obtained in this study than in earlier studies.

Characteristics of the expressed proteins

The amino acid composition, amino acid sequence at the N-terminus and molecular mass of the recombinant mMT-I and α-α mutants were determined to confirm that the desired gene products were recovered from cell lysates. The molecular mass of the recombinant mMT-I and

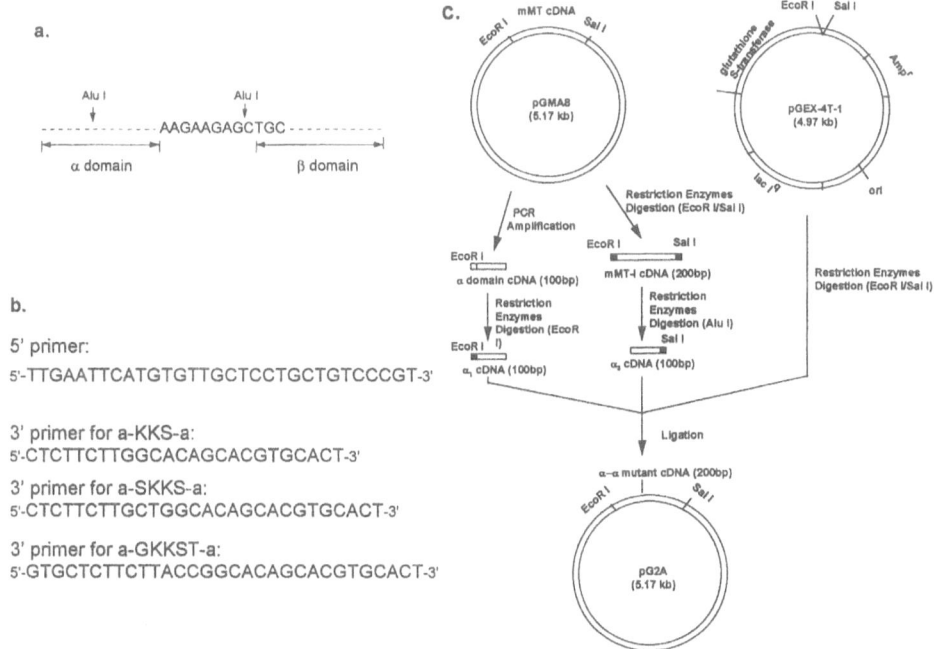

Figure 3. Construction of expression plasmid containing α-α domain mutant gene (a) The two recognition sites of Alu I in the mMT-I cDNA; (b) Different primers for amplification of α domain gene; (c) The α-α domain mutant gene was synthesized by PCR and inserted into the expression vector pGEX-4T-1 to generate expression plasmid pG2A.

α-α mutants are 6635, 6583, 6674 and 6739 Da. They are greater than that of native mMT-I because of the short peptide prior to methionine, which is due to the digestion with thrombin from the glutathione S-transferase.

The contents of bound Cd and thiol groups of recombinant mMT-I and the three α-α mutants were measured and the result show that divalent metals bind to the mutants at almost the same ratio of 8:1 as we expected. The pH at half-dissociation of Cd and Zn from the proteins were estimated to be in agreement with those from *in vitro* reconstituted rabbit MT-I (Tab. 1). The same results show that the recombinant MT purified from *E.coli* is a mammalian MT as desired, and the α-α mutants also have the same characteristics as MT.

The ability to bind Cd was determined by reaction with DTNB and the results were as follows: α-KKS-α > recombinant mMT-I > α-SKKS-α > α-GKKST-α (Fig. 4). The α-KKS-α has the strongest ability to bind heavy metal. The results also indicate that the length and the amino acid composition of the linker region are important in the complete molecule.

Table 1 Half metal dissociation pH of the recombinant mMT-I and α-α mutants isolated from *E.coli* cells

Protein	Cd	Zn	Cu
Reconstituted rabbit MT-I	3.67	5.48	1.83
Recombinant mMT-I	3.57	5.20	1.40
α-KKS-α	3.58	5.40	ND
α-SKKS-α	3.80	5.95	ND
α-GKKST-α	3.70	5.19	ND

About 50 µg protein was incubated with 2 ml buffer at different pH (from 1.0 to 9.0) and then applied to a Sephadex G25 column equilibrated with the same buffer. Metal ions eluting in void volume and total volume were quantified. The pH at which half metal were dissociated from the protein were calculated from the curve of the percent of bound metal versus pH.

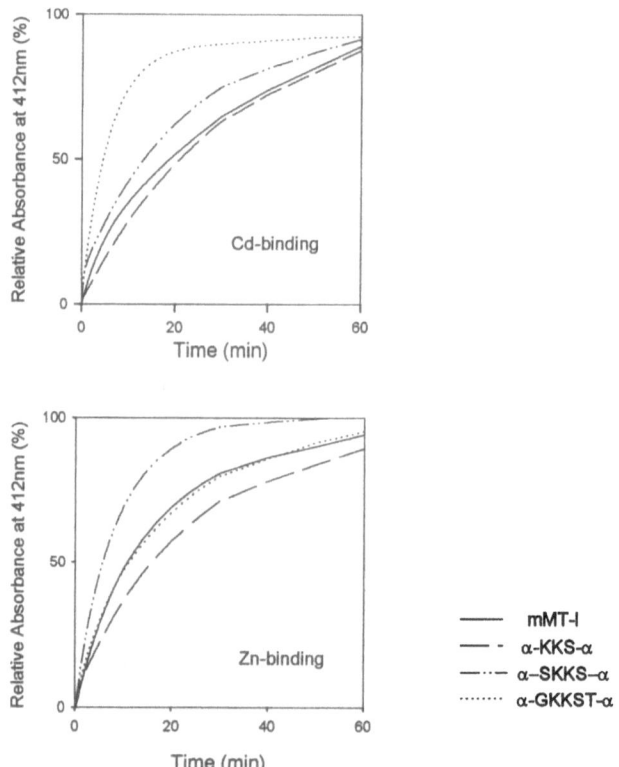

Figure 4. Reaction of the α-α mutants and mMT-I with DTNB. 10 nmol of the purified protein was dissolved in 100 µl of degassed 10 mM phosphate buffer, pH7.5, and the reaction was started by adding 25 nmol of DTNB. The absorbance at 412 nm was recorded on a Shimadzu UV-240 spectrophotometer at 10 s intervals for 90 min at room temperature, beginning at 20 s after adding DTNB.

Structure of the expressed proteins

The atomic force microscopy profiles of recombinant mMT-I and the three α-α mutants exhibit the dumbbell-like shape containing two global clusters and a hinge region. In similar ultraviolet absorption spectra of the expressed proteins, high cadmium-thiolate absorption at 250 nm and low absorbance at 280 nm are observed at pH8.0, but at pH 2.0 they both disappear. The CD spectra of the proteins are also very similar, as shown in Figure 5. The absorption band at 258 nm, corresponding to the characteristic Cd-thiolate cluster at pH8.0, diminished when pH was adjusted to 2.0. The bands at 258 nm in CD spectra must be attributed to the metal-thiolate chromophore because of the absence of aromatic acids and disulfide bridges [22]. The

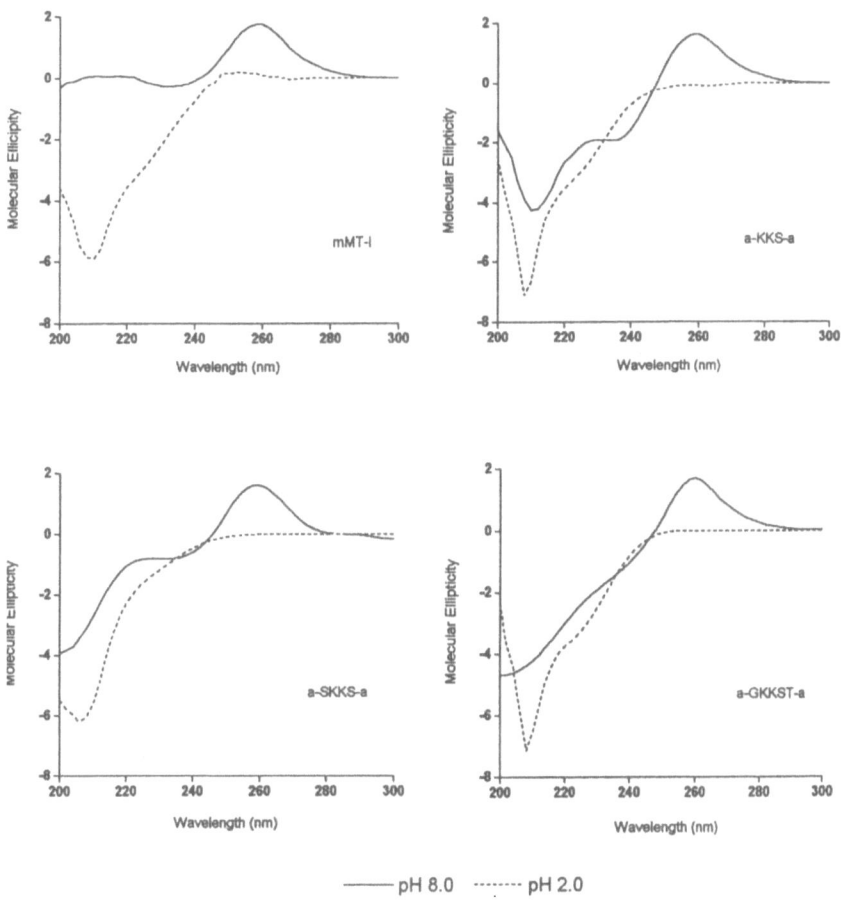

Figure 5. Circular dichroism (CD) spectra of the α-α mutants and mMT-I. Protein concentration is 6.5×10^{-5} M.

cotton effects, which disappear when lowering the pH to 2.0, refers to the dissociation of metal-thiolate cluster as all the cysteines are protonated.

The preliminary results show that the recombinant mMT-I and three α-α mutants have the same metal-thiolate structure. The α domain in the mutants is able to form metal-binding cluster properly as in the MT, which confirms the independent function of the domain.

Conclusions

(1) An efficient expression system has been set up to produce mouse MT-I in *E.coli* as we wanted; (2) The studies of the three α-α mutants demonstrate the independent structure and function of the α domain; (3) The importance of the length and the amino acid composition of the linker region between the two domains has been confirmed; (4) A α-α mutant – α-KKS-α – .was determined to have a higher ability to bind heavy metals than MT while studied *in vitro*.

Acknowledgment
This work is supported by Chinese National 863 High Technology Grant. It is a pleasure to acknowledge Dr. Richard D. Palmiter for the plasmid pBX/MT. Mr. Wanyu Zhang assisted in the atomic absorption spectrophotometry. We are also grateful to Dr. Yong Dai for his valuable suggestions and contributions to the circular dichroism studies.

References

1. Otvos JD and Armitage IM (1980) Structure of the metal clusters in rabbit liver metallothionein. *Proc Natl Acad Sci USA* 77: 7094–7098.
2. Winge DR and Miklossy KA (1982) Domain nature of metallothionein. *J Biol Chem* 257: 3471–3476.
3. Braun WM, Wagner G, Worgotter E, Vašák M, Kägi JHR and Wuthrich K (1986) Polypeptide fold in the two metal clusters of metallothionein-2 by nuclear magnetic resonance in solution. *J Mol Biol* 187: 125–129.
4. Schultze P, Worgotter E, Braun WM, Wagner G, Vašák M, Kägi JHR and Wuthrich K (1988) Conformation of [Cd7]-metallothionein-2 from rat liver in aqueous solution determined by nuclear magnetic resonance spectroscopy. *J Mol Biol* 203: 251–268.
5. Arseniev A, Schultze P, Worgotter E, Braun WM, Wagner G, Vašák M, Kägi JHR and Wuthrich K (1988) Three-dimensional structure of rabbit liver [Cd7]-metallothionein-2a in aqueous solution determined by nuclear magnetic resonance. *J Mol Biol* 201: 637–657.
6. Messerle BA, Schaffer A, Vašák M, Kägi JHR and Wuthrich K (1992) Comparison of the solution conformations of human [Zn7]-metallothionein-2 and [Cd7]-metallothionein-2 using nuclear magnetic resonance spectroscopy. *J Mol Biol* 225: 433–443.
7. Kägi JHR and Schaffer A (1988) Biochemistry of metallothionein. *Biochemistry* 27: 8509–8515.
8. Nielson KB and Winge DR (1984) Preferential binding of copper to the β domain of metallothionein. *J Biol Chem* 259: 4941–4946.
9. Zelazowski AJ, Szymanska JA, Law AYC and Stillman MJ (1984) Spectroscopic properties of the alpha fragment of metallothionein. *J Biol Chem* 259: 12960–12963.
10. Yu J, Zhou YJ and Ru BG (1997) The preparation and analysis of rabbit liver MT β domain. *Chinese Biochem J* (in Chinese) 13(4): 451–459.
11. Okada Y, Ohto N, Yagyu M, Min KS, Onosaka S and Tanaka K (1985) Synthesis of a nonacosapeptide (β-fragment) corresponding to the N-terminal sequence 1–29 of human liver metallothionein II and its heavy metal-binding properties. *FEBS Lett* 183(2): 375–378.
12. Matsumoto S, Nakayama S, Nishiyama Y, Okada Y, Min KS, Onosaka S and Tanaka K (1992) Amino acids and peptides. XXXIV. Synthsis of mouse metallothionein I. (1). Synthesis of dotriacontapeptide corresponding to C-terminal sequence 30–61 (α-fragment) of mouse metallothionein I and related peptides and examination of their heavy metal-binding properties. *Chem Pharm Bull* 40(10): 2694–2700.
13. Masaaki K, Tadasu E, Ana RLA, Masashi O, Futoshi Y, Shinji O and Yutaka K (1996) Independent self-assem-

bly of cadmium-binding α-fragment of metallothionein in Escherichia coli without participation of β-fragment. *Protein Eng* 9(12): 1173–1180.

14. Kille P, Lees WE, Darke BM, Winge DR, Dameron CT, Stephens PE and Kay J (1992) Sequestration of cadmium and copper by recombinant rainbow trout and human metallothioneins and by chimeric (mermaid and fishman) proteins with interchanged domains. *J Biol Chem* 267(12): 8042–8049.
15. Pan AH, Tie F, Yang MZ, Luo JC, Wang ZX, Ding X, Li LY, Chen ZL and Ru BG (1993) Construction of multiple copies of α-domain gene fragment of human liver metallothionein I_A in tandem arrays and its expression in transgenic tobacco plants. *Protein Eng* 6(7): 755–762.
16. Pan AH, Tie F, Duan ZW, Yang MZ, Wang ZX, Li LY, Chen ZL and Ru BG (1994) α-Domain of human metallothionein I_A can bind to metals in transgenic tobacco plants. *Mol Gen Genet* 242: 666–674.
17. Xiong Y and Ru BG (1997) Purification and characteristics of recombinant mouse metallothionein-I from *Escherichia coli*. *J Biochem* 121: 1102–1106.
18. Pande J, Vašák M and Kägi JHR (1985) Interaction of lysine residues with the metal thiolate clusters in metallothionein. *Biochemistry* 24: 6717–6722.
19. Templeton DM and Cherian MG (1984) Chemical modifications of metallothionein. Preparation and characterization of polymers. *Biochem J* 221: 569–575.
20. Cody CW and Huang PC (1993) Metallothionein detoxification function is impaired by replacement of both conserved lysines with glutamines in the hinge between the two domains. *Biochemistry* 32: 5127–5133.
21. Pan AH, Ru BG, Li LY, Shen T, Tie F and Wang WQ (1994) Purification andIdentification of metallothioneins from rabbit liver induced by zinc. *Chinese J Biochem Biophys* 24(6): 509–516.
22. Rupp H and Weser U (1978) Circular Dichroism of metallothioneins: a structural approach. *Biochim Biophys Acta* 533: 209–226.

Metallothionein in non-mammalian tissues

Involvement of metallothionein in female squirrelfish reproduction

E.D. Thompson[1], P.-E. Olsson[2], G.D. Mayer[1], C. Haux[3], P. Walsh[4] and C. Hogstrand[1,4]

[1]T.H. Morgan School of Biological Sciences, 101 Morgan Building, University of Kentucky, Lexington, Kentucky 40506-0225, USA
[2]Department of Cellular and Developmental Biology, Umeå University, S-901 87 Umeå, Sweden
[3]University of Göteborg, Zoological Institute, Department of Zoophysiology, Medicinaregatan 18, S-41390 Göteborg, Sweden
[4]RSMAS/MBF University of Miami, 4600 Rickenbacker Causeway, Miami, Florida 33149, USA

Introduction

The squirrelfish family (*Holocentridae*) presents a unique system for studying the involvement of metallothionein in zinc metabolism and reproductive physiology. Since zinc is an essential micronutrient, virtually all organisms need only maintain minimal amounts of zinc for normal cellular function. The squirrelfish (*Holcentrus rufus*) has been found to be an exception to this rule. While zinc concentrations in the liver of most vertebrates fall in the range of 15–40 µg/g wet mass [1,2], zinc hepatic concentrations of squirrelfish have been measured up to an average of 1,778 µg/g (Tab. 1). In the same fish, hepatic metallothionein levels were found to be 16,900 µg/g wet mass. This phenomenon is not exclusive to *Holocentrus rufus* but, in fact, is observed in all studied species of the *Holocentridae* family (Tab. 1). However, there is high variability between species and individuals within the same species and population. The family *Holocentridae* is made up of fish which are associated exclusively with coral reefs in tropical areas all over the world. Such a wide geographical dispersal of species makes it unlikely that high hepatic levels of zinc and metallothionein can be attributed to any local environment.

Table 1. Metallothionein (MT) and zinc levels in livers of squirrelfish collected off Bermuda[a] and in four Australian Holocentrid species[b]

Species	MT	Zinc
squirrelfish[a]	16900 (10957)	1778 (1419)
spotfin squirrelfish[b]	7444 (7147)	447 (480)
sabre squirrelfish[b]	889 (444)	67 (35)
lattice soldierfish[b]	19502 (11939)	1014 (534)
blotcheye soldierfish[b]	3458 (5585)	159 (238)

Concentrations are expressed in µg g^{-1} wet wt with standard deviations in parentheses.
[b]data from Hogstrand and Haux (1996).

Metallothionein and zinc levels in squirrelfish liver are closely correlated. This would be expected because metallothionein is involved in zinc metabolism [3,4] and metallothionein transcription can be induced by increased activity of zinc as well as other metals within groups IB and IIB of the periodic table [5]. Other metals such as copper and cadmium are found to be associated with metallothionein in squirrelfish liver but their levels are not nearly as high as that of zinc. Because of the consistently high content of zinc in squirrelfish metallothionein and the relatively low and quite variable contents of copper and cadmium, we believe that zinc is the native metal of metallothionein in all *Holocentridae* species and that copper and cadmium are inadvertently sequestered in the liver of squirrelfish due to the high binding capacity offered by metallothionein.

Is zinc accumulation a result of metallothionein structure?

After the discovery of unusual levels of zinc and metallothionein in the livers of squirrelfish, it became interesting to attempt to discern the reasons for this massive accumulation. To this end, the amino acid sequence of squirrelfish metallothionein was determined to see if there were any crucial differences in the protein itself. This was accomplished by isolating poly-A RNA from longjaw squirrelfish (*Holocentrus marianus*) and producing a cDNA library with the ZAP Express cDNA synthesis kit. The library was screened in *E. coli* with riboprobe derived from perch MT-cDNA and positive plaques were obtained and subsequently reintroduced into *E.coli*. The plasmids were recovered and purified in *E. coli* and finally amplified by the polymerase chain reaction and sequenced. This sequence was then compared to that of European perch (*Perca fluviatilis*), a relative of the squirrelfish with normal levels of zinc and metallothionein. The sequences were found to be very similar with only four differing amino acids, none of which were the cysteine residues involved in binding of metals. Therefore, the radical accumulation of zinc does not appear to be dependent on the structure of squirrelfish metallothionein.

Female-specific accumulation, subcellular distribution and seasonal variations of zinc and metallothionein

At least part of the variability in the levels of metallothionein and zinc could be explained by the fact that only females exhibit high levels, while male levels of metallothionein and zinc were normal in comparison with other fish. In one study, longjaw squirrelfish were captured and sacrificed by an overdose of MS222, sexed, and tissue samples were taken from each fish for acid digestion [6]. These samples included liver, gonads, scales, kidney, brain, muscle, blood cells, plasma, and retina. Tissue samples were ultimately analyzed for zinc content by atomic absorption spectroscopy. In this particular collection of fish, the females had 13-fold higher hepatic zinc levels than did the males [6]. The only other gender-specific accumulation occurred in the gonads. Thus, squirrelfish ovaries contain considerably higher zinc concentrations than ovaries from other species [6]. As in other vertebrates, there were high levels of zinc in the retina [7], but there was no sex-linked difference in the squirrelfish. It was considered that the increased zinc levels in females could have been a result of differences in diet between males and females. However, upon analysis of intestinal contents and experimentation involving food-deprivation, it was shown that females do not need a zinc-rich food source to maintain high levels of zinc [6].

Interspecies variability in hepatic metallothionein and zinc concentrations is seasonal. Longjaw squirrelfish captured in February, May, and June were sacrificed and analyzed for hepatic zinc content. Zinc concentrations dropped drastically from February to May and June, suggesting that there is some seasonal variation in the amount of zinc required by female squirrelfish (Fig. 1). The liver somatic index (LSI) remained relatively constant indicating that the decreasing hepatic zinc concentration could not be solely attributed to increasing liver size. The zinc concentration in the liver of males showed no significant seasonal variation. The hepatic zinc concentration of females in May and June, although greatly reduced, was still ten times greater than that of males at the same time of year.

In accordance with the hypothesis that metallothionein is regulating zinc levels in squirrelfish, we would expect metallothionein levels to drop from February to May and June in accordance with zinc levels. In fact, we found this to be the case. The same longjaw squirrelfish liver samples analyzed for zinc content were fractionated by differential centrifugation and the subcellular distributions were analyzed for metallothionein content by western analysis. Upon examination and comparison of typical western blots taken from both time periods, the February samples [6] had much more metallothionein than did samples taken in May (Fig. 2). Also, there

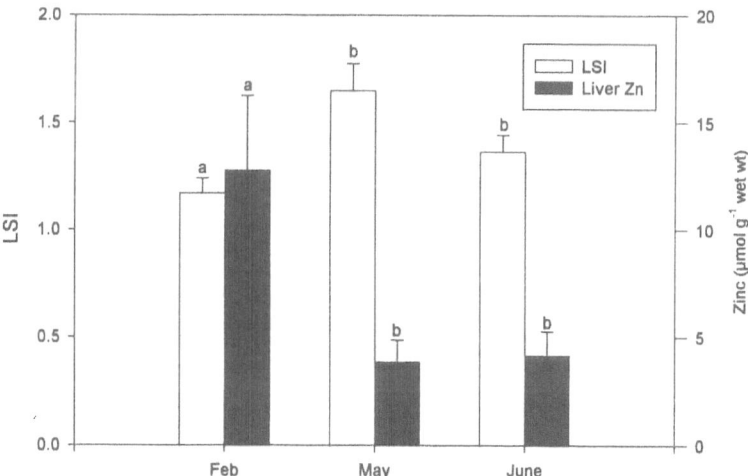

Figure 1. Seasonal variation of LSI and zinc concentrations in February, May, and June. The liversomatic index (LSI) is a measure of liver growth. LSI = liver mass/body mass × 100. Data was tested statistically by 2-way ANOVA (bars with like letters are not significantly different; p ≤ 0.05). n = 11 for February and May and n = 13 for June.

seemed to be a shift in the subcellular localization of metallothionein from February to May. In February, metallothionein was found predominantly in the nuclear fraction of females and small amounts of metallothionein were detected in the cytosolic fractions of both males and females [6]. No metallothionein could be detected in nuclear fractions from males. In May, metallothionein in females shifted from being almost entirely in the nuclear fraction to an on average equal distribution between the nuclear and cytosolic fractions. The subcellular distribution of metallothionein in the males remained unchanged. These results suggest that this seasonal variation could be the result of some female-specific factor, possibly sex hormones, and could further be related in some way to the reproductive season. Reproductive patterns of squirrelfish are relatively unknown but have been characterized to a certain extent. As stated previously, squirrelfish are tropical fish that settle on coral reefs. Upon hatching, squirrelfish pass through a series of larval stages into a postlarval rhynchichthys stage [8]. The transition from rhynchichthys to juvenile is bridged by a pelagic, prejuvenile meeki stage [9]. At the subsequent benthic juvenile stage of development, fish settle onto coral reefs. The reproductive activity of squirrelfish is dependent upon this reef settlement. Previous studies of squirrelfish reproductive ecology have revealed that greater proportions of settled juveniles were found in February than in the summer months [10]. As reef settling is believed to be coupled to sexual maturation, these findings support our suggestion that the higher accumulation and nuclear localization of metallothionein in females collected during February as compared with May and June is somehow related to the reproductive season.

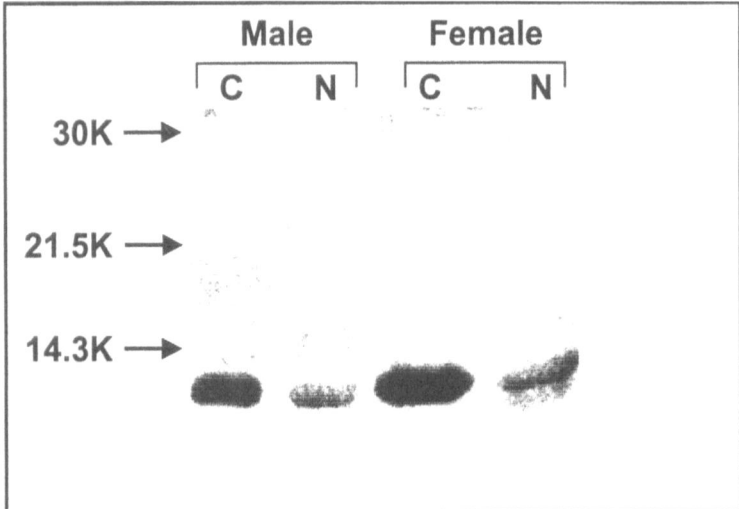

Figure 2. Western blots portraying subcellular distribution of metallothionein in male and female squirrelfish hepatocytes. C, cytosolic fraction; N, nuclear fraction. Arrows indicate the mobility of molecular weight standards (kDa).

Metallothionein and sexual maturation

In order to determine if the induction of synthesis of metallothionein is somehow related to sexual maturity and reproduction, we measured the levels of MT-mRNA to relate activation of the MT gene to sexual maturity. There was a hyperbolic relationship between gonadosomatic index (GSI) in females and concentration of MT-mRNA (Fig. 3). Levels of MT-mRNA in immature females were the same as those found in males but, as the females reach sexual maturity, the MT-mRNA levels increase rapidly. This finding suggests that the transcription of the MT gene is upregulated in females during sexual maturation.

What possible connections exist between gonadal development and metallothionein production in the liver? It is known that the liver plays a major role in the development of oocytes in non-mammalian vertebrates. In fish, the ovaries produce and secrete estradiol in response to Gonadotropic Hormone I secreted by the pituitary gland [11]. The liver cells have estradiol receptors which, when activated, can trigger the induction of several egg proteins, including vitellogenin and vitelline envelope ("egg shell") proteins [11]. The question is to determine the role of metallothionein in this process. It would be interesting to learn if estradiol also induces the production of metallothionein in squirrelfish liver.

Such an experiment to determine whether estradiol induces female-specific patterns for metallothionein and zinc can be easily designed in fish because gender is not strictly genetically determined. Males have functional estrogen receptors and can easily be feminized by treatment with female sex hormones such as estradiol. Likewise, genes encoding for "female proteins"

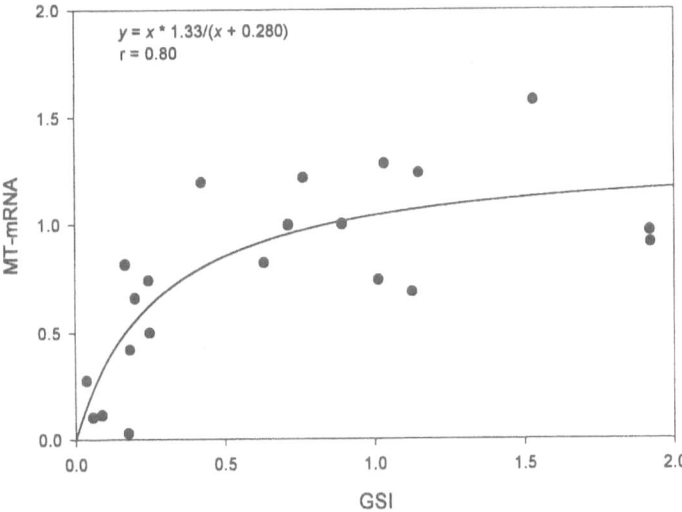

Figure 3. Levels of MT-mRNA as a function of sexual maturation. The gonadosomatic index (GSI) is a measure of sexual maturity. GSI = gonad mass/body mass × 100. GSI is plotted against MT-mRNA to illustrate the relationship. n = 21

are present and functional in males. In this manner, estrogenic effects on the regulation of fish "female proteins" can be conveniently studied in males. Utilizing this strategy, male squirrelfish were treated with 17 β-estradiol. The males were divided into a control group and an estradiol-treated group, which were injected with estradiol (5 mg/kg body weight) on days 0, 2, and 4. Both groups were then further subdivided by day of sacrifice, either day 5, 6, or 10. Both the control and estradiol-treated groups were then analyzed for both metallothionein and zinc content (Fig. 4). Liver size, expressed as the liver somatic index (LSI), can be used as an indication of increased protein production following estradiol injection. Estradiol stimulation of protein synthesis in male liver was confirmed by a rapid and pronounced increase in LSI. The effect of estradiol peaked at day 6, two days after the last estradiol injection, and subsided at day 10 when LSI values returned to normal. Thus, the estradiol treatment induced a transient production of "female proteins" in treated males. Northern analysis showed, however, that the levels of MT-mRNA remained constant (data not shown). Furthermore, zinc levels in the livers of estradiol-treated males were approximately the same as that of control males regardless of the time span of treatment (data not shown). Since zinc is the primary metal bound to metallothionein in squirrelfish, a constant zinc level is additional evidence that no metallothionein is being produced. For these reasons, we can conclude that transcription of metallothionein is probably not induced by estradiol in squirrelfish.

Estradiol did, however, have an intriguing effect on metallothionein, but the effect was not on metallothionein synthesis. Rather, the subcellular distribution of metallothionein was affected. Liver samples of control and estradiol-treated males were fractionated by differential centrifugation and cytosolic and nuclear fractions were obtained for all fish. Each sample was sub-

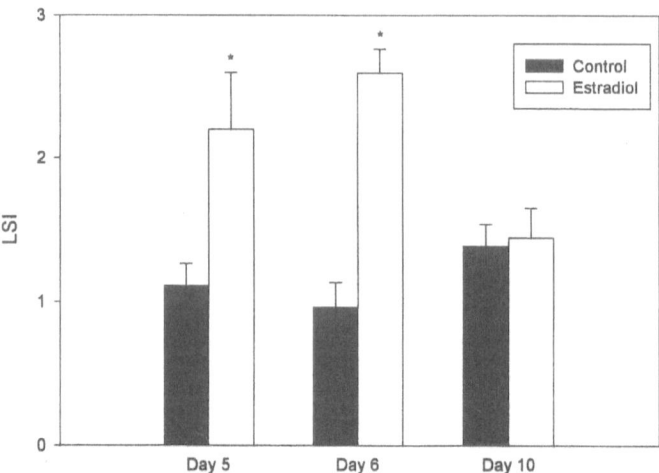

Figure 4. The effect of estradiol treatment on the LSI of male squirrelfish. Control fish are plotted alongside estradiol-treated fish at days 5, 6, and 10 following estradiol injection. n = 5–7.

jected to SDS-PAGE electrophoresis and quantified via western analysis for metallothionein (Fig. 5). In control males, there were small amounts of metallothionein in the nuclear fraction. However, in estradiol-treated males, the amount of "nuclear" metallothionein was significantly higher. In both control and estradiol-treated males, the amount of cytosolic metallothionein did not significantly increase. Therefore, the data suggests that estradiol is triggering localization of metallothionein into the nuclear fraction. Also, the level of "nuclear" metallothionein

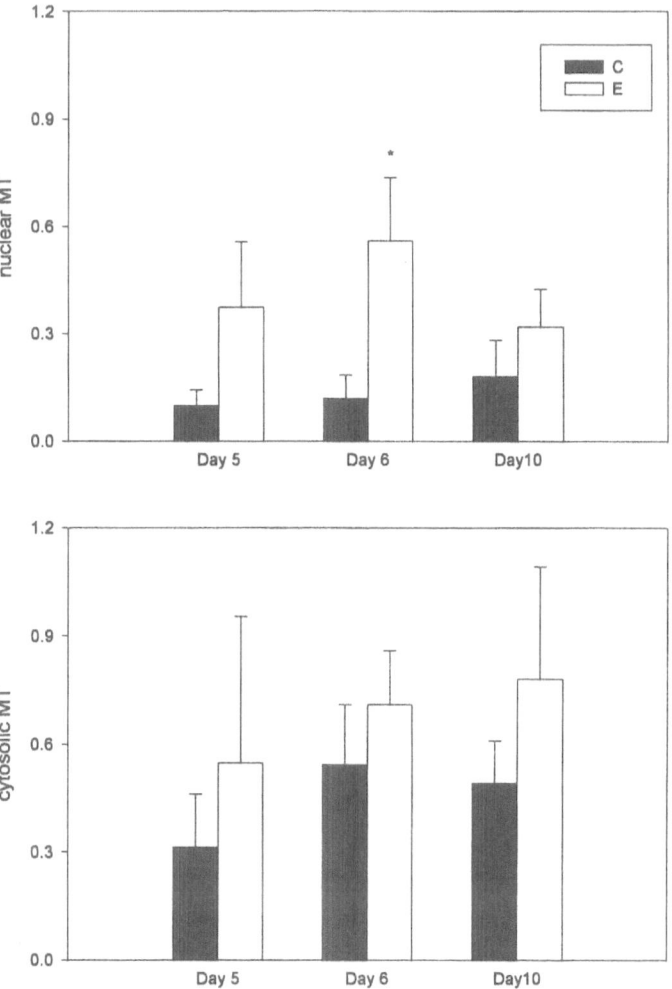

Figure 5. Subcellular distribution of metallothionein in estradiol-treated male squirrelfish. The levels of metallothionein in nuclear and cytosolic fractions of control and estradiol-treated males at days 5, 6, and 10 following estradiol injection. n = 2 for estradiol day 5, n = 3 for estradiol day 10, n = 4 for control day 10 and n = 5 for all others

peaked at day 6, the sampling point at which the highest LSI values were recorded, and both "nuclear" metallothionein and LSI returned to normal levels at day 10. This co-variation suggests that the effects of estradiol on "female protein" production and subcellular distribution of metallothionein may be related. Interestingly, total metallothionein content (nuclear + cytosolic fractions) was significantly increased by estradiol, but this increase was probably not the result of increased transcription.

Significance of nuclear metallothionein in squirrelfish

Our studies present evidence that metallothionein may be localized into the nucleus of squirrelfish hepatocytes but many questions on the mechanisms of this nuclearization still remain. How could metallothionein be directed into the nucleus? The nuclear envelope is composed of outer and inner nuclear membranes transversed by a nuclear pore complex. This complex allows the transport into the nucleus of water-soluble proteins that would normally be unable to cross lipid bilayers of the nuclear envelope. Small molecules (approximately 10,000 daltons or less) can diffuse relatively freely through the nuclear pore complex while larger proteins are actively transported into the nucleus. Also, proteins imported into the nucleus usually have a nuclear localization signal to allow for selectivity by the nuclear pore complex. This signal is a short sequence of four to eight amino acids and is typically rich in positively charged amino acids such as arginine and lysine. The signal is often split into two blocks of two to four amino acids separated by approximately ten amino acids and the blocks are thought to form a loop within the conformational structure of the protein. A mechanism for the nuclear transport of metallothionein can be proposed using these criteria. Metallothionein in squirrelfish has a molecular weight of approximately 6,500 daltons, assuming seven zinc atoms/molecule of metallothionein. Due to the size of this protein, it is possible that metallothionein could diffuse into the nucleus. A review of the amino acid sequence reveals that there are no stretches of amino acids indicative of a nuclear localization signal. This may be expected because the small size of metallothionein should make such a signal unnecessary. However, the rapid diffusion of small proteins such as metallothionein means that, to accumulate in the nucleus, these proteins would need to be bound somewhere within the nucleus to prevent subsequent diffusion back through the nuclear pore complex to the cytosol. It has been suggested that metallothionein polymerizes in order to remain in the nucleus [12], but chromatography filtration experiments reveal that this is probably not the case in squirrelfish [13]. The mechanism for localization of metallothionein into squirrelfish hepatocyte nuclei should be an interesting focus for future research.

What is the significance of nuclear metallothionein in squirrelfish hepatocytes? As noted above, the onset of reproduction in female fish and most other non-mammalian species means an increased hepatic production of many proteins such as vitellogenin. It is possible that metallothionein is needed for the increased transcription levels necessary for the massive production of these proteins. Since metallothionein binds zinc, which is involved in a number of aspects of the cell cycle such as cell proliferation, metallothionein could act as a zinc donor or zinc chelator for enzymes such as DNA polymerase. Studies in mammalian systems have shown that metallothionein is accumulated in the nucleus of a variety of fetal and tumor cells [12], and, similarly, in regenerating hepatocytes in which proliferation is stimulated by cytokines [14]. Although metallothionein seems to be involved in the replication of DNA associated with rapid cell growth, we believe that nuclear metallothionein in squirrelfish may have a more important role in the increased transcription of proteins. An alternative explanation is that metallothionein is used as a vehicle to remove zinc from the cytosol. It is also possible that metallothionein is not actually in the nucleus but is associated in some way with the nuclear envelope or other structures that co-precipitate with nuclei during sub-cellular fractionation. Species of squirrelfish were found to contain, on average, 16.9 mg metallothionein/g hepatocyte. Since the nucleus is approximately 10% of the cell, there would be 169 mg metallothionein/g hepatocyte nuclei if metallothionein were sequestered in the nucleus. This indicates that nearly one-fifth of the nucleus would be made up of metallothionein. This seems overwhelming but, assuming the volume of the nucleus remains the same upon the addition of metallothionein, the concentration of nuclear metallothionein would be 26 μM. This number is large but not entirely impossible. Experiments involving immunogold electron microscopy are under way. Hopefully this work will enable us to determine if such large amounts of metallothionein are indeed in the nucleus.

Acknowledgements
We are grateful to the numerous people that in different ways have contributed to the work described herein. Original research presented was funded by NSF grant # IBN-9631441 to CH.

References

1. Underwood EJ (1977) *Trace Elements in Human and Animal Nutrition*, 4th edition. Academic Press, New York.
2. Hogstrand C, Wood CM (1996) The physiology and toxicology of zinc in fish. *In*: EW Taylor (ed.): *Toxicology of Environmental Pollution – Physiological, Molecular and Cellular Approaches*, Society for Experimental Biology Seminar Series. Cambridge University Press, Cambridge, pp. 61–84.
3. Hogstrand C, Haux C (1991) Binding and detoxification of heavy metals in lower vertebrates with reference to metallothionein. *Comp Biochem Physiol* 100C: 137–141.
4. Vallee BL, Falchuk KH (1993) The biochemical basis of zinc physiology. *Physiol Rev* 73: 79–118.
5. Olsson P-E (1993) Metallothionein gene expression and regulation in fish. *In*: PW Hochachka, TP Mommsen (eds): *Moleclar Biology Frontiers*. Elsevier Science Publishers, Amsterdam, London, 259–278.
6. Hogstrand C, Gassman NJ, Popova B, Wood CM, Walsh PJ (1996) The physiology of massive zinc accumulation in the liver of female squirrelfish and its relationship to reproduction. *J Exp Biol* 199: 2543–2554.
7. Weitzel G, Strecker F, Roester U, Buddecke E, Fretzdorff A (1954) Zink in Tapetum lucidum. *Hoppe-Seyler's Z Physiol Chem* 296: 19–30.
8. Valenciennes A (1829) Vol. 3 *in*: G Cuvier, A Valciennces (eds): *Histoire naturelle des poissons*. Paris, 1–500.
9. Tyler JC, Johnson GD, Brothers EB, Tyler DM, Smith CL (1993) Comparative early life histories of Western Atlantic squirrelfishes (Holocentridae): age and settlement of rhynchichthys, meeki, and juvenile stages. *Bull Mar Sci* 53(3): 1126–1150.
10. Munro JL, Gaut VC, Thompson R, Reeson PH (1973) The spawning seasons of Caribbean reef fishes. *J Fish Biol* 5: 69–84.
11. Hyllner SJ, Oppen-Bernsten DO, Helvik JV, Walther BT, Haux C (1991) Oestradiol-17β induces the major vitelline envelope proteins in both sexes in teleosts. *J Endocrinol* 131: 229–236.
12. Cherian MG (1994) The significance of the nuclear and cytoplasmic localization of metallothionein in human liver and tumor cells. *Environ Health Perspect* 102: 131–135.
13. Hogstrand C, Haux C (1996) Naturally high levels of zinc and metallothionein in liver of several species of the squirrelfish family from Queensland, Australia. *Mar Biol* 125: 23–31.
14. Tsujikawa K, Imai T, Kakutani M, Kayamoni Y, Mimura T, Otaki N, Kimura M, Fukuyama R, Shimizu N (1991) Localization of metallothionein in nuclei of growing primary cultured adult rat hepatocytes. *FEBS Lett* 283: 239–242.

Metallothionein IV
C. Klaassen (ed.)
© 1999 Birkhäuser Verlag Basel/Switzerland

Accumulation of untranslated metallothionein mRNA in antarctic hemoglobinless fish (icefish)

Vincenzo Carginale[1], Rosaria Scudiero[1], Antonio Capasso[1], Clemente Capasso[1], Gianluca Passaretti[1], Guido di Prisco[1], Peter Kille[2] and Elio Parisi[1]

[1]CNR Institute of Protein Biochemistry and Enzymology, via Marconi 10, I-80125 Naples, Italy
[2]School of Molecular and Medical Biosciences, University of Wales, Museum Avenue, Cardiff, CF1 3US, UK

Summary. Icefish (Notothenioidei, Channichthyidae) are a group of Antarctic fish displaying unique phenotype characteristics. Besides the reported lack of hemoglobin, icefish have a very low metallothionein (MT) content compared to Antarctic red-blooded fish. Despite the low amount of MT protein in icefish, cDNA encoding two MT isoforms could be produced from total hepatic RNA by RT-PCR. The mRNA 5' end fragments were amplified by 5'-RACE. Steady-state mRNA levels were assessed in both icefish and red-blooded fish by high-stringency hybridization of the MT probe with total RNA. The results showed the presence of large amounts of MT mRNA in icefish liver. A further increase in MT mRNA was observed in Cd-treated icefish which was accompanied by the appearance of large amounts of MT protein. A comparison of the sequences of the 5' ends of the mRNAs showed the lack of a short motif in the 5'-UTR of only one of the two MT isoforms. On the basis of these results, we suggest that translation of the mRNA encoding an MT isoform is regulated in icefish by specific motifs in the 5'end.

Introduction

Ancient lakes are usually old, deep and isolated. Fishes living in these habitats exhibit mono-phyly, endemism and speciosity; they are useful model systems for studying evolution in isolated environments. Unlike other marine habitats, the shelf and the upper slope of Antarctica very much resemble ancient lakes; in fact, the Antarctic Ocean is equivalent to a closed basin, isolated from other areas by the Antarctic Polar Front, the oceanic frontal system currently running between 50°S and 60°S, developed approximately 25 My ago following the opening of the Drake Passage [1, 2].

The Antarctic Region provides an excellent opportunity for studying an unusual fish fauna in which molecular evolution, organismal adaptation and environmental conditions can be directly linked. Antarctica supports a low-diversity fish fauna, composed mostly of liparids and notothenioids. The latter dominate the fish biomass of the Antarctic shelf and are highly cold-adapted and remarkably stenothermal.

Antarctic fish have evolved adaptation mechanisms not found in temperate marine species, the most remarkable being the presence of antifreeze glycoproteins and the reduction of the hematocrit value. Channichthyidae (also known as icefish) with 16 species is an endemic family characterized by the total lack of hemoglobin in the colorless blood [3].

The present report describes the results of our studies on metallothionein accumulation and expression in two Antarctic fishes: the red-blooded *Trematomus bernacchii* and the hemoglobinless icefish *Chionodraco hamatus*.

MT in red-blooded and hemoglobinless fishes

Antarctic fishes are characterized by a large difference in MT content. Figure 1 shows the hepatic MT levels in red-blooded and hemoglobinless fishes determined by silver saturation assay
[4]. While an appreciable level of MT is present in *T. bernacchii*, a very low MT amount is
detectable in icefish. Consequently, no Zn-thionein was found in icefish liver extracts chro-
matographed on a Sephadex G-75 column [5]. However, as shown in Figure 1, the amount of
hepatic MT dramatically increases in icefish injected with repeated doses of Cd^{2+}.
    The icefish Cd-thionein was purified by combined gel-permeation, anion-exchange chro-
matography and reverse-phase HPLC, and finally resolved in two isoforms by capillary electrophoresis (CE). From the height of the CE peaks, we deduced that the two isoforms are synthesized in almost equimolar amounts (unpublished results).

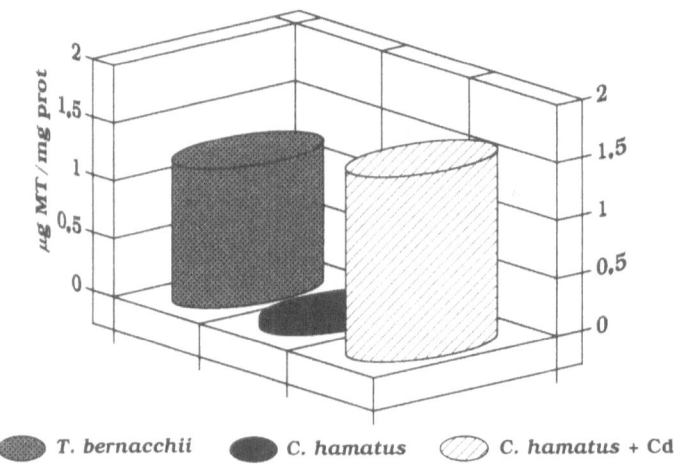

Figure 1. MT content in red-blooded and hemoglobinless Antarctic fish. Extracts from *T. bernacchii* and *C. hama-*
tus livers were prepared as described in [6]. MT content was determined by silver saturation assay [4].

RT-PCR amplification and cloning of MT specific cDNAs

MT cDNAs were generated by RT-PCR using icefish hepatic RNA and a primer derived from
the N-terminal amino acid sequence of trout MT [6]. The electrophoresis of the PCR mixtures
showed bands of 350 bp. These fragments were recovered from the gels and ligated into the
plasmid vector pGEM-T. The constructs were cloned in *E. coli* in the presence of ampicillin.
Double strand sequencing was performed using not less than 13 distinct clones for each RNA
sample. The sequences of the coding regions and of the 3'UTRs of these cDNAs can be found

in the EMBL nucleotide sequence database with the following accession numbers: Y12580 (*C. hamatus* MTa) and Y12581 (*C. hamatus* MTb).

Both sequences encode MTs of class I differing in a single amino acid substitution (T vs K) which can account for the observed difference in electrophoretic mobility. Major differences between the two cDNAs were found at the level of the 3'-UTR.

The sequence of the 5'-UTR was established following a 5' RACE protocol [7] using as reverse primer an oligonucleotide complementary to the segment comprised between nucleotide 52 and nucleotide 75 of the icefish MT cDNA. The primary structures and the predicted secondary structures of the leader regions of the two MT isoforms MTa and MTb, are shown in Figure 2. The sequences of the leader regions (Fig. 2a) differ substantially for the presence of the pentanucleotide GACAA at the 5' end of the 5'-UTR of MTb; in addition, to the sequence AGAUA at positions 30–34 of the 5'-UTR of MTb corresponds the sequence AAAGA in the 5'-UTR of MTa. The AGAUA motif is the most important in determining the differences in the predicted secondary structure shown in Figure 2b. These differences consist in the presence of hairpin loops in the leader region of MTb. It is noteworthy to mention that these structures were derived at the temperature of 0 °C which is very close to that of the environment where icefish live.

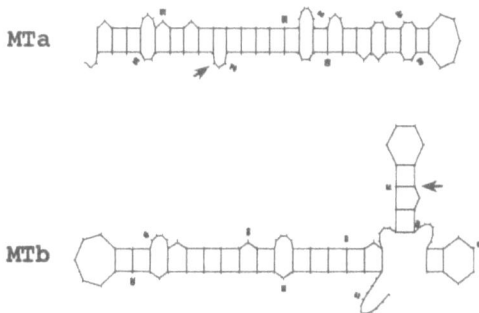

Figure 2. Structures of the 5' ends of *C. hamatus* MT isoforms. Primary structures (a) were determined by 5' RACE. The predicted secondary structures (b) were obtained from RNA sequences by using a program of the GCG Wisconsin Package.

Expression of MT genes in control and Cd-treated icefish

The levels of MT gene expression were determined by Northern blot analysis of total RNA from both *T. bernacchii* and *C. hamatus* hybridized with a homologous cDNA probe labelled

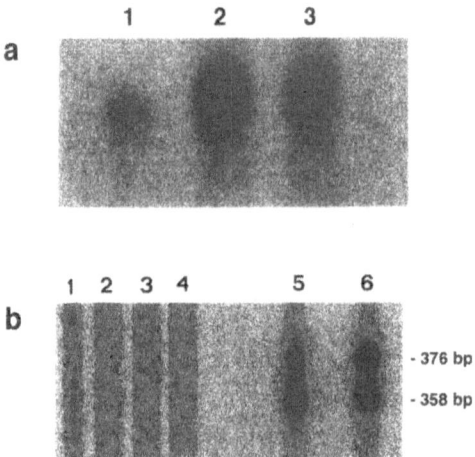

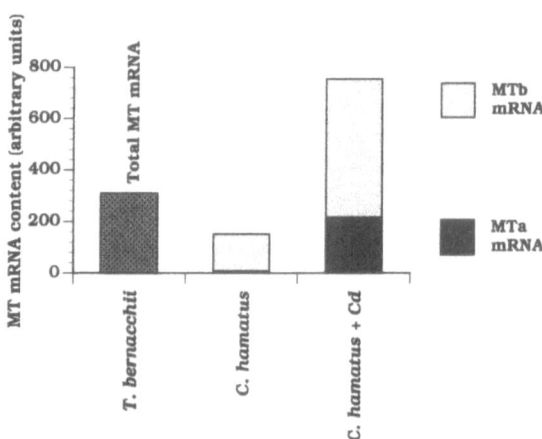

Figure 3. Expression of MT genes in Antarctic fish. a) Northern blot analysis of *C. hamatus* (lane 1), Cd-injected *C. hamatus* (lane 2) and *T. bernacchii* (lane 3) hepatic RNA. b) Polyacrylamide gel electrophoresis of radiolabeled cDNAs obtained from *C. hamatus* (lane 5) and Cd-injected *C. hamatus* (lane 6) RNA by RT-PCR. Lanes 1 to 4 are the DNA sequence ladder.

with ^{32}P (Fig. 3a). The MT mRNA levels were then estimated by densitometric analysis carried out on the Northern blot.

Figure 4. MT mRNA levels in Antarctic fish. Total MT mRNA levels were obtained by the results of the Northern blot. The MTa and MTb RNA levels were inferred from the data in Figure 3b (see text for further details) using the following relations:

$x_a + x_b = t$ where x_a = amount of MTa mRNA
$x_a/x_b = r$ x_b = amount of MTb mRNA
                                         t = total MT mRNA
                                         r = ratio of MTa mRNA/MTb mRNA

The ratio between the MTa and MTb transcripts in control and Cd-treated icefish was deduced by running RT-PCR reactions in the presence of ^{32}P labelled dCTP: the radioactive products were then separated by electrophoresis on a sequence gel of polyacrylamide (Fig. 3b), and the radioactivity present in the two bands was measured by liquid scintillation counting. The cpm ratio in the two cDNA bands, giving an estimate of the relative abundance of the MTa over MTb transcript, was found to be 0.06 for control and 0.4 for Cd-treated icefish. By combining the amounts of total MT mRNA obtained from Northern blot with this cpm ratio, the MTa and MTb mRNA levels could be algebraically calculated. The results summarized in Figure 4 show that the amount of the MT mRNA in control icefish is about 50% of the amount present in the red-blooded fish, with a 5-fold increase in MT mRNA content in Cd-treated icefish with respect to untreated icefish. However, according to our calculations, cadmium treatment brings about a 27-fold increase in the MTa transcript against a 4-fold increase only in the MTb transcript.

Conclusions

Our results show that in untreated icefish there is a discrepancy between MT and MT mRNA amounts: indeed, the MT content is only 5% of the amount found in *T. bernacchii*, whereas the MT mRNA is approximately 50% of that in the red-blooded species. This indicates the presence of a large pool of untranslated MT mRNA (mostly MTb) in icefish liver.

A marked increase in MT occurs in Cd-treated icefish: under these conditions, the expression of the gene encoding the MTa isoform is preferentially induced. Differential mechanisms of regulation of MT isoforms have been reported in a number of systems [8–10]. However, in Cd-treated icefish the two MT isoforms are synthesized in almost equimolar amounts only after cadmium induction. These results suggest that as in other systems [11, 12] icefish MT synthesis can be regulated by heavy metals at both transcriptional and post-transcriptional levels. Such a conclusion is strenghtened by the differences found in the leader regions of the two transcripts encoding MTa and MTb: in fact, the 5'-UTR of the MTb transcript, which is the most abundant in control fish, carries hairpin loops, and it has been reported that hairpin loops may affect mRNA translation [13]. A well known example of translational control exists for ferritin mRNA [14]. We propose that MTb production is post-transcriptionally regulated according to the model similar to that described for ferritin production. We postulate that MT mRNA translation is inhibited by the interaction of a metal-binding protein with a hairpin loop formed in the 5'-UTR of the MTb mRNA. When the intracellular concentration of metal increases, the inhibitory protein is dislodged from the loop following interaction with the metal, thus making mRNA translation possible.

Acknowledgments
This research is in the framework of the Italian National Programme for Antarctic Research (PNRA). One of us (R. S.) attended the 1996 S-301 Biology Course held at McMurdo Station (USA) in Antarctica, and was supported by a student fellowship awarded by the US National Science Foundation.

References

1   Barker PF, Burrell J (1977) The opening of the Drake Passage. *Mar Geol* 25: 15–34.
2   Kennett JP (1977) Cenozoic evolution of Antarctic glaciation, the circum-Antarctic ocean and their imapct on global paleoceanography. *J Geophys Res* 82: 3843–3876.
3   Ruud JT (1954) Vertebrates without erithrocytes and blood pigment. *Nature* 173: 848–850.
4   Scheuhammer AM, Cherian MG (1991) Quantification of metallothionein by silver saturation. *Methods Enzymol* 205: 78–83.
5   Scudiero R, De Prisco PP, Camardella L, D'Avino R, di Prisco G, Parisi E (1992) Apparent deficiency of met-allothionein in the liver of the Antarctic icefish *Chionodraco hamatus*. Identification and isolation of a zinc-containing protein unlike metallothionein. *Comp Biochem Physiol* 103B: 201–207.
6   Scudiero R, Carginale V, Riggio M, Capasso C, Capasso A, Kille P, di Prisco G, Parisi E (1997) Difference in hepatic metallothionein content in Antarctic red-blooded and haemoglobinless fish: undetectable metal-lothionein levels in haemoglobinless fish is accompained by accumulation of untranslated metallothionein mRNA. *Biochem J* 322: 207–211.
7   Scudiero R, Capasso C, carginale V, Riggio M, Capasso A, Ciaramella M, Filosa S, Parisi E (1997) PCR amplification and cloning of metallothionein complementary DNAs in temperate and Antarctic sea urchin characterized by a large difference in egg metallothionein content. *Cell Mol Life Sci* 53: 472–477.
8   Lehman-McKeeman LD, Andrews GK, Klaassen CD (1988) Mechanisms of regulation of rat hepatic metal-lothionein-I and metallothionein-II levels following administration of zinc. *Toxicol Appl Pharmacol* 92: 1–9.
9   Shworak NW, O'Connor T, Wong NC, Gedamu L (1993) Distinct TATA motifs regulate differential expres-sion of human metallothionein I genes MT-IF and MT-IG. *J Biol Chem* 268: 24460–24466.
10 Mididoddi S, McGuirt JP, Sens MA, Todd JH, Sens DA (1996) Isoform- specific expression of metalloth-ionein mRNA in the developing and adult human kidney. *Toxicol Lett* 85: 17–27.
11 McCormick CC, Salati LM, Goodridge AG (1991) Abundance of hepatic metallothionein mRNA is increased by protein-synthesis inhibitors. Evidence of transcriptional activation and post-transcriptional regulation. *Biochem J* 273: 185–188.
12 Vasconcelos MH, Tam SC, Beattie JH, Hesketh JE (1996) Evidence for differences in the post-transcription-al regulation of rat metallothionein isoforms. *Biochem J* 315: 665–671.
13 Kozak M (1994) Features in the 5' non-coding sequences of rabbit α and β-globin mRNAs that affect trans-lational efficiency. *J Mol Biol* 235: 95–110.
14. Theil EC (1994) Iron regulatory elements (IREs): a family of mRNA non-coding sequences. *Biochem J* 304: 1–11.

Metallothionein IV
C. Klaassen (ed.)
© 1999 Birkhäuser Verlag Basel/Switzerland

Structure and function of metallothionein isoforms in terrestrial snails

Reinhard Dallinger[1], Burkhard Berger[1], Peter Hunziker[2] and Jeremias H.R. Kägi[2]

[1]*Institut für Zoologie u. Limnologie der Universität Innsbruck, Technikerstraße 25, A-6020 Innsbruck, Austria*
[2]*Biochemisches Institut der Universität Zürich, Winterthurerstraße, Bau 44, CH-8057 Switzerland*

Introduction

Terrestrial gastropods have long been known to avidly accumulate trace elements such as copper, zinc and cadmium in their soft tissues [1,2]. It has been suggested that this extraordinary accumulation potential of terrestrial snails may be related to their being forced to avoid any substantial loss of water which might occur upon excretion of excessive amounts of trace elements [3]. As a probable consequence of such physiological constraints, land snails have developed efficient mechanisms of retaining and detoxifying trace elements within their body. Consequently, some snail and slug species belong to those animals exhibiting the highest concentration factors for cadmium ever recorded among soil invertebrates, which lead to the discovery of metallothioneins (MTs) in these organisms [3–6].

Organ-specific metal accumulation in terrestrial gastropods

Several studies have revealed that the major storage site for some important essential and nonessential trace elements in the snail's body is the midgut gland [1–3]. In Roman snails (*Helix pomatia* L.) from metal-contaminated areas, for instance, cadmium concentrations in the midgut gland can amount to more than 100 µg/g (dry weight) [7], and metal-exposed snails in the laboratory can reach even higher concentrations, with 85–90% of total cadmium body burden accumulating in this organ [2]. The essential trace element zinc also predominantly accumulates in the snail midgut gland, which can contain up to 70% of the animal's body burden of this metal [2]. Once the two metals have entered the midgut gland, cadmium and zinc are retained within the tissue over extended periods of time. However, the two metals seem to follow different cellular pathways: Whereas zinc is, at least in part, precipitated within calcium granules of basophilic cells [3], cadmium is mainly associated with a cytosolic protein which has eventually been identified as an MT [8]. Variants of this MT isoform were also characterized from the midgut gland of a related species, *Arianta arbustorum* [9]. In contrast to cadmium and zinc, copper is predominantly accumulated in the snail's foot and mantle. Within the mantle tissue, a considerable proportion of the metal is associated with an MT isoform whose primary structure significantly differs from the one detected in the midgut gland [10].

Structural properties of snail MTs

A comparison (Fig. 1) reveals that snail MTs are small molecules with 64 to 66 non-aromatic amino acid residues and molecular weights between 6.2 (mantle isoform of *Helix pomatia*) and 6.6 K (midgut gland isoforms). All of them have N-terminal acetylated serines. MT variants differing only in acetylation of their N-terminal amino acids have been reported from oyster MTs, and it has been shown that acetylated variants exhibit a slightly different time pattern of induction compared to the non-acetylated forms [11]. In contrast to mammalian MTs, snail iso-forms contain less Cys residues (18 versus 20 in mammalian MTs) [12] which are typically arranged in several Cys-X-Cys motifs (with "X" denoting any amino acid except Cys). Moreover, it can be seen that mantle and midgut gland isoforms from the same species (*Helix pomatia*) differ much more from each other than midgut gland variants between *Helix pomatia* and *Arianta arbustorum*. (Fig. 1). In fact, midgut gland and mantle isoforms display a recip-rocal homology of only 58% in their primary structure, including Cys residues. Interestingly, several amino acid positions have been replaced by asparagine residues in the mantle isoform when compared to the midgut gland MT [10]. While the significance of these replacements still remains unknown, the high degree of divergence in the primary structure between the two iso-forms already suggests significant differences in their functions [13].

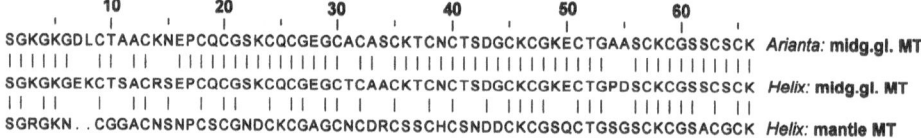

Figure 1. Primary structures of terrestrial snail Mts, showing amino acid sequences (without N-terminal acetyla-tion) from midgut gland (midg.gl.) of *Arianta arbustorum* (*Arianta*) and *Helix pomatia* (*Helix*), and from mantle of *Helix pomatia*. Sequences were aligned by the program Clustal W. [10]. Gaps are shown by dots and identical amino acids are marked by vertical lines.

So far little is known about the tertiary structure of snail MTs. Circular dichroism (CD) spec-troscopic studies prove, however, that the spatial organization of terrestrial gastropod MTs must be similar to that found in mammalian MTs. This is especially indicated by the biphasic signal obtained from the midgut gland (Cd)-holo-MT of *Helix pomatia* upon CD spectroscopy (Fig. 2), displaying two positive ellipticity lobes at 220 and 256 nm, and one negative lobe at 242 nm. After removal of cadmium by acidification, the characteristic CD signal vanishes (Fig. 2), but can be restored after reconstitution of the protein *in vitro*. The biphasic ellipticity signal of snail (Cd)-MT strongly resembles CD spectra displayed by mammalian MTs [12], and can be interpreted as a manifestation of the molecule's cadmium thiolate cluster structure [14]. It thus suggests the existence of two protein domains in snail MTs with each bearing one such cluster [14].

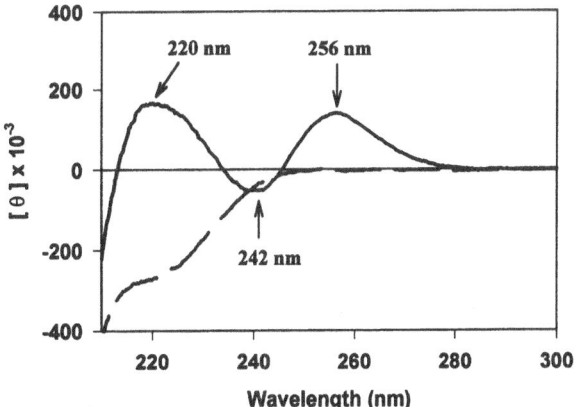

Figure 2. Circular Dichroism (CD) spectrum of native (Cd)-holo-MT from midgut gland of *Helix pomatia* (bold line) over a wavelength range from 210 to 300 nm with positive and negative ellipticity bands at 220, 256 and 242 nm (arrows), and CD spectrum of the Cd-free apo-MT (hatched line) obtained after acidification of the holo-MT with HCl.

Metal-specific functions of midgut gland and mantle isoforms

The midgut gland and mantle MT isoforms of *Helix pomatia* do not only differ in their primary structure, but also show clear preferences for different metals *in vivo*.

The native Roman snail midgut gland MT isoform, for instance, nearly exclusively contains cadmium and only traces of zinc and copper, with a molar metal ratio of Cd: Zn: Cu = 100: 6.6:2 [13]. A similarly high affinity to cadmium ions has also been assessed for MT variants isolated from the midgut gland of *Arianta arbustorum* [9]. Spectroscopic titration studies have revealed that the cadmium-binding isoform from *Helix pomatia* displays a metal stoichiometry of six cadmium ions per protein molecule, thus clearly differing from the known 1: 7 stoichiometry of mammalian MTs [12].

On the other hand, the native mantle isoform from the same species predominantly binds copper, exhibiting a molar metal ratio of Cu: Zn: Cd = 100: 6:1 [13]. Its stoichiometry has approximately been determined with six equivalents of copper ions per mol of protein [10].

The preponderant association of snail midgut gland MT with cadmium *in vivo*, as well as the increase of the protein's concentration with rising cadmium levels in the tissue with a constant molar ratio of 1: 6 (Fig. 3) proves that all the cadmium entering the organ is readily chelated by the increasing amounts of the newly synthesized MT isoform. It is therefore suggested that the main function of this protein is devoted to the detoxification of cadmium [13]. In fact, it was shown earlier that induction of synthesis of the cadmium-binding isoform in the snail midgut gland can already occur within a few hours after exposure of animals to increased metal concentrations in their diet [15], a circumstance which might be expected to be true if the protein really had to serve detoxification. Finally, the detoxification hypothesis is also supported by the fact that snails are able to survive cadmium loads in their tissues up to very high con-

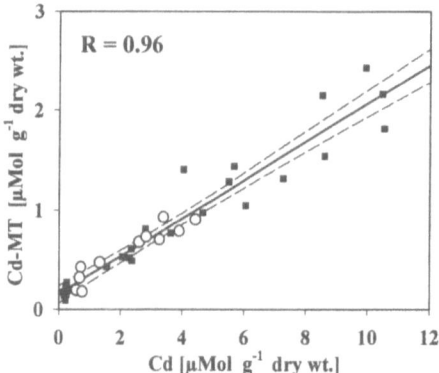

Figure 3. Linear relationship (bold line) with 95% confidence limits (hatched lines) between molar concentrations of midgut gland cadmium and midgut gland (Cd)-MT, both referred to tissue dry weight. Bold squares represent values obtained from animals fed on a cadmium-rich diet (varying concentrations) in the laboratory, whereas open circles refer to values from snails colleted at metal-polluted sites in the field. The regression coefficient (R) is shown as inset.

centrations [16]. In general, cadmium detoxification is one of the basic functions ascribed to MTs, and experimental evidence for this has been proven in cadmium-exposed fruit flies [17], and transgenic cell cultures and mice [18, 19].

In contrast to the midgut gland cadmium-binding MT, the copper-binding isoform from the mantle of *Helix pomatia* has no detoxifying function at all. In fact, when Roman snails are fed on a cadmium-rich diet, significant proportions of cadmium are diverted into the mantle tissue. However, virtually none of the metal becomes bound to the mantle MT isoform whose concentration and copper content remain unaffected by the metal insult [13] (Fig. 4). Instead, the

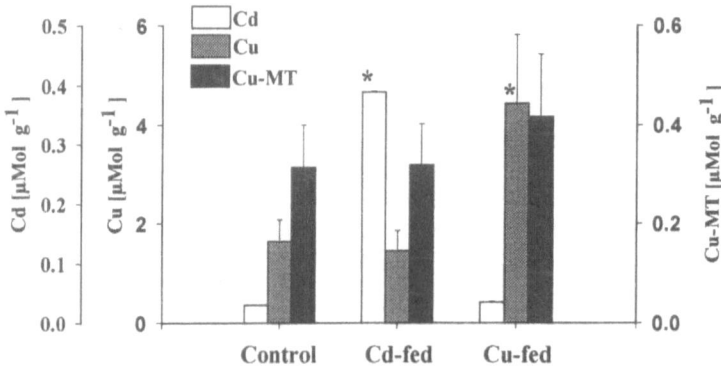

Figure 4. Molar concentrations of Cd, Cu, and (Cu)-MT (bars), referred to tissue dry weight, in the mantle of *Helix pomatia* (n = 7), with control snails (Control) and individuals fed on a cadmium-rich (Cd-fed) and a copper-rich (Cu-fed) diet. Asterisks above bars indicate significant differences (P < 0.01) from controls (Student's t-test).

absorbed cadmium is readily re-distributed from the mantle to the midgut gland where the metal becomes chelated by the midgut gland MT isoform. In a similar manner, copper which enters the mantle tissue after feeding does not interfere with the existent (Cu)-MT isoform [13]. It is therefore concluded that the copper-binding mantle isoform of Roman snails cannot be involved in any kind of detoxification of excessive amounts of cadmium or copper. Alternatively, it has to be assumed that the protein would have to serve an essential function in favor of copper metabolism. One possible task in this respect might be the interference with hemocyanin, a respiratory protein which is responsible for the snail's oxygen supply. In fact, when mantle homogenates from Roman snails are fractionated by gel chromatography, a separate, high-molecular weight, copper-binding component can always be detected besides the known (Cu)-MT isoform, and it is assumed that this component represents hemocyanin [4]. This molecule has been reported to be synthesized in the so-called pore cells of the snail's mantle [20]. One possible function which MT could perform in this respect, is copper donation during hemocyanin synthesis. Such a task has also been suggested for (Cu)-MT in marine decapods [21].

Conclusion

Until now the performance of different functions of MT in metal metabolism have been explained by the domain-specific preference of the MT molecule towards different metal species, making the protein able to serve several metal-specific functions simultaneously [22]. The existence of distinct MT isoforms devoted to cadmium detoxification and copper metabolism in terrestrial snails allows the animals to perform copper-specific metabolic tasks even under conditions of environmental cadmium insult [13], thus providing a valuable alternative model for explaining multifunctionality of these proteins.

References

1. Coughtrey PJ, Martin MH (1976) The distribution and speciation of cadmium in the terrestrial snail, *Helix aspersa*. *Oecologia* 23: 315–322.
2. Dallinger R, Wieser W (1984) Patterns of accumulation, distribution and liberation of Zn, Cu, Cd and Pb in different organs of the land snail *Helix pomatia*. L. *Comp Biochem Physiol* 79C(1): 117–124.
3. Dallinger R (1993) Strategies of metal detoxification in terrestrial invertebrates. *In*: R Dallinger, PS Rainbow (eds): Ecotoxicology of metals in invertebrates. Lewis Publishers, Boca Raton, pp. 245–289.
4. Dallinger R (1996) Metallothionein research in terrestrial invertebrates: synopsis and perspectives. *Comp Biochem Physiol* 113C (2), 125–133.
5. Cook M, Jackson A, Nicless G, Roberts DJ (1979) Distribution and speciation of cadmium in the terrestrial snail, *Helix aspersa*. *Bull Environ Contam Toxicol* 23: 445–451.
6. Dallinger R, Wieser W (1984) Molecular fractionation of Zn, Cu, Cd and Pb in the midgut gland of *Helix pomatia* L. *Comp Biochem Physiol* 79C (1), 125–129.
7. Berger B, Dallinger R (1993) Terrestrial snails as quantitative indicators of environmental metal pollution. *Environ Monit Assess* 25: 65–84.
8. Dallinger R, Berger B, Hunziker PE, Birchler N, Hauer CR, Kägi JHR (1993) Purification and primary structure of snail metallothionein – Similarity of the N-terminal sequence with histones H4 and H2A. *Eur J Biochem* 216: 739–746.
9. Berger B, Hunziker PE, Hauer CR, Birchler N, Dallinger R (1995) Mass spectrometry and amino acid sequencing of two cadmium-binding metallothionein isoforms from the terrestrial gastropod *Arianta arbustorum*. *Biochem J* 311: 951–957.
10. Berger B, Dallinger R, Gehrig P, Hunziker PE (1997) Primary structure of a copper-binding metallothionein

from mantle tissue of the terrestrial gastropod *Helix pomatia. Biochem J*; *in press*

11. Roesijadi G, Vestling MM, Murphy CM, Klerks PL, Fenselau CC (1991) Structure and time-dependent behavior of acetylated and non-acetylated forms of a molluscan metallothionein. *Biochim Biophys Acta* 1074: 230–236.

12. Kägi JHR, Schäffer A (1988) Biochemistry of metallothionein. Biochem. 27: 8509–8515.

13. Dallinger R, Berger B, Hunziker PE, Kägi JHR (1997) Metallothionein in snail Cd and Cu metabolism. *Nature* 388: 237–238.

14. Willner H, Vašák M, Kägi JHR (1987) Cadmium-thiolate clusters in metallothionein: spectrophotometric and spectropolarimetric features. *Biochemistry* 26: 6287–6292.

15. Berger B, Dallinger R, Thomaser A (1995) Quantification of metallothionein as a biomarker for cadmium exposure in terrestrial gastropods. *Environ Toxicol Chem* 14: 781–791.

16. Berger B, Dallinger R, Felder E, Moser J (1993) Budgeting the flow of cadmium and zinc through the terrestrial gastropod *Helix pomatia* L. *In*: R Dallinger, PS Rainbow (eds): Ecotoxicology of metals in invertebrates. Lewis Publishers, Boca Raton, pp. 291–313.

17. Maroni G, Wise J, Young JE, Otto E (1987) Metallothionein gene duplications and metal tolerance in natural populations of *Drosophila melanogaster. Genetics* 117: 739–744.

18. Lee DK, Fu K, Linag LC, Dalton T, Palmiter RD, Andrews GK (1996) Transgenic mouse blastocysts that overexpress metallothionein-I resist cadmium toxicity *in vitro. Mol Reprod Dev* 43 (2), 158–166.

19. Liu YP, Liu J, Iszard MB, Andrews GK, Palmiter RD, Klaassen CD (1995) Transgenic mice that overexpress metallothionein-I are protected from cadmium lethality and hepatotoxicity. *Toxicol Appl Pharmacol.* 135 (2), 222–228.

20. Sminia T, Vlugt van Daalen JE (1977) Haemocyanin synthesis in pore cells of the terrestrial snail *Helix aspersa. Cell Tissue Res* 183: 299–301.

21. Brouwer M, Whaling P, Engel DW (1986) Copper-metallothioneins in the American lobster, *Homarus americanus*: Potential role as Cu(I)-donors to apohemocyanin. *Environ Health Perspect* 65: 93–100.

22. Li H, Otvos JD (1996) Cd[III] NMR-studies of the domain specificity of Ag[+] and Cu[+] binding to metallothionein. *Biochemistry* 35: 13929–13936.

Metallothionein IV
C. Klaassen (ed.)
© 1999 Birkhäuser Verlag Basel/Switzerland

Functional and regulatory aspects of teleost metallothionein

P.-E. Olsson[1], M. Gerpe[1, 2] and P. Kling[1]

[1]Department of Cellular and Developmental Biology, Umea University, S-901 87 Umea, Sweden
[2]CONICET and Marine Sciences Department, FCEyN, University of Mar del Plata, Funes 3350, (7600) Mar del Plata, Argentina

Summary. The regulation of metallothionein (MT) has been studied in different teleost species in order to determine MTs involvement in metal-regulation, free radical scavenging, as well as, in reproduction and development. The primary sequences of MTs have been found to be highly conserved among teleosts. The regulatory regions of three species studied have shown the presence of multiple functional metal-regulatory elements. In the rainbow trout MT-A gene, several functional AP-I elements have also been found. In several studies, we have shown that MT is regulated by both metals and free radicals and that induction of MT reduces the toxic effects caused by both metals and free radicals. MT has been shown to be regulated during sexual maturation and early embryonic development. While the regulation of MT during sexual maturation appears to be coupled to zinc utilisation, the function of MT during early embryonic development remains unclear. Further work is being carried out to determine the regulatory pathways for MT during fish development.

Introduction

With more than 20 000 different species, fish are the most numerous vertebrates. The teleosts, bony fish, constitute the largest group of fish and have adapted over many hundred million years to a multitude of different habitats. This has resulted in a high degree of physiological variation among teleosts.

Fish live in highly variable environments and in many cases must be able to withstand drastic changes in temperature, oxygen and salinity. This requires adaptation of regulatory systems, including the control of trace metal metabolism. Besides natural fluctuations, the aquatic environment is being polluted with toxic heavy metals as a consequence of industrialisation. In this context, MT has been of particular interest as it is the main cellular heavy metal binding protein. MT has been cloned from a number of different teleost species [1] and so far, all studied species express MT.

Our work has been aimed at determining the structure, regulation and function of MT in teleosts. Of primary interest has been the conservation of metal-regulation of MT among teleosts, regulation of MT during reproduction and development, and the possible involvement of MT in scavenging free radicals.

Primary structure of MT in fish

The primary sequence of MT is highly conserved among vertebrates, and it is the presence and spatial conservation of cysteines in MT that dictate its metal-binding capacity. Therefore, to have a conserved function in metal homeostasis, it is important that the cysteines are conserved among teleosts. To investigate this we have cloned and sequenced MT from a number of teleosts, inhabiting a variety of environments. These studies have revealed that the cysteine residues are perfectly conserved in number and position in all studied teleosts, except for pupfish (*Cyprinodon nevadensis*) MT where an extra cysteine has replaced a serine at position 27 [1]. The pupfish is interesting in that it inhabits desert streams and is able to withstand high fluctuations in daily temperature (from up to 40 °C during the day down to 5 °C at night). The effect of this extra cysteine on the metal-binding capacity and specificity is presently being studied. However, the conservation of cysteines among teleost species indicate that the metal-binding characteristics of most teleost MTs are highly similar.

Regulation of MT gene expression

While the proteins sequence determines the metal-binding capacity of MT, the regulation of the different MT genes is also important for understanding the conservation of MT functionality. The regulatory regions of MT genes have been characterised from rainbow trout (*Oncorhynchus mykiss* [2–4]), sockeye salmon (*Oncorhynchus nerka* [5]), pike (*Esox lucius* [6]), stone loach (*Noemachelius barbatulus* [6]) and carp (*Cyprinus carpio*; K.M. Chang unpublished). All these gene promoters contain functionally active metal-response elements (MREs). In the teleost MT promoters there are generally 2–3 MREs located within 300 bp of the TATA box. Additional MREs are found further upstream, in the distal region of the promoters. We have analysed the functionality of the rainbow trout MT-A, pike MT and stone loach MT promoters and shown that both the distally located and proximal MREs are functional and required for maximum activity [4, 7].

In the rainbow trout MT-A promoter there are four distally located AP-I elements and one NF-IL6 element. AP-I elements are also found in both the pike and the stone loach MT promoters. In transfection experiments we have shown that these distal AP-I elements mediate free radical response of the gene [4]. So far we have not been able to achieve activation through the NF-IL6 element (unpublished observations). Although we have shown the functionality of the AP-I elements in the rainbow trout MT-A gene, the significance of this regulatory pathway for fish MT remains to be determined.

The MT levels in different fish tissues are regulated by a number of substances for which no cis-acting elements have been observed in the genes. Thus, glucocorticoids, progesterone and noradrenalin have been shown to elevate the MT levels in cell lines [8]. Estradiol-17β (E2) blocks metal induction of MT [9] and this is likely due to competition for divalent trace metals. It is not known by which secondary mechanisms the other compounds influence MT gene transcription and translation.

MT and free radical regulation

The discovery of multiple AP-I elements in the rainbow trout MT-A gene demonstrated that free radicals may directly regulate MT by changing the cellular redox state. Transfection experiments have shown the functionality of the AP-I elements and studies using RTG-2 cells have shown upregulation of endogenous MT gene expression by free radicals [10]. In order to determine the involvement of MT in cellular protection against free radical challenge, we have compared the response of RTG-2 cells to that of CHSE-214 cells, following combined metal and hydrogen peroxide exposure. The CHSE-214 cell line contains a heterogeneous population of cells, of which a minor subpopulation expresses MT (unpublished observations). However, since most of the cells do not express MT it remains possible to utilise this cell line as a negative control when studying the involvement of MT in cell survival experiments. Exposure of RTG-2 cells to Cd or Zn results in raised cell survival following challenge with hydrogen peroxide [10]. The same treatment of CHSE-214 cells does not provide protection against free radical challenge (Fig. 1). The equal efficiency of Cd and Zn in protecting the RTG-2 cells and the absence of effect on the CHSE-214 cells, indicates that MT is the major factor responsible for the difference between the two cell lines and that GSH does not significantly contribute to the observed effects. These studies show that MT is regulated by, and involved in the cellular protection against free radicals. It remains to be shown that MT is regulated by free radicals *in vivo*. The presence of multiple functional AP-I elements appears to be unique for teleost MT, but the functional significance of this regulatory pathway has yet to be determined.

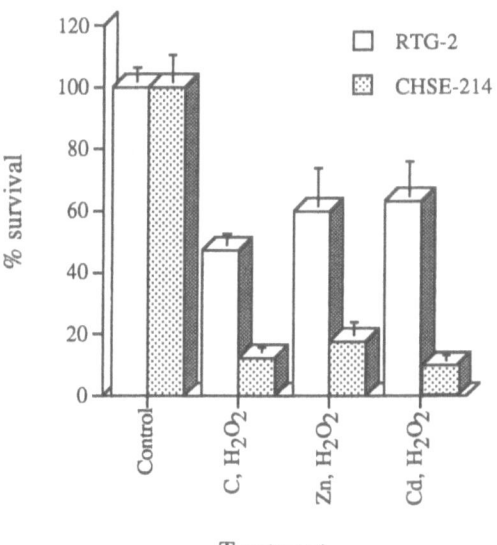

Figure 1. Effects of zinc and cadmium on hydrogen peroxide induced cell mortality. An MT expressing cell line, RTG-2, was compared to a cell line were the MT genes are methylated, CHSE-214, and therefore not expressed. The cells were treated with 10 μM Cd or 150 μM Zn for 96 h prior to challenge with 150 μM H_2O_2 for 24 h. The significantly higher % cell survival in RTG-2 than in CHSE-214 indicates that MT is involved in free radical scavenging

MT and metal-regulation

As mentioned above, teleost MTs contain both proximally and distally located functional MREs (Fig. 2A; [7]). The investigation into the functionality of MT promoters from both the cadmium sensitive rainbow trout and cadmium tolerant species (pike and stone loach) has not indicated any major differences in function (Fig. 2B; [7]). Although MT regulation may contribute to the metal sensitivity of a species, a number of other factors may also be involved. Thus, the tissue distribution of metals, as well as, the inducibility of MT in different tissues may contribute to the relative sensitivity of a teleost to heavy-metal exposure. We have studied the distribution and inducibility of MT in different tissues of arctic char (*Salvelinus alpinus* [11]). Analysis of the general distribution of MT mRNA in juvenile arctic char showed that the highest basal levels were found in the liver and gonads. Intermediate levels of MT mRNA were observed in kidney, brain, heart, eyes and blood cells while the stomach, small intestine and large intestine contained low levels of MT mRNA. High levels of MT mRNA in immature

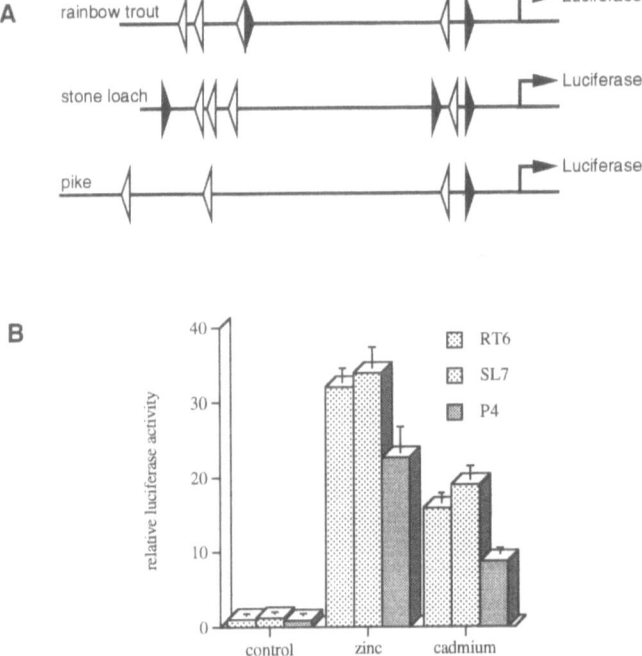

Figure 2. Comparison of MREs in the promoters of rainbow trout MT-A (RT6), stone loach MT (SL7) and pike MT (P4). (A) Localisation and orientation of MREs in the three MT gene promoters. (B) Comparison of the zinc (150 μM) and cadmium (10 μM) inducibility of the full length promoters. The MT promoters (SL7, P4) from the Cd tolerant species were not more inducible than the rainbow trout MT promoter. On the contrary the pike promoter is less sensitive to metals, due to a lower number of MREs, then the RT6 promoter, thus indicating that MT is not primarily responsible for metal tolerance.

gonads is consistent with previous studies [12]. However, the MT mRNA levels did not reflect the levels of MT in these tissues. While the MT mRNA levels were low in the stomach and intestine, the MT levels in these organs were the highest of all studied tissues (unpublished observation). The reason for these differences in MT and MT mRNA levels are unknown but may reflect differences in metal metabolism in the different tissues.

MT and hormonal regulation

While MT has been shown to be involved in trace metal homeostasis, very little is known about its role in tissue specific metal regulation in teleosts. In the liver, we have shown that the metal-inducibility of MT is inhibited as a response to elevated E2 levels [9]. Both the hepatic MT mRNA and MT levels are low during the period of vitellogenin synthesis in female fish [13]. Once the period of vitellogenesis is over, there is a redistribution of zinc, an induction of MT mRNA and a concomitant increase in MT levels. Following E2 injection of rainbow trout the vitellogenin mRNA levels increase drastically within 24 h [14]. The MT mRNA levels do not increase until approximately 14 days post E2 injection, when the liver ceases to produce vitel-logenin, the vitellogenin mRNA levels drop, and there is a redistribution of zinc from high molecular weight proteins to MT. These studies show that E2 is not a primary inducer of MT and indicate that the increase in MT at the end of vitellogenesis is induced by an increased concentration of free zinc.

It has been shown in several studies that fish are more sensitive to metal exposure during the period of vitellogenin production [13, 15]. We have shown that this is due to indirect inhibition of MT synthesis by E2 in combination with inhibition of vitellogenesis by Cd. We suggest that the effect of E2 is due to an increased need for zinc as co-factor in the transcriptional and translational machinery needed for vitellogenin production and that this results in sequestration of zinc from MT, a reduction in free metal and thereby down regulation of MT mRNA synthesis. When Cd enters the system during vitellogenesis it is sequestered by zinc requiring systems, probably disrupting their functionality. Since MT is already down regulated it can not bind Cd and the fish is rendered more sensitive to metal exposure.

We have shown that the inhibitory effect of E2 is tissue specific [16]. While Cd fails to induce hepatic MT during an E2 challenge, it is still capable of inducing renal MT mRNA transcription. When injected into rainbow trout, Cd normally accumulates in the liver and bone [16]. However, following E2 treatment, injected Cd is relocated from bone and liver to the blood and subsequently to gill, gut and muscle. E2 appear to decrease bone mineralization as indicated by the decrease in Cd accumulation in bone following E2 exposure. This implies that induction of vitellogenesis by E2 decreases the growth rate in order to redirect the animals metabolic activities toward reproduction. Thus the excess Cd, as a result of decreased bone formation, accumulate in neighbouring organs. Further research is needed to determine the importance of this redistribution for the increase in Cd susceptibility. In particular, the redistribution to the gills may prove to be harmful since Cd will interfere with Ca transport across the gill epithelium.

MT and environmental pollution

One of the major roles for MT is protection against toxic heavy metals, such as Cd and Hg. In light of the effects of E2 on MT regulation it appears that this protective function of MT may be diminished by exposure to environmental chemicals with estrogenic activity. To test this we injected 5 mM of E2 or three different PCBs (PCB#190, PCB#104 and 2',4',6'-trichloro-3-biphenyl (OH-PCB)) into juvenile arctic char followed by injection of 0.2 mg Cd two days later. On day three the fish were sacrificed and the liver, kidney and brain MT mRNA levels were determined. The OH-PCB is capable of inducing estrogen receptor transcription and is thus a functional environmental estrogen. The PCB#104 may upon hydroxylation in the body be metabolised into an environmental estrogen while PCB#190 does not have such a potential. Of the tested PCBs, only OH-PCB and PCB#104 were found to inhibit Cd induced MT mRNA transcriptional activation, although to a lesser extent than E2 (Fig. 3). This indicates that environmental estrogens, such as OH-PCB have the potential to disrupt metal metabolism and thereby cause increased metal-sensitivity in fish. Besides the implications for metal toxicity, this also indicates that MT may serve as a marker of deleterious effects caused by environmental estrogens in fish.

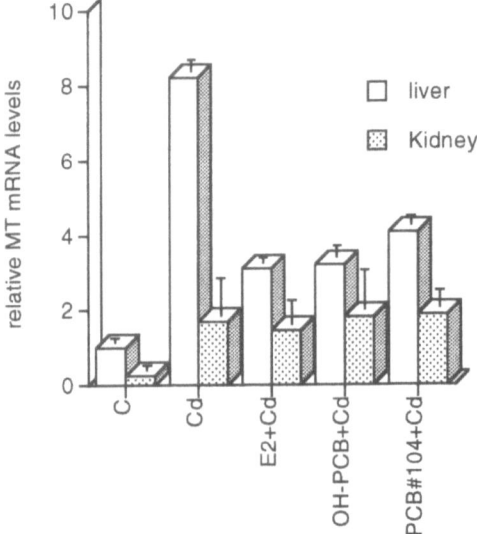

Figure 3. MT mRNA levels in arctic char liver and kidney following injection of Cd (0.2 mg/kg) alone or in combination with E2, OH-PCB or PCB#104 (1 µmol/kg each). The liver MT mRNA level in control s was arbitrarily set to 1.0. Injection of Cd resulted in approximately 8 fold inductin in both organs. Coinjection of E2, OH-PCB or PCB#104 led to reduced MT mRNA levels in liver but not in kidney, demonstrating the organ specific effect of estrogenic compounds on MT transcription.

Summary

The role of MT in homeostatic control of metal metabolism remains one of the main functions of this protein. However, recent advances suggest that it may also be important in the regulation and scavenging of free radicals. These functions, in combination with the hormonal effects on MT, demonstrate the complexity of MT regulation. While metal-regulation appears to govern the effects observed in response to E2 it still remains to be determined how MT is regulated during embryonic development. There is ample data on the function of MT during female sexual maturation in fish, but much less is known on the role that MT plays in male fish, or during early embryonic development. The role of free radical regulation of MT in fish remains unknown and will be a challenging area for future research.

Acknowledgements
These studies were supported financially by grants from the Swedish Natural Science Research Council and the Swedish Environmental Protection Agency.

References

1. Morris CA, Stürzenbaum S, Nicolaus B, Morgan J, Harwood JL and Kille P (1998) Identification and characterization of metallothioneins from envrionment, indicators as potential biomonitors; *this volume.*
2. Zafarullah M, Bonham K and Gedamu L (1988) Structure of the rainbow trout metallothionein B gene and characterization of its metal-responsive region. *Mol Cell Biol* 8: 4469–76.
3. Murphy MF, Collier P, Koutz P and Howard B (1990) Nucleotide sequence of the trout metallothionein A gene 5'-flanking region. *Nucl Acid Res* 18: 4622.
4. Olsson P-E, Kling P, Erkell LJ and Kille P (1995) Structural and functional analysis of the rainbow trout (*Oncorhynchus mykiss*) metallothionein-A gene. *Eur J Biochem* 230: 344–49.
5. Chan WK and Devlin RH (1993) Polymerase chain reaction amplification and functional characterization of sockeye salmon histone H3, metallothionein-B, and protamin promoters. *Mol Mar Biol Biotechnol* 2: 308–18.
6. Kille P, Kay J and Sweeney GE (1993) Analysis of regulatory elements flanking metallothionein genes in Cd-tolerant fish (pike and stone loach). *Biochim Biophys Acta* 1216: 55–64.
7. Olsson P-E and Kille P (1997) Functional comparison of the metal-regulated transcriptional control regions of metallothionein genes from cadmium-sensitive and tolerant fish species. *Biochim Biophys Acta* 1350: 325–334.
8. George SG and Olsson P-E (1994) Metallothionein as indicators of trace metal pollution. *In*: KJM Kramer (ed.): *Biomonitoring of Coastal Waters and Estuaries*. CRC Press Inc, pp. 151–178.
9. Olsson P-E Kling P, Petterson C and Silversand C (1995) Interaction of cadmium and oestradiol-17b on metallothionein and vitellogenin synthesis in rainbow trout (*Oncorhynchus mykiss*). *Biochem J* 307: 197–203.
10. Kling P, Erkell LJ, Kille P and Olsson P-E (1996) Metallothionein induction in rainbow trout gonadal (RTG-2) cells during free radical exposure. *Mar Environ Res* 42: 33–36.
11. Gerpe M, Kling P and Olsson P-E (1997) Metallothionein gene expression in arctic char (*Salvelinus alpinus*) following metal and PCB exposure. *Mar Environ Res*; *in press.*
12. Olsson P-E, Zafarullah M, Foster R, Hamor T and Gedamu L (1990) Developmental regulation of metallothionein mRNA, zinc and copper levels in rainbow trout, *Salmo gairdneri. Eur J Biochem* 193: 229–235.
13. Olsson P-E, Haux C and Forlin L (1987) Variations in hepatic metallothionein, zinc and copper levels during an annual reproductive cycle in rainbow trout, *Salmo gairdneri. Fish Physiol Biochem* 3: 39–47.
14. Olsson P-E, Zafarullah M and Gedamu L (1989) A role of metallothionein in zinc regulation after oestradiol induction of vitellogenin synthesis in rainbow trout, *Salmo gairdneri. Biochem J* 257: 555–559.
15. Povlsen AF, Korsgaard B and Bjerregaard P (1990) The effect of cadmium on vitellogenin metabolism in estradiol-induced flounder (*Platichtys flesys*, L.) males and females. *Aquat Toxicol* 17: 253–263.
16. Valencia R, Gerpe M, Trimmer J, Buckman T, Mason AZ and Olsson P-E (1997) The effect of estrogen on cadmium distribution in rainbow trout (*Oncorhynchus mykiss*). *Mar Environ Res*; *in press.*

Metallothionein IV
C. Klaassen (ed.)
© 1999 Birkhäuser Verlag Basel/Switzerland

Metallothionein induction in mussels exposed to a metal mixture

M.J. Bebianno[1] and W.J. Langston[2]

[1]U.C.T.R.A.-University of Algarve, Campus de Gambelas, 8 000 Faro, Portugal
[2]Plymouth Marine Laboratory – Citadel Hill, Plymouth PL1, 2PB, UK

Summary. The uptake of metals such as Cd, Cu and Zn by mussels and its accumulation in their tissues is related to the binding of these metal ions to specific ligands where the metals have higher affinity. One of these ligands are the metallothioneins.
Metallothionein induction was followed in the whole soft tissues of *Mytilus edulis* after exposure to a metal mixture of Cd (100 µg·l^{-1}), Cu (50 µg·l^{-1}) and Zn (50 µg·l^{-1}). When the mussels were exposed to the same metal concentrations individually Cd and Cu induced metallothionein, while Zn exposure did not. MT induction in the whole soft tissues of the mussels exposed to the metal mixture is also observed but the induction is smaller than expected if there was a synergistic effect. This suggests that when metals are present jointly there is competition between Cd and Cu for binding MT. Subsequently MT induction is triggered by Cu and the excess of Cd in the heat-treated cytosol is bound to the newly formed MT. Data on metal concentrations suggests that different metal binding characteristics, observed specially for Cu, might be related to the difference observed on MT induction.

Introduction

Molluscs, and particularly mussels, have developed a number of subcellular systems for accumulation, regulation and immobilisation of essential and pollutant metals. Among these are the heat-stable, cysteine-rich low molecular weight cytosolic protein known as metallothionein (MT).

The functions of MT are known to include elements of homeostasis and detoxification. The latter has been considered to be the primary role of MT due to its unique status as a Cd-MT protein. However Cd-MT seems to be originated as a Cu and Zn-MT in unstressed animals supporting the concept of dual functionality (Langston and Bebianno, 1998).

MT has been reported for some 50 different aquatic invertebrates, three-quarters of which are molluscs (Langston and Bebianno, 1998). In the mussels *Mytilus edulis* and *Mytilus gallo-provincialis* induction of MT occurs after exposure to Cd or Cu (Bebianno and Langston, 1991, 1992; Viarengo *et al.*, 1989).

The induction of MT in marine organisms is often presumed to occur as a result of exposure to environments contaminated with heavy metals. Consequently, the present research was aimed to understand more comprehensively the function of MT in the metabolism of metals such as Cd, Cu and Zn. For this purpose mussels *M. edulis* were exposed to a metal mixture of Cd (100 µg/l), Cu (50 µg/l) and Zn (50 µg/l). Besides MT concentrations, total metal and metal-binding levels, among various cytosolic components separated by gel-exclusion chromatography, were also determined in an attempt to identify mechanisms behind metal-metal interactions in mussels.

Materials and methods

Groups of 50 mussels, *M. edulis* (shell length 4–6 cm) were exposed to a mixture of Cd (100 µg/l), Cu (50 µg/l) and Zn (50 µg/l) for 41 days. Groups of controls were maintained in parallel throughout the experiments. Individual samples of six control and six metal-exposed mussels were removed at different time intervals and MT and metal concentrations were determined in the whole soft tissues of individual mussels after being removed from the shells. Tissues were weighted and homogenised in three volumes of 0.02 M Tris-HCl buffer (pH 8.6) in an ice-bath. Subsamples were taken for wet:dry weight ratio and for the determination of metal (Cd, Cu and Zn) analysis. An aliquot of the homogenate was centrifuged at 30 000 g for 1 h at 4°C. The supernatant (cytosol) was separated from the pellet, maintained at 80°C for 10 min, to precipitate the heat-denatured high molecular weight proteins (HMW), subsequently centrifuged at 30 000 g for 1 h at 4°C.

Gel-chromatography (Sephadex G-75) was used to separate metal-binding components and to determine the thiolic content in the heat-treated cytosolic extracts of molluscs during exposure experiments.

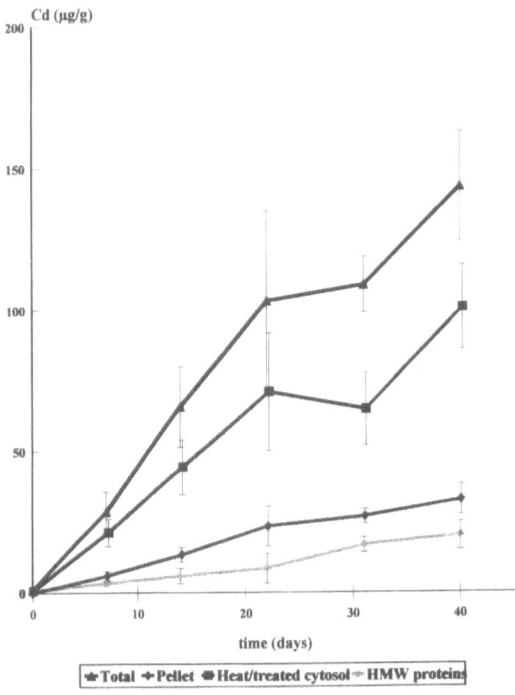

Figure 1. *M. edulis*. Cd accumulation and subcellular distribution in the mussels whole soft tissues.

MT analysis and the thiolic content of each chromatographic fraction were performed on subsamples of the heat-treated cytosol (50–250 µg/l) using Differential Pulse Polarography, according to the method described by Bebianno and Langston (1989). MT concentrations are expressed as mg/g of dry weight tissues.

Metal (Cd, Cu and Zn) analysis were performed on dried, HNO_3-digested samples of the tissues, using Atomic Absorption Spectrophotometry. Reference materials were used to validate metal determinations. Metal concentrations are expressed as µg/g of the dry weight tissue homogenate.

Statistical analysis of the data were performed using ANOVA.

Results

Metal (Cd, Cu and Zn) concentrations accumulated in the whole soft tissues of mussels revealed major differences in metal binding. Cd is mainly found in the cytosol bound to MT while Cu is almost equally divided between cytosolic and insoluble fractions (Fig. 1 and 2). On the other

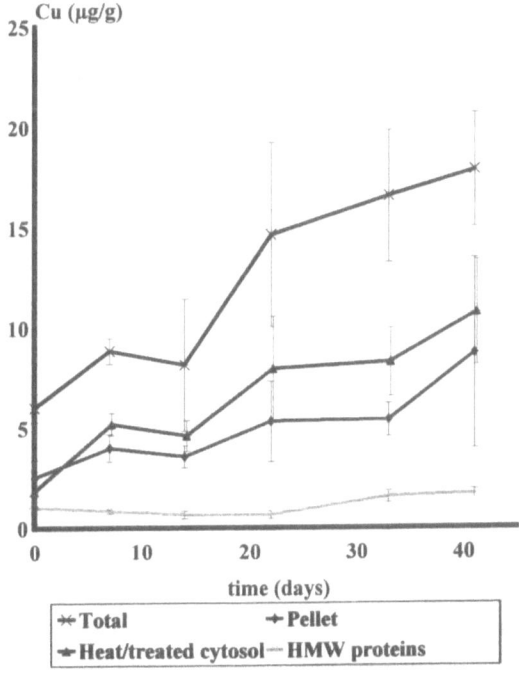

Figure 2. *M. edulis*. Cu accumulation and subcellular distribution in the mussels whole soft tissues.

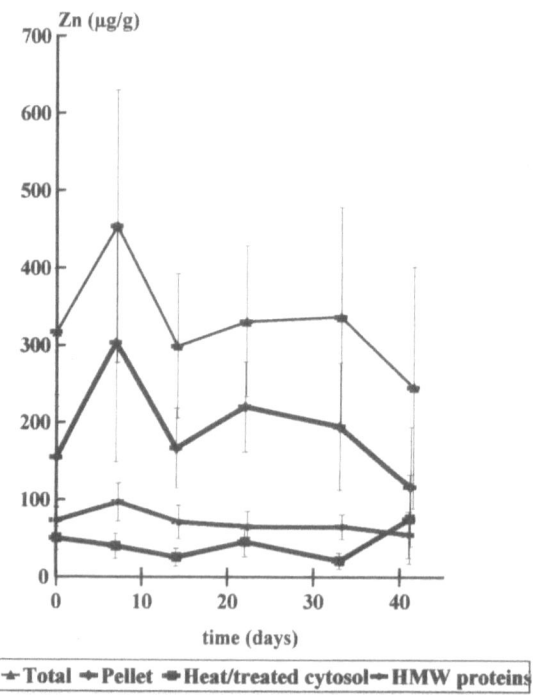

Figure 3. *M. edulis*. Zn accumulation and subcellular distribution in the mussels whole soft tissues.

hand most of the Zn (Fig. 3) is found predominantly bound in the insoluble pellet and to a lesser extent associated with the high molecular weight cytosolic proteins.

Total Cd concentrations and Cd bound to MT increased linearly during exposure to the metal mixture (3.4 $\mu g \cdot g^{-1} \cdot d^{-1}$ ($P < 0.001$) and 2.9 $\mu g \cdot g^{-1} \cdot d^{-1}$ ($P < 0.001$) (Tab. 1). Cd bound to MT. represented 66% of total Cd (Fig. 1). Cd in the other subcelullar fractions (pellet and HMW proteins) also increased linearly with the time of exposure representing 0.8 and 0.5 $\mu g \cdot g^{-1} \cdot d^{-1}$ ($P < 0.001$), respectively. Cd in the pellet represented 22% of total Cd (Fig. 4).

As for Cd, Cu increase linearly in the whole soft tissues of mussels with the time of exposure 0.16 $\mu g \cdot g^{-1} \cdot d^{-1}$ ($P < 0.001$) (Tab. 1 and Fig. 2) while Cu in control mussels did not change significantly (5.9 ± 0.1 $\mu g \cdot g^{-1}$) during the experiment. Cu accumulated in the heat-treated cytosol followed the same pattern as total Cu and represents 56% of total Cu. The accumulated Cu is greater in the heat-treated cytosol 0.08 $\mu g \cdot g^{-1} \cdot d^{-1}$ ($P < 0.001$), followed by the pellet 0.05 $\mu g \cdot g^{-1} \cdot d^{-1}$ ($P < 0.001$) (Tab. 1). However, Cu bound to the HMW content did not change significantly while the mussels were exposed to the metal mixture (Fig. 2).

In contrast to the Cd and Cu accumulated by the whole soft tissues of the mussels exposed to the metal mixture there is no significant change between Zn concentrations in the whole soft tissues of control (321 ± 82 $\mu g \cdot g^{-1}$) and of mussels exposed to the metal mixture. Subcellular Zn distribution is also different from Cd and Cu (Fig. 3). The pellet is the subcellular fraction

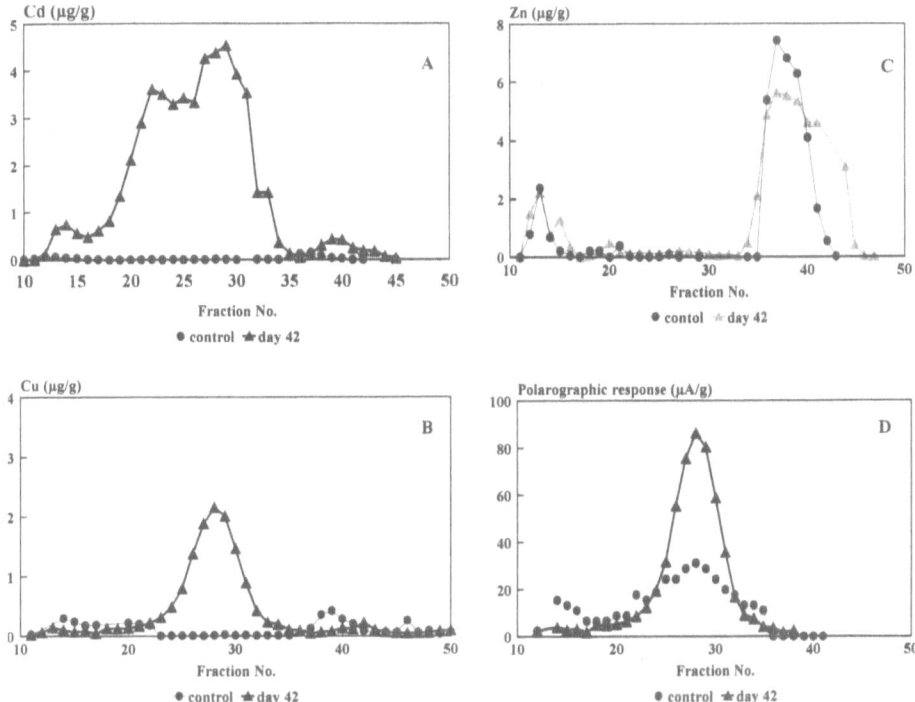

Figure 4. *M. edulis*. Gel-filtration chromatography of Cd (A), Cu (B), Zn © and thiolic contents (D) of the heat-treated cytosol of the mussels whole soft tissues.

with the highest Zn concentrations, followed by the HMW protein fraction and the heat-treated cytosol (Fig. 3). When the mussels were exposed to the metal mixture Zn only increase linearly in the HMW protein pool ($0.24\ \mu g\cdot g^{-1}\cdot d^{-1}$, $P < 0.05$) and in the heat-treated cytosol ($0.27\ \mu g\cdot g^{-1}\cdot d^{-1}$ ($P < 0.05$) (Tab. 1).

Metal-binding in the heat-treated cytosol after being separated by exclusion chromatography also revealed significant differences ($P < 0.05$) between control and contaminated mussels.

Table 1. Metal uptake rates ($\mu g\cdot g^{-1}\cdot d^{-1}$) in the whole soft tissues and subcellular fractions of the mussels exposed to a misture of Cd (50 µg/l), Cu (50 µg/l) and Zn (100 µg/l)

	Cd	Cu	Zn
Total	3.4*	0.16*	NS
Heat-treated cytosol	2.9*	0.08*	0.27**
Pellet	0.8*	0.05*	NS
HMW proteins	0.5*	NS	0.24

* $P < 0.001$; ** $P < 0.05$; NS – not significant.

Chromatographic elution profiles for Cd, Cu, Zn and the thiolic groups of the heat-treated cytosol of the mussels whole soft tissues are presented in Fig (4 A-D). For Cd (Fig. 4A) the concentrations in the whole soft tissues of control mussels are very small. After 42 days of the exposure to the metal mixture Cd increase significantly and most of the Cd was bound to the fractions with an apparent molecular weight from 10 000 to 20 000 daltons (Fractions 20–35), which corresponds to the MT pool. A double peak of Cd is observed (between 10 000 and 20000 daltons) suggesting that a MT isoform might be present.

Similarly two peaks were detected for Cu in the chromatographic profile (Fig. 4B) in the heat-treated cytosol of exposed mussels. One between fraction 13–18 which corresponds to a high molecular weight pool (where enzymes and haemocyanin appears) with a low Cu content. Another peak (fractions 20–35) which corresponded to the MT pool, with higher Cu concentrations (80% of Cu in the heat-treated cytosol) in exposed mussels. In controls this pool represented only 20% of Cu in the heat-treated cytosol (Fig. 4B).

For Zn, however (Fig. 4C), less than 15% is found in the heat-treated cytosol and even in this fraction most of the Zn is found bound to the very high molecular weight proteins (fraction 10–15 - > 50 000 daltons) and very low molecular weight weight protein fractions (35–45

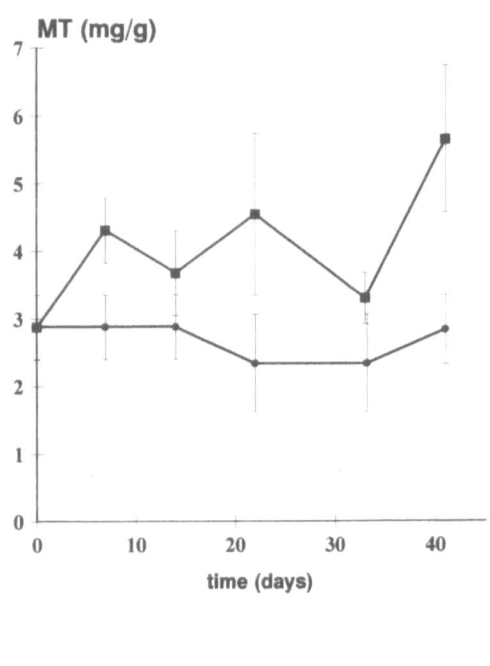

control   Cd+Cu+Zn

Figure 5. *M. edulis*. MT concentrations in the whole soft tissues of control and metal exposed mussels.

- < 5 000 daltons) (Fig. 4C). Zn is virtually absent from the MT pool, which means that in the presence of a metal mixture Zn does not induce MT.

Similarly the thiolic content of the proteins present in the heat-treated cytosol also increase with the increase of metal accumulated in the mussels (Fig. 4D). As can be seen from Figure 4D, the thiolic content in control mussels the polarographic response was low and only present between fractions 20–35. After 42 days of exposure to the metal mixture, the polarographic response in the mussels heat-treated cytosol increase significantly in the MT pool. In conclusion Cd, and Cu concentrations increase in the MT pool after 42 days of exposure and this increase is related to the increase of the polarographic response in the same pool.

MT concentrations determined in control and metal contaminated mussels revealed that there was a significant increase in MT levels ($P < 0.05$) in the whole soft tissues of mussels exposed to the metal mixture (Fig. 5). MT levels increase to 5.7 ± 1.1 mg/g which represented a two fold increase in MT concentrations when compared with controls (2.8 ± 0.5 mg/g).

For the mussels exposed to the metal mixture a direct relationship was detected between MT and Cd concentrations and between with MT and Cu. MT concentrations increased significantly ($P < 0.05$) with the increase in Cd and Cu concentrations (Fig. 6). This increase can be

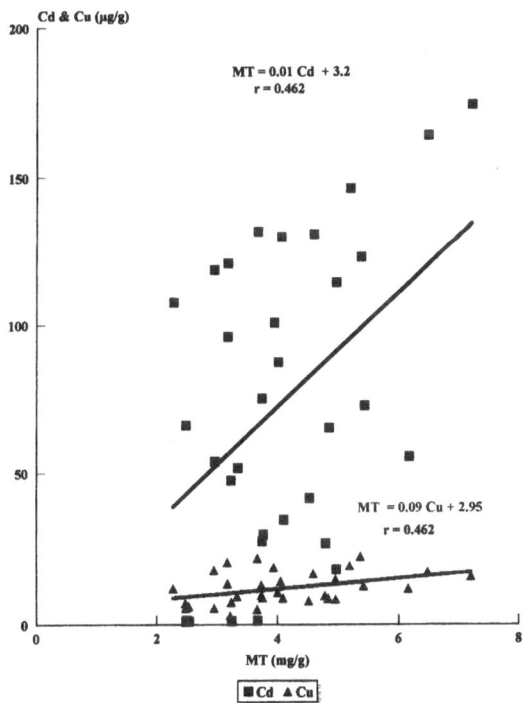

Figure 6. *M. edulis*. Relationship between Cd and Cu and MT concentrations in the mussels whole soft tissues.

expressed by the following linear equations [MT] (mg/g) = 0.01 [Cd] (μg/g) + 3.2 (r = 0.462) and [MT] (mg/g) = 0.09 [Cu] (μg/g) + 2.95 (r = 0.462).

Discussion

The acumulation of Cd and Cu in the whole soft tissues of mussels are in agreement with data obtained by other authors (Bebianno and Langston, 1991, 1992; Harrison et al., 1886;, Sholz, 1980; Viarengo et al., 1980, 1985). For Zn mussels seem to be able to regulate Zn influx.

In the presence of metal mixture a situation which is more likely to be found in the marine environment, Cd and Cu both bind to MT and seem to induce MT synthesis in the whole soft tissues of mussels. When the mussels were exposed to the same metal concentrations individually, Cd and Cu induced metallothionein concentrations, while Zn exposure did not (Bebianno and Langston, 1991; Bebianno, own data). MT induction, however is smaller than expected if there was a synergistic effect among these metals. When mussels were exposed to the same Cd concentration, MT induction was higher (Bebianno, unpublished data). When metals are present jointly there is competition between Cd and Cu for binding to MT. In fact Zn is displaced by Cd when mussels are exposed only to Cd (Bebianno and Langston, 1991). The results presented in this paper suggests that MT induction is triggered by Cu and the excess of Cd in the heat-treated cytosol is bound to the newly formed MT. Data on metal subcellular distribution suggests that different metal binding characteristics occur among these three metal ions, and the ones observed specially for Cu, might be related to the difference obtained for MT induction.

References

Bebianno MJ and Langston WJ (1989) Quantification of metallothioneins in marine invertebrates using differential pulse polarography. *Portugaliæ Electrochim Acta* 7: 59–64.

Bebianno MJ and Langston WJ (1991) Metallothionein induction in *Mytilus edulis* exposed to cadmium. *Mar Biol* 108: 91–96.

Bebianno MJ and Langston WJ (1992) Cadmium induction of metallothionein synthesis in *Mytilus galloprovincialis*. *Comp Biochem Physiol* 103C: 79–85.

Harrison FL, Lam JR, Novacek J (1988) Partitioning of metals among metal-binding proteins in the bay mussel *Mytilus edulis. Mar Environ Res* 24: 167–170.

Langston WJ and Bebianno MJ (eds) (1998) *Metal Metabolism in Aquatic Environment*. Chapmann and Hall, London 448pp.

Scholz N (1980) Accumulation, loss and molecular distribution of cadmium in *Mytilus edulis. Helgoländer Meeresunters* 33: 68–78.

Viarengo A, Pertica M, Mancinelli G, Zanicchi G and Orunesu M (1980) Rapid induction of copper-binding proteins in the gills of metal exposed mussels. *Comp Biochem Physiol* 67C: 215–218.

Viarengo A, Palmero S, Zanicchi G, Capelli R, Vaissiere R and Orunesu M (1985) Role of metallothioneins in Cu and Cd accumulation and elimination in the gill and digestive gland cells of *Mytilus galloprovincialis* Lam. *Mar Environ Res* 16: 23–36.

Viarengo A, Pertica M, Canesi L, Mazzucotelli A, Orunesu M and Bouquegneau JM (1989) Purification and Biochemical characterisation of a lysosomal copper-rich thionein-like protein involved in metal detoxification in the digestive gland of mussels. *Comp Biochem Physiol* 93C(2): 389–395.

Metallothionein IV
C. Klaassen (ed.)
© 1999 Birkhäuser Verlag Basel/Switzerland

Synthesis and properties of class III metallothioneins of *Schizosaccharomyces pombe* in response to cadmium and zinc

Mary M. Lauro and Donald J. Plocke

Department of Biology, Boston College, Chestnut Hill, Massachusetts 02167, USA

Introduction

Metal-binding peptides derived from glutathione and having the structure $(\gamma\text{Glu-Cys})_n\text{Gly}$ [$(\gamma\text{EC})_n\text{G}$] were first described by Murasugi *et al.* [1] in studies of *Schizosaccharomyces pombe* growing in the presence of added cadmium ions. These peptides were subsequently found to be synthesized in a wide variety of plant species [2] and have been designated as phytochelatins, gamma-glutamyl peptides, and class III metallothioneins. The enzyme responsible for the stepwise addition of the γEC moiety to γECG (glutathione) or an already formed $(\gamma\text{EC})_n\text{G}$ has been isolated from *Silene cucubalis* and partially characterized. The enzyme is a dipeptidyl transferase that has been given the trivial name phytochelatin synthase and has been reported to be constitutive in plants [3].

Here we report on $(\gamma\text{EC})_n\text{G}$ synthesis in *S. pombe* (strain 972 h⁻) growing in both complex (yeast extract-glucose-peptone) and Edinburgh defined minimal medium (EMM) in the presence and absence of cadmium ions. Additionally, we describe the isolation of a stoichiometrically defined zinc-peptide complex.

Results and discussion

(γEC)$_n$G peptide production in cells grown in the presence of cadmium

Substantial amounts of $(\gamma\text{EC})_n\text{G}$ peptides are present during the exponential growth phase of *S. pombe* growing in complex medium in the absence of added metal ions, to the extent of 20% of the amounts present after exposure of log phase cells for 4 h to 1.0 mM Cd^{2+} (Fig. 1a). $(\gamma\text{EC})_n\text{G}$ peptides are produced as well by cells grown in Edinburgh defined minimal medium (EMM) with no added cadmium (Fig. 1b). On exposure of cells growing in EMM to 1.0 mM Cd^{2+} for 4 h, there is a modest increase in the amount of these peptides, particularly in the case of $(\gamma\text{EC})_3\text{G}$. A marked reduction in the peptide content, particularly in the case of $(\gamma\text{EC})_2\text{G}$, is found when the zinc concentration is lowered from 1.4×10^{-6} M to 1.4×10^{-7} M (Fig. 1b). This observation is consistent with results obtained with a variety of plant species growing in metal-depleted medium, where the amounts of $(\gamma\text{EC})_n\text{G}$ peptides synthesized were proportional to the concentration of subsequently added zinc or copper [4].

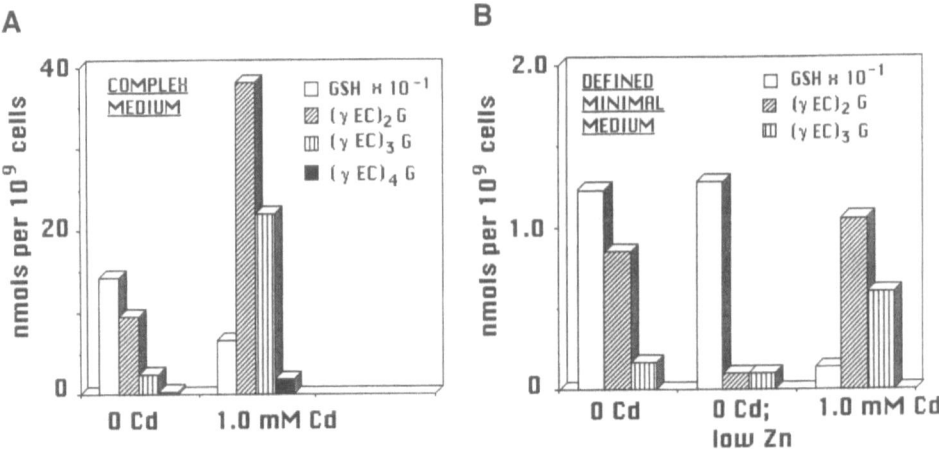

Figure 1 The GSH and (gEC)$_n$G content of mid-log phase cells grown in complex medium (1a) containing 2% D-glucose, 2% bactopeptone and 1% yeast extract and in Edinburgh minimal defined medium (1b) [6] in the absence of added metal ions and after a 4 h exposure to 1 mm Cd^{2+}. Cells were harvested by centrifugation and disrupted by shaking with glass beads, followed by acidification of the cell homogenate and centrifugation. The resulting supernatant was applied to a Nucleosil C-18 column, with separation of peptides effected by the use of a gradient of 0–20% acetonitrile in 0.05% H$_3$PO$_4$, followed by post-column derivatization with Ellman's reagent and quantitation by measurement of absorbance at 405 nm, using GSH as a standard. ESMS was employed to confirm the assignment and designation of the elution peaks.

Production of GSH and (ΥEC)$_n$G as a function of cell growth

Measurements of glutathione (GSH) and (γEC)$_n$G in cells grown in defined minimal medium with no added cadmium and harvested at various stages of cell revealed that the GSH content decreased over 20-fold as the cells proceeded from early exponential into stationary phase (from 55 to 2.3 nmol per 10^9 cells). (γEC)$_2$G decreased 5-fold (from 1.5 to 0.3 nmol per 10^9 cells). (γEC)$_3$G, on the other hand, showed a slight increase as the cells entered stationary phase, while the amounts of (γEC)$_4$G, present initially in much smaller amounts than (γEC)$_2$G, remained at a constant low level until the cells entered stationary phase. The fact that (γEC)$_n$G peptides are synthesized by S. pombe both in complex medium in the absence of added metals and in defined minimal medium containing the normal complement of added metal ions suggests that these peptides may have functions in the cell additional to their documented role in metal detoxification.

Production of aggregates in cells grown in the presence of cadmium

Exposure of S. pombe grown in defined minimal medium to 1.0 mM cadmium for 48 h resulted in the production of large aggregates containing embedded cells. Figure 2 shows representative scanning electron micrographs of two of these aggregates, and displays their typical

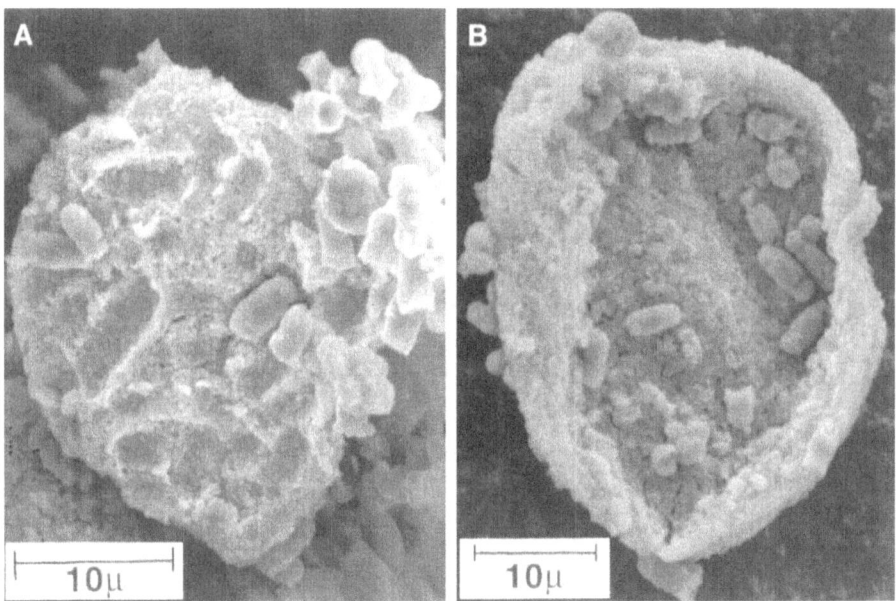

Figure 2 (a) and (b): Scanning electron micrographs of representative aggregates produced on addition of Cd^{2+} to cells growing in Edinburgh defined minimal medium (EMM). 1.0 mM Cd^{2+} was added at $OD_{590} = 1.0$; cells were harvested after a 48 h exposure to Cd^{2+}. The aggregates were separated from the cell suspension by the use of a step gradient of 35%–70% sucrose and centrifugation in a tabletop centrifuge (45 min at 12 **g**). The material was prepared for SEM using osmium tetroxide and thiocarbohydrazide, followed by lyophilization and examination in the Amray model 1610 scanning electron microscope.

appearance as hollow invaginated spheres ranging from 25–45 µm in diameter and present to the extent of approximately 1×10^4 per liter of cells. Large numbers of yeast cells appear to be embedded in the aggregate matrix in addition to cells which appear to be more loosely associated. Of the various agents employed to solubilize these aggregates, only the metal chelator pentaethylenehexamine (penten) proved to be effective. Hence it seems likely that metal ions (presumably cadmium) are involved in maintaining the integrity of the aggregate matrix. Preliminary data on the solubilized material (after removal of the intact cells by filtration) indicated the presence of as yet unidentified thiols or possibly sulfide ions. $(\gamma EC)_n G$ peptides were not detected.

These aggregates bear at least a superficial resemblance to the spherical mineral deposits observed in *Microcoleus chtonoplastes*, described as "hollow spheres, formed directly in the cell mass" and containing appreciable amounts of silicon, zinc, sulfur, calcium, and iron on the outer surface, whereas the inner surface contained no sulfur, and showed a "distinct predominance of zinc over calcium" [5]. However, beyond the remarkable superficial resemblance, there is nothing to suggest similarity of origin for the aggregates in the two instances.

Isolation of a zinc-containing complex from cells grown in complex medium with added zinc

A zinc-$(\gamma EC)_nG$ complex containing equimolar amounts of $(\gamma EC)_3G$ and Zn^{2+} was isolated from cells grown in complex medium following a 3 h exposure to tetramethylthiuram disulfide and 1.0 mM zinc sulfate. A cell extract from these cells was applied to a Sephadex G-50 column and the fractions containing $(\gamma EC)_nG$ peptides, which eluted at a range of Mr between 1000 and 2000, were further fractionated on a DEAE column. The elution pattern, shown in Figure 3(a), revealed a single thiol and zinc-containing peak that eluted from DEAE on application of 0.4 M NaCl and had a thiol:zinc ratio of 3:1. Analysis of the peak fraction for zinc (atomic absorption) and peptide (HPLC) content revealed the presence of a single peptide species, namely $(\gamma EC)_3G$ (Fig. 3b). The probable stoichiometry of the complex is: $[(\gamma EC)_3G]_2 \cdot [Zn^{2+}]_2$, corresponding to Mr = 1674, consistent with the G-50 elution results.

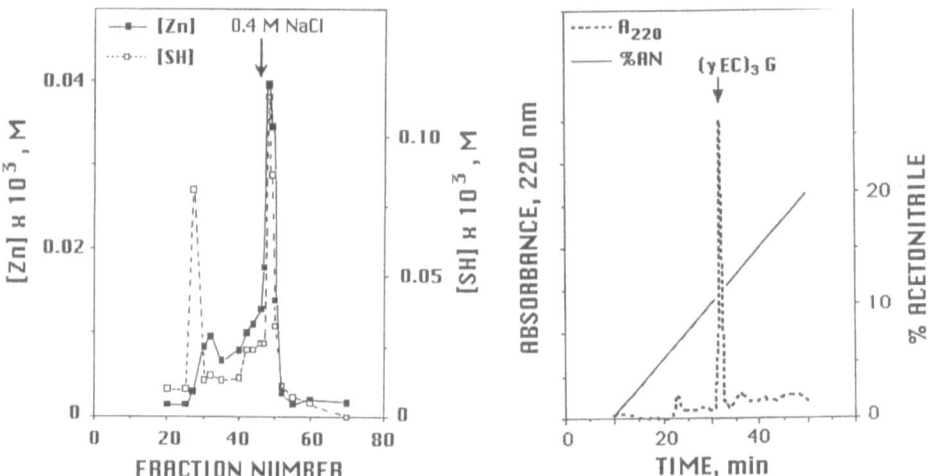

Figure 3: Isolation of a zinc-$(\gamma EC)_nG$ complex. (a) DEAE chromatography of $(\gamma EC)_nG$-containing fractions from Sephadex G-50. *S. pombe* was grown in complex medium to mid-log phase ($OD_{590} = 1.6$) followed by 3-hour exposure to 1.0 mM Zn^{2+} and 1.0 mg per ml TMTD (tetramethylthiuram disulfide); cells were harvested after 3 h at a final $OD_{590} = 5.4$. Cells were ruptured by shaking with glass beads; a cell extract was prepared by centrifugation (30 min at 12,000 **g**), followed by lyophilization and centrifugation at 40,000 **g** for 60 min and lyophilization of the resulting supernatant, which was taken up in a small volume and applied to a Sephadex G-50 column (2.0×100 cm). Fractions were analyzed for [Zn], [SH] and $(\gamma EC)_nG$ content. $(\gamma EC)_nG$ peptides were detected by HPLC in fractions corresponding to an M_r between 1000 and 2000. This peptide-containing material was pooled, applied to a DEAE column (DE-52) and eluted with 0.2 M and 0.4 M NaCl in 0.0 M Tris (pH 7.6) as shown. Elution with 0.4 M NaCl resulted in a single peak containing a thiol to zinc ratio of 3:1. (b) HPLC analysis of DEAE Fraction 48. Conditions of chromatography as indicated in caption for Figure 1, except that identification of peptides was by measurement of absorbance at 220 nm.

References

1. Murasugi A, Wada C, Hayashi Y (1981) Cadmium-binding peptide induced in fission yeast, *Schizosaccharomyces pombe*. *J Biochem* 90: 1561–1564.
2. Grill E, Winnacker EL, Zenk MH (1987) Phytochelatins, a class of heavy-metal-binding peptides from plants, are functionally analogous to metallothioneins. *Proc Natl Acad Sci USA* 84: 439–443.
3. Grill E, Löffler S, Winnacker EL, Zenk MH (1989) Phytochelatins, the heavy-metal-binding peptides of plants, are synthesized from glutathione by a specific γ-glutamylcysteine dipeptidyl transpeptidase (phytochelatin synthase). *Proc Natl Acad Sci USA* 86: 6838–6842.
4. Grill E, Thumann J, Winnacker ELZenk MH (1988) Induction of heavy-metal binding phytochelatins by inoculation of cell cultures in standard media. *Plant Cell Rep* 7: 375–378.
5. Golubev SNGerasimenko LM (1990) Biomineralization in Cyanobacteria cultures. *Mikrobiologiya* 58: 963–968.
6. Mitchison, 6.Mitchison JM (1970) Physiological and cytological methods for*Schizosaccharomyces pombe*. *In*: DM Prescott (ed.): *Methods in Cell Biology Vol. IV*. Academic Press, New York, pp. 131–165.

Metallothionein IV
C. Klaassen (ed.)
© 1999 Birkhäuser Verlag Basel/Switzerland

Phytochelatins (class III metallothioneins) and their desglycyl peptides induced by cadmium in root cultures of *Rubia tinctorum* L.

Tamio Maitani, Hiroki Kubota, Kyoko Sato and Takashi Yamada

National Institute of Health Sciences, Kamiyoga 1-18-1, Setagaya, Tokyo 158-8501, Japan

Summary. Phytochelatins (PCs) are class III metallothioneins (MTs) and the general structures are (γ-Glu-Cys)$_n$-Gly (n $\geq$ 2). We studied the induction of PCs by various metals in the root cultures of *Rubia tinctorum* L. (madder) and found that the desglycyl peptides of PCs were also induced (both PCs and their desglycyl peptides are abbreviated as CIIIMTs). To study the induction mechanism of desglycyl PCs, it is essential to analyze two precursors, glutathione (GSH) and γ-Glu-Cys. However, the separation of GSH and γ-Glu-Cys was poor in postcolumn derivatization HPLC used for CIIIMT analysis. Therefore, we applied high-performance capillary electrophoresis (HPCE) for the separation of the two compounds along with CIIIMTs. Using the HPCE method accompanied by sample pretreatment with EDTA and ABD-F (a colorimetric reagent for thiol), GSH, γ-Glu-Cys, and respective CIIIMT species were detected separately, and the analysis was performed within 6 min. Consequently, the HPCE method was found to be very useful for the study of CIIIMTs.

Introduction

By exposure to excess heavy metals, plants induce thiol (SH)-rich peptides called phytochelatins (PCs). The general structures of PCs are (γ-Glu-Cys)$_n$-Gly, and those with n = 2~11 have so far been described. They are classified as class III metallothioneins (MTs). The role of PCs is proposed in metal detoxification and metal homeostasis. Therefore, PCs have been used recently as indicators of forest decline caused by heavy metal pollution [1].

High-performance liquid chromatography (HPLC) with postcolumn derivatization is most frequently used for analyzing class III metallothioneins. By using this method, we studied the induction of PCs by various metals in the root cultures of *Rubia tinctorum* L. (madder) and found that the desglycyl peptides of PCs are also induced (hereafter, both PCs and their desglycyl peptides are abbreviated as CIIIMTs). Furthermore, we applied high-performance capillary electrophoresis (HPCE) for the separation of two precursors, glutathione (GSH) and γ-Glu-Cys, along with respective CIIIMT species.

Root cultures of *Rubia tinctorum* L.

Many attempts have so far been made to produce useful secondary metabolites of plants commercially by tissue culture techniques. The roots of *R. tinctorum* produce anthraquinone (AQ) pigments [2]. They have been used from ancient times as a source of natural dyes. We established normal adventitious roots [3] and transformed hairy roots [4] of *R. tinctorum* to study the production mechanism of AQ pigments and the controlling factors.

The production of useful secondary metabolites such as pigments is often augmented by exposing plants to heavy metals. However, plants also induce PCs against the exposure. Therefore, we are also interested in the induction of PCs by exposing the root cultures to various metals.

Induction of desglycyl PCs

Although the structures of PCs are conservative in a wide range of plants, we have demonstrated by postcolumn derivatization HPLC and electrospray ionization-mass spectrometry (ESI-MS) that the desglycyl peptides of PCs are also induced when the root cultures of *R. tinctorum* are exposed to both cadmium (Cd) and GSH, which are an effective inducer of PCs [5] and a precursor of PC, respectively.

Figure 1 shows a typical chromatogram of the postcolumn derivatization HPLC for the root cultures containing CIIIMTs. The root cultures were exposed to both Cd (1 mM) and GSH (2 mM) in Murashige-Skoog medium for 5 days. CIIIMTs were analyzed according to the method of Grill et al. [6] with some modifications [3]. The treated normal root cultures were milled in 2 vol. of 1 M NaOH containing 0.1% $NaBH_4$ with an agate mortar. After centrifugation (11,000 × g, 5 min), the supernatant fraction was acidified with 3.6 M HCl. The precipitates

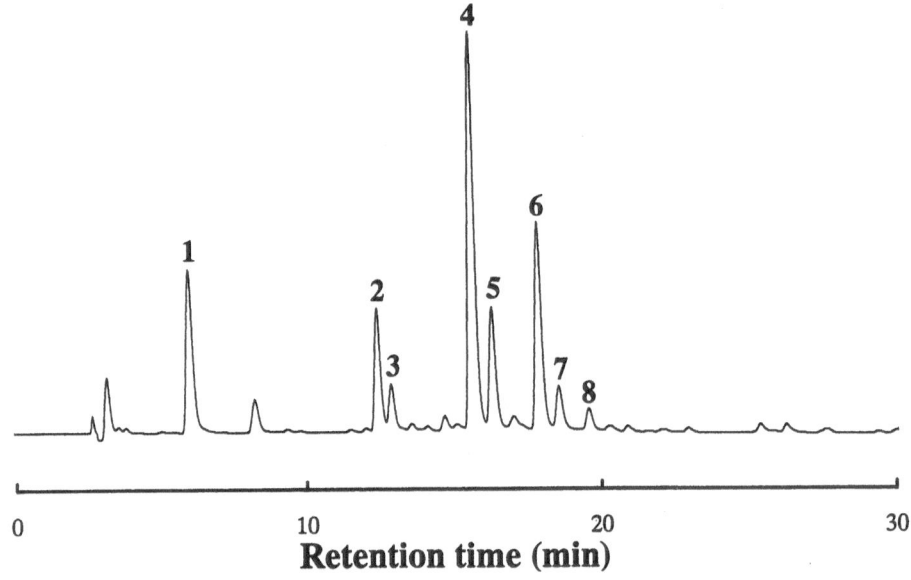

Figure 1. A typical HPLC chromatogram generated by postcolumn derivatization with DTNB. The root cultures were exposed to $CdCl_2$ (1 mM) and GSH (2 mM) for 5 days. The induced CIIIMTs were analyzed as SH-containing peptides with DTNB. 1, GSH and γ-Glu-Cys; 2, (γ-Glu-Cys)$_2$-Gly; 3, (γ-Glu-Cys)$_2$; 4, (γ-Glu-Cys)$_3$-Gly; 5, (γ-Glu-Cys)$_3$; 6, (γ-Glu-Cys)$_4$-Gly; 7, (γ-Glu-Cys)$_4$; 8, (γ-Glu-Cys)$_5$-Gly.

were separated by centrifugation and the supernatant fraction was applied to an HPLC apparatus equipped with a C_{18} column (Inertsil ODS-80A, 4.6 mm I.D. × 250 mm, GL Sciences Inc., Tokyo). Postcolumn derivatization was performed with 75 mM 5,5'-dithiobis(2-nitrobenzoic acid) (DTNB), and SH-containing substances were detected at 410 nm.

Peak 1 was assigned to the mixture of GSH and γ-Glu-Cys, because both compounds were not separated. Peaks 2 ~ 8 were identified as (γ-Glu-Cys)₂-Gly, (γ-Glu-Cys)₂, (γ-Glu-Cys)₃-Gly, (γ-Glu-Cys)₃, (γ-Glu-Cys)₄-Gly, (γ-Glu-Cys)₄, and (γ-Glu-Cys)₅-Gly, respectively, by ESI-MS [3]. Peaks 3, 5, and 7 are the desglycyl PCs.

Induction of CIIIMTs by various metals

The induced amounts of CIIIMTs were compared among various metals. To study the inducibility of metal itself, GSH was not added to the medium. As test-metals (including metalloids), soft metals and metals that have an electronic configuration of d^{10} were selected. Namely, the salts of Ag(I), As(III), As(V), Cd(II), Cu(II), Ga(III), Hg(II), In(III), Ni(II), Pb(II), Pd(II), Se(IV), and Zn(II) were used. The dose was 10, 100, or 1000 µM depending on the toxicity of the respective metals. The metals that induced CIIIMTs over 15 nmol γ-Glu-Cys/g fresh weight of cultures after 3 day-exposure were Ag(I) (dose, 100 µM) > Cd(II) (100 µM) > Pb(II) (1000 µM) > Hg(II) (10 µM) > As(III) (100 µM) > Cu(II) (100 µM) > As(V) (100 µM) > Zn(II) (1000 µM) [5]. (γ-Glu-Cys)₄-Gly was the most abundant CIIIMT species induced by Ag(I) and Cd(II). Even (γ-Glu-Cys)₅-Gly was clearly detected in Ag(I). Thus, Ag(I) and Cd(II) were effective inducers of CIIIMTs as well as those of mammalian MTs (class I MTs), while Zn(II) was not so effective in the induction of CIIIMTs.

HPCE analysis of CIIIMTs along with GSH and γ-Glu-Cys

To clarify the induction mechanism for the desglycyl PCs, the determination of GSH and γ-Glu-Cys is essential. However, the separation of these two compounds was poor on the postcolumn derivatization HPLC analysis as shown in Figure 1. We applied HPCE to determine the two compounds and respective CIIIMT species simultaneously.

The root cultures were exposed to Cd (1 mM) and GSH (2 mM) for 5 days. They were milled in 2 vol. of 0.1% NaBH₄ aqueous solution. After centrifugation (11,000 × g, 10 min), the supernatant fraction was subjected to capillary zone electrophoresis using a P/ACE 5500 system (Beckman Instruments, Fullerton, CA) equipped with an untreated fused-silica capillary (75 µm I.D. × 57 cm). As a running buffer, 100 mM borate buffer (pH 8.5) was used. Ethylenediamine-N,N,N',N'-tetraacetic acid (EDTA) disodium salt (10 mM, 5 µl) and ABD-F (4-aminosulfonyl-7-fluoro-2,1,3-benzoxadiazole, a SH-reacting agent, 20 mM, 20 µl) were added to the sample solution (20 µl) to prevent interference of Cd bound to CIIIMTs and to detect SH-group at 380 nm with a diode array detector, respectively.

CIIIMTs induced by Cd are more or less bound with Cd through SH groups [5]. Figure 2 shows the effects of EDTA on the HPCE electropherogram for the supernatant fraction of root

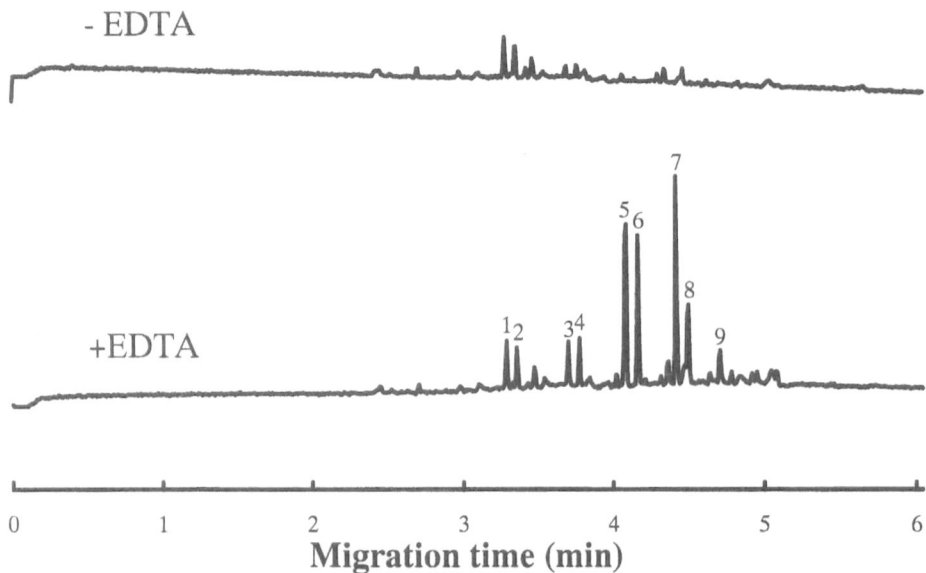

Figure 2. Effects of EDTA on the HPCE electropherogram for the supernatant fraction of root cultures containing CIIIMTs. The root cultures were exposed to Cd (1 mM) and GSH (2 mM) for 5 days. CIIIMTs, GSH, and γ-Glu-Cys were detected colorimetrically as SH-containing peptides with ABD-F before (upper) and after (lower) treatment with EDTA. 1, GSH; 2, γ-Glu-Cys; 3, (γ-Glu-Cys)$_2$-Gly; 4, (γ-Glu-Cys)$_2$; 5, (γ-Glu-Cys)$_3$-Gly; 6, (γ-Glu-Cys)$_3$; 7, (γ-Glu-Cys)$_4$-Gly; 8, (γ-Glu-Cys)$_4$; 9, (γ-Glu-Cys)$_5$-Gly.

cultures containing CIIIMTs. The peaks with migration times of 3.27 and 3.33 min were assigned to GSH and γ-Glu-Cys, respectively. Namely, the two compounds were separated well in HPCE.

The peak areas of respective CIIIMT species increased by the EDTA addition, while those of GSH and γ-Glu-Cys remained almost constant. This result may suggest that Cd ions should be removed from CIIIMTs before ABD-F (a SH-reacting agent) reacts with the CIIIMTs. Thus, by the addition of EDTA, the peaks of respective CIIIMT species were detected with good separation. Moreover, the analysis was performed within 6 min, while 20 min were required in the postcolumn derivatization HPLC analysis. Consequently, HPCE was useful for the separation of GSH and γ-Glu-Cys along with respective CIIIMT species.

Conclusion

In HPCE method accompanied by sample pretreatment with EDTA and ABD-F, GSH and γ-Glu-Cys were fully separated, and the separation of respective CIIIMT species was also good. Moreover, the analysis was performed within a short time. Consequently, the HPCE method was found to be very useful for the study of CIIIMTs.

Acknowledgments
This study was supported in part by a grant from the Japan Health Sciences Foundation.

References

1   Gawel JE, Ahner BA, Friedland AJ, Morel FMM (1996) Role for heavy metals in forest decline indicated by phytochelatin measurements. *Nature* 381 (May 2): 64–65.

2   Sato K, Kubota H, Goda Y, Yamada T, Maitani T (1997) Glutathione enhanced anthraquinone production in adventitious root cultures of *Rubia tinctorum* L. *Plant Biotechnol* 14: 63–66.

3   Kubota H, Sato K, Yamada T, Maitani T (1995) Phytochelatins (class III metallothioneins) and their desglycyl peptides induced by cadmium in normal root cultures of *Rubia tinctorum* L. *Plant Sci* 106: 157–166.

4   Maitani T, Kubota H, Sato K, Takeda M, Yoshihira K (1996) Induction of phytochelatin (class III metallothionein) and incorporation of copper in transformed hairy roots of *Rubia tinctorum* exposed to cadmium. *J Plant Physiol* 147: 743–748.

5   Maitani T, Kubota H, Sato K, Yamada T (1996) The composition of metals bound to class III metallothionein (phytochelatin and its desglycyl peptide) induced by various metals in root cultures of *Rubia tinctorum*. *Plant Physiol* 110: 1145–1150.

6   Grill E, Winnacker E-L, Zenk MH (1987) Phytochelatins, a class of heavy-metal-binding peptides from plants, are functionally analogous to metallothioneins. *Proc Natl Acad Sci USA* 84: 439–443.

Metallothionein IV
C. Klaassen (ed.)
© 1999 Birkhäuser Verlag Basel/Switzerland

Chemical and spectroscopic properties of copper metallothionein from the copper resistant fungus *Beauveria bassiana*

Satomi Kameo[1], Peter Faller[2], Milan Vašák[2] and Yutaka Kojima[1]

[1]*Department of Environmental Medicine and Informatics, Graduate School of Environmental Earth Science, Hokkaido University, Sapporo, 060, JAPAN*
[2]*Biochemisches Institut der Universitat Zürich, CH-8057 Zürich, SWITZERLAND*

Introduction

While copper is an essential trace element and plays important roles in living organisms, excess uptake of copper is known to be toxic to the organisms. However, the accumulation of heavy metals in fungi is a well-known phenomenon and has been subject of numerous investigatoins.

It was reported that incorporated copper in fungi was bound to low molecular weight ligands in the mycelia. [1, 2, 3]. An ascomycete *Neurospora crassa* accumulated copper with a concomitant synthesis of copper-binding protein, consisted of a single polypeptide chain of 25 amino acid residues and exclusively binding copper [1]. *Neurospora crassa* copper-binding protein contained 7 cysteine residues and no aromatic amino acids [1, 2].

From a mushroom *Agaricus bisporus*, a copper-binding protein was isolated and the chemical characterization of the protein was reported [3]. It consisted of 25 amino acids containing 7 cysteine residues and no aromatic amino acid. Both copper-binding proteins from *Neurospora crassa* and *Agaricus bisporus* were reported to belong to metallothionein(MT) group [1, 3]. The amino acid sequences of MTs have been elucidated from a wide variety of sources [4, 5, 6]. In contrast to vertebrate MTs, which bind different metal ions, fungal MTs were reported to contain exclusive copper ions [3].

From a mushroom *Grifola frondosa*, different type of copper-binding ligand was found [7]. A peptide was an acidic peptide of molecular weight 2,240 Da in which four kinds of amino acids (aspartic acid, glutamic acid, serine, and glycine) occupied about 84% of total residues and contains no cysteine.

In this study, we successfully isolated a copper resistant fungus,which was identified as *Beauveria bassiana*, grown on the medium containing high copper concentration (6 mM). The purposes of this study are to investigate the properties of a copper resistant fungus *Beauveria bassiana* and the properties of a copper-binding protein from the fungus.

The differences and similarlities on fungal copper-binding proteins were discussed.

Isolation and identification of a copper resistant fungus

A copper resistant fungus was isolated, which was grown on the medium containing high copper concentration (6 mM). By the taxonomic investigation, the fungus was identified as a strain of *Beauveria bassiana* (Balsamo) Vuillemin, which is a group of Deuteromycotina, Hyphomycetes.

Assay of resistance of the fungus towards copper

In order to evaluate the copper resistance, growths of the fungus *Beauveria bassiana* were assayed on the medium containing copper in several concentrations, compared to those of several fungi (*Aspergillus flavus, Cladosporium herbarum, Rhizopus nigricans, Trichoderma viride*).

 Beauveria bassiana and the other fungi were grown very well on the control medium (YGP medium; 0.2%(w/v) yeast extract, 2% peptone, 10% glucose and 2% of bact agar with PIPES buffer, pH 6.8) without additional copper. Whereas, according to the increasing copper concentration (as $CuCl_2$), differences were observed in the growth of *Beauveria bassiana* from that of the other fungi. As the results of the assays, it was clear that *Beauveria bassiana* was resistant to copper up to 10 mM, whereas the growth of the other fungi were observed on the medium only to 3 mM of copper concentration.

Assay of resistance of the fungus towards other heavy metals

Additional resistances towards zinc, cadmium and lead were also established. On the medium containing of zinc, cadmium or lead, *Beauveria bassiana* was able to grow up to 70 mM zinc content, 10 mM cadmium content or 50 mM lead content. On the other hand, on the medium over 20 mM of zinc, lead content or containing 10 mM of cadmium, the growth of the mycelia of these fungi were completely inhibited.

Copper incorpolation and isolation of copper-binding protein from the fungus

In order to elucidate the origin of copper resistance, *Beauveria bassiana* was exposed to copper (1 mM). Copper was incorporated and accumulated in the mycelia. Incorporated copper was eluted in low molecular weight region by the gel filtration, Sephadex G-50 (5 × 100 cm) in 10 mM Tris – 5 mM HCl. The low molecular weight copper containing fractions were pooled and subjected to DE-52 anion exchange chromatography. Elution profile of DE-52 chromatography shows 2 separate peaks

 (Fig. 1). The profile revealed the existence of a low molecular weight, copper-binding protein. Fractions of peak1 were pooled for the following analyses. Before the analyses, the samples were desalted by passaging over Sephadex G-10 (1 × 30 cm) in 1 mM potassium phosphate, pH 7.5.

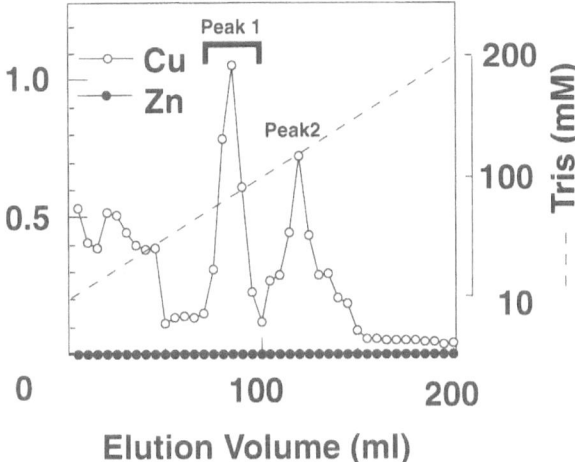

Figure 1. Elution profile of DE-52 anion exchange chromatography of the pooled Sephadex G-50 fractions. The column was eluted with a liner gradient by using 100 ml each of 10 mM Tris – 5 mM HCl and 200 mM Tris – 100 mM HCl. Fractions of peak 1 were pooled. —O—, copper; —●— , zinc; – – –, Tris gradient.

Amino acid composition of copper-binding protein (Cu-BP) from *Beauveria bassiana* and copper-binding proteins from several fungi

Table 1 shows the amino acid composition of the purified copper-binding protein from the fungus *Beauveria bassiana* and of copper-binding proteins from several fungi. The amino acid analysis of copper-binding protein was determined by a method described by Kojima and Hunziker [8, 9].

The amino acid analyses indicated, this protein from the fungus *Beauveria bassiana* is a cysteine-rich protein lacking aromatic residues, Arg, His, Ile and Leu. The amino acid composition of this protein shows very high similarity with metallothionein from *Neurospora crassa* and *Agaricus bisporus* [1, 3].

However, the amino acid composition differs substantially from that of the copper-binding peptide of *Grifola frondosa* lacking cysteines, whereas contained high aspartic acids [7].

Spectroscopic properties of Cu-BP of *Beauveria Bassiana*

In the Figure 2, spectroscopic properties of copper-binding protein from the fungus *Beauveria bassiana* are shown. This absorption spectrum with shoulder at 250 and 320 nm was typical of Cu(I)-thiolate coordination. At neutral pH, the protein had a dinstict shoulder around 250 nm. Upon removal of metal ions by lowing the pH with hydrochloride (HCl), the spectrum indicated a clear loss of absorption shoulder.

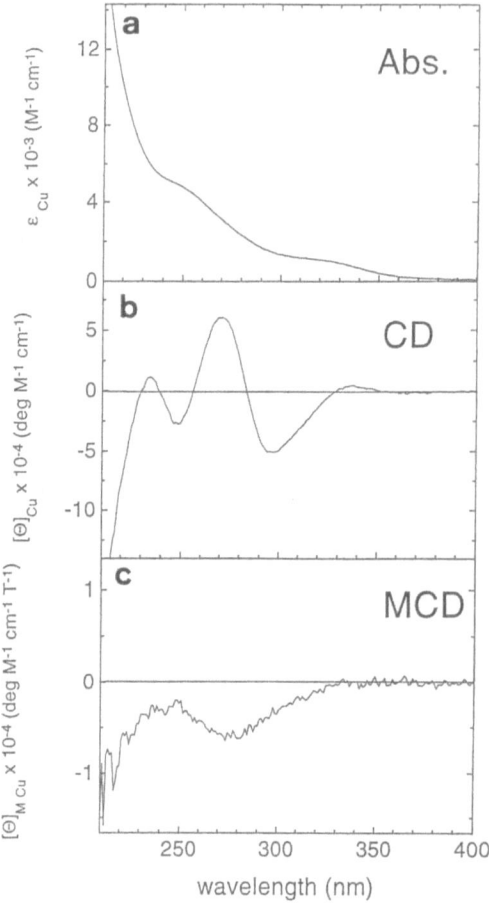

Figure 2. Spectroscopic properties of copper-binding protein from the fungus *Beauveria bassiana*. Absorption Spectrum (a), CD (circular dichroism) Spectrum (b), MCD Spectrum (c). Units based on copper concentration.

It was known that the feature above 250 nm in both absorption and CD spectra are normaly absent in mononuclear Cu(I)-thiolate complex. Thus, the feature of the existence of 250 nm in both absorption and CD spectra of the protein indicates the presence of a copper-thiolate cluster structure in the protein from this fungus. Similar Absorption and CD properties have also been found for Cu-MT from *Neurospora crassa* and *Agaricus bisporus* [2, 3].

Luminescence properties with an emission band at about 565 nm are typical for Cu(I)-thiolate complex (Fig. 3). Similar feature has also been reported for fungus *Neurospora* and *Agaricus* Cu-MT [10, 11]. EPR spectra also indicated existence of Cu(I) in the protein (data not shown).

Table 1. Amino acis compositions of fungal copper-binding proteins

	Species			
	B	N	A	G
Amino Acid		Residues per mol		
Ala	2	1	2	1
Asx	4			
Asp		1	1	7
Asn		2	0	0
Arg	0	0	0	0
Cys	5	7	7	0
Glx	1			
Glu		0	0	4
Gln		0	1	0
Gly	5	6	6	5
His0	0	0	0	
Ile	0	0	0	0
Leu	0	0	0	0
Lys	1	1	1	0
Met	0	0	0	0
Phe	0	0	0	0
Ser	3	7	5	4
Thr	1	0	0	1
Tyr0	0	0	0	
Val	0	0	0	0
Pro			1	
Total	22	25	25	23

Cys and Met were determined as cysteic acid and methionine sulfone, respectively. The numbers of amino acid residues listed are based on 1.0 mol of Lys per one mol. Species; B: *Beauveria bassiana*, N: *Neurospora crassa* [1], A: *Agaricus bisporus* [3], G: *Grifola frondosa* [7].

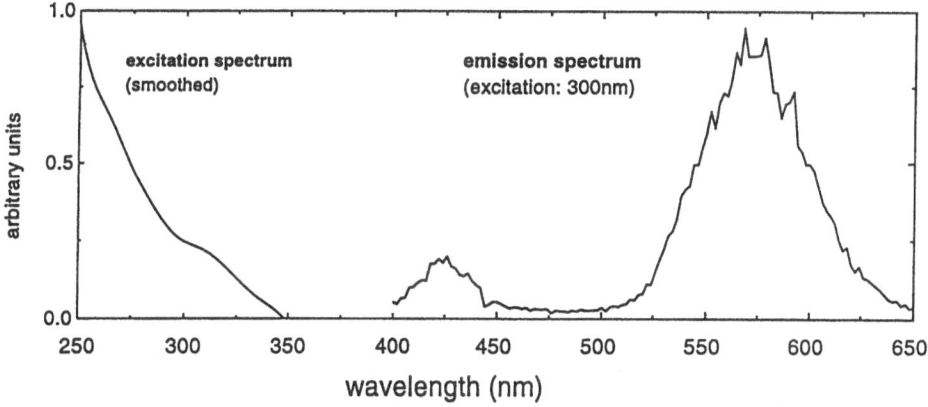

Figure 3. Luminescence Properties of Copper-binding Protein from the fungus *Beauveria bassiana*. Observed excitation and emission spectra of *Beauveria bassiana* copper-binding protein: (Luminesence at 77K) excitation at 300 nm; emission at 565 nm.

Several spectroscopic properties of the copper-binding protein from the fungus *Beauveria bassiana* suggests the existence of a copper(I)-thiolate cluster, with spectral features similar to the Cu(I)6-Cys7 cluster in metallothionein of *Neurospora crassa*.

Conclutions

Properties of *Beauveria bassiana*
* copper resistant
* additional resistant towards zinc, cadmium and lead
* copper accumulation and copper bound to a low molecular weight protein.

Properties of the copper-binding protein of *Beauveria bassiana*
* a cysteine-rich protein lacking aromatic residues nor histidine.
* the existence of a copper(I)-thiolate cluster, with spectral features similar to that of Cu-MT of *Neurospora crassa* and *Agaricus bisporus*. From these properties, it was confirmed that the copper-binding protein from the fungus *Beauveria bassiana* was a menber of Metallothioneins [12].

Acknowledgments
The authors thank Prof. Dr. Jeremias. H. R. Kägi, Dr. Peter Hunziker Biochemisches Institut der Universität Zürich, Switzerland for helpful suggestions.
The authors were also indebted to Dr. T. Yoshida for his help of identification of fungus, and Dr. F. Tomita for supply with fungal strains, Dept. Applied Micro Biology, Hokkaido Univ, Japan.

References

1. Lerch K (1980) Copper metallothionein, a copper-binding protein from *Neurospora crassa*. *Nature* 284: 368–370.
2. Beltramini M, Lerch K (1983) Spectroscopic Studies on *Neurospora* Copper Metallotionein. *Biochemistry* 22: 2043–2048.
3. Munger K, Lerch K (1985) Copper Metallotionein from the Fungus *Agaricus bisporus*: Chemical and Spectroscopic Properties. *Biochemistry* 24: 6751–6756.
4. Kägi JHR, Vašák M, Lerch K, Gilg DEO, Hunziker PE, Berhard WR, Good M (1984) *EHP, Environ Health Perspect* 54: 93–103.
5. Roesijadi G, Fowler BA (1991) Purification of invertebrate metallothionein. *Meth Enzymol* 205: 263–273.
6. Binz PA, Kägi JHR (1997) Metallothionein families: In *SWISS-PROT Protein Sequence Data Bank*, University of Geneva.
7. Shimaoka I, Kodama J, Nishino K, Itokawa Y (1993) Purification of a copper binding peptide from the mushroom *Grifola frondosa* and its effect on copper absorption. *J Nutr Biochem 4: 33–38*
8. Kojima Y, Hunziker PE (1991) Amino Acid Analysis of Metallothioneins. *Meth Enzymol* 205: 419–421.
9. Hunziker PE (1991) Cysteine Modification of Metallotioneins. *Meth Enzymol* 205: 399–400.
10. Beltramini M, Lerch K (1981) Luminescence properties of *Neurospora* copper metallothionein *FEBS Lett* 127: 201–203.
11. Beltramini M, Munger K, Germann UA, Lerch K (1987) Luminescence emission from the Cu(I) complex in metallothioneins *Experientia, Suppl* 52: 237–241.
12. Kojima Y (1991) Definitions and Nomenclature of Metallotioneins. *Meth Enzymol* 205: 8–10.

Transcription factors

Metallothionein facts and frustrations

Richard D. Palmiter

Howard Hughes Medical Institute and Department of Biochemistry, University of Washington, Box 357370, Seattle WA 98195, USA

Summary. In the years since the last metallothionein meeting, transgenic gain-of-function and loss-of-function experiments have been performed to examine the potential functions of MT-I & II and MT-III in the mouse. The most important conclusion is that none of these MT isoforms are essential for viability under normal conditions. The MT-I and MT-II-null mice are remarkably normal in all respects, except for their sensitivity to heavy metal toxicity. Thus, these MTs are not required for the biosynthesis of metalloproteins, for homeostasis of essential metals (Zn and Cu), or for protection against oxidative insults that are incurred under normal conditions. Mice that cannot make MT-III also appear normal; even two-year-old mice reveal no signs of Alzheimer's disease, consistent with our observations that MT-III levels are not reduced in people with Alzheimer's disease. However, MT-III null mice are more sensitive to kainate-induced seizures and they manifest more neuronal damage. Surprisingly, ectopic expression of MT-III, where MT-I is normally expressed, causes pancreatic necrosis, whereas expression of a comparable amount of MT-I in the pancreas has no effect. This result indicates that acinar cells are especially sensitive to MT-III and that MT isoforms are not biologically equivalent.

The transcription factor, MTF-1, is essential for the induction of MT by metals yet the mechanism by which it is regulated by metals is unclear. Our experiments suggest that all the different metals that induce MTF-1 probably do so by increasing intracellular zinc pools, but MTF-1 is not the zinc sensor; instead, we believe that MTF-1 is normally kept in check by another molecule that is the zinc sensor.

Mouse MT gene cluster

The four mouse MT genes are clustered within 50 kb on mouse chromosome 8 [1]. The 5' most gene in the cluster (MT-IV) is expressed in certain stratified epithelia after the basal cells begin to differentiate [2]. The next adjacent gene (MT-III) is expressed predominantly in neurons of the CNS that sequester zinc in their synaptic vesicles [3]; whereas MT-II and MT-I genes are expressed in most organs [4]. Surprisingly, all four MT genes are expressed in the maternal placenta [5]. The DNA sequences responsible for cell-specific regulation of MT-IV and MT-III are unknown. However, transgenic experiments suggest that the sequences required for appropriate expression of MT-I and MT-II lie within a 25 kb region that includes these two genes [6,7].

Regulation of MT-I and MT-II gene induction

The MT-I and MT-II genes are coordinately induced by glucocorticoids, cytokines and a variety of metals [4], whereas the other two MT genes are relatively unresponsive to these inducers [2, 8]. Cis-acting elements that allow induction of MT-I and MT-II by cytokines, although undefined at the sequence level, are found in both the proximal promoter of the MT-I gene [9] as well as in more distal flanking regions [10]. A pair of adjacent glucocorticoid response elements that regulate both MT-I and MT-II lie ~ 1 kb upstream of the MT-II gene [10]. Both MT-I and MT-II genes have several metal response elements (MREs) within 200 bp of their transcription start sites [4, 11]. The transcription factor MTF-1, that is essential for both basal expression and induction by metals, binds to the promoter-proximal MREs [12]. MTF-1 has six zinc-fingers that are responsible for DNA binding [13]. Although transcription of MT-I and MT-II is induced by many different metals (Cd, Zn, Cu, Co, Bi, Ag, Hg, Ni) it is unlikely that all of these metals act directly [14]. It is possible that zinc is the only physiological regulator of MTF-1. A major question concerns the mechanism by which MTF-1 responds to zinc [14]. Is MTF-1 itself the zinc sensor, or does it respond to some other molecule that is sensitive to zinc? Our experiments (Palmiter and Findley, unpublished observations) are consistent with the latter hypothesis. By using stable transfection of MTF-1 (and derivatives) into a cell line with a zinc-sensitive reporter gene (MRE-βGeo) as a test system (Fig. 1A). along with mobility shift assays to measure the relative abundance of MRE-binding proteins, we have come to the following conclusions:

1. Cycloheximide induces MRE-βGeo expression (14) without changing total cellular zinc levels (Fig. 1A).
2. A small increase in total MTF-1 expression (e.g. equal to endogenous MTF-1) greatly increases basal expression of MRE-βGeo in the absence of added zinc (Fig. 1B).
3. Expression of the zinc fingers (ZFs) alone, competes with induction of MRE-βGeo by endogenous MTF-1 (Fig. 1C).
4. Expression of the ZF fused to an activation domain (VP16) leads to high constitutive expression of MRE-βGeo that is independent of zinc. The level of expression can exceed that achieved by optimal activation of endogenous MTF-1 by zinc (Fig. 1D) or transfection with excess MTF-1 (Fig. 1B).
5. Expression of an MTF-1 from which the ZFs have been deleted, inhibits induction via endogenous MTF-1 at low concentrations of zinc (Fig. 1E).
6. Expression of a construct in which the ZFs of MTF-1 are replaced with ZFs from the yeast transcription factor ADR1, results in high constitutive expression that is unaffected by zinc when assayed with ARE-βGeo, a reporter gene with cis-acting elements that are recognized by the ZFs of ADR1 (Fig. 1F).

Figure 1. Regulation of MRE-βGeo by MTF-1 and various derivatives. BHK cells were transfected with MRE-βGeo and a stable clone (3038) was selected in G418; see [14] for details. (A) (left panel) The 3038 cells show zinc-dependent induction of β-galactosidase; half-maximal induction occurs with a 30% increase in Zn; the cells die when Zn increases > 2-fold; (right panel) Cylcoheximide induces MRE-βGeo mRNA without affecting total cellular zinc. (B) (left panel) Stable clones of 3038 cells that were transfected with murine MTF-1 have high basal β-Gal activity in absence of added zinc, (right panel) even though the total amount of MRE-binding activity measured by mobility shift assay is only slightly increased. (C) Stable transformants of 3038 cells with a construct expressing MTF-1 zinc fingers (ZFs) alone inhibits induction of β-Gal probably due to competition. (D) Stable transformants of 3038 cells with low (TG1) or higher amounts (TG2) of ZFs from MTF-1 fused to a heterologous activation domain (VP16) produced high basal activity in the absence of added zinc that was insensitive to added zinc (after subtracting the contribution from endogenous MTF-1). Maximal activity was greater than achieved with intact MTF-1 suggesting that VP16 activation domains are more effective than those in MTF-1. (E) Stable transformants of 3038 cells with MTF-1 from which the ZFs had been deleted prevented induction at low concentrations of Zn (0–30 μM) but not at higher concentrations, suggesting that it has a dominant-negative effect at low Zn. (F) Stable transformants of BHK cells with three ADR1 response elements driving β-Geo have low, zinc-insensitive activity. Transfection of these cells with an MTF-1 construct in which the ZFs are replaced by those from ADR-1 results in high β-Gal activity that is insensitive to Zn.

These results indicate that the ZFs are not the zinc sensor (point 4) and the rest of the molecule does not confer zinc-sensitivity to heterologous ZFs (point 6). Thus, if MTF-1 is the zinc sensor, then ZFs must interact with the rest of the molecule. If the latter were true, then MTF (without ZFs; point 5) might be predicted to inhibit expression ZF-VP16 (point 4) when expressed

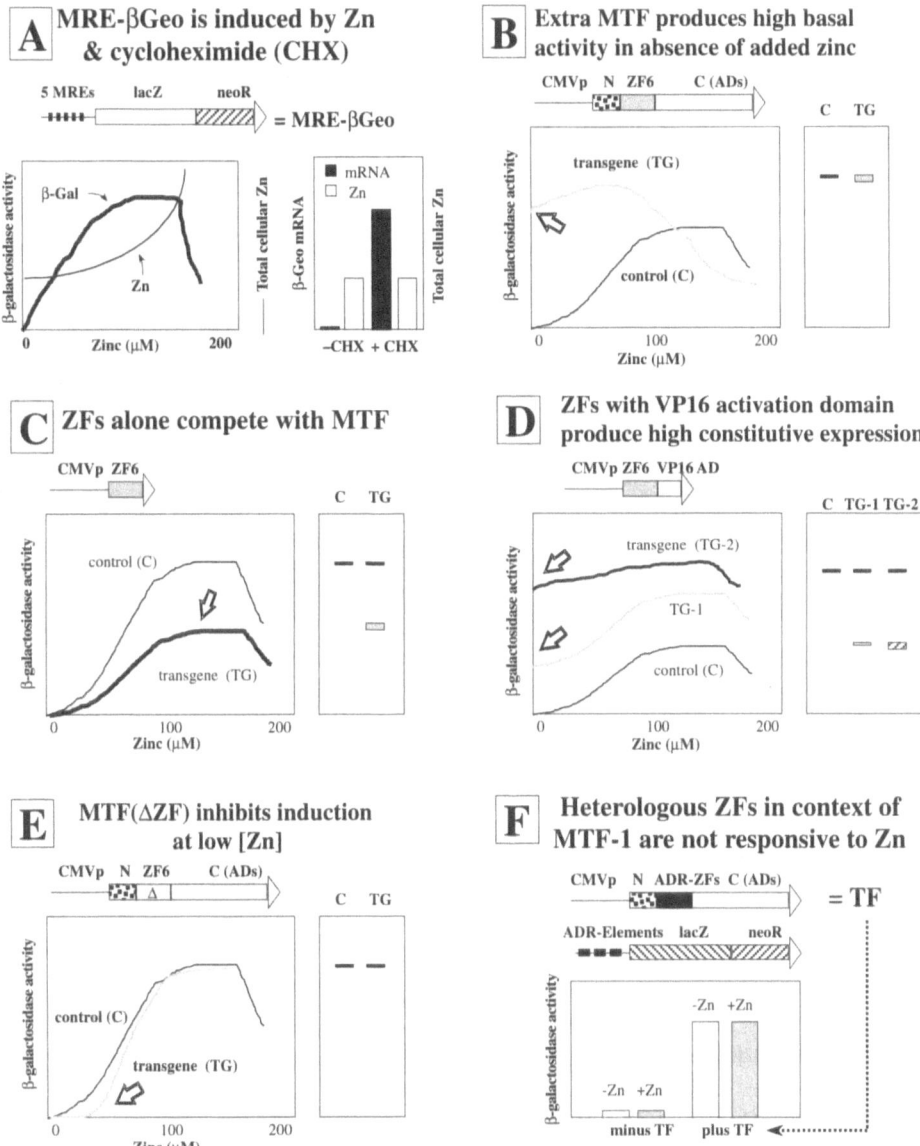

together in the same cell, but this has not been observed. In contrast, points 1 and 5 are most easily explained by assuming that some other molecule is the zinc sensor and that it interacts with MTF-1 to prevent transcription when cytoplasmic zinc levels are low.

Furthermore, we isolate fully active MTF-1 (measured by mobility shift assay) from cells that are not induced by zinc. We can inhibit induction *in vivo* or mobility shift *in vitro* with the zinc chelator (TPEN) in a reversible manner. But inactivation of MTF-1 *in vitro* is paralleled by inactivation of Sp1, another ZF-transcription factor that is not regulated by Zn, suggesting that this effect is probably due to nonspecific removal of zinc from the ZFs. The burden of proof for the existence of a distinct zinc sensor that regulates MTF-1 lies in its molecular characterization; however, this has not been achieved.

Search for MT-I and MT-II functions

The functions of MT-I and MT-II in mice has been examined by inactivating both genes [15, 16] and by over-expressing MT-I [6,7]. The most important conclusion is that these ubiquitously expressed isoforms are not essential for viability under normal conditions. The MT-null mice are remarkably normal in all respects, except for their sensitivity to heavy metal toxicity [15, 16]. There is no evidence for compensation by other MTs. Thus, MTs are not required for the biosynthesis of metalloproteins, for homeostasis of essential metals (Zn and Cu), or for protection against oxidative insults that are incurred under normal conditions. Furthermore, our attempts to create stressful conditions under which a requirement for MT might be revealed have not been very fruitful – with one exception. MT-null mice are born with less hepatic zinc and mild kidney dysplasia that resolves with age. The kidney defects are more severe when MT-null mice are reared on a zinc-deficient diet, but kidney function is still in the normal range [17]. The MT-null mice respond to 95% oxygen normally [Kelly, Horowitz and Palmiter, unpublished observations]. They can tolerate 25 mM zinc in their drinking water, but their pancrata are slightly more sensitive to massive amounts of injected zinc [17]. Their tolerance to injected copper is indistinguishable from wildtype mice [Kelly and Palmiter, unpublished observations]. However, in the absence of a functional copper efflux transporter (ATP7A) the absence of MT results in embryonic lethality [18]. The absence of MT also results in increased sensitivity to zinc in cells that lack a zinc efflux transporter [19]. Thus, it appears that MT provides an important defense against metal toxicity only when metal efflux transporters are inactive (Zn and Cu) or nonexistent (Cd). MT-I over-expression in mice has no obvious phenotype except that it provides extra protection against cadmium toxicity [20], consistent with the fact that selection for cadmium resistance in cultured cells invariably leads to MT gene amplification [21].

Search for MT-III function

MT-III-null and MT-III over-expressing mice have also been produced [22, 23]. The MT-III null mice have ~ 10% less zinc in the brain [22] and MT-III over-expressors have ~ 15% more

[23]. There is no compensation by other MTs. Histology of the brain in regions that are normally enriched in MT-III is normal. The MT-III null mice show normal response to injected Zn and Cd – total brain Zn barely changes and very little Cd gets into the brain [22]. Because MT-III was originally implicated in Alzheimer's disease [24], we assessed the learning and memory capabilities of young and 2-year-old mice and they were normal [22]. The only difference we noted was that MT-III-null mice are more sensitive to kainate-induced seizures and there is more neuronal damage when comparing mice that have undergone similar seizure episodes [22]. Furthermore, MT-III over-expressors show less neuronal damage in response to kainate-induced seizures [22]. We have no evidence that MT-III promotes Alzheimer-like symptoms in mice [22] in agreement with our observations that MT-III is not down-regulated in humans with Alzheimer's disease [25].

Expression of MT-III in all the organs where MT-I is expressed results in pancreatic atrophy, whereas expression of similar amounts of excess MT-I has no effect [26]. We do not know why MT-III is toxic to pancreatic acinar cells, but this experiment provides biological evidence that MT isoforms have different properties that probably relate to differences in they way they interact with other cellular components rather than their metal-binding properties [27].

Several attempts to delete the entire known MT locus were unsuccessful. The function of MT-IV has not been examined by the gene knockout approach.

The conundrum of MT function

Overall, the genetic analysis of MT function has been very disappointing. MT appears to be a secondary defense against extreme conditions that are not likely to be encountered normally. It seems remarkable that genes that are present in multiple copies in most organisms, that are expressed at all developmental stages in both somatic and germ cells, and that are induced by a wide range of stressful conditions, do not have an obvious essential function. The suggestion that MTs provide a secondary defense against metal toxicity and epileptic damage is not very satisfying. Perhaps we are blind to the critical conditions that led to their preservation during evolution.

The mechanisms by which cells detect changes in zinc (or any other heavy metal), regulate zinc transport out of the cell [28], to various cellular compartments [19, 29], or induce the synthesis of metallothioneins to sequester zinc [14], remains a fascinating problem. Identification of molecule(s) that sense changes in intracellular "free" zinc is likely to be the key to understanding this problem. We can be encouraged that transcription factors [12, 30] that respond either directly or indirectly to zinc have been identified since the last metallothionein meeting.

Acknowledgments
This work was supported in part by NIH grant CA 61268. I especially thank Seth Findley who helped with the cloning and analysis of MTF-1 function.

References

1. Palmiter RD, Sandgren EP, Koeller DM, Findley SD, Brinster RL (1993) Metallothionein genes and their regulation in transgenic mice. *In*: KT Suzuki, N Imura, M K Imura (eds): *Metallothionein III: Biological Roles and Medical Implications*. Birhäuser Verlag, Basel 399–406.
2. Quaife CJ, Findley SD, Erickson JC, Froelick GJ, Kelly EJ, Zambrowicz BP, Palmiter RD (1994) Induction of a new metallothionein isoform (MT-IV) occurs during differentiation of stratified squamous epithelia. *Biochemistry* 33: 7250–7259.
3. Masters BA, Quaife CJ, Erickson JC, Kelly EJ, Froelick GJ, Zambrowicz BP, Brinster RL, Palmiter RD (1994) Metallothionein III is expressed in neurons that sequester zinc in synaptic vesicles. *J Neurosci* 14: 5844–5857.
4. Palmiter RD (1987) Molecular biology of metallothionein gene expression. *Experientia Suppl* 52: 63–80.
5. Liang L, Fu K, Lee DK, Sobieski RJ, Dalton T, Andrews GK (1996) Activation of the complete mouse metallothionein gene locus in the maternal decidium. *Mol Reprod Dev* 43: 25–37.
6. Palmiter RD, Sandgren EP, Koeller DM, Brinster RL (1993) Distal regulatory elements from the mouse metallothionein locus stimulate gene expression in transgenic mice. *Mol Cell Biol* 13: 5266–5275.
7. Iszard MB, Liu J, Liu Y, Dalton T, Andrews GK, Palmiter RD, Klaassen CD (1995) Characterization of metallothionein-1 transgenic mice. *Toxicol Appl Pharm* 133: 305–312.
8. Palmiter RD, Findley SD, Whitmore TE, Durnam DM (1992) MT-III, a brain-specific member of the metallothionein gene family. *Proc Natl Acad Sci USA* 89: 6333–6337.
9. Durnam DM, Hoffman JS, Quaife CJ, Benditt EP, Chen HY, Brinster RL, Palmiter RD (1984) Induction of mouse metallothionein-I mRNA by bacterial endotoxin is independent of metals and glucocorticoid hormones. *Proc Natl Acad Sci USA* 81: 1053–1056.
10. Kelly EJ, Sandgren EP, Brinster RLPalmiter RD (1997) A pair of adjacent glucocorticoid response elements regulate expression of two mouse metallothionein genes. *Proc Natl Acad Sci USA* 94: in press.
11. Searle PF, Davison BL, Stuart GW, Wilkie TM, Norstedt G, Palmiter RD (1994) Regulation, linkage and sequence of mouse metallothionein I and II genes. *Mol Cell Biol* 4: 1221–1230.
12. Heuschel R, Radtke F, Georgiev O, Stark G, Aguet M, Schaffner W (1994) The transcription factor MTF-1 is essential for basal and heavy metal-induced metallothionein gene expression. *EMBO J* 13: 2870–2875.
13. Radtke F, Georgiev O, Muller H-P, Brugnera E, Schaffner W (1995) Functional domains of the heavy metal-responsive transcription factor MTF-1. *Nucl Acid Res* 23: 2277–2286.
14. Palmiter RD (1994) Regulation of metallothionein genes by heavy metals appears to be mediated by a zinc-sensitive inhibitor that interacts with a constitutively active transcription factor MTF-1. *Proc Natl Acad Sci USA* 91: 1219–1223.
15. Masters BA, Kelly EJ, Quaife CJ, Brinster RLPalmiter RD (1994) Targeted disruption of metallothionein I and II genes increases sensitivity to cadmium. *Proc Natl Acad Sci USA* 91: 584–588.
16. Michalska AE, Choo AKH (1993) Targeted germline transmission of a null mutation of the metallothionein I and II loci in mouse. *Proc Natl Acad Sci USA* 90: 8088–8092.
17. Kelly EJ, Quaife CJ, Froelick GJ, Palmiter RD (1996) Metallothionein I and II protect against zinc deficiency and zinc toxicity in mice. *J Nutr* 126: 1782–1790.
18. Kelly EJ, Palmiter RD (1996) A murine model of Menkes disease reveals a physiological function of metallothionein. *Nature Genet* 13: 219–222.
19. Palmiter RD, Cole TB, Findley SD (1996) ZnT-2, a mammalian protein that confers resistance to zinc by facilitating vesicular sequestration. *EMBO J* 15: 1784–1791.
20. Liu Y, Liu J, Iszard MB, Andrews GK, Palmiter RD, Klaassen CD (1995) Transgenic mice that overexpress metallothionein-I are protected from cadmium lethality and heptotoxicity. *Toxicol Appl Pharm* 135: 222–228.
21. Durnam DM, Palmiter RD (1987) Analysis of the detoxification of heavy metal ions by metallothionein. *Experientia Suppl* 52: 457–463.
22. Erickson JC, Hollopeter G, Thomas SA, Froelick GJ, Palmiter RD (1997) Disruption of the metallothionein-III gene in mice: analysis of brain zinc, behavior, and neuron vulnerability to metals, aging and seizures. *J Neurosci* 17: 1271–1281.
23. Erickson JC, Masters BP, Kelly EJ, Brinster RL, Palmiter RD (1995) Expression of human metallothionein III in transgenic mice. *Neurochem Int* 27: 35–41.
24. Uchida Y, Takio K, Titani K, Ihara Y, Tomonaga M (1991) The growth inhibitory factor that is deficient in Alzheimer's disease brain is a 68 amino acid metallothionein-like protein. *Neuron* 7: 337–347.
25. Erickson JC, Sewell AK, Jensen LT, Winge DR, Palmiter RD (1994) Enhanced neurotrophic acitivity in Alzheimer's disease cortex is not associated with down-regulation of metallothionein III. *Brain Res* 649: 297–304.
26. Quaife CJ, Kelly EJ, Masters BA, Brinster RL, Palmiter RD (1997) Ectopic expression of metallothionein-III causes pancreatic aciner cell necrosis in transgenic mice. *Toxicol Appl Pharmacol* 148: 148–157.
27. Sewell AK, Jensen LT, Erickson JC, Plamiter RD, Winge DR (1995) Bioactivity of metallothionein-3 correlates with its novel b domain sequence rather than metal binding properties. *Biochemistry* 34: 4740–4747.
28. Palmiter RD and Findley SD (1995) Cloning and functional characterization of a mammalian zinc transporter that confers resistance to zinc *EMBO J* 14: 639–649.

29. Palmiter RD, Cole TB, Quaife CJ, Findley SD (1996) ZnT-3, a putative transporter of zinc into synaptic vesicles. *Proc Natl Acad Sci USA* 93: 14934–14939.
30. Zhao H (1997) Zap1p, a metalloregulatory protein involved in zinc-responsive transcripitonal regulation in *Saccharomyces cerevisiae*. *Mol Cell Biol* 17: 5044–5052.

Liver degeneration and embryonic lethality in mouse null mutants for the metal-responsive transcriptional activator MTF-1

Çagatay Günes[1], Rainer Heuchel[1], Oleg Georgiev[1], Karl-Heinz Müller[1], Silvia Marino[2], Adriano Aguzzi[2], Horst Blüthmann[3] and Walter Schaffner[1]

[1]*Institut für Molekularbiologie, Abteilung II der Universität Zürich, Winterthurstrasse 190, CH-8057 Zürich, Switzerland*
[2]*Neuropathologie USZ, Schmelzbergstrasse 12, CH-8091 Zürich, Switzerland*
[3]*Hoffmann La-Roche PRTB, CH-4002 Basel, Switzerland*

Introduction

The metal regulatory transcription factor 1 (MTF-1) is a highly conserved transcriptional activator (Radtke et al., 1993; Heuchel et al., 1994; Radtke et al., 1995). We previously cloned mouse and human MTF-1, which share a 93% identity in amino acid sequence (Radtke et al., 1993; Brugnera et al., 1994). MTF-1 contains six zinc fingers of the C_2H_2-type as a DNA binding domain and at least three distinct domains responsible for transcriptional activation (Fig. 1). It binds to a number of metal responsive elements (MREs) present in the promoter regions of metallothionein genes I and II (MT-I and MT-II) and regulates their expression.

Expression of metallothionein genes MT-I and MT-II is enhanced by a great number of stimuli, most notably by adverse conditions such as an excess of heavy metal cations, reactive oxygen intermediates (ROI), UV- and X-irradiation (Kaegi, 1991; Heuchel et al., 1995). Consequently, these proteins have been implicated in heavy metal detoxification, metal homeostasis and radical scavenging (Masters et al., 1994; Thornally and Vašák, 1985). Since constitutive MT-I and MT-II expression is very high in fetal liver, it has been suggested that these MTs might play a pivotal role in fetal liver development. However, two groups have shown independently that mouse strains with combined deletion of the MT-I and MT-II genes, as a result of targeted gene disruption, show no altered phenotype under normal conditions. But, such null mutant mice were more sensitive to cadmium exposure, an observation which sup-

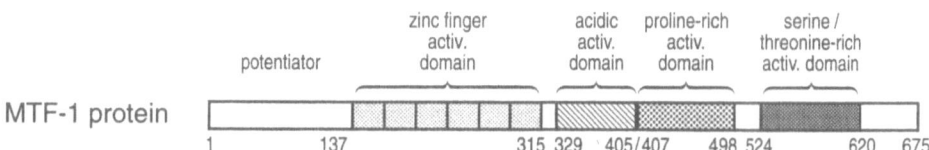

Figure 1. Schematic drawing of mMTF-1

ports the proposed role of metallothioneins in metal detoxification (Michalska and Choo, 1993; Masters et al., 1994).

Results

We have previously shown that MTF-1 is essential for basal and heavy metal-induced transcriptional activation of MT-I and MT-II genes in cultured cells (Heuchel et al., 1994). To investigate the function of MTF-1 *in vivo*, we generated mice lacking functional MTF-1 by targeted disruption of the MTF-1 genomic locus. Genotype analyses of embryos from MTF-1$^{+/-}$ intercrosses revealed no viable homozygous mutant embryo older than E14.5 days (0 of 25). Before day E13.5, homozygous mutant embryos were obtained at about the expected Mendelian frequency (10 of 41), indicating that implantation and early post-implantation development were not impaired. At E13.5, only one of 14 MTF-1 homozygous mutant embryos was found dead. Conversely, at E14.5, only one homozygous mutant embryo was found alive (Tab. 1).

Histological analysis of MTF-1$^{-/-}$ embryos at E13.5 showed variable degrees of liver degeneration. The liver structure of MTF-1$^{-/-}$ embryos at this stage was disrupted with enlarged, congested sinusoids and dissociation of the epithelial compartment. Furthermore, although individual epithelial cells of MTF-1$^{-/-}$ liver at E13.5 appeared morphologically normal in conventional histology, immunohistochemical analysis showed that these cells had a reduced cytokeratin expression which was specific for liver, since cytokeratin expression in the gut from the same embryo was not affected. At E14.5, the only null mutant embryo still alive showed severe liver necrosis, diffuse bleeding and generalized oedema as well as an almost complete lack of cytokeratin expression in the liver.

We could not detect any significant increase in the frequency of apoptotic cells in the liver of MTF-1$^{-/-}$ embryos, neither morphologically nor with the TUNEL *in situ* staining for DNA fragmentation. We also found that the cell proliferation rate in the MTF-1 homozygous mutant was not affected since bromodeoxyuridine incorporation at E13.5 revealed no decrease when compared to wild type or heterozygous littermates. Living E13.5 homozygous mutants showed not only no macroscopic color change indicative of anemia, but also no change in the number

Table 1. Lethality of MTF-1 homozygous mutants

Stage	Total tested	Live MTF-1$^{-/-}$	Dead* MTF-1$^{-/-}$
E10.5-E12.5	41	10 (24%)	0
E13.5	54	13 (24%)	1 (2%)
E14.5	24	1 (4%)	3 (13%)
E15.5-E16.5	25	0	0
postnatal	41	0	-

E, embryonic day; *dead embryo or empty decidua. No homozygous mutants were found among the born pups so far. Most of the homozygous mutants die after E13.5 and before E14.5. No living MTF-1$^{-/-}$ embryos older than E14.5 were observed. Before E13.5, living MTF-1$^{-/-}$ embryos were present in the expected Mendelian frequency.

or appearance of erythrocytes or hematopoietic precursor cells observable in histological sections, arguing against a defect in hematopoiesis. Live homozygous MTF-1$^{-/-}$ embryos at stages E13.5 and E14.5 showed no reduced liver size. Also, the placenta of MTF-1$^{-/-}$ embryos but did not reveal any abnormalities.

At the molecular level, we analyzed expression of the metallothionein I and II genes in 13.5-day old MTF-1$^{-/-}$ embryos by RNAse protection assay and found no detectable MT-I and MT-II expression whereas expression of the unrelated transcription factor Sp1was not affected. As mentioned before, MTF-1 is essential for basal and heavy metal induced expression of MT-I and MT-II genes. Since mice lacking the metallothionein I and II genes are viable (Michalska and Choo, 1993; Masters et al., 1994) and other MT isoforms cannot account for redundancy in MT-I and II knockout mice because of their temporal and spatial expression (Palmiter et al., 1992; Quaife et al., 1994) we considered it most likely that deregulated expression of at least one additional target gene, perhaps in combination with loss of MT-I and MT-II expression, is responsible for the liver damage phenotype of MTF-1$^{-/-}$ mice.

Recently, MREs have been described in the promoter of the gamma-glutamylcysteine syntethase heavy chain (γ-GCShc) gene (Mulchay and Gipp, 1995). The enzyme gamma-glutamylcysteine syntethase is essential for glutathion biosynthesis. We found additional, putative MREs in the promoter of γ-GCShc and tested them for MTF-1 binding. By electrophoretic mobility shift assays (EMSA), we observed a strong, MTF-1-specific complex with one of these MREs (located at position -114) and a specific, albeit weak interaction with the MRE located at position +296. No binding to other putative MREs could be observed in EMSA assays. We then inserted a 1.8 kb fragment of the γ-GCShc promoter upstream of a β-globin reporter gene and tested it in transient transfection assays in the human hepatoma cell line HepG2, which expresses MTF-1 endogenously. Cotransfections with mouse or human MTF-1 expression vectors yielded an increased transcription of the reporter gene. Moreover, induction with Zn or Cd further elevated reporter gene expression. More importantly, by RT-PCR analysis we found reduced γ-GCShc gene expression in 13.5-day old MTF-1 knockout embryos compared to wild type embryos. In contrast, expression of the gene encoding the light chain subunit (regulatory subunit) of gamma GCS (γ-GCSlc) or the gene for glutathione syntethase (GSH-Syn), the second enzyme involved in glutathione biosynthesis, was not altered.

Discussion

The metal-regulatory transcription factor MTF-1 is essential for normal liver development and a lack of this protein leads to embryonic lethality with liver decay in mice between E13.5-E14.5. Since MTF-1 is ubiquituosly expressed during mouse gestation and in adult mice (MTF-1 transcripts can be detected as early as embryonic day E8.5 and in all embryonic and adult tissues tested so far: Ç. Günes, unpublished results), we suggest that MTF-1 regulates expression of genes essential for normal liver development and function.

In MTF-1 null mutant embryos no expression of the known target genes MT-I and MT-II could be detected. In addition to these known MTF-1 target genes, we observed reduced transcription level of another gene that codes for the γ-GCS heavy chain. γ-GCS is the first and the rate-limiting enzyme in the biosynthesis of the glutathione (GSH) *de novo* pathway. Glutathione is essential to maintain the cellular redox-balance and is a scavenger of reactive oxygen intermediates (ROI).

We assume that MTF-1 plays additional roles in protection against cytotoxic agents and that the null mutant phenotype is caused by aberrant expression of other genes besides MT-I and MT-II, such as those involved in glutathione synthesis. Thus, loss of MT expression and γ-GCShc downregulation in MTF-1 knockouts could result in disturbed metal homeostasis and insufficient ROI scavenging. Perhaps with the contribution of other genes, MT-I, MT-II and γ-GCShc may act in concert to ensure normal liver development during embryogenesis.

References

Brugnera E, Georgiev O, Radtke F, Heuchel R, Baker E, Sutherland GR and Schaffner W (1994) Cloning, chromosomal mapping and characterisation of the human metal-regulatory transcription factor MTF-1. *Nucl Acid Res* 22: 3167–3173.

Günes Ç, Heuchel R, Georgiev O, Müller K-H, Lichtlen P, Blüthmann H, Marino S, Aguzzi A and Schaffner W (1998) Embryonic lethality and liver degeneration in mice lacking the metal-responsive transcriptional activator MTF-1. *EMBO J* 17: 2846–2854.

Heuchel R, Radtke F, Georgiev O, Stark G, Aguet M and Schaffner W (1994) The transcription factor MTF-1 is essential for basal and heavy metal-induced metallothionein gene expression. *EMBO J* 13: 2870–2875.

Heuchel R, Radtke F and Schaffner W (1995) Transcriptional regulation by heavy metals, exemplified at the metallothionein genes. *In: Inducible Gene Expression* 1, Birkhäuser Boston, Vol I, 206–240.

Kaegi JHR (1991) Overview of metallothionein. *In: Methods Enzymol.* New York: Academic Press.

Masters BA, Kelly EJ, Quaife CF, Brinster RL and Palmiter RD (1994) Targeted disruption of metallothionein I and II genes increases sensitivity to cadmium. *Proc Natl Acad Sci USA* 91: 584–588.

Michalska AE and Choo KHA (1993) Targeting and germ-line transmission of a null mutation at the metallothionein I and II loci in mouse. *Proc Natl Acad Sci USA* 90: 8088–8092.

Mulchay RT and Gipp JJ (1995) Identification of a putative antioxidant response element in the 5'-flanking region of the human γ-glutamylcysteine syntethase heavy subunit gene. *Biochem Biophys Res Commun* 209(1), 227–233.

Quaife CJ et al (1994) Induction of a new metallothionein isoform (MT-IV) occurs during differentiation of stratified squamous epithelia. *Biochemistry* 33: 7250–7259.

Palmiter RD, Findley SD, Whitmore TE and Durnam DM (1992) MT-III, a brain-specific member of the metallothionein gene family. *Proc Natl Acad Sci USA* 89: 6333–6337.

Radtke F, Heuchel R, Georgiev O, Hergersberg M, Gariglio M, Dembic Z and Schaffner W (1993) Cloned transcription factor MTF-1 activates the mouse metallothionein I promoter. *EMBO J* 12: 1355–1362.

Radtke F, Georgiev O, Müller H-P Brugnera E and Schaffner W (1995) Functional domains of the heavy metal-responsive transcription regulator MTF-1. *Nucl Acid Res* 23: 2277–2286.

Thornally PJ and Vašák M (1985) Possible role for metallothionein in protection against radiation-induced oxidative stress. Kinetics and mechanism of its reaction with superoxide and hydroxyl radicals. *Biochim Biophys Acta* 827: 36–44.

Metallothionein IV
C. Klaassen (ed.)
© 1999 Birkhäuser Verlag Basel/Switzerland

New insights into the mechanisms of cadmium regulation of mouse metallothionein-I gene expression

G.K. Andrews, D. Bittel, T. Dalton, N. Hu, W. Chu, M. Daggett, Q. Li and J. Johnson

University of Kansas Medical Center, Kansas City, KS, USA

Introduction

Mouse MT-I and MT-II gene transcription is induced rapidly by heavy metals (especially Zn & Cd) [1]. Metal response elements (MRE) are essential for this induction, and these elements are present in multiple copies in the proximal promoters of these MT genes. MREs were initially shown to mediate transcriptional response of MT genes to Zn and Cd [2–4], and more recently to oxidative stress (5, 6). The Zn-finger transcription factor MTF-1 (MRE-binding transcription factor-1) binds specifically to MREs and transactivates MT gene expression [7, 8]. The Zn-fingers of MTF-1 are of the Cys_2His_2 family and we have shown that the DNA-binding activity of MTF-1 is reversibly regulated by Zn interactions with the Zn-finger domain [9]. In contrast with some Zn-finger proteins, including Zn-finger transcription factors, which can bind Zn with pM to nM disassociation constants [10, 11], MTF-1 is regulated by μM concentrations of this metal. Thus, MTF-1 may serve as a sensor for "free" Zn within the cell.

How MTF-1 participates in the regulation of MT gene expression by transition metals other than Zn is not as well understood. Manipulation of MTF-1 expression by targeted deletion of both genes in embryonic stem cells [12] or by expression of antisense MTF-1 [13] eliminates responsiveness to all transition metals of transfected MRE-driven reporter genes. Thus, MTF-1 is thought to be essential for activation of MRE activity by all of the transition metals that have been examined. Although MTF-1 may play a role in activating MT gene expression in response to several transition metals, the nature of the interaction between MTF-1 and metals other than Zn has not been examined. In order to further elucidate the mechanisms of activation of MTF-1 by transition metals, we examined the DNA-binding activity of MTF-1 *in vivo* in cells treated with Cd We also we utilized whole cell extracts prepared from mouse and human cells and recombinant mouse and human MTF-1, transcribed and translated *in vitro,* to study effects of transition metals on MTF-1 DNA-binding activity. We found that MTF-1 DNA-binding activity is poorly activated *in vivo* by Cd, and that it is activated *in vitro* by Zn but not by any of the other transition metals tested. These data suggest that transition metals, other than Zn, activate MT gene expression through mechanisms independent of a significant increase in DNA-binding activity of MTF-1.

Transcriptional activation of gene expression by Cd involves multiple signal transduction pathways [14–17], including those leading to oxidative stress [18, 19]. Therefore, cooperation of promoter elements may lead to maximal activation of MT-I gene transcription by Cd.

Transcription of the heme oxygenase-I (HO) gene [20] is induced by Cd and this response has been mapped to a distal enhancer region [21] that contains antioxidant response elements (ARE) (also called electrophile response element), and to an upstream stimulatory factor (USF) binding site in the proximal promoter region [22]. AREs mediate induction of the rat glutathione-S-transferase Ya subunit and the rat quinone reductase genes in response to redox cycling xenobiotics and H_2O_2 [23, 24]. USFs are members of the bHLH-bZip protein superfamily and bind to DNA as homo- or heterodimers [25]. USFs interact with other transcription factors to regulate gene expression [26], including Maf, which can interact with ARE sequences [27, 28].

We previously reported that ARE consensus sequences are present in MT promoters from several species, including the mouse (MT-I at -98 to -89-bp) [5]. In the mouse MT-I promoter the ARE consensus sequence overlaps a previously identified USF binding site (−101 to −94) [29]. The USF/ARE element alone (four copies) can direct response to oxidative stress in transient transfections [5]. We have reexamined the role of USF in regulating basal as well as induced expression of the mouse MT-I gene. We discovered that the USF/ARE is dramatically responsive to Cd (2 to 10 µM), but not to Zn, and that USF is essential for this induction. The results of our recent studies are narrated below and are currently in press in *Nucleic Acids Reserach*, or were recently published in *JBC* (273: 7127, 1998).

Results

The DNA-binding activity of MRE-binding transcription factor-1 (MTF-1) is activated in vivo *and* in vitro *by Zn, but not by Cd.*

We examined the DNA-binding activity of mouse and human MTF-1 in whole cell extracts from cells cultured in medium containing Zn or Cd and from untreated cells after the *in vitro* addition of Zn or Cd, as well as using recombinant MTF-1 transcribed and translated *in vitro* and treated with various transition metals. Incubation of human (HeLa) or mouse (Hepa) cells in medium containing Cd (5 to 15 µM) did not lead to a significant increase (< 2-fold) in the amount of MTF-1 DNA-binding activity, whereas Zn (100 µM) led to a 6- to 15-fold increase within 1 h. MTF-1 binding activity was low, but detectable, in control whole cell extracts and was increased (>10-fold) after the *in vitro* addition of Zn (30 µM) and incubation at 37 °C for 15 min. In contrast, addition of Cd (6 or 60 µM) did not activate MTF-1 binding activity. Recombinant mouse and human MTF-1 were also dependent on exogenous Zn for DNA-binding activity. Cd did not facilitate activation of recombinant MTF-1, but instead inhibited the activation of the recombinant protein by Zn. Interestingly, glutathione (1 mM) protected recombinant MTF-1 from inactivation by Cd, and allowed for activation by Zn. Of the several transition metals (Zn, Cd, Ni, Ag, Cu and Co) examined, only Zn facilitated activation of the DNA-binding activity of recombinant MTF-1. These data suggest that transition metals, other than Zn, that activate MT gene expression may do so by mechanisms independent of an increase in the DNA-binding activity of MTF-1.

Identification of the USF/ARE as a Cd-responsive element in the mouse metallothionein-I gene promoter

To further explore mechanism of Cd activation of mouse MT-I gene expression we studied the functions of the USF/ARE. Mouse Hepa cells were stably transfected, using a fusion gene containing four copies of the USF/ARE in front of a minimal promoter and a β-gal reporter gene. Analysis of several stably transfected clones revealed that the USF/ARE directs response to Cd, as well as to hydrogen peroxide, but not to Zn or *tert*-butylhydroquinone. The response to Cd (10 μM) was detectable by 4 h and dramatic (about 30-fold) by 16 h of treatment. Transient transfection assays in Hepa cells demonstrated that deletion of the USF/ARE from the intact MT-I promoter (−150-bp) attenuated (50%) Cd-induction of a luciferase reporter gene. A single copy of the USF/ARE in a minimal promoter-luciferase construct, was dramatically responsive to Cd, but not to Zn, and mutation of either the USF or the ARE attenuated the Cd induction. Furthermore, co-transfection with a dominant negative USF expression vector dramatically reduced Cd responsiveness of the USF/ARE-luciferase reporter gene. Electrophoretic-mobility shift assays detected specific USF and ARE-binding protein complexes using nuclear proteins from untreated Hepa cells. However, the amount of the ARE-specific complex increased after Cd-induction, whereas USF binding activity remained unchanged. Cotransfection of dominant negative USF attenuated Cd induction of the MT promoter of the USF/ARE. In summary, induction of the mouse MT-I gene by Cd in Hepa cells is mediated not only by MTF-1 but also by USF and as yet unidentified ARE-binding proteins. These results also suggest that protein interactions between USF and ARE-binding proteins may be important for Cd responsiveness.

Discussion

The Zn-finger transcription factor MTF-1 may play an essential role in induction of MT gene expression by several transition metals [12, 13]. Among these transition metals, Zn and Cd are the most effective inducers of MT gene expression in mouse cells. The dose-response for induction of MT genes and MRE-driven reporter genes by Cd is 5 to 10-fold lower than that for Zn. We recently reported that Zn can reversibly and directly activate the DNA-binding activity of mouse MTF-1 [9], and the amount of MTF-1 DNA-binding activity, measured *in vitro*, is dramatically and rapidly increased after treatment of cells with Zn or oxidative stress-inducing agents [6]. This increased MTF-1 activity correlates with increased occupancy of MREs in the MT-I promoter [6]. Interestingly, the efficacy of induction by many transition metals, of MRE-driven reporter gene expression in transfected cells, is diminished when the concentration of Zn in the culture medium is reduced. Therefore, it has been suggested that Zn mediates induction of MT gene expression by other transition metals [13]. Furthermore, genomic footprinting suggests that Cd increases the occupancy of MREs in MT promoters in cultured cells [30] which implies that Cd activates the DNA-binding activity of MTF-1 *in vivo*. Therefore, we examined MTF-1 DNA-binding activity in cells treated with Zn or Cd, and of recombinant MTF-1 synthesized in a coupled transcription-translation system and exposed to various transition metals *in vitro*.

In contrast with oxidative stress and Zn [6, 9], Cd has little effect on the amount of DNA-binding activity of MTF-1 extracted from Hepa cells cultured in medium containing Cd. Furthermore, the modest increase in MTF-1 activity that we reported previously [9] was detected later than induction of MT-I gene expression by Cd in these cells. The results reported herein confirm and extend our previous conclusion that Cd, in concentrations that rapidly and efficiently induce MT gene expression, does not cause the rapid activation of MTF-1 DNA-binding activity in mouse, human and dog cells. These results suggest that Cd does not simply cause the redistribution of Zn which, in turn, rapidly activates MTF-1 to bind to DNA. Our results cannot exclude the possibility that Cd causes an increase in the binding affinity of a small amount of MTF-1. This might explain the increased occupancy of MREs detected *in vivo* after treatment of cultured cells with Cd [30]. Due to the difficulty in detecting the small amount of MTF-1 binding activity in the Cd-treated cells, it was not possible to accurately compare the affinity of binding of this transcription factor in Cd-treated versus Zn-treated cells.

Recombinant human and mouse MTF-1 synthesized *in vitro* in a coupled transcription-translation system is not competent to bind to DNA, but can be activated by low micromolar concentrations of Zn [2 to 3 μM] in a temperature dependent manner. Examination of the ability of several transition metals, including Cd, to activate recombinant MTF-1 *in vitro* revealed that none of these metal ions, other than Zn, directly activated this transcription factor to bind to DNA. A wide range of metal concentrations was examined in these experiments. These results are consistent with the hypothesis that transition metals that activate MT gene expression do so by indirect mechanisms independent from those used by Zn. These transition metals may make use of the activated MTF-1 that preexists in the cell to increase MT gene expression. This notion is indirectly supported by the observation that Zn concentration in the culture medium influences the extent of induction of an MRE-driven reporter gene by other transition metals, including Cd [13].

It is possible that transition metals alter the transactivation potential of MTF-1. We speculate that reversible phosphorylation of MTF-1 may play a role in modulating the activities of this transcription factor. Cd can stimulate myosin light chain kinase [14], affect calmodulin activity in the brain [15], and evoke inositol polyphosphate formation [16]. Cd induction of *c-Myc* may involve the activation of protein kinase C [17]. Cd may modulate gene expression by interfering with normal cellular signaling mechanisms at the levels of receptors, calcium and Zn homeostases, protein phosphorylation and modification of transcription factors [31]. Mouse MTF-1 contains a serine/threonine rich transactivation domain [32], and several potential sites of phosphorylation.

The proximal promoter of the mouse MT-I gene contains a complex array of transcriptional activation elements, and although it has been shown that MTF-1 is essential for basal level expression of MT, as well as heavy metal induction [12, 13], it seems likely that MTF-1 may function in conjunction with other transcription factors to regulate MT gene expression. We identified one of those transcription factors as USF and found that interactions between USF and unidentified ARE-binding proteins may be required for maximal Cd-responsiveness. Unlike the MRE, the USF/ARE mediates response to Cd, but not Zn. The MREs and the

USF/ARE also contribute to induction of the mouse MT-I gene by oxidative stress [5]. Cd induced toxicity and mutagenesis have been suggested to involve the rapid generation of reactive oxygen species inside the cell [18, 33]. Oxidative stress, in turn, can enhance tyrosine phosphorylation of the EGF receptor [34], phosphorylation of IkB [35], and Jun kinase [MAPK] activity [36]. Interestingly, the mouse HO-I gene is also inducible by oxidative stresses and Cd, but not by Zn [21, 37–41]. Cd responsiveness of the rat HO-I gene involves USF interactions with the proximal promoter and recent studies of the human HO-I promoter demonstrated that ultraviolet light causes altered protein-DNA interactions within a region overlapping an USF binding site [42]. Therefore, USF may play a central role in the mammalian cells transcriptional response to stress.

Our studies, and those cited above, also suggest that ARE-binding proteins are components of the oxidative stress and Cd-responses of MT genes. The identity of the factors that may bind to AREs is unknown. Constitutive ARE binding activity has been detected in cell extracts [23, 43], and an ARE binding protein has been isolated recently [44]. If the ARE core sequence bears homology with an AP-1 binding site, then Fos/Jun can bind *in vitro* [43]. AP-1 is thought to regulate mouse GST Ya [45] and human Qr [46] gene expression via induced binding to the ARE/TRE, but a functional ARE from the rat GST Ya gene does not bind AP-1 [43]. The terminal GC bases in the ARE consensus sequence are essential for induction by oxidative stress, and GST Ya ARE-binding activity is not induced during oxidative stress [43]. We have detected USF and ARE-specific complexes [6] in control Hepa cell nuclear extracts and the amounts of the ARE binding complex was slowly yet dramatically induced after Cd treatment. The ARE-complex was not predominantly AP-1, based on supershift assays and competition with a TRE [6]. The identity of the ARE binding proteins in Cd-treated Hepa cells remains to be determined. It has recently been shown that Maf recognizes an extended ARE-like sequence [27, 28] and that Maf can form heterodimers with Fos and Jun [27], as well as with USF [47]. In addition, Nrf1 and Nrf2 have been suggested to positively regulate ARE activity, whereas Fos and Fra1 may negatively regulate its activity. Further studies of the proteins that interact at the USF/ARE in the MT-I promoter are required.

Acknowledgements
This work was supported, in part, by an NIH grants to GKA (ES05704) and J.J. (ES08089). D. Bittel was supported, in part, by a National Research Service Award (postdoctoral fellowship ES05753), and T. Dalton was supported, in part, by a postdoctoral fellowship award from the Marion Merrell Dow Scientific Education Partnership. We are indebted to Dr. Lashitew Gedamu and Dr. Susan Samson of the University of Calgary for discussing with us their recent corroborative findings on Cd regulation of MTF-1, and to Jim Geiser and Steve Eklund for excellent technical assistance.

References

1. Andrews GK (1990) *Prog Food Nutr Sci* 14: 193–258.
2. Stuart GW, Searle PF, Chen HY, Brinster RL and Palmiter RD (1984) *Proc Natl Acad Sci USA* 81: 7318–7322.
3. Samson SL-A and Gedamu L (1995) *J Biol Chem* 270: 6864–6871.
4. Foster R and Gedamu L (1991) *J Biol Chem* 266: 9866–9875.
5. Dalton TP, Palmiter RD and Andrews GK (1994) *Nucl Acid Res* 22: 5016–5023.
6. Dalton TP, Li QW, Bittel D, Liang LC and Andrews GK (1996) *J Biol Chem* 271: 26233–26241.

7. Radtke F, Heuchel R, Georgiev O, Hergersberg M, Gariglio M, Dembic Z and Schaffner W (1993) *EMBO J* 12: 1355–1362.
8. Brugnera E, Georgiev O, Radtke F, Heuchel R, Baker E, Sutherland GR and Schaffner W (1994) *Nucl Acid Res* 22: 3167–3173.
9. Dalton TD, Bittel D and Andrews GK (1997) *Mol Cell Biol* 17: 2781–2789.
10. Shi Y, Beger RD and Berg JM (1993) *Biophys J* 64: 749–753.
11. Berg JM (1990) *J Biol Chem* 265: 6513–6516.
12. Heuchel R, Radtke F, Georgiev O, Stark G, Aguet M and Schaffner W (1994) *EMBO J* 13: 2870–2875.
13. Palmiter RD (1994) *Proc Natl Acad Sci USA* 91: 1219–1223.
14. Chao SH, Bu CH and Cheung WY (1995) *Arch Toxicol* 69: 197–203.
15. Vig PJS and Nath R (1991) *Biochem Int* 23: 927–934.
16. Smith JB, Dwyer SD and Smith L (1989) *J Biol Chem* 264: 7115–7118.
17. Tang N and Enger MD (1993) *Toxicology* 81: 155–164.
18. Yang JL, Chao JI and Lin JG (1996) *Chem Res Toxicol* 9: 1360–1367.
19. Amoruso MA, Witz G and Goldstein BD (1982) *Toxicol Lett* 10: 133–138.
20. Maines MD (1988) *FASEB J* 2: 2557–2568.
21. Alam J (1994) *J Biol Chem* 269: 25049–25056.
22. Maeshima H, Sato M, Ishikawa K, Katagata Y and Yoshida T (1996) *Nucl Acid Res* 24: 2959–2965.
23. Favreau LV and Pickett CB (1993) *J Biol Chem* 268: 19875–19881.
24. Jaiswal AK (1994) *Biochem Pharmacol* 48: 439–444.
25. Sirito M, Walker S, Lin Q, Kozlowski MT, Klein WH and Sawadogo M (1992) *Gene Expr* 2: 231–240.
26. Meier JL, Luo X, Sawadogo M and Straus SE (1994) *Mol Cell Biol* 14: 6896–6906.
27. Kataoka K, Noda M and Nishizawa M (1994) *Mol Cell Biol* 14: 700–712.
28. Inamdar NM, Ahn YI and Alam J (1996) *Biochem Biophys Res Commun* 221: 570–576.
29. Carthew RW, Chodosh LA and Sharp PA (1987) *Genes Dev.* 1: 973–980.
30. Andersen RD, Taplitz SJ, Oberbauer AM, Calame KL and Herschman HR (1990) *Nucl Acid Res* 18: 6049–6055.
31. Beyersmann D and Hechtenberg S (1997) *Toxicol Appl Pharmacol* 144: 247–261.
32. Radtke F, Georgiev O, Müller H-P, Brugnera E and Schaffner W (1995) *Nucl Acid Res* 23: 2277–2286.
33. Stohs SJ and Bagchi D (1995) *Free Radical Biol Med* 18: 321–336.
34. Gamou S and Shimizu N (1995) *FEBS Lett* 357: 161–164.
35. Meyer M, Schreck R and Baeuerle PA (1993) *EMBO J* 12: 2005–2015.
36. Karin M (1995) *J Biol Chem* 270: 16483–16486.
37. Keyse SM, Applegate LA, Tromvoukis Y and Tyrrell RM (1990) *Mol Cell Biol* 10: 4967–4969.
38. Keyse SM and Tyrrell RM (1987) *J Biol Chem* 262: 14821–14825.
39. Keyse SM and Tyrrell RM (1989) *Proc Natl Acad Sci USA* 86: 99–103.
40. Applegate LA, Luscher P and Tyrrell RM (1991) *Cancer Res* 51: 974–978.
41. Keyse SM and Tyrrell RM (1990) *Carcinogenesis* 11: 787–791.
42. Nascimento AL, Luscher P and Tyrrell RM (1993) *Nucl Acid Res* 21: 1103–1109.
43. Nguyen T, Rushmore TH and Pickett CB (1994) *J Biol Chem* 269: 13656–13662.
44. Liu SX and Pickett CB (1996) *Biochemistry* 35: 11517–11521.
45. Bergelson S and Daniel V (1994) *Biochem Biophys Res Commun* 200: 290–297.
46. Li Y and Jaiswal AK (1992) *J Biol Chem* 267: 15097–15104.
47. Kurschner C and Morgan JI (1997) *Biochem Biophys Res Commun* 231: 333–339.

Metallothionein IV
C. Klaassen (ed.)
© 1999 Birkhäuser Verlag Basel/Switzerland

Activation of mouse metallothionein I promoter by cadmium in human neuroblastoma cells

Jeffrey A. Johnson[1], Jeffrey D. Moehlenkamp[1], Waihei Chu[1], Douglas Bittel[2] and Glen K. Andrews[2]

Departments of [1]Pharmacology, Toxicology and Therapeutics, and [2]Biochemistry and Molecular Biology, University of Kansas Medical Center, Kansas City, KS 66160-7417, USA

Summary. Metallothioneins (MTs) are cysteine-rich proteins that bind metal ions with high affinity. In all cell types that have been examined, mouse MT-I (mMT-I) and MT-II are up-regulated by zinc (Zn), cadmium (Cd), as well as pro-oxidants (i.e. tert-butyl-hydroquione, tBHQ). The 5'-flanking region of the mMT-I gene contains multiple cis-acting elements that bind the trans-acting factors Sp1, MRE-binding transcription factor-1 (MTF-1) and upstream stimulatory factor (USF). The promoter also includes an antioxidant responsive element (ARE) which overlaps the USF binding site. Transcriptional activation of mMT-I has been extensively studied, however there is little data on regulation of the mMT-I promoter in neuronal cells. Truncations of the mMT-I promoter were inserted into a luciferase (luc) reporter construct. IMR-32 cells were cotransfected with the mMT-I luciferase constructs and a RSV-galactosidase constitutive reporter, and treated with tBHQ (10 µM), Cd (0.5 µM), and Zn (100 µM). The –250 mMT-I-luc, –150 mMT-I-luc, –150 mMT-IΔUSF/ARE-luc (USF/ARE deleted) and MREd$_5$-luc were activated by Cd. Remarkably, these same constructs were not activated by Zn or tBHQ. Electrophoretic mobility shift assays suggest that a functional MTF-1 is not present in the IMR-32 cells. Thus, the absence of MTF-1 could account for the lack of a Zn response. To test this hypothesis, IMR-32 cells were cotransfected with increasing concentrations of a mammalian expression vector for mouse MTF-1 (CMV-mMTF-1) and the -250 mMT-I-luc reporter construct. These conditions regenerated a Zn-mediated induction of the mMT-I reporter gene. Notably, activation this construct by Cd was unaffected by overexpression of mMTF-1. These data suggest that Cd activates the MRE by a MTF-1-independent mechanism in human IMR-32 neuroblastoma cells.

Introduction

Metallothioneins (MT) constitute a conserved family of cysteine-rich heavy metal binding proteins (Kägi, 1991). In the mouse, MT-I and MT-II display a wide tissue distribution and have been demonstrated to participate in detoxification of cadmium (Masters et al., 1994; Michalska et al., 1993), zinc homeostasis (Dalton et al., 1996a) and protection against oxidative stress (Lazo et al., 1995). A hallmark of MT-I and MT-II genes is their transcriptional induction by heavy metals (i.e. zinc and cadmium) (Andrews, 1990). Metal response elements (MRE) are essential for this induction, and these elements are present in multiple copies in the proximal promoters of MT genes. MREs were initially shown to mediate transcriptional response of MT genes to zinc and cadmium (Stuart et al., 1984), and more recently to mediate, in part, the transcriptional response to oxidative stress in mouse hepatoma cells (Dalton et al., 1996b; Dalton et al.,1994).

A protein that binds specifically to MREs and that transactivates MT gene expression has been identified in mouse and human, and is termed MTF-1 (Brugnera et al., 1994; Radtke et al., 1993). MTF-1 is a Zn-finger transcription factor in the Cys2His2 family. Targeted disruption of both MTF-1 alleles in mouse embryonic stem cells demonstrated its essential role for basal as well as heavy metal-induced MT gene expression (Heuchel et al., 1994.). The 5'-flanking region of the mouse MT-I gene (mMT-I) also contains cis-acting elements that bind the trans-acting factors Sp1, and upstream stimulatory factor (USF). An antioxidant responsive element (ARE) overlaps the USF binding site (Dalton et al.,1994), but the protein(s) that bind to the ARE have not been identified. This composite element was shown to contribute to the induction of mMT-I by pro-oxidants in Hepa cells (Dalton et al.,1994).

Transcriptional activation of mMT-I has been extensively studied, however there is little data on regulation of the mMT-I promoter in neuronal cells. Thus, the purpose of this investigation was to determine if the mMT-I promoter is transcriptionally activated by metals and oxidative stress in IMR-32 human neuroblastoma cells, and to identify trans-acting factors involved in the up-regulation of mMT-I promoter.

Materials and methods

Cell culture, transfections and plasmids

IMR-32 human neuroblastoma cells and Hepa mouse hepatoma cells were cultured, transfected, treated and assayed as previously described (Johnson and Nathanson, 1994; Dalton et al., 1996b). Truncations of the mMT-I promoter (–250mMT-I, –150mMT-I, and –42mMT-I) were inserted into a luciferase (luc) reporter construct. A –150ΔUSF/ARE-luc construct (Dalton et al., 1996a) was used to assess the role of the USF/ARE within the context of the intact promoter, whereas insertion of $MREd_5$ (5 copies) inserted upstream to the -42 mMT-I minimal promoter-luciferase reporter and USF/ARE_1 (1 copy) into a TATA-Inr minimal promoter-luciferase reporter (Wasserman and Fahl, 1997) was used to address the enhancer function of these elements. IMR-32 cells and Hepa cells were cotransfected with the mMT-I luciferase constructs and a RSV-galactosidase constitutive reporter. IMR-32 cells were treated with tBHQ (5.0 and 10 µM), Cd (0.5 and 1.0 µM), and Zn (50 and 100 µM). Hepa cells were treated with tBHQ (50 and 100 µM), Cd (5.0 and 10 µM), and Zn (50 and 100 µM). The expression plasmid CMV-MTF-1 was created by inserting the mouse MTF-1 cDNA, into the Not I site of a CMV expression vector kindly provided by Dr. James Smith (Baylor College of Medicine, Houston, TX).

Electrophoretic mobility shift assay

Electrophoretic mobility shift assay (EMSA) were performed using whole cell extracts prepared as described previously (Zimarino and Wu, 1987), with modifications (Dalton et al., 1996b). Proteins from whole cell extracts (20 µg in 1 µl) or a MTF-1 in vitro transcription/translation reaction (1 µl of a 50 µl reaction) were incubated in buffer containing 12 mM HEPES (pH 7.9), 60 mM KCl, 0.5 mM DTT, 12% glycerol, 5 mM $MgCl_2$, 0.2 µg dI/dC/µg protein, 2–4 fmol end-labeled double stranded oligonucleotide (5000 cpm/fmol) in a total volume of 20 µl for 20 min on ice (Fried, 1989). The oligonucleotide sequence used was as follows; bold bases denote the functional core.

MRE-s (Radtke et al., 1993):

5' GATCCAGGGAGCTC**TGCAC**ACGGCCCGAAAAGTA
     GTCCCTCGAGACGTGTGCCGGGCTTTTCATCTAG 3'

Protein-DNA complexes were separated at 4 °C using 4% polyacrylamide gel (acrylamide:bisacrylamide/80:1) electrophoresis (PAGE) at 15 V/Cm. The gel was polymerized and run in buffer consisting of 0.19 M glycine (pH 8.5), 25 mM Tris, 0.5 mM EDTA. After electrophoresis, the gel was dried and labeled complexes were detected by autoradiography.

Results and discussion

The -250 mMT-I-luc was transcriptionally activated by Zn (10-fold), Cd (18-fold) and tBHQ (4- to 7-fold) in Hepa cells. Cd (7- to 9-fold) also induced expression of the -250 mMT-I-luc in IMR-32 human neuroblastoma cells. Remarkably, this same construct was not activated by Zn or tBHQ in the neuroblastoma cell line. Activation of the mMT-I promoter by heavy metals and oxidative stress in Hepa has been completely or partially attributed to interaction between MTF-1 and MRE (Stuart et al., 1984; Dalton et al., 1996b; Dalton et al.,1994). Thus, the absence of MTF-1 in the IMR-32 cells could account for the lack of a Zn response.

The hypothesis that this human neuroblastoma cell line has no functional MTF-1 was evaluated by EMSAs on whole cell lysates from IMR-32 cells compared with Hepa cells. IMR-32 cellular lysates from control, tBHQ, Cd and Zn treated cells did not contain MRE-s-binding activity, and this activity was not activated by Zn *in vitro*. In contrast, Hepa cells treated with Zn and tBHQ had MRE-s-binding activity and this activity is activated *in vitro* by Zn. Interestingly, Hepa cells treated with Cd had little MRE-s-binding activity, and treatment with Zn *in vitro* could activate MRE-s-binding in the Cd-treated lysates. These data imply that Cd may be activating mMT-1 transcription by mechanism(s) not involving MTF-1 and MRE.

To test this hypothesis, IMR-32 cells were cotransfected with increasing concentrations of a mammalian expression vector for mouse MTF-1 (CMV-mMTF-1) and the -250 mMT-I-luc reporter construct. These conditions regenerated a Zn-mediated induction of the mMT-I reporter gene. Notably, activation of this construct by Cd was unaffected by overexpression of mMTF-1 suggesting that Cd activated the MRE by a MTF-1-independent mechanism in human IMR-32 neuroblastoma cells.

A possible alternative site of regulation by Cd is the USF/ARE. This composite element was shown to contribute to the induction of mMT-I by pro-oxidants in mouse hepatoma cells (Dalton et al.,1994). The mMT-I-USF/ARE$_1$-luc fusion construct was not activated by Zn, Cd or tBHQ into IMR-32 cells, and deletion of the USF/ARE element from the intact promoter (–150ΔUSF/ARE-luc) did not effect Cd-mediated induction. In Hepa cells, expression of the USF/ARE$_1$-luc reporter was induced 60-fold by Cd. Thus, the USF/ARE contributes significantly to the activation of mMT-I promoter by Cd in Hepa cells, but plays no role in Cd-mediated induction in IMR-32 cells.

Finally, the mMT-I-MREd$_5$-luc was transfected into IMR-32 cells to determine if Cd could be activating the mMT-I promoter through the MRE in the absence of MTF-1. The MREd$_5$-luc reporter was strongly activated by Cd (16- to 18-fold), slightly activated by Zn (2-fold) and not activated by tBHQ. These data suggest that Cd activates the mMT-I-MRE by a MTF-1-independent mechanism in IMR-32 human neuroblastoma cells.

Studies herein demonstrate that there are at least two alternative mechanisms of mMT-I induction by Cd. In Hepa cells, the USF/ARE predominates, while in IMR-32 cells the major cis-acting element is the MRE. However, MTF-1, the trans-acting factor known to associate with the MRE does not appear to mediate the Cd response in IMR-32 cells. Since the IMR-32 human neuroblastoma cells respond to Cd and not Zn, this cell line can be utilized to further elucidate the mechanism(s) by which Cd regulates the expression of MTs.

Acknowledgements
Supported by NIH Grants ES 08089 to J.A. Johnson and ES 05704 to G.K. Andrews, the Burroughs Wellcome New Investigator in Toxicological Sciences to J.A. Johnson. J.D. Moehlenkamp was supported by NIEHS Toxicology Program Predoctoral Fellowship T32 ES07079. An NRSA fellowship provided support for D. Bittel (F32 ES05753).

References

Andrews GK (1990) Regulation of metallothionein gene expression. *Prog Food Nutr Sci* 14: 193–258.
Brugnera E, O Georgiev F Radtke R Heuchel E Baker GR Sutherland and W Schaffner (1994) Cloning, chromosomal mapping and characterization of the human metal-regulatory transcription factor MTF-1. *Nucl Acid Res* 22: 3167–3173.
Dalton, T.P., K. Fu, R.D. Palmiter, and G.K. Andrews. (1996a) Transgenic mice that over-express metallothionein-I resist dietary zinc deficiency. *J Nutr* 126: 825–833.
Dalton, T.P., Q.W. Li, D. Bittel, L.C. Liang, and G.K. Andrews. (1996b) Oxidative stress activates metal-responsive transcription factor-1 binding activity – Occupancy *in vivo* of metal response elements in the metallothionein-I gene promoter. *J Biol Chem* 271: 26233–26241.
Dalton TP, RD Palmiter and GK Andrews (1994) Transcriptional induction of the mouse metallothionein-I gene in hydrogen peroxide-treated Hepa cells involves a composite major late transcription factor/antioxidant response element and metal response promoter elements. *Nucl Acid Res* 22: 5016–5023.
Fried Fried MFried Fried MG (1989) Measurement of protein-DNA interaction parameters by electrophoresis mobility shift assay. *Electrophoresis* 10: 366–376.
Heuchel R, F Radtke O Georgiev G Stark M Aguet and W Schaffner (1994) The transcription factor MTF-I is essential for basal and heavy metal-induced metallothionein gene expression. *EMBO J* 13: 2870–2875.
Johnson JA and NM Nathanson (1994) Differential Requirements for p21ras and Protein Kinase C in the Regulation of Neuronal Gene Expression by NGF and Neurokines. *J Biol Chem* 269: 18856–18863.
Kägi, J.H.R. (1991. Overview of metallothionein. *Methods Enzymol* 205: 613–626.
Lazo JS, Y Kondo D Dellapiazza AE Michalska KHA Choo andLazo JS, Y Kondo D Dellapiazza AE Michalska KHA Choo and BR Pitt (1995) Enhanced sensitivity to oxidative stress in cultured embryonic cells from transgenic mice deficient in metallothionein I and II genes. *J Biol Chem* 270: 5506–5510.
Masters BA, EJ Kelly CJ Quaife RL Brinster and RD Palmiter (1994) Targeted disruption of metallothionein I and II genes increases sensitivity to cadmium. *Proc Natl Acad Sci USA* 91: 584–588.
Michalska AE and KHA Choo (1993) Targeting and germ-line transmission of a null mutation at the metallothionein I and II loci in mouse. *Proc Natl Acad Sci USA* 90: 8088–8092.
Radtke, F., R. Heuchel, O. Georgiev, M. Hergersberg, M. Gariglio, Z. Dembic, and W. Schaffner. (1993. Cloned transcription factor MTF-1 activates the mouse metallothionein I promoter. *EMBO J* 12: 1355–1362.
Stuart GW, PF Searle HY Chen RL Brinster and RD Palmiter (1984) A 12-base-pair DNA motif that is repeated several times in metallothionein gene promoters confers metal regulation to a heterologous gene. *Proc Natl Acad Sci USA* 81: 7318–7322.
Wasserman WW and WE Fahl (1997) Functional antioxidant responsive elements. *Proc Natl Acad Sci USA*; in press.
Zimarino V and C Wu (1987) Induction of sequence-specific binding of Drosophila heat shock activator protein without protein synthesis. *Nature* 327: 727–730.

Transcription factors involved in heavy metal regulation of the human metallothionein-II$_A$ gene

Shinji Koizumi[1], Yasumitsu Ogra[2], Kaoru Suzuki[1] and Fuminori Otsuka[3]

[1]*Division of Hazard Assessment, National Institute of Industrial Health, 6-21-1, Nagao, Tama-ku, Kawasaki 214, Japan and*
[2]*Laboratory of Toxicology and Environmental Health, Faculty of Pharmaceutical Sciences, Chiba University, 1-33, Yayoi-cho, Inage-ku, Chiba 263, Japan*
[3]*Department of Environmental Toxicology, Faculty of Pharmaceutical Sciences, Teikyo University, Sagamiko, Kanagawa 199-01, Japan*

Transcription of mammalian metallothionein (MT) genes is activated by a variety of metals including zinc, cadmium, copper, mercury, silver and so on [1, 2]. Analysis of control sequences located in the 5'-flanking region of MT genes revealed that a short DNA motif can mediate metal responsiveness [3, 4]. This DNA element, called metal responsive element (MRE), contains a conserved core sequence TGCRCNC (R = purine; N = any base) and a less conserved GC-rich region [5, 6]. The core sequence was reported to be particularly important for mediating metal response [6]. A number of MRE-binding proteins that are assumed to be transcriptional regulators have been described [7]. Recently, cDNA that encodes an MRE-binding factor, mouse MTF-1(mMTF-1), has been isolated [8], and this protein was shown to be essential for metal-induced transcriptional regulation [9].

Mammalian MT genes so far reported have several MREs in their 5'-control region [10], and cooperation of those MRE copies seems important for metal regulation. However, it is not well known how specific regulator proteins interact with these multiple MREs to achieve metal regulation. In this work, we studied correlation between transcriptional activity and regulatory factor binding of individual MREs of the hMT-II$_A$ gene, which encodes the major human metallothionein isoform.

Activity of individual MREs of the hMT-II$_A$ gene to mediate metal response

In the upstream region of the hMT-II$_A$ gene, there are seven sites that contain the MRE core consensus sequence (Fig. 1). To examine the ability of these potential MREs to mediate metal response, reporter gene expression from promoters each containing one of these MREs was estimated. Four tandem MRE repeats were placed upstream of the herpes simplex virus-thymidine kinase gene promoter linked to the chloramphenicol acetyltransferase (CAT) reporter gene, and plasmids carrying these fusions were introduced into HeLa cells. After incubation of cells with or without heavy metals, CAT protein expression was monitored. These experiments showed that metal response is mediated only by four MREs including MREa, MREb, MREe and MREg, although levels of zinc-induced transcription largely differed between different MREs. Remaining MREs, MREc, MREd and MREf had relatively high basal activity, but were not responsive to zinc. Except the hMT-II$_A$ gene, the mouse MT-I (mMT-I) gene is the only example for which the transcriptional activity of each MRE was determined by a similar approach [5]. Comparison of MREs of these two genes revealed that the distribution of metal-responsive MREs over the promoter region is quite different. For example, in the hMT-II$_A$ gene MREa located most proximal to the cap site has the highest activity, whereas in the mMT-I gene MREd located about 150 bp upstream of the cap site is most potent [5]. These observations indicate that the arrangement of active MREs reported for the mMT-I gene is not always essential for the metal response, together with the fact that both the hMT-II$_A$ and mMT-I genes efficiently respond to multiple heavy metal species in a variety of tissues [1, 11].

Comparison of hMT-II$_A$ and mMT-I MREs showed that the sequences of two strong MREs, hMT-II$_A$ MREa and mMT-I MREd are highly homologous. Other metal responsive MREs of these two genes are less homologous to the sequence common to the strong MREs. Although these MREs share some bases in addition to the conserved bases within the MRE core, these are not always shared by metal-responsive MREs of the mMT-I gene [5].

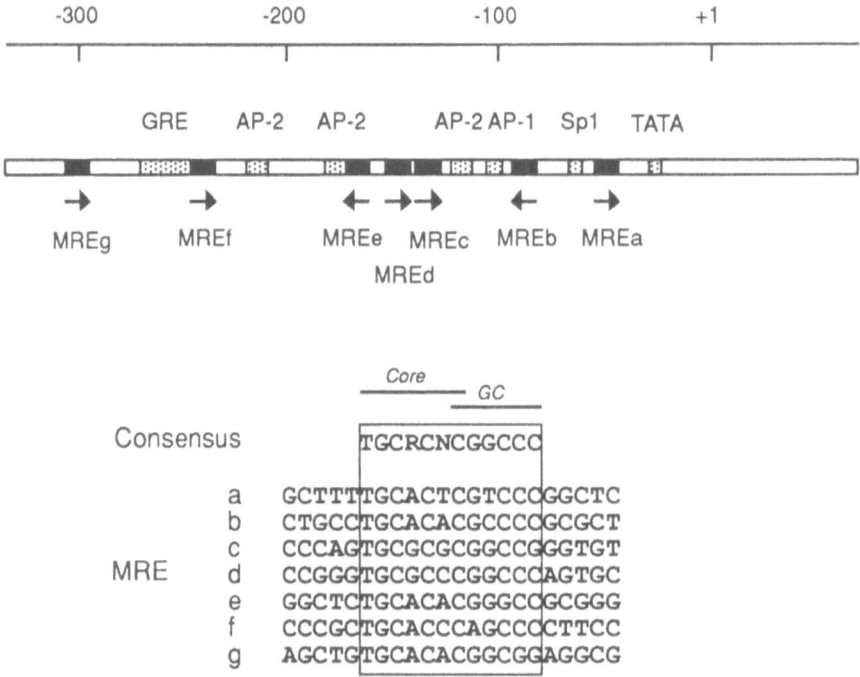

Figure 1. MREs of the hMT-II$_A$ gene. (Upper panel) transcriptional control elements in the hMT-II$_A$ gene upstream region. MREs are indicated by black boxes; arrows indicate the orientation of MRE. Other elements are indicated by gray boxes. (Lower panel) sequences of MREs.

Interaction of hMT-II$_A$ MREs with transcription factor hMTF-1

In our previous work, we purified a human MRE-binding factor named ZRF from HeLa cell nuclei [12]. Amino acid sequence analysis of ZRF fragments suggested that this protein is the human counterpart of mMTF-I [13]. Analysis of cDNA cloned based on the peptide sequence information confirmed the identity of ZRF and hMTF-I. Using a 2,000-fold purified ZRF/hMTF-1 preparation, we examined interactions between this factor and hMT-II$_A$ MREs. Competitive mobility shift assays revealed that the four MREs capable of mediating metal response preferentially bind hMTF-1, suggesting that the interaction between hMTF-1 and these selected MREs plays the central role in metal-induced transcription. MREd and MREf also bound hMTF-1 but only weakly, and MREc showed no affinity to hMTF-1. Transcriptional activity and hMTF-1 binding of hMT-II$_A$ MREs are summarized in Table 1.

Table 1. Transcriptional activity and regulatory factor binding of hMT-II$_A$ MREs

MRE	MTF-1[1]	Sp1[2]	Zinc response[3]	Induced activity[4]
a	+	−	+	high
b	+	+	+	low
c	−	−	−	
d	+	+	−	
e	+	−	+	medium
f	+	+	−	
g	+	−	+	medium

[1] hMTF-1 binding in protein-DNA binding assay *in vitro*.
[2] Sp1 binding in protein-DNA binding assay *in vitro*.
[3] Zinc responsiveness in transfection assay.
[4] Levels of zinc-induced activity in transfection assay.

Search and analysis of other MRE-binding regulatory proteins

Although MREb can mediate response to zinc, its activity is much less than other metal-responsive MREs. On the other hand, MREd and MREf show no response to metals despite that they can interact with hMTF-1 weakly. We speculated that some other proteins interacting with MREs are responsible for such discrepancy. To search additional nuclear factors that can interact with MREs, we analyzed HeLa cell crude nuclear extracts by mobility shift assays using oligonucleotide probes containing each of the seven hMT-II$_A$ MREs. In this analysis, we observed that MREb, MREd and MREf strongly bound a protein distinct from hMTF-1. Competition experiments suggested that an identical protein binds to these MREs, and that this factor recognizes the GC-rich region rather than the MRE core which is essential for hMTF-1 binding. We then noticed that these three MREs contain sequences homologous to the recognition sequence of the transcription factor Sp1 [14]. Supershift analysis using an Sp1-specific antibody demonstrated that Sp1 actually binds to all of the three MREs (Tab. 1). Furthermore, binding of Sp1 to the sequence overlapping MREb was confirmed by DNase-1 footprinting experiments using recombinant Sp1.

Sp1 probably acts as a negative modulator of MRE-mediated transcription

These results suggest a possibility that Sp1 competes with hMTF-1 for binding to these MREs, thereby modulating metal response of the hMT-II$_A$ gene. It was found difficult to see whether these two factors compete by *in vitro* DNA-binding assay, since these two complexes were similar in electrophoretic mobility, and the band intensity of the Sp1 complex was much greater than that of the hMTF-1 complex, hampering estimation of levels of the latter. We therefore examined the transcriptional activity of MREb variants with reduced affinity to Sp1. Two MREb mutants each carrying a single base substitution within the Sp1 recognition site were used. It was confirmed by competitive mobility shift assays that these mutations dramatically reduce Sp1 binding, but do not affect hMTF-1 binding. When examined by CAT assay, both of these MREb mutants could mediate transcriptional activation much more efficiently than wild type MREb. Thus our work suggests that Sp1 can act as a negative modulator in metal-induced transcription by competing with hMTF-1 for binding to certain MREs.

Acknowledgment
This work was supported in part by a grant-in-aid from the Science and Technology Agency, Japan.

References

1. Durnam DM, Palmiter RD (1981) Transcriptional regulation of the mouse metallothionein-I gene by heavy metals. *J Biol Chem* 256: 5712–5716.
2. Durnam DM, Palmiter RD (1984) Induction of metallothionein-I mRNA in cultured cells by heavy metals and iodoacetate: evidence for gratuitous inducers. *Mol Cell Biol* 4: 484–491.
3. Stuart GW, Searle PF, Chen HY, Brinster RL, Palmiter RD (1984) A 12-base-pair DNA motif that is repeated several times in metallothionein gene promoters confers metal regulation to a heterologous gene. *Proc Natl Acad Sci USA* 81: 7318–7322.
4. Karin M, Haslinger A, Holtgreve H, Richards RI, Krauter P, Westphal HM, Beato M (1984) Characterization of DNA sequences through which cadmium and glucocorticoid hormones induce human metallothionein-II$_A$ gene. *Nature* 308: 513–519.
5. Stuart GW, Searle PF, Palmiter RD (1985) Identification of multiple metal regulatory elements in mouse metallothionein-I promoter by assaying synthetic sequences. *Nature* 317: 828–831.
6. Culotta VC, Hamer DH (1989) Fine mapping of a mouse metallothionein gene metal response element. *Mol Cell Biol* 9: 1376–1380.
7. Koizumi S, Otsuka F (1993) Factors involved in the transcriptional regulation of metallothionein genes. *In*: KT Suzuki, N Imura, M Kimura (eds): *Metallothionein III*. Birkhäuser Verlag, Basel, 457–474.
8. Radtke F, Heuchel R, Georgiev O, Hergersberg M, Gariglio M, Dembic Z, Schaffner W (1993) Cloned transcription factor MTF-1 activates the mouse metallothionein I promoter. *EMBO J* 12: 1355–1362.
9. Heuchel R, Radtke F, Georgiev O, Stark G, Aguet M, Schaffner W (1994) The transcription factor MTF-1 is essential for basal and heavy metal-induced metallothionein gene expression. *EMBO J* 13: 2870–2875.
10. Koizumi S, Otsuka F (1995) Transcriptional regulation of the metallothionein gene: metal responsive element and zinc regulatory factor. *In*: B Sarkar (ed.): *Genetic Response to Metals*. Marcel Dekker, New York,

397–410.

11. Heguy A, West A, Richards RI, Karin M (1986) Structure and tissue-specific expression of the human met-allothionein I$_B$ gene. *Mol Cell Biol* 6: 2149–2157.

12. Otsuka F, Iwamatsu A, Suzuki K, Ohsawa M, Hamer DH, Koizumi S (1994) Purification and characterization of a protein that binds to metal responsive elements of the human metallothionein II$_A$ gene. *J Biol Chem* 269: 23700–23707.

13. Brugnera E, Georgiev O, Radtke F, Heuchel R, Baker E, Sutherland GR, Schaffner W (1994) Cloning, chromosomal mapping and characterization of the human metal-regulatory transcription factor MTF-1. *Nucl Acid Res* 22: 3167–3173.

14. Briggs MR, Kadonaga JT, Bell SP, Tjian R (1986) Purification and biochemical characterization of the promoter-specific transcription factor, Sp1. *Science* 234: 47–52.

Transcriptional regulation of the gene encoding mouse metallothionein-3 and its expression in the organs of the reproductive system

Pierre Moffat[1], Raffaella Faraonio[1], Olivier LaRochelle[1], Isabelle Delisle[1], René Saint-Arnaud[2] and Carl Séguin[1*]

[1]*Centre de recherche en cancérologie, CHUQ, Pavillon Hôtel-Dieu de Québec, Québec, Québec, Canada G1R 2J6; and Département de physiologie, Faculté de médecine, Université Laval, Québec, Québec, Canada G1K 7P4*
[2]*Genetics Unit, Shriners Hospital for Crippled Children, Montréal, Québec, Canada H3G 1A6*

Introduction

The survival of, as well as neurite formation by rat cortical neurons is enhanced when they are cultured in presence of Alzheimer's disease (AD) brain extracts [1]. Metallothionein-3 (MT-3) was initially discovered by its ability to antagonize this positive effect of AD brain extracts and was called "Growth Inhibitory Factor" (GIF) [2]. Because MT-3 levels appear to be reduced in AD brains, MT-3 may be involved in the pathogenesis of AD [2, 3]. However, MT-3 down-regulation in AD brains was not observed in all cases [4]. MT-3 may normally play a role in preventing the neuronal sprouting and the development of neurofibrillar tangles associated with the disease [2].

Contrary to other *MT* genes which are highly inducible by metals and which are expressed in most tissues, *MT-3* gene expression fails to respond to metals *in vivo* and was reported to be restricted to the brain. Immunohistochemical and *in situ* localization studies showed that within the normal brain, MT-3 is found in subsets of astrocytes [2, 5–9] and neurons [7–12], and is abundantly present in specialized neurons of the hippocampus that sequester Zn^{2+} in synaptic vesicles within their terminals (Zn-ergic neurons) [10].

The transcription of the *MT-3* gene is not inducible by any of the stimuli (metal ions, hormones, lipopolysaccharides, etc) that normally increase *MT-1* and *MT-2* gene transcription in liver and brain [6, 13, 14], although metal ions, dexamethasone, ethanol, interleukin-6, and kainic acid have been reported to lower MT-3 mRNA levels by 30–60% [5, 8, 12, 15].

The mechanisms that govern the specific expression of *MT-3* in normal brain, its down- regulation in the AD brain and its repression in other organs remain unknown. Our interest in the *MT* gene system led us to study the transcriptional regulation of the *MT-3* gene. In the course of these transcriptional studies on *MT-3*, we found that MT-3 mRNA, in addition to being present in the brain, is present in other organs particularly in those of the reproductive system.

Mouse *MT-3* gene transcription

The mouse *MT-3* gene contains all the *cis*-acting regulatory elements which have been shown to be required and sufficient [16] for constitutive and metal ion-regulated expression, including three metal regulatory elements (MREs), and putative binding sites for many transcription factors, including Sp1, NF-1, and AP-2, as well as a CTG repeat which has been reported to function as a silencer element contributing to *MT-3* gene negative regulation in non-permissive tissues [17] (Fig. 1). However, such a repeat is not found in the human *MT-3* gene [13]. To study the mechanisms and identify the *cis*-acting regulatory elements involved in the control of *MT-3* gene transcription in the brain and its repression in other tissues, we performed transient transfection experiments in cell cultures.

We found that *MT-3* gene transcription is induced in P19 embryonal carcinoma cells [18] induced to differentiate with retinoic acid (RA). A portion (600 bp) of the mouse *MT-3* gene promoter was inserted into LUC or CAT reporter plasmids, and various deletion mutants were generated and transfected into HepG2 cells and into RA-treated P19 cells. Reporter plasmids containing hybrid promoter sequences were also constructed by ligating the upstream promoter sequences of the mouse *MT-1* (–600 to –35) to a mouse *MT-3* minimal promoter DNA fragment (–35 to +20) and conversely, i.e. the upstream *MT-3* promoter was fused to a minimal *MT-1* promoter (Fig. 2). We found that 180 bp of *MT-3* promoter DNA are sufficient for the appropriate expression of *MT-3* in P19 cells. However, this promoter DNA fragment was poorly active in HepG2 cells, generating CAT activity barely over the background level (Fig. 2). This suggests that *MT-3* gene transcriptional inhibition in HepG2 cells is partially due to differences in the *cis*-acting control sequences of the *MT-3* gene itself, but an organization of the gene into a separate chromosomal domain cannot be excluded. Furthermore, our results do not support the contention that the CTG repeat is a silencer element since its deletion did not restore transcription in HepG2 cells (data not shown). The shuffling experiment showed that the minimal *MT-3* gene promoter is functional in HepG2 cells, while the region upstream of the TATA box is very weak (Fig. 2). These results, as well as deletion mutant analyses (data not shown), suggest that the region between -180 and -35 may contain transcriptional regulatory elements. This 145-bp DNA fragment contains three MREs, and binding sites for the transcription fac-

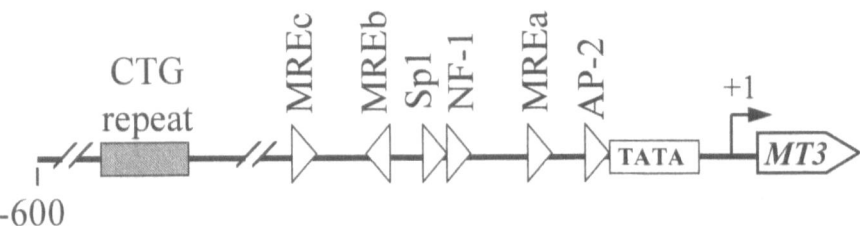

Figure 1. Map of the mouse *MT-3* promoter showing the putative binding sites of some transcription factors and the CTG repeat.

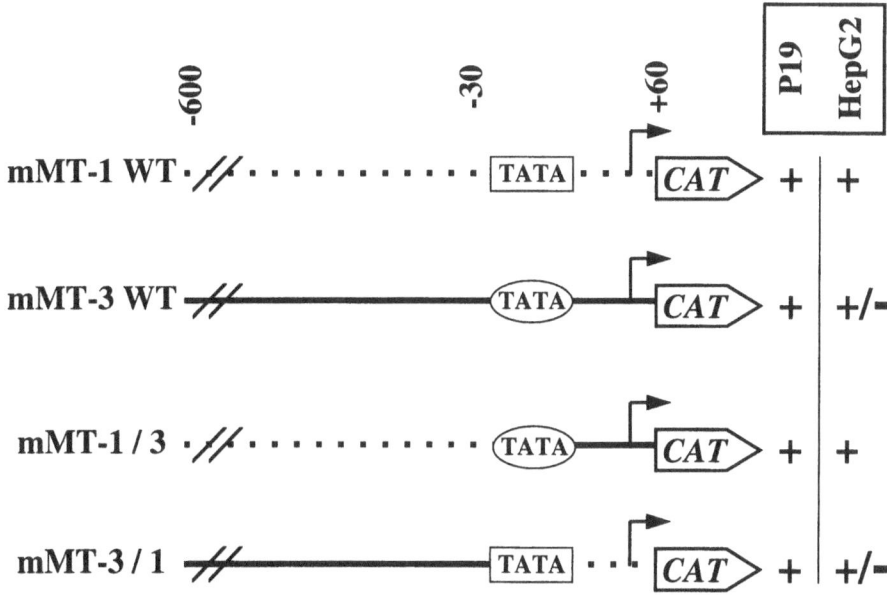

Figure 2. Plasmid construction and CAT assay. A portion of the 5'-UTR of mouse MT-1 (mMT-1) and MT-3 (mMT-3) genes, including the transcription start site (tsp) (bent arrow) were inserted into the pCATbasic (Stratagene) reporter plasmid. The plasmids mMT-1/3 and mMT-3/1 are hybrid promoters in which a portion of the mouse *MT-1* promoter (-600 to -35) was fused to a mouse *MT-3* minimal promoter DNA fragment (-35 to +20), and *vive versa*, i.e. the same portion of the mouse *MT-3* distal promoter fused to the minimal mouse *MT-1* promoter. Broken line: MT-1 promoter sequences; solid line: MT-3 promoter sequences. The resulting plasmids were transfected into P19 or HepG2 cells, and CAT activity was measured. The numbers indicate nucleotide positions relative to the *tsp*. The TATA box is indicated. Presence (+) or absence (-) of basal CAT activity is indicated at the right.

tors Sp1, NF-1 and AP-2 (Fig. 1). DNAseI footprinting analyses showed that nuclear extracts prepared from liver and brain contained proteins binding to the NF-1/Sp1, and MREb sites (data not shown). No major differences were evident in the footprints generated by the two extracts, apart from DnaseI hypersensitive sites in the MREa/AP-2 region which were present when the liver extract was used.

The insertion of two strong MREs into the *MT-3* gene promoter partially improved metal ion induction of the reporter plasmid (data not shown). To further characterize this *MT-3* promotor region, site directed mutagenesis experiments were performed on different sites. The results showed that the Sp1, NF-1 and AP-2 sites are not involved in *MT-3* gene transcriptional repression (data not shown). It is worth noting that the proximal region of the *MT-3* promoter is highly conserved between mouse and rat, suggesting that this region is important for *MT-3* transcriptional regulation. We also found that, in P19 cells, the AP-2 site, and to a lesser extent the Sp-1 and NF-1 sites, are important for maintaining optimal transcription activity (Fig. 3).

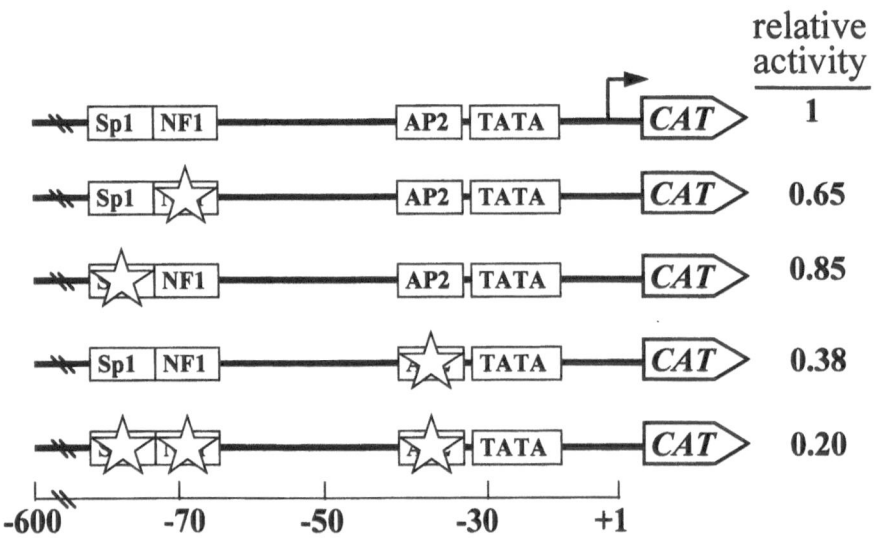

Figure 3. Transfection studies performed with mutant MT-3 promoter CAT constructs. Plasmids were transfected into RA-treated P19 cells, and CAT activity was measured. CAT activities relative to the wild type MT-3 promoter taken as 100% are indicated on the right. Stars indicate mutations introduced into the different sites. The numbers at the bottom indicate nucleotide positions relative to the *tsp* (bent arrow).

MT-3 gene expression in organs other than the brain

MT-3 mRNA is present in other organs than the brain. Northern hybridization analyses showed that in the rat, MT-3 mRNA is present in brain, testis, prostate, uterus, ovary, and seminal vesicles (Fig. 4A), and also in epididymis, tongue, stomach, heart, and kidney. Compared to the brain, MT-3 mRNA levels were approximately 4.5-fold lower in testes, epididymis, prostate and tongue, 25-fold lower in ovaries, uterus, and stomach, and at least 50-fold lower in bladder, heart and kidney. While $CdCl_2$ and LPS led to an appreciable increase of MT-1 mRNA in liver and brain, MT-3 mRNA levels remained unchanged in brain, testis, and prostate. In mice, MT-3 transcripts were also found in testis, ovary, and uterus (Fig. 4B). However, mouse MT-3 mRNA levels were lower in these organs than those present in the same organs in the rat, in relation to MT-3 mRNA levels in the brain. MT-3 transcript levels were approximately 20-fold lower in the testis than in the brain, 25-fold lower in the ovary and lower still in the uterus. A very faint signal was detected in RNA from mouse kidney and small intestine. In humans, testicular and ovarian MT-3 mRNA levels were lower still than those in mouse when compared to brain. MT-3 transcripts were readily detected in testis, but the hybrization signal was very faint in ovary, and barely over the background level in prostate.

To identify the cell types expressing *MT-3* in the human testis, *in situ* hybridization was performed. MT-3 transcript localized nonuniformly to Leydig cells and seminiferous tubules [20]. Not all Leydig cell clusters were labeled. Similarly, seminiferous tubules were not all labeled with the same pattern. Some tubules had patchy labeling while others were more uniformly

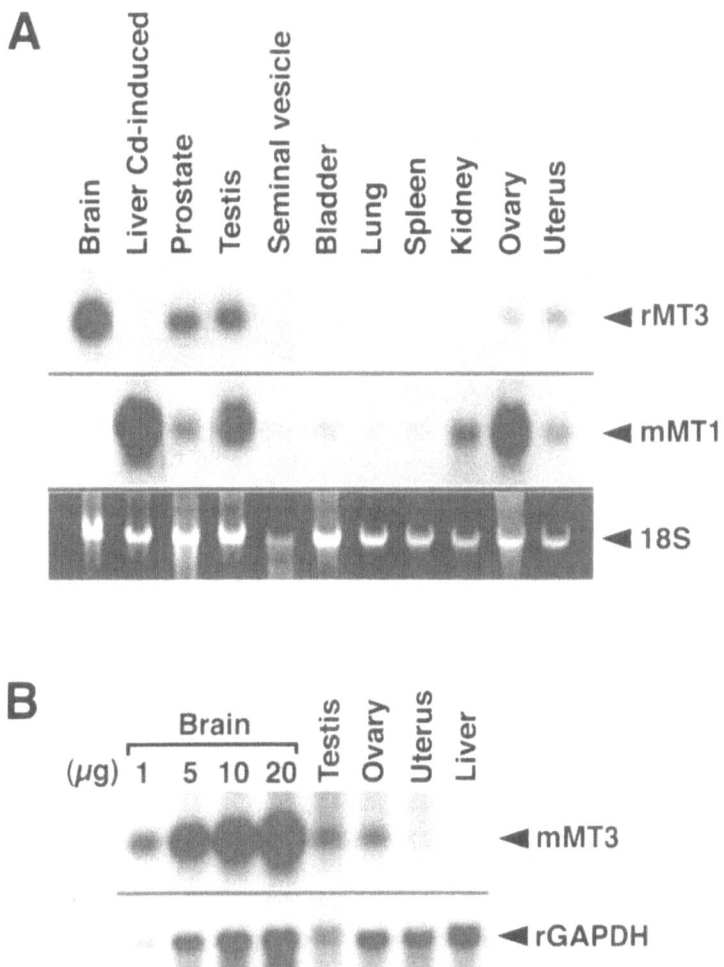

Figure 4. Northern analysis of MT-3 in adult rat (A) and mouse (B) tissues. All organs were obtained from control rats and mice. The rat liver was from an animal injected with a single dose of CdCl$_2$ (5 mg/kg) and killed 6 h later. Total RNA (20 µg, except where indicated) was electrophoresed on 1.1% agarose/2.2 M formaldehyde gels, transferred to Hybond-N membranes, and hybridized with ^{32}P-labeled cDNA probes corresponding to the rat (rMT-3) or the mouse (mMT3) MT-3 3'UTR. For the rat, the blot was washed and re-hybrydized with a mouse MT-1 (mMT1) cDNA probe. Comparable loadings were confirmed by ethidium bromide staining of 18S ribosomal RNA (18S) (A), and hybridization with a rat GAPDH (rGAPDH) cDNA probe (B).

labeled. The spermatogonial cell layer appeared mostly devoid of labeling. Heterogeneity of the labeling among Leydig cells and seminiferous tubules may reflect functional or physiological differences among the different cell populations.

The temporal expression profile of *MT-3* was analyzed in postnatal rat testis and brain. In brain, MT-3 mRNA was easily detectable in one-day rats, it increased until day 22, and then leveled off to adult levels [20]. In contrast, in testis, *MT-3* was expressed at constant levels during postnatal development. Indeed, in 6-day animals, MT-3 mRNA levels were similar to those present in adult animals.

To determine whether *MT-3* expression is controlled by androgens, male rats orchiectomized three weeks previously were treated with 5α-dihydrotestosterone (DHT) twice a day for three days or with the vehicle alone, and MT-3 mRNA levels were determined in the prostate and brain. In castrated rats, the prostate weight decreased to 10% of that of control animals. Regression of the prostate was accompanied by a 5- to 10-fold decrease of both MT-3 and MT-1 mRNAs compared to control animals [20]. DHT injections in castrated animals restored prostate mass to approximately 20% of the controls, and led to a complete recovery of MT-1 and MT-3 mRNA levels.

Taken together, these results show that the *MT-3* gene is expressed in organs other than the brain, and that species-specific differences exist in the level at which MT-3 mRNA accumulate in the organs of the reproductive system. The highest levels were found in the rat, intermediate levels were present in the mouse while the lowest levels were found in the human.

Conclusion

We have shown here that *MT-3* gene transcription is induced in P19 cells differentiated with RA, and that the AP-2 site, and to a lesser extent the Sp1 and NF-1 sites, are important for maintaining optimal transcription activity. P19 cells are developmentally pluripotent murine cells. Treatment of P19 cells with RA induces their differentiation into neurons, astrocytes and other cell types derived from neuroectoderm [18]. The P19-derived neurons express a variety of neuronal markers such as neurofilament proteins, and Wnt-1 [19], and are heterogenous, both with respect to morphology and to the neurotransmitters they synthesize. Because P19-derived neurons are able to form synapses and acquire the electrophysiological properties of mature neurons, they are an attractive cell culture system for investigating the role of MT-3 in neural cell biology.

The absence of *MT-3* gene transcription in liver cells appears to be partially due to differences in the *cis*-acting control sequences of the *MT-3* gene itself. However, other mechanisms may also be involved.

We also showed that in rats, mice and humans, *MT-3* in addition to being expressed in the brain, is expressed in organs of the reproductive system, namely testis and ovary. Furthermore, in rats, detailed Northern analyses revealed that MT-3 mRNA is also present in epididymis, prostate, tongue, uterus, seminal vesicles, kidney, stomach, and bladder.

Whether the function of MT-3 in the organs of the reproductive system is similar to that in the brain is unknown. Although MT-3 deficient mutant mice mate and reproduce normally, this does not preclude an important role of MT-3 in the gonads under particular conditions of stress.

These results show that *MT-3* tissue-specific gene expression is broader than previously reported. Although the biological significance of the presence of MT-3 mRNA remains to be elucidated, the present study provides new experimental systems and set the basis for further experiments aimed at defining the function and the action of MT-3 in cellular homeostasis.

Acknowledgements
We are grateful to A. Anderson for critical reading of the manuscript, R.D. Palmiter for providing the mouse MT-3 genomic clone, and B. Têtu for the human tissue-sections. This study was supported by National Institutes of Health Grant R01-CA61261 and Medical Research Council of Canada Grant MT-12468 (to C.S.). R.F. held a fellowship from the Programme québécois de bourses d'excellence, P.M. from the Fédération québécoise des sociétés Alzheimer, and O.L. held a studentship from the Fonds pour la formation des Chercheurs et l'aide à la recherche du Québec.

References

1. Uchida Y, Ihara Y, Tomonaga M (1988) Alzheimer's disease brain extract stimulates the survival of cerebral cortical neurons from neonatal rats. *Biochim Biophys Acta* 150: 1263–1267.
2. Uchida Y, Takio K, Titani K, Ihara Y, Tomonaga M (1991) The growth inhibitory factor that is deficient in the Alzheimer's disease brain is a 68 amino acid metallothionein-like protein. *Neuron* 7: 337–347.
3. Tsuji S, Kobayashi H, Uchida Y, Ihara Y, Miyatake T (1992) Molecular cloning of human growth inhibitory factor cDNA and its down-regulation in Alzheimer's disease. *EMBO J* 11: 4843–4850.
4. Erickson JC, Sewell AK, Jensen LT, Winge DR, Palmiter RD (1994) Enhanced neurotrophic activity in Alzheimer's disease cortex is not associated with down-regulation of metallothionein-III (GIF). *Brain Res* 649: 297–304.
5. Zheng H, Berman NEJ, Klaassen CD (1995) Chemical modulation of metallothionein I and III mRNA in mouse brain. *Neurochem Int* 27: 43–58.
6. Kramer KK, Liu J, Choudhuri S, Klassen CD (1996) Induction of metallothionein mRNA and protein in murine astrocyte cultures. *Toxicol Appl Pharmacol* 136: 94–100.
7. Yamada M, Hayashi S, Hozumi I, Inuzuka T, Tsuji S, Takahashi H (1996) Subcellular localization of growth inhibitory factor in rat brain: Light and electron microscopic immunohistochemical studies. *Brain Res* 735: 257–264.
8. Belloso E, Hernandez J, Giralt M, Kille P, Hidalgo J (1996) Effect of stress on mouse and rat brain metallothionein I and III mRNA levels. *Neuroendocrinology* 64: 430–439.
9. Hozumi I, Inuzuka T, Ishiguro H, Hiraiwa M, Uchida Y, Tsuji S (1996) Immunoreactivity of growth inhibitory factor in normal rat brain and after stab wounds – An immunocytochemical study using confocal laser scan microscope. *Brain Res* 741: 197–204.
10. Masters BA, Quaife CJ, Erickson JC, Kelly EJ, Froelick GJ, Zambrowicz BP, Brinster RL, Palmiter RD (1994) Metallothionein III is expressed in neurons that sequester zinc in synaptic vesicles. *J Neurosci* 14: 5844–5857.
11. Choudhuri S, Kramer KK, Berman NEJ, Dalton TP, Andrews GK, Klaassen CD (1995) Constitutive expression of metallothionein genes in mouse brain. *Toxicol Appl Pharmacol* 131: 144–154.
12. Kramer KK, Zoelle JT, Klaassen CD (1996) Induction of metallothionein mRNA and protein in primary murine neuron cultures. *Toxicol Appl Pharmacol* 141: 1–7.
13. Palmiter RD, Findley SD, Whitmore TE, Durnam DM (1992) MT-III, a brain-specific member of the metallothionein gene family. *Proc Natl Acad Sci USA* 89: 6333–6337.
14. Dalton T, Pazdernik TL, Wagner J, Samson F, Andrews GK (1995) Temporalspatial patterns of expression of metallothionein- I and III and other stress related genes in rat brain after kainic acid-induced seizures. *Neurochem Int* 27: 59–71.
15. Hernandez J, Molinero A, Campbell IL, Hidalgo J (1997) Transgenic expression of interleukin 6 in the central nervous system regulates brain metallothionein-I and -III expression in mice. *Mol Brain Res* 48: 125–131.
16. Imbert J, Culotta VC, Fürst P, Gedamu L, Hamer DH (1990) Regulation of metallothionein gene transcription by metals. *Adv Inorg Biochem* 8: 139–164.
17. Imagawa M, Ishikawa Y, Shimano H, Osada S, Nishihara T (1995) CTG triplet repeat in mouse growth inhibitory factor/metallothionein III gene promoter represses the transcriptional activity of the heterologous promoters. *J Biol Chem* 270: 20898–20900.
18. Bain G, Ray WJ, Yao M, Gottlieb DI (1994) From embryonal carcinoma to neurons: the P19 pathway. *Bioessays* 16: 343–348.
19. St-Arnaud R, Craig J, McBurney MW, Papkoff J (1989) The int-1 proto-oncogene is transcriptionally activated during neuroectodermal differentiation of P19 mouse embryonal carcinoma cells. *Oncogene* 4: 1077–1080.
20. Moffat P, Séguin C (1998) Expression of the gene encoding metallothionein-3 in organs of the reproductive system. *DNA Cell Biol* 17: 501–510.

Metallothionein IV
C. Klaassen (ed.)
© 1999 Birkhäuser Verlag Basel/Switzerland

Dynamic regulation of yeast metallothionein gene transcription via metalloregulatory signaling factors

D.J. Thiele, M.M.O. Peña and K.A. Koch

The University of Michigan Medical School, Ann Arbor, MI, USA

Yeast cells have provided an excellent model for understanding the functional roles of MT proteins, the transcriptional regulation of MT genes in response to metal and other stresses and how MT gene expression interfaces with cellular metal physiology. As a consequence of exposure to high levels of potentially toxic metal ions, eucaryotic cells must rapidly sense and respond to the presence of metals by rapidly activating the expression of metallothionein genes. In the baker's yeast, *S. cerevisiae*, one MT isoform is encoded by the *CUP1* gene, which is transcriptionally activated by Cu ions through the Ace1 metalloregulatory transcription factor, a Cu-activated DNA binding protein which is inducibly bound to metal responsive elements and which delivers a potent activation domain to the transcriptional machinery [1]. In the opportunistic pathogenic yeast *Candida glabrata*, a large MT gene family is transcriptionally activated by the functionally homologous Amt1 protein [2, 3]. Although the ACE1 gene is constitutively expressed and the Ace1 protein found in the nucleus, the AMT1 gene is subject to rapid positive autoregulation that is critical to mount an appropriate detoxification response in the presence of high copper ion concentration [4]. The efficient access of Amt1 to the single metal response element (MRE) in the AMT1 promoter requires both major and minor groove contacts on the DNA [5] and the use of a specially formed chromatin structure leads to dramatic defects in AMT1 autoregulation and copper ion detoxification [6]. Interestingly, in response to chronic Cu administration, the *CUP1* gene is rapidly and robustly activated by Ace1, yet *CUP1* mRNA levels diminish to a new low basal level with no diminution of Ace1 steady state levels. The construction and analysis of Ace1 DNA binding domain fusions to the *Herpes simplex* virus VP16 protein activation domain show that the *CUP1* transient expression is specific to the Ace1 Cu-activated DNA binding domain. Furthermore, *in vivo* footprinting experiments demonstrate that loss of Ace1 binding to the metal response elements correlates with the severe reduction in *CUP1* mRNA levels. Furthermore, we demonstrate that transient expression of the *CUP1* gene is dependent on the proper expression of the *S. cerevisiae* Cu transport machinery. These studies demonstrate that, although the *CUP1* MT plays an important role in Cu detoxification in yeast cells, there is a dynamic interplay between the Cu detoxification pathway and the Cu acquisition machinery, to insure that sufficient yet non-toxic levels of Cu are maintained in the cell.

References

1. Zhu Z and Thiele DJ (1996) Toxic Metal-Responsive Gene Transcription. *In*: U Feige, RI Morimoto, I Yahara and B Polla (eds): *Stress-Inducible Cellular Responses*. Birkhäuser Verlag, Basel.

2. Zhou P and Thiele (1991) Isolation of a metal-activated transcription factor gene from *Candida glabrata* by complementation in *Saccharomyces cerevisia. Proc Natl Acad Sci USA* 88: 6112–6116.
3. Zhou P, Szcypka MS, Sosinowski T and Thiele (1992) Expression of a yeast metallothionein gene family is activated by a single metalloregulatory transcription factor. *Mol Cell Biol* 12: 3766–3775.
4. Zhou P and Thiele (1993) Rapid transcriptional autoregulation of a yeast metalloregulatory transcription factor is essential for high level copper detoxification. *Gene Develop* 7: 1824–1835.
5. Koch KA and Thiele (1996) Autoregulation by a *Candida glabrata* copper metalloregulatory transcription factor requires critical minor groove contacts. *Mol Cell Biol* 16: 724–734.
6. Zhu Z and Thiele (1996) A specialized nucleosome modulates the kinetics of transcriptionfactor binding to a *Candida glabrata* metal responsive promoter *in vivo. Cell* 87: 459–470.

Metallothionein IV
C. Klaassen (ed.)
© 1999 Birkhäuser Verlag Basel/Switzerland

Metal regulation of mammalian metal responsive element (MRE) binding proteins

L. Gedamu, S.L.-A. Samson, S. Schieman and W. Paramchuk

Department of Biological Sciences, The University of Calgary, Calgary, Alberta, Canada, T2N 1N4

Summary. The cloning of the mammalian Metal Responsive Element (MRE) binding factor, MTF, has been an important advance toward understanding metallothionein (MT) gene regulation as mediated by MREs. However, there is little information on the *in vivo* regulation of this factor by metals other than zinc. Here, our hypothesis was that, if MTF is the only MRE binding metal regulatory factor in the cell, then other metals (i.e. cadmium) should be able to activate MRE binding. We detect single MRE specific complexes in both human (HeLa and Hep G2) cell and mouse (L) cell *nuclear* extracts using electrophoretic mobility shift assays (EMSAs). These complexes comigrate with *in vitro* translation products of the mouse and human MTF cDNAs. In untreated cell extracts, MRE binding is activated by the addition of zinc ions to reactions. Because no metal chelators were employed during extract preparation or in the reactions, this result indicates that MTF may be present in the nucleus in inactive form and it is available for binding by zinc. Cadmium addition *in vitro* cannot activate the factors suggesting they are specific for zinc. We also have observed that treatment of cells with zinc for 4 h in culture activates MRE binding without requiring zinc *in vitro*. These complexes are abolished by treatment with the chelator 1, 10-phenanthroline. Since cadmium cannot activate MRE binding activity directly *in vitro*, we investigated whether cadmium indirectly activates the MRE binding factors *in vivo*. However, *in vivo* treatment with cadmium does not activate MRE binding activity to detectable levels in spite of the strong induction of MT transcript accumulation during the treatment. Since the addition of increasing cadmium concentrations *in vitro* actually decreases MRE binding of zinc activated by MTF, it is possible that cadmium can displace zinc from MTF physically but not functionally. Together, the above results bring into question the role of MTF in mediating cadmium induction of MRE activity.

Introduction

Metallothioneins (MTs) are small molecular mass (6 to 7 kDa) non-enzyme proteins which act as biological chelators of heavy metals through formation of metal-thiolate bonds with their numerous cysteine residues [1]. Although the precise cellular roles of MTs have not been defined, MTs have been proposed to have a variety of cellular roles including metal ion homeostasis, detoxification and cytoprotection during the acute phase response [2]. Consistent with a diverse array of cellular functions, MTs are transcriptionally induced by numerous agents [3] and display developmentally regulated expression patterns [4, 5] as well as cell-type specific regulation [6,7]. Metals are the common and most potent inducers of all MT isoforms. Although many MT promoter cis-acting elements are unique to one gene or species, the single common motif from invertebrates to vertebrates is the Metal-Responsive Element (MRE) which is always present in multiple copies and mediates induction by metals. The MRE consists of a conserved seven b.p. core sequence (underlined) surrounded by semi-conserved flanking sequences: CTN*TGC(G/A)CNC*GGCCC (8–11). With the MRE consensus defined, the next logical step toward understanding of metal regulation is the identification and characterization of the factors that bind to MREs. Unfortunately, with numerous investigators attempting to identify, purify, and clone MRE binding proteins, the result has been a confusing array of factors with apparently different MRE sequence requirements and metal responses *in vitro* or *in vivo* [12, 13]. By far, the most significant advance in understanding MRE-directed transcription has been the isolation of cDNAs encoding mouse and human MTF-1 [14, 15].

In vitro and *in vivo* activation of human and mouse MTF

In our previous studies, we have assayed the functional contributions of the MRE sequences to MT promoter activity and have observed that increased MT expression with *in vivo* zinc treatment was paralleled with increased MRE -specific binding activity in nuclear extracts [16, 17]. To further understand the MRE binding regulation of the factor(s) detected in these studies, it was important to establish if MRE binding activity was accessible to zinc *in vitro* and how other metal ions, which are known to activate MRE activity, affect factor binding.

Using South-western analysis, we have consistently identified a 125 kDa nuclear protein from Hep G2 cells that bound to the human MREa and trout MREa sequences [16, 17]. Both of these sequences also bind to an *in vivo* zinc induced factor in electrophoretic mobility shift assays (EMSA). Throughout the course of many assays, the tMREa sequence was found to detect the human MRE binding protein more consistently and was employed to study mammalian MRE binding proteins. The upper strand of the tMREa and tMREa mutant (tMREa/-60A) sequences is as follows:

tMREa   5'-GACTGTTTT**TGCACAC**GGCACC-3'
tMREa mut   5'-GACTGTTTT**AGCACAC**GGCACC-3'

Although a large complex is formed with untreated Hep G2 cell extracts in the absence of zinc, the addition of 100 µM $ZnCl_2$ to tMREa binding reactions causes the formation of an additional slower mobility MRE-protein complex as is observed with *in vivo* zinc treated Hep G2 cell extracts (Fig. 1A). This confirms that, although the Hep G2 factor complex is rarely detectable in control nuclear extracts without added metal ions, it is present in the nucleus at the time of extract preparation and it is available for DNA binding activation by zinc *in vitro*. Zinc addition to *in vivo* zinc treated Hep G2 cell extracts consistently does not increase the amount of zinc responsive complex that is observed suggesting that the factor is fully activated by zinc in culture. The *in vivo* zinc activated MRE binding activity can be abolished by incubation in the presence of 500 µM 1,10-phenanthroline, which chelates zinc ions (Fig. 1B). Similar results were obtained when HeLa cells were used during our studies. The zinc responsive tMREa complexes observed with Hep G2 and HeLa cell extracts appear to involve equivalent MRE-binding factors. The amount of MRE-binding factor complex is observed to be increased in zinc treated HeLa cell extract compared to the control. However, in contrast to the Hep G2 factor, the amount of endogenous ZRF complex formed because of *in vivo* treatment can be increased further with addition of zinc to DNA binding reactions (Fig. 1C). This indicates that, in HeLa cells, the factor is not fully activated by zinc treatment in culture. The *in vivo* zinc induced HeLa complex can be abolished by chelation with as little as 500 µM 1,10-phenanthroline (Fig. 1D) as was observed for the Hep G2 factor. Therefore, the activation of the MRE-binding activity *in vivo* requires constitution of the factor with Zn^{2+}, as it does *in vitro*, rather than a protein modification. Further, the Zn^{2+} bound to the factor is accessible to the chelator.

Since cadmium is a potent inducer of MT gene expression, we tested whether it will activate MRE-binding activity *in vitro*. In HeLa nuclear extracts, the addition of increasing amounts of Zn^{2+} to the binding reactions cause increased formation of a slower mobility tMREa com-

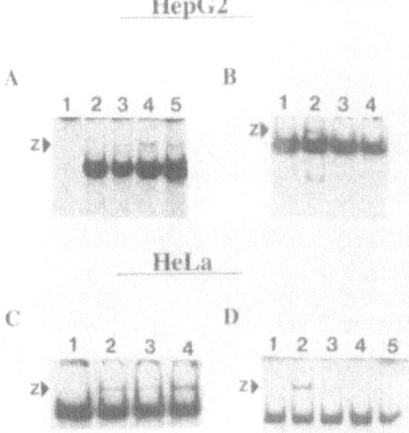

Figure 1. The effect of zinc on tMREa interactions with Hep G2 and HeLa nuclear factors. The zinc responsive mobility shift complex is indicated by 'z'. (A) *Lane 1*, no protein; *lane 2*, control extract; *lane 3*, control extract with 100 μM ZnCl2 in the binding reaction; *lane 4*, extract from cells treated for four hours with 100 μM ZnCl2; *lane 5*, zinc treated cell extract with 100 μM ZnCl2 in the binding reaction. (B) *Lane 1*, control extract; *lane 2*, extract from zinc treated cells; *lane 3*; zinc treated cell extract with 500 μM 1,10-phenanthroline in the binding reactions; *lane 4*, zinc treated cell extract with 5 mM 1,10-phenanthroline in the binding reaction. (C) *Lane 1*, control extract; *lane 2*, control extract with 100 μM ZnCl2; *lane 3*, extract from HeLa cells treated with 100 μM ZnCl2 for four hours; *lane 4*, zinc treated cell extract with 100 μM ZnCl2. (D) *Lane 1*, control extract; *lane 2*, control extract with 100 μM ZnCl2; *lane 3*, zinc treated HeLa cell nuclear extract; *lane 4*, zinc treated extract with 500 μM 1,10-phenanthroline; *lane 5*, zinc treated extract with 5 mM 1,10-phenanthroline.

plex (Fig. 2, upper panel). However, Addition of cadmium to binding reactions at concentrations ranging from 1 μM to 200 μM $CdCl_2$ does not induce the complex (Fig. 2, upper panel). The tMREa/-60A mutant oligonucleotide does not form the complex in the presence of 100 μM $ZnCl_2$ confirming that it is an MRE sequence specific interaction (Fig. 2, upper panel).The *in vitro* zinc specificity of the endogenous HeLa factor observed here is consistent with the MRE-binding properties of purified HeLa cell ZRF (generously provided for our studies by Dr. Shinji Koizumi, National Institute of Industrial Health, Kawasaki, Japan). Purified ZRF forms a complex with tMREa in the presence of zinc, but not cadmium. The tMREa-ZRF complex comigrates with the endogenous HeLa and Hep G2 zinc responsive complexes [18]. The similarities in the molecular masses (from South-western analysis), zinc dependence, and DNA complex migration suggest that the Hep G2 factor is equivalent to ZRF.

As with human ZRF/MTF, the factor is present in a sub-saturated form in untreated mouse L cell nuclear extracts since specific tMREa binding can be activated by the addition of zinc *in vitro*. However, cadmium addition *in vitro* does not promote formation of the tMREa complex (Fig. 2, lower panel). The metal response of the endogenous mouse L factor parallels the behavior of *in vitro* translated mouse MTF, which is activated by increasing zinc *in vitro* but not cadmium. The tMREa complex formed with *in vitro* translated mouse MTF comigrates with mouse L cell and mouse Hepa 1B cell zinc responsive complexes supporting the possibility that the factor detected with tMREa has identity with MTF [18].

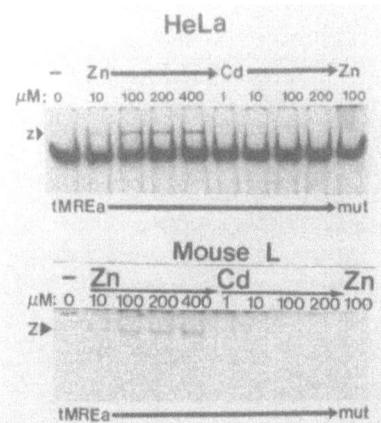

Figure 2. The effect of increasing concentrations of Zn^{2+} and Cd^{2+} on the interactions of tMREa with HeLa and Mouse L cell nuclear factors. The zinc responsive mobility shift complex is indicated by 'z'. (*Upper panel*) Untreated HeLa cell nuclear extracts were used for all reactions. The reactions contained the tMREa probe (*tMREa*) or the tMREa/-60A mutant probe (*mut*) as indicated under each lane for each panel. (*Lower panel*) Untreated mouse L cell nuclear extracts were used for all reactions.

Since the human MTF/ZRF and mouse MTF are not activated by cadmium *in vitro*, it was important to investigate if *in vivo* treatment could induce complex formation similar to zinc.Thus, treatment of Hep G2 cells with 10 µM $CdCl_2$ for four hours failed to cause formation of the tMREa complex (Fig. 3A) even though the chosen cadmium concentration and time course cause the accumulation of MT message to a level which rivals that caused by zinc treatment (Fig. 3B; [19]). Similar observations are made with HeLa cell extracts which display increased tMREa binding activity with zinc in culture while cadmium treatment of cells has no effect compared to control extracts (Fig. 3C). Again, the cadmium treatment used for HeLa cells induces the accumulation of MT message similar to zinc (Fig. 3D). In mouse L cells, the MTF-tMREa complex is not increased by the treatment of the cells with cadmium even though it is induced by *in vivo* zinc treatment (Fig. 4A). Copper treatment also has no positive effect on complex formation (Fig. 4B). However, in each of the metal treated extracts MTF-tMREa complex formation is still clearly inducible confirming that the extracts are active (Fig. 4B). At the time point of extract preparation, the mouse L MT message is accumulating from each of these *in vivo* metal treatments confirming that metal regulatory transcription factors are active for induction of MT gene expression (Fig. 4C). Furthermore, Dr. Glen Andrews laboratory have made similar observations about the lack of *in vivo* cadmium inducibility of mouse and human MTF and lack of *in vitro* response to several transition metals other than zinc [20, 21].

We further asked whether cadmium could affect the ability of zinc to activate MTF. Thus, for this study, we have utilized recombinant mouse and human MTF-1 (full length and the region representing the amino-terminus to the end of the zinc fingers), transcribed and translated *in vitro*. The DNA-binding activity of MTF-1 in the presence of zinc was reduced upon

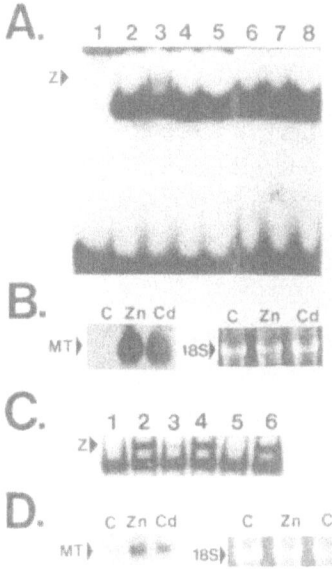

Figure 3. The effect of metals on the interaction of human MRE binding factors with tMREa. (A) The Hep G2 nuclear zinc responsive complex with tMREa is indicated by 'z'. The binding reactions contained tMREa (*lanes 1* to *5*) or the tMREa/-60A mutant (*lanes 6* to *8*) and the following: *lane 1*, no extract; *lane 2*, control extract; *lanes 3* and *6*, nuclear extract from cells treated with 100 μM ZnCl2 for four hours; *lanes 4* and *7*, control extract with 10 μM CdCl2; *lanes 5* and *8*, extract from cells treated with 10 μM CdCl2 for 4 h. (B) Northern blot analysis of total Hep G2 RNA probed with the human MT-II processed pseudogene. The hybridized MT message is indicated by 'MT'. The RNA was isolated in parallel with nuclear extracts from untreated cells (*C*), cells treated with 100 μM ZnCl2 for four hours (*Zn*), or cells treated with 10 μM CdCl2 for four hours (*Cd*). (C) The HeLa nuclear zinc responsive complex with tMREa is indicated by 'z'. The binding reactions contained no added metal (*lanes 1, 3,* and *5*) or 100 μM ZnCl2 (*lanes 2, 4* and *6*) as well as the following: *lanes 1* and *2*, control extract; *lanes 3* and *4*, extract from cells treated with 100 μM ZnCl2 for four hours; *lanes 5* and *6*, extract from cells treated with 10 μM CdCl2 for four hours. (D) Northern blot analysis of total HeLa cell RNA probed with the human MT-II processed pseudogene. The hybridized MT message is indicated by 'MT'. The RNA was isolated in parallel with nuclear extracts from untreated cells (*C*), cells treated with 100 μM ZnCl2 for four hours (*Zn*), or cells treated with 10 μM CdCl2 for four hours (*Cd*).

addition of cadmium suggesting that cadmium inactivates the factor possibly by displacing zinc from the fingers [22]. Furthermore, if MTF is reacted with cadmium first, zinc is able to bind MTF suggesting that zinc is displacing cadmium from the factor and that cadmium did not denature or inactivate the factor irreversibly [22]. These results also suggest that cadmium interacts with the zinc fingers of MTF-1 and forms an inactive complex. Similar results were obtained from our collaborative work with Dr. Glen Andrews [21]. Taken together, these data suggest that cadmium and other transition metals, with the exception of zinc, might activate MT gene expression by mechanisms independent of an increase in the DNA-binding activity of MTF-1.

Our results are difficult to reconcile with the idea that MTF is the only factor involved in metal regulation [23, 24]. Also, the human MT genes are differentially regulated by metals in Hep G2 cells [19] which is also at odds with the concept of a single factor. For example, zinc,

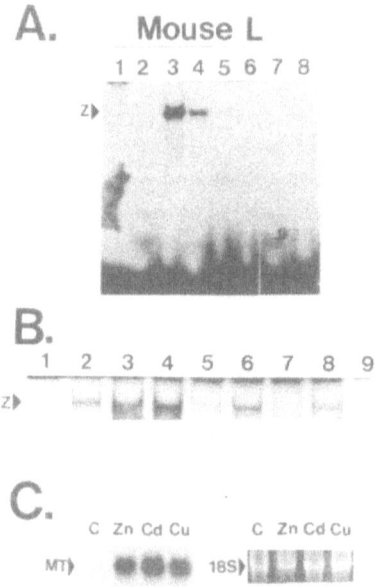

Figure 4. The effect of *in vivo* metal treatment on the formation of the tMREa complex with mouse L cell factors. The mouse L cells were untreated (control) or were treated with 100 µM ZnCl2, 10 µM CdCl2, or 400 µM CuCl2 in the media for 4 h. (A) The zinc responsive complex with tMREa is indicated by 'z'. The binding reactions contained the tMREa probe (*lanes 1* to *6*) or the tMREa/-60A mutant probe (*lanes 7* to *8*) and the following nuclear extracts: *lane 1*, no extract; *lanes 2* and *7*, control extract; *lanes 3* and *8*, extract from zinc treated cells; *lane 4*, control extract with 100 µM ZnCl2; *lane 5*, extract from cadmium treated cells; *lane 6*, control extract with 10 µM CdCl2. (B) Each reaction contained either no metal (*lanes 1, 3, 5,* and *7*) or 100 µM ZnCl2 (*lanes 2, 4, 6, 8,* and *9*) with the tMREa probe (*lanes 1* to *8*) or the tMREa/-60A mutant probe (*lane 9*). The nuclear extracts added to each binding reaction were from: *lanes 1* and *2*, control cells; *lanes 3, 4* and *9*, zinc treated cells; *lanes 5* and *6*, cadmium treated cells; *lanes 7* and *8*, copper treated cells. (C) Northern blot analysis of total RNA from control mouse L cells (C) or mouse L cells treated with zinc (Zn), cadmium (Cd), or copper (Cu). The human MT-II processed pseudogene was used as a probe.

cadmium and copper induce nearly equivalent levels of human MT-IIA transcript accumulation [19]. However, the MT-IF gene responds poorly to copper while the MT-IG gene responds poorly to cadmium [19]. If only one type of promoter element, MREs, and a single MRE binding factor is responsible for induction by these three metals, then each metal should have the same order of potency for inducing every MT gene within one cell type. Although the existence of auxiliary metal regulatory factors is tenuous at the moment, they cannot be ruled out completely, based on the current literature on MRE binding factors. MTF plays an essential role in the activation of MT gene transcription [23, 24], but auxiliary factors (for example, cadmium responsive) could be present in the cell which modulate the transcription response depending on the metal inducer. Such factors might require interaction with a basal level of active MTF to affect MT transcription. On the other hand, it is also possible that transition metals alter the transactivation potential of MTF-1 by phosphorylation. Both mouse and human MTF-1 contain a serine/threonine rich transactivation domain and several phosphorylation sites [25]. Thus, the mechanism(s) by which heavy metals initiate MT gene activation remained to be determined.

Acknowledgements
Studies from this laboratory was supported from the Natural Sciences and Engineering Council and in part from the Medical Research Council of Canada to L. Gedamu. The Alberta Heritage Foundation for Medical Research has kindly provided studentships to S.L.-A.Samson and S. Schieman. We also thank Dr. Walter Schaffner for the generous gift of the mouse and human MTF-1.

References

1. Kägi JHR, Schaffer A (1988) *Biochemistry* 27: 8509–8515.
2. Thiele DJ (1992) *Nucl Acid Res* 20: 1183–1191.
3. Hamer DH (1986) *Annu Rev Biochem* 55: 913–951.
4. Andrews GK, McMaster MT, De S, Paria BC, Dey SK (1993) *In*: KT Suzuki, N Imura, M Kimura (eds): *Metallothionein III*. Birkhäuser Verlag, Basel, 351–362.
5. Andrews GK, Adamson ED, Gedamu L (1984) *Dev Biol* 103: 294–303.
6. Jahroudi N, Foster R, Price-Haughey J, Beitel G, Gedamu L (1990) *J Biol Chem* 265: 6506–6511.
7. Price-Haughey J, Bonham K, Gedamu L (1987) *Biochim Biophys Acta* 908: 158–168.
8. Imbert J, Culotta VC, Furst PF, Gedamu L, Hamer DH (1990) *Adv Inorg Biochem* 8: 140–150.
9. Searle PF, Davison BL, Stuart GW, Wilkie TM, Norstedt G, Palmiter RD (1984), *Mol Cell Biol* 4: 1221–1230.
10. Searle PR, Stuart GW, Palmiter R (1987) *In*: *Metallothionein II*. Experentia (Basel) 52 (suppl.), 407–414.
11. Culotta VC, Hamer DH (1989) *Mol Cell Biol* 9: 1376–1380.
12. Koizuma S, Otsuka F (1993) *In*: KT Suzuki, N Imura, M Kimura (eds): *Metallothionein III*, Birkhäuser Verlag, Basel, 457–474.
13. Samson SL-A, Gedamu L (1998) *In*: K Moldave (eds): *Progress in Nucleic Acid Research and Molecular Biology*, Vol. 59, Academic Press, pp. 257–287.
14. Radtke F, Heuchel R, Georgiev O, Hergersberg M, Gariglio M, Dembic Z, Schaffner W (1993) *EMBO J* 12: 1355–1362.
15. Brugnera E, Georgiev O, Radtke F, Heuchel R, Baker E, Sutherland GR, Schaffner W (1994) *Nucl Acid Res* 22: 3167–3173.
16. Samson SL-A, Gedamu L (1995) *J Biol Chem* 270: 6864–6871.
17. Samson SL-A, Paramchuk WJ, Shworak NW, Gedamu L (1995) *J Biol Chem* 270: 25194–25199.
18. Samson SL-A (1996) Ph.D. Dissertation, Department of Biological Sciences, University of Calgary.
19. Sadhu C, Gedamu L (1988) *J Biol Chem* 263: 2679–2684.
20. Dalton TD, Bittel D, Andrews GK (1997) *Mol Cell Biol* 17: 2781–2789.
21. Bittel D, Dalton TD, Samson SL-A, Gedamu L, Andrews GK (1997) *J Biol Chem* 273: 7127–7133.
22. Schieman S, (1997) Ph.D. Dissertation, Department of Biological Sciences, University of Calgary.
23. Heuchel R, Radtke F, Georgiev O, Stark G, Aguet M, Schaffner W, *EMBO J* 13: 2870–2875.
24. Palmiter RD (1994) *Proc Natl Acad Sci USA* 91: 1219–1223.
25. Radtke F, Georgiev O, Muller H-P, Brugnera E, Schaffner W (1995) *Nucl Acid Res* 23: 2277–2286.

Metallothionein IV
C. Klaassen (ed.)
© 1999 Birkhäuser Verlag Basel/Switzerland

Metallothionein and hormone responsiveness

James Koropatnick, Janice M. DeMoor and Olga M. Collins

Department of Oncology, University of Western Ontario, London Regional Cancer Centre, 790 Commissioners Road East, London, Ontario, Canada N6A 4L6

Metallothioneins (MTs) bind both essential and toxic metals. Not surprisingly, metal ions are also classical inducers of MT expression. However, changes in MT level and intracellular localization are also associated with events that are not caused by exposure to extracellular metal ions. The events include differentiation, proliferation, cell cycle stage, tumour cell initiation and progression, and the acquisition of chemotherapeutic drug and radiation resistance. The multiple circumstances under which MTs are expressed has fuelled speculation about their function(s). They may participate in control of intracellular redox potential and help regulate thiol metabolism and/or play a role in homeostasis of essential metals such as zinc and copper. It has been suggested that MT may participate in essential metal trafficking to and from the many metal-dependent proteins, including signal-transducing macromolecules, zinc-binding transcription factors, and metal-dependent enzymes. In support of this hypothesis, we have previously shown that MT is induced by LPS-induced activation of human monocyte/macrophages [1]. Furthermore, antisense downregulation of MT prior to LPS treatment abolishes activation in the absence of toxicity [2]. We have also demonstrated that treatment with low, non-toxic levels of zinc, cadmium, or mercury to induce MT expression alters the ability of these cells to be activated by bacterial lipopolysaccharide (LPS) or phorbol myristate acetate (PMA), and to increase expression of IL-1β and MT-2 genes normally upregulated by activation [1, 3]. We report here that transient downregulation of metallothionein expression inhibits the capacity of cells to respond to glucocorticoid induction of gene expression, especially under low zinc conditions. This suggests that metallothionein participates *in vivo* in zinc-associated glucocorticoid signal transduction (possibly by mediating zinc-binding to the zinc-requiring glucocorticoid receptor), and provides the only evidence to date of a role for MT in regulating the activity of zinc transcription factors in living cells.

Antisense downregulation of MT inhibits cellular response to glucocorticoid

To investigate the role of MT in mediating the activity of a well-characterized transcription factor, we used a derivative of the C127 mouse mammary tumour cell line (denoted 2305) that harbours a stable episomal bovine papilloma virus (BPV)-based vector containing a chloramphenicol acetyltransferase (CAT) reporter gene under the control of the mouse mammary tumour virus (MMTV) promoter [4, 5] (kindly supplied by Dr. Trevor Archer, Dept. of Biochemistry, Univ. of Western Ontario, Canada). This promoter responds to dexamethasone

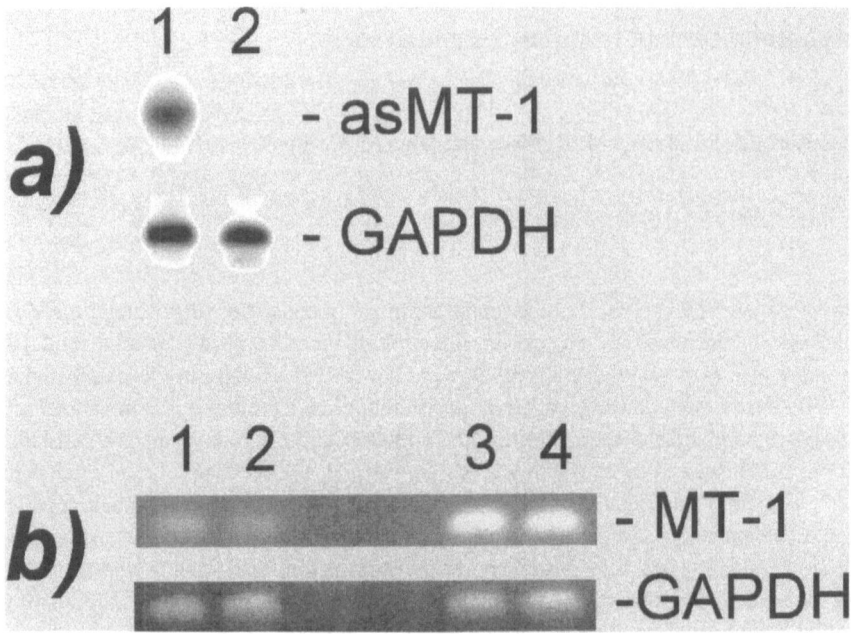

Figure 1. Inhibition of MT-1 mRNA expression by antisense MT-1 RNA in zinc-deficient mouse 2305 cells. A 322 bp mouse MT-1 cDNA was generated by polymerase chain reaction (PCR) from a full-length 400 bp mouse MT-1 cDNA template using primers containing non-complementatry *NotI* restriction enzyme sequences, and cloned into the *NotI* site of the eukaryotic expression vector pRC/CMV (Invitrogen) in reverse orientation. The resulting pRC/asMT vector, and an empty control vector (pRC/CMV), were separately transiently transfected into mouse C127 mammary carcinoma cells containing a GR-responsive CAT expression vector. Transfection efficiency was greater than 90% with the cationic liposome method used. Cells recovered for 18–20 h in medium containing zinc (approximately 4 µM), and were then transferred to medium supplemented with fetal bovine serum that had been depleted of zinc by treatment with Chelex 100 resin (BioRad). Dexamethasone (DEX, 1×10^{-7} M) was added 24 h after transfer to zinc-depleted medium, and the cells were incubated for a further 24 h in low zinc conditions in the presence of DEX. Cells were then collected and frozen for CAT assay and mRNA analysis. (a) Expression of antisense MT-1 RNA in cells transiently transfected with the pRC/asMT expression vector was assessed by specific RT-PCR treatment. cDNA was generated from antisense MT-1 RNA from cells transfected with pRC/asMT (Lane 1) or control pRC/CMV vector (Lane 2). Reverse transcription was primed by a specific oligonucleotide that would produce cDNA from antisense MT-1 RNA, but not MT-1 mRNA. A specific primer to allow production of glyceraldehyde phosphate dehydrogenase (GAPDH) cDNA from GAPDH mRNA was included in the reverse transcriptase reaction mix. PCR of the resulting antisense MT-derived cDNA and GAPDH cDNA yielded products of 344 and 752 base pairs, respectively, which Southern-blotted and hybridized to MT-1 and GAPDH cDNA probes. Hybridization was visualized by phosphorimage analysis. *Lane 1:* Cells transfected with pRC/asMT-1 vector. *Lane 2:* Cells transfected with control pRC/CMV vector. (b) Downregulation of MT-1 mRNA expression by pRC/asMT-1. cDNA from GAPDH mRNA and MT-1 mRNA was generated by reverse transcription of RNA isolated from cells transfected with pRC/asMT-1 (Lanes 1 and 2) or control pRC/CMV (lanes 3 and 4). The reverse transcription reaction was primed by specific oligonucleotides that produced cDNA from GAPDH mRNA and MT-1 mRNA, but not antisense MT-1 RNA. The ethidium bromide-stained MT-1 cDNA (344 bp) and GAPDH cDNA (752 bp) bands are shown. Products from two separate RT-PCR reactions from the same pool of zinc-depleted cells transfected with pRC/asMT-1 vector (lanes 1 and 2) or control pRC/CMV vector (lanes 3 and 4) are shown, indicating an approximately 60% decrease in MT-1 mRNA in cells transfected with the antisense MT-1 RNA expression vector.

signals transduced to the nucleus by the glucocorticoid receptor (GR), which is a zinc finger DNA-binding protein. Nuclear GR interacts with enhancer elements in the MMTV promoter to induce CAT mRNA, protein, and enzyme activity [6]. We transiently transfected into these cells a vector containing a full length mouse MT-1 cDNA designed to express RNA complementary to mouse MT-1 mRNA (pRC/asMT). Orientation and identity of the MT cDNA insert was confirmed by DNA sequence analysis. Control cells were transfected with vector that did not contain MT-1 DNA sequences. Transfected pRC/asMT vector expressed asMT-1 RNA and decreased MT-1 mRNA levels by 60%, as assessed by RT-PCR (Fig. 1).

asMT downregulation of MT expression inhibits glucocorticoid-induced CAT activity

We surmised that the approximately 4 µM endogenous zinc [7] in the fetal bovine serum used to supplement nutrient medium might be an alternative source of zinc and diminish the importance of MT in delivering zinc to macromolecules. Therefore, we grew 2305 cells in medium containing fetal bovine serum that had been depleted of zinc by chelation. Growth in this zinc-depleted medium did not significantly decrease cell growth over a 48 h period. However, CAT expression in response to dexamethasone induction was reduced by 30% in asMT-transfected cells, compared to 2305 cells transfected with control vector when grown in zinc-depleted medium (Fig. 2). The reduction in dexamethasone responsiveness in cells grown in medium containing 4 µM zinc, while significant, was only 10% (data not shown). This suggested that MT participated in glucocorticoid responsiveness to a minor degree under conditions where zinc was freely available, but played a much greater role under conditions where zinc supplies were limited.

Induction of MT by zinc enhances glucocorticoid response

To assess the effect of metal induction and resulting MT expression on glucocorticoid responsiveness, 2305 cells grown under normal conditions (without zinc depletion and without transfected vectors) were treated with 20, 40, 80, or 100 µM zinc chloride for 24 h. Only 80 and 100 µM Zn significantly induced MT levels, as assessed by ELISA (Fig. 3B). Cells were then assessed for CAT expression in response to induction with DEX for 24 h, in the presence of the inducing zinc. DEX-induced CAT expression was enhanced by zinc in a stepwise manner (Fig. 3A) in close correlation with the amount of MT protein induced (Fig. 3C). Lower levels of added zinc (20 and 40 µM), in spite of being 5- and 8-fold higher than the 4 µM zinc normally present in cell culture medium due to the contribution of fetal bovine serum [7, and unpublished data], did not significantly induce MT expression or enhance glucocorticoid responsiveness, suggesting that MT, and not only zinc ions, were critical.

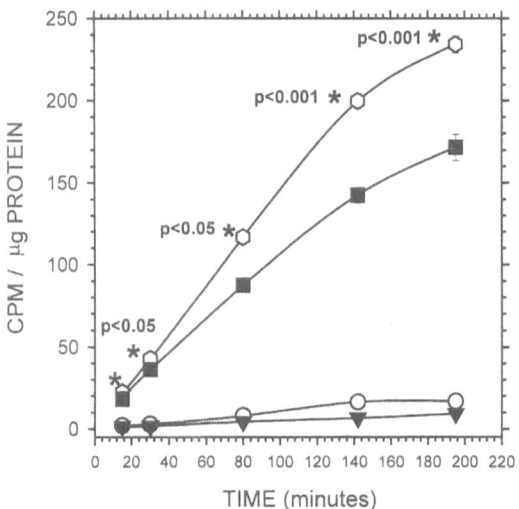

Figure 2. Mouse mammary tumour (2305) cells, stably transfected with a glucocorticoid-responsive CAT expression vector, were maintained in low zinc medium (less than 0.3 µM Zn) for 24 h. All following steps, including transfection, took place in low zinc medium. Cells were then transiently transfected for 8 h using cationic liposomes with an antisense MT-1 expression vector (pRC/CMV-asMT) to downregulate MT expression, or an empty control vector (pRC/CMV). Two days after the beginning of transfection, cells were induced with dexamethasone or ethanol solvent (1:1000, v/v) for 24 h and cumulative CAT activity assessed CAT enzyme activity assessed by a phase extraction assay (a modification of that described by Seed and Sheen, 1988). Over the course of 3 experiments, CAT activity was inhibited by antisense MT-treated cells by approximately 30%. The asterisk indicates a significant difference (p < 0.05, Student t-test). —O—Control vector (pRC/CMV), induved with dexamethasone (10^{-7} M); —O—Control vector (pRC/CMV); —▼—Antisense MT vector (pRC/CMV-asMT), induved with dexamethasone (10^{-7} M); —■—Antisense MT vector (pRC/CMV-asMT), induved with dexamethasone (10^{-7} M)

Summary

Expression of antisense RNA results in a significant decrease in MT-1 mRNA levels within cells. The status of MT-2 mRNA has not been assessed, but the high degree of similarity between mouse MT-1 and mouse MT-2 (greater than 85%), and the fact that expression of RNA antisense to mouse MT-1 has the capacity to downregulate human MT-2 mRNA and protein in a human monocytic leukemia cell line [2] suggests that MT-2 might be similarly downregulated. Reduced MT expression results in reduced capacity of a transfected reporter gene to respond to induction by glucocorticoid, particularly when zinc supplies are limited. The suggested role for MT in regulating zinc supplies within cells [8], and the requirement of the glucocorticoid receptor for zinc, suggests that MT may regulate the availability of zinc under low zinc conditions. Although direct evidence to support this postulated mechanism is not yet available, our observations are consistent with reports that insertional inactivation (gene knockout) of MT-1 and MT-2 genes in mice is not lethal under normal conditions, but leads to increased sensitivity to cadmium toxicity [9, 10] and zinc toxicity and deficiency [11], and enhanced sensitivity to cadmium and a wide range of toxic agents in cultured MT-1/MT-2 gene knockout fibroblasts in vitro [12]. We suggest that cells lacking MT grow and function normally under conditions

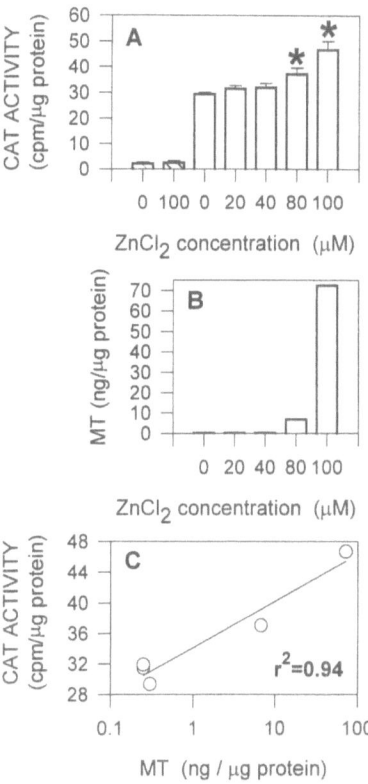

Figure 3. Zinc treatment induces enhanced responsiveness to glucocorticoid induction. 2305 cells were induced with 20, 40, 80, or 100 μM ZnCl$_2$ for 24 h. (A) ZnCl$_2$ treatment alone, at concentrations as high as 100 μM, did not induce CAT activity (shaded bars). However, dexamethasone induction following 24 h 40 or 80 μM zinc treatment, but not 20 or 40 μM zinc treatment, resulted in significant enhancement of dex-induced CAT activity (open bars). (B) Only 80 and 100 μM zinc induced significant MT protein, as measured by a competitive solid phase ELISA. (C) Zinc-induced enhancement of CAT activity correlated with induced MT expression. Asterisks (*) indicate values significantly different from controls ($p < 0.05$).

where zinc is freely available, both in culture and in whole animals. However, under stress conditions where the requirement for zinc is high or where zinc supplies are limited, some zinc-requiring proteins (including glucocorticoid receptors) would depend more heavily upon metallothionein to supply them with zinc. Lack of MT would conceivably result in a deficient transcriptional response to stress and enhanced cell death in the presence of toxins. The fact that metallothionein is produced under conditions of stress (including glucocorticoid induction), and is associated with resistance to stress-inducing toxins (including reactive oxygen intermediates and chemotherapeutic agents) is consistent with this hypothesis.

Induction of MT expression by zinc correlates with enhanced response to glucocorticoid. It is not clear whether enhancement of response is due to increased MT, or other gene products induced by zinc. The presence of zinc itself is unlikely to be the cause of enhanced response, since increasing zinc levels in serum up to eightfold (40 μM) without significantly increasing MT protein levels had no effect on glucocorticoid responsiveness. Only zinc levels that significantly induced MT (80 and 100 μM) zinc appeared to significantly enhance responsiveness to dexamethasone. Neither of these higher zinc concentrations had a deleterious effect on cell survival (as measured by a growth assay).

This novel role for MT in responsiveness to extracellular signals, and transcription factor activity, is being explored by measuring hormone responsiveness in MT-1/MT-2 gene knockout fibroblasts in comparison with parental cells with competent MT-1/MT-2 genes, in both normal and zinc-depleted nutrient medium. We are also assessing the effect of zinc induction on glucocorticoid signaling in the MT-deficient cells. Finally, investigation of altered physiological consequences of glucocortoid signaling after addition or depletion of zinc, or by antisense downregulation of metallothionein (for example, glucocorticoid-induced apoptosis in T lymphocytes) are warranted to explore the role of MT in influencing cellular responses to extracellular signals.

Acknowledgments
This work was supported by a grant from the Medical Research Council of Canada

References

1. Leibbrandt MEI, Koropatnick J (1994) Activation of human monocytes with endotoxin induces metallothionein expression and is diminished by zinc. *Toxicol Appl Pharmacol* 124: 72–81.
2. Leibbrandt MEI, Khokha R, Koropatnick J (1994) Antisense down-regulation of metallothionein in a human monocytic cell line alters cell adherence, invasion and the respiratory burst. *Cell Growth Differ* 5: 17–25.
3. Koropatnick J, Zalups RK (1997) Effect of non-toxic mercury and cadmium pretreatment on the capacity of human monocytes to undergo lipopolysaccharide-induced activation. *Brit J Pharmacol* 120: 797–806.
4. Mymryk JS, Berard D, Hager GL, Archer TK (1995) Mouse Mammary tumor virus chromatin in human breast cancer cells is constitutively hypersensitive and exhibits steroid hormone-independent loading of transcription factors *in vivo*. *Mol Cell Biol* 15: 26–34.
5. Charron J, Richard-Foy H, Berard DS, Hager GL, Drouin J (1989) Independent glucocorticoid induction and repression of two contiguous responsive genes. *Mol Cell Biol* 9: 3127–31.
6. Lapointe MC, Baxter JD (1989) *In:* Anti-inflammatory Steroid Action, Academic Press, San Diego, pp 3–23.
7. Palmiter RD (1995) Constitutive expression of metallothionein-3 (MT-3), but not MT-1, inhibits growth when cells become zinc-deficient. *Toxicol Appl Pharmacol* 135: 139–146.
8. Koropatnick J, Leibbrandt MEI (1995) Effects of Metals on Gene Expression. *In:* RA Goyer and MG Cherian (eds): *Handbook of Experimental Pharmacology (Toxicology of Metals – Biochemical Aspects)*, vol. 115, Springer-Verlag, pp. 93–120.
9. Michalska AE, Choo KH (1993) Targeting and germ-line transmission of a null mutation at the metallothionein 1 and 2 loci in mouse. *Proc Natl Acad Sci USA* 90: 8088–92.
10. Masters BA, Kelly EJ, Quaife CJ, Brinster RL, Palmiter RD (1994) Targeted disruption of metallothionein 1 and 2 genes increases sensitivity to cadmium. *Proc Natl Acad Sci USA* 91: 584–8.
11. Kelly EJ, Quaife CJ, Froelick GJ, Palmiter RD (1996) Metallothionein I and II protect against zinc deficiency and zinc toxicity in mice. *J Nutr* 126: 1782–1790.
12. Kondo Y, Woo ES, Michalska AE, Choo KH, Lazo JS (1995) Metallothionein null cells have increased sensitivity to anticancer drugs. *Cancer Res* 55: 2021–3.
13. Seed B, Sheen J-Y (1988) A simple phase extraction assay for chloramphenicol acetyl-transferase activity. *Gene* 67: 271–277.

Metallothionein IV
C. Klaassen (ed.)
© 1999 Birkhäuser Verlag Basel/Switzerland

Synergistic activation of mouse metallothionein-I gene by interleukin-6 and glucocorticoid

Norio Itoh, Keiko Kasutani, Masako Kanekiyo, Tomoki Kimura, Norio Muto and Keiichi Tanaka

Faculty of Pharmaceutical Sciences, Osaka University, Yamada-oka, Suita, Osaka 565, Japan

Summary. Metallothionein (MT)-inducing activity of interleukin (IL)-6 depends on the presence of glucocorticoid in hepatic cells. The synergistic action of IL-6 and glucocorticoid was observed in the transcriptional activation of the mouse MT (mMT)-I gene. A 281 bp promoter was sufficient for IL-6 and glucocorticoid stimulation. The putative type 1 and 2 IL-6 responsive elements (REs) are present in this region. Functional analyses of these regions were performed, and it was observed that the type 2 IL-6RE exerted the major response to the IL-6 signal. A glucocorticoid responsive element (GRE) was also required for the synergistic activation by IL-6 and glucocorticoid. The type 2 IL-6RE or GRE alone did not show this transcriptional synergism. Interestingly, this synergism was not observed when the type 2 IL-6RE and the GRE were kept apart. Therefore, the synergistic activation of the mMT-I gene by IL-6 and glucocorticoid may require not only binding of signal transducers and activators 3 (Stat3) and the glucocorticoid receptor (GR) to their respective responsive elements, but also interaction of Stat3 and the GR with one another.

Introduction

MT is known to be directly induced by heavy metals [1] and glucocorticoid [2]. Endotoxin (lipopolysaccharide; LPS) and turpentine, both of which initiate inflammatory responses, also induce hepatic MT synthesis [3–5]. These compounds stimulate MT production indirectly. We have shown that serum from LPS-treated mice contains factor(s) that stimulate MT induction, and that MT-inducing activity in the serum is mediated by IL-6 [6]. In addition, we also have shown that the expression of MT-inducing activity is dependent on the presence of glucocorticoid [6].

Cytokines mediate acute-phase responses in the liver [7]. The effect of some, but not all types of acute-phase responses mediated by cytokines are potentiated by glucocorticoid [8]. Even among such acute-phase proteins (APP) the response to IL-6 of which is modified by glucocorticoid, the extent of dependence on glucocorticoid is varied [9]. Therefore, the synergism between IL-6 and glucocorticoid observed in the induction of MT cannot be explained by mechanisms that glucocorticoid changes the responsibility to all of IL-6 inducible genes. Rather, the specific cooperation of two independent signal receivers on the MT gene appears to be taking place.

The 5' flanking regions of many APP genes have been studied in detail to identify the regulatory elements mediating the action of cytokines. Two types of IL-6 responsive elements (REs), designated types 1 and 2 have been found in APP genes [8, 9]. Type 1 IL-6REs are present in the promoter regions of C-reactive protein, haptoglobin, hemopexin, and α1-acid glycoprotein (AGP) genes. This element binds members of the CCAAT enhancer-binding protein

(C/EBP) family, of which nuclear factor IL-6 (NF-IL6) (also called C/EBPβ, AGP/EBP, LAP, or IL-6DBP) and NF-IL6β (also called C/EBPδ) have been shown to be involved in the regulation of APP genes by IL-6 [10–12]. In contrast, type 2 IL-6REs are found in the α2-macroglobulin, fibrinogen, and serine protease inhibitor genes. Signal transducers and activators of transcription 3 (Stat3) (also called APRF) bind to type 2 IL-6REs [13, 14].

In the glucocorticoid receptor (GR) pathway, the receptor interacts with its steroid hormone ligand and translocate into nucleus. Then the GR binds to specific DNA-response elements (glucocorticoid responsive elements; GREs) and modulate transcription [15, 16]. GREs have been identified at the long terminal repeat of the murine mammary tumor virus, the rat tyrosine amino transferase gene, and the human MT-IIA gene.

For MT genes, functional IL-6REs have not yet been defined in any species. It is reported that the region which allows the LPS induction of the mMT-I gene lies within the -350 to -185 bp region of the 5' flanking region of the mMT-I gene, since LPS-administered-trangenic mice harboring recombinant genes made by fusing the mMT-I promoter to the herpes simplex virus thymidine kinase structural gene revealed the response to LPS [5]. However, a more detailed analysis of this region has not been reported. The primary purpose of the present study is to identify the functional IL-6RE(s) in the promoter region of the mMT-I gene and to show the contribution of the Stat3 and NF-IL6 binding elements in the induction of target promoter(s).

Results and discussion

Mapping of functional responsive elements to IL-6 and glucocorticoid in the mMT-I promoter

A 592 bp fragment of the 5' flanking region of the mMT-I gene was cloned into the reporter plasmid encoding the firefly luciferase gene and transiently transfected into H4IIEC3 cells. The treatment of transfected cells with corticosterone or a combination of IL-6 and corticosterone resulted in a marked increase in luciferase activity, whereas IL-6 alone had no effect. Glucocorticoid alone caused an about 2-fold induction of luciferase activity, while the treatment with a combination of both factor produced a 4 to 5-fold induction. Thus, the synergistic action of IL-6 and glucocorticoid on MT induction is a transcriptional event. Within the 281 bp region of the mMT-I promoter, a GRE located at -242 to -229 bp has already been identified in transient transfection experiments [17], while no IL-6RE has been found yet. We examined the homologous site to the IL-6RE consensus sequence within the region where the GRE was identified. Two putative type 1 IL-6REs and one putative type 2 IL-6RE were located in the 281 bp fragment. They possess one mismatch each to their respective consensus sequences. Within the -259 to -251 bp region, the type 1 and 2 IL-6REs overlapped. From the results of the deletion experiment (Fig. 1), it appeared that the IL-6RE(s) located upstream (-259 to -251) on the mMT-I promoter exert the major response to the IL-6 signal.

To evaluate the contribution of the IL-6REs and GRE in the IL-6- and glucocorticoid-induced response of the mMT-I gene, we introduced point mutations. The mutation introduced into the overlapping region (-259 to -251) of the type 1 and 2 IL-6REs destroyed one of the possible binding sites but did not change the consensus for the other binding site. The mutated

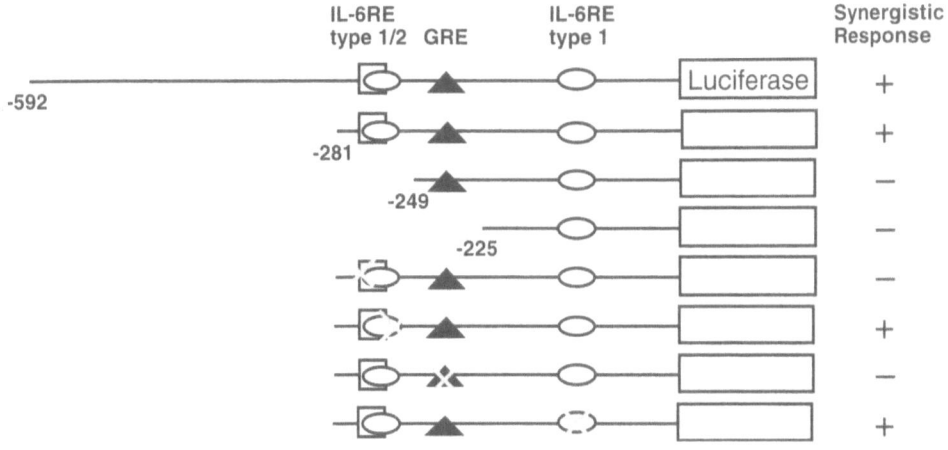

Figure 1. Functional analysis of the mMT-I promoter and its mutants. The results from deletion mapping and functional analysis of putative IL-6 and glucocorticoid responsive elements in the mMT-I promoter are summarized.

constructs were transfected into hepatoma cells for a functional analysis (Fig. 1). The mutation of the putative type 2 IL-6RE resulted in a total loss of response to the synergism of IL-6 and glucocorticoid. In contrast, the mutations of the two putative type 1 IL-6REs located upstream or downstream did not abolish the response to IL-6 and glucocorticoid. The disruption of the GRE also abolished the synergism of IL-6 and glucocorticoid. Therefore, not only the type 2 IL-6RE but also the GRE is essential for the synergistic activation of IL-6 and glucocorticoid on mMT-I gene.

Contribution of NF-IL6 and GR to the synergistic activation of the mMT-I gene by IL-6 and glucocorticoid

To determine whether NF-IL6 also plays a part in the mMT-I promoter activity by IL-6 and glucocorticoid, an NF-IL6 expression vector was cotransfected with the reporter plasmids containing various deletion mutants of the mMT-I promoter into the hepatoma cells. NF-IL6 did not activate transcription on all constructs, and synergism with glucocorticoid was not observed. In addition, we were persuaded by the results of a cotransfection experiment using an NF-IL6 responsible reporter that NF-IL6 was expressed as the functional form in our cotransfection experiment.

    To examine the contribution of the GR to the synergistic activation of the mMT-I gene by IL-6 and glucocorticoid, we used RU38486 and GR expression vectors. The synergistic activation of IL-6 and glucocorticoid on the mMT-I promoter was clearly inhibited by RU38486.

The GR(*wt*) overexpression showed a marked increase in its ability to transactivate the reporter plasmid containing the mMT-I promoter when treated with the combination of IL-6 and glucocorticoid. The synergistic activation by IL-6 and glucocorticoid was not observed in the overexpression of N525 (a mutant of GR), unlike GR(*wt*). In the overexpression of the GR(*wt*), RU38486 also inhibited the synergistic activation of the mMT-I gene. Therefore, the synergistic effect of glucocorticoid on the activation of the IL-6RE by IL-6 requires a functional GR.

The Stat3-binding site and the GRE must be closely located for the synergistic activation by IL-6 and glucocorticoid

It was demonstrated that Stat3 is involved in the signaling pathway of IL-6 *via* the type 2 IL-6RE [14]. In addition, the results described above make it clear that the GR-binding to the GRE of the mMT-I promoter is required for the synergistic activation by IL-6 and glucocorticoid. The Stat3-binding site and the GRE located closely on the mMT-I promoter prompted us to examine the possibility that the synergistic transactivation by IL-6 and glucocorticoid requires not only that Stat3 and the GR each bind to their responsive elements, but also that Stat3 and the GR physically interact with one another. Thus, a mutant was constructed in which a nonsense 16 bp fragment was inserted between the IL-6RE type 2 and the GRE. The results of the assays of this promoter are shown in Figure 2. Increasing the distance between these elements

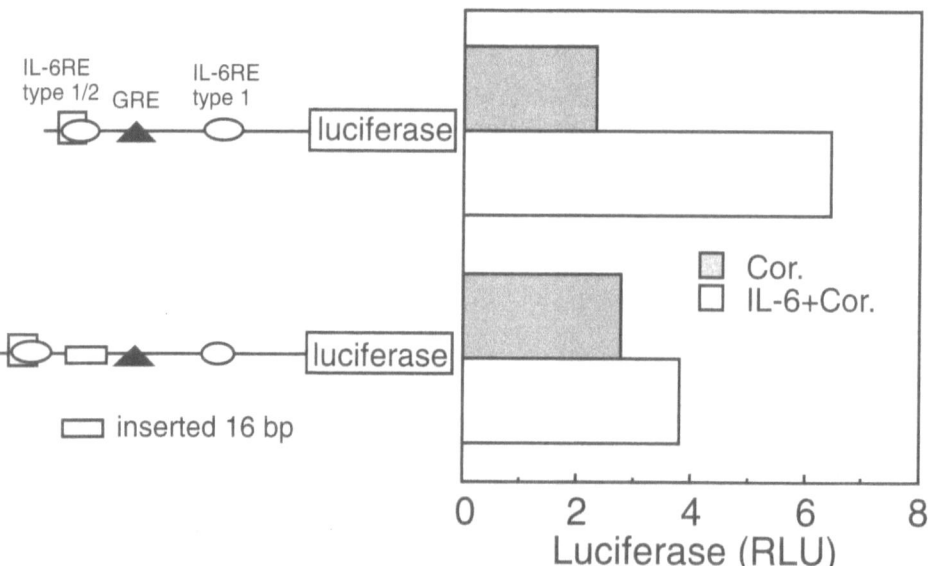

Figure 2. Abrogation of synergistic activation by insertion of a nonsense 16 bp fragment between IL-6RE and GRE in the mMT-I promoter. A nonsense 16 bp fragment was inserted between type 2 IL-6RE and GRE in the mMT-I promoter. Cells were stimulated with IL-6 and/or corticosterone.

failed to activate synergistically by IL-6 and glucocorticoid. In the signaling pathway through Stat5, which is a member of the Stat family (as is Stat3), it has been reported that the GR can act as a transcriptional co-activator for Stat5 and enhance Stat5-dependent transcription, and that Stat5 forms a complex with the GR which binds to DNA independently of the GRE [18]. We speculated a direct interaction (binding) between the two transcription factors Stat3 and the GR, for their synergistic activation of the mMT-I gene.

The difference in the requirement of GRE for gene activation by the type 2 IL-6RE between some APPs and mMT-I

Some APP genes regulated *via* the type 2 IL-6RE, it was reported that their fragments which contain the type 2 IL-6RE on their promoter were capable of response to combined stimulation of IL-6 and glucocorticoid, in spite of the absence of an obvious consensus GRE [19, 13]. To examine whether the type 2 IL-6RE core site on the mMT-I promoter function without GRE, we inserted several copies of this element in the front of the SV40 promoter linked to the luciferase gene. Contrary to the case of some APPs, these constructs were unable to response to IL-6 and glucocorticoid. Notable synergistic activation by IL-6 and glucocorticoid was observed with the trimer of the element (-281 to -206 bp) containing the type 2 IL-6RE and the GRE. These results led us to the conclusion that both the type 2 IL-6RE and the GRE of the mMT-I promoter are necessary for the synergistic response to IL-6 and glucocorticoid.

The observations in this study are significant from the viewpoint that the expression of the MT genes involved in host defense is activated by the cooperative interaction of the independent IL-6 and glucocorticoid signaling pathways involved in the inflammatory reaction of host defense.

Acknowledgements
We thank Chugai Pharmaceutical Co. for the recombinant human IL-6, ROUSSEL-UCLAF for the glucocorticoid antagonist RU38486, Dr. Palmiter for genomic clone of mMT. We also thank Dr. M. Imagawa for helpful advice and discussions. This work was supported in part by a grant from the Houansha Foundation, Osaka, Japan.

References

1. Searle PF (1987) Metallothionein gene regulation. *Biochem Soc Trans* 15: 584–586.
2. Mayo KE, Palmiter RD (1981) Glucocorticoid regulation of metallothionein-I mRNA synthesis in cultured mouse cells. *J Biol Chem* 256: 2621–2624.
3. Sobocinski PZ, Canterbury WJJr Mapes CA, Dinterman RE (1978) Involvement of hepatic metallothioneins in hypozincemia associated with bacterial infection. *Amer J Physiol* 234: E399–406.
4. DiSilvestro RA, Cousins RJ (1984) Mediation of endotoxin-induced changes in zinc metabolism in rats. *Amer J Physiol* 247: E436–441.
5. Durnam DM, Hoffman JS, Quaife CJ, Benditt EP, Chen HY, Brinster RL, Palmiter RD (1984) Induction of mouse metallothionein-I mRNA by bacterial endotoxin is independent of metals and glucocorticoid hormones. *Proc Natl Acad Sci USA* 81: 1053–1056.
6. Itoh N, Kasutani K, Muto N, Otaki N, Kimura M, Tanaka K (1996) Blocking effect of anti-mouse interleukin-6 monoclonal antibody and glucocorticoid receptor antagonist, RU38486, on metallothionein-inducing activity of serum from lipopolysaccharide-treated mice. *Toxicology* 112: 29–36.

7. Andus T, Geiger T, Hirano T, Kishimoto T, Tran-Thi TA, Decker K, Heinrich PC (1988) Regulation of synthesis and secretion of major rat acute-phase proteins by recombinant human interleukin-6 (BSF-2/IL-6) in hepatocyte primary cultures. *Eur J Biochem* 173: 287–293.
8. Akira S, Kishimoto T (1992) IL-6 and NF-IL6 in acute-phase response and viral infection. *Immunol Rev* 127: 25–50.
9. Baumann H, Prowse KR, Marinkovic S, Won KA, Jahreis GP (1989) Stimulation of hepatic acute phase response by cytokines and glucocorticoids. *Ann N Y Acad Sci* 557: 280–296.
10. Akira S, Isshiki H, Sugita T, Tanabe O, Kinoshita S, Nishio Y, Nakajima T, Hirano T, Kishimoto T (1990) A nuclear factor for IL-6 expression (NF-IL6) is a member of a C/EBP family. *EMBO J* 9: 1897–1906.
11. Poli V, Mancini FP, Cortese R (1990) IL-6DBP, a nuclear protein involved in interleukin-6 signal transduction, defines a new family of leucine zipper proteins related to C/EBP. *Cell* 63: 643–653.
12. Kinoshita S, Akira S, Kishimoto T (1992) A member of the C/EBP family, NF-IL6 beta, forms a heterodimer and transcriptionally synergizes with NF-IL6. *Proc Natl Acad Sci USA* 89 1473–1476.
13. Hocke GM, Barry D, Fey GH (1992) Synergistic action of interleukin-6 and glucocorticoids is mediated by the interleukin-6 response element of the rat alpha 2 macroglobulin gene. *Mol Cell Biol* 12: 2282–2294.
14. Wegenka UM, Buschmann J, Lutticken C, Heinrich PC, Horn F (1993) Acute-phase response factor, a nuclear factor binding to acute-phase response elements, is rapidly activated by interleukin-6 at the posttranslational level. *Mol Cell Biol* 13: 276–288.
15. Beato M, Herrlich P, Schutz G (1995) Steroid hormone receptors: many actors in search of a plot. *Cell* 83: 851–857.
16. Mangelsdorf DJ, Thummel C, Beato M, Herrlich P, Schutz G, Umesono K, Blumberg B, Kastner P, Mark M, Chambon P, Evans RM (1995) The nuclear receptor superfamily: the second decade. *Cell* 83: 835–839.
17. Plisov SY, Nichiporenko MG, Shkapenko AL, Kumarev VP, Baranova LV, Merkulova TI (1994) The immediate vicinity of metallothionein-I gene contains two sites conferring glucocorticoid inducibility to the heterologous promoter. *FEBS Lett* 352: 339–341.
18. Stocklin E, Wissler M, Gouilleux F, Groner B (1996) Functional interactions between Stat5 and the glucocorticoid receptor. *Nature* 383: 726–728.
19. Kordula T, Travis J (1996) The role of Stat and C/EBP transcription factors in the synergistic activation of rat serine protease inhibitor-3 gene by interleukin-6 and dexamethasone. *Biochem J* 313: 1019–1027.

Metallothionein IV
C. Klaassen (ed.)
© 1999 Birkhäuser Verlag Basel/Switzerland

Hammerhead ribozymes mediated down-regulation of rat metallothionein mrnas expression

Kai-Fai Lee and Shuk-Mei Ho

Biology Department, Tufts University, Medford, MA 02155, USA

Summary. Ribozymes (Rzs) are small RNA molecules with endoribonuclease activity that can inhibit gene expression by cleaving a particular mRNA in a highly sequence-specific fashion. Two hammerhead ribozymes (Rz1-2 and Rz4-9) targeted against the 3'-coding regions of rat metallothionein-I (MT) and -II RNAs were generated. The ribozymes mediated RNA cleavage in the presence of Mg^{2+} at a broad temperature range (25°C to 50°C) *in vitro*. Rz1-2 and Rz4-9 ribozymes are specific for rat MT-I and MT-II messages, respectively, while no *in vitro* cleavage activities were observed for two mutated RNA molecules, Rz2-6 and Rz3-3, derived from Rz1-2 and Rz4-9, respectively. Using primer extension assay, we confirmed the cutting sites on the MT messages were as predicted. The catalytic efficiencies of Rz1-2 on MT-I RNA cleavage was 678 $M^{-1}s^{-1}$ and of Rz4-9 on MT-II RNA was 372 $M^{-1}s^{-1}$. A tandem ribozyme construct (Rz1-2/Rz4-9) driven by a strong promoter (CMV) was generated and used to transfect a rat ventral prostate epithelial cell line, NbE-1. Several stable transfectants were selected. In the selected clones, Rz-2 and Rz-3, both MT-I and -II message were reduced to non-detectable levels. Importantly, these clones were shown to exhibit increased susceptibility to cadmium cytotoxicity. These findings therefore provide strong evidence in support of a role played by MT in protecting prostatic epithelial cells against cadmium toxicity.

Introduction

Metallothioneins (MTs) are found in most tissues of a wide variety of vertebrates and invertebrate species [1–3]. These polypeptides are characterized by their low molecular weights, high cysteine content (25–30%), lack of aromatic residues, and affinities for a range of heavy metal ions including zinc, copper, mercury and cadmium (Cd). Although the precise biological functions of these proteins remain obscure, it has been suggested that MTs are involved in detoxification of certain heavy metals, homeostasis of essential trace element such as zinc (Zn) and copper, scavenger activity of free radicals, and protection against alkylating agents [4–7]. Additionally, they may confer anti-cancer drug resistance to a variety of cancer cells [8, 9].

One of the most commonly postulated function of MTs is that enhanced expression of these proteins protect against Cd-cytotoxicity/genotoxicity [1]. In most tissues, MT expression is inducible by Cd exposure. However, in the rat ventral prostate, MT was found to be expressed at low to non-detectable levels and appeared to be non-inducible by Cd[10, 11]. It has therefore been postulated that the lack of expression and inducibility of MTs by this heavy metal ion is the underlying cause of the rat ventral prostate's hypersensitivity to Cd-toxicity and/or -carcinogeniety [10]. However, evidence in support of this notion was mostly derived from correlative studies and evidence implicating a direct role played by MTs in conferring cellular resistance to Cd is still lacking. In this study, we successfully generated hammerhead ribozymes (Rzs) that exhibited sequence-specific cleavage activities on MT-1 and MT-2 transcripts. These Rzs were cloned into expression vectors and used to transfect a rat ventral prostate epithelial cell line, the NbE-1 [12]. Using highly sensitive quantitative RT-PCR protocols, recently developed in our laboratory, we demonstrated down-regulation of both MT-1 and MT-2 mRNAs in stable transfectants expressing the Rzs. In addition, expression of Rzs in these stable transfectants were shown to correlate with an increase in susceptibility of these cells to cadmium-induced cytotoxicity. These findings thus provide direct evidence in support of MT's role in protecting prostatic cells against cadmium toxicity.

Results

Development of highly sensitive, sequence-specific RT-PCR protocols for quantification of MT-1 and MT-2 transcripts

The first part of this study focused on the development of two highly-sensitive, sequence-specific, quantitative RT-PCR protocols for the detection and quantification of MT-1 and MT-II transcripts in NbE-1 cells. These assays have sensitivities in the range of 5 to 10 amole per ng RNA and require only 0.05 to 0.1 µg of total tissue RNA. When compared to Northern and solution hybridization [10, 11], they are at least 100- to 1,000-fold more sensitive and require 200–400 times less sample. The use of RNA mimics as internal competitors in these reactions offers control over both the reverse transcription and PCR steps. Message contents can thus be quantified in absolute values making comparisons among different samples more quantitative. Additionally, the sequence-specificity of these protocols permit simultaneous quantification of both gene transcripts, thus offering a effective means to compare MT-I and MT-II transcript levels in the same sample.

MT-I and MT-II cRNA sequences, containing the entire coding regions, were generated by RT-PCR from hepatic total RNA preparations. These fragments were separately cloned into pBluescript SK+ vector and the sequences analyzed by sequencing. Two insertion mutants were constructed by splicing a fragment of lambda DNA into the MT sequences to generated chimeric molecules (Fig. 1). For the construction of the MT-I chimeric molecule, an internal *Nar*I (GG↓CGCC) site in the MT-I sequence was used to accommodate a *Msp*I (C↓CGG) cut lambda fragment (nt 6,936–7,066). Similarly, an internal *Bam*HI (G↓GATCC) stie in the MT-II sequence was used to splice in a *Sau*3AI (↓GATC) cut lambda fragment (nt 15582–15801). A poly (A) tail was then attached to each chimeric molecule to facilitate subsequent priming with oligo-d(T)$_{12-18}$ in the RT reaction. MT-I and MT-II cRNA mimics were synthesized by T3 promoter-mediated *in vitro* transcription using *Hind*III linearized pBSMT-IA and pBSMT-IIA, respectively, as templates. Each cRNA mimic therefore possesses the same primer templates as the targeted MT sequence but will generate a PCR product of larger size than that derived from the native transcript. For quantitation of MT transcript contents, 0.1 µg of total RNA from a sample was added to reaction tubes containing serially diluted cRNA mimic ranging from 0.5 to 5×10^3 attomole. The reverse transcription reaction was conducted in the presence of SuperScript II RNaseH⁻ reverse transcriptase (Gibco/BRL, Gand Island, NY) and oligo-d(T)$_{12-18}$ (Pharmacia, Uppsala, Swedan). PCR primers for rat MT-I and MT-II are as follow: MT-I (forward primer: 5'-GAA TTC CGT TGC TCC AGA TTC ACC AGA TC-3' and reverse

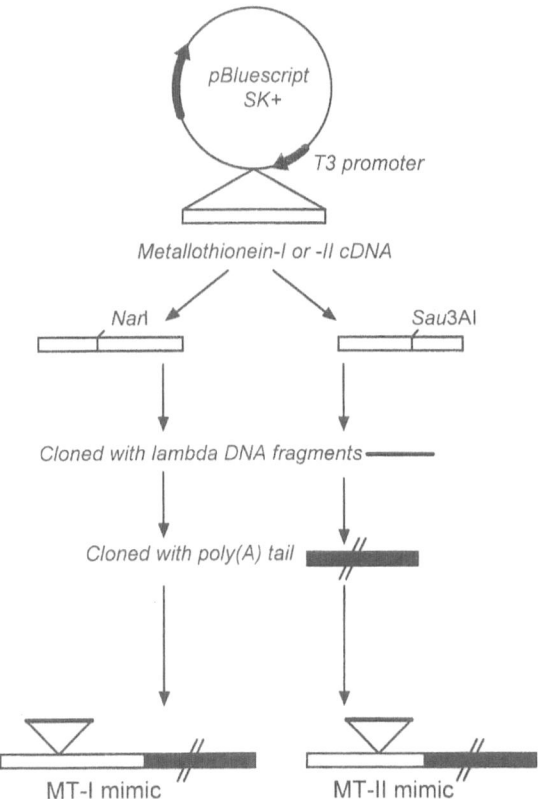

Figure 1. Schematic diagram illustrated the construction of the pBSMT-IA and pBSMT-IIA constructs for the generation of cRNAs. The MT-I or MT-II coding sequences were amplified by RT-PCR from liver samples and cloned individually into pBluescript SK+ vector. A 132 bp and 219 bp lambda DNA fragment was inserted into the MT-I and MT-II sequences to generate pBSMT-I* and pBSMT-II*, respectively. A poly (A) tail was added downstream of the MT sequences to generate pBSMT-IA and pBSMT-II A. The T3 promoter was used to *in vitro* transcribe the cRNA molecules.

primer: 5'-GAA TTC TCA CAT GCT CGG TAG AAA ACG G-3') and MT-II (forward primer: 5'-TAG ATC TCC ACC TGC CGC CTC CA-3' and reverse primer: 5'-TAG ATC TAC ACC ATT GTG AGG ACG CCC-3'). PCR used 30 cycles of 94 °C for 1 min, 60 °C for 30 s and 72 °C for 30 s. The final extension step was at 72 °C for 10 min. The PCR products were resolved with 1.5% agarose gel yielding products from mimic and target sequences of expected sizes of 462 bp an 330 bp for MT-I and 470 bp and 250 bp for MT-II, respectively. The relative intensities of the target and mimic products were quantitated, corrected for their molecular weights to derive molar values, and data plotted as log molar ratio of target/mimic against log mimic added in attomole (Fig. 2). When the amount of target RNA in the sample equaled to that of the mimic added to the reaction tube, the log target/mimic ratio would have a value equaled to zero.

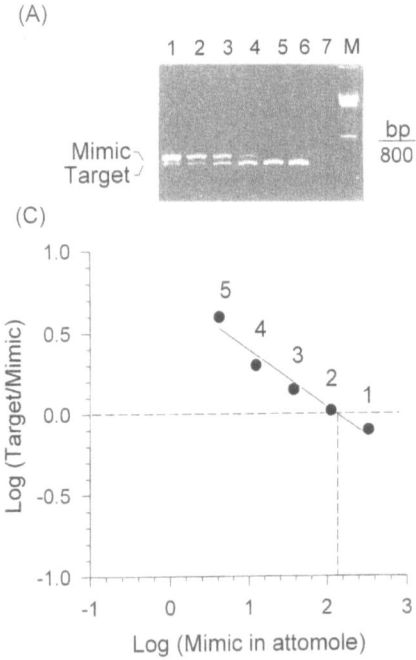

Figure 2. Titration of a MT-I mimic with a constant amount of total RNA (0.1 µg/tube) prepared from NbE-1 cells. Upper panel: Lane 1–7, a serial dilution of competitor cRNA was added to each RT-PCR tubes containing 0.1 µg total RNA from an unknown sample as follows: lane 1 (333 amol), lane 2 (111 amol), lane 3 (37.1 amol), lane 4 (12.3 amol), lane 5 (4.1 amol), lane 6 (1.4 amol) and lane M (100 bp molecular weight marker, Pharmacia). Lower panel: Following electrophoresis, signal intensities of bands corresponding to a MT target and its mimic were determined by densitometry. The data are plotted as a function of the log target band intensity/mimic band intensity (Log target/mimic) vs log amol of mimic added (Log mimic).

Generation of hammerhead ribozymes for of MT-I and MT-II mRNA and characterization of their in vitro *activities*

The basic sequence requirements of hammerhead Rzs have been discussed extensively in recent reviews [13, 14]. Briefly, a hammerhead Rz has three essential regions. The 3' (stem I) and 5' (stem III) ends of the Rz hybridize with the substrate or the targeted transcript, thus orienting the unhybridized nucleotides (cleavage site) of the substrate over the catalytic core (stem II) of the Rz. The non-base paired nucleotides flanking both sides of stem II make up the catalytic core. They contain the essential sequences, CUGAUGA on the 5' end and the GAAA on the 3' end. These sequences together with the high GC content in the stem stabilizes the hammerhead structure. The cleavage site on the target RNA is a nucleotide triplet and one that has the highest rate of cleavage is GUC while GUA and AUA also have good catalytic efficiency.

We have constructed plasmids for the generation of ribozymes for MT-I and MT-II (Fig. 3A). In our design of the Rz for rat MT-I mRNA (designated Rz1-2) a GUC recognition site was

Table 1. Rat MT sequences alignment at the ribozymes cutting sites

rat MT-I	131	GTGCCCAGGG CTGT*GTC*TGC AAAGGTGCCT
rat MT-II		GCTCCCAGGG CTGC*ATC*TGC AAAGAGGCTT

chosen and that for MT-II mRNA (designated Rz4-9) a AUC recognition site was used (Fig. 3B). Both Rzs have two ten base antisense regions flanking their respective cleavage site. Two catalytically inactive ribozymes Rz2-6 and Rz3-3 for MT-I and MT-II transcripts, respectively, were also generated by modification of two nucleotides in the stem II region of Rz1-2 and Rz4-9 (Tab. 1). These catalytically inactive ribozymes served as negative controls in our studies of the active ribozymes. A plasmid containing both Rz1-2 and Rz4-9, designated Rz(4-9/1-2), was also constructed by linking the two Rzs in tandem (Fig. 3A). The Rz-induced cleavage sites on the MT-I and MT-II transcripts were confirmed by primer extension analysis (data not shown). Rz1-2 cleaved rat MT-I mRNA into a 260- and a 175-nucleotide products while Rz4-9 cut MT-II into a 260- and 90-nucleotide fragments. Optimal enzyme activities for Rz1-2 and Rz4-9 were determined under cell-free conditions. Incubation of either Rz with its target substrate for 16 h at 37 °C gave maximal yield. Incubation at higher temperatures or for longer incubation times yield less products, probably due to degradation or inactivation of the Rzs. Enzyme kinetics analyses, determined under single turnover conditions [15], revealed that the catalytic efficiencies of Rz1-2 on MT-I mRNA was 678 $M^{-1}s^{-1}$ and of Rz4-9 on MT-II mRNA was 372 $M^{-1}s^{-1}$. Thus, Rz1-2 is a more efficient Rz than Rz4-9, probably due to the fact that it utilized a high efficiency cleavage site.

Ribozyme-mediated down-regulation of MT-I and MT-II mRNA in stably transfected NbE-1 cells, clones Rz-2 and Rz-3, is correlated with an increase in susceptibility to Cd-induced cytotoxicity

An expression vector carrying Rz(4-9/1-2) driven by a strong promoter (CMV) was generated and used to transfect an immortalized ventral prostate epithelial cell line, NbE-1 [15], using the lipofactAMINE reagents (Life Technologies, Gaithersburg, MD). Three stable transfectants Rz-1, Rz-2 and Rz-3 were selected and cloned. Total RNA was isolated from these clones using standard protocols. MT-I and MT-II transcript levels as well as the levels of Rz(4-9/1-2) expression were analyzed by RT-PCR. Results (not shown) indicated strong expression of Rz(4-9/1-2) and lack of MT-I and M-II mRNA expression in clones Rz-2 and Rz-3. In contrast, clone Rz-1 expressed low levels of Rz (4-9/1-2) and substantial levels of MT-I and M-II transcripts. A cytotoxicity experiment was conducted to compare the sensitivities of the stable transfectants to Cd-cytotoxicity to that of the parent NbE-1 cells. Cells were exposed to 1–400 µM of Cd and numbers of cells survived after 4 days of exposure were determined (Fig. 4). Our data demonstrated that Rz-2 and Rz-3 cells were more susceptible to Cd-cytotoxicity than Rz-1 and

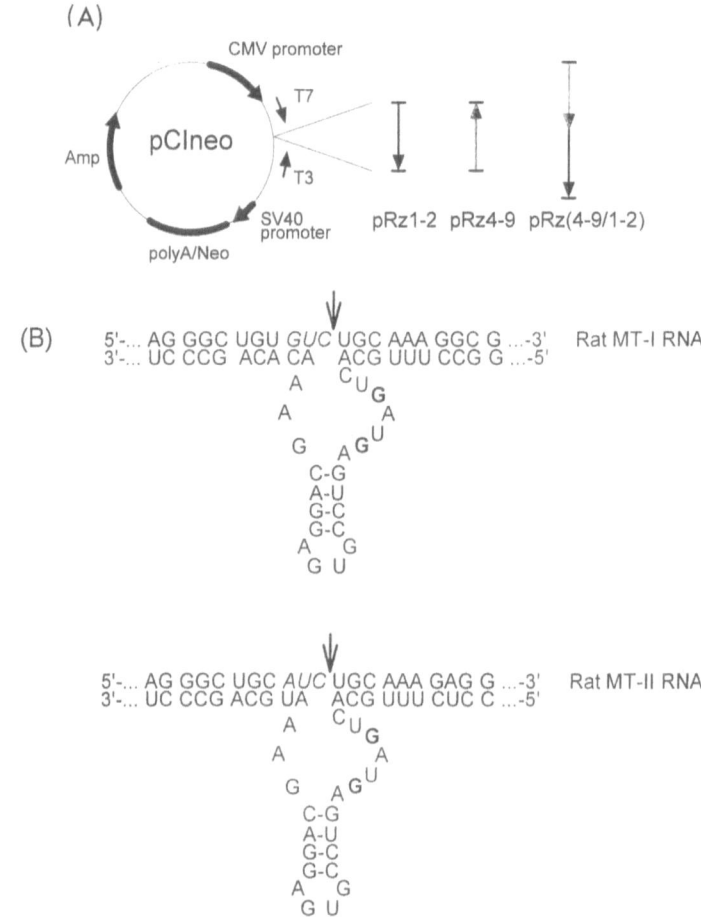

Figure 3. (A) Construction of plasmid Rz1-2, Rz4-9 and Rz(4-9/1-2). Ribozyme recognition sites were chosen from MT messages at positions containing the GUC or AUC triplet. Two ten base antisense regions flanking the cleavage site were included. The ribozymes against MT I and MT-II were designated Rz1-2 and Rz4-9, respectively. The oligonucleotide sequences were as follows: Rz1-2 forward, 5'- CCGAATTCGCGCCTTTGCACT-GATGAGTCCGTGAGGACGAA-3' and reverse, 5'-GCTCTAGAGCGTGTGTGTGTAGGGCTGTGTT-TCGTCCTCACGGAC-3'; Rz4-9 forward, CATTTCGTCCTCACGGAC-3' and reverse, 5'-GCTCTAGAGCC-CTCTTTGCACTGATGAGTCCGTGAGGACGAA-3'. Mutant oligonucleotide (5'-CGCCTTTGCACTAATG-GGTCCGTGAGGACGAA-3') was used to generate catalytically inactive ribozymes Rz2-6 and Rz3-3 for MT-I and MT-II, respectively. The extended double-stranded DNA products were subcloned into pCIneo vector (Promega). (B) Mapping of the cleavage site of the hammerhead ribozyme in the MT-I and MT-II RNAs. Primer extension analysis was performed in the presence of $[\alpha\text{-}^{32}P]$-dATP and primer specific to the 3'-end of MT-I (5'-TCGCATGCTCGGTAGAAAACGG-3') or MT-II (5'- AGGACGCCC-3'). The reaction was started by the addition of 10 U SuperScript II RNaseH⁻ reverse transcriptase and the primer extension was performed at 42 °C for 1 h. The products were loaded directly onto 6% polyacrylamide denaturing gel, next to a DNA sequencing ladder prepared from the cloned MT-I or MT-II cDNA using the same primer and the Thermo Sequenase Kit (Amersham). The figure is a schematic drawing of the MT-I RNA and MT-II RNA cut sites, as determined by primer extension analyses, that were induced by the hammerhead ribozymes Rz1-2 and Rz4-9, respectively.

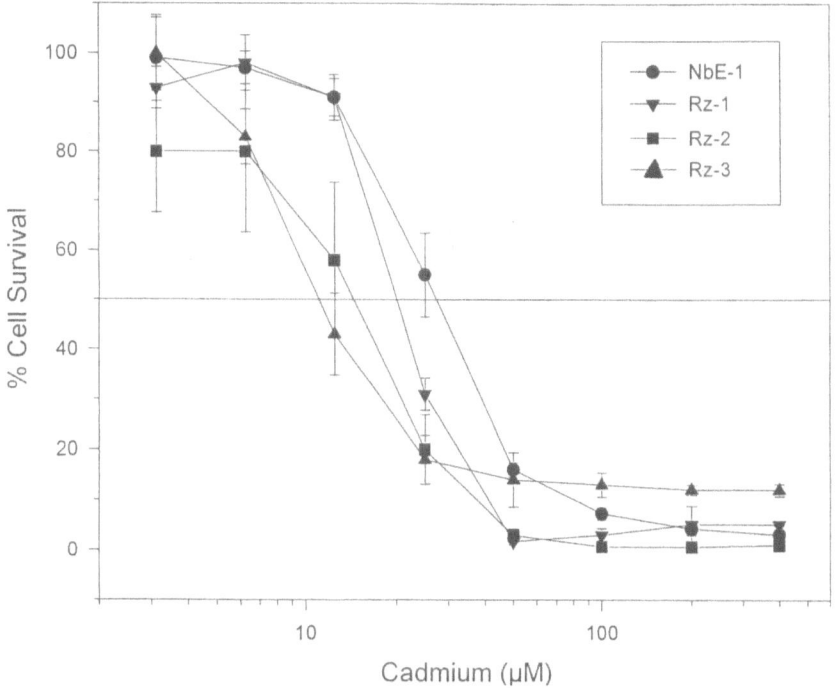

Figure 4. Cytotoxicity studies on the Rz(4-9/1-2) expressing clones. Cells (5×10^3) were plated on 96-well plate per well and subjected to Cd treatment for 4 days. Cell viability are determined by MTT assay Kit (Boehringer-Mannheim).

parent NbE-1 cells. Taken together these data suggest that susceptibility to Cd-cytotoxicity is increased in clones expressing the Rzs which presumably cleaved the endogenous MT-I and MT-II transcripts. It is therefore likely that expression of MTs in prostatic cells protects against Cd. These results strongly support the notion that MT expression protects prostatic cells against Cd cytotoxicity.

Conclusions

Two highly sensitive, sequence-specific, quantitative RT-PCR protocols were established to quantify rat MT-I and MT-II messages. Two hammerhead ribozymes (Rz1-2 and Rz4-9) specific for the rat MT-I and MT-II messages were generated and shown to be catalytically active *in vitro*. The ribozymes were found to be active under a broad temperature range (25–50 °C) and achieved maximal cleavage of the substrates after a 16 h incubation at 37 °C. Primer extension analyses revealed that the ribozymes cleaved at the expected sites on MT messages. The catalytic efficiencies of Rz1-2 and Rz4-9 on rat MT-I and MT-II RNA were estimated to be 678 and 372 $M^{-1}s^{-1}$, respectively. An expression vector carrying Rz1-2 and Rz4-9 in tandem (Rz4-9/1-2) driven by the CMV promoter was generated and used to transfect a rat prostatic epithelial cell line (NbE-1). Two transfectant clones (Rz-2 and Rz-3) were found to stably incorporated the Rz4-9/1-2 plasmid and expressed negligible levels of MT messages. A third clone, Rz-1, on the contrary, expressed little Rz4-9/1-2 and close to normal levels of the two MT transcript. Increase susceptibility to Cd was noted in Rz-2 and Rz-3 but not in Rz-1. These findings strongly suggest that MTs play a role in protecting prostatic epithelial cells against Cd cytotoxicity.

Acknowledgments
This work was supported in part by NIH Grant CA-62269. We thank Dr. Leland Chung at the University of Virginia for providing the NbE-1 cells.

References

1. Kägi JH, Schaffer A (1988) Biochemistry of metallothionein. *Biochemistry* 27: 8509–8515.
2. Vallee BL (1995) The function of metallothionein. *Neurochem Int* 27: 23–33.
3. Bremner I (1991) Nutritional and physiologic significance of metallothionein. *Methods Enzymol* 205: 25–35.
4. Nath R, Kambadur R, Gulati S, Paliwal VK, Sharma M (1988) Molecular aspects, physiological function and clinical significance of metallothionein. *CRC Crit Rev Food Sci Nutr* 27: 41–85.
5. Lohrer H, Robson T (1989) Overexpression of metallothionein in CHO cells and its effect on cell killing by ionizing radiation and alkylating agents. *Carcinogenesis* 10: 2279–2284.
6. Templeton DM, Cherian GM (1991) Toxicological significance of metallothionein. *Methods Enzymol* 205: 11–24.
7. Ebadi M, Leuschen MP, El Refaey H, Hamada FM, Rojas P (1995) The antioxidant properties of zinc and metallothionein. *Neurochem Int* 29: 159–166.
8. Kondo Y, Kuo SM, Watkins SC, Lazo JS (1995) Metallothionein localization and cisplatin resistance in human hormone-independent prostatic tumor cell lines. *Cancer Res* 55: 474–477.
9. Satoh M, Cherian GM, Imura N, Shimizu H (1994) Modulation of resistance to anticancer drugs by inhibition of metallothionein synthesis. *Cancer Res* 54: 5255–5257.
10. Coogan TP, Shiraishi N, Waalkes MP (1995) Minimal basal activity and lack of metal-induced activation of the mtallothineine gene correlates with lobe-specific sensitivity to the carcinogenic effects of cadmium in the rat prostate. *Toxicol Appl Pharmacol* 132: 164–1731.
11. Ghatak S, Oliveria P, Kaplan P, Ho SM (1996) Expression and regulation of metallothionein in mRNA levels in the prostate of noble rats: lack of expression in the ventral prostate and regulation by sex hormones in the dorsolateral prostate. *Prostate* 29: 91–100.
12. Chang SM, Chung LWK (1989) Interaction between prostatic fibroblast and epithelial cells in culture: Role of androgen. *Endocrinology* 125: 2719–2727.
13. Birikh KR, Heaton PA, Eckstein F (1997) The structure, function and application of hammerhead ribozymes. *Eur J Biochem* 245: 1–16.
14. Scott WG, Klug A (1996) Ribozymes: structure and mechanism in RNA catalysis. *Trends Biochem Sci* 21: 220–224.
15. Heidenreich O, Eckstein F (1992) Hammerhead ribozyme-mediated cleavage of the long terminal repeat RNA of human immunodeficiency virus type 1. *J Biol Chem* 267: 1904–1909.

Metallothionein IV
C. Klaassen (ed.)
© 1999 Birkhäuser Verlag Basel/Switzerland

Inactivation of metal-induced metallothionein gene expression by protein kinase C inhibitor

Je-Hsin Chen, Chih-Wen Yu and Lih-Yuan Lin

Department of Life Science, National Tsing Hua University, Hsinchu, Taiwan, Republic of China

Summary. The involvement of protein kinase C (PKC) in the induction of metallothionein (MT) gene expression by non-metallic compounds has been reported. However, whether PKC participates in the metal-induced MT gene transcription remains unclear. We used PKC inhibitor, H7 and cherelythrine, to treat CHO Cd^R and GH_3 cells, and found that both Cd- and Zn-induced MT gene expressions were blocked. Protein kinase A (PKA) was apparently not involved in the induction since PKA inhibitor, HA1004, did not affect the expression of MT gene. By using constructs with a reporter gene linked to different regions of MT promoter, we observed that transcription factor AP1 was not associated with the induction. The inactivation of MT gene expression by PKC inhibitor was not due to block of Cd transport since cellular Cd content was not affected by the inhibitor. However, the PKC inhibitor dramatically reduced cellular Zn accumulation when stimulated by Zn ions. Further analyses of MT transcriptional factor, MTF-1, by semi-quantitative reverse transcriptase-polymerase chain reaction indicated that MTF-1 gene expression was not changed by the PKC inhibitor. We also found that vanadate, a phosphatase inhibitor, increased both basal and induced level of MT gene expression. These results suggest that MT gene expression induced by metal ions involves a PKC mediated phosphorylation pathway.

Introduction

The cysteine-rich, metal binding proteins, metallothioneins (MTs), are commonly found in various organisms although their structures may vary extensively among species [1]. Four MT isoforms have been identified in the mammals. Among them, MTI and MTII are inducible proteins which can be found in all of the tissues. MTIII is constitutively and specifically expressed in the brain [2], and MTIV is primarily present in stratified squamous epithelium cells [3]. Except for MTIII, whose function may be involved in inhibiting neuron survival and outgrowth, the major biological roles for MT isoforms are associated with metal detoxification and homeostasis [4].

Transcriptional regulations for MTI and MTII genes have been studied extensively. It is well established that metal responsive elements (MREs) are present in the promoter region as the *cis*-acting elements, which are responsible for the stimulation of transcription by metal ions [5]. MT transcription factor (MTF-1) is the *trans*-acting factor which interacts with MREs and activates the MT promoter [6, 7]. MTF-1 contains six Zn finger domains and reacts only with Zn to exert its DNA binding ability [7]. It is, therefore, postulated that MTF-1 serves as a Zn sensor that monitors the fluctuation of cellular Zn concentration and modulates the activity of MT promoter [7]. Other transition metals (e.g. Cd, Cu) are speculated to stimulate MT promoter by replacing Zn from cellular components and the displaced free Zn ions activate MTF-1 [8]. However, other studies also indicate that there may be an inhibitor present in the cells [8, 9, 10]. The inhibitor binds MTF-1 when cells are not stimulated by metal ions. Upon induction, the inhibitor interacts with Zn and is released from MTF-1; the MTF-1 thus becomes active and binds MREs [8].

For non-metallic MT inducers, the mechanism of induction by glucocorticoid is the best characterized one. When the hormone moves into the cells, it binds with glucocorticoid receptor. The hormone-receptor complex then interacts with glucocorticoid responsive element in the MT promoter region and activates transcription [11]. For other chemicals, the molecular mechanisms involved in the induction are not clear at present. However, some of the chemicals (e.g. phorbol ester, calcium ionophore and heme-hemopexin) have been shown to induce MT through pathways involving PKC [12–14]. We therefor investigated whether the activity of PKC is also related to MT gene expression induced by metal ions.

We addressed the possibility by adding a PKC inhibitor, H7, to a cadmium resistant CHO cell line (Cd^R). A pronounced reduction in MT expression (approximately 70% with 50 μM H7) was observed after induction by 5 μM Cd [15]. The inhibitory effect was not metal-specific since stimulation by Zn was also blocked by H7. A similar effect was observed when another PKC inhibitor, chelerythrine was employed. The inhibitory effect was not related to the activity of PKA since the PKA inhibitor, HA1004, did not affect the MT induction by metal ions. We also examined whether the block of MT gene expression by PKC inhibitor is cell type specific. A rat thyroid cell line (GH_3) was used for this purpose. As shown in Figure 1, MT induction in GH_3 cells was also inhibited by H7, suggesting a common inhibitory effect of the inhibitor.

The role of AP1 in the PKC mediated MT induction was also investigated since AP1 activity can be regulated by PKC [16] and AP1 recognition sequences are present in MT promoters. We fused CHO MTII promoter with (−253 to +62) or without (−142 to +62) the AP1 binding site with a chloramphenicol acetyltransferase (CAT) gene [17] and used the constructs to study the role of AP1. When transfected into cells, both constructs were able to express CAT

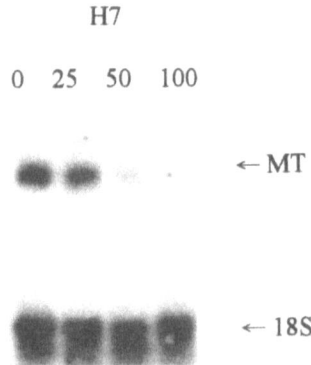

Figure 1. Inavtivation of metal-induced MT gene expression by protein kinase C inhibitor. GH3 cells were pretreated with various concentrations of H7 for 2 h and followed by the stimulation of 5 μM Cd for 10 h. MT mRNAs were determined by Northern hybridization. 18S rRNA was also probe and used as an internal control for the amount of RNA loaded in each lane. Number above each lane indicates the amount (μM) of H7 added to the cells.

activities under the control of metal ions. However, H7 effectively blocked the CAT expression even when the AP1 binding site was absent. This result indicates that AP1 is not involved in the PKC-mediated MT gene expression induced by metal ions [15].

The stimulation of MT promoter requires the presence of metal ions and MTF-1. The possibility that the reduced induction of MT gene expression resulted from the block of either metal transport or MTF-1 gene expression, was investigated by determining the effect of H7 on cellular metal accumulation and the level of MTF-1 mRNA. The results showed that Cd accumulation was not affected by H7. However, cellular Zn uptake was dramatically reduced when treated with Zn [15]. By semi-quantitative reverse transcriptase-polymerase chain reaction [17], the MTF-1 mRNA level was demonstrated to be unaffected by H7. These findings suggest that H7 may reduce the MT gene expression by inactivating PKC activity (i.e. Cd) or by blocking metal transport (i.e. Zn). Nevertheless, it cannot be ruled out that MT induced by Zn is also mediated by the activation of PKC.

In addition to kinase inhibitors, we also employed vanadate, a phosphatase inhibitor [18], to treat the cells. The phosphotyrosine phosphatase inhibitor increased both basal and Cd-induced level of MT gene expression (Fig. 2). This finding indicates that inhibition of dephosphorylation process elevates the MT mRNA level. Since activation of PKC requires phosphorylation at tyrosine residue of the enzyme [19, 20], prolongation of PKC activity by vanadate may be one of the explanations for the increase of MT gene transcripts. Taken together with the PKC inhibitor results, it appears that phosphorylation through PKC is required for the metal-induced MT gene expression. However, our results do not demonstrate how exactly PKC is involved in the MT gene expression. Although there is no direct evidence to show that MTF-1 is phosphorylated during MT induction by metal ions, several consensus PKC recognition sequences [21] can be identified in the Zn finger domains of this protein. In addition, there is a Ser/Thr

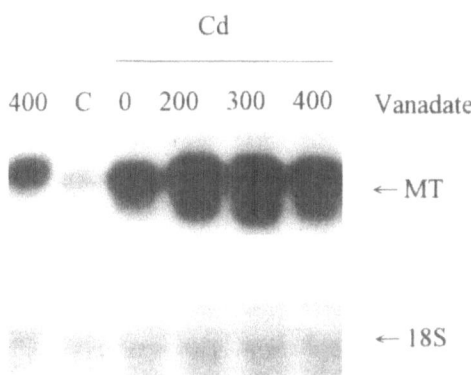

Figure 2. Activation of MT gene expression by phosphatase inhibitor. Cd^R cells were pre-treated with various concentrations of H7 for 2 h and followed by the stimulation of 5 μM Cd for 10 h. MT mRNAs were determined by Northern hybridization. 18S rRNA was also probe and used as an internal control for the amount of RNA loaded in each lane. Number above each lane indicates the concentration (μM) of vanadate added to the cells. Samples subjected to Cd treatment were also indicated. C: sample without adding both Cd and vanadate.

rich domain in the MTF-1 where there can be a potential phosphoylation site. It is also noted that nuclear extracts prepared from Zn-treated cells are more efficient in binding MREs than those of untreated cells even though equivalent amount of Zn was added to the extracts [7]. Furthermore, cells treated with hydrogen peroxide also revealed an increase of binding between MTF-1 and MREs [21]. Therefore, those results suggest that modifications of the MT transcriptional factor may occur when MT gene expression is induced.

Acknowledgment
The authors would like to thank the National Science Council of the Republic of China for financial support of this work under contract No. NSC87-2311-B-007-002-B12.

References

1. Hamer DH (1986) Metallothionein. *Annu Rev Biochem* 55: 913–951.
2. Uchida Y, Takio K, Titani K, Ihara Y, Tomonaga M (1991) The growth inhibitory factor that is deficient in the Alzheimer's disease brain is a 68 amino acid metallothionein-like protein. *Neuron* 7: 337–347.
3. Quaife CJ, Findley SD, Erickson JC, Forelick GJ, Kelly EJ, Zambrowicz BP, Palmiter RD (1994) Induction of a new metallothionein isoform (MT-IV) occurs during differentiation of stratified seuamous epithelia. *Biochemistry* 33: 7250–7259.
4. Master BA, Kelly EJ, Quaife CJ, Brinster RL, Palmiter RD (1994) Targeted disruption of metallothionein I and II genes increases sensitivity to cadmium. *Proc Natl Acad Sci USA* 91: 584–588.
5. Stuart GW, Searle PF, Palmiter RD (1985) Identification of multiple metal regulatory elements in mouse metallothionein-I promoter by assaying synthetic sequences. *Nature* 317: 828–831.
6. Radtke F, Heuchel R, Georgiev O, Hergersberg M, Gariglio M, Dembic Z, Schaffner W (1993) Cloned transcription factor MTF-1 activates the mouse metallothionein I promoter. *EMBO J* 13: 1355–1362.
7. Heuchel R, Radtke F, Georgiev O, Stark G, Aguet M, Schaffner W (1994) The transcription factor MTF-1 is essential for basal and heavy metal-induced metallothionein gene expression. *EMBO J* 13: 2870–2875.
8. Palmiter RD (1994) Regulation of metallothionein gene by heavy metals appears to be mediated by a zinc-sensitive inhibitor that interacts with a constitutively active transcription factor, MTF-1. *Proc Natl Acad Sci USA* 91: 1219–1223.
9. Mayo KE, Palmiter RD (1981) Glucocorticoid regulation of metallothionein-I mRNA synthesis in culture mouse cells. *J Biol Chem* 256: 2621–2624.
10. McCormick CC, Salati LM, Goodridge AG (1991) Abundance of hepatic metallothionein mRNA is increased by protein-synthesis inhibitor- Evidence for transcriptional activation and post-transcriptional regulation. *Biochem J* 273: 185–188.
11. Karin M, Beato M, Haslinger A, Holtgreve H, Krauter P, Richards RI, Westphal HM (1984) Characterization of DNA-sequences through which cadmium and glucocorticoid hormones induce human metallothionein-Iia gene. *Nature* 308: 513–519.
12. Imbra RJ, Karin M (1987) Metallothionein gene expression is regulated by serum factors and activators of protein kinase C. *Mol Cell Biol* 7: 1358–1363.
13. Arizono K, Peterson KL, Brady FO (1993) Inhibitors of Ca+2 channels, calmodulin and protein kinases prevent A23187 and other inductions of metallothionein mRNA in EC3 rat hepatoma cells. *Life Sci* 53: 1031–1037.
14. Yafei R, Smith A (1995) Mechanism of metallothionein gene regulation by heme-hemopexin-role of protein kinase C, reactive oxygen species and cis-acting elements. *J Biol Chem* 270: 23988–23995.
15. Yu C-W, Chen J-H, Lin L-Y (1997) Metal-induced metallothionein gene expression can be inactivated by protein kinase C inhibitor. *FEBS Lett* 420: 69–73.
16. Boyle WJ, Smeal T, Defize LH, Angel P, Woodgett JR, Karin M and Hunter T (1991) Activation of protein kinase C decreases phosphorylation of c-jun at sites that negatively regulate its DNA-binding activity. *Cell* 64: 573–584.
17. Yu C-W, Chen H-C, Lin L-Y (1998) Transcription of metallothionein isoform promoters is differentially regulated in cadmium-sensitive and -resistant. *J Cell Biochem* 68: 174–185.
18. Swarup G, Cohen S, Garbers DL (1982) Inhibition membrane phosphotyrosyl-protein phosphatase- activity by vanadate. *Biochem Biophys Res Commun* 107: 1104–1109.
19. Liu F, Roth RA (1994) Insulin-stimulated tyrosine phosphorylation of protein-kinase-C—evidence for direct interaction of the insulin-receptor and protein kinase C in cells. *Biochem Biophys Res Commun* 200: 1570–1577.

20. Denning MF, Dlugosz AA, Threadgill DW, Magnuson T, Yuspa SH (1996) Activation of the epidermal growth-factor receptor signal-transduction pathway stimulates tyrosine phosphorylation of protein kinase C-δ. *J Biol Chem* 271: 5325–5331.
21. Kennelly PJ, Krebs EG (1991) Consensus sequences as substrate specificity determinants for protein kinases and protein phosphatases. *J Biol Chem* 266: 15555–15558.
22. Dalton TP, Li Q, Bittel D, Liang L, Andrews GK (1996) Oxidative stress activates metal-responsive transcription factor-1 binding activity-occupancy *in vivo* of metal response elements in the metallothionein-I gene promoter. *J Biol Chem* 271: 26233–26241.

Cadmium-induced metallothionein expression and metallothionein mRNA in mice

Keiji Suzuki[1], Kyoumi Nakazato[1] and Katsuyuki Nakajima[2]

[1]Division of Pathology, Department of Laboratory Sciences, Gunma University School of Health Sciences, 3-39-15 Maebashi, Gunma 371, Japan
[2]Otuka America Pharmaceutical, Inc. 2440 Research Boulevard, Suite 5300 Rochville, Maryland 20850

Metallothionein (MT) isoforms I and II are different slightly in amino acid sequence and they have been identified in several organs. These proteins are thought to be involved in the homeostatic control or protection from toxic metals. MT mRNA is present in murine organs and can be induced in many organs by small amount of heavy metals. However, the functional differences between MT-Iand MT-II are poorly understood. The present study examined MT immunohistochemically using rabbit polyclonal anti-MT antibody and streptavidin biotin peroxidase kit (Nichirei Co., Japan). MT mRNAs were studied using *in situ* hybridization and northern blot hybridization, particularly in the livers and kidneys of mice, but also in the other organs as well. Results were noted at 3, 6, 12, and 24 h after intraperitoneal cadmium (3 mg/kg) treatment. The mouse MT-Iand MT-II cDNAs were provided by Dr. R.D. Palmiter (University of Washington, Seattle, WA).

Immunohistochemically, various organs showed strong MT expression after 3 h of Cd treatment, and the expression was still apparent 24 h later.

Northern blotting of MT-I mRNA and MT-II mRNA of the liver and kidney (Fig. 1) showed the highest level after 3 h of Cd treatment. Liver mRNAs gradually decreased to a level lower

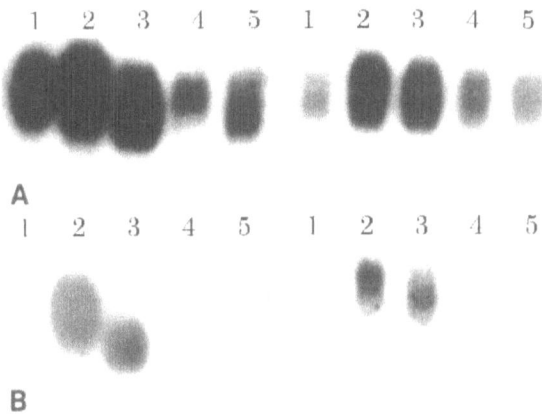

Figure 1. Northern blotting of MT-I (A. left: liver, right: kidney) and II (B. left: liver, right: kidney) mRNAs. 1: Control, 2: 3 h, 3: 6 h, 4: 12 h and 5: 24 h after Cd treatment.

than the control level but kidney mRNAs remained at a higher level than that prior to Cd treatment. MT-II mRNA in the kidney showed same decrease as the liver, but decrease of MT-I mRNA in both liver and kidney was mild and a relatively high level of MT-I mRNA was found at 24 h after Cd treatment. The level of MT-I mRNA in both organs was higher than MT-II mRNA at every hour after Cd treatment.

In situ hybridization histochemistry of non treated mice detected MT-I and II mRNA expression in liver cells, tubular epithelium of both cortex and medulla of the kidney, transitional epithelium of the urinary bladder, prostatic epithelium, Leydig cells, cells of the seminiferous tubules, epithelium of the salivary gland, esophageal epithelium, mucosal epithelium in the stomach, small and large intestines, epithelium of the exocrine and endocrine pancreas, bronchial epithelium, cells in alveolar septa of the lung, cartilage cells, and the pia mater and choroid plexus of the brain. mRNA in various organs was more strongly expressed than MT-II mRNA, and both MT-I mRNA and MT-II mRNA exhibited strong expression 3 h after Cd treatment. The strong expression of MT-I and II in these organs gradually decreased after Cd treatment, but the decrease in MT-I mRNA expression was mild.

These results demonstrate that various mouse organs express both MT-I and II genes, and that MT-I is primarily responsible for protection against Cd toxicity.

Acknowledgment
We are grateful for the kindness of Dr. R. D. Palmiter who gave us MT cDNA.

Role of metallothionein in reproduction, development and cell proliferation

Metallothionein IV
C. Klaassen (ed.)
© 1999 Birkhäuser Verlag Basel/Switzerland

The nuclear-cytoplasmic presence of metallothionein in cells during differentiation and development

M. George Cherian, John C. Lau, Margarita D. Apostolova and Lu Cai

Departments of Pathology, and Pharmacology and Toxicology, University of Western Ontario, London, Ontario N6A 5C1, CANADA

In early biochemical studies, metallothionein (MT) was mainly isolated from cytoplasm in most tissues. However, in 1982 it was reported from two different laboratories [1, 2] by immuno-histochemical localization technique that MT can be present in both cytoplasm and nucleus of epithelial cells in liver and kidney, especially after its induced synthesis by metals. In subsequent studies, it was demonstrated that the nuclear presence of MT occurred in human fetal liver and in both fetal and neonatal rat liver [3, 4]. The transitional presence of MT in nucleus of hepatocytes has been confirmed after partial hepatectomy in rats [5]. Both cytoplasmic and nuclear expressions of MT have been demonstrated in various human tumours [6]. The mechanism and significance of the nuclear-cytoplasmic localization of MT in hepatocytes and its functions are not clearly understood.

In this report, we describe some of the recent studies in progress in our laboratory using differentiating cell lines of myoblast and MT-1 isoform overexpressing transgenic mice (MT-1*) to understand more about the nuclear localization of MT, and the changes in zinc and copper levels during development in MT-1 and -2 isoforms knockout (MT-null) mice.

Developmental changes in zinc, copper and MT

Studies undertaken in MT-1*, MT-null and control C57BL/6 mice during development showed that the highest concentration of hepatic MT was found in MT-1* during late gestation to 3 days of age (about 900 µg/g), and declined to adult level (about 250 µg/g) by 21 days of age. In control mice, MT levels were about 260 µg/g during late gestational and early neonatal stage and was later declined to the adult level of about 10 µg/g (Fig. 1). In immunohistochemical localization, MT was present mainly in cytoplasm in fetal and adult livers in both strains of mice. However, MT localization in nucleus of hepatocytes was observed in newborn mice of one to three days of age. Although the adult MT-1* mice had hepatic MT levels as high as in the livers of newborn control mice, MT was localized mainly in cytoplasm. These results suggest that the localization of MT in nucleus of cells is not dependant on its concentration but is due to other factors.

Concomitantly, the hepatic concentration of zinc was highest (80 µg/g) in MT-1* and lowest in MT-null (15.9 µg/g) mice. In control mice, the high hepatic zinc levels (32 µg/g) declined to adult levels (10 µg/g) while in MT-null mice, the low hepatic zinc levels in neonates increased to an adult level of 22 µg/g of zinc. In contrast, the hepatic copper levels increased during the first week after birth and then decreased to an adult hepatic level of 5 µg/g in all groups of mice. These results are shown in Figure 1. From these results, it was possible to calculate the hepatic zinc concentration not bound to MT in newborn mice of MT-1* and control C57BL/6 strain, and compare them to that in newborn MT-null mice. The concentration of hepatic zinc not bound to MT were estimated to be 20.3 and 14.3 µg/g for MT-1* and control mice, respectively. In this calculation an assumption was made that each mol of MT contains 7 g atoms of zinc. The total hepatic zinc concentration for MT-null mice was about 15.9 µg/g throughout devel-

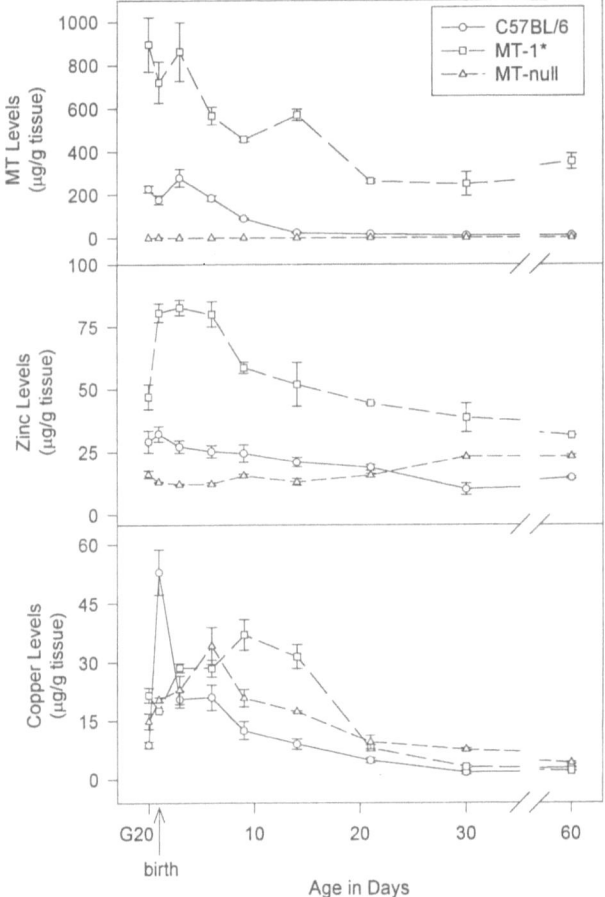

Figure 1. Hepatic MT, zinc and copper levels in developing mice. Livers from gestational day 20 to 60 days of age in MT-1*, C57BL/6 and MT-null mice were analyzed for MT, zinc and copper. Values represent mean ± SEM with a minimum of 5 animals from different litters. Hepatic MT levels (top panel) were determined by the [109]Cd-Heme assay and both hepatic zinc (middle panel) and hepatic copper (bottom panel) levels were measured by atomic absorption spectrophotometry using an air-acetylene flame.

opment. From these data, it appears that the essential hepatic concentration of zinc required is about 14.3 to 20.3 µg/g which is similar to the hepatic concentration of zinc found in MT-null newborn mice. These results suggest that the role of MT in newborn mice is to store excess zinc in the liver and this may be utilized during rapid development and growth.

The transitional nuclear localization of MT in hepatocytes of mice during the neonatal period is similar to previous reports in rat [4] and it occurs when the hepatic MT concentration is maximal in both MT-1* and control mice. However, in adult MT-1* mice, which has high hepatic MT concentration as compared to control, MT is mainly in the cytoplasm, suggesting that some other cellular factors other than high expression of MT may determine the nuclear localization of MT.

Nuclear localization of MT in cultured cells during differentiation

The significance of nuclear MT in hepatocytes during development and also in certain tumour cells is not understood but may be related to the interactions of metal ions (e.g. zinc) with various nuclear constituents during the cell cycle. The metals that bind to MT can also bind to histones, nuclear RNA and nuclear acidic proteins, and may modify gene expression and phosphorylation of regulatory proteins. For example, MT in the nucleus could effectively donate zinc to several enzymes or chelate zinc from transcription factors containing zinc fingers. These protein-metal interaction in the nucleus can modify various cellular processes, including enzymatic activity, gene transcription and apoptosis. In addition, since cell growth and differentiation are affected by both generation of free radicals and antioxidant defence mechanisms, MT and zinc may play a major role in these processes.

DNA flow cytometry studies demonstrated the localization of MT in the nucleus of both diploid and aneuploid cells during the G1/S phase of the cell cycle of adenocarcinoma of breast and colon [7]. Studies on expression of MT in tumours also showed that less differentiated cells express more MT than well differentiated cells [6]. In order to look at the direct relationship between cellular differentiation and MT expression, studies are initiated in the two fast differentiating myoblast cell lines L6 and H9C2. These myoblast cell lines can differentiate into myotubes within 7 days of growth in controlled culture media. The preliminary results suggest that myoblast cell lines contain higher levels of MT in the cell nucleus than in myotubes. During the formation of myotubules, within 5 days of growth, the MT levels decrease along with its nuclear presence. The study also suggest that in order to sustain MT in the nucleus, the cellular energy level in myoblasts should be maintained. Depletion of glucose may decrease the nuclear MT level in these cells. Similarly arresting the cell cycle at G1/S interphase with aphidicolin can also affect the presence of nuclear MT.

Role of MT during spermatogenesis

Spermatogenesis, including various stages of seminiferous testicular tubules, consists of different differentiated cells, such as stem spermatogonia, spermatocytes, spermatids and spermatozoa. The normal spermatogenesis is controlled by hormones from Leydig and Sertoli cells in cell proliferation and degeneration. The germ cell degeneration has been identified as programmed cell death (Apoptosis). Spontaneous apoptosis during spermatogenesis has been considered as a defence mechanism to remove damaged germ cells and prevent birth defects. To investigate the role of MT in spermatogenesis, the MT immunohistochemical staining was performed in the testis of MT-1*, MT-null and control mice. In MT-1* mouse testis, MT was found in the nucleus of few spermatocytes except for in Leydig and Sertoli cells, while in control and MT-null mice no MT staining was found. Observation of spontaneous apoptosis among these three strain mice indicated low spontaneous apoptosis in the testis of MT-null mice with high relative testis weight as compared to control and MT-1* mice. These results suggest the possible involvement of MT in regulation of the spontaneous apoptosis during normal spermatogenesis.

Thus, studies in our laboratory show that MT is mainly expressed in certain epithelial cells during cell proliferation. The nuclear localization of MT is associated with rapid growth and cell differentiation but is not related to the high concentration of MT in the cell. The retention of MT in cell nucleus may require energy.

References

1   Banerjee D, Onosaka S, Cherian MG (1982) Immunohistochemical localization of metallothionein in cell nucleus and cytoplasm of rat liver and kidney. *Toxicology* 24: 95–105.
2   Danielson KG, Ohi S, Huang PC (1982) Immunohistochemical detection of metallothionein in specific epithelial cells of rat organs. *Proc Natl Acad Sci USA* 79: 2301–2304.
3   Nartey NO, Banerjee D, Cherian MG (1987) Immunohistochemical localization of metallothionein in cell nucleus and cytoplasm of fetal human liver and kidney and its changes during development. *Pathology* 19: 233–238.
4   Panemangalore M, Banerjee D, Onosaka S, Cherian MG (1983) Changes in intracellular accumulation and distribution of metallothionein in rat liver and kidney during postnatal development. *Dev Biol* 97: 95–102.
5   Tohyama C, Suzuki JS, Hemelraad J, Nishimura N, Nishimura H (1993) Induction of metallothionein and its localization in the nucleus of rat hepatocytes after partial hepatectomy. *Hepatology* 18: 1193–1201.
6   Cherian MG (1994) The significance of the nuclear and cytoplasmic localization of metallothionein in human liver and tumour cells. *Environ Health Perspect* 102: 131–135.
7   Haile Meskel H, Cherian MG, Martinez VJ, Veinot LA, Frei JV (1993) Metallothionein as an epithelial proliferative compartment marker for DNA flow cytometry. *Modern Pathol* 6: 755–760.

Metallothionein IV
C. Klaassen (ed.)
© 1999 Birkhäuser Verlag Basel/Switzerland

Metallothionein and male genital organs

Chiharu Tohyama[1], Junko S. Suzuki[1], Hisao Nishimura[2] and Noriko Nishimura[1]

[1]Environmental Health Sciences Division, National Institute for Environmental Studies, 16-2, Onogawa, Tsukuba, Ibaraki 305-0053, Japan
[2]Department of Human Sciences, Aichi Mizuho University, Toyota-shi, Aichi 470-0394, Japan

Introduction

Male genital organs, such as testis and prostate are susceptible to acute and chronic cadmium insult, resulting in necrotic changes in germ cells and tumorigenesis, respectively (1). It was proposed that the susceptibility was probably due to an apparent lack of MT in these organs [2]. Since the male genital organs, testis, prostate and other auxiliary organs consist of heterogenous types of cells, we thought that it would be important to identify what kind of cells harbor or induce MTs, and studied localization of MT mRNA and protein in these tissues. Next, we studied whether and/or how testosterone affects the induction of metallothionein in the above male genital organs. Finally, we sequenced RT-PCR products obtained from the rat testicular RNA and also purified and isolated Cd-binding proteins from the rat testis, and showed that MT-I and II exist in this organ.

Localization of metallothionein in male genital organs of the rat

We first studied localization of MT in various genital organs of male rats, i.e. testis, epididymis, seminal vesicle, prostate, coagulating gland and ejaculatory duct, by utilizing antibody specific to rat MTs in immunohistochemical technique [3]. In each seminiferous tubule of the testis, MT was found to localize in Sertoli cells, the supporting cells, and all the spermatogenic cells including spermatozoa. In the seminiferous tubules it takes approximately 48 days for spermatogenesis through a mitosis and meioses, portions of each seminiferous tubule harbor different stages of spermatogenesis along its long axis, and thus a cross-section of seminiferous tubules of tissue preparation showed various differentiation stages of spermatogenesis. We found that there was only weak MT staining in all the spermatogenic cells in the seminiferous tubules where sperms were in the lumen whereas strong MT staining was observed in these cells of the seminiferous tubules that were in the process of spermatogenesis. No MT staining was observed in the basement membrane, connective tissue and Leydig cells. Metallothionein concentrations differed markedly among the male genital organs of intact rats. The highest average MT level (473 ng/mg protein) was found in dorsolateral lobes compared to other tissues such as coagulating gland (126 ng/mg protein) and the ventral lobe (non-detectable level).

Growing testes from 1 to 3 weeks old rats showed that MT is first localized in spermatogonia located peripherally and then spermatocytes in the central part of tubules. This topographical change in MT staining with testicular development may reflect an involvement of MT with cell differentiation as well as continuous processes of spermatogenesis. Changes in distribution pattern and intensity of MT were also confirmed to vary tubule to tubule in the neonatal rat testis as seen in the adult. In addition, identical types of cells did not show the same degree of intensity of MT staining, which may reflect the different stage of spermatogenesis. Appearance of MT only in the leptotene spermatocytes suggest that MT would take part in physiological roles at the earlier stage of cell division. Since no significant disturbance in spermatogenesis was found in the testis from MT-null transgenic mice, MT may be compensated by other yet unknown factors or may not be essential for this reproductive process.

The prostate gland was most intensely stained for MT antigenicity among the male genital tissues examined, but the intensity and localization of MT staining considerably differ between the ventral and dorsolateral lobes. In the ventral lobe almost no staining was observed in the mucous columnar epithelium. In contrast, MT was most strongly stained in the lateral lobe among the three lobes, and the presence of MT was shown mainly in the cytoplasm of the epithelium and partly in the nucleus. In the dorsal lobe, MT staining was observed in the cytoplasm and nucleus of the columnar epithelial cells and some cells appeared to harbor MT near their free surface facing to the lumen, in which MT staining was present with other apocrine secretions. MT immunostaining was also found in the columnar epithelial cells of the coagulating gland, seminal vesicle, ductus efferentes, epididymis and ejaculatory duct.

Another interesting observation is the presence of very strong MT staining in basal cells of the seminal vesicle, coagulating gland and ejaculatory duct. The basal cells are speculated to be a progenitor of the epithelium, and involved in transport and supply of indispensable substances for the survival of the epithelium. No MT staining was found in the connective tissue of the male genital organs examined in our studies.

Since MT has been speculated to act as store and donor of Zn for other macromolecules, presence of MT not only in the basal cell but also in Sertoli cell, the supporting cell that nourishes spermatogenetic cells, may be relevant to this physiological role since Zn ions are required for various enzymes responsible for cell divisions. This speculation is supported by a few lines of evidence that the secreted materials in these lobes were positively stained for MT, that semen contained large amount of Zn and that nuclei of actively growing basal cells contain a high level of Zn [4] and MT.

Localization of metallothionein mRNA in the rat testis and prostate

To localize MTmRNA in the testis and prostate of the rat by *in situ* hybridization, we utilized a digoxigenin-labeled riboprobe in a bright-field optics so that we could clearly visualize tissue structure and MTmRNA localization together [5]. The present study clearly proved that *in situ* hybridization technique using non-radiolabeled riboprobes is extremely useful and that the utilization of both *in situ* hybridization technique and immunohistochemical staining on the same tissue specimens show MT gene expression in specific cell types in the male genital organs. In the rat testis, MT mRNA was conspicuous in the cytoplasm of primary and secondary spermatocytes and spermatids, but not detected in spermatogonia, spermatozoa, Sertoli cells or Leydig cells. On the other hand, as described above, MT protein was present in these spermatogenic cells as well as in spermatozoa, and Sertoli cells. It should be noted that primary spermatocytes of some seminiferous tubules harbored remarkable amounts of MT mRNA whereas secondary spermatocytes and spermatids of other tubules had more MT mRNA than primary spermatocytes. These results suggest the possibility that MT gene expression is associated with stages with spermatogenesis. In the prostate, MT mRNA was found predominantly in the epithelium of the dorsolateral lobes, but not in the ventral lobe, which is in agreement with the observed localization of MT protein as described above.

Testosterone-dependent induction of MT in the dorsolateral prostate not in the ventral prostate

Constitutive expression of MT gene was found higher in the testis and dorsolateral prostate than other organs. Although the biosynthesis of MT was reported to be regulated by steroid hormones like glucocorticoid and female sex hormones *in vivo* and/or *in vitro*, it was not clear how MT biosynthesis in the male auxiliary organs was regulated by testosterone. It should be pointed out that cadmium administration to rodents suppressed the synthesis of testosterone [6]. We thus investigated effects of testosterone on MT gene expression in male genital organs of castrated rats [7]. In this study plasma testosterone was depleted to approximately 10% of the intact level by orchidectomy and an injection of testosterone to orchidectomised rats replenished the plasma testosterone level by 24 h after injection. MT mRNA and protein were found to be most abundant in the dorsolateral lobe of the prostate, some in the coagulating gland but not detectable in the ventral lobe. Orchidectomy caused involution of the prostate (both ventral and dorsolateral lobes) and coagulating gland, and amounts of both MT mRNA and MT in the dorsolateral gland were considerably decreased or became undetectable. Plasma corticosterone, a marker of physiologic stress, was within the physiological concentration range after orchidectomy. An injection of testosterone to orchidectomised rats remarkably restored not only size of these organs, but also MT mRNA and MT, particularly in the epithelium of the dorsolateral lobe and coagulating gland, but not in the ventral prostate. In the dorsolateral lobe, no selective uptake of zinc preceding the increase in MT was observed, suggesting that zinc ions did no trigger the increased expression of MT gene. These results suggest that the dorsolateral lobe and coagulating gland but not ventral lobe contain MT, the biosynthesis of which is regulated by testosterone.

Glucocorticoid responsive elements are identified in promotor regions of human MT-II and mouse MT-I genes [8]. Since the plasma testosterone concentration was markedly elevated after testosterone administration and the structure of steroid hormone receptors are very similar to one another among different animal species [9], there is the possibility that the glucocorticoid response element reacted with the androgen receptor and up-regulated MT gene transcription. However, this possibility is considered to be minimal because other male auxiliary organs such as the ventral lobe of the prostate responded to testosterone for growth but not for MT biosynthesis, and because the liver, which has glucocorticoid hormone receptors did not respond to testosterone under the present experimental conditions.

Waalkes and Perantoni [2] reported that the prostate of rats contains a cadmium-binding protein, which is different from MT in terms of amino acid composition, and proposed that the absence of MT in this organ may be relevant to the susceptibility of the prostate to cadmium tumorigenicity. Although our present result on the absence of MT in the ventral lobe may support this notion, more importantly, it confirms earlier observations on the presence of MT in the dorsolateral lobes of prostate [3, 10, 11]. Furthermore, it is reasonable to suppose that a decrease in testosterone biosynthesis caused by cadmium exposure aggravates the toxic effects of this metal on male auxiliary organs.

Existence of MT-I and -II in the rat testis

Despite the above findings on localization and induction of MT mRNA and protein in the male genital organs, we consider it essential to solve this argument by clarifying the DNA and protein sequences by more direct methodology than has been used in the earlier studies.

It was necessary to clarify these MT-I and -II exist in the rat testis because Waalkes and coworkers reported that the Cd- or Zn-binding proteins purified as MT completely differ in amino acid composition [12, 13]. First, we isolated total RNA from the rat testis, performed RT-PCR analysis, sequenced the PCR products and found that MT-I and -II genes are actually expressed. We also performed purification of Cd-binding proteins from the rat testis, by sequential purification by using gel filtration chromatography and anion HPLC, from which we obtained Cd-binding protein 1 and 2 which were subjected to protein sequence analysis. From the cDNA and protein sequences the two forms of Cd-binding proteins in the rat testis were found to have the same primary structure as that of MT-I and -II. Finally, the experimental evidence that MT-I and -II isoforms are present at least in the rat testis supports the results from earlier reports the localization of MT in the male genital tissues [22].

The reasons why MT is not inducible by heavy metals and other factors such as lipopolysaccharide are still unsolved and were discussed earlier [14]. In brief, one of the plausible explanation could be that Cd cannot be accumulated in this tissue much less than in other tissues such as liver after Cd administration, and even such small amounts of Cd can destroy spermatogenic cells. Second, hypermethylation status of MT may be responsible for uninducibility of MT gene in the rat testis since DNA methylation was reported to control MT-I gene expression in murine lymphoid cells. However, if this is the case, we do not know why a relatively large amount of MT protein as well as that of MT mRNA is present in the testis without any specific exogenous stimuli. Since the transgenic mice that had the interferon gene under the control of the MT promotor were found to express high levels of interferon mRNA in the testes without a specific inducer [15], there may be yet unknown tissue specific induction machinery in this organ.

In addition to the rat testis, other male auxiliary organs [3, 10] and female reproductive tissues [16–18] were reported to show immunohistochemical staining for MT. As described above, cell- and tissue-specific gene expression of MT was exhibited in the rat prostate: in the ventral lobe, neither MT mRNA nor MT protein was detectable whereas the dorsolateral lobe constitutively harbored MT mRNA and the protein [3, 5, 10]. In contrast, amino acid composition of Cd- or Zn-binding protein purified from the prostate of rodents and primates and that from the ovary of Syrian hamsters were reported to be markedly distinct from that of MT [2, 19, 20]. Since these studies utilized the essentially same purification and assay protocols as the one used for testicular Cd-binding protein, it would be plausible to consider that MT also exists in these male and female reproductive organs as well.

We believe that we have provided sufficient experimental evidence of the presence of MT-I and -II in the male genital organs of the rat. Furthermore, during the course of our purification study on the rat testis, we found several other protein peaks which could react with antibody against MTs. Since the testis from MT-null mice whose MT-I and -II genes were knocked out by homologous recombination may contain other proteins which interacts with anti-MT antibodies [21], we would speculate that there may be other isoforms of MT in the testis.

Acknowledgement
Norika Nishimura was supported by Science and Technology Agency to study at NIES.

References

1. Waalkes MP, Coogan TP, Barter RA (1992) Toxicological principles of metal carcinogenesis with special emphasis on cadmium. *Crit Rev Toxicol* 22: 175–201.
2. Waalkes MP, Perantoni A (1989) Apparent deficiency of metallothionein in the Wistar rat prostate, *Toxicol Appl Pharmacol* 101: 83–94.
3. Nishimura H, Nishimura N, Tohyama C, (1990) Localization of metallothionein in the genital organs of the male rat. *J Histochem Cytochem* 38: 927–933.
4. Chandler JA, Timms BG (1977) The effect of testosterone and cadmium on the rat lateral prostate in organ culture. *Virchows Arch* 25: 17–31.
5. Tohyama C, Nishimura N, Suzuki JS, Karasawa M, Nishimura H (1994) Metallothionein mRNA in the testis and prostate of the rat detected by digoxigenin-labeled riboprobe. *Histochemistry* 101: 341–346.
6. Favino A, Baillie AH, Griffiths K (1966) Androgen synthesis by the testes and adrenal glands of rats poisoned with cadmium chloride. *J Endocrinol* 35: 185–192.
7. Tohyama C, Suzuki JS, Homma S, Karasawa M, Kuroki T, Nishimura H, Nishimura N (1996) Testosterone-dependent induction of metallothionein in genital organs of male rats. *Biochem J* 317: 97–102.
8. Hamer DH (1986) Metallothionein. *Annu Rev Biochem* 55: 913–951.
9. Kumar V, Green S, Staub A, Chambon P (1986) Localisation of the oestradiol-binding and putative DNA-binding domains of the human oestrogen receptor. *EMBO J* 5: 2231–2236.
10. Umeyama T, Saruki K, Imai K, Yamanaka H, Suzuki K, Ikei N, Kodaira T, Nakajima K, Saitoh H, Kimura M (1987) Immunohistochemical demonstration of metallothionein in the rat prostate. *Prostate* 10: 257–264.
11. Bataineh ZM, Heidger PM Jr, Thompson SA, Timms BG (1986) Immunocytochemical localization of metallothionein in the rat prostate gland. *Prostate* 9: 397–410.
12. Waalkes MP, Perantoni A (1986) Isolation of a novel metal-binding protein from rat testes. Characterization and distinction from metallothionein. *J Biol Chem* 261: 13097–13103
13. Waalkes MP, Chernoff SB, Klaassen CD (1984) Cadmium-binding proteins of rat testes. Characterization of a low-molecular-mass protein that lacks identity with metallothionein. *Biochem J* 220: 811–818.
14. Tohyama C, Suzuki JS, Homma ST, Nishimura N, Nishimura H (1993) Regulation of metallothionein biosynthesis in genital organs of male rats. *In*: KT Suzuki, N Imura, M Kimura (eds): *Metallothionein III. Biological roles and medical implications*, Birkhäuser, Basel, 443–454.
15. Iwakura Y, Asano M, Nishimune Y, Kawade Y (1988) Male sterility of transgenic mice carrying exogenous interferon-beta gene under the control of the metallothionein enhancer-promoter. *EMBO J* 7: 3757–3762.
16. De SK, McMaster MT, Dey SK, Andrews GK (1989) Cell-specific metallothionein gene expression in mouse decidua and placentae. *Development* 107: 611–621.
17. Hazelhoff Roelfzema W, Tohyama C, Nishimura H, Nishimura N, Morselt AF (1989) Quantitative immuno-

histochemistry of metallothionein in rat placenta. *Histochemistry* 90: 365–369.
18. Nishimura N, Nishimura H, Tohyama C (1989) Localization of metallothionein in female reproductive organs of rat and guinea pig. *J Histochem Cytochem* 37: 1601–1607.
19. Waalkes MP, Perantoni A, Palmer AE (1988) Isolation and partial characterization of the low-molecular-mass zinc/cadmium-binding protein from the testes of the patas monkey (Erythrocebus patas). Distinction from metallothionein. *Biochem J* 256: 131–137.
20. Waalkes MP, Rehm S, Perantoni A (1988) Metal-binding proteins of the Syrian hamster ovaries: apparent deficiency of metallothionein. *Biol Reprod* 39: 953–961.
21. Tohyama C, Satoh M, Kodama N, Nishimura H, Choo A, Michalska A, Kanayama Y, Nagamuna A (1996) Reduced retention of cadmium in the liver of metallothionein-null mice. *Environ Toxicol Pharmacol* 1: 213–216.
22. Suzuki JS, Kodama N, Molotokov A, Aoki E, Johyama (1998) Isolation and identification of metallothionein isoforms (MT-1 and MT-2) in the rat. *Biochem J* 334: 695–701.

Metallothionein IV
C. Klaassen (ed.)
© 1999 Birkhäuser Verlag Basel/Switzerland

Induction of metallothionein-like cadmium binding protein in testis and its protective role against cadmium toxicity

H. Ohta, H. Tanaka, S. Asami, Y. Seki and H. Yoshikawa

Department of Occupational Health and Toxicology, School of Allied Health Sciences, Kitasato University, 1-15-1 Kitasato, Sagamihara, Kanagawa, 228 Japan

Summary. Despite the numerous studies conducted on the testicular Cadmium (Cd) binding protein (Cd-BP) or metallothionein (MT), there is still no consensus on whether the testicular Cd-BP is the same as MT. Moreover, it is unclear whether the testicular Cd-BP is inducible. Therefore, the possible induction of MT-like Cd-BP was investigated in rat testis after oral Cd administration and Cd injection.

These results indicate that testicular MT-like Cd-BP, assumed to be MT and be hardly inducible by Cd, is an inducible protein and correlates with the increased Cd accumulation in the testis in the absence of damage by Cd toxicity after oral Cd administration.

The existing mysterious problem concerning MT-like Cd-BP in the testis and the need for further studies are discussed.

Cadmium toxicity in the testis and its protection mechanism

The testis is very sensitive to acute Cd toxicity. It is well known that Cd causes testicular damage such as hemorrhagic inflammation, atrophy, and necrosis or mortality [1–5]. Moreover, it is also known that the toxic effects of Cd can be prevented by simultaneous selenium (Se) administration with Cd [6, 7] or pretreatment with small dose of Cd before the challenge Cd administration [4, 5, 8]. Furthermore it has been thought that the testicular damage mentioned above and the inhibitions of enzyme activity in the testis are protected remarkably as a result of Cd binding by MT in the testis [6, 7]. Namely, the high molecular weight complex containing Cd and Se (Cd-Se HMWC) in the plasma and testis play a role in the protection against Cd toxicity. However, Cd in the Cd-Se HMWC was unstable and decreased rapidly with time. In contrast with the decreased Cd from the Cd-Se HMWC in plasma and testis, Cd in liver and testis increased gradually. And Cd in the testis and liver was found in the MT fraction. The increased Cd in the MT fraction in the testis was related to the decrease of Cd in the HMWC fraction of the testis homogenate. It was thought that the shift of Cd from the HMWC fraction to MT fraction is an important protection mechanism.

Consequently, the secondary induction of MT in the testis corresponding to the decrease of Cd-Se HMWC containing Cd in the testis and plasma is suggested to be an important mechanism of protection against the acute Cd toxicity in the testis by simultaneous Se administration with Cd [6, 7, 24].

However, a problem is raised for MT-like Cd-BP and MT in testis, by the disagreement in amino acid composition of both proteins [20–24]. Namely, there is still no consensus on whether the testicular MT-like Cd-BP is the same as MT. Moreover, it is unclear whether this testicular Cd-BP is inducible.

MT like-metal binding proteins and MT in the testis

MT, a low molecular weight metal binding protein and containing high cysteine content (about 30% in amino acid composition), is thought to play a role in the homeostatic control of zinc (Zn) and copper (Cu) metabolism, and has the role of detoxification of the toxicity of certain heavy metals such as Cd [9]. It has been suggested reasonably that MT participates the protection mechanism in the testis against Cd toxicity. The amounts of MT-like Cd-BP as well as MT were quantified by Cd-Hem method [10] in various tissues including testes from rodents [11, 12]. Onosaka and Cherian [11] have reported the interesting finding that MT in the testis decreased markedly in inverse proportion to the increase in Cd accumulation. In general, MT is induced in proportion to increased Cd accumulation in various organs including liver and kidney; the testis is an exception. The possibility of MT existing in the testis has been supported mainly by the results such as the immunohistochemical localization and the expression of MT mRNA [13–16]. The possibility of local synthesis of MT in the testis has been suggested, because it is thought that not only liver cells but also other cells can induce MT [17, 18].

Although several studies described above have indicated the presence and the possibility of induction of MT in testis, it also has been indicated that the major Cd-BPs in testis are not MT, but rather are MT-like Cd-BP because this Cd-BP in testis is not similar to MT of liver and kidney, as evidenced by the amino acid composition, u.v.-absorption spectra, and decreased induction by Cd injection [15, 19–25]. Furthermore, it has been reported that the purified Zn-BP in the testis has a different amino acid composition to that of MT, and the antiserum reacting against this protein does not cross-react with MT from the liver and kidney in humans [15].

Despite numerous studies on testicular Cd-BP or MT, there is still no consensus whether the testicular Cd-BP is MT or not. Furthermore it is not clear why MT and MT mRNA are expressed highly in the testis under physiological conditions [16, 26–28]. Whether this testicular protein is MT or not, it is important to clarify its possible induction and physiological role. Beacuse it has been suggested that there may be other factors which affect the assay specificity of MT method, MT-like Cd-BP was determined by Cd binding method in our study. It is quite a problem whether Cd-Hem method is suitable or not for the determination of MT including all isoproteins in the testis, though this method is useful for the determination of MT in the liver and kidney.

However, in this review, we used the term MT-like Cd-BP for the testicular Cd-BP, considering its different amino acid composition from MT [15, 19–25], although this Cd-BP is similar to MT in its behavior on Sephadex G-75 gel filtration, heat stability, metal binding, and assay by Cd-hem method [9–18, 22].

Although the MT level in the testis is higher than those of liver and kidney, it may be thought that the ability of induction of MT in the testis by Cd is not high compared to the cases of liver and kidney (Fig. 1) [11]. For this reason, it has been thought that the testis is an extremely susceptible organ to Cd toxicity. Cd causes damages such as hemorrhagic inflammation, atrophy, and necrosis. Spermatogenetic cells undergo degeneration and necrosis, caused by very small amounts of Cd [15]. It has been thought that MT is hardly induced in testicular cells. In fact, in our study, the parenchymal tissue of testis was damaged severely by accumulated Cd following Cd injection (Fig. 2-B). However, the parenchymal tissue of testis given Cd orally was

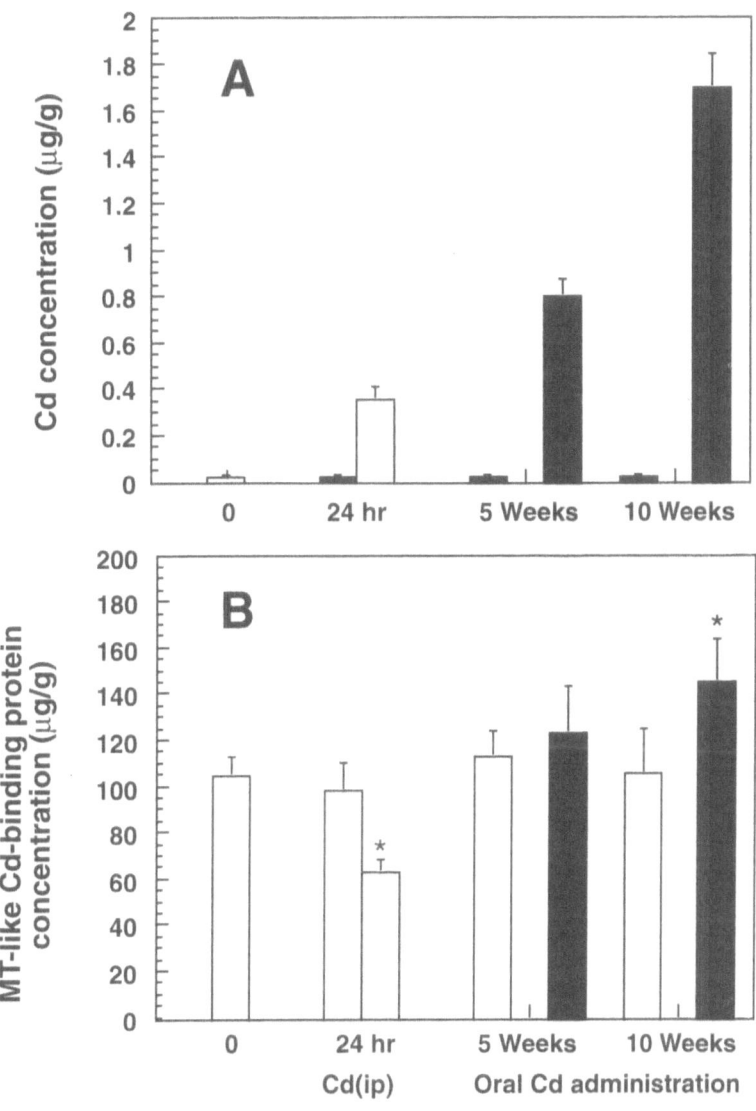

Figure 1. Cadmium (A) and metallthioneine-like cadmium-binding protein (B) concentrations in the testis after the intraperitoneal injection and oral administration of cadmium. □ – Control 0 untreated, 24 h, 5, 10 week; distilled water administration; □ – 2 mg Cd/kg ip injection; ■ – 20 mg Cd/kg oral administration; * – significantly different from the control at p < 0.05 (n = 5). (Ohta et al., Ind Health 35: 96–103, 1997)

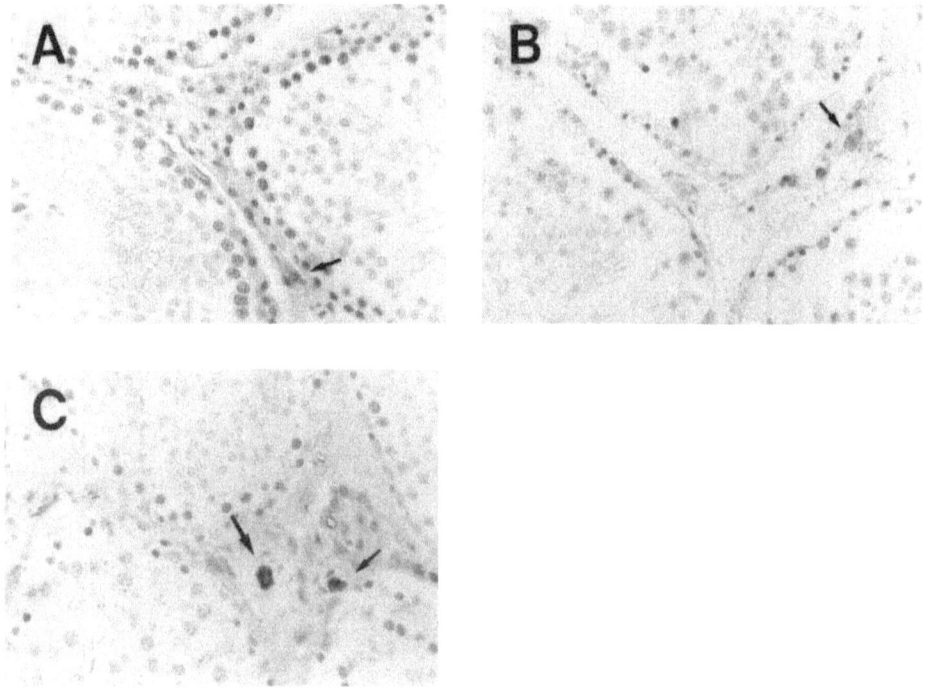

Figure 2 Metallothionein localization in the testis after the intraperitoneal injection and oral administration of cadmium. Experimental groups: (A) Control group, (B) 24 h after Cd (ip) injection, (C) 10 weeks after oral Cd administration The positive brown stain (arrow) of MT immunohistochemical staining was mainly observed in blood vessel and Leydig cells of the interstitial tissue and Sertoli cells with weak staining of testis in mice given Cd orally. In contrast, weak staining was observed in the testis of Cd (ip) injection group with tissue damages.

not damaged by accumulated Cd in the testis, as shown by the inhibited enzyme activity and decreased glutathione in the testis (Fig. 2-C, unpublished data). We also observed that the strong positive reaction of immunohistochemical MT staining by antiserum to rat liver MT was found in the testis given Cd orally; subjectively, the staining gradually increased in proportion to the increased MT-like Cd-BP determined by Cd-hem method (Figs 1, 2A–C). From these results, it was thought that when Cd accumulated in the testis without damage, the MT-like Cd-BP reaction to MT antiserum can be induced by the accumulated Cd. Consequently, we could show that if Cd accumulated in the testis without any damage, MT-like Cd-BP is induced in the testis according to increased Cd accumulation in the testis. Waalkes and associates [20, 21] have reported that the male genital organs such as testis and prostate lack MT. They have also speculated that the severe Cd toxicity found in the testis as well as tumors at injection site in the prostate is caused partly by the absence of MT, which is known to lessen metal toxicity [15]. Chellman et al. [25] reported that the Cd in MT-like Cd-BP of Cd resistant (A/J) mouse to Cd-induced testicular necrosis, is bound three times more to an MT-like protein fraction obtained from the testis of susceptible (129/J) mouse, though a similar level of Cd was accumulated in the testis of both strains. Their data suggest that MT-like Cd-BP is related to the protection to Cd toxicity.

There are some reports in which MT was found mainly in Sertoli cells and Leydig cells but not in spermatogenetic cells [29, 30]. Tohyama and associates [15, 26] have also reported that the positive MT staining in the testis was found in spermatogenetic cells, spermatozoa and Sertoli cells of adult rats under physiological conditions, but not in the interstitial cells. They have suggested that the spermatogenetic cells, which constitutively express MT gene, are so susceptible to Cd toxicity that degenerative changes by Cd occur in these cells, at levels at which Cd can induce MT in less vulnerable cells such as Leydig and Sertoli cells. It was also reported that the MT mRNA was localized mainly in spermatocytes and spermatids. In isolated Sertoli cells a very low level of MT is found [15].

From the results described above, it can be thought that the positive immunological MT staining in the Sertoli cells, Leydig cells and interstitial tissue of the testis increased gradually in proportion to the increased MT-like Cd-BP, correlating with the accumulated Cd in the testis. (Fig. 2C). Shaikh [13] reported that the low-molecular weight Cd-binding protein in the testis that reacted with the antibody of MT, is MT and that the high-molecular weight species is an MT polymer.

Recently, it has been reported that the induction of a Cd-BP in R2C cells (rat testicular Leydig cell line) did occur, as determined by Cd-hem assay, but did not induce tolerance to Cd induced single-strand DNA damage [32]. Wang et al. [34] reported that MT gene is functional and inducible, and is associated with the tolerance of testicular cells such as Leydig and Sertoli cells to Cd toxicity. Tohyama and associates [35] reported that MT mRNA in the rat testis was found predominantly in primary spermatocytes and also in secondary spermatocytes and spermatids, but not in the spermatogenetic cells, Sertoli cells, and Leydig cells. On the contrary, MT protein was present in these spermatogenetic cells as well as in spermatozoa and Sertoli cells.

Although there is evidence, described above, for the existence of MT and MT mRNA in the testis, the marked difference between the testicular Cd-BP and the liver MT is in the amino acid composition. If the Cd- or Zn-binding proteins in the testis truly are different from MT, problems will arise with aspects of not only analysis techniques but also biological significance concerning their roles in Zn metabolism and the metal detoxification in the testis [15].

Conclusion

Whether or not Cd-BP is MT, we found that this Cd-BP in the testis is an inducible protein, correlating with increasing Cd accumulation. Furthermore, we observed that Sertoli cells, Leydig cells, and stromal cells of testis were stained strongly positive for immunohistochemical staining of MT.

From the results, it is strongly suggested that the MT-like Cd-BP in the testis is MT. However, from the experimental evidence of amino acid analysis so far obtained, it is difficult to infer that the Cd-BP in the testis is MT. Therefore, it is thought that analysis of amino acid composition needs to be repeated for testicular Cd-BP. Moreover, further detailed studies are needed to clarify whether testicular Cd-BP is MT or not, including the clarification of the amino acid sequence and gene expression of MT.

References

1. Gunn SA, Gould TC, Anderson WAD (1965) Strain differences in susceptibility of mice and rats to cadmium-induced testicular damage. *J Reprod Fertil* 10: 273–275.
2. Gunn SA, Gould TC (1970) Specificity of the vascular system of the male reproductive tract. *J Reprod Fertil Suppl* 10: 75–95.
3. Parizek J, Zahor Z (1956) Effect of cadmium salts on testicular tissue. *Nature* 1956; 177: 1036.
4. Yoshikawa H (1970) Preventive effect of pre-treatment with low dose of metals on the acute toxicity of metals in mice. *Ind Health* 8: 184–191.
5. Yoshikawa H (1973) Preventive effects of pretreatment with cadmium on acute cadmium poisoning in rats. *Ind Health* 11: 113–119.
6. Ohta H, Imamiya S (1986) Selenium protection against acute cadmium toxicity in testis. Kitasato *Arch Exp Med* 59: 27–36.
7. Ohta H, Imamiya S, Yoshikawa H (1988) The protective effect of simultaneous selenium administration on acute cadmium toxicity and metallothionein. *Jpn J Ind Health* 30: 451–458.
8. Yoshikawa H, 8.   Yoshikawa H, Ohta H (1982) Interaction of metals and metallothionein. *In*: Foulkes EC (ed.): *Biological roles of metallothionein*. Elsevier/North-Holland, New York, 11–23.
9. Cherian MG (1995) Metallothionein and its interaction with metals. *In*: Toxicology of Metals. *In*: Goyer RA, Cherian MG (eds): *Biochemical Aspects*. Handbook of Experimental Pharmacology, Vol. 115, Springer-Verlag, Berlin, 121–137.
10. Onosaka S, Tanaka K, Doi M, Okahara K (1978) A simplified procedure for determination of metallothionein in animal tissue. *Eisei Kagaku* 24: 128–133.
11. Onosaka S, Cherian MG (1981) The induced synthesis of metallothionein in various tissues of rat in response to metals. I.Effect of repeated injection of cadmium salts. *Toxicology* 22: 91–101.
12. Waalkes MP, Klaassen CD (1985) Concentration of metallothionein in major organs of rats after administration of various metals. *Fund Appl Toxicol* 5: 473–477.
13. Shaikh ZA (1991) Radioimmunoassay for metallothionein in body fluids and tissues. *In*: Riordan JF, Vallee BL (eds) *Meth Enzymol* 205: Academic Press, 120–130.
14. Nakajima K, Suzuki K, Otaki N, Kimura M (1991) Detection of metallothionein in brain. *In*: Riordan JF, Vallee BL (eds) *Meth Enzymol* 205 Academic Press 387–395.
15. Tohyama C, Suzuki JS, Homma ST, Nishimura N, Nishimura H (1993) Regulation of metallothionein biosynthesis in genital organs of male rats. *In*: Suzuki KT, Imura N, Kimura M (eds) *Metallothionein III*, Birkhäuser, Verlag Basel/Switzerland, 1993; pp 443–456.
16. Shiraishi N, Hayashi H, Hiraki Y, Aono K, Itano Y, Kosaka F, Noji S, Taniguchi S (1989) Elevation in metallothionein messenger RNA in rat tissues after exposure to X-irradiation. *Toxicol Appl Pharmacol* 98: 501–506.
17. Lucis OJ, Shaikh ZA, Embil JA (1970) Cadmium as a trace element and cadmium binding components in human cells. *Experientia* (Basel) 26: 1109.
18. Nordberg GF (1971) Effects of acute and chronic cadmium exposure on the testicles of mice. *Environ Physiol* 1: 171–181.
19. Singh K, Nath R, Chakravarti RN IJT, Whanger PD (1985) Properties of cadmium-binding proteins in rat testes: characteristics unlike metallothionein. *Biochem J* 231: 279–283.
20. Waalkes MP, Chernoff SB, Klaassen CD (1984a) Cadmium-binding proteins of rat testes: Characterization of a low-molecular-mass protein that lack identity with metallothionein. *Biochem J* 220: 811–818.
21. Waalkes MP, Chernoff SB, Klaassen CD (1984b) Cadmium-binding proteins of rat testes:Apparent source of the protein of low molecular mass. *Biochem J* 220: 819–824.
22. Waalkes MP, Perantoni A, Palmer AE (1988) Isolation and partial characterization of the low-molecular-mass zinc/cadmium-binding protein from the testes of the patas monkey (Erythrocebus patas) Distribution from metallothionein. *Biochem J* 256: 131–137.
23. Waalkes MP, Perantoni A (1986) Isolation of a novel metal-binding protein from rat testes: Characterization and distribution from metallothionein. *J Biol Chem* 261: 13097–13103.
24. Ohta H, Seki Y, Imamiya S (1988) Metallothionein-like cadmium binding protein in rat testes administered with cadmium and selenium. *Bull Environ Contam Toxicol* 41: 195–200.
25. Chellman GJ, Shaikh ZA, Baggs RB, Diamond GL (1985) Resistance to cadmium- induced necrosis in testes of inbred mice: Possible role of a metallothionein like cadmium-binding protein. *Toxicol Appl Pharmacol* 79: 511–523.

26. Nishimura N, Nishimura H, Tohyama C (1990) Localization of metallothionein in the genital organs of the male rat. *J Histochem Cytochem* 38: 927–933.
27. Compere SJ, Palmiter RD (1981) DNA methylation controls the inducibility of the mouse metallothionein-I gene in lymphoid cells. *Cell* 25: 233–240.
28. De SK, Enders GC, Andrews GK (1991) High levels of metallothionein messenger RNAs in male germ cells of the adult mouse. *Mol Endocrinol* 5: 628–636.
29. Danielson KG, Ohi S, Huang PC (1982) Immunochemical detection of metallothionein in specific epithelial cells of rat organs. Pro. Natl. *Acad Sci USA* 79: 2301–2304.
30. Nolan CV, Shaikh ZA (1986) *Toxicol Appl Pharmacol* 85: 135–144.
31. Durnam DM, Palmiter RD (1981) *J Biol Chem* 256: 5712–5716.
32. Shiraishi N, Hochadel JF, Coogan TP, Koropatonick J, Waalkes MP (1995) Sensitivity to cadmium-induced genotoxicity in rat testicular cells is associated with minimal expression of the metallothionein gene. *Toxicol Appl Pharmacol* 130(2): 229–236.
33. Wang SH, Chen JH, Lin LY (1994) Functional integrity of metallothionein gene in testicular cell lines. J Cellular *Biochemistry* 55(4): 486–495.
34. Tohyama C, Nishimura N, Suzuki JS, Karasawa M, Nishimura H (1994) Metallothionein mRNA in the testis and prostate of rat detected by digoxigenin- labeled riboprobe. *Histochemistry* 101(5): 341–346.
35. Ohta H, Nakakita M, Tanaka H, Seki Y, Yoshikawa H (1997) Induction of metallothionein-like cadmium-binding protein in the testis by oral cadmium administration in rats. *Ind Health* 35: 96–103.

Zinc homeostasis during pregnancy in metallothionein-null mice on a low zinc diet

A.M. Rofe, J.C. Philcox, M. Sturkenboom and P. Coyle

Institute of Medical and Veterinary Science, Adelaide, SA, Australia

Introduction

Zinc (Zn) is an essential intracellular cation necessary for the structural integrity and functional capacity of a wide range of macromolecules. It is therefore not surprising that a deficit in Zn has profound effects on metabolism, growth, and development [1, 2]. The ability of metallothionein (MT) to respond to Zn and to cause the cellular sequestration of Zn following induction of MT by hormones or inflammatory mediators implicates a key role for metallothionein in Zn homeostasis.

Gestation causes large progressive increases in MT levels within certain tissues, coinciding with increased Zn demand. The transfer of Zn from placenta to embryo is dependent upon a number of factors, including the maternal plasma Zn concentration and fetal demand as expressed by the fetal hepatic MT. The liver is formed early in embryonic life and is thought to be responsible for Zn homeostasis during development [3]. Fetal MT-gene expression is coordinately programmed during embryogenesis, presumably to maximise the Zn availability to the fetus. In humans and laboratory animals, fetal hepatic MT levels are high during late gestation and in the first post-natal week, falling to adult levels at weaning [3, 4]. There is a high correlation between MT and Zn in the cytosol of fetal and neonatal livers [3].

Less is known about the organ-specific changes in maternal MT and associated changes in Zn homeostasis that occur during the perinatal period. High levels of MT expression have been found in the deciduum early in gestation [4]. Perinatal elevations in maternal liver MT mRNA have been shown [5] but corresponding changes in MT protein have received little attention [6]. It is not known whether the increase in maternal liver MT is more important in providing a Zn supply to the developing fetus or in building a Zn reserve to cope with the very large Zn and metabolic demands of late pregnancy and lactation.

The MT-null (MT–/–) mouse offers the opportunity to answer questions on the contribution of MT to the regulation of maternal/fetal Zn balance. Recent work by Kelly et al. [7] using MT–/– mice has shown impaired renal development in pups when dams were placed on a Zn deficient diet. It was also noted that MT–/– pups failed to accumulate Zn in bone, even when dams were consuming Zn-adequate diets. A significant finding in their study was that MT–/– pups were born with lower hepatic Zn levels; this has also been our observation (see below). Mice that overexpress MT have been shown to be more resistant to the effects of Zn deficiency, with fewer fetal growth abnormalities [8]. MT–/– mice grow and reproduce normally on Zn replete diets. Furthermore, preliminary observations from our laboratory and those of others indicate that gestation does not proceed to term when severe Zn restriction is imposed. Our aims in this study were three fold; a) to establish a dietary regimen that would limit Zn intake but still allow gestation to term b) to examine the effect of the absence of MT on maternal and fetal Zn homeostasis and c) to examine the relationship between any alterations in Zn home-ostasis and growth and development of the fetus.

Materials and methods

MT-I and MT-II null mice were produced by A.Michalska and K.H.A. Choo of the Murdoch Institute, Royal Childrens Hospital, Victoria [9]. The mice were in a mixed genetic background of OLA129 and C57BL6 strains. Control mice (MT+/+) were C57BL6 supplied by the Animal Resources Centre, Canning Vale, Western Australia. All mice were housed in plastic cages with wire mesh floors in an animal house at 22 °C. Female mice were fed a low Zn diet containing (in g/kg); casein, 180; sucrose, 152; cornflour, 514; cellulose, 50; corn oil, 54 as macronutri-ents and Zn, 8 mg/kg (confirmed by atomic absorption spectroscopy following acid digestion) and Cu 30 mg/k as well as essential vitamins and minerals. After 2 months on the diet, the mice were mated. Studies were conducted at gestational day 18. The mice were anaesthetised with halothane and 1 ml of blood withdrawn by cardiac puncture. The mice were then killed by cer-vical luxation and portions of organs and fetus were weighed and retained for Zn and MT anal-ysis by methods previously described [10, 11]. In ^{65}Zn absorption studies, pregnant MT+/+ and MT–/– mice were given ^{65}Zn bound to albumin (1uCi, 50ug ZnSO$_4$, 2% fatty acid free albu-min in 500 µl of saline) by oral gavage and tissue radioactivity was measured at 16 h. The results are expressed as the mean ± sem with differences between groups assessed using Student's t test.

Results

Zinc uptake studies

The MT+/+ mice were able to absorb and retain more ^{65}Zn from the oral gavage than their MT–/– counterparts (Tab. 1). Over two thirds of the extra absorption could be accounted for by increased hepatic uptake. Measurement of non-radiolabelled Zn pool and MT concentrations at the relevant gestational stages demonstrated that the incremental Zn absorption was almost entirely incorporated into increasing maternal hepatic MT. A further experiment demonstrated a fall in fetal hepatic Zn from g.d. 18 to g.d. 21 to half-normal concentrations in the MT–/– pups at birth. The latter finding is in accord with recently published data [7].

Effect of diet

When fed a Zn replete diet, 20 g MT+/+ and MT–/– mice grow at a rates of 1–1.2 g every 10 days. When fed the 8 mg/kg diet, the growth rates were 40–60% of the normal rate (significant

Table 1. Uptake of ^{65}Zn at gestational day 18 (100 mg/kg Zn diet) ^{65}Zn as % of oral dose

Tissue	MT +/+	MT –/–
Blood	0.40 ± 0.03	0.34 ± 0.06 *
Liver	18.4 ± 1.7	4.6 ± 0.7 *
Kidney	1.6 ± 0.2	0.8 ± 0.1 *
Muscle	4.3 ± 0.4	3.0 ± 0.4
Skin	2.5 ± 0.2	1.5 ± 0.3 *
Uterus	10.1 ± 2.0	9.6 ± 3.5
Total (gut excluded)	33.4 ± 3.0	13.1 ± 1.1 *

Mean ± sem (n = 6) *p < 0.05

effect of diet but not significantly different between groups). Plasma Zn concentrations of both genotypes were decreased by 1–2µM on this diet (not significant).

Zinc limiting diet (8 mg/kg) and pregnancy

Feeding a Zn limiting diet for 2 months prior to, as well as throughout gestation reduced the MT response to pregnancy in the dam's liver by 50% (i.e. from a 20-fold to 10-fold stimulation) but did not decrease MT concentrations in the placenta and fetal liver (Tab. 2).

Table 2. Tissue metallothionein in MT+/+ mice on 100 mg/kg and 8 mg/kg Zn diet

	Metallothionein (nmol Cd bound/g wet wt)	
	100 mg Zn/kg	8 mg Zn/kg
Maternal liver	71 ± 6	35 ± 8 *
Placenta	47 ± 4	41 ± 10
Fetal liver	114 ± 13	153 ± 26

Mean ± sem (n = 5) p < 0.05

Tissue zinc levels

The Zn concentrations in the dam's liver, placentas and particularly the fetal livers were decreased in the MT–/– mice (Tab. 3). This was mainly attributable to the absence of MT as 40% of the fetal hepatic Zn was associated with MT in the normal pups. The total Zn pool size in the dam liver, placenta and fetal liver was even lower (i.e. decreased by 37%, 58% and 67% respectively) in the MT–/– mice due to smaller maternal livers and smaller litter size (an aver-

Table 3. Tissue Zn levels in MT+/+ and MT–/– mice on the 8 mg/kg Zn diet

	Zinc	MT +/+	MT –/–
Maternal liver	nmol/g wet wt.	494 ± 28	431 ± 5
	nmol/total	875 ± 75	555 ± 35*
Placenta	nmol/g	240 ± 29	172 ± 19
	nmol/total	211 ± 39	88 ± 22 *
Fetal liver	nmol/g	373 ± 35	169 ± 12 *
	nmol/total	101 ± 31	33 ± 1

Mean ± sem (n = 5) $p < 0.05$

age of 9 pups in MT+/+ mice at birth versus 3.5 in the MT–/– mice). The reduction in fetal number in the MT–/– mice was attributed to fetal resorption. Of special note was the appearance of multiple developmental abnormalities in an MT–/– litter including spina bifida, micropthalmia, micrognathia, microcephaly and herniation of the viscera. MT–/– mice on 100 mg Zn/kg diet had normal sized litters (i.e. 9–10 pups) with no apparent deformities.

Discussion

The availability of MT–/– mice is leading to rapid advances in our understanding of the functions of MT, especially in the area of Zn homeostasis. Our initial studies with these mice clearly established the primary role of hepatic MT in Zn sequestration following inflammation [10, 12] or Zn administration [11]. From early studies with these mice, it was evident that they could grow and reproduce normally on standard diets (Zn in the range of 30–100 mg/kg). This has led to the general view that if MT has an essential role it is only likely to be revealed when Zn supply is limiting or where there are increased metabolic demands for Zn. Pregnancy is a situation where Zn supply is of paramount importance and recent experience with MT transgenic mice (knockout and overexpressors) [7, 8] has highlighted this as an area warranting closer study.

Our investigations using an 8 mg/kg Zn diet have shown marked differences between MT+/+ and MT–/– mice in zinc homeostasis and fetal development during pregnancy. While the diet decreased growth rates by half in both MT+/+ and MT–/– mice, this level of Zn intake did allow conception and gestation through to parturition. An important feature of the 8 mg/kg Zn casein-based diet was that mice did not enter a pattern of cyclical feeding, nor did they reduce their food intake as did mice subjected to more extreme Zn deficiencies. Thus there is no protein-calorie shortfall to complicate the interpretation of results. The 8 mg/kg Zn diet is also more closely aligned with the clinical setting of chronic-marginal rather than acute-severe Zn deficiency.

Our results clearly demonstrate that in the absence of MT there are severe adverse consequences to the fetus associated with a marked limitation in the amount of Zn that can be accumulated in the livers of both the dam and the fetuses. One impact of this was the appearance of multiple developmental abnormalities as previously mentioned. The small litter size observed at late gestation in MT–/– mice was attributed to fetal resorption in late gestation as the number of fetuses had been observed to be higher earlier in gestation and resorption sites were noted. This is in keeping with the predicted effects of a limitation in Zn supply.

The indication is that without MT, there is difficulty for the maternal Zn supply to support fetal demand as the fetus grows from 0.5 to 1.0 g over the last 3–4 days of gestation. When MT–/– dams are fed a normal Zn diet the pups are born in apparent good health, despite their half-normal hepatic Zn concentrations. They then continue to grow normally until weaned. An 8 mg/kg Zn diet, which is well tolerated in pregnancies of MT+/+ mice is lethal to over half the MT–/– fetuses during late gestation. Many that reached parturition were so developmentally retarded as to be unlikely to survive more than a few days.

These findings suggest that in mice, MT has a important role in late fetal development, assuming critical importance when dietary Zn is low. In this situation, the balance between maternal liver MT, and fetal liver MT, was more heavily biased in favour of the fetus to ensure adequacy of fetal Zn reserves at birth. The MT –/– mice were both unable to build sufficient liver Zn reserves and, nearing parturition, lacked the mechanism to redistribute Zn from the dam to the fetuses leading to adverse consequences. In normal mice on normal Zn diets, the maternal liver MT and Zn stores at parturition are approximately 2.5 times higher than liver stores in non-pregnant mice and are probably necessary to provide Zn to the pups during lactation, reducing the loss of Zn from other tissues, particularly bone. This situation may not be as severe in larger animals where late fetal growth is relatively much slower. The mouse is however a very sensitive to the Zn needs of late fetal development and gross changes in mice may reflect more subtle but nevertheless important changes in other species including man.

References

1. Mills CF, ed (1989) *Zinc in Human Biology*. Springer-Verlag, London.
2. Valee BL and Falchuk KH (1993) The biochemical basis of zinc physiology. *Physiol Rev* 73: 79–118.
3. Klein D, Scholz P, Drasch GA, Muller-Hocker J and Summer KH (1991) Metallothionein, copper and zinc in fetal and neonatal human liver: Changes during development. *Toxicol Lett* 56: 61–67.
4. De SK, Dey SK and Andrews GK (1990) Cadmium teratogenicity and its relationship with metallothionein gene expression in midgestation mouse embryos. *Toxicology* 64: 89–104.
5. Quaife C, Hammer RE, Mottet NK and Palmiter RD (1986) Glucocorticoid regulation of metallothionein during murine development. *Dev Biol* 118: 549–555.
6. Reis BL, Keen CL, Lonnerdal B and Hurley LS (1988) Mineral composition and zinc metabolism in female mice of varying age and reproductive status. *J Nutr* 118: 349–361.
7. Kelly EJ, Quaife CJ, Froelick GJ and Palmiter RD (1996) Metallothionein I and II protect against zinc deficiency and zinc toxicity in mice *J Nutr* 126: 1782–1790.
8. Dalton T, Fu K, Palmiter RD and Andrews GK (1996) Transgenic mice that overexpress metallothionein-I resist dietary zinc deficiency. *J Nutr* 126: 825–833.
9. Michalska A and Choo KHA (1993) Targeting and germ-line transmission of a null mutation at the metallothionein I and II loci in mouse. *Proc Natl Acad Sci USA* 90: 8088–8092.
10. Philcox JC, Coyle P Michalska A, Choo KHA and Rofe AM (1995) Endotoxin-induced inflammation does not cause hepatic zinc accumulation in mice lacking metallothionein gene expression. *Biochem J* 308: 543–546.
11. Coyle P, Philcox JC and Rofe AM (1995) Hepatic zinc accumulation in metallothionein null mice following zinc challenge: *in vivo* and *in vitro* studies. *Biochem J* 309: 25–31.
12. Rofe AM, Philcox JC and Coyle P (1996) Trace metal, acute phase and metabolic response to endotoxin in metallothionein-null mice. *Biochem J* 314: 793–797.

Metallothionein IV
C. Klaassen (ed.)
© 1999 Birkhäuser Verlag Basel/Switzerland

Restriction of cadmium transfer to egg from laying hen exposed to cadmium: Involvement of metallothionein in the ovaries

Shin Sato, Masashi Okabe, Tadasu Emoto, Masaaki Kurasaki and Yutaka Kojima

Department of Environmental Medicine and Informatics, Graduate School of Environmental Earth Science, Hokkaido University, Sapporo 060-0810, Japan

Introduction

Cd is a toxic environmental contaminant affecting the reproduction of humans and animals by producing teratogenicity and embryotoxicity [1]. Cd accumulation in embryos and fetuses is generally limited to development after maternal exposure to Cd [2, 3]. Moreover, the presence of metallothionein (MT) bound to Cd in the placenta [4, 5] and the expression of MT genes in the placenta during embryonic development has been reported [6].

In avian species, previous reports indicated that Cd content in eggs was very low after Cd administration, although Cd accumulated in the liver of laying hen [7] and Japanese quail [8]. Yolk substances, such as vitellogenine, are known to form in the liver and be transported from the blood across the follicle walls before depositing in the follicle yolks [9]. So far, in birds, it has been unclear whether Cd would accumulate in the yolks when maternal animals are exposed to high amounts of Cd. To our knowledge, little is known whether MT is associated with the Cd transfer from laying hens to eggs.

In this study, we demonstrate that Cd accumulates in the follicle walls of the ovaries but not in the follicle yolks, and that Cd-binding MT was biosynthesized in the follicle walls, when laying hens were administered with high amounts of Cd. On the results, a possible mechanism of the restriction of Cd transfer to the egg is discussed.

Cadmium adminitration to laying hens

Twelve-month old White Leghorn laying hens were used. The birds were injected *i.p.* with saline or Cd as $CdCl_2$ in saline solution at dose-levels of 0.5, 1.5, 2.5, 3.5, 5.5 and 7.5 mg/kg body weight as total dosage. The Cd concentrations in the maternal blood, livers, ovaries and eggs were measured as previously described [10].

Cd concentration in follicle walls and follicle yolks

The Cd concentration in the maternal blood and the livers increased with dose of Cd (Tab. 1), and Cd concentration in the yolks of egg laid during the period of injections with Cd ranged

Table 1. Cd Concentrations in Blood and Liver

Group	Total dose	Blood	Liver
	mg/kg	μg/g wet weight	%
A	0	ND[a]	0.3 ± 0.09
B	0.5	0.04 ± 0.00	12.0 ± 1.32[*]
C	1.5	0.12 ± 0.01[‡]	35.2 ± 3.23[*]
D	2.5	0.25 ± 0.06[‡]	53.4 ± 3.54[*]
E	3.5	0.39 ± 0.03[‡]	70.5 ± 3.70[*]
F	5.5	0.65 ± 0.15[‡]	100.1 ± 8.30[*]

Each value represents the mean ± S.E. for n = 3–5. *p < 0.05 vs. Group A as control, ‡p < 0.05 vs. Group B (0.5 mg/kg treatment).[a]ND means not detectable.

from < 0.001 μg/g wet weight to 0.04 μg/g wet weight. In the albumen of the eggs, Cd was also less than 0.001 μg/g wet weight. Although no egg was laid in the highest injected group (7.5 mg Cd/kg) on the day after the final injection, egg production was observed again after 12 to 23 days. The Cd concentration in the yolks of the first eggs laid after egg production ranged from 0.02–0.03 μg/g wet weight (Tab. 2).

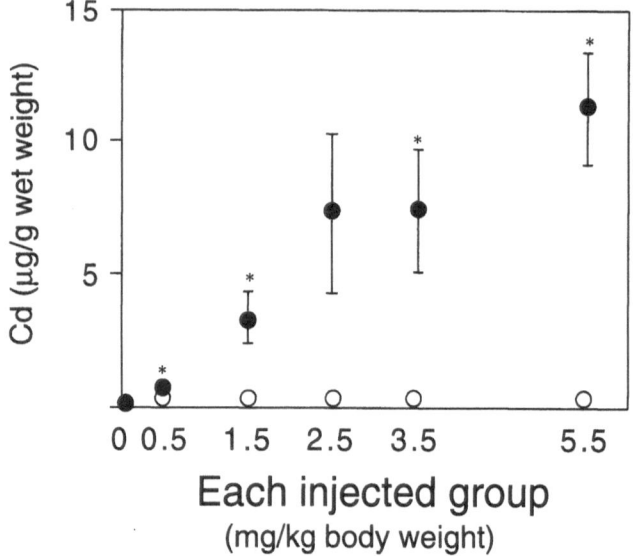

Figure 1. Cd concentration in the follicle walls (closed circle) and follicle yolks (open circle). Cd concentration in the follicle walls and follicle yolks removed 24 h after the final injection were measured by atomic absorption. Each value represents the mean ± SE (n = 3–5). *p < 0.05 vs. Cd dose-matched the follicle walls.

Table 2. Cd Concentrations in Yolks and Albumen of Laid Eggs, and Liver in Group G

Sacrifice	Egg laid		Liver (n)
	Yolk (n)	Albumen (n)	
		$\mu g/g$ *wet weight*	
after 1 day	- [a]	-	114.9 ± 4.5 (4)
after 12–23 days	$0.02–0.03$ (3)[b]	ND[c] (3)	138.7 ± 12.5 (4)

Each value represents the mean ± S.E.
[a] No egg was laid.
[b] Range of Cd concentrations.
[c] Not detectable.

Furthermore, the Cd concentration in the follicle walls of ovaries increased during the period of Cd administration, and was 13- to 52-fold higher than that in the follicle yolks (Fig. 1), when the Cd concentration in the blood was raised. These results indicate that Cd accumulates in the follicle walls rather than in follicle yolks. The Cd concentration in the follicle yolks was lower than in the follicle walls.

Effect of Cd on chick hatching

To estimate whether hatching success is affected by Cd in the eggs laid by the Cd-injected laying hens in this study, a total of 120 fertilized White Leghorn hen eggs were used, and 0, 0.3, 1.2, 4.8 and 19.2 µg Cd in 100 µl of sterilized saline was injected directly into the 20 fertilized egg yolks in each group on day 0 of incubation. The number of hatching success was as follows: 8 in saline injected group, 9 in 0.3 µg Cd, 9 in 1.2 µg Cd, 3 in 4.8 µg Cd, and only one in 19.2 µg Cd/egg-injected groups. The hatched chicks in the Cd- and saline-treated groups had no external malformations in appearance. This is supported by the report that significant mortality was found when dose greater than 15 µg Cd were injected into surviving embryos of White Leghorn hens [11]. These results suggest that the chick embryos might hatch even if low amounts of Cd accumulate in the eggs.

MT in the follicle walls

The cytosol obtained from the follicle walls was applied to a Sephadex G-75 column and the eluent was measured for Cd and Zn concentrations. Approximatly 85% of Cd in the cytosol eluted was found as a major peak in the low molecular weight fractions containing Cd with a relative elution volume (Ve/Vo) of 1.86 (Fig. 2). In control animals, no low molecular weight fraction containing Cd was found from the follicle walls. When the Cd-binding protein from the follicle walls was further purified by a column of DEAE-Sephadex A-25, only a single peak

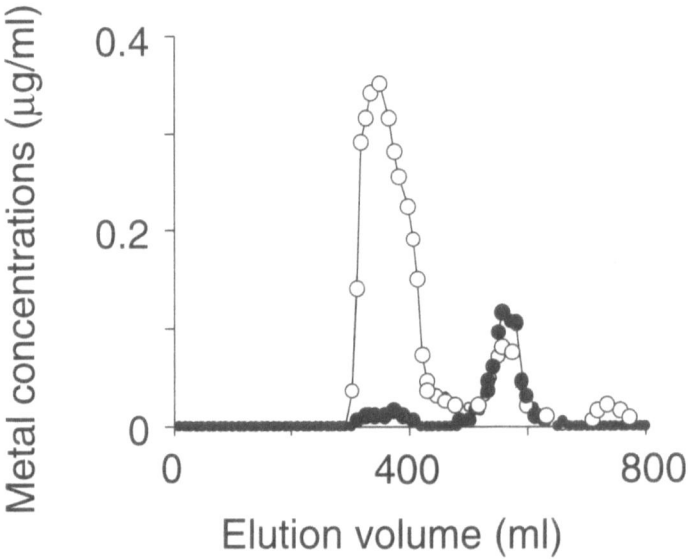

Figure 2. Sephadex G-75 gel filtration of cytosol obtained from the follicle walls of the laying hens injected with Cd. Cytosol was applied to a column (3 × 100 cm) of Sephadex G-75 equilibrated with 10 mM Tris-5 mM HCl, and eluted with the same buffer. The eluent was collected in 10 ml fractions. The elution profile of Cd (closed circle) and Zn (open circle) is indicated.

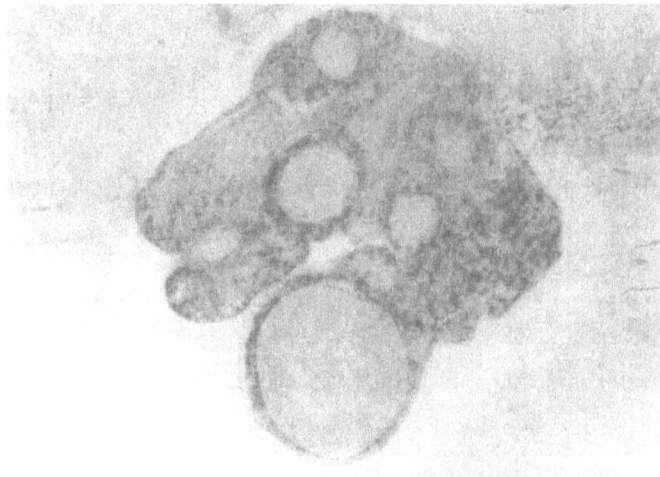

Figure 3. Localization of MT mRNA in the ovary of laying hens treated with Cd by *in situ* hybridization.

was obtained. The protein contained 27% cysteine residues as the predominant amino acid. The composition of the Cd-binding protein from the follicle walls agreed well with that of hepatic MT. The ultraviolet absorption spectrum of the protein shows a shoulder at 245–255 nm at neutral pH. When acidified, the spectrum changed with a clear loss of the shoulder, which was due to the Cd-thiolate complex [12]. Thus, the Cd-binding protein present in the follicle walls was identified as MT.

Expression of MT mRNA

To estimate whether biosynthesis of MT occurs in the follicle walls, we carried out *in situ* northern hybridization study as previously described [13]. In the peripheral parts of the follicles, which corresponded to the follicle walls, positive signals of MT mRNA were observed (Fig. 3), indicating that MT mRNA was synthesized in the follicle walls of Cd-treated laying hens. No specific signals of MT mRNA was detected in the follicle yolks of the control animals.

Possible mechanism of Cd metabolism in the follicle walls

In this study, eggs accumulated low concentration of Cd, even when 140 µg/g wet tissue of Cd had accumulated in the maternal livers. Our results are consistent with previous reports [7, 14]. It is noted that the concentration of Cd in eggs laid after the resumption of egg production remained low, even though the Cd concentration in the liver remained high (Tab. 2). Thus, our findings suggest the presence of a protective mechanism restricting the transfer of Cd between maternal birds and their eggs.

Similar results have been reported in mammals. After Cd exposure in mice [15] and rats [2, 3], Cd concentration was extremely low in the fetus. Accumulated Cd was shown in the placenta. It was believed that the placenta acts as a barrier to protect the fetus from Cd [16, 17]. In contrast with the small amount of Cd in eggs, Zn and Pb contents in egg yolks from laying hens treated with each metal, were higher than for yolks from control [18, 19]. A considerable amount of Hg was found in the yolks of eggs laid when laying quails were treated with [203]Hg [20]. From our findings, as well as from previous reports, it is reasonable to assume that not only in mammals, but also in birds, there may be a system which regulates the transfer of Cd from the maternal animals to the egg or fetus. In view of this, we propose that this system be designated the "matri-filial barrier".

It is interesting that Cd accumulated in the follicle walls was bound to MT. The functions of MT in the follicle walls remain as speculation at present. A possibility is that MT may serve to sequester Cd to restrict the Cd transfer from the maternal blood to the follicle yolks, because it has been considered that MT can detoxify Cd by sequestration [21]. With mammals, MT has been found in the placenta [4, 5, 22], and it has been suggested that the MT plays a role in protecting the fetus [4]. Although the structure and functions within the follicle walls are different from those within the placenta, MT in both tissues might play an important role in protecting the yolks or the fetus by sequestering Cd during the deposition of yolk substances or the development of the fetuses in "matri-filial barrier".

In conclusion, Cd transfer from laying hen to the eggs is restricted following exposure of the maternal bird to Cd. Furthermore, Cd bound with MT accumulates in the follicle walls of ovary and Cd-MT is biosynthesized in those tissues, suggesting that MT in the follicle walls may play a role in protecting the follicle yolks against Cd toxicity.

References

1. Barlow SM, Sullivan FM (1982) Cadmium and its compounds. *In*: Barlow SM, Sullivan FM (eds): *Reproductive Hazards of Industrial Chemicals*. Academic Press, London, 136–177.
2. Ahokas RA, Dilts PVJr (1979) Cadmium uptake by the rat embryo as a function of gestational age. *Amer J Obstet Gynecol* 135: 219–222.
3. Sowa B, Steibert E, Gralewska K, Piekarski M (1982) Effect of oral cadmium administration to female rats before and/or during pregnancy on the metallothionein level in the fetal liver. *Toxicol Lett* 11: 233–236.
4. Hanlon DP, Specht C, Ferm VH (1982) The chemical status of cadmium ion in the placenta. *Environ Res* 27: 89–94.
5. Roelfzema WH, Tohyama C, Nishimura H, Nishimura N, Morsell AFW (1989) Quantitative immunohisto-chemistry of metallothionein in rat placenta. *Histochemistry* 90: 365–369.
6. De SK, McMaster MT, Dey SK, Andrews GK (1989) Cell-specific metallothionein gene expression in mouse decidua and placentae. *Development* 107: 611–621.
7. Leach Jr RM, Wang KWL, Baker DE (1979) Cadmium and the food chain: the effect of dietary cadmium on tissue composition in chicks and laying hens. *J Nutr* 109: 437–443.
8. Nishimura M, Sakuta M, Okamoto K, Urakawa N (1974) Distribution and excretion of [115m]cadmium and its transfer to egg and bone in laying female and estrogenized male Japanese quail. *Jpn J Vet Sci* 36: 133–143.
9. Nickel R, Schummer A, Seiferle E (1977) Sex organs. *In*: Nickel R, Schummer A, Seiferle E (eds) *Anatomy of the domestic birds*. Verlag Paul Parey, Berlin, 72–84.
10. Sato S, Okabe M, Emoto T, Kurasaki M, Kojima Y (1997) Restriction of cadmium transfer to eggs from laying hens exposed to cadmium. *J Toxicol Environ Health* 51: 15–22.
11. Narbaitz R, Riedel KD, Kacew S (1983) Induction of feather malformations in chick embryos by cadmium; protection by zinc. *Teratology* 27: 207–213.
12. Vašák M, Kägi JHR, Holmquist B, Vallee BL (1981) Spectral studies of cobalt (II)- and Nickel (II)-metal-lothionein. *Biochemistry* 20: 6659–6664.
13. Kurasaki M, Emoto T, Linde Arias AR, Okabe M, Yamasaki F, Oikawa S, Kojima Y (1996) Independent self-assembly of cadmium-binding α-fragment of metallothionein in Escherichia coli without participation of β-fragment. *Protein Eng* 9: 1173–1180.
14. Sell JL (1975) Cadmium and the laying hen: Apparent absorption, tissue distribution and virtual absence of transfer into eggs. *Poultry Sci* 54: 1674–1678.
15. Berlin M, Ullberg S (1963) The fate of Cd[109] in the mouse. An autoradiographic study after a single intra-venous injection of Cd[109]Cl$_2$. *Arch Environ Health* 7: 686–693.
16. Dencker L, Danielsson B, Khayat A, Lindgren A (1983) Disposition of metals in the embryo and fetus. *In*: Clarkson TW, Nordberg GF, Sager PR (eds): *Reproductive and developmental toxicity of metals*. Plenum Press, New York, 607–631.
17. Kuhnert PM, Kuhnert BR, Bottoms SF, Erhard P (1982) Cadmium levels in maternal blood, fetal cord blood, and placenta tissues of pregnant women who smoke. *Amer J Obstet Gynecol* 142: 1021–1025.
18. Stahl JL, Cook ME, Greger JL (1988) Zinc, iron and copper contents of eggs from hens fed varying levels of zinc. *J Food Comp Anal*. 1: 309–315.
19. Mazliah J, Barron S, Bental E, Reznik I (1989) The effect of chronic lead intoxication in mature chickens. *Avian Dis* 33: 566–570.
20. Nishimura M, Urakawa N (1972) Effect of estrogenization on the distribution of radiomercury in quail. *Jpn J Pharmacol* 22: 535–543.
21. Cherian MG (1980) The synthesis of metallothionein and cellular adaptation to metal toxicity in primary rat kidney epithelial cell cultures. *Toxicology* 17: 225–231.
22. Waalkes MP, Poisner AM, Wood GW, Klaassen CD (1984) Metallothionein-like proteins in human placenta and fetal membranes. *Toxicol Appl Pharmacol* 74: 179–184.

Metallothionein IV
C. Klaassen (ed.)
© 1999 Birkhäuser Verlag Basel/Switzerland

Metallothionein expression during wool follicle development in foetal sheep

Hisao Nishimura[1,] Noriko Nishimura[2, 3], Chiharu Tohyama[4], Graham R. Cam[3] and David L. Adelson[3]

[1]School of Human Sciences, Aichi Mizuho University, Toyota, Aichi 470-03 Japan
[2]Department of Veterinary Anatomy and Pathology, University of Sydney, NSW 2006, Australia
[3]Division of Animal Production, CSIRO, Prospect, NSW 2149, Australia
[4]Environmental Health Sciences Division, NIES, Tsukuba, 305, Japan

Metallothionein in proliferating tissues

Metallothionein (MT) is known to detoxify heavy metals such as cadmium and mercury by binding them. Other roles have been suggested because the protein is present in various tissues and is found to change in the concentration and immunohistochemical localization under different physiological conditions. We demonstrated that the staining intensity of MT increased with developmental stage and decreased with a morphological and functional maturation of developing rat kidney. In contrast to the low MT immunoreactivity restricted to the cytoplasm of liver cells following laparotomy, strong MT immunofluorescence was found in the nucleus and cytoplasm of hepatocytes in the regenerating liver after partial hepatectomy, which induces active cell proliferation [1, 2]. Furthermore, spermatogenic cells in the testicular seminiferous tubules showed a strong signal for MT protein and MT mRNA in the rat testis [3, 4], which suggests a possible role for MT in cellular proliferation and differentiation. The hair follicle is well-known as having one of the highest rate of cell proliferation among various tissues, and the follicle cells differentiate to form hair fibers and other accessory cell types. We reported previously that increased intensity of MT immunostaining was found in mouse skin after topical application of cholera toxin or phorbol ester, which stimulates cell division resulting in hyperplasia. Further, the elevated expression of MT mRNA was demonstrated by Northern blot analysis in hyperplastic epidermis, suggesting that MT has a possible association with the proliferation and differentiation of mouse epidermal keratinocytes [5]. Hanada et al showed MT mRNA expression in the cultured keratinocytes and a significant increase in cells treated with 1,25-dihydroxyvitamin D_3 [6]. Recently, a new metallothionein isoform, MT-IV, was found in stratified squamous epithelia associated with oral epithelia or neonatal skin, suggesting that MT-IV plays a special role in regulating zinc metabolism during the differentiation of stratified epithelia [7].

Here, we briefly describe how MT and MT-mRNA expression patterns correlate with cell proliferation or differentiation in developing wool follicles. An earlier study [8] showed that organ- and developmentally regulated expression of mRNA of four MT isoforms in the sheep. MT-II, MT-Ia and MT-Ib were found at 120–132 days of gestation in foetal lamb lung. MT-Ic mRNA was not detected at any developmental stage. These 4 MT genes were expressed in foetal and newborn liver. In the normal sheep liver, mRNA expression of the MT isoforms was most abundant for MT-Ia mRNA, followed in decreasing order, by MT-II, MT-Ic and MT-Ib [9]. Of the two MT mRNAs examined, MT-Ia and MT-II mRNAs were detected in kidney, duodenum and skin of normal sheep [10].

Localization of MT protein and mRNA in the sheep skin

In order to study the relationship of MT with hair follicle development, we used immunohistochemistry and *in situ* hybridization for MT protein and mRNA, respectively. 5-bromo-2'-deoxyuridine (BrdU) was injected into the jugular vein of pregnant ewes two hours before euthanasia to identify dividing cells. Immunodetection of MT protein and BrdU was performed with a polyclonal rabbit antiserum to rat MT-I and BrdU. At 71 days of gestation, immunostaining for MT protein was observed in epidermal cells which in many cases also contained BrdU. At 81–95 days of gestation, MT protein was detected in the presumptive outer root sheath (ORS) and follicle rudiment lumen but not in the presumptive dermal papilla cells (Fig. 1a). At 105 days of gestation, outer root sheath cells, the keratogenous zone and the follicle bulb matrix showed a uniform pattern of MT protein localization but did not co-localize with the BrdU-proliferative zone surrounding the dermal papilla (DP). Although MT protein was found in the ORS and the basal cell layer of sebaceous glands and sweat glands, its distribution was changed to an asymmetric pattern in the follicle bulb by 120 days of gestation (Fig. 1b).

In situ hybridization with digoxigenin-labelled sheep MT-Ia, MT-Ib, MT-Ic, and MT-II riboprobes, revealed that the MT-Ib gene was expressed at the highest level in sheep skin, followed by MT-Ia [11]. Neither MT-Ic nor MT-II probes gave hybridization signals in sheep skin. The epidermal cells, primarily in the spinous cells, expressed MT-Ib mRNA at 71 days of gestation. At 81–90 days of gestation, MT-Ib mRNA was concentrated at the tip of the follicle rudiment, but not in the presumptive dermal papilla (Fig. 2a). At 112 days of gestation, the MT-Ib gene was actively expressed in a uniform pattern in the follicle bulb matrix. However, very little MT-Ib mRNA was detected in the matrix cell layers adjacent to dermal papilla. At 130 days of gestation, intense cytoplasmic staining for MT-Ib mRNA was observed in the follicle bulb and keratogenous zone (Fig. 2b).

Metallothionein and cell proliferation/differentiation in the hair follicles

In order to investigate the physiological function of MT, the expression of MT mRNA and MT protein were studied in developing wool follicles of sheep since the follicles were thought to be an appropriate model for very active cell proliferation and differentiation. Since MT mRNA and MT protein were localized in actively dividing cells and their staining intensity increased with the developmental stage of the wool follicle. MT could be involved in cell proliferation or differentiation in the developing wool follicle. Nevertheless, there may be no direct relationship with cellular proliferation since localization of MT protein and mRNA did not always coincide with BrdU incorporation.

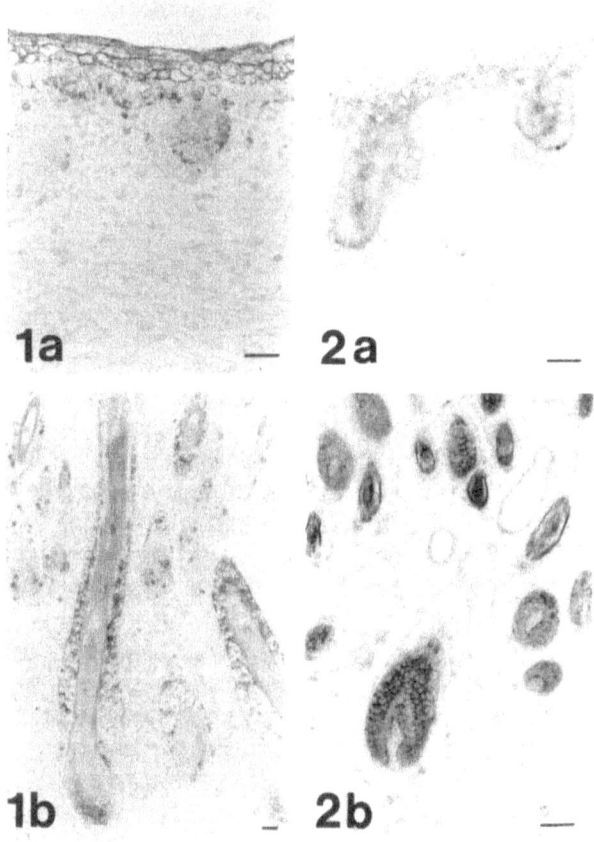

Figure 1. Immunohistochemical detection of MT protein in developing wool follicle at 81 days (1a) and 120 days (1b) of gestation. Bar, 50 μm.

Figure 2. Expression of MT-1b mRNA in developing wool follicle at 81 days (2a) and 130 days (2b) of gestation. Bar, 50 μm.

In situ hybridization of MT mRNA indicated that MT-Ib was a major MT isoform in developing sheep skin. In addition, MT-Ib mRNA was expressed in the keratogenous zone where MT protein was absent at 120 days of gestation, suggesting that the translation of MT-Ib mRNA was suppressed during cell differentiation. We have previously reported this apparent translational regulation in wool follicles in adult sheep skin [11]. As far as the potential role for MT in proliferating/developing cells, any role MT may play in cell division is still speculative. One such speculation is that they are involved in the supply of zinc ions to metalloenzymes and zinc-finger transcription factors. In view of the viability of MT-I and MT-II gene knock-out mice [12] it is difficult to envision essential roles of MT in cell division or cell differentiation unless other MT genes (such as MT-IV) are capable of compensating for the lack of MT-I and MT-II functions.

References

1. Nishimura H, Nishimura N, Tohyama C (1989) Immunohistochemical localization of metallothionein in developing rat tissues. *J Histochem Cytochem* 37: 715–722.
2. Tohyama C, Suzuki SK, Hemelraad J, Nishimura N, Nishimura H (1993) Induction of metallothionein and its localization in the nucleus of rat hepatocytes after partial hepatectomy. *Hepatology* 18: 1193–1201.
3. Nishimura H, Nishimura N, Tohyama C (1990) Localization of metallothionein in the genital organs of the male rat. *J Histochem Cytochem* 38: 927–933.
4. Tohyama C, Nishimura N, Suzuki JS, Karasawa M, Nishimura H, (1994) Metallothionein mRNA in the testis and prostate of the rat detected by digoxigenin- labeled riboprobe. *Histochemistry* 101: 341–346.
5. Karasawa M, Nishimura N, Nishimura H, Tohyama C, Hashiba H, Kuroki T (1991) Localization of metallothionein in hair follicles of normal skin and the basal cell layer of hyperplastic epidermis: Possible association with cell proliferation (1991) *J Invest Dermatol* 97: 97–100.
6. Hanada K, Sawamura D, Nakano H, Hashimoto I (1995) Possible role of 1,25-dihydroxyvitamin D_3 -induced metallothionein in photoprotection against UVB injury in mouse skin and cultured rat keratinocytes. *J Dermtol Sci* 9: 203–208.
7. Quaife CJ, Findley SD, Erickson JC, Froelick GJ, Kelly EJ, Zambrowicz BP, Palmiter RD (1994) Induction of a new metallothionein isoform (MT-IV) occurs during differentiation of stratified squamous epithelia. *Biochemistry* 33: 7250–7259.
8. Pitt BR, Brookens MA, Steve AR, Atlas AB, Davies P, Kuo SM, Lazo JS (1992) Expression of pulmonary metallothionein genes in late gestation lams. *Paediat Res* 4: 424–430.
9. Petersons GM, Mercer JFB (1988) Differential expression of four linked sheep metallothionein genes. *Eur J Biochem* 174: 425–429.
10. Lee J, Triller BAP, Grace NOD (1994) Metallothionein and trace element metabolism in sheep tissues in response to high and sustained zinc dosages. I. Expression of metallothionein mRNA. *Aust J Agric Res* 45: 321–332.
11. Nishimura N, Cam GR, Nishimura H, Tohyama C, Saitoh Y, Adelson DL (1996) Evidence for developmentally regulated transcriptional, translational and post- translatioal control of metallothionein gene expression in hair follicles. *Reprod Fertil Dev* 8: 1089–1096.
12. Michalska AE, Choo KHA (1993) Targeting and germ-line transmission of a null mutation of the metallothionein I and II loci in mouse. *Proc Natl Acad Sci USA* 90: 8088–8092.

Metallothionein IV
C. Klaassen (ed.)
© 1999 Birkhäuser Verlag Basel/Switzerland

A role of hepatic metallothionein on mercury distribution in fetal guinea pigs after in utero exposure to mercury vapor

Minoru Yoshida[1], Hiroshi Satoh[2] and Yawara Sumi[1]

[1]*Department of Chemistry, St. Marianna University School of Medicine, Kawasaki 216-8511, Japan*
[2]*Department of Environmental Health Sciences, Tohoku University Graduate School of Medicine, Sendai 980-8575, Japan*

Summary. Ethanol, which is an inhibitor of catalase, prevents the oxidation of mercury vapor (Hg^0) into ionic mercury (Hg^{2+}). Consequently, ethanol causes penetration of a large amount of Hg^0 to the fetus after exposure of pregnant animals to Hg^0. The fate of mercury in the feto-placenta unit of pregnant guinea pigs treated with ethanol prior to exposure to Hg^0 was investigated. Ethanol pretreatment of the dams resulted in transfer of more mercury to the fetuses and led to a marked increase of mercury in the liver of the fetuses. Interestingly, mercury concentrations in the fetal brain and kidney were split into two levels. In a subgroup, mercury concentrations in the fetal brain and kidney were much higher than another subgroup (about 7 times in the brain and 3.5 times in the kidney). Determination of metallothionein in the fetal liver showed that metallothionein concentrations were lower in the subgroup with higher mercury concentrations in the brain and kidney than another subgroup with lower mercury concentrations in the brain and kidney. Further, the Sephadex G-75 chromatography of the cytosol of fetal liver of the subgroup with lower mercury concentrations in the brain and kidney revealed that most mercury was bound to metallothionein like-protein. On the other hand, in the subgroup with higher mercury concentrations in the brain and kidney, small amount appeared in fractions corresponding metallothionein. These findings suggest that fetal hepatic metallothionein plays a significant defense role against mercury penetrating through the placental barrier and is involved in regulating for the mercury distribution in the fetus.

Introduction

Metallothionein is known to show unique ontogenic changes during the perinatal period of mammals. Hepatic concentrations of metallothionein are elevated during the late gestational and early neonatal periods. The level of metallothionein peaks at term or in the early postnatal stage of the developing organism, thereafter decreasing during postnatal development (Bakka and Webb, 1981). Although the physiological roles of metallothionein have not been fully understood, it is believed that metallothionein has a function in regulating metabolism of trace elements, such as zinc and copper, which are essential for the growth of developing animals (Cherian and Goyer, 1978). Elevated concentrations of metallothionein, particularly in the liver and kidney, are also observed in adult animals after exposure to heavy metals (Cherian and Goyer, 1978) and environmental stresses (Oh et al., 1978). The elevated concentrations of metallothionein are due to enhanced synthesis induced by exposure to environmental stresses and play an important role in detoxification of the toxic metals, in the case of heavy metal exposure, (Klaassen and Wong, 1982). Fetal Minamata disease in Japan as a result of *in utero* exposure to methylmercury showed that the fetuses were more sensitive than their mothers. Thereafter the placental transfer of various mercury compounds and the uptake of mercury by fetal tissue have been investigated (Suzuki et al., 1968; Greenwood et al., 1972; Satoh et al., 1981). Experimental exposure of pregnant animals to mercury vapor (Hg^0) showed that substantial amount of Hg^0 penetrated the placental barrier and deposited in the fetus, though Hg^0 is rapidly oxidized into ionic mercury (Hg^{2+}) in maternal tissue (Clarkson et al., 1972; Khayat and Shaikh, 1982; Yoshida et al., 1986). Khayat and Dencker (1982) and we (1986) also reported that the Hg^0 transferred to the fetus was oxidized to Hg^{2+} by liver catalase and accumulated in the liver. Furthermore, gel chromatography showed that a large amount of Hg^{2+} was bound to fetal hepatic metallothionein (Yoshida et al., 1987). It is considered that mercury which passes through the placental barrier displaces zinc and copper, which are present in the fetal liver associated with metallothionein (Bell et al., 1979; Bakka and Webb, 1981), by binding to metallothionein existing at high levels during late gestation (Day et al., 1984; Yoshida et al., 1987). Therefore, fetal hepatic metallothionein appears to have some reserve capacity and thus may play an important role in protecting the fetus from mercury poisoning.

It has been clearly shown that ethanol inhibits the oxidation of Hg^0 by catalase both *in vivo* (Kudsk, 1969) and *in vitro* (Magos et al., 1973) resulting in much higher levels of Hg^0 in the blood during exposure periods. Khayat and Dencker (1982) reported that ethanol pretreatment caused a marked increase in fetal mercury uptake in experiments on pregnant mice. However, it is not known to what extent fetal hepatic metallothionein is involved in protecting the tissues when a large amount of Hg^0 is transferred to the fetus after ethanol treatment.

Mercury concentrations in the blood and major organs after exposure to mercury vapor with pretreatment of ethanol

In a whole-body autoradiography study in mice, Khayat and Dencker (1982) reported that ethanol preadministration to the dams caused a marked increase in fetal uptake of mercury, particularly in the liver. In our experiment using guinea pigs, the mercury concentrations in the maternal blood of the ethanol-treated (Hg^0+EtOH, given 2 g/kg of ethanol 30 min prior to mercury vapor exposure approximately at 10 mg/m^3 for 150 min) group were lower than in the Hg^0 group without ethanol pretreatment (Tab. 1). The animals in Hg^0+EtOH group were split into two subgroups according to the mercury concentrations in the blood of their fetuses: subgroup 1 (Nos. 1–3 in Tab. 1) with the concentrations similar to those in the Hg^0 group and subgroup 2 (Nos. 4–6 in Tab. 1) with much lower concentrations. Mercury concentrations were significantly lower in the kidney of dams in the Hg^0+EtOH group than in the Hg^0 group ($p < 0.01$ by t-test). In the liver, however, mercury concentration of the Hg^0+EtOH group was about 3.5 times higher than that of the Hg^0 group ($p < 0.01$ by t-test). On the brain mercury levels of dams, there were no difference between the Hg^0 and ethanol-pretreated group.

Mercury concentrations in fetal organs fell into two groups again. The fetuses in subgroup 1 had markedly higher mercury concentration in the brain and kidney than the fetuses in the Hg^0 group. In subgroup 2 (Nos. 4–6), of which treatment was the same as in subgroup 1, mercury concentrations in the brain and kidney were similar to that of the Hg^0 group. However, the mercury concentrations in the liver of all fetuses (Nos. 1–6) pretreated with ethanol were markedly higher than in the Hg^0 group ($p < 0.01$ by t-test), similarly to the mothers'. There was no significant difference between liver mercury concentrations of fetuses between subgroups 1 and 2.

Table 1. Mercury concentrations in the blood and major organs of guinea pigs and their fetuses after exposure to mercury vapor

| | Exposure levels | Mercury concentration (ng Hg/g wet wt.) | | | |
	(mg/m^3)	Blood	Brain	Kidney	Liver
Control group					
Mothers (n = 5)	-	5 ± 1	29 ± 3	25 ± 5	24 ± 3
Fetuses (n = 15)		7 ± 1	11 ± 3	13 ± 3	38 ± 5
Hg^0 group					
Mothers (n = 4)	7–10	283 ± 29	253 ± 37	14050 ± 540	414 ± 108
Fetuses (n = 15)		25 ± 9	21 ± 9	31 ± 10	549 ± 83
Hg^0 + EtOH group					
Subgroup 1					
1 Mothers	10	70	233	3390	1170
Fetuses (n = 3)		32 ± 4	121 ± 4	145 ± 14	1760 ± 230
2 Mothers	11	87	195	3450	956
Fetuses (n = 3)		26 ± 8	288 ± 33	180 ± 31	944 ± 155
3 Mothers	10	127	171	2650	1170
Fetuses (n = 5)		22 ± 1	159 ± 6	118 ± 0	1130 ± 54
Subgroup 2					
4 Mothers	9	69	173	2230	1640
Fetuses (n = 4)		9 ± 1	32 ± 8	29 ± 2	1950 ± 331
5 Mothers	9	150	263	3490	1490
Fetuses (n = 3)		8 ± 1	23 ± 3	20 ± 11	3370 ± 153
6 Mothers	10	205	314	4930	2290
Fetuses (n = 3)		3 ± 1	32 ± 8	38 ± 14	1150 ± 254

Pregnant guinea pigs were exposed to mercury vapor for 150 min, except control animals. Pregnant guinea pigs were given ethanol 2 g/kg orally 30 min before exposure to mercury vapor. Values are mean ± standard deviation of the values obtained in each animals. The numbers of animals are given in parentheses.

The results indicate that pretreatment of dams with ethanol prior to mercury vapor exposure caused greater concentrations of mercury in the fetal kidney and liver, with marked accumulation in the maternal liver and less in the maternal kidney. An interesting finding in this study is that the same pretreatment with ethanol differently affected accumulations of mercury in the brain and kidney of the fetuses.

Metallothionein concentrations in the fetal liver

Metallothionein levels in the fetal liver were similar to or slightly higher than that in the maternal liver in all groups (Fig. 1). However, the average fetal liver metallothionein levels in subgroup 2, which had markedly lower mercury levels in the brain, was higher than that in subgroup 1, and the difference was statistically significant ($p < 0.01$, by t-test). Maternal liver

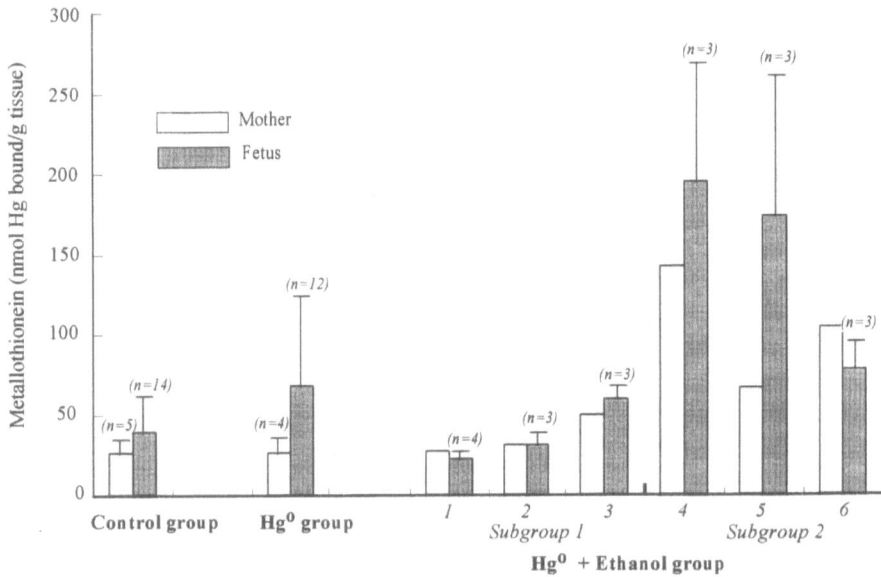

Figure 1. Metallothionein concentration in the liver of guinea pigs and their fetuses after mercury vapor exposure. Values are mean ± standard deviation of the values obtained in each animals. The numbers of animals are shown in parentheses.

metallothionein in subgroup 1 was also significantly higher than that in subgroup 2 ($p < 0.05$, by t-test). We previously reported that hepatic metallothionein concentrations vary widely in fetal guinea pigs in late gestation (Yoshida et al., 1987; Yoshida et al., 1990). Waalkes et al. (1984) have reported that acute ethanol administration exceeding the dose of 5 g/kg induces metallothionein synthesis in the rat liver. Since the dose of ethanol used in the experiment was 2 g/kg, the higher hepatic metallothionein concentrations observed in the fetuses of the dams pretreated with ethanol were probably attributable to endogenous factors, not ethanol administration.

Sephadex G-75 chromatography of the soluble liver fractions

There are high levels of metallothionein associated with copper and zinc in the fetal liver (Bell et al., 1979; Bakka et al., 1981). We showed that Hg^0 transferred through the placental barrier was oxidized to Hg^{2+} and accumulated in the fetal liver, and that a large amount of the mercury retained in the fetal liver was bound to metallothionein by displacing of zinc and copper (Yoshida et al., 1987). Gel chromatography of soluble fractions of the fetal liver in subgroup 1, which showed high mercury uptake in the brain, showed that small amount of mercury was bound to protein corresponding to metallothionein (Fig. 2). In contrast, subgroup 2, with lower

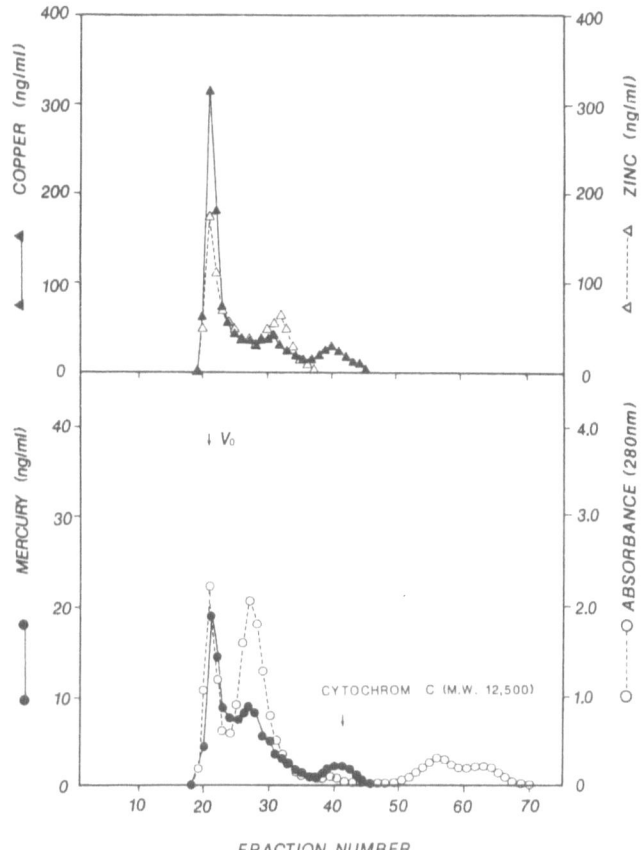

Figure 2. Sephadex G-75 gel chromatograms of the soluble fractions of the liver of the fetuses in subgroup 1 (Nos,1–3), which exhibited high mercury uptake by the brain and who mother had been pretreated with ethanol prior to exposure to mercury vapor. The upper chromatogram shows the concentrations of copper (—▲—) and zinc (---△---) in each fraction. The lower chromatograms show the mercury concentration (—●—) and the absorbance at 280 nm (---○---). Void volume was determined by blue dextran

mercury levels in the brain, a large portion of the mercury deposited in their liver was bound to a protein with a molecular weight similar to that of metallothionein (Fig. 3). Thus gel chromatograms on Sephadex G-75 columns of the soluble liver fractions were different between fetal subgroups 1 and 2. Theses findings suggest that the high mercury uptake by the brain in the fetuses in subgroup 1 may have been related to less amount of metallothionein in the fetal liver. However, since brain uptake of mercury occurs presumably in the form of Hg^0, precise mechanism is to be elucidated with investigations of possible effects on the fetal brain, which is rapidly developing and highly vulnerable.

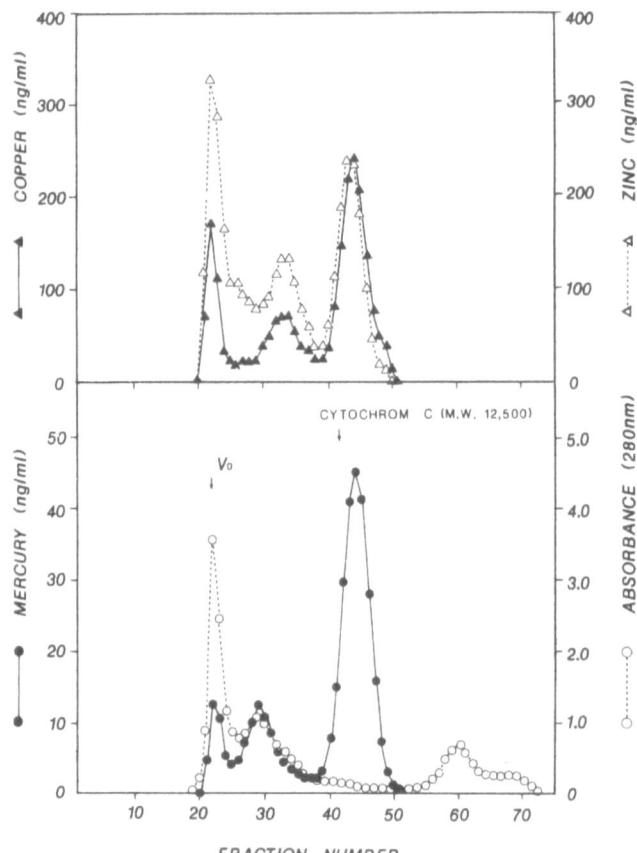

Figure 3. Sephadex G-75 gel chromatograms of the soluble fractions of the liver of the fetuses in subgroup 2 (Nos. 4–6), which exhibited less mercury uptake by the brain, although whose mother had been similarly pretreated with ethanol prior to exposure to mercury vapor. The upper chromatogram shows the concentrations of copper (—▲—) and zinc (---△---) in each fraction. The lower chromatograms show the mercury concentration (—●—) and the absorbance at 280 nm (---○---).

References

1. Bakka A and Webb M (1981) Metabolism of zinc and copper in the neonate: Changes in the concentrations and contents of thionein-bound Zn and Cu with age in the livers of the newborn of various mammalian species. *Biochem Pharmacol* 30: 721–725.
2. Bell JU (1979) A metallothionein-like protein in the hepatic cytosol of the term rat fetus. *Toxicol Appl Pharmacol* 48: 139–144.
3. Cherian MG and Goyer RA (1978) Metallothioneins and their role in the metabolism and toxicity of metals. *Life Sci* 23: 1–10.
4. Clarkson TW, Magos L and Greenwood MR (1972) The transport of elemental mercury into fetal tissue. *Biol Neonate* 21: 239–244.
5. Day FA, Funk AE and Brady FO (1984) *In vivo* and *ex vivo* displacement of zinc from metallothionein by cadmium and by mercury. *Chem Biol Int* 50: 159–174.

6. Grennwood MR, Clarkson TW and Magos L (1972) Transfer of metallic into fetus. *Experientia* 28: 1455–1456.
7. Khayat A and Dencker L (1982) Fetal uptake and distribution of metallic mercury vapor in the mouse: Influence of ethanol and aminotriazole. *Biol Res Pregnancy* 3: 38–46. 1.
8. Khayat AI and Shaikh ZA (1982) Dose-effect relationship between ethyl alcohol pretreatment and retention and tissue distribution of mercury vapor in rats. *J Pharmacol Exp Ther* 233: 649–653.
9. Klaassen CD and Wong K-L (1982) Cadmium toxicity in the newborn rat. *Can J Physiol Pharmacol* 60: 1027–1036.
10. Kudsk FN (1969) Factors influencing the *in vitro* uptake of mercury vapor in blood. *Acta Pharmacol Toxicol* 27: 161–172.
11. Magos L, Clarkson TW and Greenwood MR (1973) The depression of pulmonary retention of mercury vapor by ethanol: identification of the site of action. *Toxicol Appl Pharmacol* 26: 180–183.
12. Oh SH, Dlagen JT, Whanger PD and Weswig PH (1978) Biological function of metallothionein. V. Its induction in rats by various stresses. *Amer J Physiol* 234: E282-E285.
13. Satoh H, Suzuki T, Nobunaga T, Naganuma A and Imura N (1981) Effect of sodium selenite on distribution and placental transfer of mercuric mercury in mice of late gestational period. *J Pharmcol Dyn* 4: 191–196.
14. Suzuki T, Miyama T and Katsunuma H (1968) Placenta transfer of mercury chloride, phenyl mercury acetate and methyl mercury acetate in mice. *Ind Health* 5: 149–155.
15. Yoshida M, Yamamura Y and Satoh H (1986) Distribution of mercury in guinea pig offspring after in utero exposure to mercury vapor during late gestation. *Arch Toxicol* 58: 225–228.
16. Yoshida M, Aoyama H, Satoh H and Yamamura Y (1987) Binding of mercury to metallothionein-like protein in fetal liver of the guinea pig following in utero exposure to mercury vapor. *Toxicol Lett* 34: 1–6.
17. Yoshida M, Satoh H, Kojima S and Yamamura Y (1990) Retention and distribution of mercury in organs of neonatal guinea pigs after in utero exposure to mercury vapor. *J Trace Elem Exp Med* 3: 219–226.
18. Waalkes MP, Hjelle JJ and Klaassen CD (1984) Transient induction of hepatic metallothionein following oral ethanol administration. *Toxicol Appl Pharmacol* 74: 230–236.
19. Webb M and Cain K (1982) Function of metallothionein. *Biochem Pharmacol* 31: 137–142.

Metallothionein IV
C. Klaassen (ed.)
© 1999 Birkhäuser Verlag Basel/Switzerland

Cell differentiation alters metallothionein expression in avian growth plate chondrocytes

Glenn R. Sauer, Licia N. Y. Wu, Daotai Nie and Roy E. Wuthier

University of South Carolina, Dept. of Chemistry & Biochemistry, Columbia, SC, 29208, USA

Endochondral bone formation represents the primary mechanism of vertebrate skeletal development and is the means by which longitudinal bone growth occurs in the epiphyseal growth plate [1]. During this process growth plate chondrocytes progress through a series of distinct developmental stages, each of which is characterized by the expression of specific marker proteins and changes in electrolyte metabolism. This progression proceeds from an early resting stage, through a proliferative stage characterized by active cell division and matrix synthesis, a hypertrophic stage and finally, an apoptotic terminally differentiated stage [2, 3]. In the later stages of this developmental cascade, extracellular matrix vesicles are released from the cell membrane [4] and become the initial sites of calcium phosphate mineral formation thereby triggering matrix calcification [5]. Studies in our laboratory have shown that zinc is essential to normal chondrocyte growth and differentiation [6] and plays an important role in matrix vesicle mineralization [7]. In order to gain a better understanding of metal ion metabolism in these calcifying cells, we examined the expression of the intracellular Zn-binding protein metallothionein (MT) in avian growth plate tissue and chondrocyte cell cultures.

When chickens were exposed to Cd (1 mg/kg body weight) or Zn (10 mg/kg body weight) and the growth plate cartilage tissue fractionated by Sephadex G-75 chromatography, a protein with high metal content, high cysteine content, low UV absorbance and apparent molecular weight 10,000 was induced [8]. A MT standard and MT from liver of the same birds eluted at the same position. These findings confirmed the presence of inducible MT in growth plate tissue. The Cd-hemoglobin affinity assay [9] was next used to quantitate the levels of MT in the growth plate and compare the values with MT levels in other tissues [8]. Basal levels of MT in the growth plate of control animals were low (0.06 µg MT/mg protein) relative to MT levels in liver (0.97 µg MT/mg protein) or kidney (0.99 µg MT/mg protein). Exposure to metal ions resulted in a 25-fold induction by Cd of growth plate MT and a 5-fold induction of MT by Zn. MT levels in the growth plate increased as the duration of metal exposure increased.

Elemental analysis of tissue slices from chicken epiphyseal growth plates revealed significant differences in the distribution of Zn. Zinc concentrations were highest in the proliferative zone (278 µg Zn/g tissue), lowest in the hypertrophic zone (85 µg Zn/g tissue) and rose steadily through the calcifying zone (145 µg Zn/g tissue) to the bone tissue (203 µg Zn/g tissue). These regional differences in the distribution of Zn in the growth plate prompted us to ask whether there could also be phenotype-specific differences in chondrocyte MT expression. To study this question we used a serum-free primary culture system which mimics both morphologically and biochemically, the changes that are known to occur in the growth plate [10].

Induction of MT in growth plate chondrocyte primary cultures was observed for media Cd concentrations of 0.1 μM and greater and Zn concentrations of 100 μM and greater. Both Cd- and Zn-treatment also resulted in increased levels of MT mRNA in the cell cultures as determined by RT-PCR analysis [8]. Basal and inducible levels of MT were highest in the early days of culture corresponding to the proliferative stage of chondrocyte development and declined steadily throughout the culture period. Lowest MT levels were observed in the terminally differentiated hypertrophic chondrocytes of late culture periods that actively undergo mineralization (Fig. 1). Late stage cultures were also more sensitive to Cd toxicity than early stage cultures. Treatment of chondrocytes with Zn prior to Cd exposure resulted in a protective induction of MT [8]. On the other hand, pre-treatment of chondrocyte cultures with dexamethasone resulted in reduced MT levels and greater metal toxicity when the cells were subsequently exposed to Cd. The finding that dexamethasone treatment reduced chondrocyte MT is unusual since it has been shown to be a strong inducer of MT in many other cell types [11]. This apparent reduction in cellular levels of MT in response to dexamethasone occurred even though the levels of MT mRNA increased in these cells [8].

The unusual dexamethasone findings prompted us to conduct a further examination of MT induction in different chondrocyte phenotypes. These experiments were conducted using primary cultures of avian chondrocytes at early stages of culture (10–14 day) representative of the proliferative phenotype and late stages of culture (21–28 day) representative of the hypertrophic phenotype [10]. For a third experimental phenotype we used Rous sarcoma virus (RSV) -transformed chondrocytes [12]. Transformation was accomplished by adding RSV stocks to 3-day old primary cultures for 4 h and passaging cells on Day 7 to ensure complete transformation. RSV-transformed cells exhibited a spindle-shaped morphology whereas normal chondrocytes have a polygonal morphology. The RSV-transformed cells also showed reduced expression of Type X collagen, proteoglycan, and alkaline phosphatase and markedly enhanced expression of matrix metalloproteinase 2 (unpublished observations). Additionally cultures of the transformed cells failed to mineralize whereas normal chondrocytes readily mineralize under our culture condition beginning between days 18–21. All cells for the MT induction experiments were cultured in serum-free Hl-1 media (BioWhitaker) supplemented with 50 μg/ml ascorbate

Table 1. Metallothionein levels (pmol Cd bound/mg protein ± SE) in cell cultures of different chondrocyte phenotypes following two day treatment with various inducing agents

Treatment[a]	Cell Type		
	Proliferative	Hypertrophic	Transformed
Control	48 ± 9	34 ± 6	73 ± 10
Cadmium	1638 ± 164	1297 ± 131	859 ± 35
Dexamethasone	54 ± 8	16 ± 6	138 ± 13
1,25-Vitamin D$_3$	109 ± 25	35 ± 7	149 ± 16
Retinoic Acid	56 ± 9	44 ± 8	257 ± 19

[a]Concentration of inducing agents: cadmium – 1 μM; dexamethasone – 1 μM; vitamin D$_3$–100 nm; retinoic acid – 100 nm.

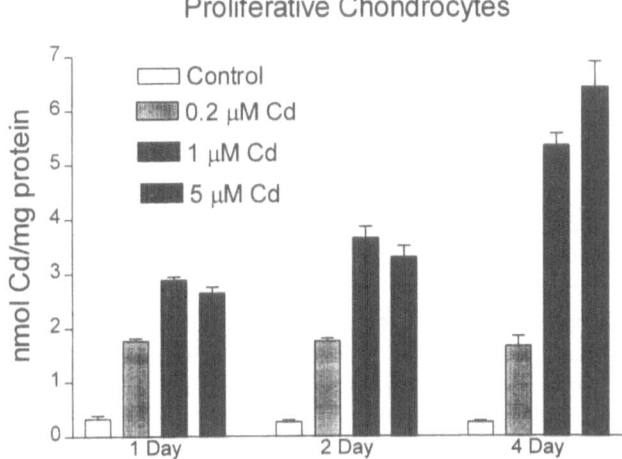

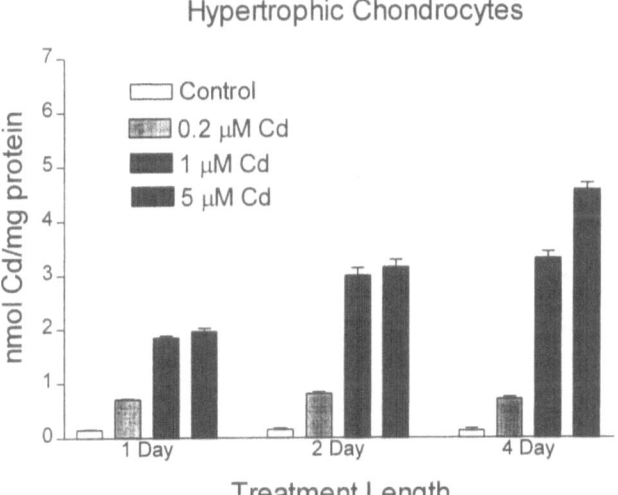

Figure 1. Metallothionein levels at different stages of growth plate chondrocyte cell cultures after 1–4 days of exposure to cadmium. Proliferative chondrocytes (13 day cultures) show higher basal and inducible levels of metallothionein than hypertrophic chondrocytes (24 day cultures). Values are mean ± SE.

as described previously [13]. In these experiments cells were exposed to various inducing agents for 2 days prior to analysis for cytosolic protein and MT.

RSV-transformed chondrocytes showed significantly higher basal levels of MT relative to normal proliferative or hypertrophic chondrocytes (Tab. 1). Despite the higher basal levels,

RSV-transformed cells were less responsive to cadmium than either of the normal chondrocyte phenotypes. It is possible that the higher basal levels of MT in RSV-transformed cells results in fewer intracellular Cd^{2+} ions available for activating MT gene transcription. As we have reported previously [8] hypertrophic chondrocytes showed lower basal and Cd-induced levels of MT relative to proliferative cells. The 3 chondrocyte phenotypes also had different responses to the non-metal inducing agents (Tab. 1). Retinoic acid treatment resulted in a strong induction of MT in RSV-transformed cells but not in normal chondrocytes. 1,25 dihydroxy Vitamin D_3 caused a significant increase in MT in proliferative and transformed chondrocytes but not hypertrophic cells. Previously we found that another metabolite, 24,25 dihydroxy Vitamin D_3 caused a modest induction of MT in late stage hypertrophic cultures [14]. Dexamethasone treatment resulted in a significant induction of MT in RSV-transformed cells while proliferative cells showed little response to dexamethasone and hypertrophic cells showed a 50% reduction in MT levels. Clearly, the response of chondrocytes to MT inducing agents is dependent upon the differentiated phenotype of the cells. If any broad generalizations can be made from the studies we have conducted so far, it is that rapidly dividing cells (RSV-transformed and proliferative chondrocytes) shower higher levels of MT than hypertrophic, calcifying cells. Additionally, agents that stimulate cell division, such as retinoic acid, increase MT levels in chondrocytes while agents that support cell hypertrophy and terminal differentiation, such as dexamethasone, result in reduced levels of MT.

As stated previously, in normal chondrocyte cultures, basal MT levels decline in response to dexamethasone in spite of an observed increase in MT mRNA expression in these cells [8]. Additionally, the induction of MT protein by Cd appears to be reduced if the chondrocytes are pre-exposed to dexamethasone (Fig. 2). In contrast, dexamethasone treatment of RSV-transformed cells resulted in enhanced expression of MT upon subsequent Cd treatment. The

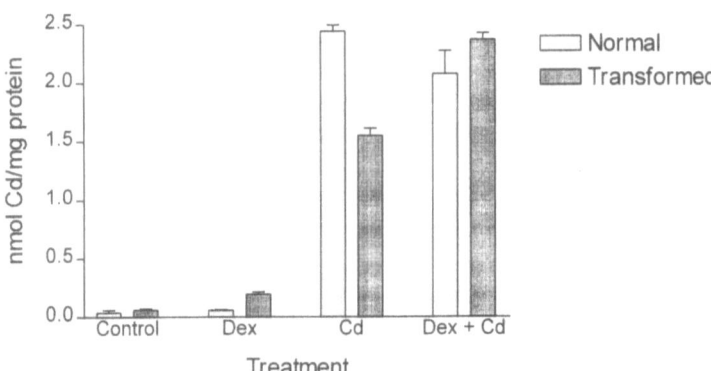

Figure 2. Metallothionein levels in normal proliferative chondrocytes (14 day cultures) and RSV-transformed chondrocytes following exposure to dexamethasone (1 μM) and cadmium (5 μM). In normal cells, dexamethasone treatment has minimal effect on metallothionein levels and lowers the observed response of cells to cadmium. In RSV-transformed cells metallothionein is induced by dexamethasone and co-induction with cadmium results in higher levels. Values are mean ± SE.

response of the RSV-transformed cells to co-induction by Cd and dexamethasone is consistent with what has been reported in other cell types [11]. The promoter region of the MT gene contains specific response elements for both metals and glucocorticoids which account for these effects [15]. The fact that we observed reduced MT levels in normal chondrocytes despite increased MT mRNA expression following dexamethasone treatment suggests an increased rate of intracellular degradation of MT in these cells. A number of recent studies have clearly shown that the terminal differentiation of hypertrophic chondrocytes is an apoptotic process [2, 3, 16]. In many cells, the production of reactive oxygen species (ROS) during the programmed cell death cascade is thought to contribute to nuclear disintegration and cytolysis [17]. It has been suggested that one of the primary physiological roles of MT in many cells is to act as an intracellular scavenger of ROS [18]. The metal-thiolate bonds of MT are sensitive to oxidation by ROS resulting in the release of bound metal ions [19]. The oxidative release of metal ions from MT in apoptotic chondrocytes may explain the elevated levels of Zn^{2+} that we observe in matrix vesicles [7] which are shed from the cells during this late stage of development. Among the many agents that have been shown to trigger programmed cell death are steroids such as dexamethasone [20]. Hence, even though dexamethasone induces MT expression in normal chondrocytes, other cellular responses that it activates or enhances may result in lower total levels of chondrocyte MT. The possible relationships between metallothionein, cell differentiation, and programmed cell death in chondrocytes is very intriguing and warrants further investigation.

Acknowledgment
This work was supported by U.S. Public Health Service grant number DE10196 from the National Institute of Dental Research.

References

1. Hunziker E (1994) Mechanism of longitudinal bone growth and its regulation by growth plate chondrocytes. *Microsc Res Techn* 28: 505–519.
2. Gibson GJ, Kohler WJ, Schaffler MB (1995) Chondrocyte apoptosis in endochondral ossification of chick sterna. *Develop Dynam* 203: 468–476.
3. Hatori M, Klatte KJ, Teixeira CC, Shapiro IM (1995) End labeling studies of fragmented DNA in the avian growth plate: evidence of apoptosis in terminally differentiated chondrocytes. *J Bone Min Res* 12: 1960–1968.
4. Anderson HC (1995) Molecular biology of matrix vesicles. *Clinical Orthopaedics* 314: 266–280.
5. Wuthier RE (1988) Mechanism of matrix vesicle-mediated mineralization of cartilage. *ISI Atlas of Science: Biochemistry* 1: 231–241.
6. Litchfield TM, Ishikawa Y, Wu LNY, Wuthier RE, Sauer GR (1997) Effect of metal ions on calcifying growth plate cartilage chondrocytes. *Calcif Tiss Int* 12: 341–349.
7. Sauer GR, Genge BR, Adkisson HD, Wuthier RE (1989) Regulatory effect of endogenous zinc and inhibitory effect of toxic metal ions on calcium accumulation by matrix vesicles *in vitro*. *Bone and Mineral* 7: 233–244.
8. Sauer GR, Nie D, Wu LNY, Wuthier RE (1997) Induction and characterization of metallothionein in chicken epiphyseal growth plate cartilage chondrocytes. *J Cellular Biochem* 68: 110–120.
9. Eaton DL, Cherian MG (1991) Determination of metallothionein in tissues by cadmium-hemoglobin affinity assay. *Meth Enzymol* 205: 83–88.
10. Wu LNY, Ishikawa Y, Sauer GR, Genge BR, Mwale F, Mishima H, Wuthier RE (1995) Morphological and biochemical characterization of mineralizing primary cultures of avian growth plate chondrocytes: Evidence for cellular processing of Ca^{2+} and Pi prior to matrix mineralization. *J Cell Biochem* 57: 218–237.
11. Klaasen CD, Liu J (1991) Induction of metallothionein in primary rat hepatocyte cultures. *Methods Enzymol* 205: 567–574.
12. Gionti E, Capasso O, Cancedda R (1983) The culture of chick embryo chondrocytes and the control of their

differentiated functions *in vitro. J Biol Chem* 258: 7190–7194.

13. Wu LNY, Sauer GR, Genge BR, Wuthier RE (1989) Induction of mineral deposition by primary cultures of chicken growth plate chondrocytes in ascorbate-containing media. *J Biol Chem* 265: 21346–21355.

14. Litchfield TM, Sauer GR (1996) Metallothionein induction in calcifying growth plate cartilage chondrocytes. *Connect Tiss Res* 35: 189–243.

15. Hamer DH (1986) Metallothionein. *Annu Rev Biochem* 55: 913–951.

16. Gibson G, Lin D-L, Roque M (1997) Apoptosis of terminally differentiated chondrocytes in culture. *Exper Cell Res* 233: 372–382.

17. Kroemer G, Petit P, Zamzami N, Vayssiere J-l Mignotte B (1995) The biochemistry of programmed cell death. *Fed Amer Soc Exp Biol J* 9: 1277–1288.

18. Lazo JS, Pitt BR (1995) Metallothioneins and cell death by anticancer drugs. *Annu Rev Pharmacol Toxicol* 35: 635–653.

19. Fliss H, Menard M (1992) Oxidant-induced mobilization of zinc from metallothionein. *Arch Biochem Biophys* 293: 195–199.

20. Schwartzman RA, Cidlowski JA (1993) Apoptosis: the biochemistry and molecular biology of programmed cell death. *Endocrin Rev* 14: 133–151.

Metallothionein in brain disease

Metallothionein IV
C. Klaassen (ed.)
© 1999 Birkhäuser Verlag Basel/Switzerland

Metallothionein in oxidative stress of Parkinson's disease

Manuchair Ebadi[1], Midori Hiramatsu[2], M.V. Ramana Kumari[2], Ruyi Hao[3] and Ronald F. Pfeiffer[3]

[1]*Department of Pharmacology, University of Nebraska College of Medicine, Omaha, NE 68198-6260, USA*
[2]*Institute of Life Support Technology, Yamagata Technopolis Fdn., Yamagata, Japan*
[3]*Department of Neurology, University of Tennessee College of Medicine, Memphis, TN, USA*

Summary. The loss of striatal dopaminergic neurons in Parkinson's disease results in enhanced turnover and metabolism of dopamine, augmenting the formation of H_2O_2 and generation of highly neurotoxic hydroxyl radicals ($\cdot OH$). In addition, the striatum of patients with Parkinson's disease has a low level of glutathione and complex I which may further predispose to oxidative stress. Neurotoxic agents causing Parkinson's disease, including 6-hydroxydopamine and 1-methyl-4-phenyl-1,2,3,6-tetrahydropyridine, generate free radicals and induce the levels of metallothionein isoforms. Moreover, metallothionein isoforms I and II are able to avert the neurotoxic effects of hydroxyl radicals, superoxide radicals, 1,1-diphenyl-2-picrylhydrazyl radicals, and the reactive oxygen species (containing singlet oxygen, superoxide and hydroxyl radicals) generated by photosensitized oxidation of riboflavin. The results of this investigation are interpreted to suggest that metallothionein isoforms, whose synthesis in the brain is induced by stress, cytokines and inflammatory processes, by being able to scavenge various free radicals, are capable of attenuating oxidative stress and of providing neuroprotection in drug-induced parkinsonism, and perhaps Parkinson's disease.

Introduction

Parkinson's disease consists of a severe reduction in the dopamine content in all components of the basal ganglia. Four separate groups of symptoms constitute the symptom complex that makes up parkinsonism. These consist of tremor, akinesia or bradykinesia, rigidity, and loss of postural reflexes. The parkinsonian tremors have been referred to as "pill-rolling," "cigarette rolling," and "to and fro" movements. These tremors are present during rest but often disappear during purposeful movements or sleep. Stress or anxiety-provoking situations aggravate the tremors and the initiation of movements becomes increasingly difficult, extremely fatiguing, and ponderously inefficient. The akinesia or bradykinesia is characterized by a poverty of spontaneous movements and slowness in initiation of movements. Rigidity or increased muscle tone occurs in response to passive movements. The loss of normal postural reflexes is a disorder of postural fixation and equilibrium [1].

The principle cytoskeletal pathology of Parkinson's disease is the occurrence of Lewy body

Parkinson's disease is characterized by degeneration of pigmented neuronal systems of the brain stem variably associated with pathology in non-nigral systems causing multiple neuromediator dysfunctions. The principle cytoskeletal pathology of Parkinson's disease is the Lewy body, which, in 85 to 100% of cases occur in many aminergic and other subcortical nuclei, spinal cord, sympathetic ganglia, and less frequently in cerebral cortex, myenteric plexuses, and adrenal medulla [2–6].

Current concepts as to the cause of Parkinson's disease suggest an inherited predisposition to environmental or endogenously produced toxic agents causing oxidative damage to nigral cells

Free radicals have been postulated to contribute to neuronal loss in Parkinson's disease [7–12] for the following reasons. The loss of striatal dopaminergic neurons accelerate the rate of synthesis of dopamine, producing H_2O_2 and hydroxyl radicals according to the following reactions.

$$L\text{-Dopa} \xrightarrow{\text{Dopa-Decarboxylase}} \text{Dopamine} + CO_2$$

$$\text{Dopamine} + O_2 + H_2O \xrightarrow{\text{MAO-B}} 3,4 \text{ Dihydroxyphenyl-acetaldehyde} + NH_3 + H_2O_2$$

$$H_2O_2 + \text{Reduced glutathione} \xrightarrow{\text{Glutathione peroxidase}} \text{Oxidized glutathione} + H_2O_2$$

$$Fe^{++} \longrightarrow Fe^{++}$$

$$^\bullet OH + OH^-$$

The level of reduced glutathione in substantia nigra is decreased [13–15; and Figure 1]. The depletion of reduced glutathione in the substantia nigra in Parkinson's disease could be the result of neuronal loss. As a matter of fact, a positive correlation has been found to exist between the extent of neuronal loss and depletion of glutathione [16, 17]. A decrease in the availability of reduced glutathione would impair the capacity of neurons to detoxify hydrogen peroxide and increase the risk of free radical formation and lipid peroxidation. Indeed, the substantia nigra contains increased levels of malondialdehyde and hydroperoxides [7, 8]. Another factor imposing increased oxidative stress is the accumulation of iron and progressive siderosis of substantia nigra [18–21]. In addition, the level of ferritin, an iron-binding protein is reduced in the substantia nigra [22] and hence aggravating the iron-induced hydroxyl radical-mediated lipid peroxidation.

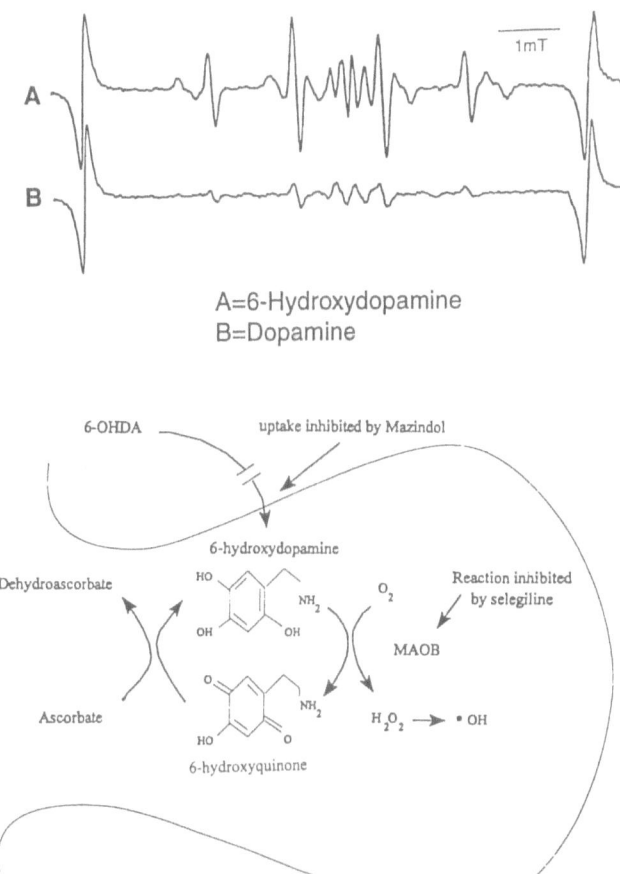

Figure 1. 6-Hydroxydopamine (6-OHDA) enters substantia nigra pars compacta via the catecholamine uptake mechanism. In the neurons, 6-OHDA autoxidizes into quinone (bottom) and generates hydroxyl radical (·OH; top panel A). The neurotoxicity of 6-OHDA is blocked by mazindol, which inhibits its uptake, or by selegiline, which inhibits its metabolism. The toxicity of 6-OHDA is decreased by deferoxamine (chelating iron) or by metallothionein (donating zinc, and scavenging free radicals). MAOB = monoamine oxidase B

Methylsalsolinol, a dopaminergic neurotoxin, increases in the cerebrospinal fluid of parkinsonian patients

Hao *et al.* [23] reported that the cerebrospinal fluid of patients with Parkinson's disease contained substance(s) that inhibited the growth and functions of dopaminergic neurons. Furthermore, selegiline, a monoamine oxidase B inhibitor, (0.125 to 0.250 μM) enhanced the number of tyrosine hydroxylase-positive neurons, augmented the high affinity uptake of dopamine, and averted the neurotoxic effects of cerebrospinal fluid of Parkinson's disease patients on rat mesencephalic neurons in culture [23]. Salsolinol, whose structure resembles

1-methyl-4-phenyl-1,2,3,6-tetrahydropyridine (MPTP), occurs in food in racemic form [24] whereas only the (R)enantiomers of both salsolinol and N-methyl salsolinol are detected in human brain [25]. The administration of N-methyl-(R)-salsolinol induces parkinsonism in rats [26, 27] and the concentration of methyl-(R)-salsolinol is increased in the cerebrospinal fluid of patients with Parkinson's disease [28].

Metallothionein protects against free radicals and oxidative stress caused by inflammation and tissue injury in the central nervous system

Metallothionein (MT) isoforms are low molecular weight zinc binding proteins consisting of 25 to 30% cysteine, with no aromatic amino acids or disulfide bonds. The areas of the brain containing high contents of zinc such as the retina, the pineal gland, and the hippocampus synthesize unique isoforms of MT on a continuous basis [29]. The four MT isoforms are thought to provide the neurons and glial elements with mechanisms to distribute, donate, and sequester zinc at presynaptic terminals; or buffer the excess zinc at synaptic junctions. A similar nucleotide and amino acid sequence has made it difficult to obtain cDNA probes and antibodies capable of distinguishing indisputably among MT isoforms. MT-I and MT-II isoforms are found in the brain and in the peripheral tissues; MT-III isoform, possessing an additional seven amino acids, is expressed mostly in the brain and to a very minute extent in the intestine and pancreas; whereas MT-IV isoform is found in tissues containing stratified squamous epithelial cells. Since MTs are expressed in neurons that sequester zinc in their synaptic vesicles, the regulation of the expression of MT isoforms is extremely important in terms of maintaining the steady-state level of zinc and controlling redox potentials [30–36].

Glutathione may participate in releasing zinc from metallothionein

Oxidative stress (or any other event shifting the balance of the glutathione/glutathione disulfide cycle to a more oxidized state) leads to zinc release from zinc-MT by glutathione disulfide and formation of thionein [37; and Figure 2]. Since the synthesis of MT is induced by interleukin-1 (IL-1) and tumor necrosis factors, MT is now considered an acute phase protein. Astrocytes do contain MT. In addition, they contain IL-1, -2, and -6, tumor necrosis factors, interferon, and granulocyte macrophage colony stimulating factor, which play a role in growth and differentiation of microglial and astroglial cells. Since the growth of astrocytes by IL-1 is associated with the production of IL-6, IL-8, tumor necrosis factor, nitric oxide synthase, and prostaglandin E_2, it is thought that the production of IL-1 by astrocytes takes place when injury or inflammation occurs in the central nervous system. IL-1 and heavy metal induce the synthesis of MT and its mRNA for a prolonged period of time. Therefore, Sawada *et al.* [38] believe that in inflammation, the level of IL-1 increases which, in turn, induces the synthesis of long-acting MT, and hence protects against free radicals and oxidative stress caused by inflammation and tissue injury in the central nervous system. Indeed lipopolysaccharide, which generates tumor necrosis factor, is more toxic in MT-null mice [39].

6-Hydroxydopamine-mediated induction of rat brain metallothionein I mRNA

6-Hydroxydopamine (6-OHDA), which utilizes the same uptake mechanism for dopamine and norepinephrine [40], was the first neurotoxin which has been shown to selectively destroy dopaminergic and adrenergic neurons [41]. Since 6-OHDA does not penetrate the blood-brain barrier, it must be given intracerebrally in order to produce a degeneration of brain dopaminergic or noradrenergic neurons. A quite selective and permanent destruction of dopaminergic nerve cells is produced by injection of 6-OHDA locally to the substantia nigra or slightly anterior to it. When given bilaterally to both substantia nigra of rodents, the animals became not only hypokinetic but also aphagic and adipsic, which require artificial tube feeding to keep the animals alive [42]. Due to this problem, a modified method was developed further by Ungerstedt [43]. We [44–46] have shown that administration of 6-OHDA generated hydroxyl radicals ($\cdot$OH; see Figure 1) and enhanced MT mRNA in some brain areas such as hippocampus, arcuate nucleus, choroid plexus, and granular layer of cerebellum, but not in the caudate putamen. The results of these studies are interpreted to suggest that zinc or MT are altered in conditions where oxidative stress has taken place. Moreover, it is proposed that areas of brain, such as striatum containing high concentrations of iron, but low levels of inducible MT are particularly vulnerable to oxidative stress [46].

Induction of metallothionein protects against neurotoxicity of 1-methyl-4-phenyl-1,2,3,6-tetrahydropyridine

Although endogenously produced neurotoxins have long been suspected of being involved in the pathogenesis of Parkinson's disease, little mechanistic evidence existed to support this concept until the neurotoxic effects of MPTP were found to induce a Parkinson-like syndrome. The discovery that MPTP produces a central nervous system pathology very similar to that observed with patients with Parkinson's disease has strengthened the endogenous neurotoxin hypothesis and provided a neuritic model for investigating the pathological process of Parkinson's disease (see Fig. 2). The first stage in the mechanism of action of MPTP appears to be its deamination by monoamine oxidase B (monoamine:oxygen oxidoreductase, EC 1.4.2.4), possibly in glial cells, which results in the formation of methyl-phenyl-tetrahydropyridinium ion (MPP$^+$). The MPP$^+$ is then selectively accumulated in dopaminergic terminals by way of a high affinity dopamine reuptake system. Neurons lacking dopamine transporter will not be affected. Once inside the nerve terminals, MPP$^+$ generates free radicals and interferes with mitochondrial respiration by inhibiting NADH CoQ reductase (Complex I) of the electron transfer complex, and consequently causing death of neurons. The activity of Complex I is reduced in the brain of Parkinson's disease patients [see 47 for a review and references]. Induction of MT protects against neurotoxic effects of MPTP [48].

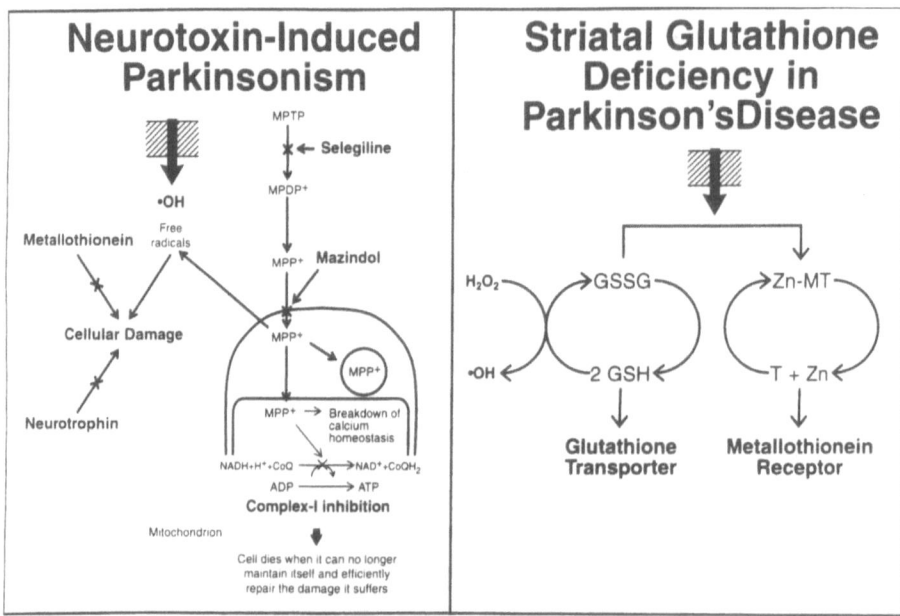

Figure 2. Neurotoxins such as 1-methyl-4-phenyl-1,2,3,6-tetrahydropyridine (MPTP) cause parkinsonism and inducers of metallothionein avert this neurotoxicity [48]. The striatal deficiency of glutathione in Parkinson's disease predisposes to oxidative stress which in part may be attenuated by induction of metallothionein isoforms. The roles of glutathione transporter and metallothionein receptors remains to be delineated. MPDP$^+$, 1-methyl-4-phenyl-2,3-dihydropyridinium ion; GSSG, glutathione disulfide; GSH, glutathione; T, thionein; MT, metallothionein; MPP$^+$, methyl-phenyl-tetrahydropyridinium ion; ADP, adenosine diphosphate; ATP, adenosine triphosphate; Zn, zinc.

Electron spin resonance spectroscopy verifies the free radical scavenging actions of metallothionein isoforms

In a group of experiments, free radicals were analyzed by an electron spin resonance spectrometer (JEOL JES-FEIXG, Tokyo, Japan). Manganese oxide was used as an internal standard and spin numbers were calculated using the ratio of signal height intensities of 2,2,6,6-tetramethyl-4-hydroxyl-piperidine-oxyl with known spin quantities as described by Hiramatsu et al. [44] and Kumari et al. [49]. The results of this study (Tab. 1) showed that although both MT-I and MT-II isoforms are capable of scavenging free radicals, MT-I appeared to be a superior scavenger of superoxide and 1,1-diphenyl-2-picrylhydrazyl radicals [50; see also Figure 2].

Conclusions

MPTP becomes metabolized by monoamine oxidase B to MPP$^+$, which generates free radicals, inhibits complex I, and causes cellular damage in the striatum. The neurotoxic effects of MPTP

Table 1. Free radical scavenging effects of metallothionein I and II isoforms: an electron resonance spectroscopic study

Free radicals studies	IC_{50} Values	
	Metallothionein I (mM)	Metallothionein II (mM)
DPPH radicals	0.025	0.14
Hydroxyl radicals (·OH)	0.145	0.14
Superoxide (·O_2)	0.16	0.40
Reactive oxygen species (containing singlet oxygen, superoxide and hydroxyl radicals)	0.004	0.0014

DPPH = 1,1-diphenyl-2-picrylhydrazyl. Data from: M. Ebadi, M.V. Ramana Kumari, and M. Hiramatsu, Soc. Neurosci. Abstract 23: 119.13, 1997.

can be blocked by selegiline inhibiting monoamine oxidase B, by mazindol preventing the uptake of MPP^+ into dopaminergic terminals, by metallothionein scavenging hydroxyl radicals, or by neurotrophin [51], rescuing and repairing dopaminergic neurons (Fig. 2). The level of glutathione is reduced in the striatum of patients with Parkinson's disease [16, 17] which hinders the ability of neurons to detoxify H_2O_2 and hence they generate hydroxyl radicals (·OH). Oxidative stress (or any other event shifting the balance of the glutathione/glutathione disulfide cycle to a more oxidized state) leads to zinc release from zinc-MT by glutathione disulfide and formation of thionein. The released zinc is thought to be made available for zinc-dependent processes [37]. The striatal status of glutathione transporter [52, 53] and MT receptor [54] in Parkinson's disease remain to be delineated.

Acknowledgments
The authors gratefully acknowledge the excellent secretarial skills of Lori Ann Clapper in typing the manuscript. This study was completed in part by a research grant provided by USPHS NS34566-04 .

References

1. Pfeiffer RF, Ebadi M (1994) Pharmacological management of Parkinson's disease. *In*: WJ Weiner, A Cohen (eds): *Interdisciplinary Treatment of Parkinson's Disease*. Demos Publications, 9–38.
2. Jellinger K (1986) An overview of morphological changes in Parkinson's disease. *Adv Neurol* 45: 1–16.
3. Jellinger K (1987) The pathology of parkinsonism. *In*: CD Marsden, S Fahn (eds): *Movement Disorders*. Butterworths, London, 124–165.
4. Hornykiewicz O (1988) Neurochemical pathology and the etiology of Parkinson's disease: Basic facts and hypothetical possibilities. *Mt Sinai J Med* 55: 11–20.
5. Forno L (1986) The Lewy body in Parkinson's disease. *Adv. Neurol* 45: 35–43.
6. Louis ED, Goldman JE, Powers JM, Fahn S (1995) Parkinsonian features of eight pathologically diagnosed cases of diffuse Lewy body disease. *Mov Disorders* 10: 188–194.
7. Dexter DT, Carter CJ, Wells FR, Javoy-Agid F, Agid Y, Lees A, Jenner P, Marsden CD (1989) Basal lipid peroxidation in substantia nigra is increased in Parkinson's disease. *J Neurochem* 52: 381–389.
8. Dexter DT, Wells FR, Lees AJ, Agid F, Agid Y, Jenner P, Marsden CD (1989) Increased nigra iron content and alterations in other metal ions occurring in brain in Parkinson's disease. *J Neurochem* 52: 1830–1836.
9. Spina MB, Cohen G (1989) Dopamine turnover and glutathione oxidation. Implications for Parkinson's dis-

ease. *Proc Natl Acad Sci USA* 86: 1398–1400.

10. Goeth ME, Freyberger A, Riederer P (1990) Oxidative stress: A role in the pathogenesis of Parkinson's disease. *J Neural Transm* 29: 241–249.

11. De Erausquin GA, Costa E, Hanbauer I (1994) Calcium homeostasis, free radical formation, and trophic factor dependence mechanisms in Parkinson's disease. *Pharmacol Rev* 46: 467–482.

12. Yoshikawa T (1993) Free radicals and their scavengers in Parkinson's disease. *Eur Neurol* 33: 60–68.

13. Perry TL, Young VW (1986) Idiopathic Parkinson's disease, progressive supranuclear palsy and glutathione metabolism in the substantia nigra of patients. *Neurosci Lett* 67: 269–274.

14. Perry TL, Godin DV, Hansen S (1982) Parkinson's disease: A disorder due to nigral glutathione deficiency? *Neurosci Lett* 33: 305–310.

15. Sofic E, Lange KW, Jellinger K, Riederer P (1992) Reduced and oxidized glutathione in the substantia nigra of patients with Parkinson's disease. *Neurosci Lett* 142: 128–130.

16. Riederer P, Konradi C, Hebenstreit G, Youdim MBH (1989) Neurochemical perspectives to the function of monoamine oxidase. *Acta Neurol Scand* 80: 41–45.

17. Riederer P, Sofic E, Rausch WD, Schmidt B, Reynolds GP, Jellinger K, Youdim MBH (1989) Transition metals, ferritin, glutathione, and ascorbic acid in Parkinsonian brain. *J Neurochem* 52: 515–520.

18. Dexter DT, Wells FR, Agid F, Agid Y, Lees AJ, Jenner P, Marsden CD (1987) Increased nigral iron content in postmortem Parkinsonian brain. *Lancet* 2: 1219–1220.

19. Dexter DT, Carayon A, Javoy-Agid F (1991) Alterations in the levels of iron, ferritin and other trace metals in Parkinson's disease and other neurodegenerative diseases affecting the basal ganglia. *Brain* 114: 1953–1975.

20. Saggu H, Cooksey J, Dexter D (1989) A selective increase in particulate superoxide dismutase activity in Parkinsonian-substantia nigra. *J Neurochem* 53: 692–697.

21. Youdim MBH, Ben-Shachar D, Riederer P (1989) Is Parkinson's disease a progressive siderosis of substantia nigra resulting in iron and melanin induced neurodegeneration? *Acta Neurol Scand* 26: 47–54.

22. Dexter DT, Carayon A, Vidailhet M (1990) Decreased ferritin levels in brain in Parkinson's disease. *J Neurochem* 55: 16–20.

23. Hao R, Ebadi M, Pfeiffer RF (1995) Selegiline protects dopaminergic neurons in culture from toxic factor(s) present in the cerebrospinal fluid of patients with Parkinson's disease. *Neurosci Lett* 200: 77–80.

24. Strolin Benedetti M, Bellotti V, Pianezola E et al (1989) Ratio of the R and S enantiomers of salsolinol in food an human urine. *J Neural Transm* 77: 47–53.

25. Deng Y, Maruyama W, Dostert P et al (1995) Determination of (R)- and (S)-enantiomer of salsolinol and N-methylsalsolinol by use of a chiral high-performance liquid chromatographic column. *J Chromatogr B Biomed Appl* 670: 47–54.

26. Naoi M, Maruyama W, Dostert P et al (1996) Dopamine-derived endogenous (1(R),2(N)-dimethyl-6,7-dihydroxy-1,2,3,4-tetrahydroisoquinoline, N-methyl-(R)-salsolinol, induced parkinsonism in rats: Biochemical, pathological and behavioral studies. *Brain Res* 709: 285–295.

27. Naoi M, Maruyama W, Dostert P, Hashizume Y (1996) Animal model of Parkinson's disease induced by naturally-occurring 1(R),2(N)-dimethyl-6,7-dihydroxy-1,2,3,4-tetrahydroisoquinoline. *Biogenic Amines* 12: 135–147.

28. Maruyama W, Abe T, Tohgi H, Dostert P, Naoi M (1996) A dopaminergic neurotoxin, (R)-N-methylsalsolinol, increases in parkinsonian cerebrospinal fluid. *Ann Neurol* 40: 119–122.

29. Ebadi M, Perini F, Mountjoy K, Garvey JS (1996) Amino acid composition, immunoreactivity, sequence analysis, and function of bovine hippocampal metallothionein isoforms. *J Neurochem* 66: 2121–2127.

30. Ebadi M (1991) Metallothionein and other zinc-binding proteins in brain. *Methods Enzymol* 205: 363–387.

31. Ebadi M, Iversen PL, Hao R, Cerutis DR, Rojas P, Happe HK, Murrin LC, Pfeiffer RF (1995) Expression and regulation of brain metallothionein. *Neurochem Int* 27: 1–22.

32. Erickson JC, Masters BA, Kelly EJ, Brinster R, Palmiter RD (1995) Expression of human metallothionein-III in transgenic mice. *Neurochem Int* 27: 35–41.

33. Zheng H, Berman NEJ, Klaassen CD (1995) Chemical modulation of metallothionein I and III mRNA in mouse brain. *Neurochem Int* 27: 43–58.

34. Masters BA, Kelly EJ, Quaife CJ, Brinster RL, Palmiter RD (1994) Targeted disruption of metallothionein-I and -II genes increases sensitivity to cadmium. *Proc Natl Acad Sci USA* 91: 584–588.

35. Masters BA, Quaife CJ, Erickson JC, Kelly EJ, Froelick GJ, Zambrowicz BP, Brinster RL, Palmiter RD (1994) Metallothionein III is expressed in neurons that sequester zinc in synaptic vesicles. *J Neurosci* 14: 5844–5857.

36. Aschner M (1996) The functional significance in brain metallothioneins. *FASEB J* 10: 1129–1136.

37. Maret W (1994) Oxidative metal release from metallothionein via zinc-thiol/disulfide interchange. *Proc Natl Acad Sci USA* 91: 237–241.

38. Sawada J, Kikuchi Y, Shibutani M, Mitsumori K, Inoue K, Kasahara T (1994) Induction of metallothionein in astrocytes by cytokines and heavy metals. *Biol Signals* 3: 157–168.

39. Rofe AM, Philcox JC, Coyle P (1996) Trace metal, acute phase and metabolic response to endotoxin in metallothionein-null mice. *Biochem J* 314: 793–797.

40. Sachs C, Jonsson G (1975) Mechanism of action of 6-hydroxydopamine. *Biochem Pharmacol* 24: 1–8.

41. Kaakkola S, Teräväinen H (1990) Animal models of parkinsonism. *Pharmacol Toxicol* 67: 95–100.

42. Ungerstedt U (1971) Adipsia and aphagia after 6-hydroxydopamine induced degeneration of the nigro-stri-

atal dopamine system. *Acta Physiol Scand* 82: 95–122.
43. Ungerstedt U (1971) Postsynaptic supersensitivity after 6-hydroxydopamine induced degeneration of the nigro-striatal dopamine system. *Acta Physiol Scand* 82: 69–93.
44. Hiramatsu M, Kohno M, Mori A, Shiraga H, Pfeiffer RF, Ebadi M (1994) An ESR study of 6-hydroxy-dopamine: Generated hydroxy radicals and superoxide anions in brain. *Neurosciences* 20: 129–138.
45. Ebadi M, Hiramatsu M, Rojas P (1996) Zinc and metallothionein isoforms in free radical-mediated striatal damage in Parkinson's disease. *In*: L Packer, M Hiramatsu, T Yoshikawa (eds): *Free Radicals in Brain Physiology and Disorders*. Academic Press, New York, 141–156.
46. Rojas P, Cerutis DR, Happe HK, Murrin LC, Hao R, Pfeiffer RF, Ebadi M (1996) 6-Hydroxydopamine-mediated induction of rat brain metallothionein I mRNA. *NeuroToxicology* 17: 323–334.
47. Ebadi M, Srinivasan SK, Baxi M (1996) Oxidative stress and antioxidant therapy in Parkinson's disease. *Prog Neurobiol* 48: 1–19.
48. Rojas P, Ríos C (1997) Metallothionein inducers protect against 1-methyl-4-phenyl-1,2,3,6-tetrahydropyridine neurotoxicity in mice. *Neurochem Res* 22: 17–22.
49. Kumari MVR, Yoneda T, Hiramatsu M (1996) Scavenging activity of "βCATECHIN" on reactive oxygen species generated by photosensitization of riboflavin. *Biochem Mol Biol Int* 38: 1163–1170.
50. Ramana Kumari MV, Hiramatsu M, Ebadi M (1998) Free radical scavenging actions of metallothionein isoforms I and II. *Free Rad Res* 29: 93–101.
51. Ebadi M, Bashir RM, Heidrick ML et al (1997) Neurotrophins and their receptors in nerve injury and repair. *Neurochem. Int.* 30: 347–374.
52. Kannan R, Kuhlenkamp JF, Ookhtens M, Kaplowitz N (1992) Transport of glutathione at blood-brain barrier of the rat: Inhibition by glutathione analogs and age-dependence. *J Pharmacol Exp Ther* 263: 964–970.
53. Fernández-Checa JC, Kannan R, Garcia-Ruiz C, Ookhtens M, Yi J-R (1996) GSH transporters: Molecular characterization and role in GSH homeostasis. *Biol Chem Hoppe-Seyler* 377: 267–273.
54. El Refaey H, Ebadi M, Kuszynski CA, Sweeney J, Hamada FM, Hamed A (1997) Identification of metallothionein receptors in human astrocytes. *Neurosci Lett* 231: 131–134.

Metallothionein IV
C. Klaassen (ed.)
© 1999 Birkhäuser Verlag Basel/Switzerland

Metallothionein-1,2 and heme oxygenase-1 are expressed in damaged brain regions following chemically-induced seizures

Mitchell R. Emerson[1], Robert S. Cross[1], Shaohua Jin[1], Fred E. Samson[2] and Thomas L. Pazdernik[1]

[1]*Department of Pharmacology, Toxicology, and Therapeutics and* [2]*Ralph L. Smith Research Center, University of Kansas Medical Center, Kansas City, KS 66160-7417 USA*

Introduction

Chemicals that induce limbic hyperactivity which generalizes into a status epilepticus cause severe neurodegeneration in limbic-related structures (piriform cortex, amygdala, and CA_1 and CA_3 regions of the hippocampus) [1–5]. Two examples are soman (O-1,2,2-trimethylpropyl methylphosphono-fluoridate), an irreversible acetylcholinesterase inhibitor, and kainic acid (KA), a glutamate analog. KA-induced seizures occur first in the ventral hippocampus and lateral septum; the areas where increases in local cerebral glucose use (LCGU) occur first after injection of KA. They are also the only areas where increases in LCGU occur when the spread of electrical activity from the initiation sites is prevented by diazepam pretreatment [6]. In contrast, when the spread of electrical activity in soman exposed rats is prevented by diazepam pretreatment, increases in LCGU, occur in the globus pallidus, ventral pallidum and substantia nigra [7]. Although the seizure-initiation sites for these two agents are different, eventually the spread of seizure activity via enhanced glutamate release within limbic-related structures leads to hyperactivity which generalizes throughout the brain causing a full-blown status epilepticus. Furthermore, the neuropathology associated with both chemical agents are similar [1–5]. LCGU dramatically increases in many brain regions, including limbic-related areas during the seizure phase induced by either soman or KA [1–5]. The increased neuronal activity causes a calcium stress leading to production of reactive oxygen species which contributes to the neuropathology [8].

We found that KA-induced seizures cause simultaneous induction of metallothionein-1 (MT-1) and heme oxygenase-1 (HO-1) mRNAs, indicating an oxidative stress during the seizure phase [9]. Thus, we hypothesize that seizures increase stress-related proteins in response to biochemical changes associated with the hyperexcited state and these stress proteins can either contribute to or counteract cell injury events [10].

MTs are a family of low molecular weight proteins (6–7 kD) with high capacity to bind metals (Zn, Cu) [11] and are principal regulators of metal disposition and Zn metalloproteins [12]. In the brain, MTs are important in regulating metals that influence glutamatergic and GABAergic neurotransmission [13]. The brain has three isoforms, MT-1, MT-2 (coordinately regulated) and MT-3. Expression of MT-1,2 is regulated by metals (Zn, Cd), glucocorticoids, endotoxin, and oxidative stress whereas MT-3 is not [13, 14]. MT-3 knockout mice are more susceptible to KA-induced seizure-related neurodegeneration [15]. Therefore, MT may be important in protection against neurodegeneration associated with status epilepticus.

Two isozymes of heme oxygenase, HO-1 and HO-2 convert heme to biliverdin which is subsequently reduced to bilirubin by biliverdin reductase [16]. HO-1, is a stress-related protein induced by heat shock, heme, H_2O_2, hyperthermia, oxidative stress, traumatic brain injury, and glutathione depletion [17–19]. Bilirubin, the end-product of heme catabolism, in micromolar concentrations, *in vitro*, scavenges peroxyl radicals [20] and *in vivo* protects against oxidative stress due to $CoCl_2$ administration [21]. Thus, by converting a pro-oxidant (heme) to antioxidants (biliverdin, bilirubin), HO-1 may provide cytoprotection against free radical damage.

The regional protein levels of MT-1,2 and HO-1 following soman or KA-induced status epilepticus are described here and discussed in relationship to adaptive responses to cell injury.

Methods

Animals

Adult male Wistar rats weighing between 250–350 g at the time of use, were purchased from Harlan Sprague Dawley, Inc., Indianapolis, IN and were kept under standard conditions with access to food and water *ad libitum*. All procedures involving animals were in accordance with the NIH Guide for the Care and Use of Laboratory Animals and were approved by the Institutional Animal Care and Use Committee (IACUC).

Experimental design

Animals were injected with a seizurogenic dose of either soman (85–90 µg/kg i.m.) or KA (14 mg/kg i.p.) dissolved in 0.9% saline. Control animals received 0.9% saline only. All subjects were rated on the severity of symptoms and only responders with tonic-clonic seizures were used. At various time points following injection, the animals were anesthetized with halothane, decapitated, and the brains removed for analysis by northern and/or western blotting.

Northern blotting

Total RNA isolation was performed using TRIzol reagent (Gibco, Grand Island, NY) according to the manufacturer's instructions. RNA preparation, agarose gel electrophoresis, transfer, and hybridization were performed according to Choudhuri, *et. al.* [14]. Hybridization was performed using a DNA oligonucleotide probe specific for MT-I (5'-TAG TAA ACT GGG TGG AGG TG-3') complementary to the 3'-untranslated region (between +949- and +968-bp position with respect to the transition start site) [11].

Western blotting

Brain tissue homogenates were prepared and clarified according to Mizzen, *et. al.* [22]. Total protein was determined by the Lowry method (Sigma, St. Louis, Mo). Samples were subjected to sodium dodecyl sulfate polyacrylamide gel electrophoresis (SDS-PAGE) using 18% gels (MT-1,2) or 15% gels (HO-1) [23]. For MT-1,2, electrophoretic transfer of the separated protein to nitrocellulose membranes at 40 V for one hour in 10 mM CAPS (3-[cyclohexylamino]-1-propanesulfonic acid), 2 mM $CaCl_2$ and 10% methanol, pH = 10.8 and subsequent glutaraldehyde treatment were performed according to Mizzen, *et. al.* [22]. For HO-1, transfer was performed similarly except at 100 V for one hour in 25 mM Tris, 192 mM glycine, 20% v/v methanol, pH 8.3. Duplicate gels were stained with Coomassie Blue to ensure equal lane loading.

BSA-blocked nitrocellulose membranes were incubated at 4 °C overnight with primary antibody (mouse monoclonal antibody to polymerized equine renal MT-1,2 (Dako, Cardinteria, CA) or a rabbit polyclonal antibody to HO-1 (Stressgen, Victoria, BC, Canada)]. The secondary antibody was either goat anti-mouse or goat anti-rabbit IgG conjugated to alkaline phosphatase (Bio-Rad, Hercules, CA) and the Immun-Star Chemiluminescent Protein Detection System (Bio-Rad, Hercules, CA) was used to develop the blots, with the signal captured on x-ray film.

Results

Responding rats injected with KA (14 mg/kg i.p.) developed a reproducible pattern of symptomatic behaviors. Initially, activity was depressed (a catatonic-like state), then approximately 30 min later, wet-dog shakes and head-bobbing occurred which progressed to limbic motor seizures by one hour post-injection. Limbic seizures are characterized by masticatory movements, forepaw tremors, rearing, and loss of postural control culminating in a full-blown status epilepticus that may persist for hours. Soman (85–90 µg/kg i.m.) caused a behavioral pattern similar to KA but with an earlier onset of symptoms: within 5–10 min of injection strong tremors occurred followed by a full-blown status epilepticus.

Soman-induced seizures caused an induction of MT-1 mRNA in the piriform cortex (Fig. 1) similar to what we previously reported for KA [9]. However in the hippocampus, a biphasic induction of MT-1 mRNA occurred after soman; a 5-fold increase at 2 h, a return to basal levels, and then a secondary 4-fold induction at 24 h (Fig. 1).

The basal protein levels of MT-1,2 were lowest in piriform cortex, whereas the hippocampus had the lowest constitutive level of HO-1 (Fig. 2). MT-1,2 levels doubled in the piriform cortex by 48 h after exposure to either soman or KA (Fig. 3). Both agents caused approximately a doubling of MT-1,2 in the hippocampus at 48 h post-injection (Fig. 4). There was a 2-fold increase in HO-1 in piriform cortex at 24 h (soman) and at 48 h (KA). HO-1 changed little in hippocampus after exposure to either soman or KA with the exception of slight decreases below basal levels at 48 h post-injection (not shown).

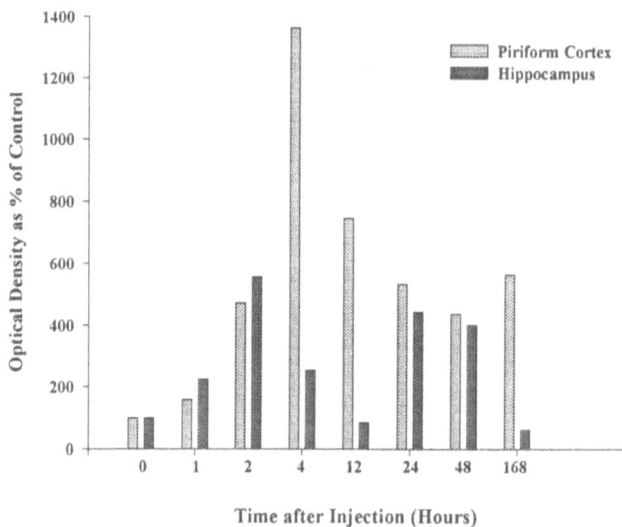

Figure 1. Quantitation of northern blot analysis assessing mRNA levels for MT-1 in piriform cortex and hippocampus at various time points following soman exposure.

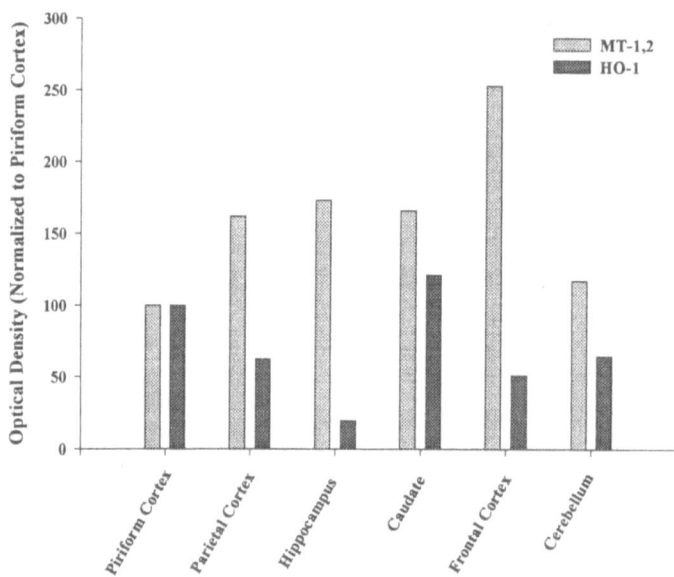

Figure 2. Quantitation of western blot analysis showing basal levels of MT-1,2 and HO-1 in various brain regions. Data is normalized to levels in piriform cortex.

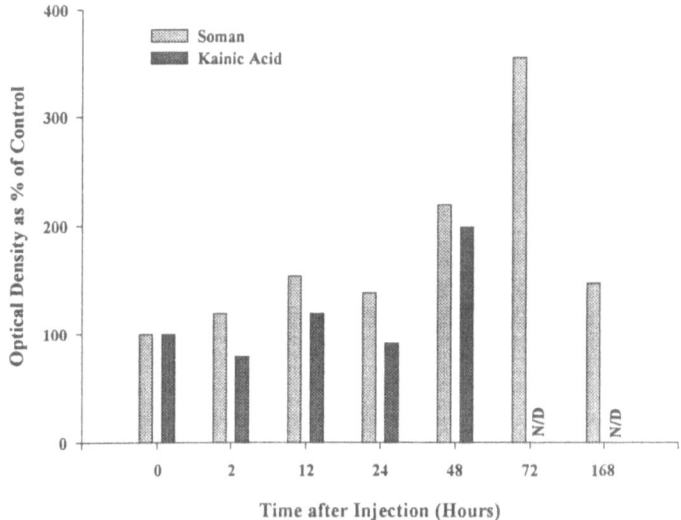

Figure 3. Quantitation of western blot analysis assessing MT-1,2 levels in piriform cortex following exposure to either soman or KA. N/D = time point was not evaluated.

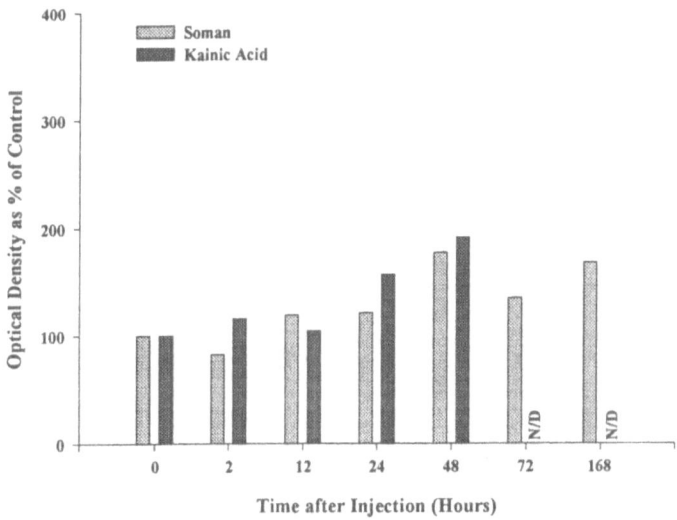

Figure 4. Quantitation of western blot analysis assessing MT-1,2 levels in hippocampus following exposure to either soman or KA. N/D = time point was not evaluated.

Discussion

These studies show that status epilepticus induced by either soman or KA increases the level of MT-1,2 and HO-1 in the brain. We speculate that these stress proteins are increased in response to zinc release and/or oxidative stress associated with status epilepticus. However, if these stress related proteins are neuroprotective their increase is too little and too late to completely prevent neurodegeneration in vulnerable brain regions.

Glutamate, the major excitatory neurotransmitter in the piriform cortex and hippocampus, is colocalized/coreleased with Zn, which in turn can regulate the excitatory actions of glutamate. The highest levels of *vesicular* Zn are found in limbic-related areas [24] and, interestingly MT-3 mRNA is also highest in these regions [12], suggesting a possible functional role for MTs in neuronal Zn disposition and release. MT-1 and HO-1 mRNAs are dramatically increased in the piriform cortex following KA [9] or soman (Fig. 1) administration. The biphasic induction of MT-1 mRNA in the hippocampus parallels the extracellular glutamate levels in the hippocampal CA_1 region following soman exposure (i.e. a transient increase of glutamate release at 10 min followed by a second increase at 50 min) [1]. The glutamate-driven process initiates a cascade of events that causes a dramatic increase in transcription of MT-1 mRNA, likely due to Zn release and possibly by oxidative stress [9, 13], and HO-1 mRNA more likely via oxidative stress [9].

The enhanced expression of stress-related proteins is an adaptive response to seizures, as previously reported [10]. Indeed, Gass, *et. al.* [25], report a correlation between the intensity of limbic seizures (KA-induced) and the induction of HSP72 in highly vulnerable neurons. Likewise, soman induces transcription of c-fos and HSP70 mRNAs in piriform cortex and hippocampus, but HSP72 protein did not increase, indicating a lack of HSP70 gene translation in areas of extensive neurodegeneration [26]. Here we show that the mRNAs for MT-1 and HO-1 increase dramatically during seizures whereas the protein levels only increase 2-fold and are not sufficient to provide protection against neurodegeneration. Recent *in vivo* and *in vitro* evidence with Hmox1-deficient mice indicates that HO-1 is crutial for iron homeostasis and HO-1 upregulation serves as an adaptive mechanism to protect cells from oxidative damage during stress [27, 28].

KA or soman exposure initiates an excitotoxic process driven by glutamate (zinc is also released) which results in conspicuous neuropathology in limbic-related brain regions. The increase in both MT-1,2 and HO-1 in limbic regions are biochemical markers of excitotoxic injury but whether or not they function in endogenous neuroprotective/reparative processes has yet to be definitively shown. The status epilepticus induced by soman or KA injection is associated with increases in mRNAs that are considerably greater than the increases in protein levels of MT-1,2 and HO-1. Since some regions (e.g. piriform cortex) have extensive damage the increases in protein levels are too little and too late to provide total neuroprotection.

Acknowledgements
Supported by DAMD 17-94C 4045 and DAAH04-95-0217.

References

1. Lallement G, Carpentier P, Collet A, Pernot-Marino I, Baubichon D and Blanchet G (1991) Effects of Soman-Induced Seizures on Different Extracellular Amino Acid Levels and on Glutamate Uptake in Rat Hippocampus. *Brain Res* 563: 234–240.
2. Sperk G (1994) Kainic Acid Seizures in the Rat. *Prog Neurobiol* 42: 1–32.
3. Pazdernik TL, Cross RS, Giesler M, Samson FE, Nelson SR (1985) Changes in Local Cerebral Glucose Utilization Induced by Convulsants. *Neuroscience* 14: 823–835.
4. Pazdernik TL, Cross RS, Geisler M, Nelson S, Samson FE, McDonough Jr J (1985) Delayed effect of soman: glucose use and pathology. *Neurotoxicology* 6: 61–70.
5. McDonough JH, Hackley BE, Cross R, Samson F and Nelson S (1983) Brain Regional Glucose Use During Soman-Induced Seizures. *Neurotoxicology* 4: 203–210.
6. Chastain JE, Samson F, Nelson S and Pazdernik TL (1987) Attenuation of Cerebral Glucose Use in Kainic Acid-Treated Rats by Diazepam. *Eur J Pharmacol* 142: 215–224.
7. Pazdernik T, Samson F and Nelson SR (1985) Soman's Actions on the Brain: A 2-[^{14}C]- deoxyglucose Study. *In*: Second International Symposium on the Synthesis and Application of Isotopically Labelled Compounds. RR Mucino (Ed.), Elsevier Science Publishers, Amsterdam, 415–420.
8. Pazdernik TL, Layton M, Nelson SR and Samson FE (1992) The Osmotic/Calcium Stress Theory of Brain Damage: Are Free Radicals Involved? *Neurochem Res* 17: 11–21.
9. Dalton T, Pazdernik TL, Wagner J, Samson F and Andrews GK (1995) Temporalspatial Patterns of Expression of Metallothionein-I and -III and Other Stress Related Genes in Rat Brain After Kainic Acid-Induced Seizures. *Neurochem Int* 27: 59–71.
10. Massa SM, Swanson RA and Sharp FR (1996) The Stress Gene Response in Brain. *Cerebrovasc and Brain Metab Rev* 8: 95–158.
11. Andersen RD, Birren BW, Taplitz SJ and Herschman HR (1986) Rat Metallothionein-1 Structural Gene and Three Pseudogenes, One of Which Contains 5'-Regulatory Sequences. *Mol Cell Biol* 6: 302–314.
12. Aschner M (1996) The Functional Significance of Brain Metallothioneins. *FASEB J* 10: 1129–1136.
13. Aschner M, Cherian MG, Klaassen CD, Palmiter RD, Erickson JC and Bush AI (1997) Metallothioneins in Brain-The Role in Physiology and Pathology. *Toxicol Appl Pharmacol* 142: 229–242.
14. Choudhuri S, McKim JM and Klaassen CD (1993) Differential Expression of the Metallothionein Gene in Liver and Brain of Mice and Rats. *Toxicol Appl Pharmacol* 119: 1–10.
15. Erickson JC, Hollopeter G, Thomas SA, Froelick GJ and Palmiter RD (1997) Disruption of the Metallothionein-III Gene in Mice: Analysis of Brain Zinc, Behavior, and Neuron Vulnerability to Metals, Aging, and Seizures. *J Neuroscience* 17: 1271–1281.
16. Tenhunen R, Marver HS and Schmid R (1968) The Enzymatic Conversion of Heme to Bilirubin by Microsomal Heme Oxygenase. *Proc Natl Acad Sci USA* 61: 748–755.
17. Maines MD (1988) Heme Oxygenase: Function, Multiplicity, Regulatory Mechanisms, and Clinical Applications. *FASEB J* 2: 2557–2568.
18. Ewing JF and Maines MD (1993) Glutathione Depletion Induces Heme Oxygenase-1 (HSP32) mRNA and Protein in Rat Brain. *J Neurochem* 60: 1512–1519.
19. Ewing JF, Haber SN and Maines MD (1992) Normal and Heat-Induced Patterns of Expression of Heme Oxygenase-1 (HSP32) in Rat Brain:Hyperthermia Causes Rapid Induction of mRNA and Protein. *J Neurochem* 58: 1140–1149.
20. Stocker R, Yamamoto Y, McDonagh AF, Glazer AN and Ames BN (1987) Bilirubin is an Antioxidant of Possible Physiological Importance. *Science* 235: 1043–1046.
21. Llesuy SF and Tomaro ML (1994) Heme Oxygenase and Oxidative Stress. Evidence of Involvement of Bilirubin as Physiological Protector Against Oxidative Damage. *Biochim Biophys Acta* 1223: 9–14.
22. Mizzen CA, Cartel NJ, Yu WH, Fraser PE and McLachlan DR (1996) Sensitive Detection of Metallothioneins-1, -2, and -3 in Tissue Homogenates by Immunoblotting: a Method for Enhanced Membrane Transfer and Retention. *J Biochem Biophys Meth* 32: 77–83.
23. Laemmli UK (1970) Cleavage of Structural Proteins During the Assembly of the Head of Bacteriophage T4. *Nature* 227: 680–685.
24. Frederickson CJ (1989) Neurobiology or Zinc and Zinc-Containing Neurons. *Int Rev Neurobiol* 31: 145–238.
25. Gass P, Prior P and Kiessling M (1995) Correlation Between Seizure Intensity and Stress Protein Expression After Limbic Epilepsy in the Rat Brain. *Neuroscience* 65: 27–36.
26. Baille-Le Crom V, Collombet JM, Burckhart MF, Foquin A, Pernot-Marino I, Rondouin G and Lallement G (1996) Time Course and Regional Expression of C-FOS and HSP70 in Hippocampus and Piriform Cortex Following Soman-Induced Seizures. *J Neurosci Res* 45: 513–524.
27. Poss KD and Tonegawa S (1997) Heme Oxygenase 1 is Required for Mammalian Iron Reutilization. *Proc Natl Acad Sci USA* 94: 10919–10924.
28. Poss KD and Tonegawa S (1997) Reduced Stress Defense in Heme Oxygenase 1-Deficient Cells. *Proc Natl Acad Sci USA* 94: 10925–10930.

Metallothionein IV
C. Klaassen (ed.)
© 1999 Birkhäuser Verlag Basel/Switzerland

Molecular cloning and expression of the metallothionein (MT)-I and MT-III in brain lesions of dogs

Akinori Shimada[1], Seiji Kojima[1], Takashi Uemura[1], Yoshiaki Yamano[2], Kinji Kobayashi[3], Takehito Morita[1] and Takashi Umemura[1]

[1]Department of Veterinary Pathology, Tottori University, Tottori 680, Japan
[2]Department of Metabolic Biochemistry, Tottori University, Tottori 680, Japan
[3]Safety Research Laboratory, Tanabe Seiyaku Co. Ltd., Yodogawa-ku, Osaka 532, Japan

Introduction

Following the discovery of brain-specific MT-III [1–3], the implications of MTs in brain pathophysiology have been strengthened. MT-III mRNA may or may not be down-regulated in patients with Alzheimer's disease [4, 5]. Limited numbers of studies have revealed that MT-III mRNA and MT-III immunostaining are localized predominantly to the astrocytes of human and rat brains [2, 3, 6, 7]. In contrast, MT-III mRNA was demonstrated exclusively in neurons of normal mouse brains [8]. Thus, physiological function(s) as well as cellular localization of MT-III is still to be determined.

Dogs are known to share a variety of pathologic brain changes with humans including age-related changes [9]. In an attempt to better understand the pathophysiological roles of MT-I &-II and MT-III, we studied the localization of MT-I &-II and MT-III at both mRNA and protein level in dog brains which contained a variety of lesions.

Results and discussion

We have isolated and determined the complete nucleotide sequence of canine metallothionein-I (MT-I) [10] and MT-III cDNA (Fig. 1). The predicted amino acid sequence of the canine MT-III showed a high homology (93%, 87% identity respectively) to that of human and mouse MT-III [1, 2, 5]. The canine MT-III had two insertions relative to known mammalian MT-I and MT-II: a threonine after the fourth amino acid and a block of six amino acids near the carboxyl terminus (Fig. 1). Following synthesis of a polypeptide specific for canine MT-III (Cys-Lys-Gly-Gly-Glu-Gly-Thr-Glu-Ala-Glu-Ala-Glu-Lys), we have generated a polyclonal antibody to canine MT-III. Expression of the canine MT-III mRNA was found exclusively in the central nervous system, where neurons in the olfactory bulb, hippocampus, thalamus and cerebral cortex showed predominant signals (Fig. 2a). The findings were also supported by the MT-III immunohistochemistry (Fig. 2b).

Double-immunohistochemistry and *in situ* hybridization techniques demonstrated expression of MT-III in the hypertrophic astrocytes in most brain lesions examined regardless of the

```
  1 AAAAGCCTCGCCACCTGCTGCCACTAGTGCATCCAGTTGC    40
 41 CCGGAGCAGCCAATTCACCTCCTCCAGCGGCGCCTCCTCT    80
 81 GGACATGGACCCTGAGACCTGCCCTTGCCCTACTGGTGGT   120
121 TCCTGTACCTGCGATGGCTCCTGCAAGTGTGAGGGATGCA   160
161 AATGCACCTCCTGCAAGAAGAGCTGCTGCTCCTGCTGCCC   200
201 TGCAGAATGTGAGAAATGCGCCAAGGATTGTGTGTGCAAA   240
241 GGCGGAGAAGGGACAGAAGCTGAGGCCGAGAAGTGTAGCTG   280
281 CTGCCAGTGAGGACATGCGCCTCCCTATGAACCAGGTGTG   320
321 AAGAGTGCCGGGTGGCACTCTCCTGTGCTGAGGCTTGTC   360
361 TGTGTGTCCCCTTCCTCTGCCAGCCCCTGGCAAGTGACAA   400
401 TAAATCCTATGAATGGCAAAAAAAAAAAAAAAAAAAAA     436
```

Canine MT-I
  MDPDCSCSTGGSCTCAGSCKCKECKCTSCKKSCCSCCPVGCAKCAQGCICKG
  ......ASDKCSCCA
Canine MT-III
  MDPETCPCPTGGSCTCDGSCKCEGCKCTSCKKSCCSCCPAECEKCAKDCVCKG
  GEGTEAEAEKCSCCQ

Figure 1. Nucleotide sequence of canine MT-III cDNA and deduced amino acid sequence of the canine MT-I and MT-III. Termination codon is in the box (□). Polyadenylation signal is underlined. Two insertions are double underlined.

type of lesions (Figs 3a and 3b); the lesions included acute to chronic changes. Confocal laser scanning microscopy also showed that MT-I and MT-III were co-localized in hypertrophic astrocytes in the lesions (data not shown).

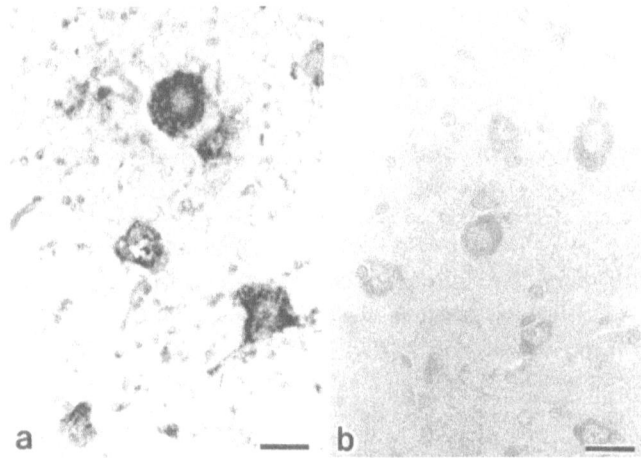

Figure 2 Predominant signals of MT-III mRNA (a) and MT-III immunolabeling (b) in neurons in the thalamus. a: 15-year-old dog, DIG-labeled antisense RNA probe (noncoding region of canine MT-III cDNA), bar = 40 μm. b: 3-week-old dog, MT-III immunohistochemistry (polyclonal antibody to canine MT-III), bar = 25 μm.

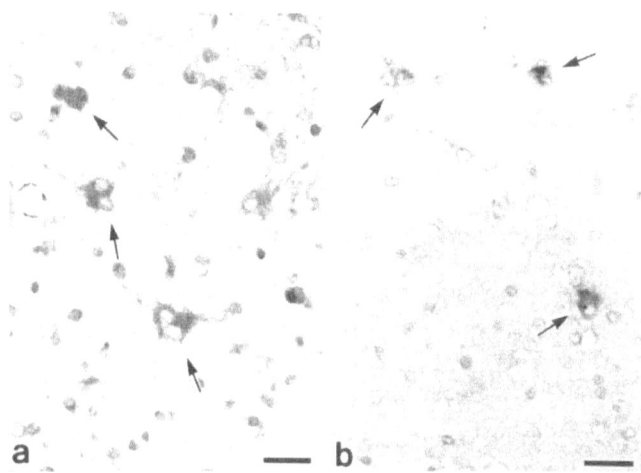

Figure 3 Hypertrophic astrocytes (arrows) in the necrotic lesion in the cerebro-cortical infarct of a 10-year-old dog. a: MT-III immunohistochemistry (polyclonal antibody to canine MT-III). b: DIG-labeled antisense RNA probe (noncoding region of canine MT-III cDNA). Bars = 25 µm.

The canine MT-III had two insertions relative to known mammalian MT-I and MT-II: a threonine after the fourth amino acid and a block of six amino acids near the carboxyl terminus. The insertions near the N-terminus are conserved among human, mouse, rat, porcine and canine MTs. The insertions near the C-terminus of MTs, however, are not conserved; human, mouse, porcine and canine MT-III have 6-amino-acid insertions (Gly-Glu-Ala-Ala-Glu-Ala, Glu-Glu-Gly-Ala-Lys-Ala, Gly-Glu-Gly-Ala-Glu-Ala and Gly-Glu-Gly-Thr-Glu-Ala, respectively) and rat MT-III has a 4-amino-acid insertion (Glu-Glu-Gly-Ala). The significance of this insertion regarding the biological function of MT-III is not known. If this insertion is critical for MT-III function, its structure may be more important than its exact sequence.

MT-III was discovered as a novel protein possessing ability to inhibit neuronal survival *in vitro* [2], although the mechanism of this inhibition and its physiological significance are not known. Based on the findings that MT-III is enriched in zinc-containing neurons in the mouse brain, Masters et al.[8] suggested the involvement of MT-III in the utilization of zinc as a neuromodulator. Thus, there is a discrepancy between the proposed roles of MT-III *in vitro* and *in vivo*.

In this study, MT-I and MT-III positive astrocytes were demonstrated in all the brain lesions regardless of the type of lesions; the changes ranged from acute to chronic. The findings imply that MT may play roles in common, fundamental processes including protection and repair in neuropathology. Following neural injury, astrocytes express genes for a variety of proteins including tissue matrix, cytokines and trophic factors [11]; zinc is used during the process of gene expression by transcription factors with zinc-finger motifs [12]. In this context, it would be plausible that MTs may serve as a provider of essential trace metals including zinc and copper, which are required for the astrocytic gene expression throughout the time course of repair of the injured neural tissues.

Acknowledgment
This study was supported by a grant-in-aid for general scientific research (08660382) from the Ministry of Education, Science and Culture, Japan.

References

1. Palmiter RD, Findley SD, Whitmore TE, Durnam DM (1992) MT-III, a brain-specific member of the metal-lothionein gene family. *Proc Natl Acad Sci USA* 89: 6333–6337.
2. Uchida Y, Takio K, Titani K, Ihara Y, Tomonaga M (1991) The growth inhibitory factor that is deficient in the Alzheimer's disease is a 68 amino acid metallothioein-like protein. *Neuron* 7: 337–347.
3. Kobayashi H, Uchida Y, Ihara Y, Nakajima K, Kohsaka S, Miyatake T, Tsuji S (1993) Molecular cloning of rat growth inhibitory factor cDNA and the expression in the central nervous system. *Mol Brain Res* 19: 188–194.
4. Erickson JC, Sewell AK, Jensen LT, Winge DR, Palmiter RD (1994) Enhanced neurotrophic activity in Alzheimer's disease cortex is not associated with down-regulation of metallothionein-III (GIF). *Brain Res* 649: 297–304.
5. Tsuji S, Kobayashi H, Uchida Y, Ihara Y, Miyatake T (1992) Molecular cloning of human growth inhibitory factor cDNA and its down-regulation in Alzheimer's disease. *EMBO J* 11: 4843–4850.
6. Anezaki T, Ishiguro H, Hozumi I, Inuzuka T, Hirasawa M, Kobayashi H, Yuguchi T, Wanaka A, Uda Y, Miyatake T et al (1995) Expression of growth inhibitory factor (GIF) in normal and injured rat brain. *Neurochem Int* 27: 89–94.
7. Yamada M, Hayashi S, Hozumi I, Inuzuka T, Tsuji S, Takahashi H (1996) Subcellular localization of growth inhibitory factor in rat brain: light and electron microscopic immunohistochemical studies. *Brain Res* 735: 257–264.
8. Masters BA, Quaife CJ, Erickson JC, Kelly EJ, Froelick GJ, Zambrowicz BP, Brinster RL, Palmiter RD (1994) Metallothionein III is expressed in neurons that sequester zinc in synaptic vesicles. *J Neurosci* 14: 5844–5857.
9. Shimada A, Kuwamura M, Umemura T, Takada K, Ohama E, Itakura C (1991) Modified Bielschowsky and immunohistochemical studies on senile plaques in aged dogs. *Neurosci Lett* 129: 25–28.
10. Kobayashi K, Shimada A, Irie M, Yamano Y, Umemura T (1997) Molecular cloning of a canine metallothionein cDNA. *J Vet Med Sci* 59: 819–823.
11. Martin PM, O'callaghan JP (1996) Gene expression in astrocytes after neural injury. *In*: M Aschner, Kimelberg H K (eds): The role of glia in neurotoxicity, CRC Press, Florida, 285–310.
12. Herdegen T, Kiessling M, Bele S, Bravo R, Zimmermann M, Gass P (1993) The KROX-20 transcription factor in the rat central and peripheral nervous systems: novel expression pattern of an immediate early gene-encoded protein. *Neuroscience* 57: 41–52.

Metallothionein IV
C. Klaassen (ed.)
© 1999 Birkhäuser Verlag Basel/Switzerland

Liver and brain metallothionein regulation in transgenic mice overexpressing interleukin-6 and in mice carrying a null mutation in the interleukin-6 gene

Juan Hidalgo[1], Javier Carrasco[1], Joaquín Hernández[1], Mercedes Giralt[1], Amalia Molinero[1], Berta González[2], Horst Bluethmann[3] and Iain L. Campbell[4]

[1]Departamento de Biología Celular y Fisiología, Unidad de Fisiología Animal, Facultad de Ciencias, and
[2]Unidad de Histología, Facultad de Medicina, Universidad Autónoma de Barcelona, Bellaterra, Barcelona,
 Spain 08193
[3]Department of Biology, Pharmaceutical Research Gene Technology, F. Hoffmann-La Roche Ltd., Basel,
 Switzerland
[4]Department of Neuropharmacology, The Scripps Research Institute, La Jolla, CA 92037, USA

Summary. Interleukin-6 (IL-6) is a major cytokine which is an essential mediator of cellular communication not only of the immune system but also of other physiological systems. Results with mice carrying a null mutation in the IL-6 gene (IL-6$^{-/-}$) demonstrate that IL-6 is an essential factor for normal liver metallothionein (MT) response not only to inflammatory stimuli (endotoxin, turpentine) but also to a basically psychogenic stress model-immobilization stress. This indicates that this cytokine is an important component of the physiological response of the organism to stress in addition to its expected role during an immunological challenge. In contrast to the liver, in the brain the IL-6 deficiency only affected significantly MT-I+II response to inflammatory stimuli. Results with transgenic mice expressing IL-6 under the regulatory control of the glial fibrillary acidic protein gene promoter (GFAP-IL6) indicate that the MT-I+II isoforms are strongly up-regulated by IL-6. The fold induction in different brain areas correlates with the associated inflammatory response, and is comparable to that of the acute-phase response gene EB22/5, suggesting that MT-I+II could be considered acute-phase proteins in the brain. MT-III expression in the GFAP-IL6 mice is elevated in some areas (e.g. cerebellum) and decreased in others (e.g. occipital cortex), while that in IL-6$^{-/-}$ mice is slightly affected in a brain area-specific manner.

Introduction

Since the pioneering work of Margoshes and Vallee on the cadmium binding proteins from horse kidney, metallothioneins (MTs) [1], an exponential number of reports have been published dealing with every aspect of these proteins, ranging from their biochemical and molecular characterization to their putative physiological functions. In the mouse, there are four known isoforms classified as MT-I, -II, -III and -IV [2, 3]. MT-I and MT-II are the most widely distributed MT isoforms, being expressed in virtually all cells, although with a wide variation in concentrations and inducibility depending on the tissue. These proteins are dramatically up-regulated in response to many factors, including metals, hormones, inflammation related stimuli, stressful agents and many others. Importantly, this upregulation occurs in a coordinate manner [4]. In contrast to MT-I and MT-II, the expression of MT-III and MT-IV is normally restricted to the central nervous system [2, 5] and to tissues containing stratified squamous epithelia [3], respectively. In pregnant mice, however, the four MT isoforms are expressed in the maternal deciduum [6]. All MT isoforms have been implicated in disparate physiological functions such as zinc and copper metabolism, protection against oxidative agents, or adaptation to stress, whereas MT-III has been involved additionally in neuromodulatory events and in the pathogenesis of Alzheimer disease. It is therefore important to characterize the factors involved in the control of the different MT isoforms in order to better understand the putative physiological functions of these proteins. Here we will discuss the role of IL-6 as a major MT regulator taking advantage of genetically modified mice.

Role of IL-6 on liver metallothionein regulation

The involvement of cytokines in liver MT regulation was suggested early by results with rats subjected to bacterial infection [7] or injected with endotoxin [8]. These pioneering studies have been confirmed by many laboratories, and it is now clear that the inflammatory response is a major physiological inducer of liver MT isoforms [9]. Many of the hallmarks of the inflammatory response are elicited by a unique group of proteins, cytokines, which are essential mediators of the immune system that also have widespread effects in many tissues and in other regulating systems such as the endocrine and nervous systems [10]. Consequently, the role of a number of cytokines as putative liver MT regulators has been studied by different approaches. Thus, the exogenous administration of IL-1α/β, IL-6, TNF-α and IFN-γ, was found to increase liver MT-I+II levels [11–13], suggesting that these cytokines might be relevant for MT-I+II regulation *in vivo*. Further support for this notion was obtained with the C3H/Hej mice, a strain that has a mutation in the *Lps* gene which causes a very low cytokine response to endotoxin. When injected with endotoxin, these mice produced less hepatic MT-I+II than controls, suggesting that cytokines are mediating the effect of endotoxin on liver MT regulation [12, 14]. A mediating role of cytokines during the inflammatory response was also suggested by studies where pretreatment with glucocorticoids, which decrease cytokine production, was given to the animals prior to the administration of the inflammatory agent [15–17].

Among the many cytokines produced during the inflammatory response, IL-6 might have a prominent role as suggested by the blocking effect of an antibody against IL-6 on *n*-Hexane-induced synthesis of hepatic MTs [18]. However, whether or not IL-6 is important in the inflammatory response elicited by other factors regarding liver MT regulation needs further studies. Furthermore, IL-6 could be important not only during an immunological insult but also in other physiological situations. This notion is supported by recent studies which demonstrate that serum IL-6 levels are strongly increased by exposure to psychological or physical stressors [19], suggesting that this cytokine could mediate liver MT induction by stress. That a number of stressors were able to increase liver MT levels was already observed in the seventies [20], which has been confirmed repeatedly. Stress, similarly to the inflammatory response, is a major physiological inducer of the hepatic MTs. However, despite many efforts, the factors responsible for the upregulation of hepatic MTs during stress are still unclear (reviewed in [21]). In this regard, the generation of mice carrying a null mutation in the IL-6 gene [22] is a unique opportunity for studying the role of IL-6 on liver MT regulation during stress. In Figure 1 a Northern blot for liver MT-I mRNA of stressed mice is shown. The results clearly demonstrate that the IL-6$^{-/-}$ mice responded significantly less than the three proper controls. In other studies, a more dramatic effect of the IL-6 deficiency was observed in mice injected with either endotoxin or

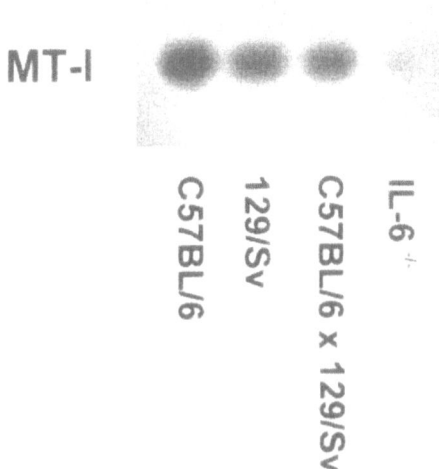

Figure 1. Northern blot analysis of the effect of IL-6 deficiency on liver MT-I mRNA response to 4 h of immobilization stress. Pooled samples from stressed animals were electrophoresed, blotted and hybridized with the mMT-I probe. It is clear that MT-I mRNA levels are lower in the IL-6–/– mice than in the three types of control mice (C57BL/6, 129/Sv, C57BL/6x129/Sv). Basal animals are not shown because of their low signal compared with the stressed ones.

turpentine (unpublished data). Thus, IL-6 is a major cytokine regarding liver MT regulation not only during an immunological insult but also during the stress response. Further work is needed to establish whether or not IL-6 is also important in other peripheral tissues, and if the acute-phase protein profile during stress is comparable to that during the inflammatory response.

Role of IL-6 on brain metallothionein regulation

There are several reports which demonstrate that the expression of the MT-I+II isoforms is up-regulated in the brain of mice and rats injected with endotoxin, suggesting that cytokines may also be involved in brain MTs regulation [2, 4, 12, 23–25]. Results with established neuronal and astrocytic cell lines suggest that IL-1 and IL-6 might be physiological regulators of MT-I+II in brain cells, but those with primary cultures are less clear [26, 27]. These discrepancies could be due on the one hand to the obvious fact that established cell lines can not in many respects be compared physiologically to primary cultures, and, on the other hand, to the likely lack of essential factors in a cell culture system for normal actions of cytokines.

As for the liver, the use of mice carrying a null mutation in the IL-6 gene [22] is a unique approach for studying the putative role of IL-6 in brain MTs regulation. In Figure 2 an *in situ* hybridization analysis for MT-I and MT-III isoforms is shown. As previously established [28],

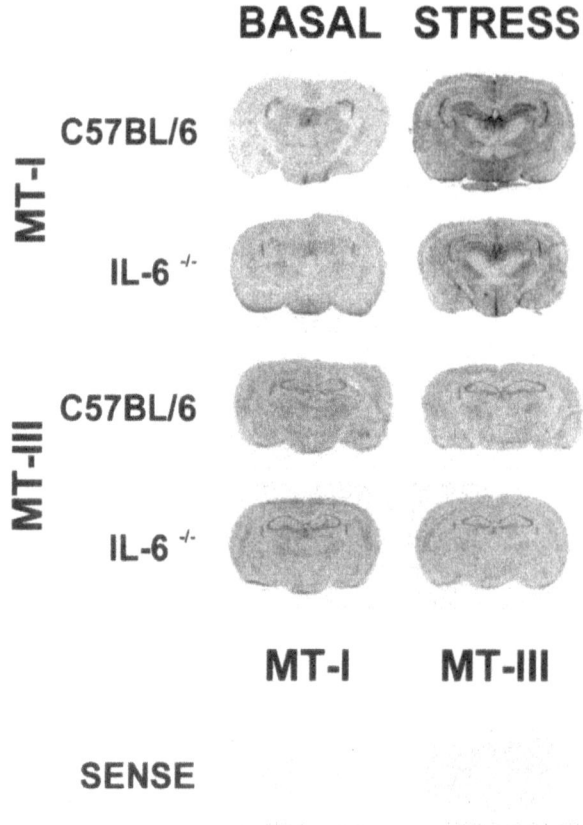

Figure 2. Representative *in situ* hybridization results of the effect of IL-6 deficiency on MT-I and MT-III responses to 4 h of immobilization stress. The sense probes produced a very weak signal compared with the antisense mMT-I and mMT-III probes. Stress had a prominent effect on MT-I but not MT-III mRNA levels throughout the brain, whereas slight but significant effects were caused by IL-6 deficiency.

the different pattern of cellular expression of these isoforms was clear: mainly neuronal for MT-III (producing therefore an intense hybridization of the MT-III probe to pyramidal neurons of the hippocampus in the CA1-CA3 regions), and mainly astrocytic for MT-I (as shown in Figure 3), although prominent signals were also observed in the choroid plexus and ependimal cells. Regarding MT-I expression, it was clear that stress had a very significant effect in most brain areas studied, in line with previous studies measuring the MT-I+II total protein (see [21] for review). An evaluation of the effect of IL-6 deficiency in both basal and stress conditions revealed only marginal effects in some brain areas, indicating that IL-6 has a distinct role in the brain compared to the liver regarding MT-I regulation during stress.

con   str   lps   tur

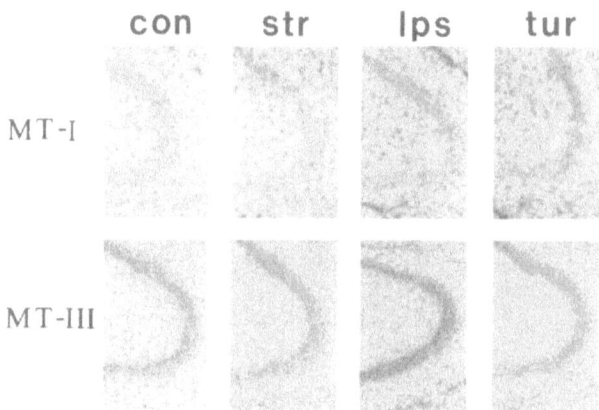

MT-I

MT-III

Figure 3. Microautoradiographies of the CA2-CA3 area of the hippocampus. Animals were either subjected to immobilization stress (STR) or injected with endotoxin (LPS) or turpentine (TUR), and killed 4 h later along control (CON) mice. MT-I expression was increased by the three treatments following a comparable pattern of induction, whereas that of MT-III remained basically unchanged. Power: ×100.

Figure 3 shows a microautoradiography of the CA2-CA3 area of the hippocampus of mice either subjected to immobilization stress or injected with endotoxin or turpentine. A very significant MT-I upregulation was observed with the three treatments, which remarkably produced a very similar pattern of MT-I induction (presumably in glial cells). In contrast, MT-III remained basically unchanged. Preliminary results indicate that in contrast to during stress, during the inflammatory response a significant effect of the IL-6 deficiency on brain MT-I regulation was observed (unpublished observations).

In order to gain further insight into the role of IL-6 on the regulation of brain MTs, we have recently undertaken studies [29] with transgenic mice expressing IL-6 under the regulatory control of the glial fibrillary acidic protein gene promoter (GFAP-IL6 mice) [30]. These mice show clear evidence of progressive neurodegeneration, astrocytosis, microgliosis, angiogenesis and up-regulation of several inflammatory and other host-response genes including IL-1 α/β, TNF-α, GFAP, ICAM-1, and the acute-phase protein EB22/5 [30, 31]. Moreover, concordant with these structural alterations, the GFAP-IL6 mice develop a progressive learning deficit [32]. Our studies in the GFAP-IL6 mice [29] indicate that brain MT-I+II regulation is also profoundly affected by the astrocyte production of IL-6. Thus, the MT-I+II levels, as measured by radioimmunoassay, in one and three month old GFAP-IL6 mice (G16-low expressor and/or G36-high expressor lines) were elevated in the cerebellum (highest induction), medulla plus pons, hypothalamus and remaining brain (lowest induction), but not in hippocampus. Figure 4 shows an *in situ* hybridization analysis of the cerebellum, which shows a strong increase of the MT-I signal in the Purkinje cell layer of the GFAP-IL6 mice, presumably in Bergmann glia cells. This notion is supported by immunocytochemical studies (Fig. 4), which show a significant MT-I+II immunostaining in the cellular processes of Bergmann glia in addition to the granular layer and

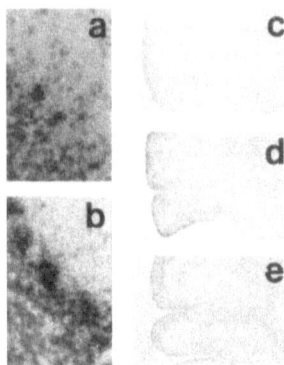

Figure 4. Effect of astrocyte-targeted expression of IL-6 in the cerebellum. An *in situ* hybridization analysis (a,b) shows a strong induction of the MT-I isoform in the GFAP-IL6 mice (b) compared to control mice (a), presumably in the Bergmann glia cells. Immunocytochemical analysis for GFAP and MT-I+II (c,d,e) revealed a strong astrocytosis as well as increased MT-I+II immunoreactivity (with a prominent staining in the Bergmann glia cell processes) in the GFAP-IL6 mice. C, GFAP+MT-I+II of control mice. D, GFAP+MT-I+II of GFAP-IL6 mice. E, GFAP of GFAP-IL6 mice. GFAP: brown. MT-I+II: black. Power: a,b ×400; c-e ×20.

white matter in the cerebellum. The MT-I+II protein and MT-I mRNA (unpublished) increases observed in the different brain areas of the GFAP-IL6 mice correlated with the associated inflammatory response caused by the transgene expression of IL-6, and, furthermore, they were comparable to the expression of the acute-phase response gene EB22/5 [30, 31], suggesting that these MT isoforms could be considered acute-phase proteins in the brain.

Little is known regarding MT-III expression in the brain and the putative role of cytokines as physiological regulators. Palmiter et al., measuring MT-III mRNA by solution hybridization, did not find a significant effect of endotoxin [2]. By Northern analysis, another study showed that endotoxin tended to decrease MT-III mRNA levels [25]; the same study also indicated that the decreasing effect was apparent in the hippocampus but not in other brain areas, as revealed by *in situ* hybridization experiments. We also observed a decreasing effect of endotoxin on MT-III mRNA levels [29]. The effect, however, is small and poorly reproducible. We have evaluated MT-III expression in basal and stress situations in both control and IL-6$^{-/-}$ mice, and the results were more complex than for MT-I, since stress and IL-6 affected MT-III mRNA levels in a brain area-specific manner (unpublished observations) which deserves further work. Nevertheless, IL-6 seems to have some role on MT-III regulation, as already suggested by the results in the GFAP-IL6 mice [29]. More recent *in situ* hybridization studies indicate that MT-III mRNA levels increase in the cerebellum and decrease in other areas (unpublished observations), highlighting again the different pattern of regulation of MT-III *versus* MT-I+II isoforms.

Acknowledgements
The financial support is gratefully appreciated: CIRIT GRQ93-2.096, DGICYT PB94-0667, and CICYT SAF96-0189 (to J.H.), DGICYT PB95-0662 (to B.G.), and USPHS Grant MH50426 (to I.L.C.). J. Carrasco is supported by a fellowship of the FI/96-2613.

References

1. Margoshes M, Vallee BL (1957) A cadmium protein from equine kidney cortex. *J Amer Chem Soc* 79: 4813–4814.
2. Palmiter RD, Findley SD, Whitmore TE, Durnam DM (1992) MT-III, a brain-specific member of the metallothionein gene family. *Proc Natl Acad Sci USA* 89: 6333–6337.
3. Quaife CJ, Findley SD, Erickson JC, Froelick GJ, Kelly EJ, Zambrowicz BP, Palmiter RD (1994) Induction of a new metallothionein isoform (MT-IV) occurs during differentiation of stratified squamous epithelia. *Biochemistry* 33: 7250–7259.
4. Searle PF, Davison BL, Stuart GW, Wilkie TM, Norstedt G, Palmiter RD (1984) Regulation, linkage, and sequence of mouse metallothionein I and II genes. *Mol Cell Biol* 4: 1221–1230.
5. Kobayashi H, Uchida Y, Ihara Y, Nakajima K, Kohsaka S, Miyatake T, Tsuji S (1993) Molecular cloning of rat growth inhibitory factor cDNA and the expression in the central nervous system. *Mol Brain Res* 19: 188–194.
6. Liang L, Fu K, Lee DK, Sobieski RJ, Dalton T, Andrews GK (1996) Activation of the complete mouse metallothionein gene locus in the maternal deciduum. *Mol Reprod Dev* 43: 25–37.
7. Sobocinski PZ, Canterbury WJ, Jr Mapes CA, Dinterman RE (1978) Involvement of hepatic metallothioneins in hypozincemia associated with bacterial infection. *Amer J Physiol* 234: E399–406.
8. Suzuki KT, Yamamura M (1980) Induction of hepatic zinc-thionein in rat by endotoxin. *Biochem Pharmacol* 29: 2260.
9. Sato M, Bremner I (1993) Oxygen free radicals and metallothionein. *Free Radical Biol Med* 14: 325–337.
10. Akika S, Hirano T, Taga T, Kishimoto T (1990) Biology of multifunctional cytokines: IL 6 and related molecules (IL 1 and TNF). *FASEB J* 5: 2860–2867.
11. Cousins RJ, Leinart AS (1988) Tissue-specific regulation of zinc metabolism and metallothionein genes by interleukin 1. *FASEB J* 2: 2884–2890.
12. De SK, McMaster MT, Andrews GK (1990) Endotoxin induction of murine metallothionein gene expression. *J Biol Chem* 265: 15267–15274.
13. Sato M, Sasaki M, Hojo H (1992) Tissue specific induction of metallothionein synthesis by tumor necrosis factor-alpha. *Res Commun Chem Pathol Pharmacol* 75: 159–172.
14. Liu J, Liu YP, Sendelbach LE, Klassen CD (1991) Endotoxin induction of hepatic metallothionein is mediated through cytokines. *Toxicol Appl Pharmacol* 109: 235–240.
15. Min KS, Terano Y, Onosaka S, Tanaka K (1991) Induction of hepatic metallothionein by nonmetallic compounds associated with acute-phase response in inflammation. *Toxicol Appl Pharmacol* 111: 152–162.
16. Min KS, Mukai S, Ohta M, Onosaka S, Tanaka K (1992) Glucocorticoid inhibition of inflammation-induced metallothionein synthesis in mouse liver. *Toxicol Appl Pharmacol* 113: 293–298.
17. Giralt M, Blanquez A, Avila J, Hidalgo J (1993) Effect of superoxide dismutase, allopurinol and glucocorticoids on liver and lung metallothionein induction by endotoxin in the rat. *Biometals* 6: 101–106.
18. Itoh N, Okamoto H, Ohta M, Hori T, Min KS, Onosaka S, Nakanishi H, Okabe M, Muto N, Tanaka K (1994) n-hexane-induced synthesis of hepatic metallothionein is mediated by Il-6 in mouse. *Toxicol Appl Pharmacol* 124: 257–261.
19. Lemay LG, Vander AJ, Kluger M (1990) The effects of psychological stress on plasma interleukin-6 activity in rats. *Physiol Behav* 47: 957–961.
20. Oh SH, Deagen JT, Whanger PD, Weswig PH (1978) Biological function of metallothionein. V. Its induction in rats by various stresses. *Amer J Physiol* 234: E282–285.
21. Hidalgo J, Gasull T, Giralt M, Armario A (1994) Brain metallothionein in stress. *Biol Signals* 3: 198–210.
22. Kopf M, Baumann H, Freer G, Freudenberg M, Lamers M, Kishimoto T, Zinkernagel R, Bluethmann H, Köhler G (1994) Impaired immune and acute-phase responses in interleukin-6-deficient mice. *Nature* 368: 339–342.
23. Itano Y, Noji S, Koyama E, Taniguchi S, Taga N, Takahashi T, Ono K, Kosaka F (1991) Bacterial endotoxin-induced expression of metallothionein genes in rat brain, as revealed by *in situ* hybridization. *Neurosci Lett* 124: 13–16.
24. Choudhuri S, Kramer KK, Berman NE, Dalton TP, Andrews GK, Klaassen CD (1995) Constitutive expression of metallothionein genes in mouse brain. *Toxicol Appl Pharmacol* 131: 144–154.
25. Zheng H, Berman NE, Klaassen CD (1995) Chemical modulation of metallothionein I and III mRNA in mouse brain. *Neurochem Int* 27: 43–58.
26. Hidalgo J, Garcia A, Oliva AM, Giralt M, Gasull T, Gonzalez B, Milnerowicz H, Wood A, Bremner I (1994) Effect of zinc, copper and glucocorticoids on metallothionein levels of cultured neurons and astrocytes from rat brain. *Chem -Biol Interact* 93: 197–219.
27. Kramer KK, Liu J, Choudhuri S, Klaassen CD (1996) Induction of metallothionein mRNA and protein in murine astrocyte cultures. *Toxicol Appl Pharmacol* 136: 94–100.
28. Masters BA, Quaife CJ, Erickson JC, Kelly EJ, Froelick GJ, Zambrowicz BP, Brinster RL, Palmiter RD (1994) Metallothionein III is expressed in neurons that sequester zinc in synaptic vesicles. *J Neurosci* 14: 5844–5857.
29. Hernández J, Molinero A, Campbell IL, Hidalgo J (1997) Transgenic expression of interleukin 6 in the cen-

tral nervous system regulates brain metallothionein-I and -III expression in mice. *Mol Brain Res* 48: 125–131.

30. Campbell IL, Abraham CR, Masliah E, Kemper P, Inglis JD, Oldstone MBA, Mucke L (1993) Neurologic disease in transgenic mice by cerebral overexpression of interleukin 6. *Proc Natl Acad Sci USA* 90: 10061–10065.

31. Chiang C-S, Stalder A, Samimi A, Campbell IL (1994) Reactive gliosis as a consequence of interleukin 6 expression in the brain: studies in transgenic mice. *Dev Neurosci* 16: 212–221.

32. Heyser CJ, Masliah E, Samimi A, Campbell IL, Gold LH (1997) Progressive decline in avoidance learning paralleled by inflammatory neurodegeneration in transgenic mice expressing interleukin 6 in the brain. *Proc Natl Acad Sci USA* 94: 1500–1505.

Effects of capsaicin on metallothionein (MT) mRNA in sensory neurons: Possible role of MT-III in nociception

Yongjiu Cai, Rubén A. Velázquez and Alice A. Larson

Department of Veterinary PathoBiology, University of Minnesota, Saint Paul, Minnesota 55108, USA

Introduction

Primary afferent C-fibers are small-diameter, unmyelinated axons that are believed to play a role in pain transmission [1]. Painful stimuli initiated in the periphery are conducted to the dorsal spinal cord through primary afferent fibers whose cell bodies reside in the dorsal root ganglia (DRG). Noxious stimulation results in an aversive response reflecting pain in humans or nociception in animals. Polymodal nociceptive primary afferent fibers are selectively excited by capsaicin, a pungent compound found in hot peppers [2]. When injected spinally (intrathecally) in the adult, capsaicin depolarizes primary afferent C-fibers causing a transient hyperalgesia followed by a prolonged analgesia, which is thought to reflect C-fibers desensitization [3]. When administered to neonates, capsaicin is relatively toxic and decreases the number of primary afferent C-fibers in rats and mice [4].

Although zinc is not historically associated with pain processing, zinc has been found to be densely localized in the dorsal spinal cord [5], an area associated with input from nociceptive afferents. Histologically reactive zinc is thought to reflect areas containing a releasable pool of zinc, concentrated in synaptic vesicles of zincergic neurons [6]. Zinc has also been found to modulate pain [7], raising the possibility that zinc released from neurons innervating the spinal cord plays a role in the modulation of nociceptive activity.

Metallothionein (MT) is a zinc-binding protein that is widely distributed in the central nervous system. Regulation of MT expressed in neurons is extremely important in maintaining the steady-state level of zinc [8]. While MT-I and -II are expressed throughout the body, zincergic neurons uniquely express MT-III [6]. In view of the association between MT-III and releasable zinc, we tested the hypothesis that primary afferent sensory neurons are zincergic by measuring MT mRNA in the lumbar DRG of rats. We further hypothesized that MT is synthesized in small, capsaicin-sensitive cells of the DRG that play a role in nociception. The experiments described here were designed to measure the contents of MT-I and -III mRNA in DRG and spinal cord of adult rats that were either treated neonatally or injected intrathecally with capsaicin.

Materials and methods

Groups of 7 rats (5 male and 2 female) were treated neonatally at 2 days of age with 50 mg/kg of capsaicin (8-methyl-*N*-vanillyl-6-nonenamide) or vehicle subcutaneously. The effectiveness of neonatal capsaicin treatment was verified histologically by decreased thiamine monophosphase staining, which is selective for a subpopulation of primary afferent C-fibers. Six male rats, weighing 250–300 g, were injected intrathecally (i.t.) with 75 µg of capsaicin in 15 µl of vehicle. The i.t. injection was made at the L5-L6 intervertebral space using a 25-gauge, 3.81-cm disposable needle attached to a 50-µl Hamilton syringe under ether anesthesia.

Tissues were sampled at 65 days of age from neonatally capsaicin-treated rats or at 6 h after i.t. injection of capsaicin. The lumbar region of the spinal cord and lumbar DRG (L1-6) were collected. Cerebral cortex and liver were also sampled as positive and negative control tissues for MT-III mRNA, respectively. All samples were stored at −80 °C until total RNA isolation, using the Total RNA Isolation System (Promega, Madison, WI).

Approximately 2-µg aliquots of total RNA were used in cDNA synthesis for each sample. The composition of PCR primers (from 5' to 3') was the followings: GAPDH (335 bp): GTG–GACATTGTTGCCATCAACGAC, and TTTCTCGTGGTTCACACCCATCAC [9]; and MT-I (180 bp): ATGGACCCCAACTGCTCCTGCTCCA and ACAGCACGTGCACTTGTCCGAGGCA [10]. PCR primers for MT-III (201 bp) were those used by Amoureux et al [11]. All PCR primers were purchased from Life Technologies (Grand Island, NY).

A master mixture of PCR components was prepared for all PCR analyses, from which an aliquot was added to a pair of specific primers for the targeted gene. Each PCR reaction mixture (25 µl) contained 1.5 µl of the synthesized cDNA. Six controls (Tab. 1) were included in each set of PCR amplification with denaturation at 94 °C for 1 min, annealing at 60 °C for 1 min, and extension at 72 °C for 1 min. Reactions were run for 20 cycles for GAPDH and 25 cycles for MT-I and -III.

The amplified products were separated by electrophoresis in a 2% agarose gel containing ethidium bromide. The DNA bands were visualized using Eagle Eye II (Stratagene, La Jolla, CA) under UV light. The intensity of bands was analyzed using the NIH Image (1.6.0 for the Macintosh). The content of mRNA was expressed as optical density units.

Data were statistically analyzed using a 2-tailed, unpaired Student's *t*-test with animals as experimental units [12]. Differences were considered significant when $P < 0.05$.

Table 1. Controls for quantitative RT-PCR

Name	Derivation	Purpose
Positive	Mouse MT cDNA	Verify amplification of MT-I or -III mRNA
Negative 1	No reverse transcriptase	Trace contaminating genomic DNA
Negative 2	No RNA template	Confirm the lack of pseudogene product in PCR signal
Brain	Cerebral cortex	Demonstrate tissue containing detectable MT
Liver	Vehicle-treated rat	Demonstrate tissue lacking MT-III
GAPDH	Housekeeping gene	Normalize quality of the synthesized cDNA

Results

Both MT-I and -III mRNA were localized in DRG as well as the spinal cord. Rat MT-I and -III mRNA were each amplified as effectively as MT cDNA of mouse (Fig. 1), the species in which the amplified mRNA fragments match the target 180 bp of MT-I and the 201 bp of MT-III, respectively. Mouse and rat tissues can be used by the PCR conditions because of the high degree of similarity of genes encoding MT-I or -III among mammalian MT genes characterized by Palmiter et al [13]. The contents of MT-I and -III mRNA amplified in the DRG and spinal cord are identical to those found in the brain of vehicle-treated rats (Fig. 2). The liver lacked MT-III mRNA, but contained a higher content of MT-I mRNA than the brain. These results indicate that our PCR technique amplifies MT-I and -III mRNA with high sensitivity and selectivity.

MT-I mRNA was increased in the DRG of rats treated neonatally with capsaicin (Fig. 3). However, neonatal treatment did not affect the contents of MT-III mRNA in the DRG. On the other hand, MT-III mRNA was decreased in the DRG by an i.t. injection of capsaicin in adult rats while MT-I mRNA was unaffected by this treatment. In addition to the increase in MT-I mRNA following neonatal capsaicin treatment, the housekeeping gene GAPDH was decreased ($P < 0.01$) in neonatally capsaicin-treated rats, but not in adults injected with capsaicin, suggesting a decreased tissue content in response to the toxic effect of capsaicin when adminis-

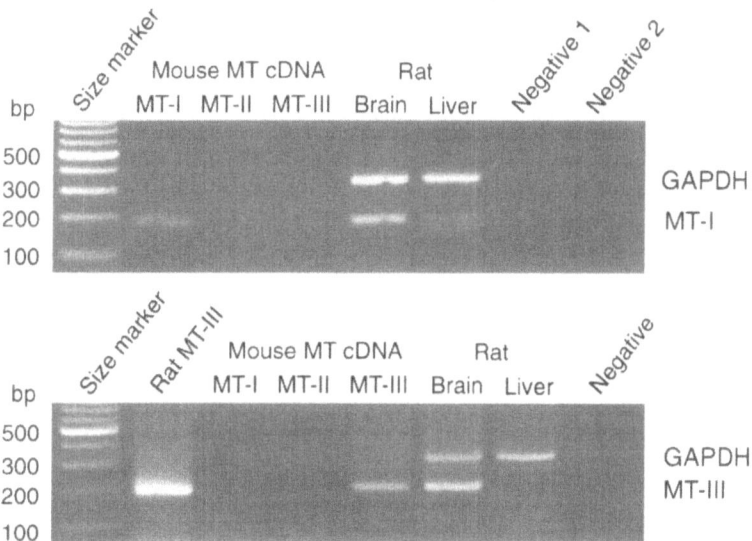

Figure 1. RT-PCR analyses of MT-I and -III mRNA in rat tissue compared with mouse MT-I, -II, and -III cDNA. Negative controls contained the same amount of all reagents as needed in the RT-PCR procedure but no reverse transcriptase (Negative 1) and no RNA template (Negative 2). The rat MT-III is a purified positive control. The negative control was a sample combined from the Negative 1 and 2.

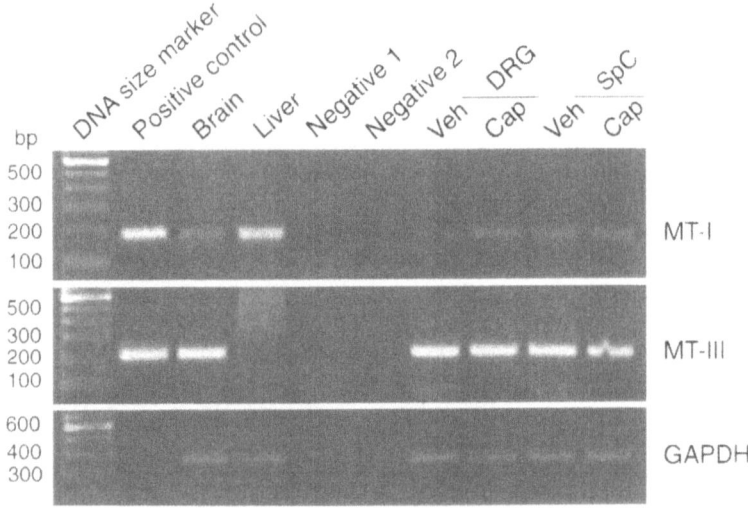

Figure 2. Representative of RT-PCR results for MT-I and -III mRNA found in DRG and spinal cord (SpC) of rats treated neonatally with vehicle (Veh) or capsaicin (Cap). Purified positive controls were derived from mouse MT-I or -III cDNA. Negative controls contained the same amount of all reagents but no reverse transcriptase (Negative 1) and no RNA template (Negative 2).

tered neonatally. The ratios of MT-I or -III to GAPDH mRNA (Fig. 3), typically used to express changes in gene expression, was similar to that expressed by absolute content (data not shown).

In the spinal cord tissue from the same animals, the mRNA contents of MT-I, -III, and GAPDH were unchanged by either neonatal capsaicin treatment or i.t. injection of capsaicin in adult rats (Fig. 3).

Discussion

The present data demonstrate that primary sensory neurons express MT mRNA. In view of the theory that MT-III has been found to be synthesized in neurons capable of concentrating and releasing zinc [6], primary afferent sensory neurons may innervate the spinal cord with terminals that contain releasable zinc. The regulation of MT isoforms expressed in neurons has been proposed to be important in terms of maintaining the steady-state level of zinc and controlling redox potentials [8]. Uchida [14] failed to visualize growth inhibitory factor (MT-III) immunoreactivity in DRG of humans. This discrepancy between the presence of mRNA (rats) and the apparent absence of protein (human) maybe due to a difference in species, a failure to synthesize MT protein from mRNA, or the tendency of neurons to transport protein to nerve terminals, not allowing it to accumulate in the soma.

Based on the ability of neonatal capsaicin to increase MT-I mRNA, MT-I may be preferentially localized in larger, capsaicin-insensitive DRG cells where it tends to be concentrated by

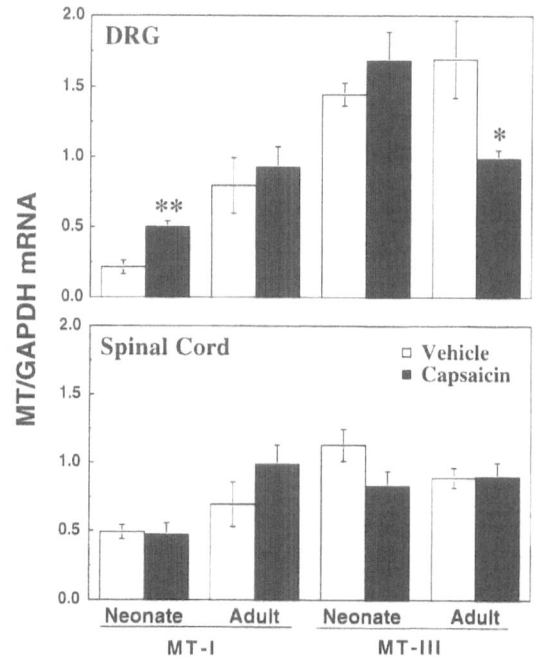

Figure 3. Effects of neonatal capsaicin treatment (Neonate) and spinal injection of capsaicin to adult rats (Adult) on MT-I and -III mRNA in the DRG and spinal cord of rats treated with vehicle or capsaicin. Effects are indicated as * $P < 0.05$ and ** $P < 0.01$, based on a 2-tailed, unpaired Student's t-test when n = 7 for Neonate and n = 6 for Adult. The relative contents of MT to GAPDH mRNA are graphed as a mean ± SEM.

neonatal capsaicin treatment. The insensitivity of MT-III mRNA synthesis in the DRG to neonatal capsaicin indicates that zincergic afferent neurons are not capsaicin-sensitive. This insensitivity does not appear to be due to a lack of efficacy of neonatal capsaicin treatment as thiamine monophosphatase stain, an enzyme that is selectively distributed in a subset of primary afferent C-fibers [15], is decreased in the spinal cords of these animals.

In contrast to the lack of effect of neonatal capsaicin on MT-III synthesis in the DRG, an i.t. injection of capsaicin, which is not as toxic yet causes antinociception, did decrease MT-III mRNA in this area. Thus, a decrease in the MT-III mRNA content may be induced by capsaicin injected in the adult and associated with antinociception. Desensitization of primary afferent C-fibers results in a decreased synthesis of many proteins localized in capsaicin-sensitive neurons, such as substance P [3], a protein thought to act a nociceptive transmitter. The relative contribution of decrease in MT-III synthesis in capsaicin-induced antinociception requires further study.

In conclusion, sensory DRG neurons appear to express MT-I and -III mRNA, suggesting that zinc is involved in sensory transmission. Neither MT-I nor MT-III mRNA appears to be preferentially localized in capsaicin-sensitive DRG cells. Yet MT-III mRNA is decreased in adult tissue desensitized by i.t. injection of capsaicin. These data are consistent with two distinct mechanisms of action of capsaicin, depending on whether it is administered neonatally or as an adult.

Acknowledgments
Authors thank Dr. Richard Palmiter at the University of Washington for providing us with mouse MT cDNA, Dr. Susan Liew Giovengo for neonatally treating rats, the United States Public Health Science for funding support NIDA07234 to Y.C., NIDA04090 to A.A.L., and NINDS-NS09882 to R.A.V.

References

1. Jessell TM, Kelly JP (1991) Pain and analgesia. *In*: ER Kandel, JH Schwartz, TM Jessell (eds): *Principles of Neural Science*. Appleton and Lange Press, Norwalk, CT, 385–399.
2. Dray A, Dickenson A (1993) Capsaicin, nociception, and pain. *In*: Wood JN (ed.): *Capsaicin in the study of pain*. Academic Press, San Diego, 239–253.
3. Gamse R (1982) Capsaicin and nociception in the rat and mouse: possible role of substance P. *Naunyn-Schmiedeberg's Arch Pharmacol* 320: 205–210.
4. Jancso G, Kiraly E, Jancso GA (1977) Pharmacologically induced selective degeneration of chemosensitive primary sensory neurons. *Nature* 270: 741–743.
5. Danscher G (1986) Do the Timm sulphide silver method and selenium method demonstrate zinc in the brain? *In*: CJ Frederickson, GA Howell, EJ Kasarskis (eds): *Neurobiology of Zinc*. Liss Press, New York, 273–288.
6. Masters BA, Quaife CJ, Erickson JC, Kelly EJ, Froelick GJ, Zambrowicz BP, Brinster RL, Palmiter RD (1994) Metallothionein III is expressed in neurons that sequester zinc in synaptic vesicles. *J Neurosci* 14: 5844–5857.
7. Larson AA, Kitto KF (1997) Manipulations of zinc in the spinal cord, by intrathecal injection of zinc chloride, disodium-calcium-EDTA, or dipicolinic acid, alter nociceptive activity in mice. *J Pharmacol Exp Ther* 282: 1319–1325.
8. Ebadi M, Iversen PL, Hao R, Cerutis DR, Rojas P, Happe HK, Murrin LC, Pfeiffer RF (1995) Expression and regulation of brain metallothionein. *Neurochem Int* 27: 1–22.
9. Fort P, Marty L, Piechaczyk M, Sabrouty S, Dani C, Jeanteur P, Blanchard JM (1985) Various rat adult tissues express only one major mRNA species from the glyceraldehyde-3-phosphate-dehydrogenase multigenic family. *Nucl Acid Res* 13: 1431–1442.
10. Andersen RD, Birren BW, Taplitz SJ, Herschman HR (1986) Rat metallothionein-1 structural gene and three pseudogenes, one of which contains 5' regulatory sequences. *Mol Cell Biol* 6: 302–314.
11. Amoureux MC, Wurch T, Pauwels PJ (1995) Modulation of metallothionein-III mRNA content and growth rate of rat c6-cells by transaction with human 5HT 1D receptor genes. *Biochem Biophys Res Commun* 214: 639–645.
12. Steel RGD, Torrie JH (1980) *Principles and Procedures of Statistcs: A Biometrical Approach*. McGraw-Hill, New York.
13. Palmiter RD, Findley SD, Whitmore TE, Durnam DM (1992) MT-III, a brain-specific member of the metallothionein gene family. *Proc Natl Acad USA* 89: 6333–6337.
14. Uchida Y (1994) Growth inhibitory factor, metallothionein-like protein, and neurodegenerative diseases. *Biol Signals* 3: 211–215.
15. Bucsics A, Sutter D, Jancso G, Lembeck F (1988) Quantitative assay of capsaicin-sensitive thiamine monophosphatase and beta-glycerophosphatase activity in rodent spinal cord. *J Neurosci Methods* 24: 155–162.

Metallothionein IV
C. Klaassen (ed.)
© 1999 Birkhäuser Verlag Basel/Switzerland

Metallothionein expression in the mammalian brain

Adrian K. West[1], Adele F. Holloway[2], Fiona A. Stennard[3] and Janine M. Miller[1]

[1]*Biochemistry, University of Tasmania, GPO Box 252-58, Hobart, Australia 7001*
[2]*Human Immunology, Hanson Centre, Frome Rd, Adelaide, Australia, 5000*
[3]*Wellcome/CRC Institute, Tennis Court Rd, Cambridge CB 1QR, UK*

The expression of metallothionein (MT) in the mammalian brain is of great current interest, and several reports localising MT immunoreactivity to specific regions of the adult brain have appeared [1–5]. These have found that MT-I and MT-II expression is widespread in the CNS and is mainly in astrocytes. It appears that both astrocytes [5] and neurons [6] are able to express MT-III, and these findings have been confirmed independently [7].

The situation in many mammals is complicated by the multiplicity of isoforms that have been identified. Four major isoforms of mammalian MT have been characterised, of which three, MT-I, MT-II and MT-III are known to be present in the brain. In the human the MT-I isoform class is encoded by at least 13 genes [8], of which eight, *MT1A, MT1B, MT1E, MT1F, MT1G, MT1H, MT1L* and *MT1X* are known to be expressed in cultured cell models at the mRNA level [e.g. 9, 10]. Likewise, sheep have one expressed MT-II gene, and at least three functional MT-I genes [11, 12]. The predicted amino acid sequences of the MT isoforms differ but it is not yet known whether this reflects specific physiological roles for each, and indeed, few of the cognate proteins for each isoform have actually been identified at the protein level in human or sheep tissues. There is good evidence from cultured cell models that the various human MT-I and MT-II isoforms are differentially expressed, for example in response to different inducers, but the situation in the living brain is less clear. Antisera which discriminate between MT-I and MT-II are not commonly available, and indeed their routine production might be extremely difficult. Therefore, the best route to identifying specific MT expression is to use a combination of molecular biological and histochemical approaches. This report summarises some recent work we have done using human and sheep models of metallothionein expression in the brain.

Human MT: Multiple isoforms contribute to human neural MT expression

We attempted to assign MT expression in the human brain to specific MT isoforms. As a first step, human brains were dissected into discrete anatomical regions. Duplicate samples were then processed for immunohistochemistry, or for RNA isolation respectively. The immunohistochemical survey was done to identify major expressing cell types, and to estimate the proportion of MT-positive cells. The RNA samples were analysed for isoform-specific expression by northern blotting using gene-specific probes.

Immunohistochemistry

In accordance with published reports in the human [3, 4], we found that MT immunoreactivity was observed in a distinct subset of cells. We used the commercially available monoclonal antibody which is thought to detect MT-I and/or MT-II protein. The cerebral cortex showed extensive staining of glial cells having a definite astrocyte morphology, and was present in the nucleus, perinuclear cytoplasm and processes. These astrocytes were observed predominantly in the grey matter, whilst the molecular layer was not stained. In contrast, staining with GFAP, a commonly used marker of a subset of astrocytes, showed that MT and GFAP staining was mutually exclusive, and in fact occupied mirror image regions of the cortex. This suggested that MT was found predominantly in the protoplasmic astrocytes, whilst GFAP was found in the fibrous astrocytes. In addition to MT staining in the astrocytes, cytoplasmic staining of choroid epithelial cells of the choroid plexus, and in the pia mater of meninges was observed. No staining of neuronal or vascular structures was observed.

Northern blotting

To derive semi-quantitative information, RNA was prepared from 18 anatomical regions of human brains, and was screened by northern blotting using a panel of general and gene specific probes. Firstly, a probe (*MT2A* cDNA) was used which is thought to hybridise to all *MT1* and *MT2* genes in the human [8]. The results obtained from this probe are indicated in Figure 1 as "*MT1 + 2*". Secondly, a series of oligonucleotide probes specific for *MT1E, MT1F, MT1G, MT1H, MT1X* and *MT2A* and *MT3* were used to screen the RNA. In each case, the RNA was standardised by co-hybridisation to a β-actin probe. In this way comparisons of absolute levels of the mRNA for each isoform can be made from region to region of the brain.

Figure 1 shows the collective expression of mRNA encoding *MT1* and *MT2* isoforms compared to the expression of *MT3*, in a single, typical brain. It is clear that *MT1/2* and *MT3* are widely expressed in the brain, and that regional levels of these two classes of isoforms are broadly similar, with high levels in the caudate, putamen and the amygdala.

The levels of mRNA of individual MT1 isoforms, *MT1G, MT1H, MT1X* and also *MT2A* are shown in Figure 1. Interestingly, no mRNA corresponding to *MT1E* or *MT1F* could be detected, although it was possible to amplify message for these genes by RT-PCR. However, mRNA for each of the other isoforms was detected, and it can be seen that there is demonstrable regional variation in levels. For example, *MT1* expression was examined in the putamen and the globus pallidus, two adjacent regions which together are referred to as the lentiform nucleus. (Fig. 1, tracks 13 and 15). Despite their anatomical proximity, the levels of expression of all isoforms are markedly different between the two regions, the expression of each *MT* gene always being higher in the putamen. In general, the expression profiles of each of the *MT1* and *MT2* genes was similar when compared from region to region. This suggests that the *MT* locus as a whole is coordinately regulated (whether directly at the locus level, or indirectly by factors such as zinc is unclear) and that some regional factor is setting the level of *MT* expression. Likewise, the strong similarity in the profiles between *MT3* and the other isoforms is intrigu-

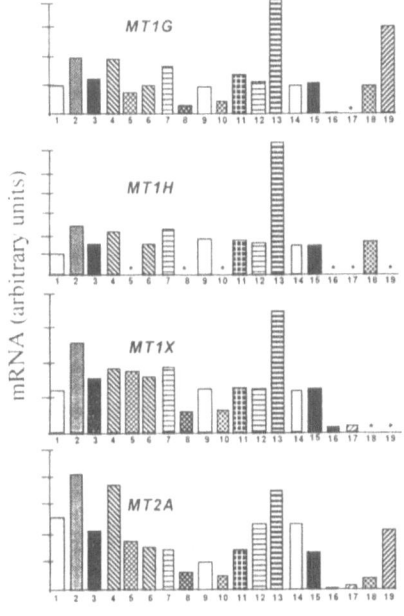
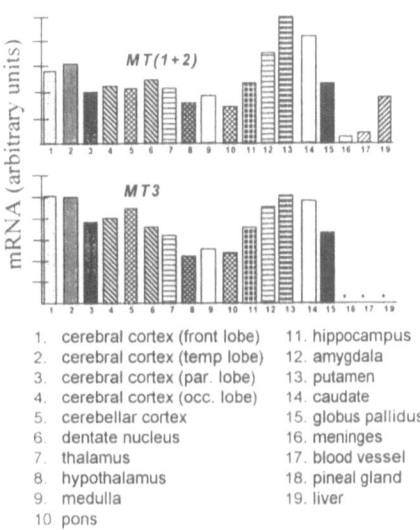

Figure 1. Relative MT isoform, *MT1 + 2* and *MT3* mRNA levels in anatomical regions of the human brain. RNA isolated from regions of the human brain (lanes 1–18) and from human liver (lane 19) were analysed with a probe which hybridises to all expressed *MT1* and *MT2* genes (MT1 + 2) or with probes specific for individual MT isoform mRNAs. MT-specific hybridisation was measured by densitometry and standardised to β-*actin* levels to give an arbitrary value for that isoform which can be compared from region to region. * indicates that no signal was detectable on the autoradiograph.

ing because it is generally assumed that *MT3* is subject to different regulatory factors than the other *MT* genes. We have analysed RNA from a number of human brains in this way with similar conclusions, but it must be noted that for human studies it is extremely difficult to control for factors such as cause of death, illness and treatment prior to death. This is especially true when investigating genes such as *MT* which are known to be responsive to physiological stress and to many pharmacological treatments.

Metallothionein expression in the sheep brain

Metallothionein immunoreactivity was examined in the developing sheep brain with the aim of determining which cell types produce MT, and whether MT synthesis is a heterogeneous event within a particular class of cells [13].

MT immunoreactivity was first observed at E73, that is, 73 days after conception and increased to a peak at E116. MT in the E80 fetal brain was localised to the ventricular zone of the diencephalon, and the molecular layer. Interestingly, there was otherwise little staining of the cerebral cortex. Cells contributing to MT immunoreactivity were primarily radial glial cells, progenitor cells and immature astrocytes.

By E116, MT immunoreactivity was widespread throughout most regions of the brain (Fig. 2). Cells which resembled astrocytes were observed in the white matter of the cerebrum and the molecular layer, and intense immunoreactivity in the glial limiting membrane was also observed, presumably due to the end-feet of astrocytes. Interestingly, a subset of cells in the cerebrum which were MT positive, GFAP negative and GalC positive (a marker of oligodendrocytes), were confirmed to be oligodendroctyes on both morphological and immunochemical criteria.

In order to make a correlation between MT expression and the proliferative status of cells, sections of the sulcus of E116 fetal sheep brain were examined for both MT and PCNA expression. This latter protein becomes reactive in the nuclei of cells in S-phase. MT and GFAP immunoreactivity appeared to be localised to the same population of cells, but these cells were not proliferating as indicated by lack of PCNA immunostaining.

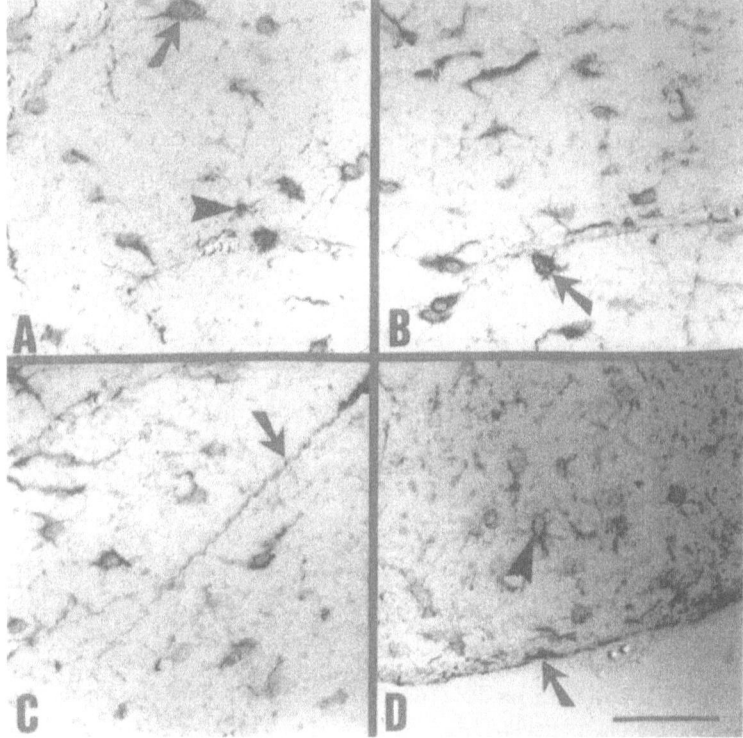

Figure 2. MT immunoreactivity in the E116 fetal sheep brain. (A) Astrocytes (arrow) and oligodendrocytes (arrow head) in the white matter of the cortex. (B) Glial cells (arrow) associated with cortical blood vessels. (C) Astrocytic fibre tracts (arrow) in the white matter of the cortex. (D) Astrocytes (arrow head) and astrocyte end feet (arrow) of the glial limiting membrane. The bar is 50 µm.

An interesting observation from the immunohistochemical survey of the sheep brain is that only a small proportion of cells express MT at high levels. For example it was possible to observe adjacent cells, one of which expressed MT at high levels and one in which no MT could be detected, but which were otherwise morphologically similar and had similar expression profiles with respect to GFAP and S100. It would seem unlikely that these cells differed in their exposure to blood-borne hormones or heavy metals, suggesting that major systemic factors are not responsible for these observations. One possibility is that MT levels correlate to cellular proliferation but we were unable to find evidence for this, either in the sheep model as mentioned [13], or in the regenerating nasal epithelia [14]. It is relevant to note that the sheep brains were obtained from "untreated animals", that is, the animals had not been exposed to exogenous heavy metals or other agents. This suggests that the basal level of MT in individual cells within an anatomical region is set by factors which are intrinsic to that individual cell.

Sheep MT-III

In order to compare the expression of MT-III in the sheep brain to the work reported above, we isolated MT-III clones from a sheep brain cDNA library. The predicted protein product was similar to previously isolated MT-III proteins, containing "inserted" amino acids in both the α and β domains and the *Cys-Pro-Cys-Pro* motif, however it also contained deletion and substitution mutations which removed three of the cysteine residues which are invariant in MT-III proteins from other species. We confirmed the structure of sheep MT-III cDNAs from a number of other sheep, and also showed by Southern blotting that there is apparently only one MT-III gene in this species. We are currently attempting to isolate the MT-III protein from sheep tissues. MT-III is expressed at the mRNA level from late fetal development, and is maximal in the same neural regions which express high levels of MT-I and II, namely pineal gland, cortex, midbrain and basal ganglia, further suggesting that MT-III and the MT-I/II isoforms are similarly regulated at a gross regional level.

Acknowledgments
We would like to thank Mrs Ros Thompson and Mr Graeme McCormack for their excellent technical assistance.

References

1. Young JK, Garvey JS and Huang PC (1991) Glial immunoreactivity for metallothionein in the rat brain. *Glia* 4: 602–610.
2. Nishimura N, Nishimura H, Ghaffar A and Tohyama C (1992) Localization of metallothionein in the brain of rat and mouse. *J Histochem Cytochem* 40: 309–315.
3. Blaauwgeers H G T, Sillevis Smitt P A E, De Jong J M B V and Troost D (1993) Distribution of metallothionein in the human central nervous system. *Glia* 8: 62–70.
4. Suzuki K, Nakajima K, Otaki N, Kimura M (1994) Metallothionein in developing human brain. *Biol Signals* 3: 188–192.
5. Uchida Y, Takio K, Titani K, Ihara Y and Tomonaga M (1991) The growth inhibitory factor that is deficient in the Alzheimer's brain is a 68 amino acid metallothionein-like protein. *Neuron* 7: 337–347.
6. Masters B A, Quaife C J, Erickson J C, Kelly E J, Froelick G J, Zambrowicz B P, Brinster R L and Palmiter R D (1994) Metallothionein III is expressed in neurons that sequester zinc in synaptic vesicles. *J Neurosci* 14: 5844–5857.

7.   Yamada M, Hayashi S, Hozumi I, Inuzuka T, Tsuji S and Takahashi H (1996) Subcellular localisation of growth inhibitory factor in rat brain: light and electron microscopic immunohistochemical studies. *Brain Res* 735: 257–264.
8.   West A K, Stallings R, Hildebrande C E, Chiu R, Karin M and Richards R I (1990) Human metallothionein genes: structure of the functional locus at 16q13. *Genomics* 8: 513–518.
9.   Stennard F A, Holloway AF, Hamilton J and West AK (1994) Characterisation of six additional human metallothionein genes. *Biochim Biophys Acta* 1218: 357–365.
10. Holloway AF, Stennard FA and West AK (1997a) Human metallothionein gene *MT1L* is expressed in several tissues but is not likely to encode a metallothionein protein. *FEBS Lett* 404: 41–44.
11. Peterson MG, Hannan F and Mercer JFB (1988a) The sheep metallothionein gene family. *Eur J Biochem* 174: 417–424.
12. Peterson MG and Mercer JFB (1988b) Differential expression of four linked sheep metallothionein genes. *Eur J Biochem* 174: 425–429.
13. Holloway AF, Stennard FA, Dziegielewska KM, Weller L and West AK (1997b) Localisation of metallothionein immunoreactivity in the adult and fetal sheep brain. *International J Develop Neurosci* 15: 195–203.
14. Skabo SJ, Holloway AF, Chuah MI and West AK (1997) Metallothioneins 1 and 2 are expressed in the olfactory mucosa of mice in untreated animals and during the regeneration of the epithelial layer. *Biochim Biophys Res Commun* 232: 136–142.

Metallothionein IV
C. Klaassen (ed.)
© 1999 Birkhäuser Verlag Basel/Switzerland

Induction of metallothionein-I (MT-I) mRNA in primary astrocyte cultures is mediated by hypotonicity

M. Aschner[1,2] and D.R Conklin[1]

[1]Department of Physiology and Pharmacology and [2]Interdisciplinary Program in Neuroscience, Bowman Gray School of Medicine, Winston-Salem, NC, USA

A brief survey of astrocytic and MT functions, and their relationship to ethanol neurotoxicity

Astrocytes occupy about 20% of the cell volume within the central nervous system (CNS). Their processes are found around synapses and in close association with nodes of Ranvier, axon tracts, and blood vessels [1, 2]. A partial list of astrocyte functions includes secretion of neurotrophic factors, control of extracellular pH, inactivation of extracellular glutamate, and concentrative uptake and metabolism of neurotransmitters [1, 3].

CNS damage in hypoxia, hypoglycemia, as well as neurodegenerative disorders and aging is thought to be partly due to excessive stimulation of neuronal glutamate-gated ion channels [4]. The origin of glutamate (and its analog, aspartate) has been tacitly assumed to be presynaptic nerve endings. However, it is known that astrocytes remove extracellular glutamate by a Na^+-dependent mechanism [5]. This transport has a likely stoichiometry of 1 glutamate and 3 Na^+ transported inwardly, and 1 K^+ transported outwardly to offset the negative charge of glutamate. In the presence of ammonia, glutamate is metabolized to glutamine by the astrocyte-specific enzyme glutamine synthetase (GS)[6], maintaining $[glutamate]_o$ at 0.3 μM [7, 8]. This represents a 10,000-fold gradient $vs.$ $[glutamate]_i$ (3 mM). This glutamate-glutamine pathway constitutes the pool of brain glutamate originally described by Berl [9]. Astrocytes also efficiently remove extracellular taurine by a Na^+-dependent mechanism [10]. This transport system generates and maintains a $[taurine]_i/[taurine]_o$ of about 10,000, and has a likely stoichiometry of 1 taurine and 2 Na^+ ions transported inwardly, to generate and maintain the observed taurine gradient [10]. Release from both the glutamate and taurine pools occurs as a result of astrocytic swelling [2, 11]. In its exaggerated form, astrocytic swelling is deleterious, and can be viewed as a pathological extension of more limited and controlled volume changes which are otherwise part of the normal homeostatic function of astrocytes. Astrocytic swelling is seen as an early event (within 1 h of injury), followed by regulatory volume decrease (RVD) in which astrocytes re-establish their pre-swelling volume by extruding ions (e.g., K^+ and Cl), and compensatory organic osmolytes (e.g., taurine, myo-inositol). Astrocytic swelling is also associated with release of both glutamate and aspartate [11, 12], but these do not appear to substantially contribute to RVD per se [11, 13].

Similar to the "housekeeping" functions ascribed to MTs in other tissues, central nervous system (CNS) MTs are implicated in metal metabolism, cellular repair processes, growth and differentiation, where they are likely to serve as the source of zinc for newly synthesized apoen-zymes, as well as regulator molecules in gene expression. Additional likely functions of MTs include control of intracellular redox potential, and metal detoxification [14, 15, 16, 17, 18, 19, 20, 21]. While ethanol (EtOH) has been reported to stimulate MT-I synthesis in the liver [22, 23], it is not well established whether it can effectively induce MT-I synthesis in the CNS. Accordingly, we have recently studied the effect of EtOH on MT-I mRNA induction in a well characterized model of neonatal rat primary astrocyte cultures. Results from these studies have been recently published [24].

Effects of EtOH on astrocytic MT-I mRNA expression

Recent studies carried out in our laboratory demonstrated increased abundance of astrocytic MT-I mRNA following 6 h treatment with isosmotic EtOH (72 mM NaCl/100 mM EtOH). The increased abundance of MT-I mRNA in cells exposed to isosmotic EtOH (72 mM NaCl/100 mM EtOH) was not apparent in monolayers exposed to hyperosmotic EtOH (122 mM NaCl/100 mM EtOH), suggesting that this effect was not directly associated with EtOH. Hyperosmotic/hypertonic exposure led to a decrease in MT-I mRNA levels compared with controls. The EtOH-independent nature of MT-I mRNA induction raised the question of whether the specificity of the effect resulted from exposure of astrocytes to hypotonic solutions. Therefore, additional cultures were exposed to hyposmotic/hypotonic EtOH-free solutions. Six hours. exposure to 72 mM NaCl bathing solution (EtOH-free) led to an increase in MT-I mRNA levels. Replacement of 50 mM NaCl with 100 mM of the non-permeable salt N-methyl-D-gluc-amine (NMDG) abolished this effect, leading to a decrease in MT-I mRNA expression.

As shown in Table 1, quantitative analysis of the same northern blots suggested that expo-sure to isosmotic EtOH led to a significant ($p < 0.001$) increase in astrocytic MT-I mRNA expression compared with controls. MT-I mRNA levels in cells treated with hyperosmotic EtOH solutions was unchanged. Hyposmotic/hypotonic EtOH-free bathing solution had a sim-ilar effect to isosmotic EtOH exposure, significantly ($p < 0.001$) increasing astrocytic MT-I mRNA levels. Replacement of EtOH with NMDG, or exposure to hyperosmotic/hypertonic solutions led to a significant ($p < 0.0002$) decrease in MT-I mRNA levels.

Table 1. Quantitative analysis of the relative abundance of MT-1 mRNA in neonatal rat primary astrocyte cul-tures

Treatment group	Means	± S.E.
Controls – (122 mM NaCl)	100	-
Isosmotic EtOH – (72 mM NaCl plus 100 mM EtOH)	139.3[*]	21.9
Hyperosmotic EtOH – (122 mM NaCl plus 100 mM EtOH)	103.7	0.1
Hyposmotic/hypotonic, EtOH-free (72 mM NaCl)	132.0[*]	17.8
Isosmotic NMDG, EtOH-free (72 mM NaCl plus 100 mM NMDG)	37.4[**]	16.0
Hyperosmotic/hypertonic EtOH-free (172 mM NaCl)	38.1[**]	8.1

[*]$p < 0.001$ compared with controls. [**]$p < 0.0002$ compared with controls.

These studies suggest that EtOH *per se* does not directly increase the expression of MT-I mRNA in primary astrocyte cultures as evident by the failure of hyperosmotic EtOH to induce MT-I mRNA expression. In contrast, isosmotic EtOH effectively increases the abundance of MT-I mRNA transcripts in the same cells. Exposure to hypotonic solution in the absence of EtOH has an identical effect. We conclude, therefore, that hypotonicity secondary to isosmotic EtOH exposure is associated with the induction of MT-I mRNA.

Ethanol, swelling, and MT: physiological significance

A variety of pathological conditions, including exposure to isosmotic ethanol [25], but not hyperosmotic ethanol [1, 25], lead to swelling of astrocytes. Astrocytic swelling, in turn, stimulates ion release by activation of ion channels at the plasma membrane [1]. Measurement of intracellular pH have recently demonstrated a pronounced acidification following astrocytic swelling [26]. While severe acidosis is an important mediator of brain infarction, a drop in pH and sustained cytosolic acidosis have been postulated to afford cytoprotective mechanisms [27]. For example, during energy depletion, cells lose control of their autolytic enzymes [28, 29], which are capable of inflicting irreversible damage to vital cellular components. However, because most lytic enzymes require a neutral or alkaline condition to optimally operate, the injured cells maintain low intracellular pH, retarding the autolysis and increasing their survival. However, when the plasma membrane is sufficiently weakened by mechanisms such as reactive oxygen species (ROS), protons leak out, the intracellular pH becomes more alkaline, and autolysis is activated [28]. Thus, while severe acidosis is a mediator of brain infarction, mild acidosis may be protective.

Although the mechanisms associated with the antioxidant properties of MT remain speculative, recent evidence supports the hypothesis that increased levels of MTs could potentially provide the "neutralizing" equivalents to oxygen reactive species [30–34]. Nitric oxide (NO) reacts rapidly with $O_2^{\cdot-}$ to produce the peroxynitrite anion (ONOO$^-$), which protonates at relevant pH (pK$_a$ 6.8) to form peroxynitrous acid (ONOOH)[35, 36]. Both ONOO$^-$ and ONOOH are potent oxidizers; ONOOH exhibits hydroxyl radical ($\cdot$OH)-like activity [36] which can initiate a chain reaction that generates numerous toxic metabolites. The rate constant of MT for its reaction with hydroxyl radicals is very high [37]. Since thiols are known targets for both ONOO$^-$ and ONOOH [35], it is plausible that MT isoforms may intercept both these oxidizers in acidotic conditions. Others have suggested that the cysteine residues of MT might serve as an expandable target for ROS [38]. Thus, increased MT levels associated with cell swelling and acidosis may function to intercept ROS, reducing the likelihood for rapid activation of autolytic processes. As such, the up-regulation of MT mRNA may represent an important adaptive response to a variety of physiological insults, such as those occurring during ischemia, anoxia, and hypoxia.

Acknowledgments
This study was partially supported by grants from the National Institute of Environmental Health Services ES 07331, and the US Environmental Protection Agency R-819210 to MA.

References

1. Kimelberg, H.K. and M. Aschner. Astrocytes and their functions, past and present. *In: National Institute on Alcohol Abuse and Alcoholism Research Monograph, Alcohol and Glial Cells*. NIH Publication No. 94-3742, Bethesda, MD, Monograph 27, 1–40, 1994.
2. Kimelberg HK and Norenberg MD (1989) Astrocytes. *Sci Amer* 260: 66–76.
3. Aschner M, Kimelberg HK (eds) (1996) *The Role of Glia in Neurotoxicity*. CRC Press, Inc., Boca Raton, FL, 1996.
4. Coyle JT, Puttfarcken P (1993) Oxidative stress, glutamate, and neurodegenerative disorders. *Science* 262: 689–695.
5. Hertz L (1979) Functional interactions between neurons and astrocytes I. Turnover and metabolism of putative amino acid transmitters. *In: Progress in Neurobiology* 13. Pergamon, Oxford, 277–323.
6. Norenberg MD, Martinez-Hernandez A (1979) Fine structural localization of glutamine synthetase in astrocytes of rat brain. *Brain Res* 161: 303–310.
7. Schousboe A, Divac I (1979) Differences in glutamate uptake in astrocytes cultured from different brain regions. *Brain Res* 177: 407–409.
8. Waniewski RA, Martin DL (1986) Exogenous glutamate is metabolized to glutamine and exported by primary astrocyte cultures. *J Neurochem* 47: 304–313.
9. Berl S, Lajtha A, Waelsch A (1961) Amino acid and protein metabolism. VI. Cerebral compartments of glutamic acid metabolism. *J Neurochem* 7: 186–197.
10. Martin DL (1992) Synthesis and release of neuroactive substances by glial cells. *Glia* 5: 81–94.
11. Vitarella D, DiRisio DJ, Kimelberg HK, Aschner M (1994) Potassium and taurine release are highly correlated with regulatory volume decrease in neonatal primary rat astrocyte cultures. *J Neurochem* 63: 1143–1149.
12. Gilles R, Hoffman EK, Bolis L, eds (1991) *In: F Gilles, EK Hoffman, L Bolis (eds): Advances in Comparative and Environmental Physiology: Volume and Osmolality Control in Animal Cells*, Springer Verlag, Berlin.
13. Mountian I, Declercq PE, Driessche WV (1996) Volume regulation in rat brain glial cells: Lack of a substantial contribution of free amino acid. *Amer J Physiol* 270: C1319-C1325.
14. Hamer DH (1986) Metallothionein. *Annu Rev Biochem* 55: 913–951.
15. Bremner I (1987) Interactions between metallothionein and trace elements. *Prog Food Nutr* 11: 1–37.
16. Kägi JHR, Kojima Y (1987) Chemistry and biochemistry of metallothionein. *Experientia (Suppl)* 52: 25–61.
17. Kägi JHR, Schäffer A (1988) Biochemistry of metallothionein. *Biochemistry* 27: 8509–8515.
18. Rising L, Vitarella D, Kimelberg HK, Aschner M (1995) Metallothionein induction in neonatal rat primary astrocyte cultures protects against methylmercury cytotoxicity. *J Neurochem* 65: 1562–1568.
19. Vallee BL, Maret V (1993) The functional and potential functions of metallothioneins: A personal perspective. *In: KT Suzuki, N Imura, M Kimura (eds): Metallothionein III: Biological Roles and Medical Implications*, Birkhäuser Verlag, Berlin, 1–27.
20. Aschner M (1996) The functional significance of brain metallothioneins. *FASEB J* 10: 1129–1136.
21. Aschner M, Cherian MG, Klaassen CD, Palmiter RD, Erickson JC, Bush A (1997) Metallothioneins in brain – the role in physiology and pathology. *Toxicol Appl Pharmacol* 142: 229–242.
22. Ebadi M, Pfeiffer RF, Huff A (1992) Differential stimulation of hepatic and brain metallothioneins by ethanol. *Neurochem Int* 21: 555–62.
23. Zheng H, Berman NE, Klaassen CD (1995) Chemical modulation of metallothionein I and III mRNA in mouse brain. *Neurochem Int* 27: 43–58.
24. .Aschner M, Conklin DR, Aschner J. Induction of Metallothionein-I (MT-I) mRNA in Primary Astrocyte Cultures is Mediated by Hypotonicity and not Ethanol (EtOH) *per se. Brain Res; in press*
25. Kimelberg HK, Cheema M, O'Connor ER, Tong H, Goderie SK, Rossman PA (1993) Ethanol-induced aspartate and taurine release from primary astrocyte cultures. *J Neurochem* 60: 1682–1689.
26. Busch GL, Wiesinger H, Gulbins E, Wagner HJ, Hamprecht B, Lang F (1996) Effect of astroglial cell swelling on pH of acidic intracellular compartments. *Biochim Biophys Acta* 1285: 212–218.
27. Wu Y, Taylor BM, Sun FF (1996) Alterations in reactive oxygen, pH, and calcium in astrocytoma cells during lethal injury. *Amer J Physiol* 270: C115-C124.
28. Harrison DC, Lemasters JJ, Herman B (1991) A pH-dependent phospholipase A_2 contributes to loss of plasma membrane integrity during chemical hypoxia in rat hepatocytes. *Biochem Biophys Res Commun* 174: 654–659.
29. Madshus IH (1988) Regulation of intracellular pH in eukaryotic cells. *Biochem J* 250: 1–8.
30. Tamai KT, Gralla EB, Ellerby LM, Valentine JS, Thiele DJ (1993) Yeast and mammalian metallothioneins functionally substitute for yeast copper-zinc superoxide dismutase. *Proc Natl Acad Sci USA* 90: 8013–8017.
31. Tamai KT, Liu X, Silar P, Sosinowski T, Thiele DJ (1994) Heat shock transcription factor activates yeast gene expression in response to heat and metallothionein gene expression in response to heat and glucose starvation via distinct signalling pathways. *Mol Cell Biol* 14: 8155–8165.
32. Murphy BJ, Laderoute KR, Chin RJ, Sutherland RM (1994) Metallothionein IIA is up-regulated by hypoxia in human A431 squamous carcinoma cells. *Cancer Res* 54: 5808–5810.
33. Schwarz MA, Lazo JS, Yalowich JC, Reynolds I, Kagan VT, Tyurin V, Kim Y-M, Watkins SC, Pitt BR (1994) Cytoplasmic metallothionein overexpression protects NIH 3T3 cells from tert-butyl hydroperoxide toxicity.

J Biol Chem 269: 15238–15243.

34. Schwarz MA, Lazo JS, Yalowich JC, Reynolds I, Kagan VT, Tyurin V, Kim Y-M, Watkins SC, Pitt BR (1995) Metallothionein protects against the cytotoxic and DNA-damaging effects of nitric oxide. *Proc Natl Acad Sci USA* 92: 4452–4456.
35. Radi R, Beckman JS, Bush KM, Freeman BA (1991) Peroxynitrite oxidation of sulfhydryls. The cytotoxic potential of superoxide and nitric oxide. *J Biol Chem* 266: 4244–4250.
36. Beckman JS, Crow JP (1993) Pathological implications of nitric oxide, superoxide and peroxynitrite formation. *Biochem Soc Trans* 21: 330–334.
37. Sato M, Bremner I (1993) Oxygen free radicals and metallothionein. *Free Radical Biol Med* 14: 325–337.
38. Lazo JS, Kondo Y, Dellapiazza D, Michalska AE, Choo KHA, Pitt BR (1995) Enhanced sensitivity to oxidative stress in cultured embryonic cells from transgenic mice deficient in metallothionein I and II genes. *J Biol Chem* 270: 5506–5510.

Metallothionein in the cerebral cortex of rats showing hereditary high ability in sidman avoidance test

H. Aikawa[1], T. Yoshida[1], H. Ohta[2] and Y. Seki[2]

[1]*Department of Environmental Health, School of Medicine, Tokai University, Bohseidai, Isehara, Kanagawa 259-11 Japan*
[2]*Department of Occupational Health and Toxicology, Kitasato University, 1-15-1 Kitasato, Sagamihara, Kanagawa, 228 Japan*

Summary. The changes in metallothionein (MT) and trace elements (Zn, Cu, Fe, Mn) in the cerebral cortex with aging were investigated for THA (Tokai High Avoider) rats and Wistar rats as control. The level of MT in the cerebral cortex of THA rats increased until 8 weeks old, and the high MT level was maintained thereafter (about 2.5 fold) compared to the levels in Wistar rats. MT in Wistar rats maintained a lower level from age 3 days to 15 weeks, and thereafter increased gradually up to age 40 weeks. In line with the change in MT level in the cerebral cortex, Zn concentration of THA rats also increased with age of rats. It was found that MT concentration in the cerebral cortex of THA rats increased significantly in response to the increasing Zn concentration from early age, compared to Wistar rats. However, the increased Zn concentration in Wistar rats did not correspond to the increase in MT level. The relationship between MT levels and trace elements, particularly Zn, in the cerebral cortex of THA rats, is described with respect to the age of rats.

Characteristics of THA rat

The Tokai High-Avoider (THA) rat was established in Tokai University, and has an innate high-avoidance ability in Sidman avoidance test. THA rat strain was derived from the JCL-Wistar strain rat [1]. The rats have been selectively bred by brother-sister mating of rats at 12–13 weeks of age for high rate of electric shock avoidance by lever pressing. The selection criterion for breeding was avoidance rate of over 95% in the Sidman avoidance schedule in the latter five sessions of the total ten sessions. All avoidance tests were performed for 60 min per day for 10 days from 7 weeks of age. THA rats exhibiting high and stable avoidance behavior with low individual variability by bar-pressing, were established at the 29th generation, and maintained as the new strain in Tokai University (Fig. 1). THA rats performed suerior learning and displayed small individual differences of learning performance, not only in the Sidman electric shock avoidance test, but also in two different tasks, namely water E-maze and two-way shuttle avoidance tests. There was no apparent difference in the development, circadian rhythm, and open-field activity (spontaneous motor activity) when compared with Wistar rat. However, the sensitivity of THA rat to electric shock or to heart stress was less than that of Wistar rat. THA rat possesses an innate ability, not acquired by sensitivity to the shock or developmental differences [2]. Also THA rats have already proven to be quite useful model rats compared to Wistar rats in assessing behavioral toxicity at subclinical doses of toxic agents, to which they had been exposed at the fetal and/or suckling period [3, 4].

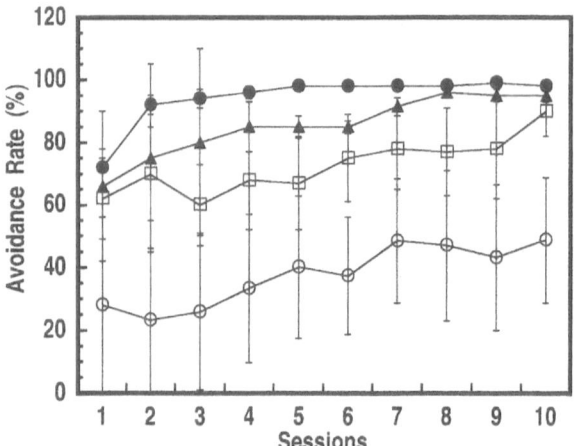

Figure 1. Avoidance rate in male rats during the latter half of the 60 min of daily Sidman avoidance tests for 10 trial days. Values are shown as mean with SD. Experimental groups: ○, Wistar rat (F1: the first generation); △, F5: the fifth generation; □, F2: the 2nd generation; ●, THA rat (F29: the 29th generation).

Metallothionein (MT) in brain

From a recent commentary review on the role of MTs in brain, a partial extract is quoted below.

Metallothionein (MT) proteins and MT-like metal binding protein are induced in various species, and in various organs including brain and testis [5, 6]. MT are proteins of low molecular weight (6–7 kDa) containing about 30% cysteines in their 60 amino acid residues, and existing as the an isoforms MT-I, MT-II, MT-III, and MT-IV. The functions attributed to MT include homeostasis of zinc and copper, and cellular cytoprotection from reactive oxygen species caused by ionizing radiation, electrophilic anti-cancer drugs and mutagen, and metals such as cadmium. Both MT-I and MT-II are found in various organs and tissues while MT-III is thought to be expressed only in brain along with MT-I and MT-II. MT-III protein can be determined by Cd-Heme method as total MT including MT-I and MT-II [7].

Purpose of this study

In this study, the changes in MT levels in the cerebral cortex with aging were investigated for THA rats, using Wistar rats as control. Also the changes in zinc, copper, manganese, and iron levels in the cerebral cortex were monitored to determine the characteristics of THA rat.

THA rats were derived from the JCL-Wistar strain rats (Japan Clea, Tokyo) according to our previous study [2]. Female and male THA rats (the 29th generation) were used in the present study. Animals were deeply anaesthetized with carbon dioxide (dry ice) and sacrificed at 3 days old, and 7, 8, 15, 20, 30, and 40 weeks old. To obtain the cerebral cortex, dissection of the brain

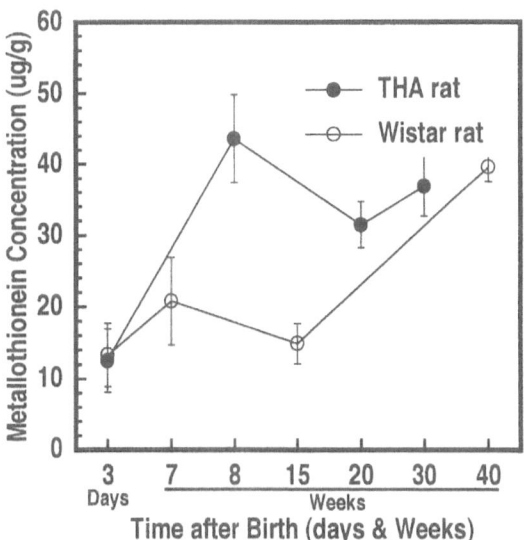

Figure 2. The time cours of metallothionein with aging in the cerebral cortex of THA and Wistar rats.

was performed according to the method described by Glowinski and Iversen [8]. MT levels were determined as total MT by the Cd-Hem assay [9]. Zn, Cu, Fe, and Mn were determined by the flame or flameless methods using Hitachi 180–80 Polarized Zeeman atomic absorption spectrophotometer after tissue digestion with nitric acid. Statistical analysis was performed by ANOVA-Scheffe, PLSD, and Student's t-test or Aspin Welch's t-test at $p < 0.05$.

Results

In the present study, we report changes in MT in cerebral cortex with aging of THA rat, using Wistar rats as control.

MT levels in the cerebral cortex of THA rats increased until age 8 weeks, and thereafter remained at the same high level (about 2.5 fold) compared to Wistar rats. The MT levels of Wistar rat remained at a lower MT level from age 3 days to 15 weeks, and thereafter increased gradually up to age 40 weeks (Fig. 2). In line with the change in MT levels, Zn levels in cerebral cortex of THA rats increased prominently with increasing age. On the other hand, the Fe level of Wistar rats also increased with increasing age. Additionally, Cu concentration increased in line with the MT level. However, although Cu and Fe showed the same tendency with Zn, there was no positive correlation with each other. Mn concentration showed no change in both THA and Wistar rats (Fig. 3).

The results show that the MT induction in cerebral cortex of THA rats occurred from an early age, and high MT levels were maintained compared to Wistar rats.

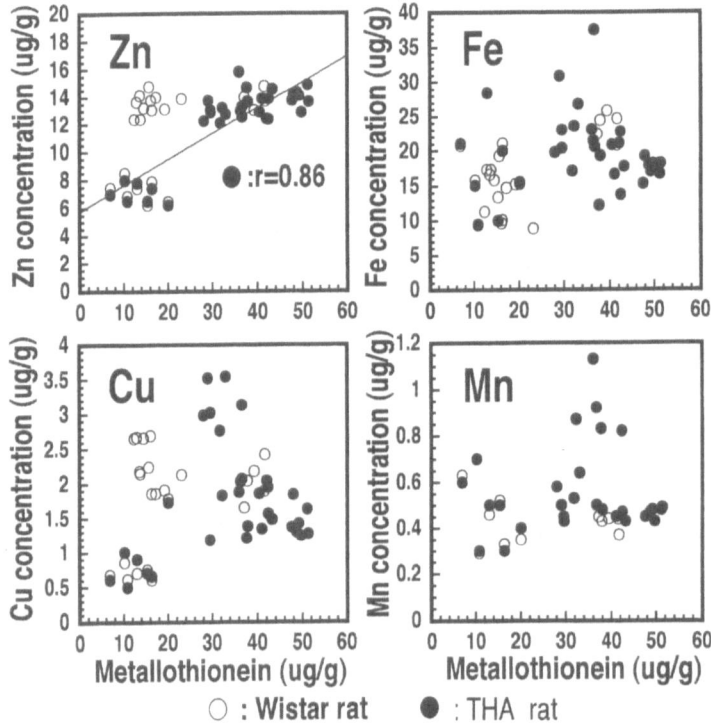

Figure 3. Relationship among metals (Zn, Cu, Fe, Mn) and metallothionein in the cerebral cortex of THA and Wistar rats.

The presence of MT-III in brain and ist effects on learning and memory

Expression of MT-III and MT-I mRNA in various regions of the mouse brain has been reported by Palmiter et al. [10]. The MT mRNA levels are almost same in those regoins except for hippocampus. MT-III mRNA level in hippocampus was significantly higher than MT-II level. Cerebral Zn metabolism and its relationship to MT homeostasis relating to Alzheimer's disease was discussed. MT-III was first isolated by Uchida et al. [5], as a growth inhibitory factor (GIF) from neuronal cultures and was later identified as MT isomer because of the structural similarities to MT-I and MT-II [11].

Preliminary studies using the antibody of MT-I and MT-II suggest that the concentration of MT in human brain may be age-dependent; MT concentration is high in aged brain. The localization of MT-III in neurons that sequester zinc in synaptic terminals, so-called zincergic neurons, suggested several possible unique functions for this MT isoform. And it has been suggested that MT-III might play a role in protecting these cells from zinc toxicity, or it might facilitate transport of zinc from the plasma membrane to synaptic vesicles, and it might protect against oxidative damage. Furthermore, because the dentate gyrus and hippocampal formation are implicated in several form of learning and memory, it has been suggested that MT-III might influence these behaviors. It is thought that the hippocampal formation and neocortex are usually the primary sites of excessive neuronal activity and neurodegeneration that characterize epileptic seizures. Thus, the role of MT-III or other MT isoforms in the brain might be revealed by exciting the zincergic neurons pharmacologically [10].

Palmiter and co-workers [10] have reported preliminary results of studies systematically testing possibilities for the participation of MT-III in learning ability and memory as follows: MT-III did not influence learning ability, based on studies using adult mice that either cannot synthesize MT-III (MT-III–/–), wild type, or over-express human MT-III (hMT-III). In their studies, animals were tested for their ability to learn and remember a simple association using a shuttle box. In another test, three groups of mice were tested in a Morris water maze, in which they had to find a submerged platform in a pool of milky water using visible cues within the room. There were no differences in their ability to learn to find the platform. Also, mice treated with zinc were tested in the water maze before and after the zinc treatment. They also tested a set of 2 year old MT-III–/– mice and their wild-type littermates in the water maze to see if MT-III has any effect on the function of zincergic neurons during the aging process. Again, there was no difference between these groups, too.

From the results described above, they concluded that even though the zincergic neurons that express MT-III are thought to be involved in learning and memory, neither the absence of MT-III nor an excess of MT-III affects these processes in young adult mice, senescent mice, or zinc-treated mice.

Total MT level in the cerebral cortex of THA rat

As mentioned above, although there are some studies reporting the presence and induction of MT and its mRNA (MT-I, MT-II, and MT-III) in brain, there are no detailed studies on changes in MT levels with aging, learning and memory. In our studies using THA rat, we found that the MT concentration in the cerebral cortex of THA rat increased significantly from an early age compared to Wistar rats. And Zn concentration in the cerebral cortex correlated with increasing MT concentration (Figs 2, 3). These results are quite interesting to understand the characteristics of THA rats, and to understand the relationship between the MT level and avoidance behavior in the Sidman test.

We have already reported some studies on the effect of toluene or lead on the behavior response in the Sidman test using THA rats. The learning impairment appearing in THA rat was caused primarily by lead [3]. Exposure of rats to 100 and 500 ppm toluene during fetal and suckling period affected the learning in the rats after termination of the growth period, and the learning disorder still remained as long as 100 days after the end of toluene exposure, although the effect decerased gradually [12]. Recently, we reported additionally that the striatum influences motor activity via changes in the amounts of neurotransmitters released during the period of high-dose (2000 ppm and 4000 ppm) toluene exposure [4].

On the other hand, THA has higher level of amino acid and acetylcholine releasehe in the cerebral cortex and the striatum. Additionally, the counts of acetylcholine esterase positive cell was higher than that of Wistar rat (our unpublished data).

Conclusion

From our preliminary study presented here, we would like to suggest that THA rat may be a useful animal model for the study of aging or diseases affecting behavior and memory.

Acknowlegement
The present study was supported in part by a grant from Kitasato university (SAHS-A-104 & B-110).

References

1. Shigeta S, Miyake K, Misawa T, Aikawa H, Yoshida T, Katoh H (1990) A new inbred rat strain "THA". *Rat News Letter* 23: 9–11.
2. Shigeta S, Misawa T, Yoshida T, Aikawa H, Momotani H, Miyake K (1989) Neurobehavioral analysis of high-rate Sidman avoidance rat strain. *Jpn J Psychopharmacol* 9: 217–244.
3. Shigeta S, Aikawa H, Misawa T, Yoshida T, Momotani H, Suzuki K (1986) Strain difference in learning impairment in rats following lead administration during brain development. *Tokai J Exp Clin Med* 11: 241–247.
4. Aikawa H, Yoshida T, Shigeta S (1997) Changes in the amounts of neurotransmitters released from the striatum and spontaneous motor activity in rats exposed to high doses of toluene. *Environ Health Prevent Med* 1: 171–177,1997.
5. Uchida Y, Takio K, Titani K, Ihara Y, Tomonaga M (1991) The growth inhibitory factor that is deficient in Alzheimer's disease is a 68 amino acid metallothionein-like protein. *Neuron* 7: 337–347.
6. Ohta H, Nakakita M, Tanaka H, Seki Y, Yoshikawa H (1997) Induction of metallothionein-like cadmium-binding protein in the testis by oral cadmium administration in rats. *Ind Health* 35: 96–103.
7. Aschner M, Cherian MG, Klaassen CD, Palmiter RD, Erickson JC, Bush AI (1997) Commentary Metallothionein in brain – The role in physiology and pathology. *Toxicol Appl Pharmacol* 142: 229–242.
8. Glowinski J, Iversen LL (1966) Regional studies of catecholamines in the rat brain-I The disposition of [3H] norepinephrine, [3H]dopamine and [3H]dopa in various regions of the brain. *J Neurochem* 13: 655–669.
9. Onosaka S, Tanaka K, Doi M, Okahara K (1978) A simplified procedure for determination of metallothionein in aminal tissue. *Eisei Kagaku* 24: 128–133.
10. Palmiter RD (1997) Genetic analysis of metallothionein III function *In*: Aschner M, Cherian MG, Klaassen CD, Palmiter RD, Erickson JC, Bush AI (1997) Commentary Metallothionein in brain – The role in physiology and pathology. *Toxicol Appl Pharmacol* 142: 233–236.
11. Cherian MG (1997) Biochemical properties of metallothionein *In*: Aschner M, Cherian MG, Klaassen CD, Palmiter RD, Erickson JC, Bush AI (1997) Commentary Metallothionein in brain – The role in physiology and pathology. *Toxicol Appl Pharmacol* 142: 229–231.
12. Shigeta S, Misawa T, Aikawa H, Momotani H, Suzuki K, Yoshida T (1986) Effects of low level toluene exposure during the developing stage of the brain on learing in high avoider rats. *Jpn J Tnd Health* 28: 445–454.
13. Shigeta S, Misawa T, Aikawa H (1980) Effects of concentration and duration of toluene exposure on Sidman avoidance in rats. *Neurobehav Toxicol* 2: 85–88.
12. Ebadi M (1991) Metallothionein and other zinc-binding proteins in brain. *Methods Enzymol* 205: 363–387.
13. Shiraishi K, Nakazawa S, Ito H (1993) Zinc enhances kainate neurotoxicity in the rat brain. *Neurol Res* 15: 113–1161415.

Role of metallothionein in copper metabolism, diabetes and autoimmune disease

Metallothionein IV
C. Klaassen (ed.)
© 1999 Birkhäuser Verlag Basel/Switzerland

Roles of metallothionein in the cytotoxicity of copper in the liver of LEC rats

Kazuo T. Suzuki[1], Ming Rui[1], Jun-ichi Ueda[2] and Toshihiko Ozawa[2]

[1]*Faculty of Pharmaceutical Sciences, Chiba University, Chiba 263-8522, Japan*
[2]*National Institute of Radiological Sciences, Chiba 263, Japan*

Summary. Copper (Cu) accumulates in the livers of Long-Evans cinnamon (LEC) rats, an animal whose genetic deficiency models Wilson disease in humans. Although excess Cu is assumed to cause hepatitis and hepatocarcinoma, the exact mechanism of Cu-induced injury is not known. Our investigation of the mechanism leading to Cu-induced hepatotoxicity (indicated by the onset of jaundice) has revealed that: 1) Cu in the livers of LEC rats is primarily bound to metallothionein (MT) before the onset of jaundice. However, the proportion of hepatic Cu not associated with MT increases just prior to the onset of jaundice. 2) Cu accumulates in the livers of Wistar rats (which have normal Cu metabolism) after repeated injections of Cu. However, it does not cause hepatitis, even at concentrations twice as high as that seen in livers of LEC rats. 3) Hydroxyl radicals are generated from hydrogen peroxide in the presence of Cu-containing MT. This is consistent with the proposal that, although Zn-containing MT acts as an anti-oxidant, Cu-containing MT functions as a pro-oxidant. 4) In liver homogenates prepared from LEC rats before the onset of jaundice, ascorbate radicals are observed first, followed by the appearance of hydroxyl radicals. After jaundice appears, however, hydroxyl radicals appear simultaneously with or without ascorbate radicals. Taken together, these data suggest that Cu,Zn-containing MT acts as an anti-oxidant to protect liver from hydroxyl radicals generated in the presence of ascorbic acid. However, Cu-MT without Zn is a pro-oxidant in the livers of LEC rats.

Introduction

Copper (Cu) is an essential heavy metal for the life and forms the active center of Cu-enzymes that catalyze redox reactions in enzymes such as Cu,Zn-superoxide dismutase (Cu,Zn-SOD), ceruloplasmin, lysyl oxidase, and others. Cu is harmful to the body, organs and cells when it exists in excess, especially when not bound to appropriate biological constituents such as Cu-enzymes and metallothionein (MT, a protein induced by excess essential heavy metals (Cu and Zn) and non-essential harmful heavy metals (cadmium (Cd), mercury (Hg), and others).

Wilson disease is a hereditary human disease involved disordered Cu metabolism. It is caused by a mutation in an autosomal recessive gene (WND or ATP7B, on chromosome 13) encoding a Cu-binding, copper-transporting ATPase [1–3], which results in Cu accumulation in the liver in an MT-bound form. The excessive accumulation of Cu is the cause of pathological consequences of Wilson disease, which progress with age.

Long-Evans cinnamon (LEC) rats harbor a mutation in a homologue of the human ATP7B gene (atp7b, on rat chromosome 16) [4–6]. They accumulate MT-associated Cu in the liver [7–11]. LEC rats suffer from acute hepatitis and/or hepatocarcinoma caused by the excess liver Cu [10–13]. However, binding of Cu to MT is believed to play a protective role in suppressing Cu toxicity by decreasing the ability of metal ions to participate in harmful intracellular interactions. As a result, Cu not bound to MT (non-MT-bound Cu) has been assumed to be the toxic form, and to appear when the metal accumulates to levels that exceed the capacity of a cell to synthesize sufficient MT [10, 11, 14].

However, it is not known how non-MT-bound Cu causes hepatitis and hepatocarcinoma (i.e. directly through interaction of (ionic) Cu with biological constituents, or indirectly by reactive oxygen species (ROS) generated by the non-MT-bound Cu participating in redox reactions). The present study was intended to explore the mechanism underlying the induction of acute hepatitis in LEC rats by Cu, and the role of MT in the production of reactive oxygen species (Scheme 1).

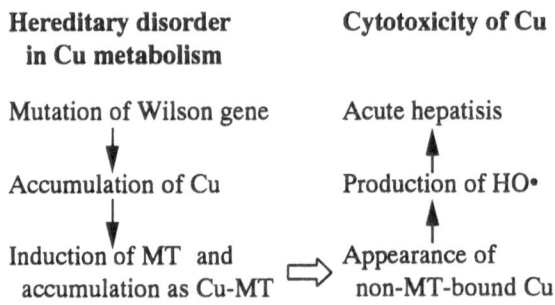

Scheme 1. Proposed mechanism underlying the occurrence of hepatitis by the hereditary disorder of Cu metabolism in LEC rats.

Induction of hepatitis by Cu bound to MT in the livers of LEC rats

Cu accumulating in the livers of LEC rats causes acute hepatitis at 16–18 weeks of age when the concentration of Cu reaches approximately 200 µg Cu/g wet weight [10, 11]. Early in life (well before jaundice appears), Cu is bound mostly to MT. However, the amount of hepatic Cu unbound to MT increases just prior to the onset of jaundice. Hepatic Cu levels decrease after the appearance of acute hepatitis and jaundice because of desquamation of damaged cells with the highest accumulated Cu levels [10, 11]. Thus, in LEC rats, hepatocytes appear to be able to sequester a maximum of approximately 200 µg Cu bound to MT per gram wet weight of liver.

Unlike LEC rats, Wistar rats with normal Cu metabolism, subjected to repeated subcutaneous Cu injection, accumulated Cu in the liver to levels greater than 400 µg Cu/g wet weight [15] without exhibiting any signs of hepatitis [15]. Hepatic Cu in Wistar rats was partly bound to MT at the beginning of the repeated injections, but, after repeated injections, was primarily in a non-MT-associated form. These results indicate that Cu present in a non-MT-bound form is not sufficient, alone, to cause hepatitis. In addition, they suggest that the accumulation of Cu as Cu-MT aggravates the development of hepatitis, as proposed in Scheme 1 and Figure 1.

Towards the onset of jaundice, Cu in the bloodstream increases followed by increase of Cu in kidneys, suggesting that Cu was released as Cu-MT from the damaged liver cells [16].

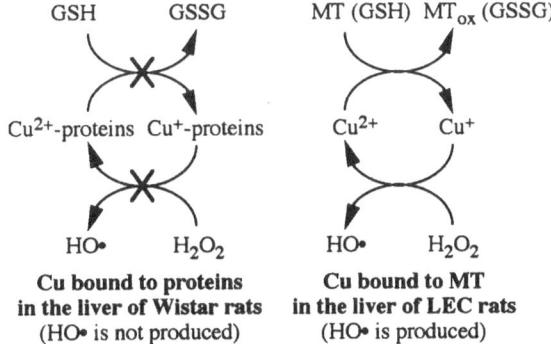

GSH   GSSG   MT (GSH) MT_{ox} (GSSG)

Cu^{2+}-proteins Cu^+-proteins Cu^{2+} Cu^+

HO• H_2O_2 HO• H_2O_2

Cu bound to proteins   **Cu bound to MT**
in the liver of Wistar rats **in the liver of LEC rats**
(HO• is not produced)   (HO• is produced)

Figure 1. Proposed mechanism underlying the occurrence of hepatitis by Cu accumulating in the liver of LEC rats but not of Wistar rats.

Production of hydroxyl and ascorbate radicals in the liver of LEC rats

Hydroxyl radicals are produced in the livers of LEC rats in the presence of hydrogen peroxide. The production of hydroxyl radicals is age-dependent (i.e. dependent on hepatic Cu levels) [17–19]. In addition, our results indicate that ascorbate radicals are produced in the presence of hydrogen peroxide in the livers of young, unjaundiced LEC rats. In older rats, just prior to the onset of jaundice, hydroxyl radicals are produced. The lack of evidence of hepatic damage in the presence of ascorbate suggests that ascorbic acid may work protectively by inhibiting lipid peroxidation caused by hydroxyl radicals [20].

Production of hydroxyl radicals *in vitro* by Cu-containing metallothionein; Roles of metallothionein as anti- and pro-oxidants depending on the presence or absence of Zn bound to metallothionein

The observation that hydroxyl radicals appear in the livers of LEC rats toward the onset of jaundice [20], and that acute hepatitis occurs in LEC rats (but not in Cu-injected Wistar rats that accumulated even higher Cu levels) [15], suggests that accumulation of Cu to high levels is not, in itself, the cause of hepatitis in LEC rats. Rather, Cu must be accumulated in an MT-bound form, and the concentration of Cu must be sufficient to result in MT that is bound primarily to Cu, and not to Zn (Scheme 1, see also Figure 4).

It is known that cadmium (Cd) accumulates in the liver in an MT-bound form, and becomes toxic when the metal accumulates to levels higher than the capacity of cells to synthesize MT, and when Cd unbound to MT appears [21]. The ratio of Cd to Zn in MT increases just prior to the point with non-MT-bound Cd becomes evident [21].

In the liver of LEC rats, the ratio of Cu/Zn in MT and the amount of non-MT-bound Cu increased close to the onset of jaundice [10, 11], suggesting that the appearance of non-MT-bound Cu is a prerequisite for the onset of jaundice. However, the presence of non-MT-bound Cu is, alone, not sufficient to cause cytotoxicity [10], and in this respect differs from the situation with Cd. As proposed in Figure 1, the appearance of both MT, and Cu unassociated with MT (i.e. non-MT-bound Cu as a result of spill-over of Cu from MT in LEC rats accumulating Cu to a level exceeding the capacity of cells to synthesize MT) seem to be required.

Cu catalyzes redox reactions in the presence of reducing agents and hydroxyl radicals are produced from hydrogen peroxide by the Fenton reaction in the presence of reducing agents as shown in Figure 2. MT is rich in cysteinyl residues and efficiently reduces cupric to cuprous ions [22], suggesting that MT is an efficient reducing agent to reduce Cu in the Fenton reaction.

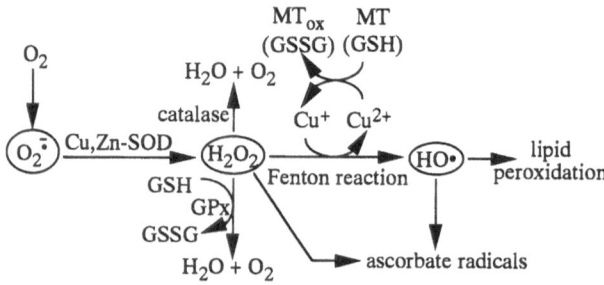

Figure 2. Production of reactive oxygen species.

Zn-MT was prepared from the livers of Wistar rats that had been injected with Zn. Zn in MT was replaced *in vitro* by adding Cu ions. Cuprous ions are known to replace Zn bound to MT without oxidizing MT, while cupric ions replace Zn after being reduced by MT, thereby oxidizing MT and forming disulfide bonds [22] as shown in Figure 3. The quantitative relationship between Cu-reduced and Zn-replaced MT was determined, and then the production of

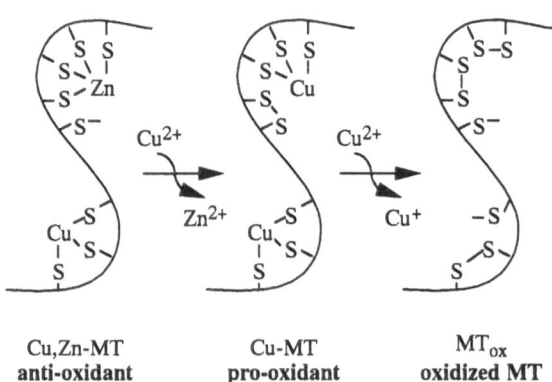

Figure 3. Roles of Cu-containing MT as anti- and pro-oxidants according to the presence or absence of Zn as a co-existing metal in MT.

hydroxyl radicals in the presence of hydrogen peroxide plus Zn-MT, Cu,Zn-MT, or Cu-MT were determined [23].

Zn bound to MT was replaced by cupric ions at the ratio of Cu/Zn = 1.0. Hydroxyl radicals (measured by DMPO trapping) were not produced by Zn-MT or Cu,Zn-MT as long as Zn was present bound to MT. Instead, hydroxyl radicals were produced when cupric ions were added in excess to replace all Zn bound to MT, and then to oxidize all sulfhydryl groups in MT [23]. Theses results suggest that MT works as an anti-oxidant as long as Zn is present bound to Cu-containing MT. On the other hand, MT turns to a pro-oxidant when Zn is not present bound to Cu-containing MT, indicating that MT is working as the reducing agent to reduce cupric to cuprous ions in the Fenton reaction as represented in Figures 1, 2 and 4.

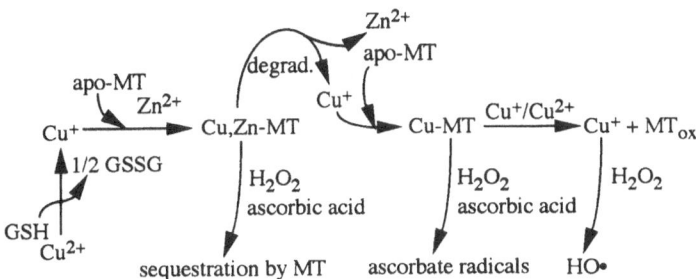

Figure 4. Production of ascorbate and hydroxyl radicals with accumulation of Cu as Cu-MT in the liver of LEC rats.

Conclusions

Summarizing our observations on the mechanisms underlying the cytotoxicity of Cu in the liver of LEC rats (i.e. the occurrence of acute hepatitis), we found that: 1) Cu accumulating in the liver owing to a defective Cu-binding, Cu-transporter protein to remove Cu from cells induces the synthesis of MT. Excess Cu accumulates in the form bound to MT. 2) Zn is present bound to Cu-containing MT as Cu,Zn-MT as long as cells retain the capacity to synthesize sufficient MT to accommodate for Cu and Zn within cells: Cu,Zn-MT functions as an anti-oxidant by sequestering excess Cu and/or scavenging radicals as long as Zn is present bound to MT (that is, as long as excess Cu is reduced and sequestered by MT by replacing zinc). MT is transformed into a pro-oxidant, however, when it reduces Cu by oxidizing its thiol groups. In this case, Cu liberated from oxidized MT is effectively reduced from cuprous to cuprous by oxidizing sulfhydryl groups on MT molecules to disulfide bonds. 3) Hepatitis is suggested to be caused by hydroxyl radicals that are produced from hydrogen peroxide by the Fenton reaction where Cu works as a catalyst and MT (and probably glutathione) as a reducing agent.

Genotoxicity of the Cu accumulating in the livers of LEC rats leading to hepatocarcinoma has to be explained by distinct mechanisms from those of the cytotoxicity.

References

1. Bull PC, Thomas GR, Rommens JM, Forbes JR, Cox DW (1993) The Wilson disease gene is a putative copper transporting P-type ATPase similar to the Menkes gene. *Nature Genet* 5: 327–337.
2. Petrukhin K, Fischer SG, Pirastu M, Tanzi RE, Chernov I, Devoto M, Brzustowicz LM, Cayanis E, Vitale E, Russo JJ, Matseoane D, Boukhgalter B, Wasco W, Figus AL, Loudianos J, Cao A, Sternlieb I, Evgrafov O, Parano E, Pavone L, Warburton D, Ott J, Penchaszadeh GK, Scheinberg IH, Gilliam TC (1993) Mapping, cloning and genetic characterization of the region containing the Wilson disease gene. *Nature Genet* 5: 338–343.
3. Tanzi RE, Petrukhin K, Chernov I, Pellequer JL, Wasco W, Ross B, Romano DM, Parano E, Pavone L, Brzustowicz LM, Devoto M, Peppercorn J, Bush AI, Sternlieb I, Pirastu M, Gusella JF, Evgrafov O, Penchaszadeh GK, Honing B, Edelman IS, Soares MB, Scheinberg IH, Gilliam TC (1993) The Wilson disease gene is a copper transporting ATPase with homology to the Menkes disease gene. *Nature Genet* 5: 344–350.
4. Sasaki N, Hayashizaki Y, Muramatsu M, Matsuda Y, Ando Y, Kuramoto T, Serikawa T, Azuma T, Naito A, Agui T, Yamashita T, Miyoshi I, Takeichi N, Kasai N (1994) The gene responsible for LEC hepatitis, located on rat chromosome 16, is the homolog to the human Wilson disease gene. *Biochem Biophys Res Commun* 202: 512–518.
5. Wu J, Forbes JR, Chen HS and Cox DW (1994) The LEC rat has a deletion in the copper transporting ATPase gene homologous to the Wilson disease gene. *Nature Genet* 7: 541–545.
6. Yamaguchi Y, Heiny ME, Shimizu N, Aoki T, Gitlin JD (1994) Expression of the Wilson disease gene is deficient in the Long-Evans cinnamon rats. *Biochem J* 301: 1–4.
7. Li Y, Togashi YSato S, Emoto T, Kang J-H, Takeichi N, Kobayashi H, Kojima Y, Une Y, Uchino J (1991) Spontaneous hepatic copper accumulation in LEC rats with hereditary hepatitis A model of Wilson's disease. *J Clin Invest* 87: 1858–1861.
8. Sakurai H, Kamada H, Fukudome A, Kito M, Takeshima S, Kimura M, Otaki N, Nakajima K, Kawano K, Hagino T (1992) Copper-metallothionein induction in the liver of LEC rats. *Biochem Biophys Res Commun* 185: 548–552.
9. Sugawara N, Sugawara C, Sato M, Katakura M, Takahashi H, Mori M (1991) Copper metabolism in LEC rats aged 30 and 80 days old: Induction of Cu-metallothionein and status of zinc and iron. *Res Commun Chem Pathol Pharmacol* 72: 353–362.
10. Suzuki KT, Kanno S, Misawa S, Sumi Y (1993) Changes of hepatic copper distributions leading to hepatitis in LEC rats. *Res Commun Chem Pathol Pharmacol* 82: 217–224.
11. Suzuki KT, Kanno S, Misawa S, Aoki Y (1995) Copper metabolism leading to and following acute hepatitis in LEC rats. *Toxicology* 97: 81–92.
12. Li Y, Togashi Y, Sato S, Emoto T, Kang J-H, Takeichi N, Kobayashi H, Kojima Y, Une Y, Uchino J (1991) Abnormal copper accumulation in non-cancerous and cancerous liver tissues of LEC rats developing hereditary hepatitis and spontaneous hepatoma. *Jpn J Cancer Res* 82: 490–492.
13. Okayasu T, Tochimaru H, Hyuga T, Takahashi T, Takekoshi YLi Y, Togashi Y, Takeichi N, Kasai N, Arashima S (1992) Inherited copper toxicity in Long-Evans cinnamon rats exhibiting spontaneous hepatitis a model of Wilson's disease. *Pediatr Res* 31: 253–257.
14. Suzuki KT (1995) Disordered copper metabolism in LEC rats, an animal model of Wilson disease roles of metallothionein. *Res Commun Mol Pathol Pharmacol* 89: 221–240.
15. Kanno S, Aoki Y, Suzuki JS, Takeichi N, Misawa S, Suzuki KT (1994) Enhanced synthesis of metallothionein as a possible cause of abnormal copper accumulation in LEC rats. *J Inorg Biochem* 56: 117–125.
16. Suzuki KT, Kanno S, Misawa S, Sumi Y (1993) Changes of copper distributions in the plasma and kidneys of LEC rats following acute hepatitis. *Res Commun Chem Pathol Pharmacol* 82: 225–232.
17. Sakurai H, Satoh H, Hatanaka A, Sawada T, Kawano K, Nagino T, Nakajima K (1994) Unusual generation of hydroxyl radicals in hepatic copper-metallothionein of LEC (Long-Evans Cinnamon) rats in the presence of hydrogen peroxide. *Biochem Biophys Res Commun* 199: 313–318.
18. Nakayama K, Okabe M, Aoyagi K, Yamanoshita O, Okui T, Ohyama T, Kasai N (1996) Visualization of yellowish-orange luminescence from cuprous metallothioneins in liver of Long-Evance Cinnamon rat. *Biochim Biophys Acta* 1289: 150–158.
19. Nakamura M, Nakayama K, Shishido N, Yumino K, Ohyama T (1997) Metal-induced hydroxyl radical generation by Cu⁺-metallothioneins from LEC rats. *Biochem Biophys Res Commun* 231: 549–552.
20. Suzuki KT, Rui M, Ueda J, Ozawa T (1997) Production of ascorbate and hydroxyl radicals in the liver of LEC rats in relation to hepatitis. *Res Commun Mol Pathol Pharmacol* 96: 137–146.
21. Suzuki KT (1984) Studies of cadmium uptake and metabolism in the kidney. *Environ Health Persp* 54: 21–30.
22. Suzuki KT, Maitani T (1981) Metal dependent properties of metallothionein: Replacement *in vitro* of zinc in zinc-thionein with copper. *Biochem J* 199: 289–295.
23. Suzuki KT, Rui M, Ueda J, Ozawa T (1996) Production of hydroxyl radicals by copper-containing metallothionein Roles as pro-oxidant. *Toxicol Appl Pharmacol* 141: 231–237.

Metallothionein IV
C. Klaassen (ed.)
© 1999 Birkhäuser Verlag Basel/Switzerland

Fate of copper and metallothionein in the liver of LEC rats

Dominik Klein[1], Josef Lichtmannegger[2], Ulrich Heinzmann[3], Josef Müller-Höcker[4] and Karl H. Summer[2]

[1]Institute of Toxicology and Environmental Hygiene, Technical University, D-80636 Munich, Germany
[2]Institute of Toxicology and [3]Institute of Pathology, GSF-National Research Center for Environment and Health, D-85764 Neuherberg, Germany
[4]Institute of Pathology, University of Munich, D-80337 Munich, Germany

Summary. The Long-Evans cinnamon (LEC) rat has a mutation homologous to the human Wilson's disease gene leading to copper induced hepatotoxicity. The mechanism of how excess copper damages the liver or what chemical form of copper is toxic is yet unclear. In liver cytosol copper levels are highest just before the onset of hepatitis and decline thereafter. In this compartment total copper was bound to metallothionein (MT). In lysosomes both copper and iron accumulate with increasing age and development of liver damage and considerable amounts of non MT-bound copper are present. In severely affected livers, large amounts of copper are associated with insoluble material of high density, which upon ultrastructural information, is derived from lysosomes of Kupffer cells. This copper-rich material is considered to consist of polymers of degradation products of copper-MT. We suggest that chronic copper toxicity in LEC rats involves the uptake of copper-loaded MT into lysosomes where it is incompletely degraded and polymerizes to an insoluble material containing reactive copper. This copper together with iron initiates lysosomal lipid peroxidation leading to hepatocyte necrosis. Subsequent to phagocytosis by Kupffer cells the reactive copper may amplify liver damage either direct or through stimulation of these cells.

The progressive accumulation of copper in hepatocytes of humans and animals may lead to hepatocellular necrosis. In Wilson's disease, an inherited disorder of copper metabolism in humans, the excretion of copper into bile is impaired. The phenotype of this disease is caused by a mutation of the ATP7B gene leading to an altered function of the gene product, a copper-transporting P-type ATPase [1]. Recently, an animal model for this disease, the Long-Evans cinnamon (LEC) rat was discovered. These animals have a mutation in the gene homologous to the human Wilson's disease gene and show many features of the disease like elevated hepatic copper levels, reduced biliary copper excretion, hemolysis, ceruloplasmin deficiency and increased hepatic iron levels [2]. The mechanism of how excess copper damages the hepatocyte or what chemical form of copper is toxic, however, is unclear.

The hepatocellular accumulation of copper is generally associated with elevated levels and increased binding of copper to metallothionein (MT). Whereas in LEC animals the binding of copper to MT is well known for the cytosol, the contribution of MT in the binding of copper in the particulate fraction has not been adequately investigated. In the following the role of lysosomes in copper accumulation and hepatotoxicity in these animals is discussed.

Pathology

LEC rats aged up to 78 days were unremarkable as compared to control animals (Tab. 1). In the LEC rat aged 84 days serum GOT activity was elevated, and histologically single cell

Table 1. Assignment of the LEC rats to the experimental groups

group (No. of animals)	age (d) range (mean)	serum GOT (U/L) range (mean)	serum Bilirubin (mg/dL) range (mean)	histopatho- logical findings in the liver
Wistar (n = 9)	55–62 (57 ± 2)	190–311 (246 ± 38)	< 0.5–0.6	negative
LEC non-diseased, 56–64 d old (n = 9)	56–64 (59 ± 3)	161–322 (238 ± 42)	< 0.5	negative
LEC non-diseased, 70–78 d old (n = 4)	70–78 (75 ± 3)	194–232 (213 ± 15)	< 0.5	negative
LEC, slightly affected liver (n = 1)	84	520	< 0.5	positive
LEC, jaundice (n = 3)	87–91 (88 ± 2)	990–1065 (1037 ± 33)	11.8–27.7 (17.1 ± 7.5)	positive

necroses, nuclei irregular in size and enlarged hepatocytes were noted, indicating that hepatitis has already started. LEC rats aged 87 and 91 days suffered from systemic jaundice. Light microscopically, these animals showed an increased number of mesenchymal cells, lymphoid celliform infiltrates, cell infiltrates in the connective tissue also with granulocytes, macrophages loaded with cereous pigments partly accompanied by lymphoid cells, single cell necroses, partly hyalin-spherical cytoplasmic inclusions, various sizes of nuclei, and partly very large cells. Ultrastructural findings in hepatic parenchymal cells of LEC rats aged 70–78 days included unremarkable mitochondria, nuclei and bile canaliculi, but notably increased numbers of lysosomes. Also, focal cytoplasmic necroses, and, in the cells of the reticulo-endothelial system an increased phagocytotic activity and numerous electron dense bodies were observed. LEC rats with systemic jaundice additionally showed dilatation of cisternae of the rough endoplasmatic reticulum. Infrequently, minimal dilated mitochondrial cristae were observed. However, in these animals an abundance of lysosomes packed with electron dense material was observed in Kupffer cells (Fig. 1).

Copper and metallothionein in subcellular compartments of the liver

Copper concentrations in the liver of LEC rats increased with age until 78 days and remained fairly constant during the progression of liver disease. In contrast, in the particulate fraction copper levels steadily increased with age and the development of liver damage.

In the cytosol, copper concentrations were highest just before the onset of hepatitis and declined thereafter (Fig. 2A). Both the concentration of total and copper-containing MT changed in a similar pattern to that of copper (Fig. 2B). As can be calculated from the concentrations of copper and copper-containing MT, and as confirmed by HPLC analysis (Fig. 6), total cytosolic copper was bound to MT, independent of age and the occurrence of liver disease (Fig. 2A).

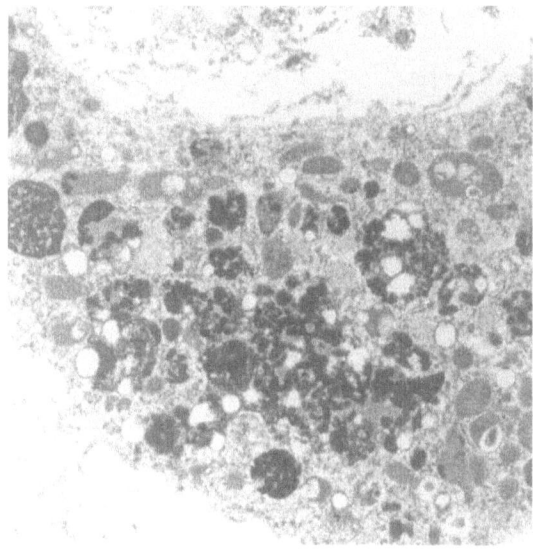

Figure 1. Electron micrograph of a Kupffer cell from a jaundiced rat with numerous electron-dense lysosomal particles (original magnification × 16,000).

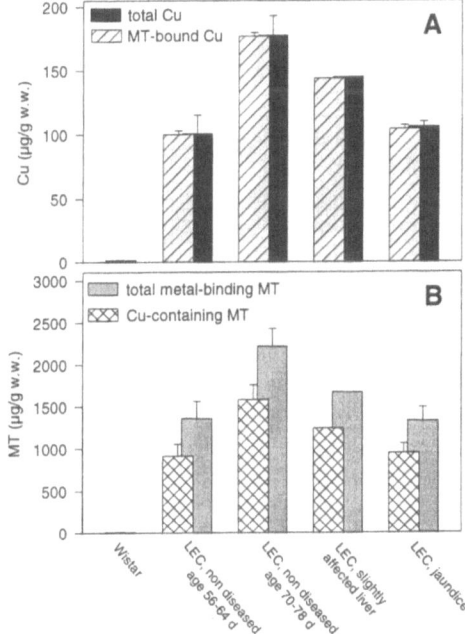

Figure 2. Copper (A) and metallothionein (MT) (B) in hepatic cytosol of non-diseased LEC rats (mean ± S.D., n = 4–9), a LEC rat with slightly affected liver (single values) and LEC rats with systemic jaundice (mean ± S.D., n = 3).

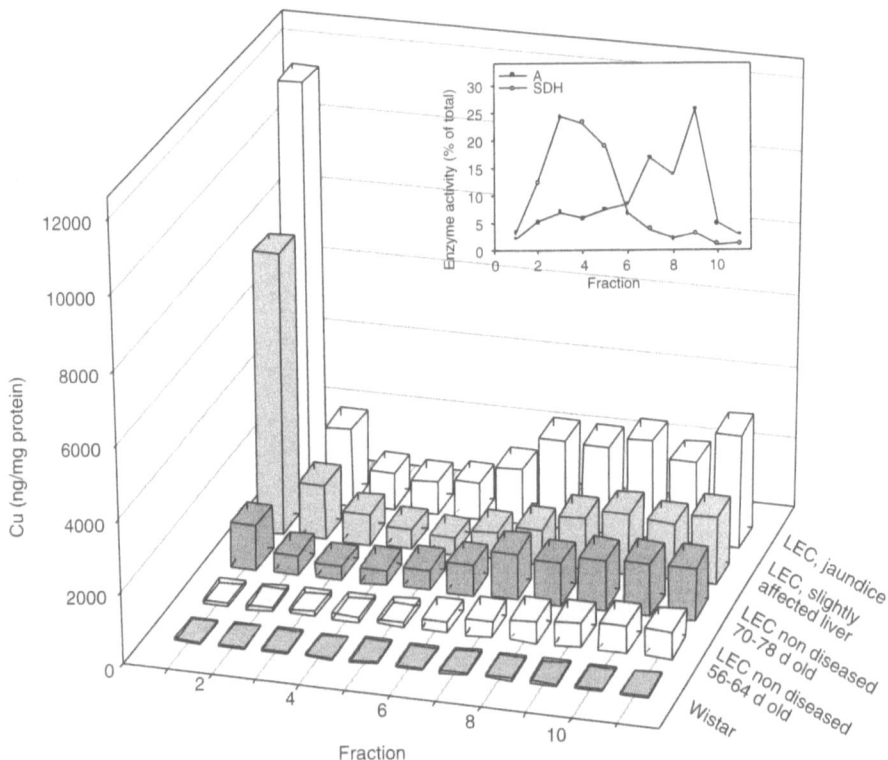

Figure 3. Copper in density fractions after separating crude lysosomes from livers of non-diseased rats (mean values, n = 4–9), a LEC rat with slightly affected liver (single values) and LEC rats with systemic jaundice (mean values, n = 3) by a density gradient centrifugation (highest density fraction No. 1). The insert shows a typical distribution profile of the lysosomal marker acid phosphatase (AP) and the mitochondrial marker succinate dehydrogenase (SDH) activities in these fractions.

The distribution of copper in density gradient fractions of crude lysosomes which were obtained through differential centrifugation of liver homogenate is shown in Figure 3. Except for fractions 1 and 2, copper was preferentially associated with lysosomes. Slightly increased levels of copper were also observed in mitochondria. In the lysosome-specific fraction No. 9 of the density gradient, both copper and iron levels increased with age of the animals and the progression of liver disease (Fig. 4A). Total MT concentrations were highest just before the onset of liver disease and remained fairly constant thereafter. Oxidized MT and copper-containing MT represented equally about 50% of total MT independent of age and liver disease (Fig. 4B). As a consequence, lysosomes of diseased animals contained about 6 times higher levels of non MT-bound copper than the non-diseased LEC rats of the younger age group (Fig. 4A).

In the rats with slightly affected liver or systemic jaundice, enormous amounts of copper were associated with insoluble material in the fraction of the highest density (fraction No. 1 of

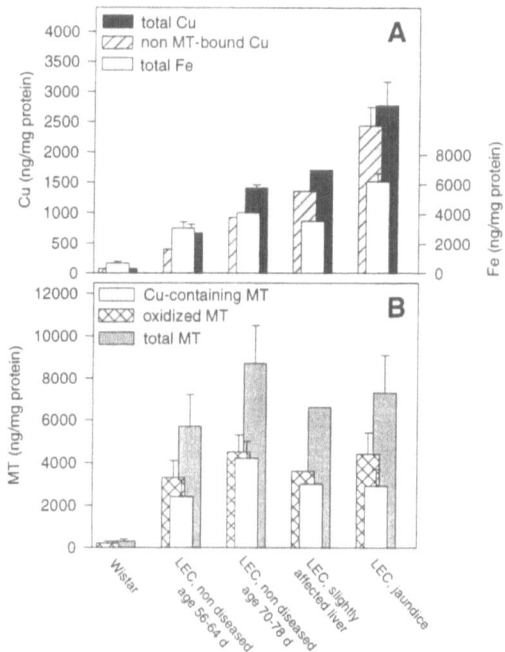

Figure 4. Copper, iron (A) and metallothionein (MT) (B) in hepatic lysosomes (fraction No. 9 of the density gradient) of non-diseased LEC rats (mean ± S.D., n = 4–9), a LEC rat with slightly affected liver (single values) and LEC rats with systemic jaundice (mean ± S.D., n = 3).

the density gradient) which was lacking lysosome- or mitochondria-specific enzyme activities (Fig. 3). Ultrastructurally, this insoluble material consisted of lysosomes containing electron-dense granules (Fig. 5A). In the more severely diseased animals granules of even higher density also sedimented with nuclei and cellular debris (not shown). Using energy dispersive X-ray microanalysis the electron-dense material could be identified as copper (Fig. 5B). The granules could be solubilized with guanidinium thiocyanate under reducing conditions. HPLC analysis revealed a copper-containing component of about 7.3 kD, corresponding to 50% of the apparent molecular weight of native MT (Fig. 6). This component cross reacted with MT-specific antibodies in an ELISA assay (not shown). The same component, although at lower concentration, could be isolated in addition to native MT from the lysosome-specific fraction No. 9 of the density gradient (Fig. 6).

Discussion

In agreement with the key role of copper in LEC rats developing hepatitis, copper accumulated in the liver with age. Strikingly, maximum copper levels were reached before the onset of

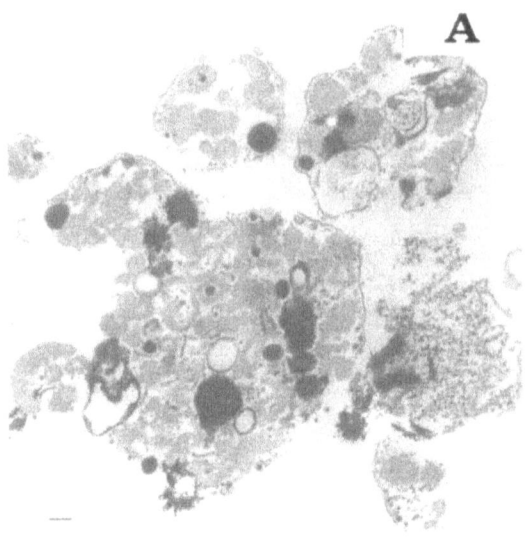

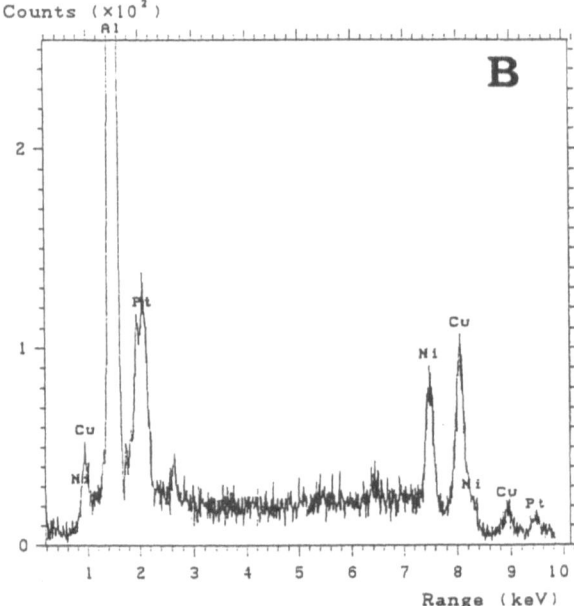

Figure 5. (A) Electron micrograph of the high density fraction of crude lysosomes from a jaundiced LEC rat exhibiting dense material predominantly inside the lysosomes and in rare cases adhering the membrane (original magnification × 52,400; bar indicates 100 nm). (B) Energy dispersive X-ray microanalysis of the lysosomal electron-dense matrix. Aluminum (Al), Nickel (Ni) and Platinum (Pt) peaks are derived from the stub, grid and the surface coating of the specimen, respectively.

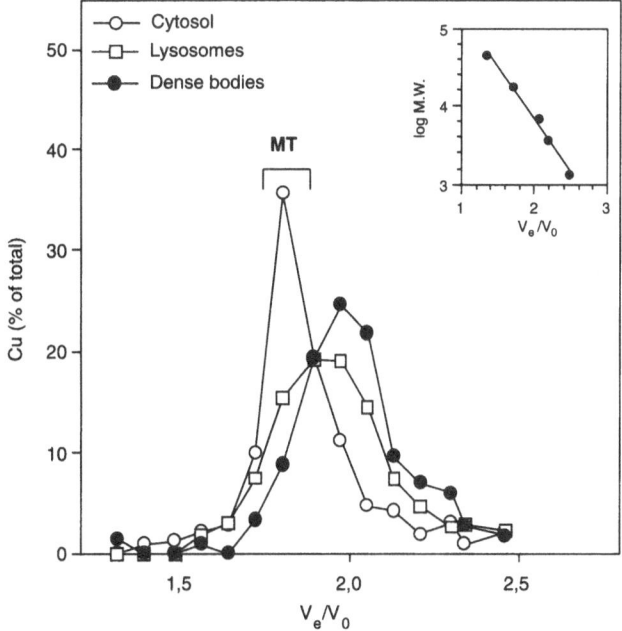

Figure 6. HPLC gelchromatography of cytosol, solubilized lysosomes (fraction No. 9 of the density gradient) and solubilized dense bodies (derived from fraction No. 1) from the liver of a jaundiced rat. The insert shows the relative elution volumes of chicken ovalbumin (M.W. 44,000), equine myoglobin (M.W. 17,000), aprotinin (M.W. 6,500), β-endorphin (M.W. 3,465), and angiotensin (M.W. 1,297). The bracket indicates the relative elution volume of native MT.

hepatitis. Thereafter copper levels increased only in the particulate fraction of the liver. According to our ultrastructural and biochemical findings, the increase in particulate copper levels may represent a shift in the distribution of copper from the hepatocyte to insoluble components of Kupffer cells rather than a change of distribution of copper within hepatocytes.

Consistent with the function of lysosomes in biliary copper excretion, copper levels were already elevated in hepatocyte lysosomes of non diseased LEC rats. Correspondingly, the accumulation of the metal in lysosomes was considered to represent a detoxification mechanism [3]. The sequestration of excess copper by lysosomes which increased in number and size with copper loading [4], may be of general importance for the maintenance of a constant concentration of copper in the cytosol of liver cells. However, findings in humans [5] and animals [6] suggested that copper cytotoxicity may be mediated by a labilisation of lysosomal membranes. Undoubtedly, the molecular association of copper within the lysosomes will be of crucial importance with regard to its toxicity. According to our MT and HPLC analyses, part of the copper in the lysosomes was bound to MT. Whereas in younger, non diseased LEC rats MT accounted for the binding of approximately 41% of lysosomal copper, this value was lowered to 12% in the liver of animals with systemic jaundice. Thus, the concentration of non MT-bound copper was higher in jaundiced LEC rats than in non-diseased animals. This copper was associated with insoluble material. High amounts of a similar copper-rich insoluble material was isolated from the liver of diseased animals. Relying on ultrastructural information, the vast majority of this high density material was derived from lysosomes of Kupffer cells. Both the parenchymal and the mesenchymal insoluble material could be solubilized under reducing conditions. The solubilized product had an apparent molecular weight of about 7.3 kD and cross reacted with MT specific antibodies. Accordingly, the insoluble material is considered to consist of polymers of degradation products of copper-containing MT. With a molecular weight of 50% of that of native MT, the degradation product is likely to represent the β-domain of MT which in copper-containing MT has been shown to be more stable than the α-domain [7].

There is considerable uncertainty on the subcellular site of injury produced by copper. From studies with Wilson's disease patients it has been postulated that hepatic copper in the cytosol which is predominant in the early stage of the disease is toxic but lysosomal copper seen in the later stage is less toxic [8]. In the liver of our LEC animals, independent of age or liver disease, the total of copper in cytosol was bound to MT. Consistent with the role of MT in detoxifying metals, this finding does not support an activity of cytosolic copper in initiating cytotoxicity in LEC rats. Therefore, the non MT-bound copper in the cytosol observed at later stages of jaundice [9] likely follows and not precedes the liver damage.

There is evidence that the hepatocytes packed with copper-loaded lysosomes are the cells that undergo necrosis [10]. The crucial role of lysosomal copper in causing hepatotoxicity is supported by our observations on the association of copper in these organelles during the development of liver injury. Since total copper in the cytosol was bound to MT, copper will be taken up by the lysosomes as highly copper-loaded MT. An important question concerns the fate of the copper-containing MT inside the lysosome. It will be mainly determined by its metal composition: due to the acidic pH of the lysosomal matrix, zinc-containing MT will easily lose the metal, and the apoprotein consequently will be degraded [11]. Since metal removal from copper-MT requires lower pH [12], the lysosomal pH is likely not low enough to remove all the copper from highly copper-loaded MT thus rendering the protein fairly stable against proteolysis. Consistently, copper-containing MT was reported to be fairly resistant against hydrolysis by lysosomal enzymes in vitro [13]. Furthermore, in vivo, the intralysosomal pH has been found to be increased after copper overload [6]. Our results suggest that the highly copper-loaded MT in lysosomes is only partly degraded. The resulting proteolysis-resistant product, presumably the β-domain of MT, might polymerize thereby accumulating as an oxidized, insoluble material which in LEC rats apparently cannot be eliminated from the hepatocyte. The polymerization process likely proceeds through a radical-mediated mechanism involving iron which was massively elevated in the hepatocyte lysosomes of LEC rats. The degradation products obviously are polymerized through disulfide bridges since the solubilization of the polymer by guanidinium thiocyanate required the presence of 2-mercaptoethanol. Therefore, it is reasonable to assume that copper associated with the polymer is loosely bound and therefore more reactive than copper bound to native MT. In agreement with its increased reactivity, copper in the copper-rich granules is histochemically stainable with rhodanine whereas copper bound to MT is not [14]. The reactive polymer-associated copper may initiate lipid peroxidation of lysosomal membranes resulting in lysosomal rupture and release of hydrolytic enzymes into the cytoplasm. The elevated levels of lysosomal iron may amplify this process. Fe(II) may reduce Cu(II) to Cu(I) which is a more potent generator of hydroxyl radicals than Fe(II). Consistent with the importance of iron in both the polymerization of degradation products of copper-containing MT and lysosomal lipid peroxidation none of LEC rats fed an iron-deficient diet died of fulminant hepatitis [15].

According to our ultrastructural findings, in the liver of jaundiced LEC rats copper is massively located in Kupffer cells. This is most likely the result of phagocytotic copper uptake from injured hepatocytes. Once within the macrophages, this material seems to further aggregate to copper-rich granules which electronmicroscopically appear as electron-dense bodies. The accumulation of high amounts of presumably reactive copper in Kupffer cells may amplify the liver damage either direct or through stimulation of these cells. Supporting this assumption, the importance of Kupffer cells in metal-induced hepatotoxicity has been reported [16].

On the basis of these results, the development of copper toxicity in LEC rats is suggested to proceed through the following steps: First, due to the basic gene defect copper accumulates in liver cytosol leading to high concentrations of highly copper-loaded MT. The second step involves the uptake of copper-containing MT into hepatocyte lysosomes and the incomplete degradation of the protein due to its relatively high stability. The third step is a likely iron-mediated polymerization of degradation products of copper-containing MT to an insoluble material containing reactive copper. In the next step, this copper together with iron initiates lysosomal lipid peroxidation finally leading to hepatocyte necrosis. Subsequent to phagocytosis of necrotic hepatocytes by Kupffer cells the reactive copper may amplify liver damage either direct or through stimulation of these cells.

References

1. Bull PC, Thomas GR, Rommens JM, Forbes JR, Cox DW (1993) The Wilson disease gene is a putative copper transporting P-type ATPase similar to the Menkes gene. *Nat Genet* 5: 327–337
2. Harris ED (1995) The iron-copper connection: the link to ceruloplasmin grows stronger. *Nutr Rev* 53: 170–173
3. Goldfischer S, Popper H, Sternlieb I (1989) The significance of variations in distribution of copper in liver disease. *Am J Pathol* 99: 715–723
4. Gross JB, Myers BM, Kost LJ, Kuntz SM, LaRusso NF (1989) Biliary copper excretion by hepatocyte lysosomes in the rat. *J Clin Invest* 83: 30–39
5. Sternlieb I (1980) Copper and the liver. *Gastroenterology* 78: 1615–1628
6. Myers BM, Prendergast FG, Holman R, Kuntz SM, LaRusso NF (1993) Alterations in hepatocyte lysosomes in experimental hepatic copper overload in rats. *Gastroenterology* 105: 1814–1823
7. Li H, Otvos JD (1996) [111]Cd NMR studies of the domain specificity of Ag+ and Cu+ binding to metallothionein. *Biochemistry* 35: 13929–13936
8. Goldfischer S, Sternlieb I (1968) Changes in the distribution of hepatic copper in relation to the progression of Wilson's disease (hepatolenticular degeneration). *Am J Pathol* 53: 883–901
9. Suzuki KT, Kanno S, Misawa S, Aoki Y (1995) Copper metabolism leading to and following acute hepatitis in LEC rats. *Toxicology* 97: 81–92
10. Kumaratilake JS, Howell JMcC (1989) Lysosomes in the pathogenesis of liver injury in chronic copper poisoned sheep: An ultrastructural and morphometric study. *J Comp Pathol* 100: 381–390
11. Mehra RK, Bremner I (1985) Studies on the metabolism of rat liver copper-metallothionein. *Biochem J* 227: 903–908
12. Hunziker PE (1991) Metal removal from mammalian metallothioneins. *Methods Enzymol* 205: 451–452
13. Bremner I, Mehra RK (1983) Metallothionein: Some aspects of its structure and function with special regard to its involvement in copper and zinc metabolism. *Chemica Scripta* 21: 117–121
14. Sumi Y, Kawahara S, Kikuchi Y, Sawada JI, Suzuki T, Suzuki KT (1993) Histochemical and immunohistochemical localization of copper, iron and metallothionein in the liver and kidney of LEC rats. *Acta Histochem*

Cytochem 26: 5–9
15. Kato J, Kobune M, Kohgo Y, Sugawara N, Hisai H, Nakamura T, Samasaki S, Sawada N, Niitsu Y (1996) Hepatic iron deprivation prevents spontaneous development of fulminant hepatitis and liver cancer in Long-Evans cinnamon rats. *J Clin Invest* 98: 923–929
16. Sauer JM, Waalkes MP, Hooser SB, Kuester RK, McQueen CA, Sipes IG (1997) Suppression of Kupffer cell function prevents cadmium induced hepatocellular necrosis in the male Sprague-Dawley rat. *Toxicology* 121: 155–164

Metallothionein IV
C. Klaassen (ed.)
© 1999 Birkhäuser Verlag Basel/Switzerland

Relationship between Cu metabolism hereditary disorders and distribution of Cu-metallothionein in kidneys

Masashi Okabe[1], Shigeru Saito[1], Mika Suzuki-Kurasaki[1], Takeshi Saito[2], Akira Hata[3], Fumio Endo[4], Koji Nagano[4], Ken-ichi Urakami[5], Ichiro Matsuda[4] and Masaaki Kurasaki[1]

[1]*Department of Environmental Medicine and Informatics, Graduate School of Environmental Earth Science, and* [2]*Department of Hygiene and Prevented Medicine, and* [3]*Department of Public Health, Hokkaido University, School of Medicine, Sapporo 060, Japan*
[4]*Department of Pediatrics, Kumamoto University Shool of Medicine, Kumamoto 860, Japan*
[5]*Terumo Corporation, Research and Development Center, Kanagawa 259-01, Japan*

Introduction

Cu is an essential trace element which requires a delicate cellular balance between necessity and toxicity. Cu accumulation in kidney has been reported in genetic disorders of Cu metabolism, such as Long-Evans Cinnamon (LEC) rats, a model for Wilson's disease [1], macular mice, a model for Menkes' disease [2], and the rat injected with Au [3]. In this study, to investigate the mechanism of Cu metabolism in mammalian kidneys, the distribution of Cu-metallothionein (Cu-MT) in the kidneys of above mentioned renal Cu accumulating rodents was revealed using auto-fluorescence, which depends on Cu-S cluster in the protein [4–8]. The fluorescent signals were in good agreement with several biochemical and immunochemical aspects of Cu-MT. Moreover, the emissions may give information such as the state of MT, e.g., partial Cu ion release or oxidation of the protein [4]. In addition, to clarify the relationship between transported MT and renal induced MT, *in situ* Northern hybridization and immuno-histochemistry using an antibody against MT were also employed. The significance of the obtained results was discussed in relation to Cu transporting ATPases.

Visualization of Cu-MT

The visualization procedures for Cu-MT were carried out according to our previous report [9]. The sections were mounted on glass slides or dry nylon membranes and thawed at 25°C. The slides were immediately immersed in acetone. Auto-fluorescence signals on glass slides were observed with an epi-fluorescence microscope using a modified U-MWU filter cube. Signals of Cu-MT on nylon membranes were also detected by illumination with UV light at 365 nm. To visualize Cu-MT in the kidneys, a barrier filter of 530 nm (cut-off range, below 530 nm) was suitable, since the spectral range of the auto-fluorescence emission of Cu-MT in the tissue was 550–700 nm. On the other hand, nuisance auto-fluorescent signals from other bio-molecules have been reported below 550 nm of emission [6]. Using a 420 nm barrier filter (cut-off range, below 420 nm), the background appeared bright-blue and the auto-fluorescent signals of Cu-MT appeared red-orange.

The localization of the signals was identical to those of MT immuno histochemistry, cysteine staining and MT mRNA signals. Moreover, the auto-fluorescent signals disappeared after Hg(II) treatment. Thus the observed signals were identified as Cu-MT [9].

Exogenously injected Cu-MT in rat kidney

The auto-fluorescent signals for Cu-MT and immuno-histochemistry of MT were predominantly detected in the renal cortex of rats administered Cu-MT (Fig. 1, A). In microscopic observation, the signals were located in the lysosome of PCT S1 and S2 adjacent to the glomeruli in the cortex (Fig. 1, B). The orange signal was gradually converted to yellow (from 24 h to 120 h after the Cu-MT injection) indicating that the Cu-MT was involved in the degradation process in lysosome by oxidation. During the time course, the level of MT mRNA increased in the cortex (Fig. 1, C and D). The level of immuno-reactivity against MT antibody was almost the same in the same region (not shown). From these results, it was thought that Cu bound to the injected MT was released in lysosome and became a new inducer of *de novo* biosynthesis of MT in the same region. The results of gel filtration also supported this inference. Zn content in the MT fraction dramatically increased in the renal cytosol of rat administered Cu-MT from 24 h to 120 h after the injection. Since it is well known that Cu-binding ability to MT is stronger than Zn-binding ability to the protein, the Cu, Zn-MT obtained from the renal cytosol 120 h after the administration was thought to be the *de novo* synthesized protein by Cu ions released from Cu-MT in lysosome.

LEC rat kidney

The interesting findings of Cu-MT and its mRNA localization are; (i) Orange fluorescent Cu-MT was located in the cytoplasm and nuclei of PST S3 of the outer stripe of outer medulla (Fig. 1, E), and MT mRNA was also localized in the same stripe (Fig. 1, F), (ii) yellow-orange fluorescent signals indicating that Cu-MT is involved in the degradation process were observed in the lysosome of PCT S1 and S2 adjacent to the glomeruli in the cortex (Fig. 1, G), but the MT mRNA was not detected in this region (Fig. 1, F).

The protein located in the cortex was estimated to be transported from other organ(s), since no MT mRNA was observed in this region. In contrast, the Cu-MT emitting the orange signals was thought to be biosynthesized in the outer stripe of outer medulla, because the distribution of MT mRNA showed identical patterns to that of immuno-reactivity for MT (not shown) and the orange signals in the outer stripe of outer medulla. We postulate that the inducer of MT in the outer stripe of outer medulla was Cu ions that were released from partially degraded Cu-MT in the lysosome in the cortex.

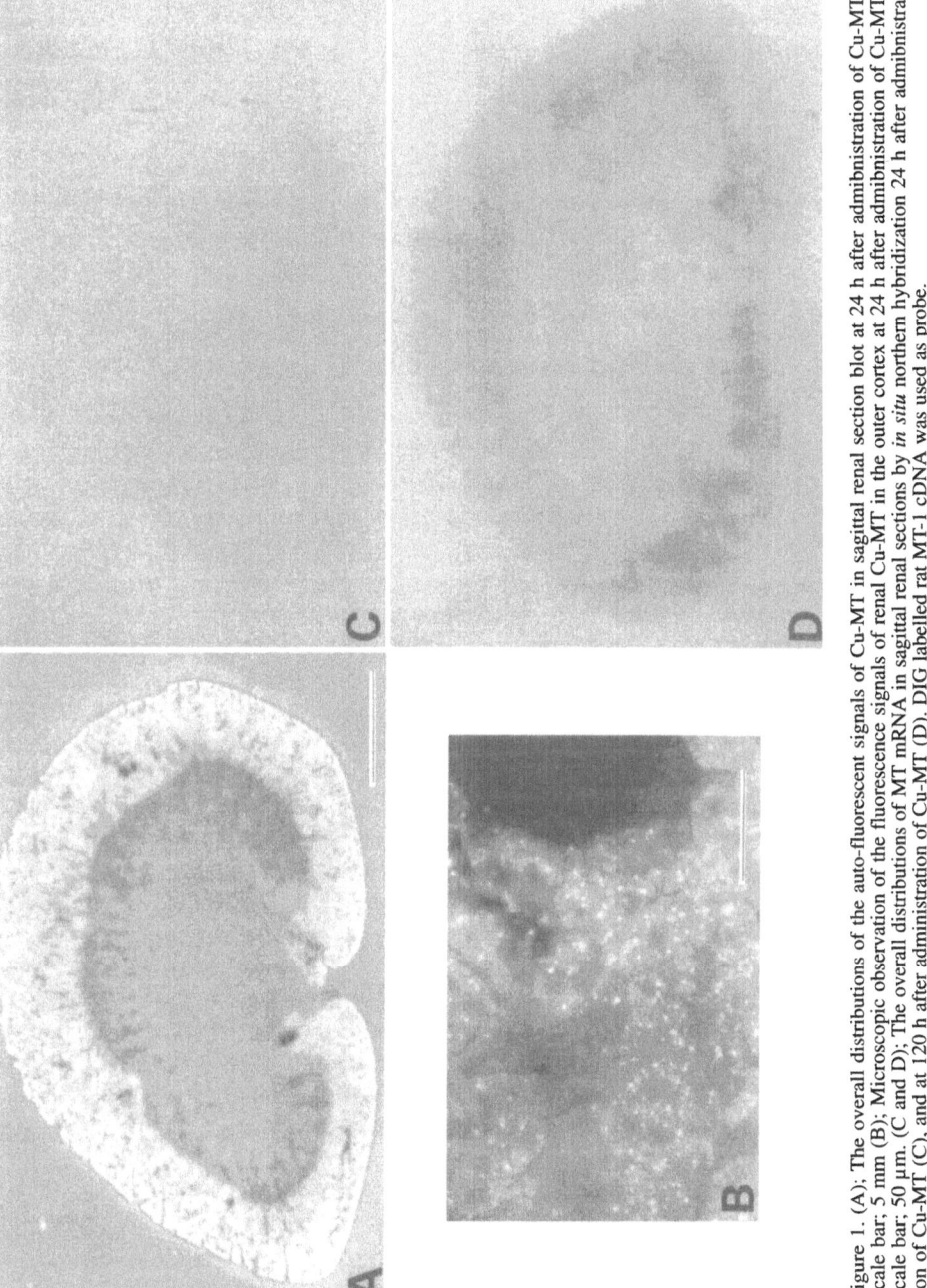

Figure 1. (A): The overall distributions of the auto-fluorescent signals of Cu-MT in sagittal renal section blot at 24 h after admibnistration of Cu-MT. Scale bar; 5 mm (B); Microscopic observation of the fluorescence signals of renal Cu-MT in the outer cortex at 24 h after admibnistration of Cu-MT. Scale bar; 50 µm. (C and D); The overall distributions of MT mRNA in sagittal renal sections by *in situ* northern hybridization 24 h after admibnistration of Cu-MT (C), and at 120 h after administration of Cu-MT (D). DIG labelled rat MT-1 cDNA was used as probe.

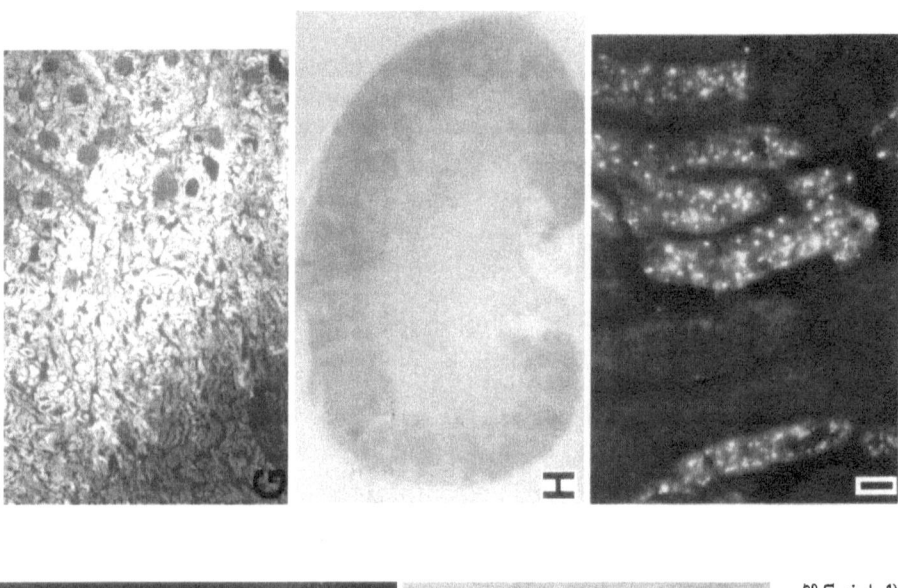

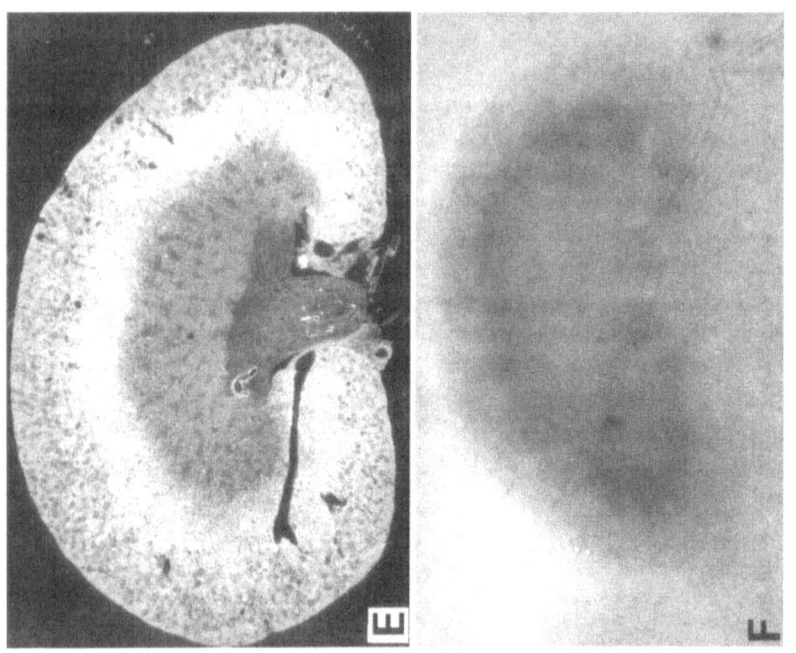

Figure 1.(continued) (E); Low power microphotographs of sagittal renal sections showing the overall distributions of the auto-fluorescence of Cu-MT of LEC rat. (F); Northern hybridization on the blot of LEC rat kidney using ^{32}P-labeled rat MT-I cDNA as probe. (G); Fine localization of the signals of Cu-MT of LEC rat. Boundary region between cortex and outer stripe of outer medulla. (H); Northern hybridization of the kidney from the Macular mice using a labeled MT-I cDNA as probe. (I); Detailed localization of the fluorescent signals of Cu-MT is shown. Cu-MT is visible as a yellow signal in the tissue

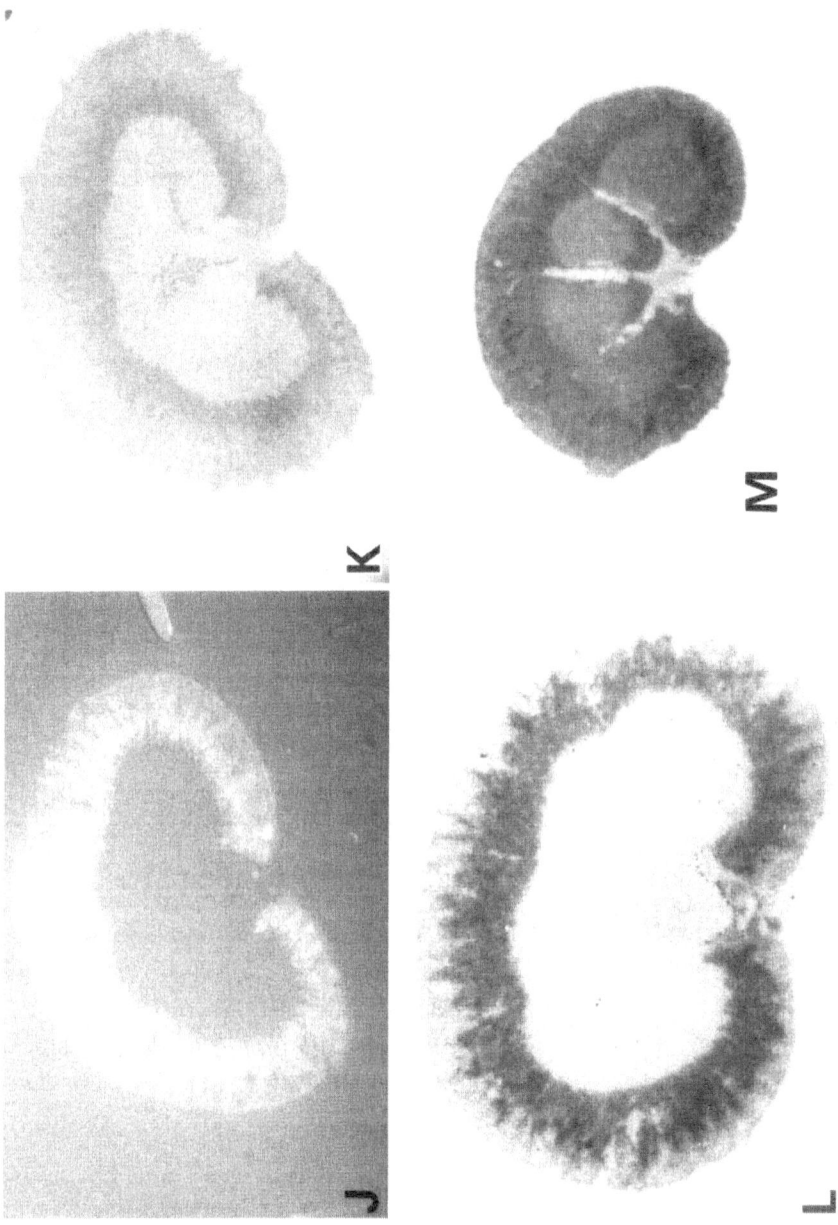

Figure 1.(continued) (J); Low-power photomicrographs of Au injected rat kidney. Renal section of Au-injected rat illuminated with UV light (312 nm). The yellow-orange autofluorescent signals with a ring shape are observed predominantly in the outer stripe of the outer medulla. (K); Localization of MT mRNA in the renal section of Au-injected rat visualized by *in situ* northern hybridization. (L); Immuno-histochemical localization of MT in the kidney of Au inject-ed rat. (M); Immuno-histochemical visualization of ATP 7B in wistar rat kidney.

Macular mouse kidney

In the kidney of Macular mice, auto-fluorescent signals both of Cu-MT and MT mRNA were found only in the cortex (Fig. 1, H). Our observations suggest that Cu-MT and/or released Cu ions were accumulated in the PCT cells (Fig. 1, I). The expression of mRNA in the renal cortex of the Macular mouse suggests that the Cu-MT in the cortex is biosynthesized in the region, but not that Cu-MT is transported to the kidney. In addition, as signals of the observed auto-fluorescence also corresponded to that of acid phosphatase as a lysosomal marker enzyme, the biosynthesized Cu-MT was degraded by lysosomal degradation in the PCT cells of the cortex. The results suggested that Cu-MT is continually synthesized in this region, especially in the S1 and S2 segments as in the case of kidneys of rats administered Cu-MT (see chapter Exogenously Cu-MT injected in rat kidney). In macular mice as well as rats administered Cu-MT, the released Cu ions induce MT mRNA to newly synthesize the protein, and the newly synthesized Cu-MT was also degraded in lysosome in the same region. The conclusion proposed as a lysosomal cycle is supported by the theory of lysosomal system for Cd-MT described by Nordberg and Nordberg [10]. In contrast, in Macular mice, Cu-MT was not found in the outer stripe of outer medulla, suggesting that the released Cu ions from the Cu-MT in the cortex diwered not transported to the outer stripe of outer medulla as in LEC rats.

Au injected rat kidney

It has been reported that the Cu accumulation in the kidney of rat was caused by Au injection, although the precise mechanisms are still unclear. In this study, the Cu-MT signals with a ring shape were observed predominantly in only the outer stripe of the outer medulla in the kidneys of Au-injected rats (Fig. 1, J). MT mRNA (Fig. 1, K) and MT protein (immunohistochemistry) (Fig. 1, L) were also detected in the same region, indicating that Cu-containing MT was *de novo* biosynthesized in only the outer stripe of the outer medulla but not transported from other organs. It is interesting that Cu contents dramatically increased in the kidneys and the Cu-binding MT was observed predominantly in the outer stripe of the outer medulla in the kidneys of Au-injected rats as well as kidneys of LEC rats. So the Au-injected rat might be used as a model for Cu metabolic disorders such as LEC rats [11].

Relationship between Cu-MT and ATP7A and B

In this study, we discovered the different renal Cu-MT distribution between LEC rat and Au-injected rat, and Macular mice and Cu-MT-injected rat. Cu-MT and a high level of MT mRNA were only observed in the outer stripe of outer medulla in the kidney of LEC rats, and rats injected with Au. On the contrary, in Macular mice and rats administered with Cu-MT, MT mRNA was observed only in the cortex of the kidneys, suggesting that the released Cu ions from the Cu-MT in the cortex were not transported into the outer stripe of outer medulla. The reason for the difference is still unclear.

Recently, two types of the gene encoding P-type cation-transporting ATPase (ATP7A and ATP7B) were shown to be responsible for Cu-transport [12–14]. ATP7A is a Cu-efflux pump [15] which is a candidate gene for Menkes disease, and the other (ATP7B) is believed to be responsible for Wilson disease. Since we postulated that the difference observed in this study, depended on function and/or dysfunction of a Cu transport substance, we carried out a study to clarify the distribution of the Cu-transporters. As shown in Figure 1, M, the ATP7B protein was observed in the outer stripe of outer medulla; so also was and its mRNA (not shown). The apparent discrepancy in Cu-MT localization between Menkes disease (and rat injected Cu-MT) and Wilson's disease (and Au-injected rats) may be due to the candidate gene. In normal organisms, it is believed that Cu ions are reabsorbed into PCT cells and the reabsorbed Cu ions are released into the vein. In the kidney of their rodents, one possibility is that the function of the ATP7B received some damage and the Cu ions were accumulated in the PCT cells in the outer stripe of outer medulla. Of course, more detailed investigations are needed to study the relationship between accumulation of Cu-MT and expression of P-type cation-transporting ATPase in the kidney.

References

1. Li Y, Togashi Y, Sato S, Emoto T, Kang JH, Takeichi N, Kobayashi H, Kojima Y, Une Y, Uchino J (1991) Spontaneous hepatic copper accumulation in LEC rats with hereditary hepatitis: A model of Wilson's disease. *J Clin Invest* 87: 1858–1861.
2. Danks DM, Stevens BJ, Campbell PE, Gillespie JM, Walker-Smith J, Blomfield J, Turner B (1972) Menkes' kinky-hair syndrome. *Lancet* 1: 1100–1103.
3. Mcvety KJ and Shaikh ZA (1987) Subchronic treatment of rats with aurothioglucose; effects on plasma, hepatic, renal and urinary zinc, copper and metallothionein. *Toxicology* 46: 295–306.
4. Beltramini M, Munger K, Germann UA, Lerch K (1987) Luminescence emission from the Cu(I)-thiolate complex in metallothioneins. *In*: Kägi JHR, Kojima Y (eds): *Metallothionein II*, Birkhäuser Verlag, Basel, 237–241.
5. Stillman MJ, Law AYC, Cai W, Zelazowski AJ (1987) Information on metal binding properties of metallothioneins from optical spectroscopy. *In*: Kägi JHR, Kojima Y (eds): *Metallothionein II*, Birkhäuser Verlag, Basel, 203–211.
6. Stillman MJ, Gasyna Z, Zelazowski AJ (1989) A luminescence probe for metallothionein in liver tissue: emission intensity measured directly from copper metallothionein induced in rat. *FEBS Lett* 257: 283–286.
7. Stillman MJ, Gasyna Z (1991) Luminescence spectroscopy of metallothioneins. *In*: Riordan JF, Vallee BL (eds): *Meth Enzymol* 205(B), Academic Press, San Diego, 540–555.
8. Stillman MJ (1992) Optical spectroscopy of metallothioneins. *In*: Stillman MJ, Show CF, Suzuki KT (eds): *Metallothioneins*, VCH Publishers, New York, 55–127.
9. Okabe M, Nakayama K, Kurasaki M, Yamasaki F, Aoyagi K, Yamanoshita O, Sato S, Okui T, Ohyama T, Kasai N (1996) Direct visualization of copper-metallothionein in kidney of LEC rat: Application of auto-fluorescence signal of copper-thionein cluster. *J Histochem Cytochem* 44: 865–873.
10. Nordberg M, Nordberg GF (1987) On the role of metallothionein in cadmium induced renal toxicity. *In*: Kägi JHR, Kojima Y (eds): *Metallothionein II*, Birkhäuser Verlag, Basel, 669–675.
11. Saito S, Okabe M, Kurasaki M (1997) Localization of renal Cu-binding metallothionein induced by gold injection to rat. *Biochim Biophys Acta* 1335: 353–358.
12. Bull PC, Thomas GR, Romens JM, Forbes JR, Cox GW (1993) The Wilson disease gene is a putative copper transporting P-type ATPase similar to the Menkes gene. *Nature Genet* 5: 327–337.
13. Mercer JFB, Livingston J, Hall B, Paynter JA, Begy C, Chandrasekharappa S, Lockhart P, Grimes A, Bhave M, Siemieniak D, Glover TW (1993) Isolation of a partial candidate gene for Menkes disease by positional cloning. *Nature Genet.* 3: 20–25.
14. Vulpe C, Levinson B, Whitney S, Pacman S, Gitschier J (1993) Isolation of a candidate gene for Menkes disease and evidence that it encodes a copper-transporting ATPase. *Nature Genet.* 3: 7–13.
15. Camakaris J, Petris MJ, Bailey L, Schen P, Lockhart P, Glover TW, Barcroft CL, Patton J, Mercer JFB (1995) Gene amplification of the Menkes (MNK: ATP7A) P-type ATPase gene of CHO cells is associated with copper resistance and enhanced copper efflux. *Hum Mol Genet* 4: 2117–2123.

Metallothionein IV
C. Klaassen (ed.)
© 1999 Birkhäuser Verlag Basel/Switzerland

Zincsulphate protects from streptozotocin-induced diabetes mellitus in mice: A function of metallothionein?

Patricia Ohly[1], Zhiyong Wang[1], Claudia Dohle[1], Josef Abel[2] and Helga Gleichmann[1]

[1]Clinical Department, Diabetes Research Institute, and [2]Division of Toxicology, at the Heinrich-Heine-University, 40225 Düsseldorf, Germany

Introduction

Human type 1 diabetes results from a chronic autoimmune destruction of the insulin-producing β-cells in the pancreatic islets of Langerhans [1]. This autoimmune process appears to be initiated by environmental agents triggering T cell-dependent inflammatory reactions with mononuclear cell infiltrates of the pancreatic islets. An animal model of autoimmune diabetes is that induced by multiple low-doses of streptozotocin (MLD-STZ), a naturally occurring diabetogen [2–4]. Since STZ activates STZ-specific Th_1 lymphocytes [3], induces infiltration of the islets with mononuclear cells including macrophages [2, 4], and generates hydrogen peroxide (H_2O_2) in islets [5], reactive oxygen species (ROS) are assumed to be also involved as non-specific mediators in the pathogenesis of MLD-STZ-induced diabetes. It is proposed that hydroxyl radicals (HO·), which are the most toxic of ROS, are generated by the Fenton reaction in the presence of adventitious iron and cause β-cell damage, which might be prevented by induced metallothionein (MT) as antioxidant [6].

Pathogenesis of STZ-induced diabetes

STZ is a bipartite molecule consisting of a glucose moiety and a methylnitrosourea moiety. The glucose moiety provides relatively specific accumulation of STZ to the β-cells, whereas toxicity is mediated by the aglycone moiety. Diabetes can be induced in laboratory animals with STZ applying two different experimental protocols which involve two different pathogenic pathways. A single injection of a high dose of STZ induces toxicity in β-cells with subsequent cell necrosis and hyperglycemia development within 48 to 72 h in rat and mice. In the MLD-STZ model 5×40 mg STZ/kg body weight are injected on five consecutive days and hyperglycemia will develop in susceptible strains of male mice two weeks after the first injection as consequence of two different STZ-exerted effects, i.e. β-cell subtoxicity followed by T cell-dependent inflammatory reactions [7]. In MLD-STZ diabetes the glucose transporter 2 (GLUT2) of β-cells was found to be a preferential, if not selective target molecule as indicated by significant reduction of protein and mRNA expression, whereas mRNA expression of glucokinase and proinsulin which are essential molecules in β-cell function and β-actin as internal control remained unaffected [8]. The reduction of the GLUT2 expression was observed already one day after the third STZ injection, i.e. in the preimmunologic phase and was not due to β-cell loss induced by the toxin. The mediators damaging the GLUT2, however, remain to be resolved. It is speculated that STZ-generated HO⋅ might be the causative oxidant (Fig. 1).

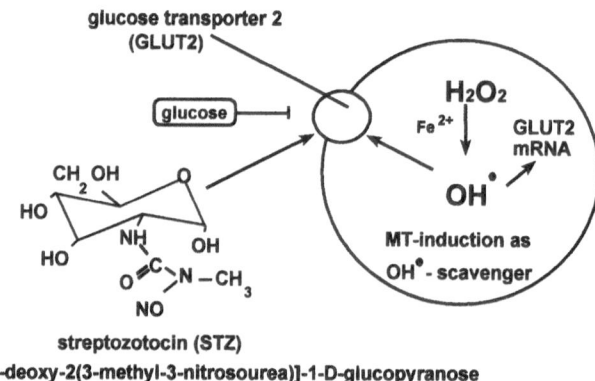

Figure 1. Hypothesis of protection of STZ-induced β-cell toxicity by MT. Toxicity may be mediated by HO⋅ produced by a Fenton reaction from STZ-generated H_2O_2 in the presence of Fe^{++}. MT induced by metal ions may prevent toxicity by scavenging HO⋅.

MT expression in pancreatic cells

High MT levels in exocrine cells

The two major isoforms MT-I and MT-II are constitutively expressed in most vertebrate major tissues such as intestine, liver, kidney, and pancreas, and MT-I is more abundant than MT-II [9]. Induction of MT synthesis is stimulated by various metal ions such as Zn^{2+}, Cu^{2+}, and Cd^{2+}, but also by numerous nonmetallic agents including glucocorticoids, interleukins, interferon [6] and STZ [10]. By immunohistochemical staining MT is predominantly localized in the cytoplasma of the exocrine cells in rat and mice [11, 12]. Remarkably high constitutive levels of MT were found in the pancreatic juice of mice, which are more induced by Zn^{2+} than by overexpression in transgenic mice [12]. Therefore, it has been suggested that the observed nonsecretory release of MT from the exocrine cells into the pancreatic juice may indicate a physiologic role in Zn^{2+} homeostasis.

MT expression in endocrine Langerhans islets

In contrast to several data on MT expression in either pancreatic exocrine cells [11, 12] or unseparated pancreas tissue [10, 13, 14], there are only few reports on MT levels in pancreatic islets, the endocrine cells. In rats, pancreatic islets stained uniformly for MT by immuno-histochemistry using an antibody with cross-reactivity for both isoforms MT-I and MT-II [11]. This staining pattern was not changed by treatment with Zn^{2+} of the donor animal. The constitutively low level of MT RNA, however, was markedly increased after treatment with Zn^{2+}. The effect of Zn^{2+} on MT induction appeared to be specific because the level of insulin-I mRNA remained unaffected.

In mice, discrepant observations have been reported. By immunohistochemistry, an antibody reacting with both MT-I and MT-II isoforms failed to stain islets of normal mice and of transgenic mice overexpressing MT-I [12]. In our laboratory, ex vivo, constitutive levels and significant (p < 0.01) induction of MT were found in islets isolated from untreated mice and mice which had been injected intraperitoneally with $ZnSO_4$, respectively [15]. In vitro, cultured isolated islets also had constitutive levels of MT which were significantly (p < 0.05) induced upon exposure to $ZnSO_4$ [16]. As shown in Figure 2, treatment of mice with $ZnSO_4$-enriched (25 mM) drinking water ad libitum was also effective in inducing significant (p < 0.05) MT induction in pancreatic islets as measured ex vivo by the [109]Cd-labelled hemoglobin-exchange assay [17]. Immunostaining for MT with a polyclonal anti-rat antiserum (kindly provided by Dr. Summer, München, Germany) localized MT in a uniform pattern in the cytoplasma of all islet cells, whereas no staining was obtained when the primary antiserum was omitted (Fig. 3 A and B). Immunostaining with an antibody against insulin identified the majority of the islet cells as insulin-producing β-cells, which remained unstained by omitting the pimary antiserum as control (Fig. 3 C and D).

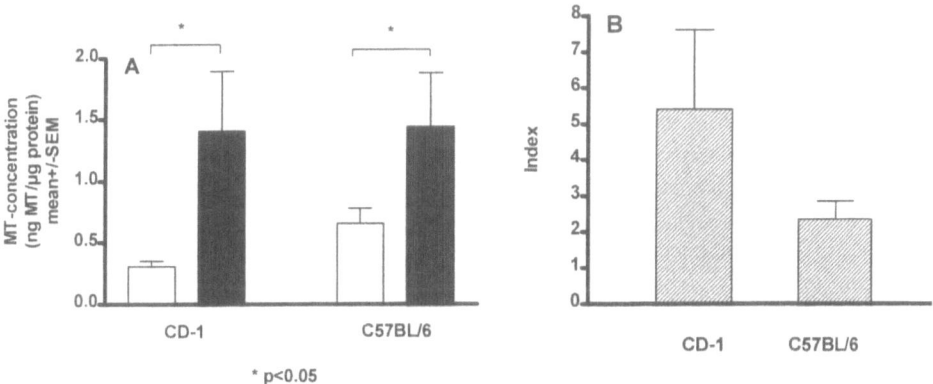

Figure 2. MT levels in isolated pancreatic islets of CD-1 and C57BL/6 mice treated with either (■) $ZnSO_4$-enriched (25 mM) drinking water for one week or (□) with the solvent. Mean values +/–SE of groups of 5 mice each are given. MT was determined by the [109]Cd-hemoglobin-exchange assay (A). Mean values ± SE of the stimulation index (B).

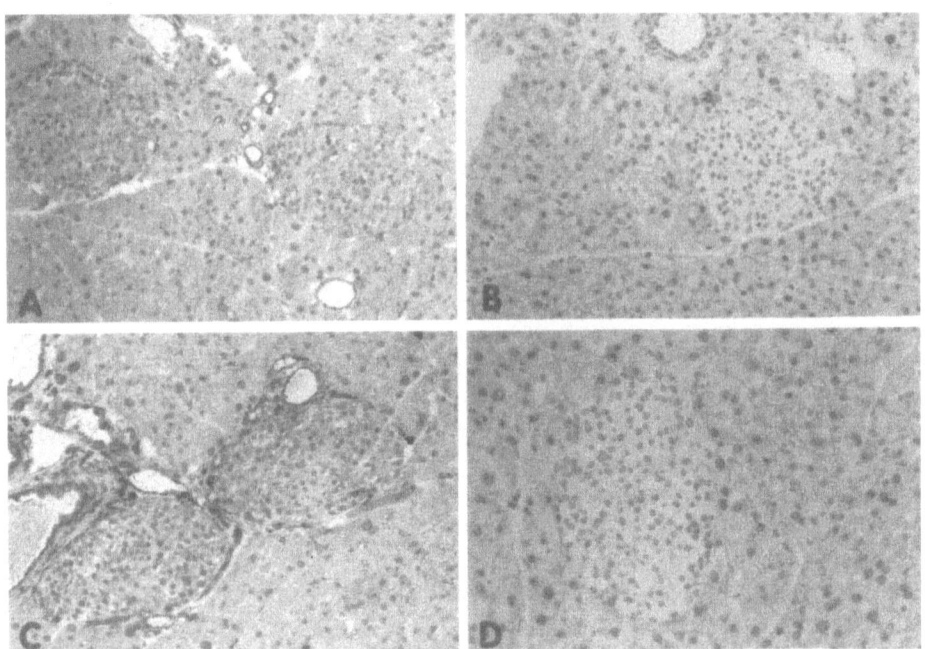

Figure 3. Immunohistochemistry of mouse pancreatic sections. (A) Uniform cytoplasmic localization of the primary anti-rat polyclonal antiserum against MT-I and MT-II in islets, weak cytoplasmic staining of exocrine cells. (B) No staining of the section when omitting the primary antiserum as control. (C) Cytoplasmic localization of the primary anti-pig insulin antiserum in β-cells. D: No staining when omitting the primary antiserum as control. Original magnification × 400.

Effect of Zn^{2+} on streptozotocin-induced diabetes

In rats, subcutaneous pretreatment with $ZnSO_4$ (10 mg/kg body weight) significantly reduced hyperglycemia induced with one single intraperitoneal injection of STZ (85 mg/kg body weight) [13]. In these experiments $ZnSO_4$ was injected 12 h before STZ. Superoxide dismutase (SOD) activities in pancreatic tissue remained unaffected by $ZnSO_4$. It is suggested that the protective effect on STZ-induced hyperglycemia may be related to MT as scavenger of HO·.

We studied the effect of treatment with $ZnSO_4$ on MLD-STZ-induced diabetes in mice. Pretreatment with intraperitoneal injections of $ZnSO_4$ (10 mg/kg body weight) 24 h before each of the five STZ injections significantly reduced hyperglycemia development and diabetes prevalence in C57BL/6 male mice [18]. Diabetes was diagnosed when a mouse had a blood glucose level $\geq$ 13.9 mM (250 mg/dl) on three consecutive weeks. In the $ZnSO_4$-treated group, 75% of the MLD-STZ recipients were euglycemic compared with only 20% in the group treated with MLD-STZ solely. Treatment with $ZnSO_4$-enriched drinking water was also effective in reducing significantly MLD-STZ hyperglycemia over time in both C57BL/6 and [C57BL/6 × SJL]F_1 hybrid mice (Fig. 4). $ZnSO_4$-enriched drinking water was given conti-

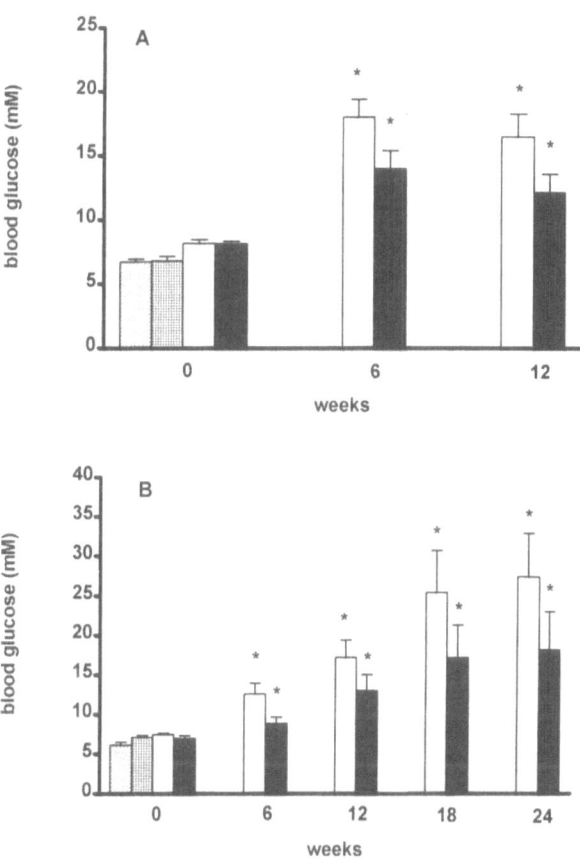

Figure 4. Blood glucose concentration in male C57BL/6 and [C57BL/6 × SJL]F$_1$ hybrid mice. (□) Treatment with MLD-STZ only; (■) treatment with ZnSO$_4$-enriched (25 mM) drinking water which was started one week before treatment with MLD-STZ; (⊞) treatment with ZnSO$_4$-enriched drinking water; (⊠) treatment with the solvent of STZ.

nously from week one before starting injections with MLD-STZ until the end of the experiments. [C57BL/6 × SJL]F$_1$ mice which had received MLD-STZ only started to develop severe hyperglycemia around week twelve after the first injection of STZ, whereas the group which had received ZnSO$_4$-drinking water as well developed only mild hyperglycemia. These results suggest that the preventive effect of treatment with Zn^{2+} may be sequelae of induced MT as scavenger of HO˙ as illustrated in Figure 1.

Our preliminary results obtained in TgN(Mt1)174Bri transgenic mice, overexpressing MT-I, however, do not appear to support the above hypothesis, because hyperglycemia with blood glucose levels of approximate 14.1 mM were found already on day 7 after starting treatment with MLD-STZ and was not influenced by treatment with $ZnSO_4$-enriched drinking water. Yet, for three reasons these observations are not rejecting a possible role of MT as antioxidant protecting β-cells from HO˙-mediated damage induced by MLD-STZ: 1) the rapid course of diabetes induction indicates acute toxicity rather than a more chronic autoimmune process which is the pathogenesis in MLD-STZ diabetes; 2) it is not known, if MT is expressed in the β-cells of the transgenic mice used by us, since in a previous investigation islets failed to immunostain for MT-I and MT-II in pancreatic sections prepared from mice derived from the original MT-transgenic breeding stock [12]; and 3) it remains to be analyzed if a protective function may be isoform-restricted.

In this context, the recent report on the effect of treatment with $ZnSO_4$ on the effect of STZ treatment in MT-I and M-II knockout (MT–/–) mice also could not unequivocally resolve the possible function of MT as protective antioxidant [19]. Pretreatment of female MT–/– recipients with one intraperitoneal injection of $ZnSO_4$ (1, 5, or 10 mg/kg body weight) 12 h prior to one intraperitoneal injection of STZ (85 mg/kg body weight) significantly reduced the blood glucose level in MT–/– mice, the mean values in the three different groups being 7.3, 6.7, and 6.2 mM, respectively, compared with the corresponding mean values of 10.0, 8.8, and 6.7 in the MT+/+ counterparts. Injection of STZ alone induced mean blood glucose levels of 9.4 and 10.4 in MT–/– and MT+/+ mice, respectively. Despite the significant differences obtained and despite being an important attempt to assay for a biological function of MT in induced diabetes, the results appear to be too incomplete for final conclusions for two essential reasons: 1) the highest mean blood glucose level induced was about 10.0 mM, which is below the generally used level of at least 11.1 mM for diagnosis of diabetes; and 2) groups of 4 mice each may be insufficient to test for unequivocal biological effects. Nevertheless, of particular interest is the observation that pretreatment with $ZnSO_4$ prevented loss of STZ-induced SOD activity. This effect may point to the possibility that Zn^{2+} are not specific MT-inducers. Thus, more experimental investigations are required with regard to dose-response kinetics and specificity of antioxidant activation.

Discussion

In laboratory animals STZ-induced diabetes was reduced by pretreatment with $ZnSO_4$ using two different experimental protocols [13, 21]. In rats, pretreatment with one intraperitoneal injection of $ZnSO_4$ significantly reduced severe hyperglycemia induced by acute toxicity with one single STZ injection [13]. In mice, we found that both pretreatment with intraperitoneal injections of $ZnSO_4$ and continous treatment with $ZnSO_4$-enriched drinking water prevented autoimmune diabetes induced by MLD-STZ [21 and present data]. Both groups of investigators conclude that the protective effect obtained with $ZnSO_4$ may be mediated by $ZnSO_4$-induced MT as scavenger of HO˙. The rationale for this assumption is based on the fact that STZ has been reported to generate H_2O_2 [5] which may lead to the generation of HO˙ in the presence of adventitious Fe^{++} by the Fenton reaction [6]. Since HO˙ are constitutively expressed in islets [20], it is likely to be increased by certain compounds such as STZ. Having observed significant MT induction with $ZnSO_4$ in pancreatic islets [15, 16, present data], it is hypothesized that MT, which has been reported to be superior to glutathione in preventing HO˙-mediated DNA damage [21] may prevent HO˙-mediated β-cell damage.

Perspectives

Our data demonstrate that treatment with $ZnSO_4$ significantly induced MT in pancreatic islets cells and significantly protected mice from MLD-STZ diabetes. Therefore, we think these animal models to be useful to analyze if the underlying mechanisms are due to MT as scavenger of HO˙ induced by MLD-STZ. Such investigations are feasible because constitutive prevalence of HO˙ has been demonstrated by electron spin resonance spectroscopy [20]. Furthermore, additional experiments in transgenic mice overexpressing MT, preferably in β-cells only, can be expected to contribute to elucidate if MT is involved in the protection of loss of β-cell function. Obviously, analysis of the response of MT-knockout mice are also expected to resolve mechanisms contributing to the sensitivity and resistance to MLD-STZ. As described above, up to date, results on both the expression of MT in pancreatic islet cells and on the possible function as cellular defense system against STZ-induced hyperglycemia are inconsistent. In order to resolve these issues, it is necessary to conduct investigations by using comparable experimental designs, since different strains of animals and different treatment protocols may lead to different results and conclusions. It is also likely that the protective function is associated with one of the MT isoforms and that treatment with metal ions to induce MT may also affect other intracellular antioxidant systems such as SOD and catalase which are known to interact in the process of HO˙ generation.

Acknowledgements
This study was supported by the Deutsche Forschungsgemeinschaft SFB 503 "Molecular and cellular mediators of exogenous noxae", project B5 and by the Bundesministerium für Gesundheit, Bonn, Germany.

References

1. Castano L and Eisenbarth GS (1986) Type I diabetes mellitus: a chronic autoimmune disease. *Annu Rev Immunol* 8: 647–679.
2. Like AA and Rossini AA (1976) Streptozotocin-induced pancreatic insulitis: new model of diabetes mellitus. *Science* 193: 415–417.
3. Klinkhammer C, Popowa P and Gleichmann H (1988) Specific immunity to streptozotocin: Cellular requirements for induction of lymphoproliferation. *Diabetes* 37: 74–80.
4. Wang Z, Dohle C, Friemann J, Green BS and Gleichmann H (1993) Prevention of high- and low-dose streptozotocin diabetes with D-glucose and 5-thio-D-glucose. *Diabetes* 42: 420–428.
5. Takasu N, Komiya I, Asawa T, Nagasawa Y and Yamada T (1991) Streptozotocin- and alloxan-induced H_2O_2 generation and DNA fragmentation in pancreatic islets. *Diabetes* 40: 1141–1145.
6. Sato M and Bremner I (1993) Oxygen free radicals and metallothionein. *Free Radical Biol. and Med.* 14: 325–337.
7. Wilson GL and Leiter EH (1990) Streptozotocin interactions with pancreatic β-cells and the induction of insulin-dependent diabetes. *In*: T Dyrberg (editor): *Current Topics in Microbiology and Immunology*. Springer Verlag, Berlin, Heidelberg, 27–54.
8. Wang Z, and Gleichmann H (1998) Glucose transporter 2 in pancreatic islets: crucial target molecule in diabetes induced with multiple-low-dose streptozotocin in mice. *Diabetes* 47: 50–56.
9. Andrews GK (1990) Regulation of metallothionein gene expression. *Prog Food Sci* 14: 193–258.

10. Baumann JW, Liu J and Liu P, Klaassen CD (1991) Increase in metallothionein produced by chemicals that induce oxidative stress. *Toxicol Appl Pharmacol* 110: 347–354.
11. Andrews GK, Kage K, Palmiter-Thomas P and Sarras Jr MP (1990) Metal ions induce expression of metallothionein in pancreatic exocrine and endocrine cells. *Pancreas* 5: 548–554.
12. De Lisle RC, Sarras Jr MP, Hidalgo J and Andrews GK (1996) Metallothionein is a component of exocrine pancreas secretion: implications for zinc homeostasis. *Amer J Physiol* 271 (*Cell Physiol* 40): C1103-C1110.
13. Yang J and Cherian MG (1994) Protective effects of metallothionein on streptozotocin-induced diabetes in rats. *Life Sci* 55: 43–51.
14. Onosaka S, Min K-S, Fujita Y, Tanaka K, Iguchi S and Okada Y (1988) High concentration of pancreatic metallothionein in normal mice. *Toxicology* 50: 27–35.
15. Zimny S, Gogolin F, Abel J and Gleichmann H (1993) Metallothionein in isolated pancreatic islets of mice: induction by zinc and streptozotocin, a naturally occuring diabetogen. Arch. *Toxicol* 67: 61–65.
16. Ohly P and Gleichmann H (1995) Metallothionein: *In vitro* induction with zinc and streptozotocin in pancreatic islets of mice. *Exp Clin Endocrinol Diabetes* 103: 79–82.
17. Eaton DL and Cherian MG (1991) Determination of metallothioenin in tissues by cadmium- hemoglobin affinity assay. *Methods Enzymol* 205: 83–88.
18. Ohly P, Wang Z, Abel J and Gleichmann H (1998) Zincsulphate induced metallothionein in pancreatic islets and protected against the diabetogenic toxin streptozotocin. *Talanta* 46: 355–359.
19. Apostolova MD, Choo KHA, Michalska AE and Tohyama C (1997) Analysis of the possible protective role of metallothionein in streptozotocin-induced diabetes using metallothionein- null mice. *J Trace Elements Med Biol* 11: 1–7.
20. Pieper GM, Felix CC, Kalyanaraman B, Turk M and Roza AM (1995) Detection by ESR of DMPO hydroxyl adduct formation from islets of Langerhans. *Free Radical Biol Med* 19: 219–225.
21. Abel J and de Ruiter N (1989) Inhibition of hydroxyl-radical-generated DNA degradation by metallothionein. *Toxicol Lett* 47: 191–196.

Metallothionein IV
C. Klaassen (ed.)
© 1999 Birkhäuser Verlag Basel/Switzerland

MT does not protect endocrine cell damage induced by alloxan in pancreas of mice

Takeshi Minami[1], Hidenori Tanaka[2], Yuko Okazaki[2], Setsuko Tohno[1] and Yoshiyuki Tohno[1]

[1]*Laboratory of Cell Biology, Department of Anatomy, Nara Medical University, 840 Shijo-Cho, Kashihara, Nara 634, Japan*
[2]*Faculty of Pharmaceutical Sciences, Kinki University, 3-4-1 Kowakae, Higashi-Osaka, Osaka 577, Japan*

Metallothionein (MT) has been proposed to have various functions, such as detoxification of heavy metals, storage of essential metals, and scavenging of free radicals. Induction of MT by zinc is reported to reduce both endocrine and exocrine damages of the pancreas induced by chemicals. However, the damage is not inhibited completely. The first question is that an essential concentration zone of zinc is too narrow to apply to organs, especially to the pancreas. It is, of course, that zinc is an essential metal. Zinc acts as an essential cofactor for many metalloenzymes with various biological functions. Zinc, itself, has antioxidant action, and prevents the production of hydroxyl and superoxide radicals. Zinc is needed for storage of insulin in the pancreas, and zinc contents in the plasma and urine are altered in diabetes. However, excess zinc may cause toxic effect in pancreas, as we observed congestion of the pancreas after zinc administration. The other question is that no MT was detected immunohistochemically in the endocrine cells after the administration of zinc. Andrews et al. [1] reported that mRNA of MT was detected in both the exocrine and endocrine cells. Ohly and Gleichmann [2] also observed MT induction in the isolated islets of pancreas by zinc and streptozotocin. MT has a strong radical scavenging action, and alloxan is used for an experimental model of type-1 diabetes. Alloxan generates superoxide radicals, and causes hyperglycemia and hypoinsulinemia. As the destruction of β-cell is mediated by superoxide and nitric oxide radicals, MT may have a protective role in the endocrine cells. However, we previously observed immunohistochemically that MT was located in the exocrine cells, but not in the endocrine cells after zinc injection to mice. In addition, Apostolova et al. [3] reported that the pretreatment of zinc inhibited the streptozotocin-induced diabetes in MT-null mice. Is MT induced in the endocrine cells by zinc? The object of this study was to elucidate the induction of MT in endocrine cells of pancreas as well as the possible role of MT in preventing oxidative damage induced by alloxan.

Toxicity of zinc in the pancreas

Mice were injected subcutaneously with various doses of zinc (0, 0.1, 1, 5, 12.5, 25, and 50 mg/kg), and after 3, 6, 10, or 24 h, blood was collected and the pancreas was removed. Plasma α-amylase activities increased in mice injected with 25 or 50 mg zinc/kg after 10 and 24 h, while 12.5 mg zinc/kg or less had no significant effect. Plasma glucose levels were unaltered by 25 mg zinc/kg or less during all periods. In contrast, mice injected with 50 mg zinc/kg had significantly decreased plasma glucose levels after 24 h. Zinc contents in pancreas of mice injected 1 mg zinc/kg increased to a maximum 10 h after injection, and returned back to the control levels after 24 h. In contrast, when 5 mg zinc/kg or more were injected to mice, pancreatic zinc levels were significantly increased at 3 h after injection in a time-dependent manner. Only mice injected with 50 mg zinc/kg exhibited lower pancreatic zinc levels at 24 h than 10 h after injection. Mice injected with either 0.5 or 1 mg zinc/kg had maximal pancreatic MT levels after 10 h, which decreased to control levels by 24 h. In mice injected with higher doses of zinc, pancreatic MT increased in a time dependent fashion. Only at the highest injected dose (50 mg zinc/kg) did pancreatic MT decline after 24 h. Pancreatic wet weights were 1.4 and 1.8 fold higher in mice injected with 25 and 50 mg zinc/kg, respectively, as compared to control mice. But, at low doses of zinc, the pancreatic weight did not change. When α-amylase activities were measured in the pancreatic cytosol fractions, they were decreased in a dose-dependent manner in the groups injected over 5 mg zinc/kg. In the group injected 50 mg zinc/kg, the activity levels in pancreas did not return to normal at 24 h after injection. Swelling and vacuolization were observed histochemically in the exocrine cells, revealing damage with increasing dose. However, no changes were found in the endocrine cells even at 50 mg zinc/kg. Additionally, MT was immunohistochemically detected in the interstitia, cytoplasm, and nuclei of exocrine cells, but not in the endocrine cells in all the groups of mice injected zinc. These data suggest that mice injected with zinc caused pancreatic exocrine damage. In addition, MT was only induced in exocrine cells of pancreas by zinc injection.

Relationships between MT and alloxan-induced endocrine damage

Next was investigated whether MT could inhibit the endocrine cell damage induced by alloxan. Mice were given water containing four different concentrations of zinc (0%, 0.05%, 0.1%, or 0.5% zinc as zinc sulfate) *ad libitum* for 18 days. On day 14 of the experiment, mice were intravenously injected with alloxan (60 mg/kg) or with 50 mM citrate buffer (pH 4.5) as the control vehicle group. Four days after alloxan injection, blood was collected and the pancreas was removed. In mice injected with alloxan, pancreatic zinc levels increased with increasing zinc contents in water. In the control vehicle group, pancreatic zinc levels were unaltered when mice were given 0.05 or 0.1% zinc water. However, when 0.5% zinc water was given, pancreatic zinc levels in control mice increased to similar levels measured in alloxan injected mice. The pancreatic MT levels increased in relation to increasing doses of zinc and were highest in both the vehicle and alloxan injected mice receiving 0.5% zinc water. Pancreatic MT levels were significantly higher after alloxan injection as compared to the vehicle in mice given 0.05 or 0.1% zinc water. In addition, alloxan injection without zinc supplementation in drinking water had no effect on pancreatic MT levels. Plasma α-amylase activity increased in both the vehicle and alloxan groups given 0.5% zinc water. Plasma glucose level increased in the all groups injected with alloxan, while the level did not increase in the vehicle groups, including the group given at 0.5% zinc water. In addition, MT was immunohistochemically observed in the exocrine cells, not in the islet's cells of both alloxan- and vehicle-administered mice, even if mice were received with 0.5% zinc water and alloxan.

All mice were given food and tap water, which contained 49 mg zinc/100g and 10 ng zinc/ml, respectively. On average daily zinc intakes were about 2.5 mg per mouse. Under the experimental conditions of 0.05% and 0.1% zinc in drinking water, mice received a total of 5 and 7.5 mg zinc, respectively. In contrast, when mice were given 0.5% zinc water, the daily drinking volume decreased one fourth in comparison with that of the control mice. In those mice, total zinc intake was 8.8 mg. As the contents of zinc in pancreas did not increase in the vehicle groups given 0.1% or less zinc water, homeostatic balance of zinc in pancreas is maintained in those groups. While, an excess of pancreatic zinc was detected in both alloxan-injected group and the vehicle group given 0.5% zinc water. Pancreatic zinc excretion may decrease in these groups compared with the uptake of zinc. After increasing zinc contents in pancreas, MT may be induced in the exocrine cells, but not in the endocrine cells. The damage induced by alloxan occurs not only in endocrine cells, but also in exocrine cells.

Conclusion

We demonstrated that zinc did not reduce the increase in plasma glucose level induced by alloxan, and that zinc contents increased in the pancreas after alloxan injection when mice were supplied with excess zinc. No MT was observed in the endocrine cells even when both zinc and alloxan were given. From these data, induction of MT by zinc may not reduce pancreatic endocrine damage induced by alloxan.

References

1. Andrews GK, Kage K, Palmiter-Thomas P, Sarras Jr MP (1990) Metal ions induce expression of metallothionein in pancreatic exocrine and endocrine cells. *Pancreas* 5: 548–554.
2. Ohly P, Gleichmann (1995) Metallothionein: *In vitro* induction with zinc and streptozotocin in pancreatic islets of mice. *Exp Clin Endcrinol* 103: 79–82.
3. Apostolova MD, Choo KHA, Michalska AE, Tohyama C (1997) Analysis of the possible protective role of metallothionein in streptozotocin-induced diabetes using metallothionein-null mice. *J Trace Elements Med Biol* 11: 1–7.

Plasma MT levels in diabetic and renal failure patients

Takeshi Minami[1], Setsuko Tohno[1], Yoshiyuki Tohno[1], Noriko Otaki[2], Masami Kimura[3] and M. George Cherian[4]

[1]Laboratory of Cell Biology, Department of Anatomy, Nara Medical University, 840 Shijo-Cho, Kashihara, Nara 634, Japan
[2]Department of Occupational Disease, National Institute of Industrial Health, Kawasaki 214, Japan
[3]Department of Molecular Biology, Keio University School of Medicine, Tokyo 160, Japan
[4]Department of Pathology, The University of Western Ontario, London, Ontario N6A 5C1, Canada

Metallothionein (MT) has various functions such as storage of essential metals, detoxification of heavy metals, and scavenging of free radicals. To investigate these functions of MT, various models of experimental animals, plants, and cultured cells are used. Recently, MT transgenic mice, MT-null and MT-overexpressed mice, are used, but the roles of MT are still unclear. Changes in MT levels in human have been studied in many countries with environmental pollution along with certain diseases and nutritional mineral deficiency. Since MT was discovered as a Cd binding protein, interactions between MT and Cd exposure have been studied. And it is confirmed that levels of urinary MT and β_2-microglobulin increased with Cd exposure. Indeed, the level of urinary MT has been used as a clinical parameter of renal dysfunction induced by Cd. And it is known that Cd-MT induces a renal dysfunction, although intracellular MT acts to reduce the toxicity of heavy metals. Smoking is also reported to increase the renal Cd and MT levels. The changes in metabolism of Zn and Cu may have a marked effect in MT levels in liver, kidney, and pancreas. Marked differences in MT content in several organs and plasma have been reported in certain diseases such as primary biliary cirrhosis, Wilson diseases, and amyotrophic lateral sclerosis. An increase in MT has been shown in human fetal liver and various types of tumors of thyroid, testis, urinary bladder, salivary gland, breast, and colorectal [1]. However, it is still unclear whether the change in MT either in tissue or biological fluids are always related to change in mineral metabolism or due to certain changes in inflammation or cancers. There are only few reports on plasma levels of MT, because sensitive methods of MT measurement such as ELISA were not readily available in most laboratories. Recent reports suggest that MT may behave as an acute phase protein in certain diseases such as myocardial infarction [2]. Analysis of plasma and urinary MT levels will provide importantly on the role of MT in various pathological status and changes in mineral metabolism. In this report, the changes in plasma and urinary MT levels in diabetes and chronic renal failure (CRF) patients are discussed.

MT and diabetes

The Plasma and urinary MT levels were measured on 10 healthy volunteers, 22 diabetes patients, 26 CRF patients and 42 CRF patients treated with hemodialysis (HD). The MT contents were measured by RIA, and the detection limit of MT was about 10 ng/ml. Diabetes is divided into two types, as type I (insulin dependent diabetes mellitus: IDDM) and type II (non-insulin dependent diabetes mellitus: NIDDM). The patients of IDDM are very few in Japan, but we measured plasma MT contents in two IDDM patients. The others were NIDDM. MT could not be found in plasmas of IDDM and NIDDM as well as the healthy people. But when CRF was complicated with diabetes, plasma MT levels were increased.

In the experimental animal model of diabetes, preinduction of MT in the pancreas is reported to inhibit damages of β-cells induced by chemicals. In addition, zinc is needed for insulin stabilization in the pancreas, and urinary excretion of zinc is increased in diabetes. We did not detect MT in plasma of diabetes patients, except patients complicated with CRF. Furthermore, no MT was observed immunohistochemically in several pancreases of diabetes patients. Therefore, it is suggested that the plasma MT is not reflected on the stage of diabetes. In contrast, the urinary MT was higher in diabetes patients than in healthy people. And there was a positive relationship between urinary MT and zinc contents in diabetes patients ($r = 0.551$, $p < 0.05$). MT was present mainly in proximal tubules of the kidney of diabetes patients. Renal failure is a major complication in diabetes. The MT produced in proximal tubules of the kidney may be excreted to the urine, complexes with zinc.

MT and chronic renal failure

Unlike in diabetes, MT was detected in plasma of 73% CRF patients, and the average of MT was detected in 33.9 ± 3.5 ng/ml (mean $\pm$ SD) of patients. MT was detected in 88% of HD patients, and the average of MT was 46.6 ± 6.2 ng/ml. Further, we divided HD patients into two groups during the period of clinical treatment. MT was detected in 84% of short-term HD patients, who had received a hemodialysis treatment during 6 ± 1 months, and the average of MT was 36.4 ± 8.0 ng/ml in the plasma. In long-term HD patients, who had reviewed a hemodialysis treatment during 70 ± 8 months, plasma MT was detected in 89% of them, and increased to 57.4 ± 9.1 ng/ml. CRF patients received a hemodialysis treatment, when the plasma creatinine level increased over 10 mg/dl and the plasma urea nitrogen level increased over 100 mg/dl.

MT is often detected in plasma samples of CRF patients. The increase of MT contents in plasmas of CRF patients may depend on the stage of CRF (Fig. 1). The plasma MT level was shown a good correlation with the urinary MT level ($r = 0.855$, $p < 0.01$). In the patients of CRF, lots of MT may be produced in proximal tubules of the kidney, and subsequently, the plasma and urinary MT contents may increase. On the contrary, the plasma MT content is not related to the plasma zinc and copper levels, similar to Vilanova et al. observed in acute myocardial infection [2]. No relationships between the plasma MT and zinc in diabetes are reported. In HD patients, the plasma MT level increased during their treatment period. MT may not pass through the dialyzing membrane, as the plasma MT level did not differ before and after a single hemodialysis treatment. It is unknown whether MT in the kidney may delay the development of renal dysfunction. Recent development of ELISA methods for MT may provide a useful tool to further investigate the role of MT in real disorders.

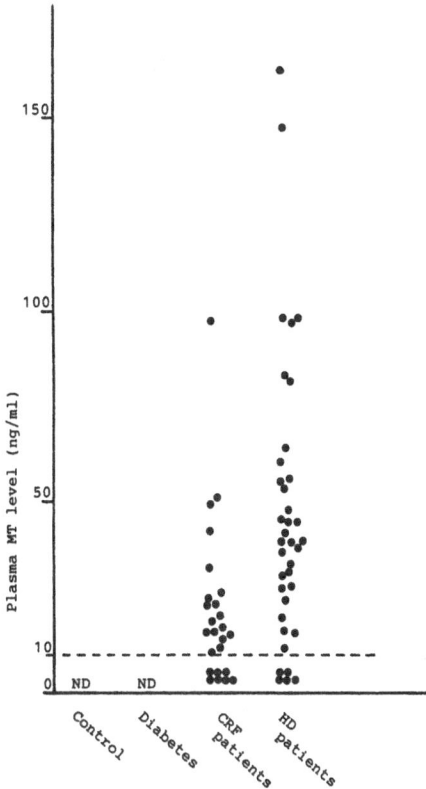

Figure 1. Plasma MT level in normal people, diabetes, and CRF patients. ND: below 10 ng/ml.

In conclusion, MT does not increase in plasma of diabetes as well as control healthy people, but urinary MT in diabetes is related to the excretion of zinc. In contrast, the plasma MT levels increased in CRF patients, and this may be due to dysfunction of proximal tubules in the kidney.

References

1. Cherian MG (1994) The significance of the nuclear and cytoplasmic localization of metallothionein in human liver and tumor cells. *Environ Health Perspect* 102 (Suppl 3): 131–135.
2. Vilanova A, Gutierrez C, Serrat N, Raga X, Paternain JL (1997) Metallothionein, zinc and copper levels: Relationship with acute myocardial infarction. *Clinical Biochem* 30: 235–238.

Metallothionein-mediated alterations in autoimmune disease processes

M.A. Lynes[1], C.A. Richardson[1], R. McCabe[2], K.C. Crowthers[1], J.C. Lee[1], J. Youn[1], I.B. Schweitzer[2] and L.D. Shultz[2]

[1]*Department of Molecular and Cell Biology, University of Connecticut, Storrs, CT 06269, USA*
[2]*The Jackson Laboratory, 600 Main St., Bar Harbor, ME 04609, USA*

Introduction

Cells under stress produce metallothionein as one way of coping with the biological consequences of that stress. A broad array of conditions are known to elicit increases in metallothionein (MT) synthesis, including exposure to various toxic chemicals [1], increased demands on biosynthetic machinery [2], irradiation [3], psychological stress [4], and inflammation [5]. In each of these circumstances, immune function may also be altered, and these changes may contribute yet further to the conditions that are stressful to cells. Metallothionein produced under these conditions may have important implications for immune function, and for the subsequent susceptibility to further exposure to stress.

Our hypothesis is that MT, synthesized as a consequence of these stressful stimuli, may contribute to the actual pattern of immune function, and that manipulations of the role(s) that MT plays may shift the consequences for immunity following exposure to stress [6]. A number of observations lend support to this conjecture. We have found that MT can stimulate lymphoproliferation *in vitro* both when added alone to cultures of splenocytes and when added to cultures simultaneously stimulated by another polyclonal activator [7]. This proliferative response has been further explored with purified subpopulations of lymphocytes. While purified B cells proliferate in response to MT alone, T lymphocytes are unresponsive unless MT is added to the culture in the context of a T cell mitogen. MT added to cultures can also stimulate some macrophage functions [8]. This stimulation by extracellular MT appears to depend upon MT interactions at the plasma membrane and is substantially reduced by 2-mercaptoethanol. Therefore, cell stimulation may be mediated in part by the thiols found in MT [9], and in part by the heavy metals associated with the MT.

Observation of these *in vitro* interactions of MT with immune cells have led us to examine the potential effects of MT on *in vivo* immune responsiveness. Exogenous MT, administered concurrently with a T-dependent antigen, was found to significantly reduce the humoral response to that antigen, while having no effect on total Ig levels in the animal. This specific suppression by MT could be eliminated by injections of UC1MT, a monoclonal anti-MT antibody [10]. Recently, we have measured the effects of MT on cytotoxic T cell (CTL) activity. In these experiments, MT was found to suppress CTL killing of allogeneic target cells, as well as to reduce proliferative responses of cytokine-dependent CTLL-2 cells. Moreover, we have found that the binding of MT to the surface of target cells can mask the expression of allogeneic MHC target molecules (J.Y., unpublished observations). Finally, we have preliminary evidence that UC1MT acts as a modifier of normal immune function in the absence of exogenous MT. These results suggest that MT can have a variety of immunomodulatory consequences, serving to activate some components of the immune response while reducing activities of other components.

MT is commonly considered to be a cytosolic molecule, largely because 90% of the MT mRNA is found in association with free polysomes [11]. However, a number of reports have described MT in serum, in urine, in broncho-alveolar spaces, and in culture media [12–15] suggesting that MT may be secreted by non-traditional means, as are other soluble immune mediators (e.g. IL-1; [16]). Alternatively, MT may be released from cells as a non-specific consequence of toxic cell death. In either circumstance, the presence of MT in the extracellular and intracellular environments suggests that there may be two different and important pools of the molecule with important implications for immune function.

One set of inducers of MT synthesis is the acute phase cytokines, including TNF-α [17], IL-1 [18], and IL-6 [19]. These cytokines are synthesized during inflammatory responses, and are carried throughout the body to initiate a systemic response to foreign agents. Under some circumstances, most notably during some forms of autoimmune disease, this acute phase response can occur in the absence of foreign agents. Thus, in some forms of autoimmune disease, MT may be synthesized, and may alter an organism's capacity to moderate the consequences of that autoimmunity. One of our interests has been to explore the role that endogenous MT might play in autoimmune disease, and the potential that manipulation of MT levels might have for the manipulation of the course of the autoimmune disease.

The autoimmune model that we have chosen for study is the "viable motheaten" mouse mutant. The mutation ($Hcph^{mev}$) is an autosomal recessive mutation in the gene encoding hematopoietic cell phosphatase [20], a cytoplasmic protein tyrosine phosphatase [reviewed in 21]. A homozygous viable motheaten animal displays a severe systemic autoimmune syndrome, with multiple organ systems that are damaged by autoimmune attack. One hallmark of the disease is severe inflammatory dermatitis, progressive alopecia, and necrosis of the extremities. These mice display a number of immunologic abnormalities, including high titers of several different autoantibodies, hypergammaglobulinemia, and pervasive immune complex deposition. Lymphopoiesis is abnormal, and results in substantial shifts in the lymphoid populations of mutant animals. Pre-B cells in the bone marrow are reduced, and an Ly-1+ population of B cells (a minor population in normal animals) predominates in peripheral blood of the mutant animals. Both T and B cell responses to antigens and mitogens are diminished in viable motheaten animals. The $Hcph^{mev}$ mutation also affects NK cell function and macrophage/monocyte function, and allows these cell populations to expand abnormally in various tissues and in cultures of cells from lymphoid tissue.

Effect of endogenous MT on immune function

In light of previous observations which showed that exogenous MT can suppress a T-dependent humoral response, we examined the role that endogenous Mt1 and Mt2 might play in the development of a humoral response to a T-dependent antigen. To do this, we injected OVA into

129/Sv-+/+ and 129/Sv-$Mt1^{\text{tm1Bri}}$ $Mt2^{\text{tm1Bri}}$ mice. Figure 1 shows that 129/Sv-$Mt1^{\text{tm1Bri}}$ $Mt2^{\text{tm1Bri}}$ produced significantly higher levels of anti-OVA antibodies throughout the course of the anti-OVA response than did the wild type control animals. The interval preceding the development of detectable levels of anti-OVA was similar in both experimental groups, but the amount of antibody produced was approximately 5 times higher at the point of greatest difference (day 21) between the two experimental groups.

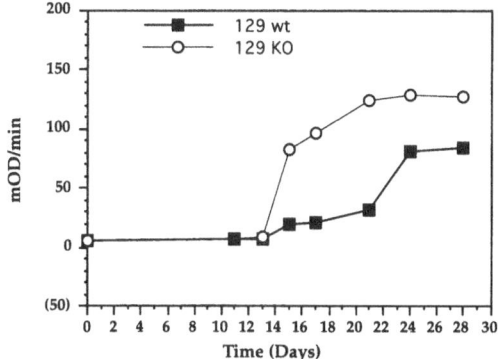

Figure 1. Comparison of the anti-OVA (IgG) response between 129/J-±/± and 129/Sv-Mt1 Mt2 knockout. Mice were injected with 200 μl of ovalbumin (OVA) i.p. on day 0 and again on day 10. Serum was collected at the times indicated and ELISA assays were performed to determine the presence of antibodies (IgG) against OVA. Results are reported as averages of four mice in each group ($p < 0.05$).

MT levels in C57BL/6J-$Hcph^{\text{mev}}$ mice

Preliminary studies suggested that the profound autoimmune and systemic inflammatory phenotype of C57BL/6J-$Hcph^{\text{mev}}$ mice was associated with the expression of metallothionein. Figure 2 shows that C57BL/6J-$Hcph^{\text{mev}}$ mice express significantly elevated levels of MT when compared to littermate controls as measured with a competition ELISA. The levels of MT in the circulation are elevated as early as serum can be collected from mutant animals. This level fluctuates over the course of the mutant animal's life, but the level always exceeds that found in normal littermate controls. Interestingly, liver homogenates prepared from the mutant and wild type animals do not differ for MT levels (data not shown) suggesting that another tissue is the source of the MT found in the circulation. Experiments underway in our laboratory are designed to identify the source of serum MT in C57BL/6J-$Hcph^{\text{mev}}$ mice.

Effects of the Mt knockout on gross pathology of the viable motheaten syndrome

129/Sv-$Mt1^{\text{tm1Bri}}$ $Mt2^{\text{tm1Bri}}$ (the MT1 and MT2-knockout region will be designated as $Mtko$) and C57BL/6J-$Hcph^{\text{mev}}$ mice were obtained from the the Jackson Laboratory. The origin of the

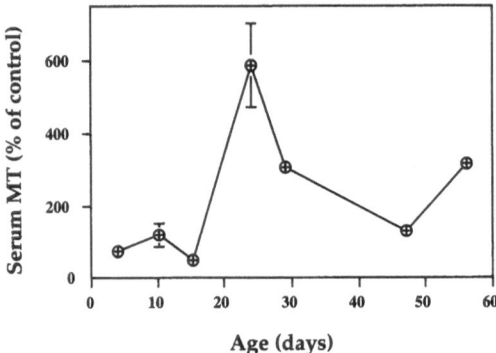

Figure 2. C57BL/6J Viable motheaten mice express high levels of MT in the serum. Data was derived from competition ELISA assays and expressed as a percent of the littermate controls.

129/Sv-$Mt1$[tm1Bri] $Mt2$[tm1Bri] strain [22] and the $Hcph$[mev] [23] mutation has been previously described. For production of the strain that is mutant at both $Hcph$ and at the $Mt1$ and $Mt2$ loci, male 129/Sv-$Mt1$[tm1Bri] $Mt2$[tm1Bri] knockout mice were crossed with ovariectomized recipients of C57BL/6J- $Hcph$[mev] (mev/mev) ovaries. Mice heterozygous at both genetic regions (+/Mt ko, +/mev) were then backcrossed to 129/Sv-$Mt1$[tm1Bri] $Mt2$[tm1Bri] mice. Homozygous $Mtko$ mice (that were also homozygous normal or heterozygous +/mev) were then genotyped by PCR to identify the mev heterozygotes and those mice were intercrossed. Subsequently, the line was maintained with the $Mtko$ construct fixed as homozygous and the mev as a segregating locus. This mating scheme requires genotyping at each generation, but produces mev/mev and littermate controls. Genotyping was done by PCR using primers specific for the neo insert, for the mutant and wild type alleles of $Hcph$, and for $Mt2$.

One of the earliest external signs of autoimmune attack in the viable motheaten syndrome are granulocytic lesions of the skin. These lesions can be observed as early as 24 h after birth. Subsequent gross changes include progressive inflammation of the extremities, including the digits. The mev homozygotes that are also homozygous for the MT knockout construct display fewer granulocytic lesions on the skin surface, and a decreased inflammation of the digits. In fact, early in life, the only visible characteristic of the $mev/mev,Mtko/Mtko$ that suggests the mev/mev genotype is the small body size that is characteristic of mev/mev animals. Examination of the offspring numbers of each genotype from the cross of double heterozygotes (+/mev, +/$Mtko$ × +/mev, +/$Mtko$) showed a marked increase ($p < 0.05$) in the survival of $Hcph$[mev] homozygotes if the progeny were also heterozygous or homozygous for the $Mtko$ gene construct, suggesting that elimination of the Mt1 and Mt2 genes was beneficial to mev/mev survival up to the time of weaning.

Flow cytometric analysis

A panel of antibodies against lineage-specific differentiation markers was used to analyze leukocyte populations in the spleen. Antibodies were obtained commercially (Pharmingen, San Diego, CA) or were purified from hybridoma cell lines as ascites and labelled with biotin, phycoerythrin, or fluorescein isothiocyanate for single and multiparameter analyses. The following panel of monoclonal antibodies was used: anti-CD3 (145-2C11, pan T cell receptor associated polypeptide); anti-B220 (RA3-6B2, B lineage cells); anti-IgLk (R5-240, B cells); anti-Mac-1 (M1/70, C3bi receptor on activated macrophages and B1 cells); and anti-Gr-1 (RB5-8C6, granulocytes).

Because of the high number of atypical nucleated cells in *mev/mev* blood, standard differential counts of blood cells from these mice are difficult to interpret. Instead, flow cytometric analysis was performed on whole blood samples to determine cell populations (Fig. 3). The percentage of lymphocytes (Mac-1-, Gr-1-) in the blood of *mev/mev* and *mev/mev, Mtko/Mtko* mice, while significantly reduced from normal littermate percentages, was similar to each other. In contrast, *mev/mev, Mtko/Mtko* had significantly more macrophages (Mac-1+, Gr-1-) and fewer granulocytes (Mac-1+, Gr-1+) than were found in *mev/mev, +/+* blood.

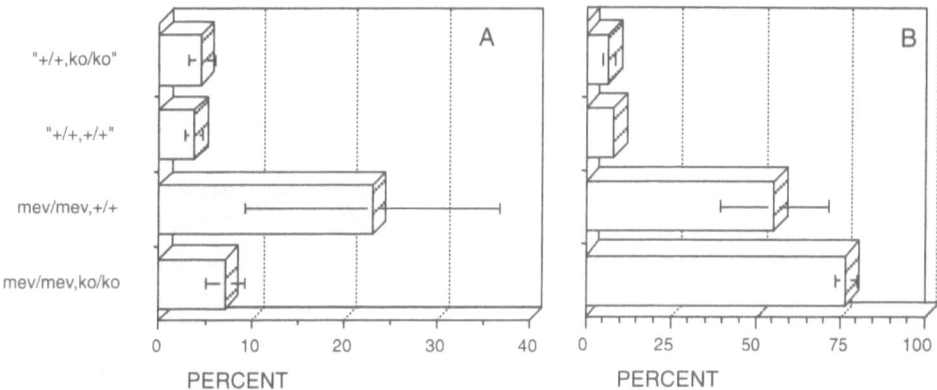

Figure 3. Blood granulocyte and monocyte populations. Comparison of granulocyte (panel A) and macrophage (panel B) populations in the peripheral blood of age matched (50 day old) mice. No effect of the *Mtko* construct is seen in either population of cells when the wild type *Hcph* allele is present. However, there appear to be reciprocal shifts in macrophage and granulocyte populations on the *Hcph*[mev] background.

Leucocyte populations in the spleen were also determined by flow cytometry. Age-matched animals were analyzed and three cell populations showed consistent differences between *mev/mev, Mtko/Mtko* mice and *mev/mev, +/+* littermates (Fig. 4). The percentage of mature splenic T-cells (CD3+) in the *mev/mev, Mtko/Mtko* mice was above that of the *mev/mev, +/+*. This appears to be a direct effect of the lack of metallothionein on the viable motheaten phenotype because the same trend is observed when comparing CD3+ cells in the +/+, *Mtko/Mtko*

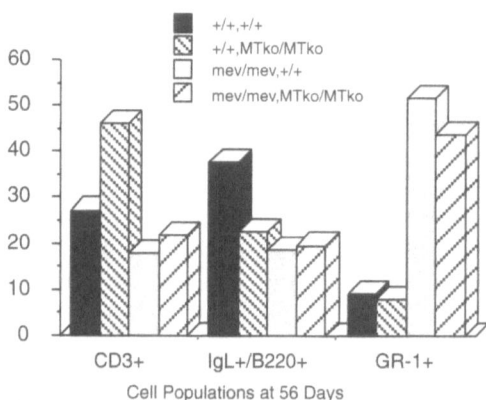

Figure 4. Spleen cell population shifts in viable motheaten and metallothionein mice. T cell (CD3⁺), B cell (Ig⁺, B220⁺), and granulocyte (Gr-1⁺) numbers were assessed in age matched (56 days old) mice. The *Mtko* construct was associated with an increase in T cells on both the wild type and mutant *Hcph* backgrounds. Granulocytes were slightly reduced on the *Hcph*^mev background at this age.

compared to normal (+/+, +/+) littermates. B-cell percentages (Ig+, B220+) in viable motheaten knockouts are also higher than viable motheaten alone despite the decreases in B cell numbers observed in +/+, *Mtko/Mtko* compared to normal (+/+, +/+). In contrast to the patterns of change in the T and B cell populations, changes in granulocyte populations (Gr-1+) appear to fluctuate with age. At 40 days of age, the *Mtko* construct has no effect, but at 56 days of age, there is a decrease in the percentage of granulocytes in *mev/mev*, *Mtko/Mtko* mice compared to *mev/mev*, and at 100 days there is an increase in the percentage of granulocytes in *mev/mev*, *Mtko/Mtko* mice compared to *mev/mev*.

Lifespan studies

Initial lifespan studies were carried out on small numbers of each viable motheaten genotype (n ≥ 10). Mean lifespan of *mev/mev* was significantly reduced in animals that were either heterozygous or homozygous for the *Mt* knockout construct (data not shown).

Discussion

Heretofore, our experiments have employed exogenous MT added to *in vitro* or *in vivo* systems to measure the impact of this protein on defined immune responses. In the series of experiments described here, we begin to explore the impact of endogenous MT on the functioning of the immune system. Our results suggest that elimination of the *Mt1* and *Mt2* gene products enhances at least some instances of specific B cell activity in normal immune function. This seems to be in accord with our previous observations that excess exogenous MT can suppress humoral responsiveness. The observation that Mt knockout animals are increasingly susceptible to CDDP-induced apoptosis [24] suggests the possibility that apoptotic events related to the immune response (e.g., CTL mediated killing, thymic T cell selection) might also be enhanced in Mt knockout animals. Assessment of splenic T cell numbers in the +/+, *Mtko/Mtko* mice suggests that the knockout construct is responsible for an increase in splenic T cells in unstimulated animals. Shifts in regulatory T cell numbers could have potent effects on T-dependent humoral immunity.

Eliminating Mt1 and Mt2 synthesis also appears to alter the course of the viable motheaten autoimmune syndrome. Our evidence suggests that knocking out Mt1 and Mt2 synthesis improves survival early in the course of the viable motheaten disease, but decreases survival as the animals age. Intriguingly, parallel experiments designed to produce a viable motheaten mouse with an excess of Mt1 transgenes has produced many fewer (than expected) offspring that are homozygous for *Hcph*^mev and that have at least one copy of the transgene construct (data not shown). This observation suggests that an excess of MT is also detrimental in autoimmune disease.

MT has been shown in a variety of ways and by many laboratories to be important as a reservoir of essential heavy metals, as a regulator of some transcription factor activity, as a scavenger of free radicals, and as an extracellular regulator of immune function. Each of these roles can have critical implications in the course of both normal immune function and autoimmune dysfunction.

Acknowledgements
This work is supported in part by NIH grants ES 07408 to MAL and CA20408 to LDS.

References

1. Klaassen C, Lehman-McKeeman L (1989) Induction of metallothionein *J Amer Coll Toxicol* 8: 1315–1321.
2. Karasawa M, Nishimura N, Nishimura H, Tohyama C, Hashiba H, Kuroki T (1991) Localization of metallothionein in hair follicles of normal skin and the basal cell layer of hyperplastic epidermis: possible association with cell proliferation *J Invest Dermatol* 97: 97–100.
3. Koropatnick J, Leibbrand M, Cherian MG (1989) Organ-specific metallothionein induction in mice by X irradiation. *Radiat Res* 119: 356–365.
4. Hidalgo J, Campmany L, Marti O, Armario A (1991) Metallothionein-I induction by stress in specific brain areas. *Neurochem Res* 16: 1145–1148.
5. Min KS, Mukai S, Ohta M, Onosaka S and Tanaka K (1992) Glucocorticoid inhibition of inflammation-induced metallothionein synthesis in mouse liver. *Toxicol Appl Pharmacol* 113: 293–298.
6. Borghesi L, Lynes MA (1996) Stress proteins as agents of immunological change: some lessons from metallothionein. *Cell Stress Chaperones* 1: 99–108.
7. Lynes MA, Garvey JS and Lawrence DA (1990) Extracellular metallothionein effects on lymphocyte activities. *Mol Immunol* 27: 211–219.
8. Youn J, Borghesi LA, Olson EA and Lynes MA (1995) Immunomodulatory Activities of Extracellular Metallothionein II. Effects on Macrophage Functions. *J Toxicol Environ Health* 45: 397–413.
9. Borghesi LA, Youn J, Olson EA, Lynes MA (1996) Interactions of Metallothionein with murine lymphocytes: plasma membrane binding and proliferation. *Toxicology* 108: 129–140.
10. Lynes MA, Borghesi LA, Youn J, Olson EA (1993) Immunomodulatory activities of extracellular metallothionein I. Metallothionein effects on antibody production. *Toxicology* 85: 161–177.
11. Shapiro SG, Cousins RJ (1980) Induction of rat liver metallothionein mRNA and its distribution between free

and membrane-bound polyribosomes. *Biochem J* 190: 755–764.

12. Thomas DG, Linton H, Garvey JS (1986) Fluorometric ELISA for the detection and quantitation of metal-lothionein. *J Immunol Meth* 89: 239–247.

13. Danielson KG, Ohi S, Huang PC (1982) Immunochemical detection of metallothionein in specific epithelial cells of rat organs. *Proc Natl Acad Sci USA* 79: 2301–2304.

14. Hart BA, Garvey JS (1986) Detection of metallothionein in bronchoalveolar cells and lavage fluid following repeated cadmium inhalation. *Environ Res* 40: 391–398.

15. Kido T, Shaikh ZA, Kito H, Honda R, Nogawa K (1991) Dose-response relationship between dietary cadmi-um intake and metallothionuria in a population from a cadmium polluted area of Japan. *Toxicology* 66: 271–278.

16. Matsushima K, Taguchi M, Kovacs EJ, Young HA, Oppenheim JJ (1986) Intracellular localization of human monocyte associated IL-1 activity and release of biologically active IL-1 by trypsin and plasmin. *J Immunol* 136: 2883–2891.

17. Sato M, Sasaki M (1992) Tissue specific induction of metallothionein synthesis by tumor necrosis factor-α. *Res Comm Chem Path Pharm* 75: 159–171.

18. Cousins R, Leinart A (1988) Tissue-specific regulation of zinc metabolism and metallothionein genes by inter-leukin -1. *FASEB J* 2: 2884–2890.

19. Schroeder J, Cousins R (1990) Interleukin 6 regulates metallothionein gene expression and zinc metabolism in hepatocyte monolayer cultures. *Proc Natl Acad Sci USA* 87: 3137–3141.

20. Shultz LD, Schweitzer PA, Rajan TV, Yi T, Ihle JN, Matthews RJ, Thomas ML, Beier DR (1993) Mutations at the murine motheaten locus are within the hematopoietic cell protein-tyrosine phosphatase (*Hcph*) gene. *Cell* 73: 1445–1454.

21. Tsui FW, Tsui HW (1994) Molecular basis of the motheaten phenotype. *Immunol Rev* 138: 185–206.

22. Kelly-EJ, Quaife-CJ, Froelick-GJ, Palmiter-RD (1996) Metallothionein I and II protect against zinc deficien-cy and zinc toxicity in mice. *J Nutr* 126: 1782–90.

23. Shultz LD, Coman DR, Bailey CL, Beamer WG, Sidman CL (1984) "Viable motheaten": a new allele at the motheaten locus. I. Pathology. *Amer J Pathol* 116: 179–186.

24. Kondo Y, Rusnak JM, Hoyt DG, Settineri CE, Pitt BR, Lazo JS (1997) Enhanced apoptosis in metallothionein null cells. *Mol Pharm* 52: 195–201.

Metallothionein IV
C. Klaassen (ed.)
© 1999 Birkhäuser Verlag Basel/Switzerland

Dietary zinc, photoimmunosuppression and metallothionein (MT)

Vivienne E. Reeve[1], Noriko Nishimura[1], Meira Bosnic[1], Andy K.H. Choo[2] and Anna E. Michalska[2]

[1]*Department of Veterinary Anatomy and Pathology, University of Sydney, NSW 2006, Australia*
[2]*Murdoch Institute, Royal Children's Hospital, Parkville, Melbourne, VIC 3052, Australia*

Introduction

Zn is an essential trace element, a deficiency resulting in skin abnormalities in particular [1]. Human skin contains approximately 20% of the body's Zn stores. It is notable that epidermal Zn is found unevenly distributed, being histochemically demonstrated to be concentrated in the stratum germinativum, an observation that has led to the suggestion that Zn may have an association with the terminal differentiation of epidermal cells, which produces the stratified epidermal structure, and thus may be involved in the regulation of epidermal proliferation [2, 3]. Zn deficiency causes characteristic hyperkeratotic changes, abnormal keratohyalin formation, and depresses dermal collagen synthesis; the skin thus seems to be particularly sensitive to the Zn status of the animal.

Recent surveys indicate that up to 85% of Australian adults have Zn intakes below the RDA, and that mild but symptomless Zn deficiency is widespread and may be exacerbated by the current trend for dietary switching from a mixed to a lactovegetarian diet [4–7]. Since one of the functions of metallothionein (MT) is to maintain Zn homeostasis in animals, it is significant that the topical application of Zn has been shown to induce skin MT [8], with similar distribution of both Zn and MT found in the epidermal basal cells. In addition, the induction of cutaneous MT is a response to a variety of oxidative or DNA-damaging agents such as cadmium [9], UVB (280–320 nm) radiation [10, 11], 1,25-dihydroxyvitamin D-3 [12] and cold stress [13], and in several models the induction of MT has been correlated with conferred protection from damage resulting from subsequent UV irradiation, measured as a reduction in sunburn cell formation [9, 12, 13]. Record et al [14] have also found that topical Zn application reduced sunburn cell formation in UV-irradiated mouse skin. The photoprotective mechanism of MT appears to be associated with its potential to reduce oxidative radicals in the skin, and a role for cutaneous MT as an endogenous antioxidant has been proposed [10]. Since reactive oxygen species contribute to phototoxic injury in UV-exposed skin, it has been of interest to examine the potential of dietary Zn to modulate the levels of epidermal MT, and thus for Zn supplementation to provide protection from other forms of photodamage.

The incidence of sunlight-induced skin cancer in Australia is the highest in the world. More than 60% of the population acquire skin cancer at some time, and the incidence continues to rise in spite of the availability of sunscreens and excellent educational awareness programs. It is important therefore to investigate alternative lifestyle methods of intervention, especially as most skin cancers are believed to be initiated by childhood sunlight exposure, and to remain latent but viable in the majority of the population for an extended latency period. Dietary interventions of other kinds [15] are currently proving promising.

Photocarcinogenesis is immunologically regulated in both man and experimental rodents [16]. UVB (290–320 nm) radiation not only initiates the tumour cell by causing DNA damage, but also suppresses specific functions of the cellular immune system, an impairment by which the tumour progenitor cell is able to evade normal immunological recognition and rejection. The defect in UV-irradiated animals is in the activity of T helper-1 cells (Th-1 cells) [17], and can be readily observed by the suppression of the delayed type, or contact hypersensitivity (CHS) response. Thus the degree of suppression of CHS in UV-irradiated mice is believed to be an indicator of the risk of tumour formation if the irradiations are continued daily until chronic. It is noteworthy that Zn is necessary for normal T lymphocyte function, and that Zn deficient mice have atrophic lymphoid tissues, including the thymus, abnormal maturation of T cells, reduced thymic hormone levels and impaired delayed type hypersensitivity reactions [18, 19].

We have therefore been interested in observing the influence of various levels of dietary Zn on the photoimmunosuppressive property of UVB radiation, using the CHS response as the assay of the relevant immunological status. Furthermore, we have also studied the relationship between dietary Zn and cutaneous MT presence, and the influence of alterations in these parameters on UVB radiation-induced hyperplasia and immunosuppression.

Dietary Zn and UVB exposure

Hairless mice of the Skh:HR-1 strain were fed semi-purified diets [20] providing normal Zn (30 mg/kg diet, or 120 µg Zn/mouse/day), 5X or 10X normal Zn (150 or 300 mg Zn/kg diet), or a diet with no added Zn (3–4 mg Zn/kg diet). Body weights were monitored weekly, but there was no difference between groups for up to 26 weeks of feeding. After 6 weeks of feeding, all mice responded equally to an exposure of $3 \times$ minimal erythemal dose (3 X MED) of UVB radiation, quantitated by the mid-dorsal skinfold thickness; this UVB exposure resulted in an approximate doubling of skinfold thickness [21]. Even after 10 weeks of feeding there was no difference in this response. Thus the low Zn diet, or the supplementation with Zn, had no obvious clinical effects on the mice at these time points.

Systemic CHS reactions were also measured after acute exposure of the mice in each dietary group, to UVB radiation [21]. After 6 weeks of feeding, all dietary groups responded similarly with normal CHS reactions in all the unirradiated mice, and approximately 60% suppression of the CHS response by exposure to 3 X MED of UVB radiation. However when the feeding was continued until 10 weeks, we observed a significant suppression of CHS in mice fed no added Zn, compared with the normal and Zn supplemented diets. This observation was consistent with other studies of the requirement of Zn for optimal T lymphocyte function, although the mice did not display overt signs of Zn deficiency. Exposure to 3 X MED of UVB radiation was found to slightly, but not significantly, further suppress the CHS reaction.

In contrast, whereas the control-fed mice were markedly immunosuppressed by 3 X MED of UVB radiation (47% suppression of CHS) as expected, mice fed 5XZn were significantly spared (31% suppression of CHS), and mice fed 10XZn were not significantly immunosuppressed at all (10% suppression of CHS). Thus we had observed a dependance of photoimmunosuppression on the dietary Zn level, with higher Zn supplementation providing effective protection from this UVB-induced defect.

Mechanism of photoimmunoprotection by dietary Zn

The possibility that MT might be involved in the mechanism was investigated using transgenic mice of the C57Bl/6-Ola129 cross bearing a null mutation of MT-I and -II isoforms [22]. These mice proved to be significantly more sensitive to UVB-suppression of the CHS response than the normal C57Bl/6-Ola129 mice, e.g. following exposure to the equivalent of 3 X 0.6MED of UVB radiation, the MT+/+ (wild type) mice retained normal CHS responses, but the MT–/– mice had a 25% suppression of CHS. At a higher UVB dose of 3 X MED, the CHS response of MT+/+ was suppressed by 43%, but the response of the MT–/– was significantly more suppressed by 51%.

Groups of MT–/– and +/+ mice were also fed the semi-purified diets supplying normal Zn or 10XZn, for 10 weeks, and their CHS responses measured after exposure to 3 X MED of UVB radiation. As previously observed with the hairless Skh mice, the 10XZn supplementation of MT+/+ mice rendered them resistant to UVB-suppression of CHS (9% suppression, compared with 51% suppression in control-fed +/+ mice). However, the CHS response in control-fed MT–/– mice was suppressed by 66% by UVB exposure, and remained suppressed (51% suppression) when MT–/– mice were supplemented with 10XZn and then UVB-exposed.

It is interesting that the MT–/– mouse appears to have a normal phenotype, to respond normally to contact sensitisers, and to maintain good health and demeanour approaching a normal mouse lifespan. Perhaps there is redundancy with other MT isoforms which as yet have not been characterised, or perhaps pathological compromise occurs in MT–/– mice only if severely challenged by an oxidative and/or mutagenic insult such as an acute UVB exposure. Further studies with these mice will be most interesting.

Localisation of MT

Immunohistochemical localisation of cutaneous MT was demonstrated with a rabbit antiserum against MT-I, known to cross-react with mammalian MT-I and -II and human MT-III, using the avidin-biotin-complex method [23] in HistoChoice-fixed skin samples embedded in paraffin. We found that supplementation of hairless mice with 10XZn enhanced the staining for MT, which was localised in the epidermal basal cell layer. UVB irradiation also induced MT staining, in both hairless and MT+/+ haired mice, first appearing in dermal fibroblasts and inflammatory cells at 24 h post-irradiation, and then strongly staining the basal cells in the epidermis at 48 h, which coincided with the appearance of epidermal hyperplasia in response to the UVB irradiation. The basal cell MT was maximum at 72 h, after which it decreased slowly, although the epidermis remained thickened. The early positive staining of dermal cells for MT, a skin stratum not normally penetrated by UVB radiation, suggests that these cells, which may be early indicators of the immunosuppressive effect of UVB radiation, were responding to cytokines released directly by UVB targets in the epidermis.

Conclusions

We found that supplementation of the diet with Zn provided Zn-dose-dependant photoim-munoprotection after prolonged feeding, implying that a gradual alteration of body Zn pools may be involved. However the immunological protection occurred without overt signs of altered Zn status, as may occur in Western populations. The Zn supplementation concurrently increased the immunopositive MT presence in epidermal basal cells, whereas the absence of MT in MT–/– mice prevented the protective effect of supplemental Zn, indicating an important mechanistic role for MT in facilitating the Zn effect. In addition to epidermal MT induction by Zn, we have observed the time course of MT induction by UVB radiation, and suggest that two epidermal UVB targets may be involved, one resulting in direct MT protein synthesis in the basal cells, the other perhaps caused by release of epidermal cytokines or signalling by the local UVB response, resulting secondarily in a dermal immune cellular response. The significance of these observations to the photocarcinogenesis process in chronically UV-irradiated mice is currently under study.

Acknowledgements
These studies were supported by the University of Sydney Cancer Research Fund. We thank Dr. C. Tohyama, NIEHS, Tsukuba, Japan, for access to the MT-antiserum.

References

1. Hsu JM, Kim KM, Anthony WL (1974) Biochemical and electronmicroscopic studies of rat skin during zinc deficiency. *Adv Exp Med Biol* 8: 347–388.
2. Greaves MW (1970) Effects of long-continued ingestion of zinc sulphate in patients with venous leg ulceration. *Lancet* 2: 889–891.
3. Kapp P, Simon F (1980) Ultrastructural cutaneous changes in zinc-depleted pigs. *Acta Vet Scand Scient Hungaricae* 28: 463–471.
4. National Dietary Survey of Adults (1983) – No.2, Nutrient Intakes. *Aust Govt Public Service*, Canberra, Australia.
5. Baghurst KI, Dreosti IE, Syrette JA, Record SJ, Baghurst PA, Buckley RA (1991) Zinc and magnesium status of Australian adults. *Nutr Res* 11: 23–32.
6. Ruz M, Cavan KR, Bettger WJ, Gibson RS (1992) Erythrocytes, erythrocyte membranes, neutrophils and platelets as biopsy materials for the assessment of zinc status in humans. *Brit J Nutr* 68: 515–527.
7. Srikumar TS, Johansson GK, Ocherman PA, Gustafsson JA, Akesson B (1992) Trace element status in healthy subjects switching from a mixed to a lactovegetarian diet for 12 months. *Amer J Clin Nutr* 55: 885–890.
8. Morgan AJ, Lewis G, Van Den Hoven WE, Akkerboom PJ (1993) The effect of zinc in the form of erythromycin-zinc complex (Zineryt lotion) and zinc acetate on metallothionein expression and distribution in hamster skin. *Brit J Dermatol* 129: 563–570.
9. Hanada K, Gange RW, Siebert E, Hasan T (1991) Protective effects of cadmium chloride against UVB injury in mouse skin and in cultured human cells: a possible role of cadmium-induced metallothionein.

Photodermatol Photoimmunol Photomed 8: 111–115.

10. Hanada K, Baba T, Hashimoto I, Fukui R, Watanabe S (1993) Possible role of cutaneous metallothionein in protection against photo-oxidative stress – epidermal localization and scavenging activity for superoxide and hydroxyl radicals. *Photodermatol Photoimmunol Photomed* 9: 209–213.

11. Anstey A, Marks R, Long C, Navabi H, Pearse A, Wynfordthomas D, Jasani B (1996) *In vivo* photoinduction of metallothionein in human skin by ultraviolet irradiation. *J Pathol* 178: 84–88.

12. Hanada K, Sawamura D, Nakano H, Hashimoto I (1995) Possible role of 1,25-dihydroxyvitamin D-3-induced metallothionein in photoprotection against UVB injury in mouse skin and cultured rat keratinocytes. *J Dermatol Sci* 9: 203–208.

13. Ota T, Hanada K, Hashimoto (1996) The effect of cold stress on UVB injury in mouse skin and cultured keratinocytes. *Photochem Photobiol* 64: 984–987.

14. Record IR, Jannes M, Dreosti IE (1996) Protection by zinc against UVA- and UVB-induced cellular and genomic damage *in vivo* and *in vitro*. *Biol Trace Element Res* 53: 19–25.

15. Black HS, Thornby JI, Wolf JEJr Goldberg LH, Herd JA, Rosen T, Bruce S, Tschen JA, Scott LW, Jaax S et al (1995) Evidence that a low-fat diet reduces the occurrence of non-melanoma skin cancer. *Int J Cancer* 62: 165–169.

16. Fisher MS, Kripke ML (1982) Suppressor T lymphocytes control the development of primary skin cancers in UV irradiated mice. *Science* 216: 1133–1134.

17. Ullrich S (1996) Does exposure to UV radiation induce a shift to a TH-2-like immune reaction? *Photochem Photobiol* 64: 254–258.

18. Fraker PJ, Haass S, Luecke RW (1977) Effect of zinc deficiency on the immune response of the young adult A/Jax mouse. *J Nutr* 107: 1889–1895.

19. Lynes MA, Borghesi LA, Youn J, Olson E (1990) Immunomodulatory activities of extracellular metallothionein I. Metallothionein effects on antibody production. *Mol Immunol* 27: 211–219.

20. Reeve VE, Matheson M, Greenoak GE, Canfield PJ, Boehm-Wilcox C, Gallagher CH (1988) Effect of dietary lipid on UV light carcinogenesis in the hairless mouse. *Photochem Photobiol* 48: 689–696.

21. Reeve VE, Boehm-Wilcox C, Bosnic M, Reilly WG (1994) Differential photoimmunoprotection by sunscreens is unrelated to epidermal cis urocanic acid formation in hairless mice. *J Invest Dermatol* 103: 801–806.

22. Michalska AE, Choo KH (1993) Targeting and germ-line transmission of a null mutation at the metallothionein I and II loci in mice. *Proc Natl Acad Sci USA* 90: 8088–8092.

23. Nishimura H, Nishimura N, Tohyama C (1990) Localisation of metallothionein in the genital organs of the male rat. *J Histochem Cytochem* 38: 927–933.

Role of metallothionein in Cd-induced nephrotoxicity and bone loss

Metallothionein IV
C. Klaassen (ed.)
© 1999 Birkhäuser Verlag Basel/Switzerland

Metallothionein-null mice are susceptible to chronic CdCl$_2$-induced nephropathy: Cd-induced renal injury is not necessarily mediated by CdMT

Curtis D. Klaassen, Jie Liu, Yaping Liu and Sultan S. Habeebu

Department of Pharmacology, Toxicology & Therapeutics University of Kansas Medical Center Kansas City, KS 66160-7417, USA

Summary. Chronic exposure to Cd results in kidney injury. It has been proposed that nephrotoxicity produced by chronic Cd exposure is via the Cd-metallothionein complex (CdMT), and not by inorganic forms of Cd. If this hypothesis is true, then MT-null mice, which cannot form CdMT, should not develop nephrotoxicity to inorganic Cd. Control and MT-null mice were injected sc with CdCl$_2$ (0.01 to 1.6 mg Cd/kg) daily, 6 times/week for up to 6 weeks, and the renal MT concentration and nephrotoxicity were examined. In control mice, renal Cd burden increased in a dose- and time-dependent manner, reaching as high as 135 µg Cd/g kidney, along with 150-fold increases in renal MT concentrations, reaching 450 µg MT/g kidney. In MT-null mice, renal Cd concentration (< 10 µg/g) was much lower. The maximum-tolerated dose of Cd in MT-null mice was approximately one-tenth that of controls. Cd-induced renal injury was more pronounced in MT-null than in control mice, as evidenced by proteinuria, glucosuria, enzymuria and blood urea nitrogen levels. CdCl$_2$ produced lesions throughout the kidney, including tubular degeneration, atrophy, interstitial inflammation, glomerular proliferation and sclerosis. These lesions were more severe in MT-null than in control mice, mirroring the biochemical analyses. These data indicate that (1) Cd-induced renal injury is not necessarily mediated through the CdMT complex, and (2) induction of MT is an important adaptive mechanism protecting against chronic Cd nephrotoxicity.

Introduction

Cadmium (Cd) is an environmental pollutant. Cd is used in batteries, electroplating, pigments, plastic stabilizers, alloys, and is a contaminant in phosphate fertilizer. Cd is non-biodegradable and environmental levels are increasing due to man's activities. In humans, Cd exposure occurs primarily via ingestion of Cd-contaminated food or via smoking of Cd-contaminated tobacco. Inhalation of Cd dusts and fumes has been an occupational hazard [1].

Acute Cd exposure produces liver injury, while chronic Cd exposure results in kidney injury. Cd-induced nephrotoxicity is the most important and the most frequently occurring ailment in humans chronically exposed to Cd [1–2]. Cd-induced nephrotoxicity is thought to be caused by the Cd-metallothionein complex (CdMT), which is synthesized in liver, released into plasma during liver damage, and is taken up by the kidney to produce nephrotoxicity. Thus, a single injection of CdMT, instead of CdCl$_2$, has been used as a model to study Cd nephropathy for years.

Using MT-transgenic and MT-null mice, we have recently demonstrated that intracellular MT plays an important protective role in acute Cd-induced lethality and hepatotoxicity [3–4]. However, these genetically-altered animals do not have a different sensitivity to acute CdMT-induced nephrotoxicity [4–5]. Thus, it is puzzling why intracellular MT does not protect against the nephrotoxicity produced by acute administration of CdMT. It is not known whether MT can provide long-term protection against Cd toxicity, as the reported protective effects of MT in animals have been done in short-term studies. Therefore, the goal of the study was to use the MT-null mouse model to critically evaluate the role of MT in Cd toxicity during a long-term exposure.

MT-null mice are susceptible to Cd-induced body weight loss and lethality

Homozygous MT-I and II knock-out mice [6] and corresponding controls (129/SvPCJ background) were given daily $^{109}CdCl_2$ at wide dose ranges (0.025 to 1.6 mg Cd/kg, 0.01 $\mu Ci^{109}Cd/kg$, sc), 6 days/week for 6 weeks.

Chronic administration of $CdCl_2$ resulted in a dose-dependent loss of body weight in both control and MT-null mice. However, MT-null mice were approximately ten-times more sensitive to Cd-induced body weight loss than control mice. In addition, control mice survived a daily dose of 1.6 mg Cd/kg for 6 weeks, while MT-null mice could not survive a daily dose of more than 0.2 mg Cd/kg.

MT-null mice are susceptible to chronic Cd-induced nephrotoxicity

Cd accumulated in kidneys of control mice in a dose- and time-dependent manner. Maximum renal Cd concentrations (>120 µg Cd/g kidney) occurred in mice receiving daily doses of 0.4 mg Cd/kg or higher for 6 weeks. In comparison, renal Cd concentrations were remarkably lower in MT-null mice, only reaching 10 µg/g kidney after 6 weeks of exposure to the maximum tolerated daily dose (0.2 mg Cd/kg) (Fig. 1)

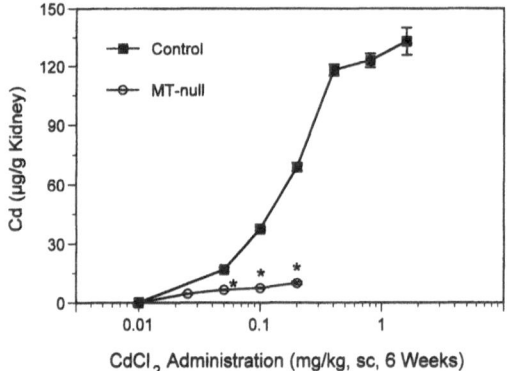

Figure 1. Renal Cd concentrations following chronic $CdCl_2$ administration to control (0.05 to 1.6 mg Cd/kg, sc, 6 times/week) and MT-null mice (0.0125 to 0.2 mg Cd/kg, sc, 6 times/week) for 6 weeks. Data are mean ± SE of 6–8 mice. *Significantly different from controls (P < 0.05).

We originally hypothesized that chronic administration of CdCl$_2$ would not produce renal injury in MT-null mice, because no CdMT complex is formed in the liver, thus, no CdMT is delivered to the kidney, and because the Cd concentration is very low in the kidney of the MT-null mice. However, this is not true. In fact, with Cd administered at a dose of 0.1 mg Cd/kg or higher, the MT-null mice excreted more urinary protein (Fig. 2a) and glucose (data not shown) than control mice. Urinary activity of N-acetyl-D-glucosaminidase (NAG), a sensitive index for proximal tubular damage, was also much higher in MT-null mice receiving the daily dose of 0.1 mg/Cd/kg or higher (Fig. 2b). Increases in blood urea nitrogen (BUN) concentrations par-

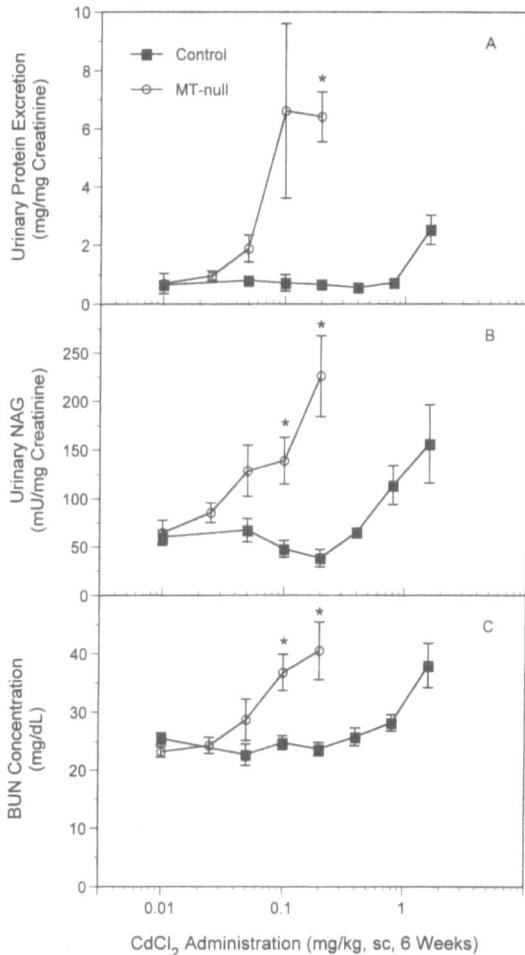

Figure 2. Urinary excretion of protein, N-acetyl-D-glucosaminidase (NAG), and blood urea nitrogen concentrations following chronic CdCl$_2$ administration to control (0.05 to 1.6 mg Cd/kg, sc) and MT-null mice (0.0125 to 0.2 mg Cd/kg, sc) for 6 weeks. Data are mean ± SE of 6–8 mice. *Significantly different from controls (P < 0.05).

alleled proteinuria and enzymuria. Again, MT-null mice had higher BUN concentrations than did control mice, indicating more renal damage (Fig. 2c). In addition, histopathological examination showed that chronic Cd administration produced more severe renal injury in MT-null than in control mice (data not shown).

MT-null mice are unable to synthesize MT during chronic Cd exposure

The renal MT concentrations following chronic Cd-exposure are markedly different between control and MT-null mice. In control mice, renal MT was increased 100-fold by Cd exposure, up to 450 µg MT/g kidney. In contrast, in MT-null mice, renal MT (Cd-binding protein concentration) remained at background level (2 µg/kg) (Fig. 3).

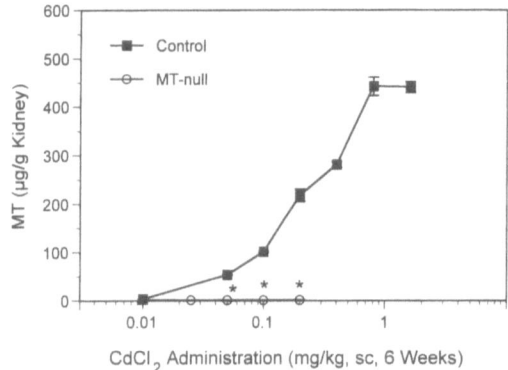

Figure 3. Renal MT concentrations following chronic CdCl$_2$ administration to control (0.05 to 1.6 mg Cd/kg, sc, 6 days/week) and MT-null mice (0.025 to 0.2 mg Cd/kg, sc, 6 days/week) for 6 weeks. Data are mean ± SE of 6–8 mice. *Significantly different from controls (P < 0.05).

Discussion

The present study shows that MT-null mice are much more susceptible than controls to chronic Cd-induced lethality and nephrotoxicity, as evidenced by proteinuria, enzymuria and blood urea nitrogen levels, as well as by histopathological examination. This is the first report to demonstrate that MT also plays an important role in protecting against chronic Cd-induced toxicity.

MT-null mice cannot synthesize functional MT protein, and cannot form the CdMT complex, yet they are susceptible to Cd nephrotoxicity. This indicates that Cd-induced renal injury is not necessarily mediated via the CdMT complex. In fact, Cd ion has been shown to be directly toxic to various functions of renal cells, such as inhibition of Na$^+$-glucose cotransport in renal cortical cells [7], disruption of intracellular junctions and actin filaments in LLC-PK1 cells (a pig proximal tubular cell line) [8], interference with DNA and protein synthesis in mesangial cells [9], and increases in LDH leakage from primary cultures of renal cortical cells [10]. Cd salts also produce acute renal injury in hamsters [11]. These observations further indicate that Cd-induced toxicity to kidney cells is not necessarily mediated via the CdMT complex.

On the basis of the present finding, the acute CdMT injection model for studying chronic Cd nephrotoxicity is questioned. We further compared the nephrotoxic potential of CdCl₂, acute CdMT and chronic CdMT. We found that the pathological features and mechanisms of renal injury from chronic Cd exposure are quite different from those produced by a single injection of CdMT. Therefore, it is proposed that acute CdMT injection is not an appropriate model for the study of chronic Cd-induced nephrotoxicity [12].

In conclusion, the present study using MT-null mice has demonstrated that (1) the renal injury produced by repeated Cd administration is not necessarily mediated via the CdMT complex, and (2) induction of intracellular MT is an important adaptive mechanism protecting against chronic Cd nephropathy.

Acknowledgement
Supported by NIH Grant ES-01142.

References

1. Goering PL, Waalkes MP and Klaassen CD (1995) Toxicology of cadmium. *In*: RA Goyer and MG Cherian (eds): *Toxicology of Metals: Biochemical Aspects*. Handbook of Experimental Pharmacology Vol. 115, pp 189–213, 1995. Springer-Verlag, New York.
2. Goyer RA and Cherian MG (1995) Renal effects of metals. *In*: RA Goyer, CD Klaassen, and MP Waalkes (eds): *Metal Toxicology*, Academic Press pp 389–412, San Diego.
3. Liu YP, Liu J, Iszard MB, Andrews GK, Palmiter RD and Klaassen CD (1995) Transgenic mice that overexpress metallothionein-I are protected from cadmium lethality and hepatotoxicity. *Toxicol Appl Pharmacol* 135: 222–228.
4. Liu J, Liu YP, Michalska AE, Choo KHA and Klaassen CD (1996) Metallothionein plays less of a protective role in CdMT-induced nephrotoxicity than CdCl₂-induced hepatotoxicity. *J Pharmacol Exp Ther* 276: 1216–1223.
5. Liu YP, Liu J, Palmiter RD and Klaassen CD (1996) Metallothionein-I transgenic mice are not protected from cadmium-metallothionein-induced nephrotoxicity. *Toxicol Appl Pharmacol* 137: 305–315.
6. Masters BA, Kelly EJ, Quaife CJ, Brinster RL and Palmiter RD (1994) Targeted disruption of metallothionein I and II genes increases sensitivity to cadmium. *Proc Natl Acad Sci USA* 91: 584–588.
7. Blumenthal SS, Lewand DL, Buday MA, Kleinman JG, Kerzoski SK and Petering DH (1990) Cadmium inhibits glucose uptake in primary cultures of mouse cortical tubule cells. *Amer J Physiol* 258, F1625-F1633.
8. Prozialeck WC, Wellington DR and Lamar PC (1993) Comparison of the cytotoxic effects of cadmium chloride and cadmium-metallothionein in LLC-PK1 cells. *Life Sci* 53, PL337–342.
9. Chin TA and Templeton DM (1992) Effect of CdCl₂ and Cd-metallothionein on cultured mesangial cells. *Toxicol Appl Pharmacol* 116: 133–141.
10. Groten JP, Luten JB, Bruggeman IM, Temmink JHM and Van Bladeren PJ (1992) Comparative toxicity and accumulation of cadmium chloride and cadmium-metallothionein in primary cells and cell lines of rat intestine, liver and kidney. *Toxicol Vitro* 6: 59–67.
11. Rehm S and Waalkes MP (1990) Acute cadmium chloride-induced renal toxicity in the Syrian hamster. *Fundam. Appl. Toxicol.* 104: 94–105.
12. Liu J, Sultan S Habeebu Liu YP and Klaassen CD (1998) Acute CdMT injection is not a good model to study chronic Cd nephropathy: Comparison of chronic CdCl₂ and CdMT exposure with acute CdMT injection in rats. *Toxicol Appl Pharmacol*; *in press*.

Cellular metallothionein: Properties of apometallothionein and the comparative toxicity of Cd^{2+} and Cd-metallothionein

David H. Petering, Munira Dughish, Meilin Huang, Susan Krezoski, Sara Krull, Donna Lewand, Amalia Muñoz, Lifen Ren, Sudha Venkatesh, Samuel Blumenthal and C. Frank Shaw III

Department of Chemistry, University of Wisconsin-Milwaukee, Milwaukee, WI 53201, USA and Department of Nephrology, Medical College of Wisconsin, Milwaukee, WI 53233, USA

Introduction

Two controversial facets of metallothionein biochemistry and toxicology center on the existence of measurable concentrations of apometallothionein in cells and the hypothesis that Cd-metallothionein is the species which causes toxicity to the kidney proximal tubule [1, 2][1] This contribution addresses both issues.

Apometallothionein: Synthesis, dynamics, and model reactions with metallo-transcription factors

Introduction

It has been thought that MT exists as metal-saturated forms in cells. Metallothionein isolated from many sources contains a full complement of metal ions. Furthermore, since apoMT has a high affinity for a variety of metal ions, apoMT should act as a thermodynamic sink for metals including Zn^{2+} which is available in plasma. Since apoMT is proteolytically hydrolyzed faster than metal-containing species *in vitro*, it has seemed unlikely that significant concentrations of apoMT could accumulate in cells [3].

An attractive hypothesis is that apoMT competes for intracellular pools of Zn during its synthesis [4]. It has also been suggested that changes in zinc occupancy in zinc finger transcription factors might control gene expression [5]. Merging these ideas, Kägi and coworkers showed that preincubation of apoMT with two Zn-finger transcription factors, TFIIIA and Sp1, prevents their specific association with DNA [4, 6].

Abbreviations: ACE1, Cu responsive transcription factor for metallothionein mRNA synthesis in *Saccharomyces cerevisiae;* CA, carbonic anhydrase, DTNB, 5,5'-dithio-bis(2-nitrobenzoate); ICR, internal control region; MT, metallothionein; Npt1, Na^+-phosphate cotransporter; RT-PCR, reverse transcriptase polymerase chain reaction; TdR, thymidine; TFIIIA, transcription factor IIIA; UAS_{CUP1}, upstream activation sequences for the CUP1 promoter; P, Zn-protein.

The presence of substantial concentrations of constitutive apoMT has been documented in a number of tumors and tumor cell lines [1]. The present report describes means of generation of cellular apoMT, its properties in cells, and model reactions addressing the hypothesis that apoMT can alter intracellular Zn or Cu distribution, including removal of Zn from Zn-finger proteins.

Constitutive and induced apometallothionein in cells

Methods to detect apoMT in cells have been described [7, 8]. Constitutive MT has been found in many tumor cells in the metal-free or unsaturated state [1]. It can also be produced in several ways. Ehrlich cells, which normally contain readily measurable concentrations of Zn-MT, lose Zn preferentially from MT either in Zn-deficient mice or Zn-depleted cell culture medium [9, 10]. In both cases apoMT of comparable concentration to Zn-MT in control cells can be measured. ApoMT has also been detected in control U-373, human glioblastoma cells. These cells respond to 10 µg/mL dexamethasone by synthesizing MT. Four hours incubation of cells with this steroid in Zn-deficient medium results in a large accumulation of apoMT. Notably, during the induction process, the concentration of Zn in the high molecular weight band remains unchanged despite the synthesis of apoMT, a new high affinity site for Zn.

Large concentrations of apoMT also stem from the incubation of kidney LL-CPK$_1$ cells with bismuth citrate. Although Bi^{3+} readily complexes with apoMT during titration experiments, it induces but does not bind to MT in cells (Fig. 1). As above, the presence of apoMT does not grossly disturb the intracellular Zn distribution. However, in contrast to induction by dexam-

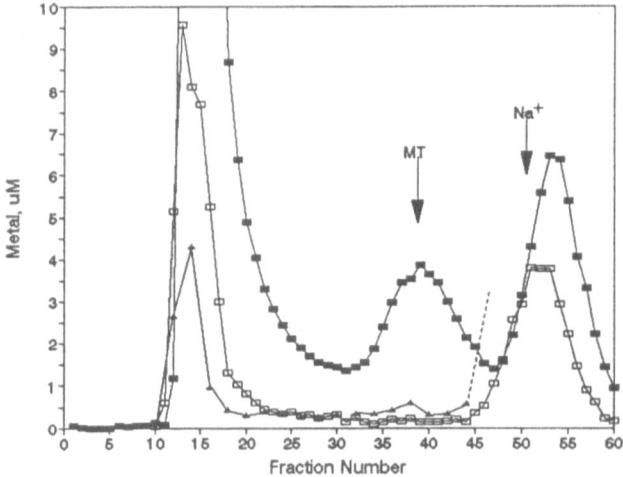

Figure 1. Induction of apoMT in LL-CPK1 cells by Bi-citrate. Sephadex G-75 profile of Zn (□) and Bi (▲) in cell supernatant of cells for 24 h with Bi-citrate and of the same supernatant incubated with Cd (■) to bind to apoMT and displace Zn from Zn-MT.

ethasone, little Zn becomes associated with the Bi-induced protein, although abundant Zn is present in the growth medium. The results of this and the previous method of apoMT synthesis show that the growth of a thermodynamic sink for Zn within the cell neither markedly disturbs Zn distribution nor necessarily draws extracellular Zn into the protein.

Dynamics of apometallothionein in cells

Exploratory experiments have probed the function of apoMT in cells. TE-671, human rhabdomyoblastoma cells contain a pool of apoMT. Incubation of these cells with thymidine to inhibit cell cycle progression leads to the saturation of apoMT with Zn^{2+} without alteration of the total protein concentration (Fig. 2). RT-PCR of the MT mRNA in cells before and after treatment with TdR reveals that mRNA concentrations are similar. Thus, it appears that only the Zn concentration changes in a fixed, steady state concentration of MT and that the underlying synthesis dynamics of MT are unaffected by TdR. The elevation in Zn content of MT is fully reversible when TdR was removed and the cells returned to their normal rate of cell division.

The following hypothesis is consistent with this experiment and the finding that *de novo* synthesis of apoMT does not alter the larger features of Zn distribution in cells (Fig. 4): MT acts as an intermediate between the extracellular source of Zn and its final sites of binding in Zn-containing proteins [8, 11]. In the model, when there is demand for Zn to form Zn-proteins as during rapid cell division, the MT pool may exist largely metal-free in the steady state. In static cell populations with a lowered requirement for Zn, the pool is transformed into Zn-MT.

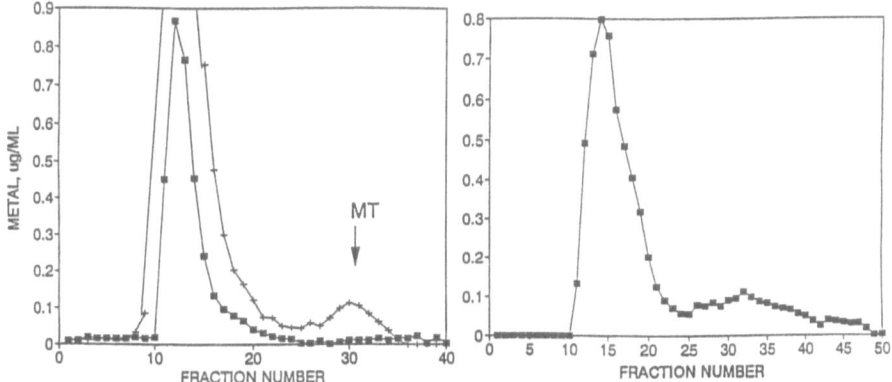

Figure 2. Influence of thymidine on Zn distribution in TE671 cells. (a) 1×10^8 control cells: Zn (■) distribution and apoMT revealed by incubation with Cd (+) as in Figure 1 and (b) Zn-distribution in 1×10^8 cells after 24 h incubation with 1.5 mM thymidine to inhibit cell proliferation 70%.

Reactivity of apometallothionein: Model ligand substitution reactions with metallo-transcription factors

The observation that the presence of apoMT does not alter the gross features of cellular zinc distribution needs to be examined in light of the large thermodynamic affinity of apoMT for metal ions and its reactivity with Zn-TFIIIA and Sp1 according to reaction 1 [4, 6].

$$\text{Zn-P} + \text{apoMT} \longrightarrow \text{Zn-MT} + \text{P} \qquad (1)$$

These latter results suggest that the entire array of zinc finger proteins which bind zinc with two cysteinyl thiolate and two histidinyl imidazole groups is vulnerable to reaction with apoMT. Yet, there is no evidence that cells containing apoMT have lost widespread control of gene expression by Zn-finger proteins. A hypothesis that reconciles these two observations is that Zn binding sites in such proteins are kinetically inert to ligand substitution reaction with apoMT when bound to their DNA recognition sites.

Direct Zn exchange between Zn_n-TFIIIA (n = 3 ± 1 or 7 ± 2) and apoMT (either MT 1 or MT 2 from rabbit liver) has been demonstrated using Sephadex G-75 chromatography to separate zinc-containing reactants and products or by employing centricon filtration in which Zn-MT but not Zn_n-TFIIIA can pass through the membrane. The reaction monitored by centricon filtration is simulated as a second-order process, first order in Zn and in apoMT binding sites, with a rate constant of $40 \pm 10 \text{ s}^{-1}\text{M}^{-1}$ (Fig. 3)

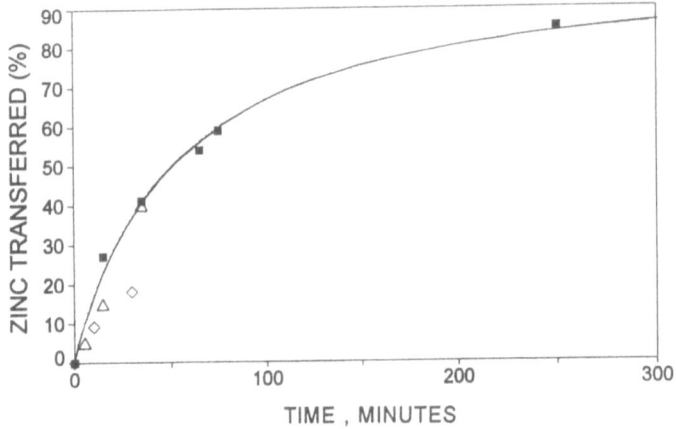

Figure 3. Reaction of apoMT with Zn-TFIIA in the absence and presence of ICR. Centricon filtration of 1.4 µM apoMT and 4.2 µM Zn as Zn_3-TFIIIA in 20 mM Hepes, pH 7.4 and 25° (□). Centricon filtration of 1.6 µM apoMT and 7.2 µM Zn as Zn_7-TFIIIA (◇) in 20 mM Hepes, pH 7.4 and 25° buffer plus 25 mM NaCl, 1% glycerol, 5 µM Zn, and 0.5 mM dithiothreitol. Gel shift assay of Zn_7-TFIIIA•ICR (7.2 µM Zn as Zn-TFIIIA preincubated with ^{32}P-labelled ICR for 30 min) and 1.6 µM apoMT (△) in 20 mM Tris, pH 7.5, 5 mM MgCl$_2$, 70 mM KCl, 1 mM DTT, 5 mM ZnSO$_4$, 10% glycerol, and 50 µg/ml poly(dI.dC).

The same reaction has been examined in the presence of the DNA binding site for Zn-TFIIIA, the internal control region of the 5S rRNA gene (Fig. 3b). The DNA binding site provides little or no protection from reaction with apoMT. By contrast, it completely eliminated reaction with EDTA over a wide range of concentrations. Evidently, apoMT is effective in competing for Zn in TFIIIA whether or not it is associated with the ICR and the hypothesis stated above is incorrect. If this is so, then the question must be considered, how can holo zinc-finger proteins exist in cells which contain apoMT? Is it possible that apoMT is not found in the nucleus?

Apometallothionein as a kinetically reactive ligand for protein bound metal ions

A survey has begun of the capacity of MT to compete for metal ions bound to other proteins. Mammalian apoMT removes Cd from Cd-substituted carbonic anhydrase. The peptide sequence 49–61 from the C-terminal end of MT reacts rapidly with Cd-CA, consistent with the hypothesis that it is this sterically less-hindered part of MT which reacts with these proteins. That the peptide contains the only 4 cysteines which coordinate to a single metal ion in the α-domain metal-thiolate cluster adds attraction to this proposed mechanism of reaction.

In an experiment similar to that shown in Figure 3, yeast apoMT was effective in competing for Cu(I) bound to ACE1, the copper responsive transcription factor for MT synthesis, in the presence its metal response element DNA binding site (Fig. 4). This result suggests that down regulation of MT induction in yeast might be controlled by the appearance of apoMT upon exhaustion of Cu in the medium which provides the original stimulus for MT synthesis. This and other results in this section demonstrate that apoMT is a remarkably effective agent in ligand substitution reactions (Rxn 1).

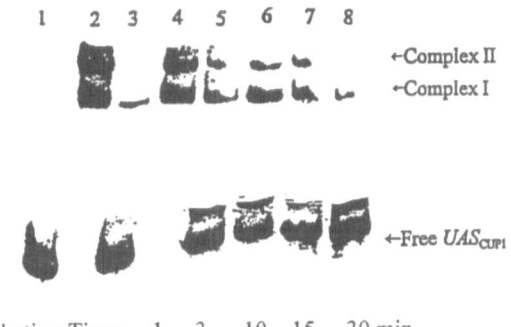

Figure 4. Reaction of apoMT with Cu-ACE1 in the presence of UAS_{CUP1}. Agarose gel electrophoresis gel shift analysis of reaction of 0.12 μM Cu-ACE1 and UAS_{CUP1} (30 min preincubation) with 0.75 μM apoMT at pH 7.4 and 25°.

Comparative effects of Cd^{2+} and Cd-metallothionein on kidney cortical cell Na^+-glucose and Na^+-phosphate cotransport

Introduction

Cadmium accumulation in the kidney leads to pervasive inhibition of sodium-dependent nutrient resorption by the proximal tubule [12]. The mechanism by which Cd causes these effects is unknown. Indeed, the species of Cd involved is in dispute. Although exposure of animals to Cd^{2+} initiates the process, it has been argued that Cd-MT released from injured hepatocytes, migrates into and is rapidly degraded by proximal tubule cells [13]. The bolus of liberated Cd^{2+} directly damages tubular function. The studies described below examine both the mechanism of Cd-induced renal dysfunction and the species of Cd that causes it.

Interaction of Cd^{2+} and Cd-metallothionein with Na^+-dependent Glucose and Phosphate Cotransport

Incubation of mouse kidney cortical cells with Cd^{2+} inhibits Na^+-glucose (α-methyl-glucoside) and Na^+-phosphate cotransport with the same concentration dependence [14]. Cells remain viable and do not display perturbed ATP or Na^+ and K^+ concentrations, indicating that inhibition of energy dependent nutrient transport is not due to derangement in oxidative phosphorylation or the Na^+,K^+ ATPase. Indeed, inhibition is correlated with apparent loss of functional transporters [14, 15].

It has now been shown that exposure of cells to Cd^{2+} produces concentration-dependent reduction in mRNA for each of the transporters (Fig. 5). The time course of changes in mRNA content of the Na^+-phosphate cotransporter, Npt1, shows that after an initial induction period of about 3 h, mRNA declines over the 24 h period of incubation with Cd^{2+}. Even 48 h after the removal of Cd^{2+}, the Npt1 mRNA does not recover.

Incubation of cells with Cd-MT prepared from rabbit MT 1 or MT 2 reveals that very little Cd from Cd-MT enters cells in culture in comparison with its facile uptake by proximal tubule in animals [15]. Nevertheless, at comparable concentrations of Cd (ca. 8 µM), Cd^{2+} inhibits Na^+-glucose cotransport, whereas, Cd-MT stimulates its activity 50% and, correspondingly, elevates its mRNA content (Fig. 6) [15].

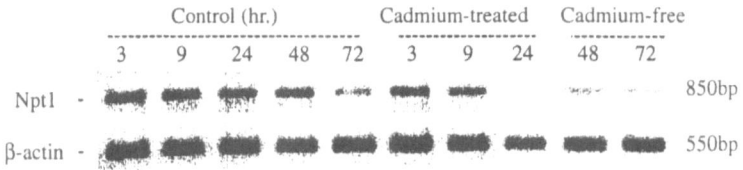

Figure 5. Na^+-dependent glucose and phosphate cotransporter mRNA concentrations. 24 h incubation of cells with 7.5 µM Cd^{2+} followed by RT-PCR.

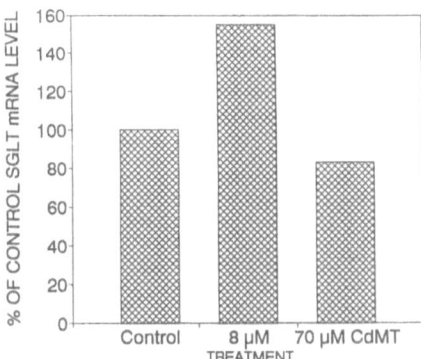

Figure 6. Na^+-glucose cotransporter mRNA concentrations. 24 h incubations of cells with 8 µM Cd as Cd_7-MT followed by RT-PCR.

These results indicate first that Na^+-dependent glucose and phosphate cotransport appear to be coordinately inhibited by Cd^{2+}. While some of the effect may be do to direct interaction between the metal ion and the transporters, reduction in net protein synthesis reflected in the lowered mRNA contents of the two transporters also appears to occur.

Cd-MT behaves differently than Cd^{2+}. Not only does it upregulate transporter mRNA levels at low concentration of Cd-MT, it also has less inhibitory effect on Na^+-glucose cotransport than Cd^{2+} when equal amounts of Cd from each source are internalized [15]. Thus, the internal pathway of Cd^{2+} metabolism seems to be different in the two cases and leads to different toxicological outcomes.

Acknowledgements
This research was supported by NIH grants ES-04026 and ES-04184.

References

1. Pattanaik A, Shaw III CF, Petering DH, Garvey J, Kraker AJ (1994) Basal metallothionein in tumors: widespread presence of apoprotein. *J Inorg Biochem* 54: 91–105.
2. Wang XP, Chan HM, Goyer RA, Cherian MG (1993) Nephrotoxicity of repeated injections of cadmium-metallothionein in rats. *Toxicol Appl Pharmacol* 119: 11–16.
3. Feldman SL, Failla ML and Cousins RJ (1978) Degradation of rat liver metallothionein *in vitro*. *Biochim Biophys Acta* 544: 638–646.
4. Zeng J, Vallee BL, Kägi JHR (1992) Zinc transfer from transcription factor IIIA fingers to thionein clusters. *Proc Natl Acad Sci USA* 88: 9984–9988.
5. O'Halloran TV (1993) Transition metals in control of gene expression. *Science* 261: 715–725.
6. Zeng J, Heuchel R, Schaffner W, Kägi JHR (1991) Thionein (apometallothionein) can modulate DNA binding and transcription activation by zinc finger containing factor Sp1. *FEBS Lett* 279: 310–312.
7. Krezoski SK, Shaw III CF and Petering DH (1991) Role of metallothionein in essential, toxic, and therapeutic metal metabolism in Ehrlich cells. *In:* JF Riordan, BL Vallee (eds): *Metallobiochemistry Part b, Metallothionein and Related Molecules, Meth Enzymol* 205. Academic Press, San Diego, 302–311.
8. Petering DH, Quesada A, Dughish M, Krull S, Gan T, Lemkuil D, Pattanaik A, Byrnes RW, Savas M, Whelan

H and and Shaw III CF (1993) Metallothionein in tumor and host: intersections of zinc metabolism, the stress response, and tumor therapy. *In*: Suzuki,KT, Imura N, Kimura M (eds): Metallothionein III, Biological Roles and Medical Implications. Birkhäuser Verlag, Basel, 329–346.

9. Kraker AJ, Krakower G, Shaw III CF, Petering DH and Garvey JS (1988) Zinc metabolism in Ehrlich cells: properties of a metallothionein-like zinc-binding protein. *Cancer Res* 48: 3381–3388.
10. Krezoski SK, Villalobos J, Shaw III CF and Petering DH (1988) Kinetic lability of zinc bound to metallothionein in Ehrlich cells. *Biochem J* 255: 483–491.
11. Otvos JD, Petering DH and Shaw III CF (1989) Structure-reactivity relationships of metallothionein, a unique metal-binding protein. *Comment Inorg Chem* 9: 1–35.
12. Friberg L, Elinger CG and Kjellström T (1992) Cadmium. World Health Organization, Geneva, 133.
13. Dudley R, Gammal L and Klaassen C (1985) Cd-induced renal injury in chronically exposed rats: likely role of hepatic Cd-MT in nephrotoxicity. *Toxicol Appl Pharmacol* 77: 414–426.
14. Blumenthal SS, Lewand D, Buday MA, Kleinman JG, Krezoski SK and Petering DH (1990) Cadmium inhibits glucose uptake in primary cultures of mouse cortical tubule cells. *Amer J Physiol* 258: F1625–1633.
15. Blumenthal S, Lewand D, Krezoski SK and Petering DH (1996) Comparative effects of Cd^{2+} and Cd-metallothionein on cultured kidney tubule cells. *Toxicol Appl Pharmacol* 136: 220–228.

Metallothionein IV
C. Klaassen (ed.)
© 1999 Birkhäuser Verlag Basel/Switzerland

Zinc-induced protection against cadmium-metallothionein nephrotoxicity depends on glutathione status

Zahir A. Shaikh, Weifeng Tang and Selma Sadovic

Department of Biomedical Sciences, College of Pharmacy, University of Rhode Island, Kingston, RI 02881, USA

Summary. The mechanism by which zinc (Zn) protects against the acute nephrotoxicity of cadmium-metallothionein (CdMT) is unknown. Since Zn offers protection even in metallothionein-null (MT-null) mice, we investigated the role of glutathione (GSH) in Zn-induced protection. Pretreatment of mice with a high dose of Zn, which elevated renal cortex, GSH levels protected against CdMT nephrotoxicity. Elevation of renal GSH was not essential for Zn-induced protection since a lower dose of Zn, which did not affect renal cortex GSH level, was also protective. Normal GSH levels, however, were required for the Zn-induced protection as inhibition of GSH synthesis by buthionine sulfoximine (BSO) in Zn-pretreated mice completely abolished the beneficial effect of Zn. Pretreatment with Zn reduced the accumulation of Cd in the renal cortex and this appears to be the mechanism by which Zn may exert its protective action. These results suggest that although Zn protects against CdMT nephrotoxicity apparently by decreasing Cd accumulation in the renal cortex, maintenance of normal GSH status is essential for Zn to be effective.

Introduction

The nephrotoxicity upon chronic Cd exposure is linked to the circulating CdMT, which is released from the liver [1–3]. Circulating CdMT is freely filtered through the renal glomeruli, efficiently absorbed by the proximal tubular epithelial cells, and rapidly degraded by lysosomal enzymes to liberate Cd^{2+} [4, 5]. It is generally agreed that within the renal epithelial cells Cd^{2+} binds to endogenous MT, however, if the MT levels are insufficient Cd interacts with sensitive sites and causes cytotoxicity [6]. Since administration of CdMT to laboratory animals produces acute renal effects similar to those observed after chronic Cd exposure, CdMT is often used to study the mechanism of Cd-induced nephrotoxicity [7, 8].

Zn is reported to protect against Cd-induced nephrotoxicity by inducing MT synthesis [9]. A recent study revealed that Zn also protects against CdMT nephrotoxicity and that this protection is offered not only in normal mice but also in MT-null mice [10]. Thus Zn protection is not mediated via MT induction.

Besides MT, GSH also modulates the lethality [11] as well as the nephrotoxicity of CdMT [12]. Administration of Zn is reported to increase not only MT but also GSH levels [13]. Therefore, we investigated the possibility that Zn protects against CdMT nephrotoxicity by raising renal GSH levels. MT-null mice were used to avoid the complication of Zn-associated MT induction.

Effect of Zn pretreatment on CdMT nephrotoxicity

The animals were injected ip with 0.3 mg Cd/kg as CdMT, with or without Zn pretreatment. The Zn pretreatment consisted of 200 µmol Zn/kg as $ZnCl_2$, given sc 48 and 24 h prior to the CdMT injection. Urinary lactate dehydrogenase (LDH) and protein used as indices of renal damage and dysfunction, respectively, established that the dose of CdMT chosen was clearly nephrotoxic (Tab. 1). Pretreatment with Zn caused a marked reduction in LDH excretion and brought the protein levels to within the normal range, confirming the protective effect of Zn.

Table 1. Modulation of CdMT nephrotoxicity in MT-null mice by pretreatment with either Zn or Zn and BSO

Pretreatment	Treatment	Urinary LDH (IU/24 h)	Urinary protein (µg/24 h)
	Saline	0.02 ± 0.01	1.71 ± 0.60
	CdMT	$0.92 \pm 0.18^*$	$5.96 \pm 1.02^*$
Zn	CdMT	$0.14 \pm 0.03^{*\#}$	$1.59 \pm 0.16^\#$
Zn & BSO	CdMT	$2.98 \pm 0.03^{*\#}$	$6.13 \pm 1.70^*$

MT-null mice were randomly divided into four groups. The first group was injected ip with saline. The rest of the groups were injected ip with 0.3 mg Cd/kg as CdMT with or without pretreatments. The Zn-pretreated group received two sc injections of 200 µmol Zn/kg/day and the Zn- and BSO-pretreated group received a sc injection of 4 mmol BSO/kg on the second day, 4 h prior to the CdMT injection. The mice were placed in plastic metabolism cages after the saline or CdMT injection and their urines were collected over ice for 24 h.
*Significantly different from the saline-injected group ($p < 0.05$).
$^\#$Significantly different from the group given only CdMT ($p < 0.05$).

Role of GSH in Zn protection

Renal cortex GSH levels of Cd-MT injected mice were not significantly different from those of the saline-injected controls 24 h after the injection (1.8 ± 0.3). Treatment with 200 µmol Zn/kg at 48 and 24 h prior to the Cd-MT injection resulted in 63% increase in renal cortex GSH levels. However, a lower dose of Zn (100 µmol/kg), which also offered the same level of protection as the higher dose, had no significant effect on GSH levels. This observation suggests that Zn protection against the nephrotoxicity of Cd-MT is not mediated via the elevation of renal cortex GSH levels. Nevertheless, maintenance of normal GSH levels was essential for the Zn protection as inhibition of GSH synthesis by buthionine sulfoximide (BSO) not only abolished the protective effect of Zn but also enhanced the nephrotoxicity of Cd-MT. Thus while GSH appears to play a vital role in protection against Cd-MT nephrotoxicity, Zn protection only requires normal GSH levels.

Modulation of renal cortex Cd levels by Zn

Since both Zn and Cd can bind to MT it is conceivable that Zn may protect against CdMT nephrotoxicity by replacing some of the Cd in the circulating CdMT and thereby reducing renal Cd uptake. To explore this possibility renal cortex Cd concentrations were determined in control and in Zn-pretreated animals 2 h after the CdMT injection. Pretreatment with 200 μmol Zn/kg X 2 caused a 39% reduction in the renal cortex Cd concentration. Thus, it seems that reduction of Cd uptake by the kidneys and not the elevation of renal GSH, is the mechanism of Zn protection against CdMT nephrotoxicity.

Conclusion

Whereas Zn appears to protect against CdMT nephrotoxicity by reducing the amount of Cd taken up by the renal cortex, Zn protection is effective only under normal or elevated GSH status.

Acknowledgement
The work described in this report was supported by a research grant no. ES03187 from NIEHS.

References

1. Shaikh ZA, Hirayama K (1979) Metallothionein in the extracellular fluids as an index of cadmium toxicity. *Environ Health Perspect* 28: 267–271.
2. Dudley RE, Gammal LM, Klaassen CD (1985) Cadmium induced hepatic and renal injury in chronically exposed rats: likely role of hepatic cadmium metallothionein in nephrotoxicity. *Toxicol Appl Pharmacol* 77: 414–426.
3. Chan HM, Zhu LF, Zhong R, Grant D, Goyer RA, Cherian MG (1993) Nephrotoxicity in rats following liver transplantation from cadmium exposed rats. *Toxicol Appl Pharmacol* 123: 89–96.
4. Cherian MG, Shaikh ZA (1975) Metabolism of intravenously injected cadmium-binding proteins. *Biochem Biophys Res Commun* 65: 863–869.
5. Squibb KS, Ridlington JW, Carmichael NG, Fowler BA (1979) Early cellular effects of circulating cadmium-thionein on kidney proximal tubules. *Environ Health Perspect* 28: 287–296.
6. Goyer RA, Miller CR, Zhu SY, Victery W (1989) Non-metallothionein-bound cadmium in the pathogenesis of cadmium nephrotoxicity in the rat. *Toxicol Appl Pharmacol* 101: 232–244.
7. Cherian MG, Goyer RA, Delaquerriere-Richardson L (1976) Cadmium metallothionein induced nephropathy. *Toxicol Appl Pharmacol* 38: 399–408.
8. Vestergaard P, Shaikh ZA (1994) The nephrotoxicity of intravenously administered cadmium-metallothionein: effect of dose, mode of administration, and preexisting renal cadmium burden. *Toxicol Appl Pharmacol* 126: 240–247.
9. Leber AP, Miya TS (1976) A mechanism for cadmium and zinc induced tolerance to cadmium toxicity: involvement of metallothionein. *Toxicol Appl Pharmacol* 37: 403–414.
10. Liu J, Liu Y, Michalska AE, Choo KH, Klaassen CD (1996) Metallothionein plays less of a protective role in cadmium-metallothionein-induced nephrotoxicity than in cadmium chloride-induced hepatoxicity. *J Pharmacol Exp Ther* 276: 1216–1223.
11. Singhal RK, Anderson ME, Meister A (1987) Glutathione, a first line of defense against cadmium toxicity. *FASEB J* 1: 220–223.
12. Suzuki CA, Cherian MG (1989) Renal glutathione depletion and nephrotoxicity of cadmium-metallothionein in rats. *Toxicol Appl Pharmacol* 98: 544–552.
13. Iszard MB, Liu J, Klaassen CD (1995) Effect of several metallothionein inducers on oxidative stress defense mechanisms in rats. *Toxicology* 104: 25–33.

Metallothionein IV
C. Klaassen (ed.)
© 1999 Birkhäuser Verlag Basel/Switzerland

Mechanism of improved cadmium-induced renal dysfunction after cessation of cadmium exposure

Kazuo Nomiyama, Hiroko Nomiyama and Naoki Kameda

Department of Environmental Health, Jichi Medical School, Tochigi-Ken 329-04, Japan

Introduction

Cadmium-induced renal dysfunction has been believed not to recover because cadmium in the renal cortex, the critical organ, hardly decreased even after the cessation of cadmium exposure [1]. Because this conclusion disagrees with our findings [2], the present experiments were carried out to help clarify this issue.

Materials and methods

Male rabbits weighing 2.5 kg were given $CdCl_2$ (0.3 mgCd/kg) sc daily until detection of renal dysfunction (two signs out of proteinuria, glucosuria and aminoaciduria were detected in consecutive 2 examinations). After cessation of cadmium exposure, plasma cadmium-thionein (CdMT), and hepatic and renal function parameters were monitored in 12 rabbits for 4 weeks.

Results

1) After cessation of cadmium exposure, 5 rabbits out of 12, died within 4 weeks. The plasma CdMT of dead rabbits exceeded 100 µgCd/L at the time of cadmium cessation. 2) Hepatic dysfunction was alleviated both in dead and surviving rabbits after cessation of cadmium exposure, with some exception. 3) Renal dysfunction was alleviated in surviving rabbits while it was aggravated in dead rabbits. 4) Hepatic function parameters correlated closely with plasma CdMT. 5) As plasma CdMT decreased, some measures for dysfunction of glomeruli and renal tubules also improved. 6) Plasma CdMT correlated closely with plasma total cadmium as well.

Discussion

Cadmium-induced renal dysfunction was reversible after cessation of cadmium exposure, only when plasma CdMT was below 100 µgCd/L. This finding accorded well with those in rabbits given oral cadmium [2]. It was also the same for rabbits given sc cadmium at 10 times higher dose level, i.e. 3 mgCd/kg: cadmium-induced renal dysfunction was reversed as plasma CdMT decreased after cessation of cadmium exposure [3]. These findings support our hypotheses on the mechanism of cadmium-induced renal dysfunction, even though they only reflect reversal in the recovery stage: upon cadmium-induced hepatic dysfunction, CdMT is released from the liver into blood stream. Then CdMT passes easily through glomeruli because of its small molecular weight, to result in injury to the brush border membrane of proximal tubules [4].

Conclusion

1) Cadmium-induced renal dysfunction in rabbits recovered after cessation of cadmium exposure only when plasma CdMT fell below 100 µgCd/L. 2) Plasma CdMT is an ideal exposure index for cadmium-induced renal dysfunction based on the mechanism. 3) Because plasma cadmium correlated well with plasma CdMT, plasma cadmium can also be used as an alternative exposure index for cadmium-induced renal dysfunction.

References

1. World Health Organization (1992) Cadmium, World Health Organization, Geneva, 143–146.
2. Nomiyama K, Nomiyama H (1984) Reversibility of cadmium-induced health effects in rabbits. *Environ Health Perspect* 54: 201–211.
3. Nomiyama K, Nomiyama H (1996) Plasma cadmium-thionein, the new and ideal biomarker for cadmium-induced renal dysfunction, based on the mechanism, Abstract I: 93, 25th Int Congress Occup Health, Stockholm.
4. Nomiyama K, Nomiyama H (1998) Cadmium-induced renal dysfunction: New mechanism, treatment and prevention. *J Trace Elem Exptl Med* 11: 275–288..

The role of metallothionein in cadmium-induced bone resorption

Maryka H. Bhattacharyya, Carmen A. Blum and Allison K. Wilson

Center for Mechanistic Biology and Biotechnology, Argonne National Laboratory, Argonne, IL 60439-4833, USA

Introduction

Cadmium (Cd) causes bone loss independent of renal dysfunction at occupational exposure levels [1–3]. *In vitro*, Cd accelerates the formation of osteoclasts, the bone resorbing cells, to cause a transient increase in their numbers [4–5] and increases their activation and/or activity [5]. The cellular mechanism for the Cd-induced increase in bone resorption is still unclear.

Metallothionein (MT) is known to bind and sequester metals, including Cd. Upon oral exposure, approximately 1% of the Cd dose is transported across the GI tract in the mouse [6] and absorbed into the blood, where it is complexed to red blood cells and serum proteins and transported to the liver, where it induces the synthesis of MT [7]. Cd bound to MT is translocated to the kidney, where it accumulates over long periods of time in the renal cortex. Renal excretion of oral Cd in mice is minimal, < 2% of the absorbed dose [8].

MT is thought to be a major mechanism for decreasing the toxicological effects of Cd. MT induction by Cd has been observed in many cells, including bone cells [9–11]. A low, nontoxic exposure to Cd or other metals can induce MT expression, conferring a protective effect against the toxicity of subsequent, higher doses of metal exposure. This protective effect is thought to be due to the binding of Cd to MT, making Cd unavailable for direct tissue damage. MT does not contain enzymatic properties, but it may work with glutathione to protect against reactive oxygen radicals [7].

MT knockout (MT-KO) and MT-normal (MT-N) mice have been used to elucidate the role of MT in Cd-induced toxicity [12–5]. This study used MT-KO mice to determine whether Cd-induced bone resorption is mediated by MT and if MT can confer protection to the bone against a second exposure.

In vivo model for the early bone response to Cd

An animal model has been used to investigate the early skeletal response to a single oral Cd exposure [3]. Several weeks prior to Cd exposure, mice are switched to a low-calcium diet (0.02% Ca and 0.40% P, Purina Mills) to increase the GI absorption of calcium such that endogenous fecal calcium excretion is equal to the total calcium excreted in the feces. For similarly-treated mice whose skeletons were labeled with ^{45}Ca, the endogenous calcium measured in feces was demonstrated to be released from bone [2]. For this mouse model, Cd is administered by a single gavage administration (200 µg Cd as CdCl₂) to nonfasted mice housed in metabolism cages. The bone response is followed by quantitating the increase in stable fecal calcium by atomic absorption spectroscopy and does not require a radioactive label.

In this model, acute Cd exposure induces a transient four-fold increase in calcium released from bone [3]. The bone response is not observed in the first 8 h, when the resorption machinery of osteoclasts is being upregulated – a time course similar to that required for the production of resorption pits by osteoclasts *in vitro* [5, 16]. Cd-induced bone calcium release drops to background levels by 104 h. Bone resorption initially occurs at blood Cd levels of 5–8 µg/L [1, 3], but, in this single-exposure model, the bone response decreases while blood Cd levels continue to rise to 10 µg/L at 104 h. These results suggest that a threshold level of Cd is required for a bone response but that chronic levels of Cd in blood do not necessarily indicate the occurrence of continuous active bone resorption. Because Cd is not quickly cleared from the blood after a single oral exposure, as is the case with intravenous exposure, the data suggest that the intestine provides a continuous Cd source to keep blood Cd levels elevated over levels of unexposed mice. Duodenal MT levels are increased to 140 µg/g at 96 h in response to the above Cd gavage, suggesting not only that Cd bound to MT provides a first line of defense against oral Cd exposure, but also that Cd bound to intestinal MT may be an important source of long-term Cd exposure.

Role of MT in the bone response to Cd

Female 129/SV mice (MT-N, 8 mice) and 129/SV mice that do not express the MT1 and MT2 genes (MT-KO, 8 mice) (Jackson Labs, Bar Harbor, ME) were housed individually in metabolism cages and switched to a low-calcium diet two weeks before Cd administration. At the start of experiment, mice were 9–12 weeks old, with a mean body weight of 19.5 ± 0.3 g (n = 16) and no significant weight difference between the strains. Half of the mice of each strain received two gavage doses of Cd (200 µg Cd as CdCl₂ per dose) separated by one week; the other half were controls and received vehicle only. This protocol utilizes the *in vivo* model to evaluate the effect of MT status on the bone response to oral Cd, and to determine whether a first Cd exposure would confer protection against a second exposure. Fecal calcium excretion decreased dramatically during the first rounds of excreta collection, as the mice increased intestinal calcium absorption and decreased endogenous calcium excretion in response to the low-calcium diet (Fig. 1, R1-R4); thereafter, fecal calcium excretion decreased very slowly and steadily through collection Round 20 for the mice not exposed to cadmium (data not shown). After Cd gavage #1, fecal calcium excretion (measure of calcium release from bone) did not rise significantly above pre-exposure levels during day 1 after the administration for either the MT-N or MT-KO mice (Fig. 1; R8 (0–8 h) and R9 (8–24 h)). However, during days 2–3 (R10-R11), excretion rates increased significantly, with a peak calcium excretion rate (µg/h) that was significantly greater in MT-KO (75 ± 15 µg/h) than in MT-N mice (45 ± 11 µg/h) (mean ± SE, n = 4). After the first gavage, the total amount of calcium excreted in response to Cd was two- to threefold greater in MT-KO mice (2580 ± 270 µg vs. 1070 ± 250 µg, R8-R13). The bone response was transient for both strains, returning to baseline levels by 96 h (R12). For both MT-

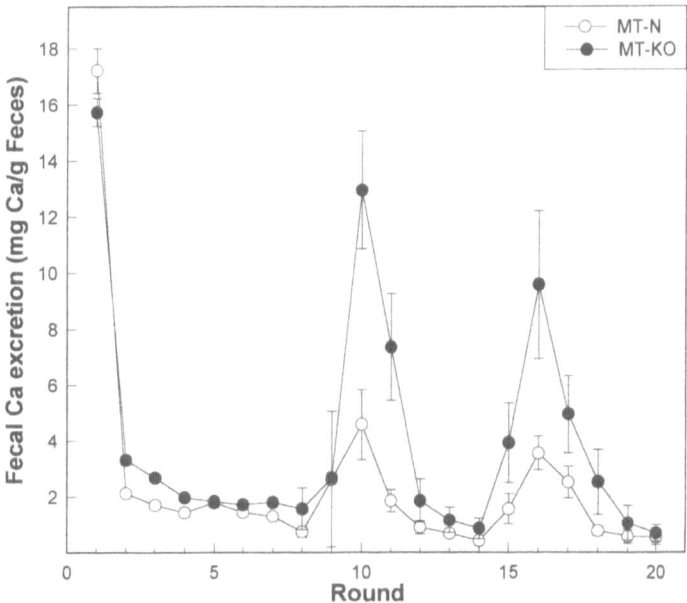

Figure 1. Effect of MT status on early bone response to Cd. Mice were switched to low-calcium diet two weeks before the first Cd administration, and calcium release from bone was monitored by determining fecal calcium excretion using flame atomic absorption spectrometry. Each mouse received two Cd gavages (200 µg Cd as CdCl$_2$ per dose), one just prior to collection Round 8 (R8) and one just prior to R14. The duration of each fecal collection round was 24 h except for the following: 72 h, R1-R4; 8 h, R8 and R14; 16 h, R9 and R15; 48 h, R12 and R20. Values are mean ± SE for 8 mice/group before Cd gavage #1 and 4 mice/group thereafter. Each mean with no apparent vertical bar has an SE value smaller than the size of the dot.

N and MT-KO mice, an identical Cd gavage one week later produced a bone response similar in magnitude and duration (Fig. 1). As with the first dose, the total amount of calcium excreted in response to Cd was about twofold greater in the MT-KO mice (1980 ± 610 µg vs. 950 ± 200 µg, R14-R19).

The smaller bone response to Cd in MT-N vs. MT-KO mice suggests that MT protects against Cd-induced bone loss after oral exposure. However, our results indicate that induction of MT in bone cells by an initial Cd exposure does not protect against a second exposure to Cd, as is the case for liver and kidney. MT may be exerting its protective effects at the level of the intestine by sequestering Cd to keep it from entering the blood. Because the bone response in MT-KO mice was only two- to threefold greater than in MT-N mice, though important, MT1 and MT2 appear not to play a highly critical role in preventing Cd-induced bone toxicity, indicating that there may be a backup system of protection in the MT-KO mice.

Acknowledgments
We would like to acknowledge the fine help of E Brako and EA Cerny in conducting the MT assays and R Santarelli for his excellent assistance with animal care. This research was made possible by grants ES04816 and ES07398 from the National Institutes of Health (NIH).

References

1. Sacco-Gibson N, Chaudhry S, Brock A, Sickles AB, Patel B, Hegstad R, Johnston S, Peterson D and Bhattacharyya M (1992) Cadmium effects on bone metabolism: Accelerated resorption in ovariectomized, aged beagles. *Toxicol Appl Pharmacol* 113: 274–283.
2. Wang C and Bhattacharyya MH (1993) Effect of cadmium on bone calcium and ^{45}Ca in pregnant mice on a calcium-deficient diet: Evidence of direct effect of cadmium on bone. *Toxicol Appl Pharmacol* 120: 228–239.
3. Wilson AK and Bhattacharyya MH (1997) Effects of cadmium on bone: An *in vivo* model for the early response. *Toxicol Appl Pharmacol* 145: 68–73.
4. Miyahara T, Takata M, Miyata M, Nagai M, Sugure A, Kozuka H and Kuze S (1991) Cadmium stimulates osteoclast-like multinucleated cell formation in mouse bone marrow cell cultures. *Bull Environ Contam Toxicol* 47: 283–287.
5. Wilson AK, Cerny EA, Smith BD, Wagh A and Bhattacharyya MH (1996) Effects of cadmium on osteoclast formation and activity *in vitro*. *Toxicol Appl Pharmacol* 140: 451–460.
6. Jonah MM and Bhattacharyya MH (1989) Early changes in the tissue distribution of cadmium after oral but not intravenous cadmium exposure. *Toxicol* 58: 325–338.
7. CDKlaassen (ed): (1996) *Casarett and Doull's Toxicology*. McGraw Hill, New York.
8. Bhattacharyya MH, Whelton BD and Peterson DP (1981) Gastrointestinal absorption of cadmium in mice during gestation and lactation 1. Short-term exposure studies. *Toxicol Appl Pharmacol* 61: 335–342.
9. Kimura M, Otaki N, Yoshiki S, Suzuki M, Horiuchi N and Suda T (1974) The isolation of metallothionein and its protective role in cadmium poisoning. *Arch Biochem Biophys* 165: 340–348.
10. Miyahara T, Yamada H, Ando R, Nemoto S, Kaji T, Mori M, Kozuka H, Itoh N and Sudo H (1986) The effects of cadmium on a clonal osteogenic cell, ML3T3-E1: Inhibition of calcification and induction of metallothionein-like protein by cadmium. *Toxicol Lett* 32: 19–27.
11. Angle CR, Thomas DJ and Swanson SA (1993) Osteotoxicity of cadmium and lead in HOS TE 85 and ROS 17/2.8 cells: Relation to metallothionein induction and mitochondrial binding. *Biometals* 6: 179–184.
12. Lazo JS, Kondo Y, Dellapiazza D, Michalska AE, Choo KH and Pitt BR (1995) Enhanced sensitivity to oxidative stress in cultured embryonic cells from transgenic mice deficient in metallothionein I and II genes. *J Biol Chem* 270: 5506–5510.
13. Liu J, Liu Y, Michalska AE, Choo KH and Klaassen CD (1996) Metallothionein plays less of a protective role in cadmium-metallothionein-induced nephrotoxicity than in cadmium chloride-induced hepatotoxicity. *J Pharmacol Exp Ther* 276: 1216–1223.
14. Liu J, Liu Y, Michalska SE, Choo KH and Klaassen CD (1996b) Distribution and retention of cadmium in metallothionein I and II null mice. *Toxicol Appl Pharmacol* 136: 260–268.
15. Zheng H, Liu J, Choo KH, Michalska AE and Klaassen CD (1996) Metallothionein-I and -II knock-out mice are sensitive to cadmium-induced liver mRNA expression of c-jun and p53. *Toxicol Appl Pharmacol* 136: 229–235.
16. Tamura T, Takahashi N, Akatsu T, Sasaki T, Udagawa N, Tanaka S and Suda T (1993) New resorption assay with mouse osteoclast-like multinucleated cells formed *in vitro*. *J Bone Miner Res* 8: 953–960.

Metallothionein IV
C. Klaassen (ed.)
© 1999 Birkhäuser Verlag Basel/Switzerland

Metallothionein induction in human proximal tubular cell cultures – lack of protection against heavy metal toxicity

Adrian T. Miles[1], Vicente Rodilla[1], Antony G. Breen[1], John Beattie[2], William Jenner[3] and Gabrielle M. Hawksworth[1]

[1]Departments of Medicine and Therapeutics and Biomedical Sciences, Polwarth Building, Foresterhill, Aberdeen AB25 2ZD, UK
[2]Rowett Research Institute, Bucksburn, Aberdeen AB21 9SB, UK
[3]Glaxo Wellcome R&D, Ware, Herts, SG12 0DP, UK

Exposure to a number of drugs and xenobiotics, including aminoglycosides, cephalosporins, cisplatin, halogenated alkenes and heavy metals, results in proximal tubular nephrotoxicity. The toxicity is specific for different regions of the tubule, depending on cell-specific transport and metabolic processes [1]. The nephrotoxicity of mercury and cadmium is well documented, their toxic effects being manifested primarily in the S_3 segments and the S_1/S_2 segments of the proximal tubule respectively, where they are known to accumulate [2, 3]. Heavy metals induce the synthesis of, and bind to, metallothionein and it has been suggested that this may ameliorate the toxicity of heavy metals *in vivo* [4]. We have previously shown that heavy metals induce metallothionein to different extents in primary cultures of human proximal tubular cells [5]. Human *in vitro* systems are potentially useful in understanding the roles of metallothionein isoforms in protection against drug- or xenobiotic-induced toxicity, since there are marked species differences in the expression and induction of the isoforms. The aim of this study was two fold: 1) to determine if bismuth, a relatively non toxic metal used in combination therapy for the treatment of peptic ulcers and as an adjunct in chemotherapy, could induce metallothioneins in human proximal tubular cell cultures and 2) to determine if the induced metallothionein protected against heavy metal or drug-induced proximal tubular toxicity.

Methods

Human proximal tubular cells were isolated from the cortex of nephrectomy specimens with enzymatic digestion followed by mechanical disaggregation of the tissue and further purification of the cell fraction by isopycnic centrifugation in Percoll. Cells were cultured in 50/50 DMEM/Ham's F12 medium supplemented with 10% fetal calf serum. Toxicity studies were carried out in defined medium (containing insulin, human transferrin, sodium selenite and hydrocortisone) as previously described [6]. Cells were seeded on 96 well plates and, three days after seeding, the medium was removed and replaced with the appropriate metal solutions in defined medium. The toxic effects of $Bi(NO_3)_3$, $HgCl_2$, $CdCl_2$ and $ZnCl_2$ were evaluated over a 96 h period using the MTT assay. Selection of the metal salts was determined by their solubility or capacity for comparing toxicity in other systems.

The induction of metallothionein was quantified at the mRNA level by Northern blotting using a 5' digoxigenin labelled 'consensus' oligonucleotide probe for human metallothionein which recognised a highly conserved region (base pairs 79–96) of human MT-Ia, MT-Ie, MT-If and MT-IIa (7). Immunocytochemical detection of metallothionein was carried out using the standard alkaline phosphatase anti-alkaline phosphatase technique with a monoclonal antibody (Clone E9, Dako Corp.).

The potential protective effects of induced metallothioneins on heavy metal or drug toxicity were evaluated in confluent proximal tubular cell cultures. Cells were incubated with bismuth, cadmium or dexamethasone (1 μM and 10 μM) for up to 96 h after which they were exposed to toxic concentrations of either metals or cisplatin and toxicity determined using the MTT assay.

Heavy metal toxicity to proximal tubular cell cultures

Toxicity occurred in a time and dose dependent manner. Mercury was the most toxic metal, followed by cadmium then zinc, resulting in IC_{50} values of 7 μM, 24 μM and 67 μM respectively after 48 h. IC_{50} values for bismuth were only determinable in two cultures at 72 h and 96 h, the other cultures being insufficiently sensitive to determine Bi toxicity (Tab. 1). None of the metals was toxic at the lowest dose of 1 μM. Considerable variation in the sensitivities towards the metals was observed both between and within cultures. For example, one culture displayed considerable resistance towards mercury, whilst being sensitive to zinc, whereas another culture was particularly sensitive to the effects of 1 μM bismuth after 24 h, in contrast to the other cultures. The toxicity of the metals overall did not differ significantly between confluent and proliferating cultures, except for mercury (50 μM), which was more toxic to subconfluent cultures. The reason for this could be related to the mechanisms by which mercury exerts its toxicity. This involves DNA damage resulting from oxidative stress and interference with spindle function due to the binding of mercury to sulphydryl groups of tubulin, both of which would be more obvious in proliferating cells [8]. An alternative explanation is that there was a greater accumulation of mercury in the subconfluent cell cultures since it has been demonstrated that subconfluent cultures of rat renal cortical cells accumulate more mercury than confluent cul-

Table 1. Mean IC_{50} values (μM) with standard deviations for mercury, cadmium, zinc and bismuth after 24, 48, 72 and 96 h treatment

Metal	24h	48h	72h	96h
Hg	13.6 (± 4.7)	7.0 (± 3.0)	4.1 (± 4.4)	5.2 (± 4.3)
Cd	41.1 (± 29.9)	24.1 (± 11.2)	21.9 (± 10.3)	26.9 (± 19.9)
Zn	52.8 (± 26.5)	66.5 (± 1.1)	48.0 (± 33.3)	51.8 (± 43.3)
Bi	-	-	56.0	42.2

n = between 4 and 6. For Bi at 72 h and 96 h, n = 2. (-) not determinable, as mean optical densities did not fall below 50% of the control values.

tures [9]. Differences in the sensitivity of individual human proximal tubular cell cultures to the toxic effects of cadmium have been reported previously [10], although no correlation was found between the sensitivity of individual cultures towards cadmium, the extent of metallothionein accumulation or cytosolic cadmium concentrations [11].

Induction of metallothioneins in human proximal tubular cell cultures

The induction of metallothionein mRNA by 1 µM Bi(NO$_3$)$_3$ and CdCl$_2$ was studied over a 96 h period using a 'consensus' oligonucleotide probe for MT1 and MT2. Both bismuth and cadmium caused a time dependent increase in MT mRNA expression over control values, although cadmium was a more effective inducer than bismuth, producing increases of 10 times the control values compared with 3.5 times for bismuth after 96 h in culture (Fig. 1). Induction increased up to 96 h. Constitutive expression of metallothionein varied between cultures and may have masked induction in some cultures (Fig. 2).

The induction of metallothionein by a non toxic concentration (1 µM) of the metals over a 96 h period was also investigated using immunocytochemical techniques. The time for maximal induction of metallothionein was both metal and culture dependent and varied between 24 and 72 h (Tab. 2). However bismuth consistently induced maximal MT expression after 48 h. Staining was localised mainly in the cytoplasm for both control and metal induced cells, consistent with previous studies [12]. Variations were also observed between cultures in the levels

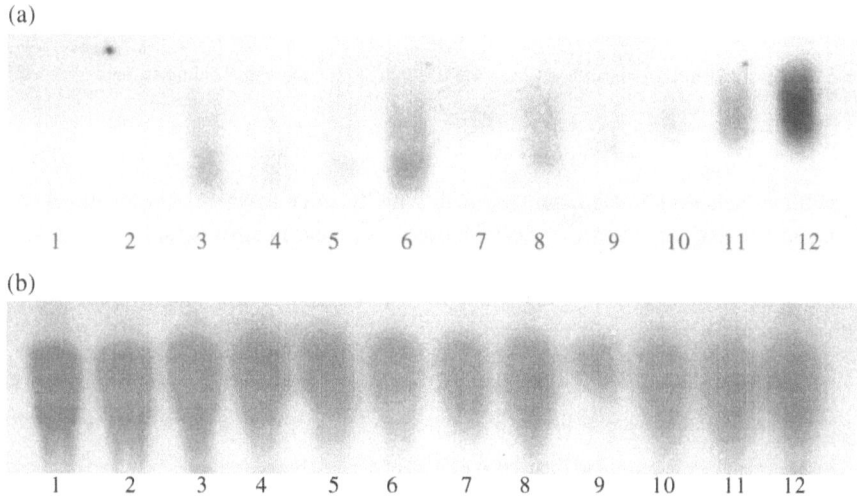

Figure 1. Northern blots for (a) metallothionein and (b)18S (control) after exposure of HPT cells to 1 µM Bi(NO$_3$)$_3$ and 1 µM CdCl$_2$ for a 96 h period. Lanes 1–3: samples taken after 24 h, lanes 4–6: samples taken after 48 h, lanes 7–9: samples taken after 72 h, lanes 10–12: samples taken after 96 h. At each time point the samples are in the following order: control sample, bismuth-treated sample, cadmium-treated sample.

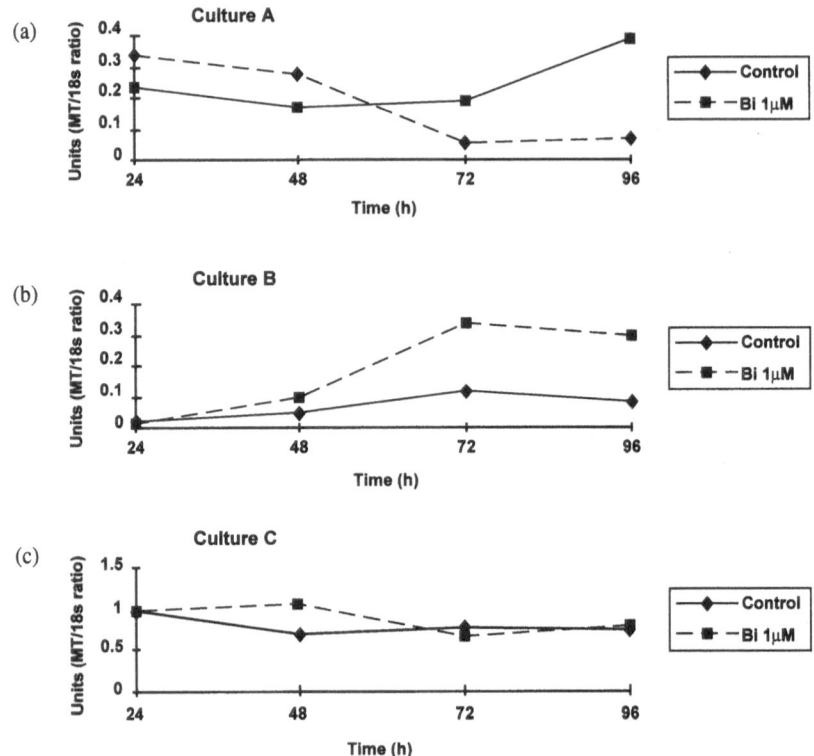

Figure 2. Profile of MT mRNA expression (using MT/18s ratios) for 3 individual cultures after exposure to 1 μM Bi(NO₃)₃ for 96 h.

of constitutive metallothionein expressed with time. In some cases constitutive levels of metallothionein remained constant over the 96 h period, whereas in other cultures, constitutive levels declined with time in culture. These data are consistent with the profile of constitutive metallothionein mRNA observed (Fig. 2). The different time course of induction at the mRNA and protein levels suggests that post-transcriptional regulation may occur.

Table 2. Time taken (h) for maximal MT induction by each of the metals as determined by immunocytochemistry

	$Bi(NO_3)_3$	$HgCl_2$	$CdCl_2$	$ZnCl_2$
Time for maximal MT induction	48 h	24–48 h	24–72 h	48–72 h

Results obtained from 4 independent cultures.

Heavy metal and drug toxicity after induction of metallothionein in human proximal tubular cell cultures

When confluent human proximal tubular cell cultures were pretreated with bismuth for 48 h prior to the addition of 50 μM or 100 μM HgCl₂, CdCl₂ or ZnCl₂, no protection against the toxic effects of these metals was seen over the following 48 h (Fig. 3). Similarly pretreatment with cadmium (1 μM) for 48 h also failed to offer any protection (Fig. 4).

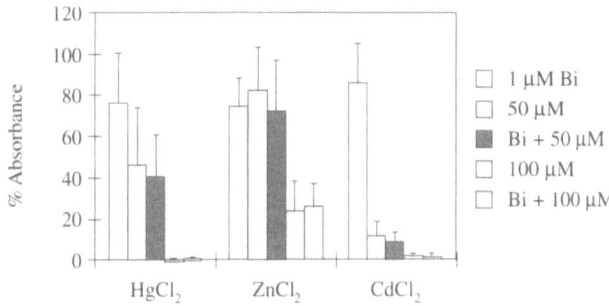

Figure 3. Effect of bismuth pretreatment for 48 h on the toxicity of metals after 48 h incubation using the MTT assay. The error bars represent the standard deviation of 5 independent experiments.

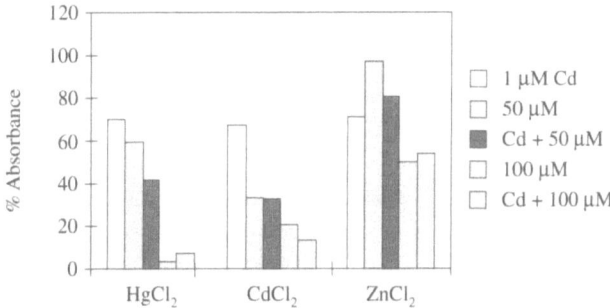

Figure 4. Effect of cadmium pretreatment for 48 h on the toxicity of metals after 24 h incubation using the MTT assay. n = 2 experiments.

Cis-diamminedichloroplatinum (II) (Cisplatin) is a potent cytostatic drug, but its use is restricted by dose-limiting nephrotoxicity [13]. Cisplatin was toxic to confluent cultures of human proximal tubular cells with an IC_{50} of 238 ± 32 μM. Pretreatment of these cultures with $Bi(NO_3)_3$ (1 or 10 μM) for 48 or 96 h did not protect against cisplatin toxicity (Fig. 5). In addition, there was slight, but not significant, protection when the cultures were treated with dexamethasone (1 or 10 μM) for 48 or 96 h (Fig. 6).

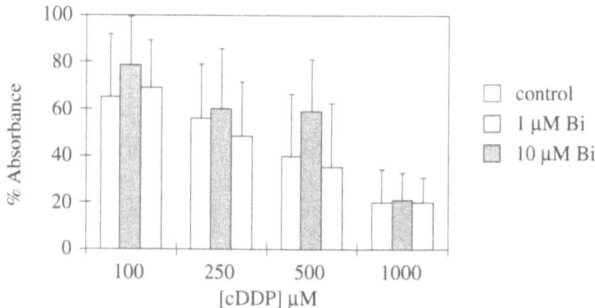

Figure 5. Effect of bismuth pretreatment for 48 h on the toxicity of cisplatin after 48 h incubation using the MTT assay. The error bars represent the standard deviation of 3 independent experiments.

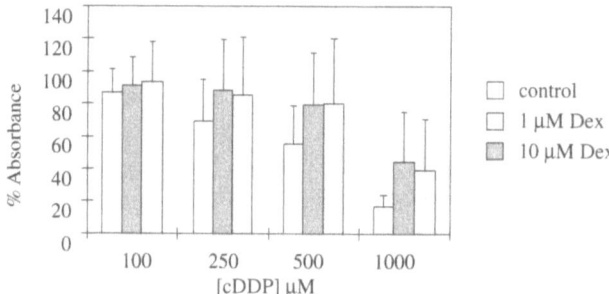

Figure 6. Effect of dexamethasone pretreatment for 48 h on the toxicity of metals after 48 h incubation using the MTT assay. The error bars represent the standard deviation of 3 independent experiments.

These results are in contrast to those obtained after pretreatment of animals with bismuth *in vivo*. Boogard *et al* [14] showed that proximal tubular cells isolated from rats receiving bismuth subnitrate (50 μmol/kg/day) for 8 d were less susceptible to the toxicity of cisplatin, $HgCl_2$, $CdCl_2$ and para-aminophenol compared with proximal tubular cells from untreated rats. These authors attributed the protection to the anti-oxidant properties of metallothionein. This confirmed an earlier study in mice where the lethal and renal toxicity of cisplatin was decreased by induction of metallothionein synthesis, without compromising its antitumour activity [15].

The data that we obtained after induction of metallothionein *in vitro* are also in contrast to studies with embryonic stem cells from transgenic mice with targeted disruption of of metallothionein 1 and 2 genes. The metallothionein null cells were more sensitive to the cytotoxic effects of cadmium, t-butylhydroperoxide and paraquat than cultured cells that were wild type or heterozygous for the loss of the metallothionein gene [16]. It is not surprising that differences in cellular sensitivity to toxicity are more marked when comparisons are made with the complete absence of metallothionein than between constitutive and induced levels of metallothionein in cells.

It has been suggested that universal overexpression of human metallothionein 2a indicates a role for this particular isoform in cisplatin resistance in human carcinoma cell lines [17]. Also it has been shown that when human MT2 is expressed in *E. coli*, the metallothionein functions in cadmium detoxification (18). In contrast Bylander *et al* [11] demonstrated using human proximal tubular cells that the expression of MT2a, 1e, 1f and 1 g mRNA and MT protein after exposure to CdCl$_2$ showed no correlation with lethality towards cadmium, whereas MT1a differed in expression and correlated with the differing lethalities of each isolate. There are several possible reasons for the lack of protection against heavy metal toxicity after exposure of human proximal tubular cells to concentrations of bismuth shown by immunocytochemistry to induce metallothionein. The induction *in vitro* may be less than that occurring *in vivo*, there may be species differences in induction or there may be isoform-specific effects. It is therefore important to determine the isoforms of metallothionein induced in primary cultures of human proximal tubular cells and to relate this to indices of toxicity after administration of heavy metals or drugs.

Acknowledgements
This work was supported by a Wellcome Trust Toxicology Fellowship (VR), Glaxo Wellcome R & D, Aberdeen Royal Hospitals NHS Trust Endowments and the European Social Fund.

References

1. Commandeur JNM, Vermeulen NPE (1990) Molecular and biochemical mechanisms of chemically induced nephrotoxicity: a review. *Chem Res Toxicol* 3: 171–194.
2. Zalups RK, Lash LH (1994) Advances in understanding the renal transport and toxicity of mercury. *J Toxicol Environ Health* 42: 1–44.
3. Goyer RA (1996) Toxic effects of metals. *In*: CD Klaassen, MO Amdur, J Doull (eds): *Cassarett and Doull's Toxicology: the basic science of poisons.* McGraw-Hill Co.,Inc, New York, 691–736.
4. Cherian MG, Goyer RA (1978) Metallothioneins and their role in the metabolism and toxicity of metals. *Life Sci* 23: 1–10.
5. Miles AT, Rodilla V, Jenner W, Hawksworth GM (1996) Relative toxicity and induction of metallothionein by heavy metals in cultured human proximal tubular cells. *Human Exp Toxicol* 15: 674.
6. Rodilla V, Hawksworth GM (1996) Isolation and culture of human renal cortical cells with characteristics of proximal tubules. *In*: GE Jones (ed.): *Methods in Molecular Medicine: Human Cell Culture Protocols.* Humana Press Inc., Totowa, NJ, 409–417.
7. Palmiter RD (1987) Molecular biology of metallothionein gene expression *In*: JHR Kägi, Y Kojima (eds): *Metallothionein II, Experientia suppl. 52* Birkhäuser Verlag, Basel, 63–80.
8. Ogura H, Takeuchi T, Morimoto K (1996) A comparison of the 8-hydroxydeoxyguanosine, chromosome aberrations and micronucleus techniques for the assessment of the genotoxicity of mercury compounds in human blood lymphocytes. *Mutation Res* 340: 175–182.
9. Endo T, Sakata M, Shaikh ZA (1995) Mercury uptake by primary cultures of rat renal cortical epithelial cells. I. Effects of cell density, temperature and metabolic inhibitors. *Toxicol Appl Pharmacol* 132: 36–43.
10. Sens MA, Hazen-Martin DJ, Bylander JE, Sens DA (1994) Heterogeneity in the amount of ionic cadmium necessary to elicit cell death in independent cultures of human proximal tubule cells. *Toxicol Lett* 70: 185–191.
11. Bylander JE, Li S, Sens MA, Sens DA (1995) Exposure of human proximal tubule cells to cytotoxic levels of CdCl$_2$ induces the additional expression of metallothionein Ia mRNA *Toxicol Lett* 76: 209–217.
12. Mididoddi S, McGuirt JP, Sens MA, Todd DH, Sens DA (1996) Isoform-specific expression of metallothionein mRNA in the developing and adult human kidney *Toxicol Lett* 85: 17–27.
13. Von Hoff DD, Schilsky R, Reichert CM, Reddick RL, Rozencweig M, Young RC, Muggia FM (1979) Toxic effects of *cis*-dichlorodiammineplatinum (II) in man. *Cancer Treat Rep* 63: 1527–1531.
14. Boogaard PJ, Slikkerveer A, Nagelkerke JF, Mulder GJ (1991) The role of metallothionein in the reduction of cisplatin-induced nephrotoxicity by Bi^{3+}-pretreatment in the rat *in vivo* and *in vitro*. *Biochem Pharmacol* 41: 369–375.
15. Naganuma A, Satoh M, Imura N (1987) Prevention of lethal and renal toxicity of *cis*-diammine-dichloro-

platinum (II) by induction of metallothionein synthesis without compromising its antitumor activity in mice. *Cancer Res* 47: 983–987.

16. Lazo JS, Kondo Y, Dellapiazza D, Michalska AE, Choo KHA, Pitt BR (1995) Enhanced sensitivity to oxidative stress in cultured embryonic cells from transgenic mice deficient in metalothionein I and II genes. *J Biol Chem* 270: 5506–5510.

17. Yang YY, Woo ES, Reese CE, Bahnson RR, Saijo N, Lazo JS (1994) Human metallothionein isoform gene expression in cisplatin-sensitive and resistant cells *Mol Pharmacol* 45: 453–460.

18. Odawara F, Kurasaki M, Suzuki-Kurasaki M, Oikawa S, Emoto T, Yamasaki F, Arias ARL, Kojima Y (1995) Expression of human metallothionein-2 in *Escherichia coli*: cadmium tolerance of transformed cells. *J Biochem* 118: 1131–1137.

Metallothionein IV
C. Klaassen (ed.)
© 1999 Birkhäuser Verlag Basel/Switzerland

Relationship of metallothionein to cadmium and to zinc in human liver and kidney

Arata Teranishi, Ruriko Ninomiya, Naoko Koizumi

Department of Public Health, Hyogo College of Medicine, Nishinomiya, Hyogo, 663 Japan

Summary. We have studied the correlations between Metallothionein(MT) and heavymetals in various species. Human liver and kidney samples were measured for Cd, Zn, and MT concentrations. We found high correlations between Zn and MT in the liver(correlation coefficient 0.884), and between Cd and MT in the renal cortex(correlation coefficient 0.799), while we found low correlations between Cd and MT in the liver(correlation coefficient -0.043). This suggests that the essential metal Zn, which is necessary for growth may be related in humans to Cd accumulation with age through MT.

Introduction

Zn and Cd concentrations in the liver and kidney of various species have been reported. And we found that Cd showed the highest concentration in the human kidney compared to other animals(1–6). Zn utilized actively in the neonatal period decreases after birth and is kept at an even level(7).On the other hand, Cd is scarcely present in the fetus and is accumulated with age in the human liver and kidney(8). Therefore, in order to evaluatethe relations of Zn and Cd with MT in the human body, we investigated the correlations between Zn, Cd, and MT in the human liver and kidney under the condition of non-exposure.

Materials and methods

The human liver and kidney samples were obtained at autopsy from 20 adult Japanese (age 39 to 83 years, 6 female and 14 male, without Cd-polluted histories) were subjected to wet ashing. The concentrations of Cd and Zn were determined by flame or flameless atomic absorption spectrophotometer using a polarized Zeeman atomic absorption spectrophotometer. For the determination of MT concentrations, about 1 g each of the liver and renalcortex was taken out at thawing, homogenized with the four or five fold amount(weight basis) of 0.01 M Tris-HCl buffer, and then centrifuged at 20,000 rpm (49,000 g)for 80 min. The resultant supernatants of the liver and renal cortex were treated with Cd-saturated hemolysate by the method of Onosaka et al. (9). The MT concentration was calculated on the assumption that seven Cd atoms were bound to thionein and 100% of the metal in the MT fraction was extracted from the organ supernatant. The correlation coefficients between Cd, Zn, and MT concentrations were calculated to investigate how they were related in the body.

Results

Table 1. shows the Cd, Zn, and MT concentrations in human kidney and renal cortex. The Cd concentration in the liver was about 10 µg/g, and in the renal cortex was about 100 µg/g. The Zn concentration in the liver was about 100 µg/g, and in the renal cortex was 60 µg/g. The MT concentration in the liver was about 80 nmol/g, and in the renal cortex was 140 nmol/g. In the human liver, the MT concentration was highly correlated with the Zn concentration(Fig. 2) but was scarcely correlated with the Cd concentration(Fig. 1). It is thus suggested that MT in the liver is, for the most part, bound to Zn, which exists in high concentrations. It seems that MT synthesis in the liver is induced at about 40 µg/g and higher. The MT concentration in the renal cortex was highly positively correlated with the Cd concentration in the renal cortex. There was a linear relationship at the Cd concentration of about 50 µg/g and higher(Fig. 3). The MT concentration in the renal cortex was also highly positively correlated with the Zn concentration in the renal cortex. There was a linear relationship at the Cd concentration of about 50 µg/g and higher(Fig. 4).

Table 1. MT, Cd, and MT concentrations in the liver and kidney

age	sex	Liver			Renal cortex		
		MT(nmol/g)	Cd(µg/g)	Zn(µg/g)	MT(nmol/g)	Cd(µg/g)	Zn(µg/g)
39	M	8.2	0.8	39.3	92.6	46	63
42	M	83.5	11	97.8	332	166	104
45	M	5.7	2.2	44.9	117	84	68.2
46	M	106	4.5	75.6			
49	M	14	4.5	54.8	90.3	86	75.7
50	F	52.3	11	62.2			
55	F	199	7.2	169	71.9	111	55.9
56	F	34	16.9	58.1	214	148	66.5
63	M	245	4.3	188			
67	M	122	2	171			
68	M	64.5	12	39.7	30.3	45	50.3
71	M	73	8.4	92.9	77.5	70	5.4
71	M	38.3	9.6	96			
72	F	78.8	25.6	99.3			
73	M	99.5	18	84.5	352	170	85
74	F		169	93	41.9		
80	M	24.9	19	50.6	76.5	131	63.1
81	M	37.5	16	67.7	81.6	100	66.5
83	M	170	21	197	148	86	59.6
83	F	32.6	16	85.9			

Discussion

Cd ranks 62nd in Clarke and its concentration in the earth's crust is extremely low,0.2 ppm. Nevertheless, Cd is positively accumulated in the livers and kidneys of such mammals as

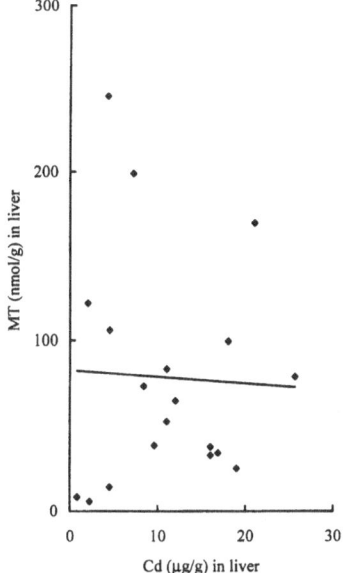

Figure 1. Correlation between MT and Cd in liver. Correlation coefficient: –0.432.

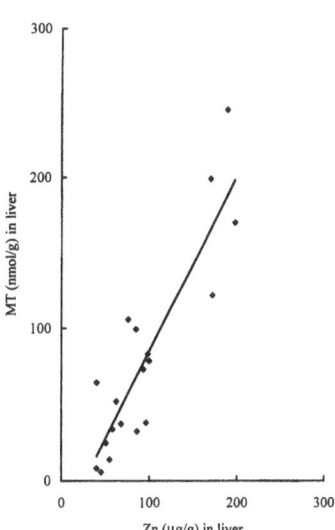

Figure 2. Correlation between MT and Zn in liver. Correlation coefficient: 0.884, p < 0.001.

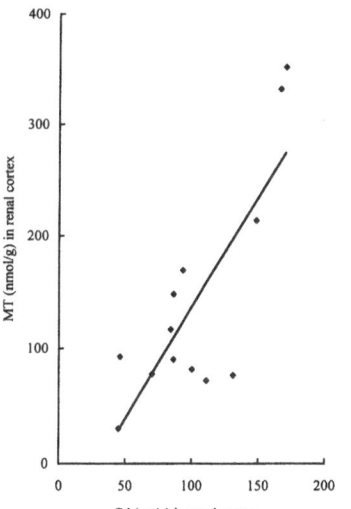

Figure 3. Correlation between MT and Cd in renal cortex. Correlation coefficient: 0.799, p < 0.01.

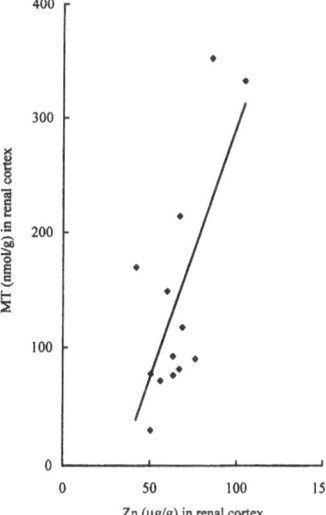

Figure 4. Correlation between MT and Zn in renal cortex. Correlation coefficient: 0.706, p < 0.01.

humans, monkeys, and horses. Such positive accumulation of Cd and its species specificity suggests that in man Cd has no detrimental effect on the body but rather some biological effect as a trace element. On the other hand, in this study, there were high correlations between Zn and MT in the liver, and between Cd and MT in the renal cortex, while we found low correlations between Cd and MT in the livers of humans. These findings differ from the data of Cd loaded animals or Cd polluted humans. Besides, Zn is actively utilized in the neonatal period and decreases with growth, while Cd is scarcely present in the fetus and is accumulated with age in humans and a few other species. Such opposite kinetic profiles of Zn and Cd during the growing period after birth suggest that the essential metal Zn, which is necessary for growth, may be related in humans to Cd accumulation with age through MT.

Acknowledgements
We are much indebted to patients for supplying specimens.

References

1. Tipton H, Schroeder HA, Perry HM, Cook MJ (1965) Trace elements in human tissue.Å@Part III. Subjects from Africa, the near and far east and Europe. *Health Phys* 11: 403–451.
2. Sumino K, Hayakawa K, Shibata T, Kitamura S (1975) Heavy metals in normal Japanese Tissues. *Arch Environ Health* 30: 487–494.
3. Koizumi N (1975) Fundamental studies on the movement of cadmium in animals and humans. *Japan J Hyg* 30: 27–34.
4. Elinder CG, Kjellstrom T, Friberg L (1976) Cadmium in kidney cortex, liver, and pancreas from Swedish autopsies. *Arch Environ Health* 31: 292–302E.
5. Nomiyama K, Nomiyama H, Nomura Y, Taguchi T, Matsui K, Yotoriyama M, Akabori F, Iwao S, Koizumi N, Masaoka T et al (1979) Effects of dietary cadmium on rhesus monkeys. *Environ Health Perspect* 28: 223–243.
6. Koizumi N, Inoue Y, Ninomiya R, Fujita D, Tsukamoto T (1989) Relationship of cadmium accumulation to zinc or copper concentration in horse liver and kidney. *Environ Res* 49: 104–114.
7. Riordan JR, Richards V (1980) Human fetal liver concentrations both zinc- and Copper rich forms of metallothionein. *J Biol Chem* 255: 5380–5383.
8. Tsuchiya K (1973) study of cadmium distribution in human organs. kankyo Hoken Repoto 19: 53–60.
9. Onosaka K, Tanaka K, Doi M, Okahara K (1978) A simplified procedure for determination of metallothionein in animal tissues. *Eiseikagaku* 24: 128–131.

Metallothionein IV
C. Klaassen (ed.)
© 1999 Birkhäuser Verlag Basel/Switzerland

Apoptosis of human kidney 293 cells is promoted by polymerized Cd-metallothionein

Tetsuo Hamada, Akihide Tanimoto, Takakazu Sasaguri, Nobuyuki Arima, Shohei Shimajiri, Ryuji Nakano and Yasuyuki Sasaguri

Department of Pathology and Cell Biology, University of Occupational and Environmental Health, School of Medicine, 1-1 Iseigaoka, Yahatanishi-ku, Kitakyushu, 807-0804 Japan

Summary. Exposure to 12.5 to 37.5 µM $CdCl_2$ induced apoptosis in transformed human kidney cells (293 cells) as confirmed by characteristic ultrastructural features, a ladder on gel electrophoresis of extracted DNA, and fragmentation of nucleosomes as detected by enzyme linked immunosorbent assay (ELISA). Higher concentrations of Cd were less effective in inducing apoptosis. Furthermore, addition of the protein extract from the serum-free medium used for Cd-exposure promoted apoptosis exhibiting the same features as that after Cd-exposure. The apoptosis induced by the protein was dose-dependent. The molecular weight of the protein (Cd-protein) was shown to be 40 kDa by gel filtration. Two-dimensional electrophoresis revealed the Cd-protein as a single spot with a molecular weight of 6 kDa and pI of 4.5. Competitive ELISA showed that the Cd-protein reacted with anti-metallothionein antibody. The present findings suggest that apoptosis is induced not only by Cd itself but also by polymerized metallothionein (MT) molecules released from cells into the medium.

Introduction

Apoptosis is inducible by various stimuli including radiation [1], heat shock [2], chemically synthesized substances [3] and physiological substances such as steroid hormones [4], and cytokines [5]. Recently we have demonstrated that Cd-induced cell death in renal tubular epithelial cells is due to apotosis by morphologic study of beagle dogs during long term administration of Cd [6] and cultured proximal tubule cells exposed to Cd [7]. In the present study, we obtained biochemical evidence for apoptosis in Cd-induced cell death and attempted to clarify the mechanism of apoptosis with reference to MT.

Ultrastructural features of cell death in renal tubular cells *in vivo* and *in vitro*

Formerly, cell death induced by Cd was regarded as a form of necrosis, an accidental cell death, for a long time even after the concept of apoptosis had been introduced to cell biology [8–11]. In 1991 we first described that Cd-induced cell death was apoptotic in nature, based on ultrastructural observations of atrophic kidneys excised from beagle dogs after long-term Cd exposure. Electron microscopic examination revealed that proximal renal tubule cells were shrunken with increased electron density and had a characteristic nucleus showing peripheral chromatin condensation; on the other hand, membrane systems such as the endoplasmic reticulum or mitochondrial cristae were found to be well preserved [12]. The morphology was identical to that given in the first description on apoptosis. In this experiment, no necrosis could be found in kidneys after various periods of exposure to Cd. Similar cytological alterations were observed in rat renal tubules in another chronic Cd-exposure experiment [6]. Subsequent Cd-exposure experiments using cultured cells demonstrated similar morphological changes in B131 cells derived from canine proximal tubule cells [7]. In the present study using transformed human kidney cells, electron microscopic examination revealed that the dead cells had the characteristic features of apoptosis (Fig. 1). After our first report, several other authors added further evidence that Cd-induced cell death was apoptotic in a human T cell line[13], porcine renal epithelial cells [14], and rat testicular tissue [15]. It seems reasonable to assume that necrosis can be induced by Cd only when damage to blood vessels feeding the organ occurs upon acute Cd exposure [16].

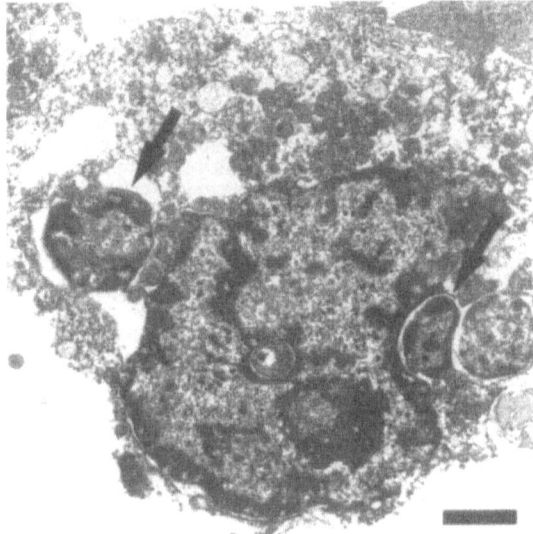

Figure 1. Electron micrographs of apoptotic 293 cells exposed to polymerized metallothionein. Ultrastructural examination revealed peripherally aggregated chromatin, fragmentation of nuclei and nuclear blebbing (arrows). Mitochondria and the membrane system were relatively well preserved. Bar: 1 μm.

Biochemical features of Cd-induced cell death

In the present study, DNA extracted from human kidney 293 cells exposed to 25 μm Cd for 48 h showed a ladder by agarose gel electrophoresis (Fig. 2). ELISA for determining nucleosomes disclosed that apoptotic cell death was induced satisfactorily at 12.5 to 37.5 μM Cd (Fig. 3). It is noteworthy that the apoptosis was not dependent on the concentration of Cd in the medium. Although a higher dose of Cd did affect 293 cells, apoptosis tended to decline at a Cd concentration exceeding 37.5 mM. This suggested that the mode of Cd-induced cell death was biphasic. Cd at a lower concentration might induce apoptosis, whereas Cd at higher con-

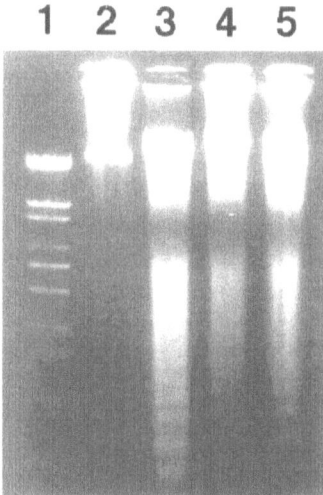

Figure 2. Agarose gel electrophoresis of DNA extracted from 293 cells exposed to Cd or polymerized metallothionein. Note distinct ladder pattern of DNA from 293 cells exposed to 25 mM $CdCl_2$ for 48 h (lane 2). Polymerized metallothionein (10 µg/ml) also produces an obscure or clear ladder pattern of DNA from 293 cells at 5 mg/ml (lane 3) or 10 µg/ml (lane 4), respectively. Lane 1 shows the electrophoretic pattern of the molecular weight marker Sty1.

centration might induce cell death through a different mechanism, perhaps necrosis. On the other hand, polymerized MT extracted from the medium induced apoptosis dose-dependently in 293 cells, as described below (Fig. 3).

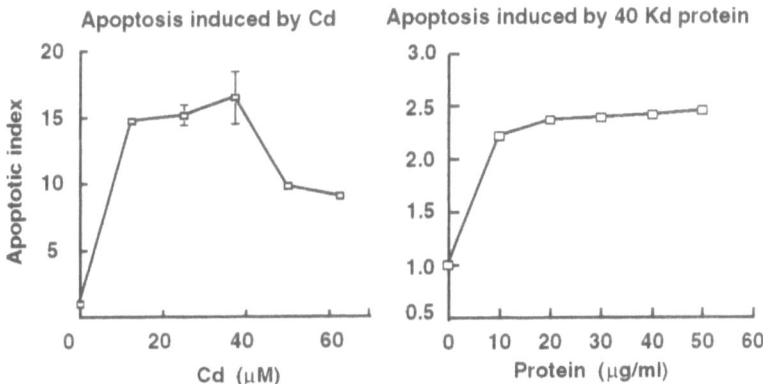

Figure 3. Quantitative analysis of apoptosis in 293 cells induced by $CdCl_2$ or polymerized metallothionein. $CdCl_2$ at 12.5 to 37.5 mM induced apoptosis in 293 cells 17 fold in comparison with control cells. Less than 10 µg/ml polymerized MT induced apoptosis and its effect was dose dependent. Apoptotic index of control cells was defined as 1.0.

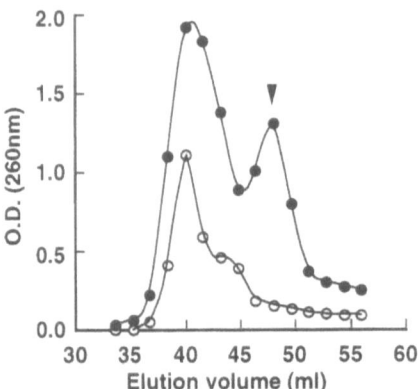

Figure 4. Gel chromatography of polymerized metallothionein. The second peak (arrow head) after void elution of protein was consistent with polymerized Cd-metallothionein in terms of molecular weight (closed circle). In control protein (open circle), no peak of polymerized metallothionein appeared in the gel filtration.

The protein extract obtained by 100% ammonium sulfate salting out from the medium used for Cd-exposure showed two major peaks by gel chromatography and monitoring of UV absorbance at 260 nm. On the other hand, only one major peak appeared upon gel chromatography of control protein (Fig. 4). The molecular weight of the second peak protein (Cd-protein) was estimated to be 40 kDa by gel chromatography. Two dimensional electrophoresis revealed the Cd-protein as a single spot after Coomassie blue staining and consisted of 6 kDa subunits with a pI of 4.5 (Fig. 5). Upon competitive ELISA for estimation of MT, the extracted 40 kDa protein reacted with anti-MT antibody (Fig. 6). The molar ratio of Cd and MT was determined

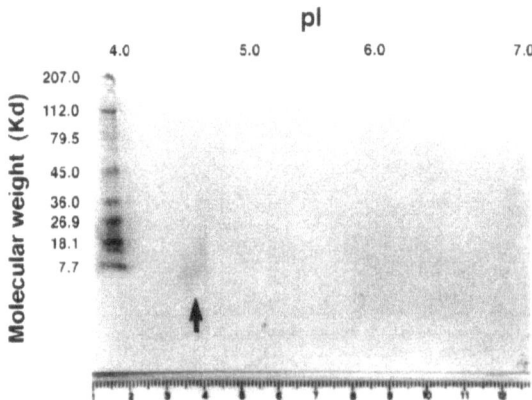

Figure 5. Electrophoretic 2-dimensional analysis of polymerized metallothionein. One spot was visible on the PVDF membrane stained by Coomassie blue solution (arrow). The electrophoretic distribution of the protein was consistent with monomeric metallothionein.

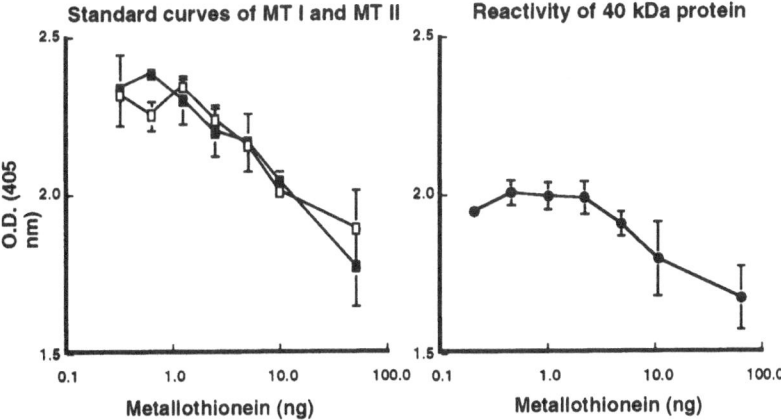

Figure 6. Standard curves of metallothionein I and metallothionein II, and reactivity of 40 kDa protein upon competitive ELISA. Cd-induced protein in culture medium showed reactivity with anti-metallothionein antibody using this method. Open square: metallothinein I, closed square: metallothionein II, closed circle: 40 kDa protein.

to be 7.9 by atomic absorbance spectrophotometry. The monomeric form had biochemical characteristics similar to human kidney MT in terms of molecular weight, pI and the molar ratio of Cd [17]. Thus the protein was proved to be polymerized MT. DNA extracted from the cells cultured in medium containing the Cd-protein (40 kDa) for 48 h also exhibited a ladder in agarose gel. Electron microscopic examination revealed that the dead cells had the characteristic features of apoptosis described above. ELISA for quantitation of apoptosis revealed that less than 10 µg/ml protein promoted apoptosis in the 293 cells. Quantitative analysis with ELISA showed that the induction of apoptosis by the protein was dose-dependent (Fig. 3).

Cell death and MT in renal tubular cells exposed to Cd

The present study on Cd-induced apoptosis added biochemical evidence to the morphologic observations, indicating a ladder upon gel electrophoresis of DNA extracted from 293 cells exposed to Cd. Although transformed cells or malignant cells have been reported to show apoptosis [18], it was considered that Cd at least promoted apoptosis of transformed 293 cells in comparison with control cells.

Although the protective effects of MT have been studied extensively both in animal experiments and cell culture systems, detoxification is by no means universally accepted as its primary function. Recent studies have pointed out that the nephrotoxicity of Cd in animal experiments is related to urinary excretion of Cd-MT rather than Cd in the renal cortex [17, 19]. Our findings that polymerized Cd-MT was an effective inducer of apoptosis supports this possibility. MT extracted from the medium in the present study showed a high molecular weight in the native form. However, the Cd-protein in a reduced state had a lower molecular weight of 6 kDa.

The Cd-protein extracted from the medium was considered to be a polymeric form of MT. Apometallothionein has been shown to polymerize rapidly by disulfide formation when adjusted to neutral pH in the absence of metal [20]. Porter showed that the polymeric form of Cu-MT was present in neonatal liver [21]. It seems reasonable to conclude that polymerized Cd-MT, which was one of the major proteins present in the conditioned medium, was liberated from 293 cells.

Up to now, the source of renal MT has been considered to be the liver and renal MT production has been neglected in many animal experiments [22]. However, a considerable amount of MT was found to be liberated from Cd-exposed 293 cells into the medium, which originally lacked exogenous MT. We consider that Cd-MT in addition to Cd itself may play an important role in inducing apoptosis of renal tubule cells upon Cd intoxication.

Genotoxicity of Cd and metallothionein

As to the genotoxicity of Cd, the metal is reported to cause DNA strand damage in cultured liver cells, which is reduced by Zn pretreatment [23]. Incubation of nuclei with different concentractions of free calcium combined with cadmium produces stronger inhibition of DNA fragmentation than zinc. Cadmium alone is able to stimulate endonuclease, and thus replace Ca^{2+} [24]. As to the effect of metallothionein on DNA, Cd/Zn-MT induces DNA strand breaks *in vitro* [25]. MT induced by Zn has a protective effect against oxidative damage to DNA [26]. Our finding that addition of extracted Cd-MT produced apoptosis in 293 cells suggests that Cd-MT activates endonuclease, and leads to DNA fragmentation.

Recent studies have indicated that MT is inducible by various stimuli other than heavy metals [27, 28]. MT expression has also been reported in various malignant tumors and its overexpression has been discussed in association with their grade of malignancy or proliferative activity [29, 30, 31]. Thus it is possible that MT may be related to apoptosis, which is the predominant type of cell death occurring in proliferating tissue.

References

1. Baxter GD, Lavin MF (1992) Specific protein dephosphorylation in apoptosis induced by ionizing radiation and heat shock in human lymphoid tumor lines. *J Immunol* 148: 1949–54.
2. Migliorati G, Nicoletti I, Crocicchio F, Pagliacci C, D'Adamio F, Riccardi C (1992) Heat shock induces apoptosis in mouse thymocytes and protects them from glucocorticoid-induced cell death. *Cell Immunol* 143: 348–56.
3. Anilkumar TV, Sarraf CE, Hunt T, Alison MR (1992) The nature of cytotoxic drug-induced cell death in murine intestinal crypts. *Brit J Cancer* 65: 552–8.
4. Billig H, Furuta I, Hsueh AJ (1993) Estrogens inhibit and androgens enhance ovarian granulosa cell apoptosis. *Endocrinology* 133: 2204–12.
5. Kessler JA, Ludlam WH, Freidin MM, Hall DH, Michaelson MD, Spray DC, Dougherty M, Batter DK (1993) Cytokine-induced programmed death of cultured sympathetic neurons. *Neuron* 11: 1123–32.
6. Tanimoto A, Hamada T, Koide O (1993) Cell death and regeneration of renal proximal tubular cells in rats with subchronic cadmium intoxication. *Toxicol Pathol* 21: 341–52.
7. Hamada T, Tanimoto A, Iwai S, Fujiwara H, Sasaguri Y (1994) Cytopathological changes induced by cadmium-exposure in canine proximal tubular cells: a cytochemical and ultrastructural study. *Nephron* 68: 104–11.
8. Payne BJ, Saunders LZ (1978) Heavy metal nephropathy of rodents. *Vet Pathol* 5: 51–87.

9. Aughey E, Fell GS, Scott R, Black M (1984) Histopathology of early effects of oral cadmium in the rat kidney. *Environ Health Perspect* 54: 153–61.
10. Kojima S, Ono H, Kiyozumi M, Honda T, Takadate A (1989) Effect of N-benzyl-D-glucamine dithiocarbamate on the renal toxicity produced by subacute exposure to cadmium in rats. *Toxicol Appl Pharmacol* 98: 39–48.
11. Rehm S, Waalkes MP (1990) Acute cadmium chloride-induced renal toxicity in the Syrian hamster. *Toxicol Appl Pharmacol* 104: 94–105.
12. Hamada T, Nakano S, Iwai S, Tanimoto A, Ariyoshi K, Koide O (1991) Pathological study on beagles after long-term oral administration of cadmium. *Toxicol Pathol* 19: 138–47.
13. el-Azzouzi B, Tsangaris GT, Pellegrini O, Manuel Y, Benveniste J, Thomas Y (1995) Cadmium induces apoptosis in a human T cell line. *Toxicology* 88: 127–39.
14. Matsuoka M, Call KM (1995) Cadmium-induced expression of immediate early genes in LLC-PK1 cells. *Kidney Int* 48: 383–9.
15. Xu C, Johnson JE, Singh PK, Jones MM, Yan H, Carter CE (1996) *In vivo* studies of cadmium-induced apoptosis in testicular tissue of the rat and its modulation by a chelating agent. *Toxicology* 107: 1–8.
16. Nolan CV, Shaikh ZA (1986) The vascular endothelium as a target tissue in acute cadmium toxicity. *Life Sci* 39: 1403–9.
17. Groten JP, Sinkeldam EJ, Luten JB, Van BP (1991) Cadmium accumulation and metallothionein concentrations after 4-week dietary exposure to cadmium chloride or cadmium metallothionein in rats. *Toxicol Appl Pharmacol* 111: 504–13.
18. Fukuda K, Kojiro M, Chiu JF (1993) Demonstration of extensive chromatin cleavage in transplanted Morris hepatoma 7777 tissue: apoptosis or necrosis? *Amer J Pathol* 142: 935–46.
19. Wang XP, Chan HM, Goyer RA, Cherian MG (1993) Nephrotoxicity of repeated injections of cadmium-metallothionein in rats. *Toxicol Appl Pharmacol* 119: 11–6.
20. Kägi JHR, Nordberg M (1979) Metallothionein: proceedings of the First International Meeting on Metallothionein and Other Low Molecular Weight Metal-binding Proteins. *Experientia* Suppl. Metallothionein 34: 64–70.
21. Porter H (1974) The particulate half-cystine-rich copper protein of newborn liver. Relationship to metallothionein and subcellular localization in non-mitochondrial particles possibly representing heavy lysosomes. *Biochem Biophys Res Commun* 56: 661–668.
22. Chan HM, Tabarrok R, Tamura Y, Cherian MG (1992) The relative importance of glutathione and metallothionein on protection of hepatotoxicity of menadione in rats. *Chem -Biol Interact* 84: 113–24.
23. Coogan TP, Bare RM, Waalkes MP (1992) Cadmium-induced DNA strand damage in cultured liver cells: reduction in cadmium genotoxicity following zinc pretreatment. *Toxicol Appl Pharmacol* 113: 227–33.
24. Lohmann RD, Beyersmann D (1993) Cadmium and zinc mediated changes of the Ca(2+)-dependent endonuclease in apoptosis. *Biochem Biophys Res Commun* 190: 1097–103.
25. Muller T, Shuckelt R, Jaenicke L (1991) Cadmium/zinc-metallothionein induces DNA strand breaks *in vitro*. *Arch Toxicol* 65: 20–26.
26. Chubatsu LS, Meneghini R (1993) Metallothionein protects DNA from oxidative damage. *Biochem J* 291: 193–198.
27. Kägi JHR (1991) *Overview of Metallothionein*. San Diego: Academic Press, Inc.
28. Luce MC, Schyberg JP, Bunn CL (1993) Metallothionein expression and stress responses in aging human diploid fibroblasts. *Exp Gerontol* 28: 17–38.
29. Bahnson RR, Banner BF, Ernstoff MS, Lazo JS, Cherian MG, Banerjee D, Chin JL (1991) Immunohistochemical localization of metallothionein in transitional cell carcinoma of the bladder. *J Urol* 146: 1518–20.
30. Fresno M, Wu W, Rodriguez JM, Nadji M (1993) Localization of metallothionein in breast carcinomas. An immunohistochemical study. *Virchows Arch A Pathol Anat Histopathol* 423: 215–9.
31. Schmid KW, Ellis IO, Gee JM, Darke BM, Lees WE, Kay J, Cryer A, Stark JM, Hittmair A, Ofner D (1993) Presence and possible significance of immunocytochemically demonstrable metallothionein over-expression in primary invasive ductal carcinma of the breast. *Virchows Arch A Pathol Anat Histopathol* 422: 153–9.

Role of metallothionein in oxidative stress

Metallothionein IV
C. Klaassen (ed.)
© 1999 Birkhäuser Verlag Basel/Switzerland

Subcellular localization mediated functioning of metallothionein: Protection against oxygen radicals and anticancer agents

Elizabeth S. Woo, Bruce R. Pitt and John S. Lazo

Department of Pharmacology, University of Pittsburgh School of Medicine, Pittsburgh, Pennsylvania 15261, USA

Summary. Our laboratories have demonstrated subcellular location-specific functionality of metallothionein, specifically showing cytoprotection from three mechanistically diverse toxins. In the case of tBH, the increased survival of the cytoplasmic MT cells may relate to their heightened ability to reduce oxygen radical formation or facilitate radical removal. Our observation that MT overexpression in SPAEC and NIH3T3 cells reduces AMVN-induced phospholipid oxidation is consistent with this hypothesis. Collectively, these studies suggest MT directly scavenges oxygen radicals or endows local target proteins with antioxidant activity.

Introduction

Metallothioneins (MT) comprise a class of intracellular thiols that confer cellular resistance to heavy metals, oxidizing agents, some electrophilic mutagens and anticancer drugs [1]. The mechanisms underlying this protective functionality are unclear but may relate to the ability of MT to exchange or donate metals to key proteins, for example transcription factor Sp1 and Cu/Zn superoxide dismutase, or its free radical scavenging activity. Irrespective of the precise mechanism, it is clear that the interaction of MT with putative targets could be facilitated or attenuated by subcellular partitioning.

In its fully metal bound state of seven copper, zinc, or cadmium atoms, MT is only 7 kDa. Interestingly, subcellular compartmentation of MT has been observed in cultured cells and in tumors, despite its small and presumably, diffusible size. Moreover, MT contains no obvious nuclear localization sequence (NLS) or other signal peptide that would suggest subcellular location specificity. Previous studies reveal an association between nuclear localized MT and neoplastic phenotypes originating from several tissue origins including bladder, prostate, and thyroid [2, 3, 4]. Nucleocytoplasmic movement of MT also has been examined with respect to the cell cycle and proliferative stimuli [5, 6, 7]. A subcellular location-specific functioning for MT has been suggested in studies showing resistance to *tert*-butylhydroperoxide (tBH) and S-nitrosoacetylpenicillamine-induced oxidant injury in NIH3T3 cells harboring cytoplasmic MT, and reversible protection against H_2O_2-induced DNA strand scission in V79 cells, where MT is nuclear [8, 9, 10].

Our laboratories have developed a definitive cell model for exploring MT-mediated cytoprotection as a function of subcellular location, and importantly, in the absence of metal or hormone induction of MT expression. Secondly, we have surveyed the National Cancer Institute (NCI) tumor cell panel comprising a comprehensive and diversified group of 53 cell lines for total MT content and subcellular localization. We have also characterized subcellular compartmentation of both endogenous MT and a fluorescent-tagged MT with respect to cellular energy requirements. Finally, utilizing a vascular endothelial cell model, we have examined the protective function of MT in response to oxidative stress and demonstrate peroxyl radical scavenging activity in intact cells.

Diversity of MT content and nucleocytoplasmic localization

As a first step in examining subcellular location-specific functioning of MT, we analyzed the NCI tumor panel of 53 human cell lines for total MT content and found a 400-fold range from 0.01 µg MT/mg protein for ovarian cell line SKOV-3 to 3.84 µg MT/mg protein for melanoma-derived LOX IMVI [1]. Confocal laser scanning microscopy was used to localize endogenous MT in the cell lines following incubation of fixed cells with an antiMT-antiserum [1] that was fluorescently-labeled with carboxymethylindocyanine (Cy3). Nuclear and cytoplasmic MT concentrations were calculated from digitized micrographs as the average pixel intensity in nuclear areas, defined by co-localization with Hoechst 33342, as well as cytoplasmic regions. The cell lines could be divided roughly in half between karyophilic and cytoplasmophilic phenotypes. Interestingly, some tissue specificity was apparent with respect to MT subcellular localization, for example, all seven breast cancer cell lines showed a cytoplasmophilic MT phenotype, while both prostate cells lines were karyophilic. Using the nuclear to cytoplasmic concentration as a subcellular localization index, we found it spanned a 10-fold range that was independent of total MT levels. Thus, considerable diversity was observed both in MT content and nucleocytoplasmic location among the NCI tumor panel.

Regulation of MT nucleocytoplasmic distribution

We employed two distinct strategies to examine the regulation of MT subcellular localization. In the first approach, we monitored endogenous MT distribution in several nuclear MT-localized human tumor cell lines in response to temperature and energy perturbations [7]. Secondly, we synthesized a fluorescently-labeled MT (MT-Cy3) and examined its distribution properties in digitonin-permeabilized cells [unpublished results]. We found nuclear retention of both cadmium-induced endogenous MT and the synthetic MT-Cy3 to be sensitive to ATP depletion by 2-deoxy-D-glucose and carbonyl cyanide p-(trifluoromethoxy)phenylhydrazone. Moreover, nuclear accumulation of MT-Cy3 was inhibited by wheat germ agglutinin and a 100-fold molar excess of unlabeled MT, but not by anti-nucleoporin antibody, mAb414, or reduced temperature. Collectively, these data suggest passive nuclear entry of MT and subsequent saturable binding to a nuclear partner. No credible endogenous binding partner has been identified for MT; however, *in vitro* thermodynamically stable binding has been demonstrated for glutathione (GSH), which in combination with MT, comprises the bulk of cellular thiols [1]. Total GSH levels did not correlate with MT levels or nucleocytoplasmic location in the NCI tumor panel [11]. Similar to MT, GSH partitions in subcellular compartments despite its small size, lacks an NLS, and its nuclear localization is sensitive to ATP depletion [14]. Co-localization studies of GSH and MT may provide important information regarding cellular protection against oxidant and electrophile-induced injury.

Functional consequences of MT subcellular localization

Because our survey of the NCI tumor panel revealed substantial heterogeneity both in total MT levels and nucleocytoplasmic localization, we devised a cell culture model 1) with a null MT background, 2) that restricted MT to either nuclear or cytoplasmic locales, and 3) whose nuclear or cytoplasmic MT expression could be regulated exogenously. Human MT IIA, the most highly and ubiquitously expressed isoform of MT, was fused to the SV40 large T antigen NLS and the bacterial *lacZ* gene, the translation of which resulted in a 120 kDa MT fusion protein [15, Figure 1]. The expressed protein could be visualized exclusively in the cell nucleus of stably transfected mouse embryo fibroblasts carrying a targeted disruption in both the MT I and II genes by light microscopy using the chromogenic substrate, 5-bromo-4-chloro-3-indolyl-—D-galactoside, and by direct immunofluorescence microscopy using our Cy3-labeled anti-MT antiserum [15]. Analogously, stable transfectants were isolated in which the NLS was deleted, and thus, expressed only cytoplasmic MT. Expression of both DNA constructs was under the control of an isopropyl-—D-thiogalactoside-regulated RSV long terminal repeat promoter, providing internal controls for fusion protein function. Both *in vitro* cadmium binding functionality and similarity in local concentrations of MT fusion protein were demonstrated for the nuclear and cytoplasmic MT cell lines. The nuclear MT cells gave modest cytoprotection against the mutagen N-methyl-N-nitro-N-nitrosoguanidine, while no protection was observed in the cytoplasmic MT cells. In contrast, increased cytoprotection against cadmium and tBH (1 h, 1 µM–1 mM) was observed for the cytoplasmic MT expressing cells, relative to either the nuclear MT expressing cells or in the absence of IPTG induction. Further examination of the

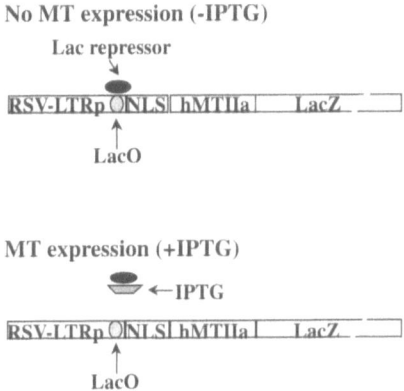

Figure 1. IPTG regulation of MT fusion protein.

tBH effect on these cells showed that the formation of the oxygen radical-sensitive dichlo-rofluorescein following tBH exposure was elevated significantly in the nuclear, but not the cyto-plasmic MT cells. These data are the first direct demonstration of location-specific functioning of MT, and suggest that essential cytotoxic targets of tBH and cadmium reside in the cytoplasm.

Antioxidant functionality of MT in pulmonary endothelial cells

The pulmonary endothelium is a primary *in vivo* target for heavy metal and oxidant-induced injury, yet remarkably little is known regarding pulmonary protective mechanisms. There is evidence that intrapulmonary MT affords some protection, for example, pretreatment of ani-mals with cadmium induces MT expression such that they are resistant to subsequent exposures of cadmium or hyperoxia [16]. In previous work, we demonstrated an antioxidant role for MT in NIH3T3 cells, specifically showing that a four-fold increase in MT levels by direct gene transfer effected a six-fold increase in resistance to tBH, relative to cells transfected with a pro-moterless expression vector [9]. Moreover, other indices indicative of resistance were evident in the MT-transfected cells, including a reduction in thiobarbituric acid-reactive substances and in DCF fluorescence, and a greater capacity to scavenge phenoxyl radicals [9]. In recent stud-ies on sheep pulmonary artery endothelial cells (SPAEC), we extended these observations and showed that overexpression of MT either by direct gene transfer or by a 24 h preincubation with 10 μM cadmium protected SPAEC against the cytotoxicity of both tBH and the peroxyl rad-ical generator, 2,2'-azobis(2,4-dimethylvaleronitrile) (AMVN) [17]. The transfectants, express-ing mouse MT-I, showed resistance to lipopolysaccharide-induced apoptosis, as determined by DNA fragmentation analysis (unpublished results), and hyperoxia (95% O_2/5% CO_2), as mea-sured by 5-HT transport [17], relative to control (sham-transfected) cells. Lipid peroxidation in the MT-transfected SPAEC was assessed in response to 0.5 mM AMVN using *cis*-pariniric acid (PnA). Oxidant-induced fluorescence of metabolically incorporated PnA was quantified in cel-lular phospholipids by HPLC and fluorescence detection. With the exception of phos-phatidylethanolamine, MT overexpression afforded complete protection against AMVN-induced oxidation of all phospholipid classes, including phosphatidylcholine, phosphatidylserine, and sphingomyelin. These data are the first demonstration of protection by MT in SPAEC in response to four distinctive forms of oxidative stress. Furthermore, our stud-ies showing an association between MT and diminution in lipid peroxidation in intact cells sup-port the role of MT in directly scavenging oxygen radicals.

Acknowledgements
Supported by NIH grant CA61299 (J.S.L) and AHA, Pennsylvania Affiliate (E.S.W)

References

1. Lazo JS and Pitt BR (1995) Metallothioneins and cell death by anticancer drugs. *Annu Rev Pharmacol Toxicol* 35: 635–653.
2. Nartey NO, Cherian MG and Banerjee D (1987) Immunohistochemical localization of metallothionein in

human thyroid tumors. *Amer J Pathol* 129: 177–182.

3. Kuo S-M, Kondo Y, DeFilippo JM, Ernstoff MS, Bahnson RR and Lazo JS (1994) Subcellular localization of metallothionein IIA in human bladder tumor cells using a novel epitope-specific antiserum. *Toxicol Appl Pharmacol* 125: 104–110.

4. Kondo Y, Kuo S-M, Watkins SC and Lazo JS (1995) Metallothionein localization and cisplatin resistance in human hormone-independent prostatic tumor cell lines. *Cancer Res* 55: 474–477.

5. Tsujikawa K, Imai T, Kakutani M, Kayamori Y, Mimura T, Otaki N, Kimura M, Fukuyama R and Shimizu N (1991) Localization of metallothionein in nuclei of growing primary cultured adult rat hepatocytes. *FEBS Lett* 283: 239–242.

6. Nagel WW and Vallee BL (1995) Cell cycle regulation of metallothionein in human colonic cancer cells. *Proc Natl Acad Sci USA* 92: 579–583.

7. Woo ES, Kondo Y, Watkins SC, Hoyt DG and Lazo JS (1996) Nucleophilic distribution of metallothionein in human tumor cells. *Exper Cell Res* 224: 365–371.

8. Chubatsu LS and Meneghini R (1993) Metallothionein protects DNA from oxidative damage. *Biochem J* 291: 193–198.

9. Schwarz MA, Lazo JS, Yalowich JC, Reynolds I, Kagan V, Tyurin V, Kim Y-M, Watkins SC and Pitt BR (1994) Cytoplasmic metallothionein overexpression protects NIH3T3 cells from *tert*-butyl hydroperoxide toxicity. *J Biol Chem* 269: 15238–15243.

10. Schwarz MA, Lazo JS, Yalowich JC, Allen WP, Whitmore M, Bergonia HA, Tzeng E, Billiar TR, Robbins PD, Lancaster JR andwich JC, Allen WP, Whitmore M, Bergonia HA, Tzeng E, Billiar TR, Robbins PD, Lancaster JR and Pitt BR (1995) Metallothionein protects against the cytotoxic and DNA-damaging effects of nitric oxide. *Proc Natl Acad Sci USA* 92: 4452–4456.

11. Woo ES, Monks A, Watkins SC, Wang AS and Lazo JS (1997) Diversity of metallothionein content and subcellular location in the National Cancer Institute tumor panel. *Cancer Chemother Pharmacol* 41: 61–68.

12. Kuo S-M, Kondo Y, DeFilippo JM, Ernstoff MS, Bahnson RR and Lazo JS (1994) Subcellular localization of metallothionein IIA in human bladder tumor cells using a novel epitope-specific antiserum. *Toxicol Appl Pharmacol* 125: 104–110.

13. Brouwer M, Hoexum-Brouwer T and Cashon RE (1993) A putative glutathione-binding site in CdZn-metallothionein identified by eequilibrium binding and molecular-modelling studies. *Biochem J* 294: 219–225.

14. Bellomo G, Vairetti M, Stivala L, Mirabelli F, Richelmi P and Orrenius S (1992) Demonstration of nuclear compartmentalization of glutathione in hepatocytes. *Proc Natl Acad Sci USA* 89: 4412–4416.

15. Woo ES and Lazo JS (1997) Nucleocytoplasmic functionality of metallothionein. *Cancer Res* 57: 4236–4241.

16. Hart BA, Voss GW, Shatos MA and Doherty J (1990) Cross-tolerance to hyperoxia following cadmium aerosol pretreatment. *Toxicol Appl Pharmacol* 103: 255–270.

17. Pitt BR, Schwarz M, Woo ES, Yee E, Wasserloos K, Tran S, Weng W, Mannix RJ, Watkins SA, Tyurina YY, Tyurin VA, Kagan VE andee E, Wasserloos K, Tran S, Weng W, Mannix RJ, Watkins SA, Tyurina YY, Tyurin VA, Kagan VE and Lazo JS (1997) Overexpression of metallothionein decreases the sensitivity of cultured pulmonary endothelial cells to oxidant injury. *Amer J Physiol* 273: L856–_865.

Obesity and hyperleptinemia in a colony of metallothionein (-I and -II) null mice

John H. Beattie[1], Anne M. Wood[1], April M. Newman[1], Ian Bremner[1], K.H. Andy Choo[3], Anna E. Michalska[3], Jackie S. Duncan[2] and Paul Trayhurn[2]

[1]*Trace Element and Gene Expression Group,* [2]*Molecular Physiology Group, Rowett Research Institute, Greenburn Road, Bucksburn, Aberdeen AB21 9SB, Scotland*
[3]*Murdoch Institute for Research into Birth Defects, Royal Children's Hospital, Flemington Road, Parkville, 3052, Australia*

Introduction

Mice with targeted disruption of metallothionein (MT)-I and -II genes are reported to have no phenotypic abnormalities and show normal reproduction and development [1, 2]. With the exception of the pancreas, adult tissue Zn levels are largely unaffected by lack of MT-I and -II [3] and retention of Cd by the liver and kidney of MT-null mice after injection of a Cd salt is reduced [4]. Nevertheless, MT-null mice are more sensitive to Cd [5], Cu [6] and Zn [7] toxicity and to oxidant stress [8]. These compounds are also more toxic to MT-null mouse embryo fibroblasts in primary culture [9] and such results are consistent with the proposed roles of MT-I and MT-II in metal detoxification and in scavenging free radicals.

We obtained MT-I and -II null mice of mixed 129/Ola-C57BL/6J genetic background [1] at the F5 generation. The original mouse colony at the Murdoch Institute had been maintained in conventional animal facilities and in order to establish a colony at the Rowett Institute, MT-null mice were re-derived by sterile embryo transfer to isolator-reared surrogate mothers. The offspring tested negative to all common viral and bacterial mouse pathogens and the resulting colony was maintained in a minimal disease unit. Four lines of mice, all with solid black coat colour, were obtained from 4 original pairs of re-derived animals and were bred separately for 1 year to verify their phenotypic similarity before interbreeding. Thereafter, male mice, which were maintained for periods of up to a further year at an environmental temperature of 23 °C with a 12 h light/dark cycle, were used in our studies. C57BL/6J mice of similar health status obtained from a commercial supplier (Harlan UK, Bicester, UK) were used for comparison. All mice were given a commercial pelleted mouse diet (CRM Diet, containing 2.4% oil, 18.1% protein, 57% carbohydrate and 3.6% fibre, in addition to essential minerals, vitamins and amino acids: Labsure, Poole, UK), and water *ad libitum*.

Growth of MT-null mice

While our intention was to focus on the putative antioxidant role of MT, survey of the MT-null mouse colony indicated that male mice were significantly heavier than male C57BL/6J mice of the same age. Indeed, over 20% of the male MT-null mice aged 22–39 weeks had a body weight of 46–59 g, with an average for the colony of 40.3 ± 6.9 g (SD; n = 58). The heavier mice had noticeably large depots of white adipose tissue and fatty livers. In contrast to preparations from smaller MT-null and C57BL/6J mice, hepatocytes prepared for primary culture from these fatty livers using a standard collagenase perfusion technique showed poor viability. This gave rise to variable results when mice of similar age were selected at random for hepatocyte preparation and the study of metal and oxidant toxicity. In growth studies from weaning to 14 weeks of age, we found that MT-null mice grew at a significantly faster rate than C57BL/6J mice and that the weight divergence was most marked between 5–7 weeks [10]. From 8–4 weeks, the growth rate of MT-null mice paralleled that of the C57BL/6J mice, thus maintaining a weight difference of about 6 g.

Study of mice aged 22–39 weeks

Three groups of older MT-null mice (6 animals/group) were selected so that 1 group (Lean Group: mean 32 g) was weight matched with the C57BL/6J mice (Control Group: mean 32 g), a second group (Average Group: mean 39 g) had a mean weight equivalent to the mean for the MT-null male colony aged 22–39 weeks and a third group was representative of the larger mice (Obese Group: mean 50 g). All animals were of similar mean age and age variance and the control mice were acclimated for 22 weeks in the same environmental conditions as the MT-null mice. Epididymal WAT (eWAT) weight and eWAT/body weight ratio were found to be significantly greater in the Average and the Obese Groups, indicating that the mice in both groups were relatively obese [10].

Body fat accretion depends crucially on the balance between energy intake and energy expenditure, which in turn can be influenced by a feedback mechanism involving leptin, the product of *obese* (*ob*) gene expression in white adipose tissue (WAT) [11]. Leptin is secreted into the circulation from WAT and there is a strong correlation between the amount of body fat and the levels of plasma leptin. The *ob* gene expression was measured in eWAT in the 3 MT-null groups and the Control Group by analysis of *ob* mRNA using northern blotting [12] and plasma leptin levels were analysed using an ELISA [13]. *ob* gene expression was significantly elevated in the Average and Obese MT-null Groups, compared to the Lean and Control Groups (Fig. 1). More strikingly, the mean plasma leptin concentration in the Obese Group was over 80 ng/mL, which was 25-fold higher than levels in the Control Group and higher even than those found in other mutant animal models of obesity such as the Zucker fatty (*fa/fa*) rat [13]. The Average Group mice, which were by comparison only moderately obese, had similar plasma leptin levels to those in the grossly obese *fa/fa* rat.

Plasma insulin levels increase in response to food intake and insulin directly promotes *ob* gene expression. On the basis that insulin could be driving leptin synthesis, we therefore measured plasma insulin concentrations in mice from each group and found that while there was no significant difference between the Average and Control Group levels, there was indeed a significant correlation between plasma insulin and plasma leptin (Fig. 2). Food intake is normally inhibited in response to high plasma leptin concentration and we therefore measured the food intake of 15 week old MT-null and C57BL/6J mice over a period of 17 days. C57BL/6J

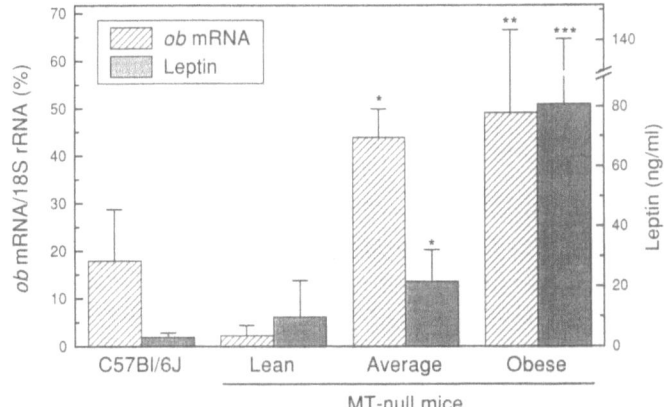

Figure 1. Leptin and *ob* mRNA levels in epididymal white adipose tissue of male C57BL/6J and MT-null mice, aged 22–39 weeks. MT-null mice were divided into 3 groups: Lean (weight-matched with C57BL/6J mice), Average (average body weight for the male MT-null mouse colony) and Obese (larger mice showing obesity). Statistically significant differences (*$P < 0.05$, **$P < 0.01$, ***$P < 0.001$) between MT-null groups and the C57BL/6J group were determined using Student's *t*-tests and a pooled estimate of error.

and MT-null mice consumed an average daily amount of 4.5 ± 0.3 g/mouse and 6.1 ± 1.0 g/mouse, respectively, a difference that was statistically significant (Student's *t-test:* $P < 0.001$). Thus, contrary to expectation, the MT-null mice actually consumed significantly more food than the C57BL/6J mice, indicating leptin insensitivity and a mechanism for accretion of fat. In addition, higher food intake may have been responsible for inducing higher insulin

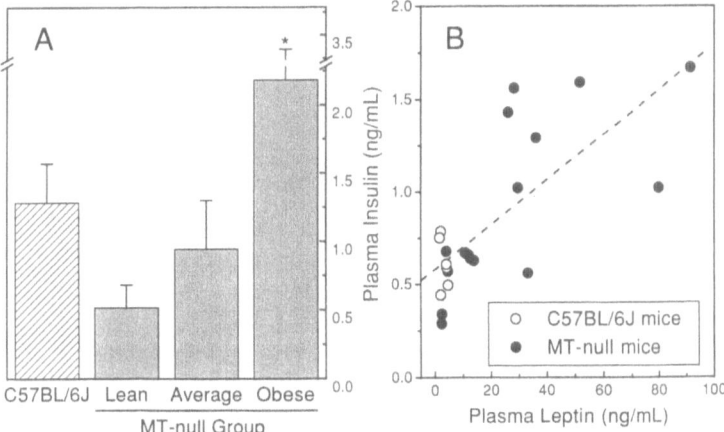

Figure 2. (A) Plasma insulin levels in 22–39 week old C57BL/6J mice and in MT-null mice of 3 body weight categories (see legend to Fig. 1) and (B) the relationship between plasma insulin and plasma leptin levels. $R = 0.66$ ($P = 0.003$) by linear regression (2 MT-null mouse datapoints off scale).

secretion. Plasma glucose and triglyceride levels were however unaffected in all MT-null groups and liver glycogen levels were reduced, though not significantly [10].

Study of mice aged 7 weeks

Since growth rate divergence between C57BL/6J and MT-null mice was first noted at 5–7 weeks of age, we investigated *ob* gene expression and plasma leptin concentration in 7-week old animals. MT-null mice contained more eWAT than C57BL/6J mice and eWAT *ob* mRNA levels were also much higher in the MT-null mice (Fig. 3). In addition, plasma leptin levels of MT-null mice were significantly higher [10], indicating that even at 7 weeks, the gain in body weight and accretion of body fat initiated some feedback regulation, which was apparently incompletely effective. Lipoprotein lipase (LPL) gene transcription is generally upregulated in obese animals and we found that eWAT LPL mRNA levels were much higher in MT-null mice than in C57BL/6J control animals (Fig. 3). Likewise, CCAAT enhancer binding protein α (C/EBP α) mRNA levels in MT-null mouse eWAT were elevated (Fig. 3), indicating an enhanced role of this transcription factor in promoting terminal differentiation of adipocytes. These results all support the proposal that our MT-null mice are moderately obese and that a propensity to obesity is evident by 7 weeks of age. Indeed, we have further (unpublished) evidence that the body weight of weanling MT-null mice can exceed that of control mice, an observation previously noted elsewhere [14].

Implications of obesity in MT-null mice

We have established that male mice in our MT-null mouse colony are obese. eWAT/body weight ratios are significantly raised and eWAT *ob* mRNA and plasma leptin levels, together with sev-

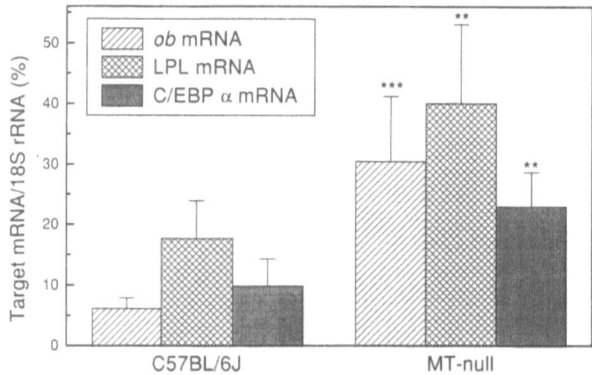

Figure 3. Levels of *ob*, lipoprotein lipase (LPL) and CCAAT enhancer binding protein α (C/EBP α) mRNA in epididymal white adipose tissue of C57BL/6J and MT-null mice aged 7 weeks, measured by northern blotting. Asterisks refer to significant differences compared to C57BL/6J mice (see legend to Fig. 1).

eral other indicators of obesity, are elevated in a large proportion of the animals. The nature of this obesity shows similarities, for example hyperleptinaemia combined with hyperphagia, and differences, such as lack of hypoglycemia in the presence of hyperinsulinaemia, to some other mutant animal models of obesity. The degree of obesity observed in MT-null mice is moderate, and we suspect that these animals also have a higher lean body mass, but this has yet to be confirmed.

MT has not previously been linked to energy balance but the presented data indicate that such a link should be considered. MT is induced by a wide variety of different stress factors in many organisms. This suggests that it may have some basic role which although not essential for survival, is nevertheless required by primitive unicellular and complex multicellular organisms alike. It is possible that MT could exert a direct effect on *ob* gene expression, but more likely, MT could influence energy utilisation or expenditure at a more fundamental level, thus affecting energy balance and consequently, *ob* gene expression. The apparent leptin insensitivity observed in our studies seems to indicate a deficiency in the hypothalamic-mediated or post-hypothalamic control of food intake or energy expenditure, but could also reflect reduced transport of leptin across the blood-brain barrier. The mice in these studies were maintained at 23 °C, below thermoneutrality for this species, and so a degree of thermogenesis, as indicated by uncoupling protein-1 (UCP-1) gene expression in interscapular brown adipose tissue (iBAT), was observed in the animals aged 7 weeks. There was, however, no difference in the level of iBAT UCP-1 mRNA levels between C57BL/6J mice and MT-null mice, indicating that the development of obesity was not obviously related to thermoregulation, at least by this mechanism (unpublished observations).

Lack of MT-I and MT-II is not the only possible explanation for the observed obesity. The MT-null mice in our colony are of mixed 129/Ola and C57BL/6J genetic background, so is it possible that the obesity observed relates simply to the cross between these two strains? While this explanation cannot be discounted, the marked growth rate of the MT-null mice, (comparable only to the most vigorous outbred strains of mice), combined with marked hyperleptinaemia, would suggest that there is another more compelling explanation. The insertion of foreign DNA sequences during targeted disruption of a gene can influence the expression of neighboring genes [15]. MT-I and MT-II genes were disrupted with a 20 bp frameshift oligomer and a positive selection marker (neomycin resistance gene), respectively [1], and so such an effect on genes neighboring the MT locus on chromosome 8 is not impossible.

Whatever the underlying mechanism involved, it is clear that our MT-null mice provide a new model of obesity characterized by moderate elevation of fat depots, hyperleptinaemia and hyperphagia. Mice of 129/SvCPJ strain with targeted disruption of the MT-I and MT-II genes [2] and appropriate control animals are available from the Jackson Laboratory (Bar Harbor, Maine, USA). Growth data on these mice are not published but studies comparing the physiological and biochemical characteristics of this strain with those of the MT-null mice in our colony are planned and should help reveal if lack of MT-I and/or MT-II is the cause of the observed obesity. Further genetic studies on our MT-null mice are also planned to evaluate whether or not this obese phenotype segregates with the null mutation.

Acknowledgments
The authors wish to thank Mr. T. Atkinson for analysis of plasma insulin and also Mrs. D. Bourke, Mrs. K. Simpson and Miss P. Dorward for helpful advice and assistance with establishment and maintenance of the MT-null mouse colony. We also appreciate useful discussions with Dr. A.K. West, Biochemistry Department, University of Tasmania, Hobart, Australia. This work was funded by the Scottish Office Agriculture, Environment and Fisheries Department, UK.

References

1. Michalska AE, Choo KHA (1993) Targeting and germ-line transmission of a null mutation at the metallothionein I and II loci in mouse. *Proc Natl Acad Sci USA* 90: 8088–8092.
2. Masters BA, Kelly EJ, Quaife CJ, Brinster RL, Palmiter RD (1994) Targeted disruption of metallothionein I and II genes increases sensitivity to cadmium. *Proc Natl Acad Sci USA* 91: 584–588.
3. Liu J, Liu Y, Michalska AE, Choo KHA, Klaassen CD (1996) Distribution and retention of cadmium in metallothionein I and II null mice. *Toxicol Appl Pharmacol* 136: 260–268.
4. Tohyama C, Satoh M, Kodama N, Nishimura H, Choo A, Michalska A, Kanayama Y, Naganuma A (1996) Reduced retention of cadmium in the liver of metallothionein-null mice. *Environ Toxicol Pharmacol* 1: 213–216.
5. Liu J, Liu YP, Michalska AE, Choo KHA, Klaassen CD (1996) Metallothionein plays less of a protective role in cadmium-metallothionein-induced nephrotoxicity than in cadmium chloride-induced hepatotoxicity. *J Pharmacol Exp Ther* 276: 1216–1223.
6. Kelly EJ, Palmiter RD (1996) A murine model of Menkes disease reveals a physiological function of metallothionein. *Nat Genet* 13: 219–222.
7. Kelly EJ, Quaife CJ, Froelick GJ, Palmiter RD (1996) Metallothionein I and II protect against zinc deficiency and zinc toxicity in mice. *J Nutr* 126: 1782–1790.
8. Sato M, Apostolova MD, Hamaya M, Yamaki J, Choo KHA, Michalska AE, Kodama N, Tohyama C (1996) Susceptibility of metallothionein-null mice to paraquat. *Environ Toxicol Pharmacol* 1: 221–225.
9. Lazo JS, Kondo Y, Dellapiazza D, Michalska AE, Choo KHA, Pitt BR (1995) Enhanced sensitivity to oxidative stress in cultured embryonic cells from transgenic mice deficient in metallothionein I and II genes. *J Biol Chem* 270: 5506–5510.
10. Beattie JH, Wood AM, Newman AM, Bremner I, Choo KHA, Michalska AE, Duncan JS, Trayhurn P (1997) Obesity and hyperleptinaemia in metallothionein (-I and -II) null mice. *Proc Natl Acad Sci USA*; *in press*.
11. White BD, Martin RJ (1997) Evidence for a central mechanism of obesity in the Zucker rat: role of neuropeptide Y and leptin. *Proc Soc Exp Biol Med* 214: 222–232.
12. Trayhurn P, Duncan JS, Rayner DV (1995) Acute cold-induced suppression of ob (obese) gene expression in white adipose tissue of mice: mediation by the sympathetic system. *Biochem J* 311: 729–733.
13. Hardie LJ, Rayner DV, Holmes S, Trayhurn P (1996) Circulating leptin levels are modulated by fasting, cold exposure and insulin administration in lean but not Zucker (fa/fa) rats as measured by ELISA. *Biochem Biophys Res Commun* 223: 660–665.
14. Duffy JY, Baines D, Keen CL, Daston GP (1997) Developmental outcome of metallothionein-null mice fed various levels of zinc during gestation. *Teratology* 55: 54.
15. Olson EN, Arnold H-H, Rigby PWJ, Wold BJ (1996) Know your neighbors: three phenotypes in null mutants of myogenic bHLH gene MRF4. *Cell* 85: 1–4.

Cardiac protection by metallothionein against ischemia-reperfusion injury and its possible relation to ischemic preconditioning

Y. James Kang and Ju-Feng Wang

Departments of Medicine, and Pharmacology and Toxicology, University of Louisville School of Medicine, 530 South Jackson St., Louisville, KY, 40202, USA

Summary. Oxidative stress is believed to play a major role in ischemia-reperfusion injury to the heart. Metallothionein (MT), a potential free radical scavenger, may function in protection against this cardiac injury. Hearts isolated from cardiac MT overexpressing transgenic mice and from the nontransgenic littermates were subjected to 50 min of warm (37°C) zero-flow ischemia followed by 90 min reflow. Compared with the nontransgenic controls, the transgenic hearts with MT concentrations about 10-fold higher than normal, showed significantly improved recovery of contractile force (73% versus 22% at the end of 90 min reperfusion, p < 0.01). Efflux of creatine kinase was reduced by more than 50% and the zone of myocardial infarction as demarcated by triphenyl-tetrazolium at the end of reperfusion was reduced by about 40% in the transgenic hearts compared with the nontransgenic controls. The second part of this study was to determine the role of MT in the ischemic preconditioning (PC). An open-chest mouse model was used to determine the effect of ischemic PC on MT synthesis in the heart, then the effect of MT induction on ischemic injury. Mouse hearts were processed four cycles of 5 min occlusion and 5 min reperfusion via ligation of left anterior descending coronary artery to produce ischemic PC. At different time points, cardiac MT was measured by a cadmium-hemoglobin affinity assay. Compared to sham-operated mice (6.6 ± 1.8 μg/g tissue), MT was significantly (p < 0.01) increased to 15.8 ± 1.8 and 17.9 ± 5.0 μg/g 12 and 24 h, respectively, after ischemic PC. Treatment with cadmium for 24 h increased cardiac MT to the same level as that induced by the ischemic PC and inhibited the subsequent injury induced by ischemia reperfusion. These results indicated that MT functions in cardiac protection against ischemia-reperfusion injury, and it may be an important factor involved in the "second window" of protection induced by ischemic PC.

Recent studies [1] have suggested that MT plays a role in the scavenging of free radicals, which are produced under various stress conditions. Zinc-MT is a very potent scavenger of hydroxyl radical *in vitro* and is more effective than glutathione (GSH) in preventing hydroxyl radical-induced DNA degradation [2]. Our recent studies using a transgenic mouse model, in which MT was overexpressed specifically in the heart, have demonstrated that MT significantly inhibited adriamycin-induced morphological changes in the myocardium and creatine kinase release from the heart [3].

Ischemia-reperfusion causes depressed myocardial function and associated deleterious morphological alterations that lead to heart failure and cell death [4]. Mechanisms by which this injury occurs are not well defined. Studies using antioxidants such as superoxide dismutase (SOD) and catalase suggest that oxidative stress and burst of free radical production are important mediators of the myocardial damage [5]. Because MT protects the heart from adriamycin-induced oxidative damage, it is possible that MT also functions in protection against ischemia-reperfusion injury in the heart. To test this hypothesis, the present study was undertaken to determine whether the cardiac MT overexpressing transgenic mice are resistant to ischemia-reperfusion injury. In this study we also determined whether MT is involved in the ischemic preconditioning (PC), a phenomenon that offers cardiac protection against ischemic injury via a brief sublethal ischemia of the heart prior to the subsequent prolonged ischemia [6].

Transgenic mice overexpressing cardiac MT about 10-fold higher than normal [3], and non-transgenic littermates of 8 wks old were anesthetized by intraperitoneal injection of sodium pentobarbital (150 mg/kg) coadministered with 100 IU of heparin. The heart was isolated and prepared for Langendorff perfusion as described previously [7]. After 30 min of preischemia equilibration, the heart was made ischemic by turning off the buffer flow for 50 min, then reperfused with the perfusion buffer (3 mL/min) for 90 min. Mechanical activity was measured throughout each perfusion experiment as described [7]. The recovery of contractile force of the transgene positive and negative mouse hearts after 50 min ischemia is shown in Figure 1. There was no significant difference in the developed contractile force between the transgenic and control hearts during the 30 min equilibration period, 0.55 ± 0.08 g and 0.56 ± 0.06 g, respectively. There was no significant difference in tension between transgene positive and transgene negative hearts during ischemia. The hearts from the transgene positive mice showed significantly better post-ischemic recovery of the suppressed contractile force ($p < 0.01$). The changes in the heart rate between transgene positive and negative hearts were not significantly different (data not shown). Creatine kinase (CK) release from the heart was measured in the effluent buffer. Samples were collected during the last minute of preischemia, and at 1, 2, 3, 5, 10, 20, 30 and 60 min of reperfusion. As shown in Figure 2, CK release from the transgene positive mouse hearts was also significantly ($p < 0.01$) suppressed as determined by a spectrophotometric method [7]. After reperfusion for 90 min post ischemia, the heart was lowered into the organ bath, and a 10% (wt/vol) solution of triphenyltetrazolium in phosphate buffer (Na_2HPO_4 88 mM, NaH_2PO_4 1.8 mM) was infused into the coronary vasculature through the side arm of

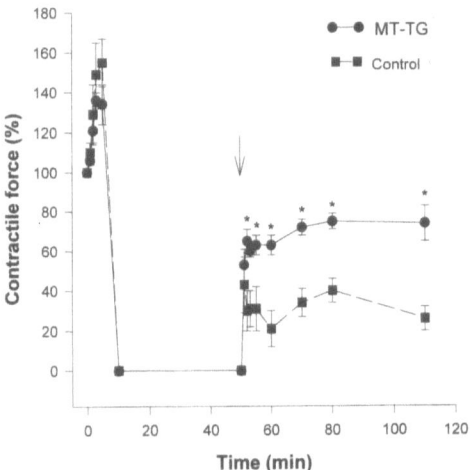

Figure 1. Effects of ischemia and reperfusion on contractile force of the hearts isolated from transgenic mice overexpressing cardiac MT about 10-fold higher than normal (MT-TG) and non-transgenic controls (Control). The isolated hearts were retrogradely perfused with Krebs-Henseleit buffer at 37^EC, made ischemic by turning off the buffer flow for 50 min after 30 min pre-ischemia equilibration, then reperfused for 60 min (starting at 50 min as indicated by the arrow). Each point represents the mean ± SD of 5 mouse hearts. *P < 0.01. Arrow indicates the time point at which reperfusion starts.

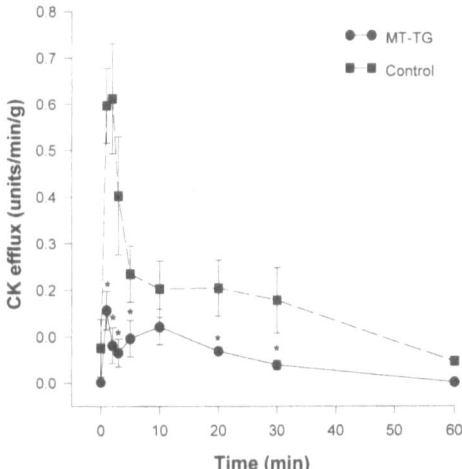

Figure 2. Creatine kinase (CK) activity in effluent buffer of the hearts isolated from transgenic (MT-TG) and non-transgenic control (Control) mice. CK release was measured at the last minute of the pre-ischemic period (0 min) and during the 60 min reperfusion. Each point represents the mean ± SD of 5 animals. *P < 0.01.

the aortic cannula. The infarct zone was examined as described previously [7]. As shown in Figure 3, the total volume of myocardial infarction was significantly greater in the transgene-negative hearts than in the transgene-positive hearts (p < 0.01).

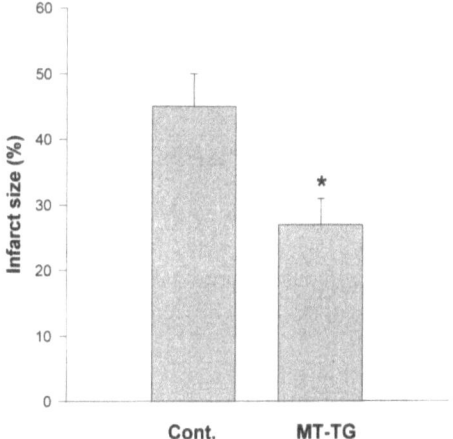

Figure 3. Myocardial infarction caused by 50 min of global ischemia and 90 min reperfusion delineated by tetra-zolium staining as described previously (7). The data presented are mean ± SD values from 5 mouse hearts of each group. *P < 0.01.

The results obtained above clearly defined the role of MT in cardiac protection against ischemia reperfusion injury. The regulation of MT expression has been well studied and oxidative stress has been shown to induce MT in multiple organ systems [1]. Because ischemia reperfusion *per se* produces oxidative stress to the heart, it is possible that MT in the heart is inducible under this stress condition. Particularly, it is important to know whether ischemic PC includes MT production. In this context, we used an open-chest mouse heart model [8]. Mouse hearts were processed four cycles of 5 min occlusion and 5 min reperfusion via ligation of left anterior descending coronary artery to produce ischemic PC. Cardiac MT was measured by the cadmium-hemoglobin affinity assay [3]. Compared to sham-operated mice (6.6 ± 1.8 µg/g tissue), MT was significantly ($p < 0.01$) increased to 15.8 ± 1.8 and 17.9 ± 5.0 µg/g 12 and 24 h, respectively, after ischemic PC.

The phenomenon "ischemic PC" was derived from a report that the amount of infarction resulting from a 40-min coronary occlusion in dogs could be marked reduced if the heart was "preconditioned" by four cycles of transient (5 min) coronary occlusion, each separated by 5 min of reperfusion [6]. Further studies have revealed that the protection offered by ischemic PC is characterized by two distinct phases; the early and transient one immediately following one or several brief period(s) of ischemia, and the delayed and long-lasting "second window of protection" occurring between about 12 h and 3 days after the brief period(s) of ischemia [9]. Several studies have demonstrated that the delayed phase offers sustained protection and is probably more important and more clinically relevant than the early phase [9].

The cellular mechanisms through which cardioprotection is manifested in the second window of protection are not clear. Several lines of evidence suggest that the protection could be related to the appearance of new proteins and/or to alterations in protein activities. It has been shown that manganese dependent superoxide dismutase (Mn-SOD) activity changes following ischemic preconditioning in canine myocardium [9]. A biphasic pattern of the enzyme activity change over a 24-h period was observed and correlated with the pattern of subsequent ischemic tolerance. Other studies have shown that myocardial content of HSP70i was elevated 24 h after preconditioning, a time when the second window of protection against ischemic injury was observed [9]. Although only these two candidate families have been identified to be altered after ischemic PC and related to the second window of protection, it is conceivable that more proteins could be involved in the protection because sublethal ischemia alters the regulation of a large number of proteins.

To determine whether the induced MT is involved in the ischemic PC, MT concentrations were elevated by treating the mice with cadmium chloride at 1.0 mg/kg for 24 h. MT concentrations in the cadmium-treated mouse hearts and in the saline-treated mouse hearts were 11.8 ± 1.6 and 5.9 ± 0.5 µg/g tissue ($p < 0.01$), respectively. These cadmium-treated mouse hearts and saline-treated controls were subjected to the Langendorff ischemia-reperfusion. The cadmium-treated heart displayed much better post-ischemic recovery of the suppressed contractile force ($p < 0.01$, data not shown), the same result as that obtained from the transgenic mouse heart (Fig. 1). CK activity in the collected perfusion effluent samples was measured. A dramatic reduction in CK release in the cadmium-treated hearts was also observed, especially during the first 5 min of reperfusion. The peak values of the CK activity in the effluent samples collected from the control and cadmium-treated hearts were 0.62 ± 0.12 and 0.13 ± 0.05 IU/min/g wet wt, respectively ($p < 0.01$).

Myocardial ischemia occurred when the perfusion flow rate of the Langendorff perfused heart was zero. The myocardial oxygen demand therefore exceeded oxygen supply under this condition. This situation resulted in cell injury as shown by the high level of CK activity measured in the effluent immediately after reperfusion and the repression of the contractile force. Reperfusion of the ischemic myocardium would restore oxygen. But it has been shown to produce another form of myocardial damage, termed "reperfusion injury". Myocardial infarction or cell death is more related to the reperfusion injury. In the present study, we have observed that MT functions in protection against both ischemia- and reperfusion-induced damage. It improved the recovery of the suppressed contractile force post ischemia and inhibited CK release from the ischemic myocardium upon reperfusion. It also reduced the size of the infarction zone produced by ischemia and reperfusion.

Possible mechanisms for MT functioning in cytoprotection against oxidative injury have been studied in vitro [1]. It has been suggested that the cysteine residues of the MT were the primary target for the reaction of hydroxyl radicals with this protein. Alternatively, the primary determinant of MT protection is the release of its associated metal, zinc, and the subsequent uptake of zinc by the membrane, since zinc protects against lipid proxidation and thereby stabilizes membranes. Another possibility is that MT chelates iron or otherwise mediates its conversion into a form that is not active as a Fenton reactant. MT may also donate an H atom to a cellular target radical on a DNA molecule, restoring it to an undamaged state. In the present study, the damage induced by ischemia-reperfusion would be more ascribed to the generation of reactive oxygen species as discussed above. The fact that MT functions as a potent scavenger of hydroxyl radical may be the primary mechanism by which this protein suppressed the ischemia-reperfusion induced detrimental effects.

It is important to note that ischemic PC induced MT synthesis in the heart and MT is likely involved in the ischemic PC evoked cardiac protection. Exploring the potential for MT protection against cardiac ischemia-reperfusion injury would therefore likely result in novel approaches to prevention of myocardial ischemic disease and would eventually be beneficial to the patient.

Acknowledgments
This study was supported in part by a National Institutes of Health Grant CA68125 and an American Heart Association Established Investigator Award (9640091N). YJK is a University Scholar of the University of Louisville.

References

1. Sato M, Bremner I (1993) Oxygen free radicals and metallothionein. *Free Radical Biol Med* 14: 325–337.
2. Abel J, de Ruiter N (1989) Inhibition of hydroxyl-radical-generated DNA degradation by metallothionein. *Toxicol Lett* 47: 191–196.
3. Kang YJ, Chen Y, Yu A, Voss-McCowan M, Epstein PN (1997) Overexpression of metallothionein in the heart of transgenic mice suppresses doxorubicin cardiotoxicity. *J Clin Invest* 100: 1501–1506.
4. Becker LC, Ambrosio G (1987) Myocardial consequences of reperfusion. *Prog Cardiovasc Dis* 30: 23–44.

5. Gross GJ, Farber NE, Hardman HF, Warltier DC (1986) Beneficial actions of superoxide dismutase and cata-lase in "stunned" myocardium of dogs. *Amer J Physiol* 25: H372-H377.
6. Murry CE, Jennings RB, Reimer KA (1986) Preconditioning with ischemia: A delay of lethal cell injury in ischemic myocardium. *Circulation* 74: 1126–1136.
7. Li G, Chen Y, Saari JT, Kang YJ (1997) Catalase overexpressing transgenic mouse heart is resistant to ischemia-reperfusion injury. *Amer J Physiol* 273: H1090–1095.
8. Michael LH, Entman ML, Hartley CJ, Youker KA, Zhu J, Hall SR, Hawkins HK, Berens K, Hallantyne CM (1995) Myocardial ischemia and reperfusion: a murine model. *Amer J Physiol* 269: H2147-H2154.
9. Baxer GF, Yellon DM (1996) Delayed myocardial protection following ischemic preconditioning. *Basic Res Cardiol* 91: 53–56.

Metallothionein and its importance relative to glutathione in cardiac protection against doxorubicin toxicity

Guang-Wu Wang, Hui-Yun Wu and Y. James Kang

Departments of Medicine, and Pharmacology and Toxicology, University of Louisville School of Medicine, 530 South Jackson St., Louisville, KY, 40202, USA

Summary. Controversial results have been reported regarding whether metallothionein (MT) functions in doxorubicin (DOX) detoxification in the heart. To determine unequivocally the role of MT in this cardiac function, a primary culture of myocardial cells was established from 3-day-old transgenic mouse hearts in which MT was overexpressed more than 40-fold higher than normal. DOX was added directly into the cultures to a final concentration of 0, 0.5, 1.0, 2.0 or 4.0 µM after the cells were cultured for 6 days. As compared to nontransgenic controls, transgenic myocardial cells displayed a significant resistance to DOX cytotoxicity as measured by the leakage of lactate dehydrogenase (LDH) 72 h after treatment, and by cell survival using a tetrazolium colorimetric (MTT) assay 12 h after treatment. To determine the importance of MT relative to glutathione (GSH) in cardiac protection against DOX toxicity, seven-week-old mice from the same transgenic line were treated with buthionine sulfoximine (BSO) by ip injection at 5 mmol/kg, two times with a 12-h interval, before being treated with DOX at a single dose of 15 mg/kg for 4 days. Cardiac GSH was depleted by about 70% in both transgenic and non-transgenic mice. The blood level of creatine kinase released from the heart, an important indicator of cardiac toxicity, was dramatically increased in the BSO-treated non-transgenic mice. This increase was completely inhibited in the BSO-treated transgenic mice. These results thus demonstrated that cardiac MT overexpressing transgenic mice are resistant to DOX cardiotoxicity, and this resistance remains the same under the condition of cardiac GSH depletion. Therefore, MT can compensate for the loss of protection from depletion of GSH in the heart. Selective modulations of decreasing DOX resistance in tumors by BSO and increasing cardioprotection by MT induction may provide an alternative approach to the improvement of DOX chemotherapeutic efficacy.

Cardiotoxicity is an important factor that limits the clinical use of doxorubicin (DOX), one of the most important anticancer agents. The proposed mechanism for DOX cardiotoxicity is the production of reactive oxygen species during its intracellular metabolism [1]. Recent studies have suggested that metallothionein (MT) plays a role in scavenging free radicals, thereby preventing oxidative injury [2]. This leads to the hypothesis that elevation of cardiac MT provides protection from DOX toxicity. However, controversial results regarding whether MT functions in DOX detoxification in the heart have been obtained from *in vivo* studies: preinduction of cardiac MT by bismuth subnitrate decreased DOX toxicity [3], but MT overexpressing transgenic mice were not resistant to DOX cardiotoxicity [4]. Our recent studies using a specific cardiac MT overexpressing transgenic mice have shown that MT significantly protected the heart from DOX-induced cardiomyopathy and creatine phosphokinase release from the heart [5]. To further define the role of MT in cardiac protection against DOX toxicity, the present study was undertaken to directly examine the effect of MT elevation on DOX toxicity using primary myocardial cell cultures established from the transgenic neonatal mice in which cardiac MT was more than 40-fold higher than normal.

Acquired drug resistance in tumor cells is another major impediment for the clinical application of DOX. Trials are ongoing to use buthionine sulfoximine (BSO) to deplete glutathione (GSH) content in tumors, whose elevation was found to contribute to the drug resistance.

However, BSO also decreases GSH content in the heart, enhancing DOX cardiotoxicity. MT may be an important factor in cardiac protection against DOX toxicity. The present study was also aimed to determine whether MT can compensate for the loss of protection from GSH depletion by BSO in the heart.

A new procedure for culturing ventricular cardiomyocytes from neonatal mice was established by modifying the methods used for neonatal rat and fetal mouse cardiomyocyte cultures. Briefly, three day old neonatal transgenic FVB mice were sacrificed by cervical dislocation. Hearts were removed and ventricles were kept in cold Hanks' buffered saline solution (HBSS without Ca^{2+} and Mg^{2+}, pH 7.4). The ventricles were washed and minced into small fragments. The cells were dissociated at 37 °C with 5% CO_2 and 95% air for 15 min with an enzyme solution (0.25% trypsin in HBSS without Ca^{2+} and Mg^{2+}, pH 7.4). The digestion was added to an equal volume of cold HBSS with Ca^{2+} and Mg^{2+} (pH 7.4) until all tissues were dissociated. After centrifugation, the cells were suspended in the FBS-MEM (MEM supplemented with 20% FBS, 100 U/ml penicillin and 100 μg/ml streptomycin). To exclude non-muscle cells, the cells were preplated in tissue culture dishes at 37 °C for 2 h in FBS-MEM under a water-saturated atmosphere of 5% CO_2 with 95% air. The suspended cells were then collected and plated at a density of 1.5×10^5 cells/ml medium and incubated under the same conditions as above. After being cultured for 24 h, almost all the cardiomyocytes attached and were spread out on the substrata of the dishes and beat spontaneously. Tissue culture medium was changed and repeated thereafter every 3 days. MT concentrations in the transgenic neonatal hearts were 190.6 ± 19.4 μg/g tissue, compared to 4.6 ± 0.6 μg/g tissue in the nontransgenic controls, measured by a cadmium-hemoglobin affinity assay as described previously [5]. GSH concentrations were not different between these two types of neonatal hearts.

DOX was added directly into the cultures to a final concentration of 0, 0.5, 1.0, or 2.0 μM after the myocardial cells were cultured for 6 days. The release of cytoplasmic enzyme lactate dehydrogenase (LDH) into the MEM from the cells and the LDH retained in the cells were determined at 0, 6, 9, 24, 48, and 72 h after DOX treatment. As shown in Figure 1, DOX of 2 μM significantly increased the LDH activity in the medium of nontransgenic cardiomyocyte cultures after 24 h treatment. This effect was inhibited in the transgenic cardiomyocytes. These changes in the extracellular LDH activity were correlated with the retention of the enzyme activity in the cells as shown in Figure 2. To further determine the effect of MT on DOX-induced toxicity in the cardiomyocytes, cell viability was examined by a short-term MTT assay. In a 96-well microplate, 2.5×10^4 cells/well were incubated in 100 μl culture medium for 36 h before DOX exposure. Following 12 h of drug treatment, cell survival was determined as described previously [6]. As shown in Figure 3, the transgenic cardiomyocytes again displayed much better survival.

These results obtained from the cultured cardiomyocytes thus provide direct evidence for the role of MT in cardiac protection against DOX toxicity. To examine whether the cardioprotective effect of MT against DOX toxicity, as assessed by changes in the serum creatine phosphokinase (CPK) activity, would be maintained under reduced cardiac GSH levels brought on by BSO treatment, transgenic mice overexpressing cardiac MT about 40-fold higher than normal were used. These animals along with non-transgenic controls were treated with BSO by ip injection at 5 mmol/kg, two times with a 12-h interval. Four hours after the second injection,

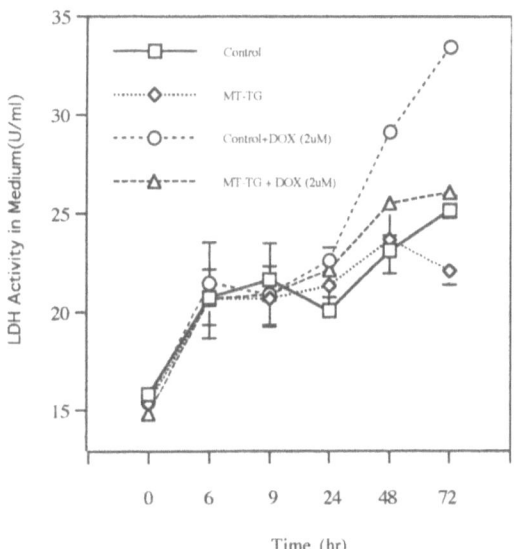

Figure 1. Time course of the effect of DOX on LDH release from the cultured MT overexpressing transgenic cardiomyocytes and nontransgenic controls. LDH activities were measured from triplicate cultures from each treatment at each time point and the data represent mean ± SD values.

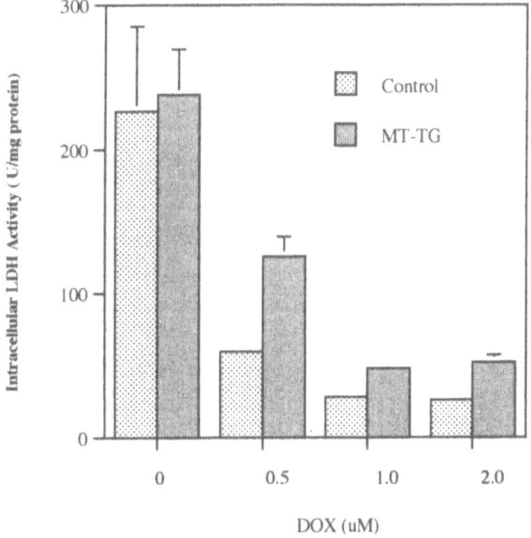

Figure 2. Dose dependent effect of DOX on LDH leakage from the cultured MT overexpressing transgenic cardiomyocytes and nontransgenic controls. Intracellular LDH activities were measured from triplicate cultures. Data represent mean ± SD values.

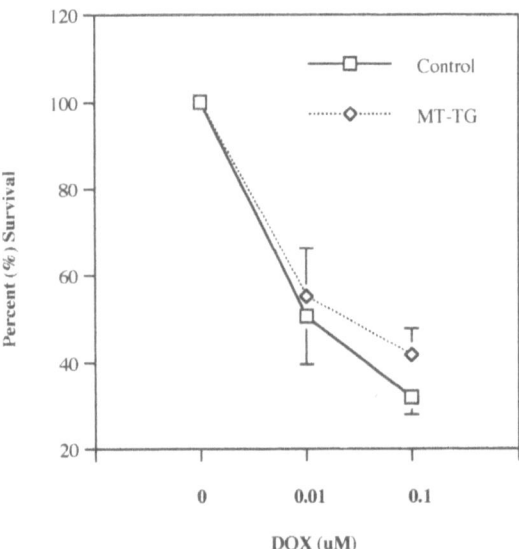

Figure 3. Effect of DOX on cell survival of the MT overexpressing transgenic cardiomyocytes and nontransgenic controls measured by MTT assay.

cardiac GSH and MT concentrations were assayed. GSH concentrations in the heart were significantly ($p < 0.01$) decreased to the same low level in both transgenic and nontransgenic mice, from 1.04 ± 0.05 to 0.28 ± 0.06 and 1.03 ± 0.13 to 0.26 ± 0.04 µmol/g tissue in transgenic and nontransgenic mouse hearts, respectively. The MT concentrations in these hearts were not altered by the BSO treatment.

At the same time (4 h after the second injection of BSO) these animals were treated with DOX by a single ip injection at 20 mg/kg. Four days after the DOX treatment, the serum CPK activity was measured. As shown in Figure 4, the MT overexpressing transgenic mouse heart was significantly ($p < 0.01$) resistant to the DOX-induced increase in the serum CPK activity. BSO treatment dramatically increased the DOX-elevated serum CPK activity in the nontransgenic mice. However, the inhibitory effect of MT elevation on this toxic effect by DOX was not altered by BSO treatment. These results suggest that MT may compensate for the loss of GSH in the heart in protection of this organ against DOX toxicity.

The results presented here, together with that obtained from our previous studies using a specific cardiac MT overexpressing transgenic mouse model [5], clearly demonstrated the important role of MT in cardiac protection against DOX toxicity. The fact that the study using MT overexpressing transgenic mice [4] failed to show a protective effect of MT on DOX cardiotoxicity, may suggest that a threshold level is required for MT to play its role. This level is clearly pharmaceutically achievable because bismuth subnitrate treated mice had induced cardiac MT levels effective for protection against DOX cardiotoxicity [3]. It is important to search for physiological inducers to increase cardiac MT in humans. If such inducers are available,

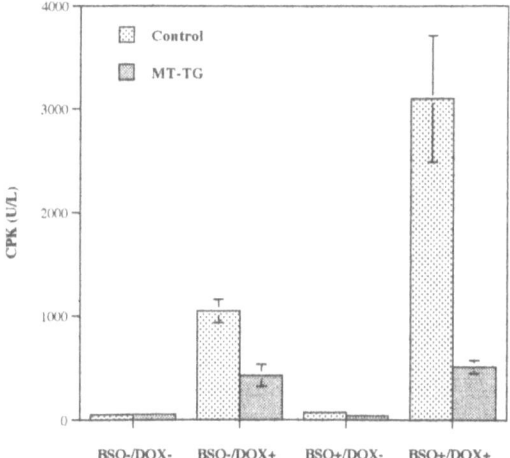

Figure 4. Effect of MT on DOX-elevated serum CPK activity in both BSO-treated and non-BSO-treated trans-genic mice in comparison with that of nontransgenic controls. Serum CPK was measured using a CK-20 kit from Sigma. DOX significantly increased serum CPK level in the non-transgenic control mice (BSO-/DOX+). MT over-expressing transgenic mouse hearts displayed significant resistance to the increase in serum CPK. BSO treatment significantly increased the DOX-elevated serum CPK activity (BSO+/DOX+), but did not alter the protective effect of MT.

selective modulations of decreasing DOX resistance in tumors by BSO and increasing cardio-protection by MT induction may be an alternative approach to enhance DOX chemotherapeu-tic efficacy.

Another important issue is how MT functions in cardiac protection against DOX toxicity *in vivo*. Studies *in vitro* have demonstrated that MT is a very potent hydroxyl radical scavenger. The rate constant for its reaction with hydroxyl radical *in vitro* is about 340-fold higher than that for GSH [7]. However, the half-life of the hydroxyl radical is extremely short, and MT can only be effective if it is located sufficiently close to the site of production of the radical to inter-act with it before reaction with other cellular components *in vivo*. A recent study using HL-60 cells has demonstrated that thiolate groups in the MT fraction were preferential attacking tar-gets of hydrogen peroxide relative to other pools of sulfhydryl groups such as GSH and pro-tein thiols [8]. This study suggests that MT reaction with hydrogen peroxide, which is much less reactive than hydroxyl radical and diffusible, may be a major protective action of MT *in vivo*. This will be investigated in our future studies.

Acknowledgments
This study was supported in part by a National Institutes of Health Grant CA68125 and an American Heart Association Established Investigator Award (9640091N). YJK is a University Scholar of the University of Louisville.

References

1. Myers CE, McGuire WP, Liss RH, Grotzinger K, Young RC (1977) Adriamycin: The role of lipid peroxidation in cardiac toxicity and tumor response. *Science* 197: 165–167.
2. Sato M, Bremner I (1993) Oxygen free radicals and metallothionein. *Free Radical Biol Med* 14: 325–337.
3. Satoh M, Naganuma A, Imura N (1988) Involvement of cardiac metallothionein in prevention of adriamycin-induced lipid peroxidation in the heart. *Toxicology* 53: 231–237.
4. DiSilvestro RA, Liu J, Klaassen CD (1996) Transgenic mice overexpressing metallothionein are not resistant to adriamycin cardiotoxicity. *Res Commun Mol Pathol Pharmacol* 93: 163–170.
5. Kang YJ, Chen Y, Yu A, Voss-McCowan M, Epstein PN (1997) Overexpression of metallothionein in the heart of transgenic mice suppresses doxorubicin cardiotoxicity. *J Clin Invest* 100: 1501–1506.
6. Hatcher EL, Alexander JM, Kang YJ (1997) Decreased sensitivity to adriamycin in cadmium-resistant human lung carcinoma A549 cells. *Biochem Pharmacol* 53: 747–754.
7. Thornalley PJ, Vašák M (1985) Possible role for metallothionein in protection against radiation-induced oxidative stress. Kinetics and mechanism of its reaction with superoxide and hydroxyl radicals. *Biochim Biophys Acta* 827: 36–44.
8. Quesada AR, Byrnes RW, Krezoski SO, Pettering DH (1996) Direct reaction of H_2O_2 with sulfhydryl groups in HL-60 cells: Zinc-metallothionein and other sites. *Arch Biochem Biophys* 334: 241–250.

Metallothionein metal composition is a determinant of its induction of the mitochondrial permeability transition

Cuthbert O. Simpkins, Tracy Lloyd, Sai Li and Samuel C. Balderman

Erie County Medical Center, Department of Surgery, State University of New York at Buffalo, Buffalo, NY 14215, USA

Introduction

Metallothionein is rapidly induced by a broad range of stress mediators, such as interleukin 1, interleukin 6, interferon gamma, steroids, and catecholamines [1]. Mitochondria play a significant role in stress through the production of ATP, oxygen free radicals [2], maintenance of intracellular calcium homeostasis [3], and the initiation of apoptosis [4]. Our initial finding was that apometallothionein was capable of transferring electrons to cytochrome c [5]. Since cytochrome c is a component of the mitochondrial respiratory chain we hypothesized that metallothionein would modulate mitochondrial function. As these experiments progressed we were encouraged by the finding of Sakurai et al that the organelle with the highest amount of metallothionein was the mitochondrion [6].

The mitochondrion has an outer and inner membrane. The space between the two membranes is the intermembrane space. The inner compartment is termed the matrix. The outer membrane is permeable to molecules of up to 10,000 Da, such as metallothionein [7]. The inner membrane contains pores which open upon stimulation and admit molecules of up to 1500 Da [8]. The maintenance of the proton gradient across the inner membrane is essential to the production of ATP and the translocation into mitochondria of cytoplasmically translated proteins. Opening of the inner membrane pore is accompanied by loss of this gradient.

Sucrose, which has a molecular weight of 342 Da, can also pass through the inner membrane pore into the matrix. The sucrose is accompanied by water leading to mitochondrial swelling. This swelling is detectable as a decrease in the absorbance at 540 nm and has been shown to occur via the opening of the inner membrane pore [8, 9]. For these experiments we utilized a buffer consisting of 5 mM HEPES, 0.1 mM phosphate, 5 mM succinate, 200 mM sucrose, pH 7.4, 25 degrees C. Mitochondria were isolated from homogenized rat liver by differential centrifugation [10]. Metallothionein 1 and 2 from rabbit liver was obtained from SIGMA.

Results

Pore Opening: We found that at physiological concentrations (6–100 μM) metallothionein 1 induced pore opening (Fig. 1). Figure 2 shows the mean % of maximal swelling ± SE caused by various concentrations of metallothionein 1 using mitochondria isolated from six rats.

Maximal swelling was defined as that caused by 100 µM calcium chloride. Overall there was no increase in swelling at concentrations beyond 50 µM metallothionein 1. The combined effect of calcium and metallothionein 1 were simply additive and not synergistic (data not shown).

We proceeded to determine whether metal composition was a factor in the effect of metallothionein. We compared metallothionein 2 which was 7% zinc and 0.5% cadmium by weight (MT2-Zn) to metallothionein 2 which was 5.3% cadmium and 0.7% zinc (MT2-Cd). We found that the initial phase of swelling of MT2-Zn terminated sooner than that for MT2-Cd. However, MT2-Cd caused a greater amount of swelling compared to MT2-Zn. Zinc or cadmium alone at concentrations as high as 10–7 M had no effect (data not shown). These concentrations far exceed those that would be present in equilibrium with metallothionein which has a low dissociation constant for binding to these metals (< 10–13) [11].

We directly compared MT2-Zn with MT2-Cd at 25 µM and 50 µM. Mitochondria from the same rat were used for each direct comparison. In each comparison the initial phase of swelling was shorter for MT2-Zn and the maximal swelling observed when all swelling had ceased was greater for MT2-Cd (Tab. 1).

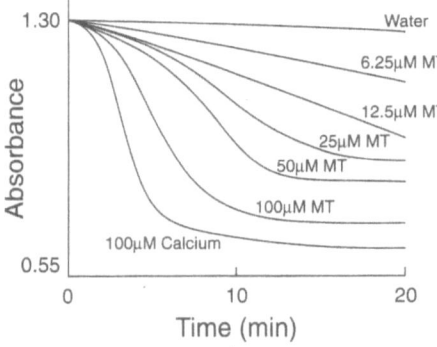

Figure 1. Representative traces of the effect of various Metallothionein 1 concentrations and 100 µM calcium chloride on mitochondrial swelling. Absorbance was at 540 nm.

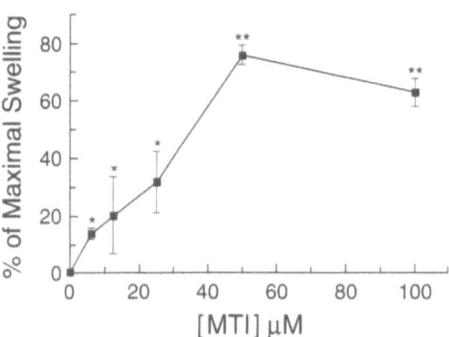

Figure 2. Effect of metallothionein 1 (0–100 µM) on mitochondrial swelling as means of six experiments ±SE. Baseline swelling induced by the addition of water and mixing is subtracted out. * = $p < 0.05$ and ** = $p < 0.01$ for significant differences from water alone, which is set at 0.

Table 1. Comparison of MT2-Cd and MT2-Zn at 25 µM and 50 µM

[MT 1]	Time for initial phase (sec)		maximum change in absorbance	
	MT2-Cd	MT2-Zn	MT2-Cd	MT2-Zn
25 µM	633	244	.458	.306
	555	256	.426	.360
	644	267	.327	.262
50 µM	467	133	.458	.262
	561	211	.480	.338
	533	233	.338	.262

Initial phase is the time before the rate of swelling became close to zero. Maximum change in absorbance is the decrease in absorbance at 540 nm that occurred by the end of the observation period

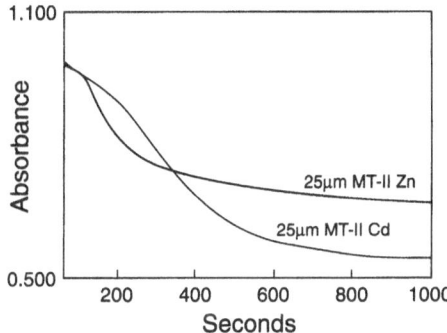

Figure 3. Differential kinetics of swelling induced by metallothinein 2 of differing metal compositions.

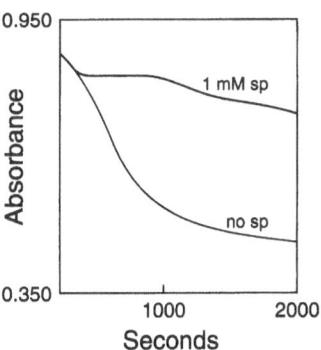

Figure 4. Inhibition of 50 µM metallothionein 1 by 1 mM spermine. The line that extends lower is the result of adding 50 µM metallothionein 1. The top line is the result of adding 50 µM metallothionein 1 100 s after the addition of 1 mM spermine.

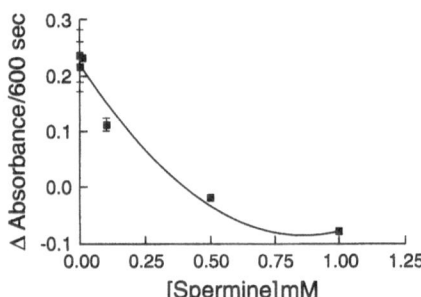

Figure 5. Concentration dependence of the inhibition by spermine of swelling caused by 50 μM metallothionein 1. Data are the result of two independent assays. The variance between the two assays for most points was too small for error bars to be distinguishable from the squares.

Spermine is an aliphatic polyamine. Polyamines are essential to the growth of mammalian cells. They carry a high positive charge and are involved in DNA replication and mRNA translation [12]. Spermine has been previously shown to inhibit calcium-induced mitochondrial swelling [13]. We found that metallothionein-induced mitochondrial swelling was also inhibited by spermine, and that this inhibitory effect was concentration dependent (Figs 4 and 5). The physiological concentration of spermine is 1 mM [14].

Conclusion

We have shown that physiological concentrations of metallothionein 1 and 2 induce mitochondrial swelling which is likely to be due to the opening of the mitochondrial inner membrane pore. We also we found that the kinetics of this differential effect of metallothionein is dependent on its metal composition. We do not yet know the mechanism of this action. Moreover, there remains the question of whether our *in vitro* observations occur in cells. Nonetheless ou r findings raise many interesting possibilities. For example it is possible that mitochondria are regulated by the balance between metallothionein and the polyamines of which spermine is one. Differential regulation of mitochondria by metallothionein may result from varied metal composition. Additional experiments have revealed that metallothionein inhibits mitochondrial oxygen consumption and that this action is synergistic with that of calcium (unpublished observations). Because of the many cellular processes involving mitochondria, the rapid inducibility of metallothionein and its long half life once induced it is possible that this fascinating metalloprotein plays a fundamental role in cellular function, especially in stress.

References

1. Kägi JHR (1991) Overview of Metallothionein. *In*: J. Riordan, B. Vallee (eds): Methods in Enzymology. Academic Press, London, Vol. 205: 613–625.
2. Miesel R, Murphy MP, Kroger H (1996) Enhanced Mitochondrial Radical Production in Patients Which Rheumatoid Arthritis Correlates with Elevated Levels of Tumor Necrosis Factor Alpha in Plasma. *Free Radical Res* 25: 161–169.
3. Kristan T, Siesjo BK (1996) Calcium-Related Damage in Ischemia. *Life Sci* 59: 357 367.
4. Zamzani N, Maarchette P, Castedo M, Zanin C, Vayssiere J, Petit P, Kroemer G (1995) Reduction in mitochondrial potential constitutes an early irreversible step of programmed lymphocyte death in vivo. *J Exp Med* 181: 1661–1672.
5. Simpkins C, Eudaric P, Torrence C, Yand Z (1993) Metallothionein I Reduction of Cytochrome c. *Life Sci* 53: 1975–1980.
6. Sakurai H, Nakajima K, Kamada H, Satoh H, Otaki N, Kimura M, Kawano K, Hagino T (1993) Copper-Metallothionein Distribution in the Liver of Long-Evans Cinnamon Rats: Studies on Immunohistochemical Staining, Metal Determintion, Gel Filtration and Eletron Spin Resonance Spectroscopy. *Biochem Biophys Res Commun* 192: 893–898.
7. Klingenberg M (1975) Energetic Aspects of transport of ADP and ATP Through the Mitochondrial Membrane. *Ciba Foundation Symposium* 31: 105–124.
8. Bernardi P, Veronese P, Petronilli V (1993) Modulation of the Mitochondrial Cyclosporin A-Sensitive Permeability Transition Pore. *J Biol Chem* 268: 1011–1016.
9. Hunter FE, Smith EE, (1967) Measurement of Mitochondrial Swelling and Shrinking-High Amplitude In RWEastabrook MEPullman (eds) (1967) Methods in Enzymology. Academic Press, London, Vol. 10: 689–696.
10. Johnson D, Lardy H (1967) Isolation of Liver or Kidney Mitochondria *In*: RW Eastabrook, ME Pullman (eds) Methods in Enzymology. Academic Press, London, Vol. 10: 94–96.
11. Li T-Y, Kraker AJ, Shaw CF, Petering DH (1980) Ligand Substitution Reactions of Metallothioneins with EDTA and Apocarbonic Anhydrase. *Proc Natl Acad Sci USA* U.S. 77: 6334–6338.
12. Molitor J, Pace W, Stunz L, Louie S, Ashman R (1994) Induction of Ornithine Decarboxylase Activity in Mouse B Lymphocytes. *Int Immunol* 6: 1777–1784.
13. Tassani V, Biban C, Toniello A, Siliprandi D (1995) Inhibition of Mitochondrial Permeability Transition by Polyamines and Magnesium. *Biochem Biophys Res Commun* 207: 661–667.
14. Chaffee RRJ, Salganicoff L, Marine R, Rochelle RH, Schultz EL 91977) Polyamine Effects on Succinate-linked and Alphaketoglutarate-linked Rat Liver Mitochondrial Respiration. *Biochem Biophys Res Commun* 77: 1009–1016.

Metallothionein IV
C. Klaassen (ed.)
© 1999 Birkhäuser Verlag Basel/Switzerland

Protective effect of metallothionein on DNA damage induced by hydrogen peroxide and ferric ion-nitrilotriacetic acid

Kyong-Son Min, Kayo Nishida, Yukiko Nakahara and Satomi Onosaka

Department of Nutrition, Kobe Gakuin University, Kobe, 651-21, Japan

Metallothionein (MT) is a low molecular weight protein in which cysteine residues comprise 1/3 of the amino acids. Heavy metals induce synthesis of this protein directly and then bind to it. It is widely believed that MT functions in the detoxification of heavy metals and in the metabolism of essential metals. Beside metals, an injection of a prooxidative agent, such as menadione, carbon tetrachloride, paraquat, and acetaminophen, induces MT synthesis [1–4]. In addition, oxidative stress, e.g., radiation and high oxygen tension, causes induction of MT synthesis [5, 6]. It has been proposed that MT might play an antioxidant role in the cell, since MT has many sulfhydryl groups as does GSH [7]. Abel and de Ruiter [8] has demonstrated that each cysteine residue in MT is 38.5-fold more effective at protecting DNA from hydroxyl radicals ($^{\cdot}$OH) attack than the GSH cysteine *in vitro*. The Cd-resistant Chinese hamster cells, which are enriched in MT and its gene [9], were significantly more resistant than parental cells to oxidative stress by extracellular H_2O_2 or a mixture of H_2O_2 and superoxide anion (O_2^-) generated by xanthine oxidase plus acetaldehyde [10]. However, transfected cells with expressing vectors containing MT genes did not become more resistant than the parental cells to the lethal effect of ionizing radiation [11]. Moreover, neither overexpression nor underexpression of MT resulted in differential resistance to the killing action of H_2O_2 [12]. The antioxidant properties of MT have not been defined in terms of whole cellular toxicity induced by H_2O_2. Chubatsu and Meneghini [12] demonstrated that, contrary to the cell-killing effect, the DNA-strand-breaking effect of oxidative stress was influenced by the cellular MT content. This suggests that its properties should be investigated in terms of molecular toxicology, because it has been shown that MT is a protein that can be found in not only the cytosol but also the nucleus [13, 14]. MT seems to have an excellent capacity for scavenging $^{\cdot}$OH in the nucleus.

Oxygen free radical species such as H_2O_2 and O_2^- are produced in mammalian cells during normal aerobic metabolism. These radical species are considered to be involved in many important biological reactions. However, excess generation of oxygen-derived species *in vivo* results in damage to many biological molecules such as DNA [15], and induces various diseases such as cancer, AIDS, chemical induced liver damage, rheumatoid arthritis, several autoimmune diseases and aging [16–8]. However, O_2^- or H_2O_2 do not modify DNA bases directly under physiologically relevant conditions [19]. It has been proposed that much of the toxicity of these species to living organisms is due to the iron-dependent generation of $^{\cdot}$OH, and/or other powerful oxidants, by what is essentially Fenton chemistry [20].

$$Fe^{2+} + H_2O_2 \longrightarrow Fe^{3+} + {}^{\cdot}OH + OH^-$$

$$Fe^{3+} + 2 H_2O_2 \longrightarrow Fe^{2+} + {}^{\cdot}OH + H^+ + H_2O + O_2$$

Inoue and Kawanishi proposed that ${}^{\cdot}OH$ could also formed by reaction of ferric ion-nitrilotri-acetic acid complex (Fe^{3+}-NTA) with H_2O_2 [21].

$$Fe^{3+}\text{-NTA} + H_2O_2 \longrightarrow Fe^{2+}\text{-NTA} + O_2^- + 2H^+$$

$$Fe^{3+}\text{-NTA} + O_2^- \longrightarrow Fe^{2+}\text{-NTA} + O_2$$

$$Fe^{2+}\text{-NTA} + H_2O_2 \longrightarrow Fe^{3+}\text{-NTA} + {}^{\cdot}OH + OH^-$$

NTA is a synthetic chelating agent that has been used in various countries as a constituent of detergents. NTA is a powerful carcinogen and promoter of carcinogenesis in experimental animals, causing acute nephrotoxicity, renal carcinoma and lipid peroxidation [22]. Incubation of the ferric ion chelate with H_2O_2 at pH 7.4 generated a reactive species able to produce chemical modifications of the bases in DNA that are very similar to those produced in DNA by the ${}^{\cdot}OH$ generating systems [23]. The pattern of base modification clearly indicates that the reactive species generated by Fe^{3+}-NTA/H_2O_2 is unlikely to be anything other than ${}^{\cdot}OH$ [23]. The high rate of ${}^{\cdot}OH$ generation would make the Fe^{3+}-NTA/H_2O_2 system a good system for DNA damage by ${}^{\cdot}OH$.

In the present study, to investigate whether MT can directly protect DNA damage induced by ${}^{\cdot}OH$, the effects of MTs on ras DNA strand scission induced incubation of Fe^{3+}-NTA with H_2O_2 at pH 7.4 (Fe^{3+}-NTA/H_2O_2) were studied *in vitro*.

Inhibition effect of MTs on deoxyribose cleavage induced by ${}^{\cdot}OH$

The ability of Fe^{3+}-NTA/H_2O_2 system to produced ${}^{\cdot}OH$ was confirmed by using the deoxyribose assay. The incubation of 2-deoxy-D-ribose with Fe^{3+}-NTA/H_2O_2 resulted in much higher rate of deoxyribose cleavage than that with Fe^{3+}/H_2O_2. Compared with other complexes used, Fe^{3+}-NTA produced by far the highest cleavage. Fe^{3+}-NTA/H_2O_2 is the best ${}^{\cdot}OH$ production system using ferric ion chelate complex.

It has been shown that this deoxyribose cleavage is inhibited by several well established scavengers of ${}^{\cdot}OH$, namely hypotaurine, mannitol, and dimethyl sulfoxide [23]. MTs and GSH as well as these scavengers inhibited the deoxyribose cleavage in Fe^{3+}-NTA/H_2O_2 system. Each cysteine residue in MT is twice as effective as GSH cysteine residues. This effect was observed in the incubation with Zn-MT and Cd-MT, not with Zn^{2+} nor Cd^{2+} at same concentration. MT II was shown to have a higher capacity to inhibit deoxyribose cleavage induced by ${}^{\cdot}OH$ than MT I regardless of metals bound to MT. These results indicate that MTs have a higher capacity for scavenging ${}^{\cdot}OH$ than GSH.

Inhibition effect of MTs on DNA strand scission induced by ˙OH

[^{32}P]-labeled Hras-1 DNA fragment (348 bp) was incubated with Fe^{3+}-NTA (molecular ratio 1:4) and 2.8 mM H_2O_2 in 10 mM KH_2PO_4-KOH buffer (pH 7.4) at 37 °C for 60 min. After incubation of DNA with Fe^{3+}-NTA/H_2O_2, DNA strand scission was observed at higher concentration of Fe^{3+}-NTA/H_2O_2 than 50 µM. The damage to the suger-phosphodiester chain was predominant over the chemical modifications of the base in DNA. The DNA strand scission was inhibited by both Zn-MT and Cd-MT (Fig. 1). MT pretreated with EDTA and N-ethylmaleimide to alkylate its sulfhydryl groups did not inhibit the DNA damage. These results indicate that MT can protect DNA from ˙OH attack and the thiol group of MT may be involved in its antioxidant properties. It is interesting to note that, contrary to MTs, GSH stimulated the DNA strand scission. This effect of GSH was contrary to the result in deoxyribose cleavage assay. It suggests that only MT, and not GSH, can be a ˙OH scavenger in the nucleus.

Protective effect of MT on the plaque formation of bacteriophage M13mp18 inhibited by Fe^{3+}-NTA/H_2O_2

After bacteriophage DNA (M13 mp18) was incubated with Fe^{3+}-NTA/H_2O_2, the DNA molecule was transfected into base-mismatch repair deficient strain *E. coli* BMH71-18 mutS. This assay

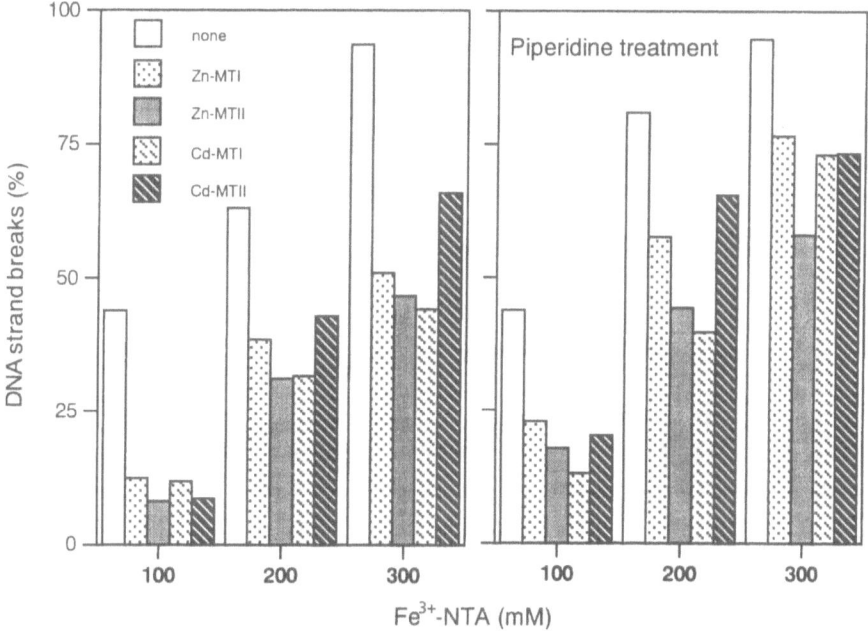

Figure 1. Protective effect of MTs on Hras-1 DNA strand scission by incubation with Fe^{3+}-NTA/H_2O_2.

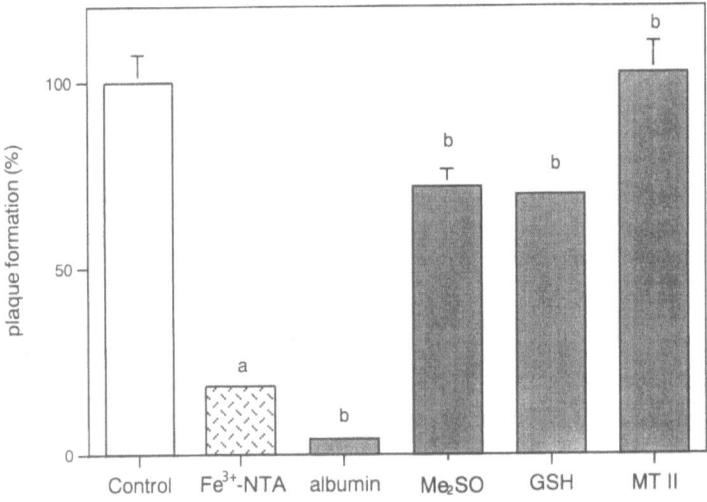

Figure 2. Effect of MT and radical scavengers on plaque formation of bacteiophage M13mp18 inhibited by Fe^{3+}-NTA/H_2O_2. [a]Significantly different from control group (p < 0.01). [b]Significantly different from Fe^{3+}-NTA group (p < 0.01).

can detect DNA damage induced by much lower concentrations of Fe^{3+}-NTA than the DNA fragment assay. Its plaque formation was inhibited according to the increasing concentrations of Fe^{3+}-NTA (1–50 μM), but no mutant plaque was detected. It has been demonstrated that 8-hydroxyguanine, an abundant form of oxidative DNA damage, caused mutation, especially the base substitutions [24]. Though the incubation with Fe^{3+}-NTA/H_2O_2 produced chemical modifications of the bases [23], the inhibition of plaque formation seem to be caused by DNA strand scission. Preincubation of MT with Fe^{3+}-NTA/H_2O_2 recovered its plaque formation to the control level (Fig. 2). MT was more effective than other scavengers such as GSH (Fig. 2). These results suggested that MT may play directly a protective role against oxidative DNA damage.

DNA strand scission by MT

Recent work has shown that Cu-containing MT might function as a prooxidant [25–27]. Sakurai et al. [25] suggest that the development of hepatitis in LEC rats is related to an unusual accumulation of Cu with MT in the liver, because generation of ˙OH is observed in the system of hepatic Cu-MT from LEC rats and H_2O_2 at pH 8.0. ˙OH were produced in the presence of H_2O_2 in proportion to the amount of cuprous ions liberated from MT [26]. Moreover, it is suggested that reactions of LEC rat liver Cu (I)-MTs with $HgCl_2$ cause the liberation of cuprous ions from MTs, followed by reaction with oxygen, leading to ˙OH formation through a Fenton-type Haber-Weiss reaction [27]. Conversely, these results indicate that only Cu-MT can generate ˙OH. However, MT, containing Cd and Zn, nicked plasmid DNA [28]. We also showed that Zn-MT or Cd-MT (7 atoms/molecule) induced DNA strand scission by itself at 100 times higher concentration than that observed to inhibit of DNA strand scission by Fe^{3+}-NTA/H_2O_2 system. GSH had a similar activity of DNA strand scission. These results indicate that SH groups of the various cysteine residues of MT may be involved in the DNA strand scission activity.

Based on our results, it is suggested that MT might have opposite effects to DNA in nuclei, independent of metals bound to MT. At low concentration, MT can be an excellent hydroxyl radical scavenger due to the SH groups of the various cysteine residues of MT and protect DNA from hydroxyl radical attack in nuclei. On the contrary, MT can induce DNA strand scission by itself at high concentration. This effect may be involved in Cd-related carcinogenesis and hepatic injury in LEC rats. These opposite effects on DNA suggests that MT may control the living cell depending on the concentration of MT in the nuclei.

References

1. Oh SH, Deagen JT, Whanger PD and Weswig PH (1978) Biological function of metallothionein. V. Its induction by various stresses. *Amer J Physiol* 234: E282–285.
2. Sato M, Ohtake K, Mizunuma H and Nagai Y (1989) Metallothionein-1 accumulation in the rat lung following a single paraquat administration. *Toxicol Lett* 45: 41–47.
3. Warmser U and Calp D (1988) Increased levels of hepatic metallothionein in rat and mouse after injection of acetaminophen. Toxicol 53: 323–329.
4. Min K, Terano Y, Onosaka S and Tanaka K (1992) Induction of metallothionein synthesis by menadione or carbon tetrachloride is independent of free radical production. *Toxicol Appl Pharmacol* 113: 74–79.
5. Shiraishi N, Aono K and Utsumi K (1983) Increased metallothionein content in rat liver induced by X irradiation and exposure to high oxygen tension. *Radiat Res* 95: 298–302.
6. Hidalgo J, Camomany L, Brass M, Gavey JS and Armario A (1988) Metallothionein response to stress in rats: Role in free radical scavenging. *Amer J Physiol* 255: E518–524.
7. Thornalley PJ and Vašák M (1985) Possible role for metallothionein protection against radiation-induced oxidative stress: Kinetics and mechanism of its reaction with superoxide and hydroxyl radicals. *Biochim Biophys Acta* 827: 36–44.
8. Abel J and Ruiter N (1989) Inhibition of hydroxyl-radical-generated DNA degradation by metallothionein. *Toxicol Lett* 47: 191–196.
9. Beach LR and Palmiter RD (1981) Amplification of the metallothionein-I gene in cadmium-resistant mouse cells. *Proc Natl Acad Sci USA* 78: 2110–2114.
10. Mello-Filho AC, Chubatsu LS and Meneghini R (1988) V79 Chinese-hamster cells rendered resistant to high cadmium concentration also become resistant to oxidative stress. *Biochem J* 256: 475–479.
11. Kaina B, Lohrer H, Karin M and Herrlich P (1990) Overexpressed human metallothionein IIA gene protects Chinese hamster ovary cells from killing by alkylating agents. *Proc Natl Acad Sci USA* 87: 2710–2714.
12. Chubatsu LS and Meneghini R (1993) Metallothionein protects DNA from oxidative damage. *Biochem J* 291: 193–198.
13. Banerjee D, Onosaka S and Cherian MG (1982) Immunohistochemical localization of metallothionein in cell nucleus and cytoplasm of rat liver and kidney. *Toxicology* 24: 95–105.
14. Tsujikawa K, Suzuki N, Sagawa K, Itoh M, Sugiyama T, Kohama Y, Otaki N, Kimura M and Mimura T (1994) Induction and subcellular localization of metallothionein in regenerating rat liver. *Eur J Cell Biol* 63: 240–246.
15. Halliwell B and Aruoma OI (1991) DNA damage by oxygen-derived species. Its mechanism and measurement in mammalian systems. *FEBS Lett* 281: 9–19.
16. Cerutti P (1985) Prooxidant states and tumor promotion. *Science* 237: 375–381.
17. Halliwell B and Cross CE (1991) Reactive oxygen species, antioxidants and acquired immunodeficiency syndrome. *Arch Intern Med* 157: 29–31.
18. Jaeschke H (1995) Mechanisms of oxidant stress-induced acute tissue injury. *Proc Soc Exp Biol Med* 209: 104–111.
19. Aruoma OI, Halliwell B and Dizdaroglu M (1989) Iron ion-dependent modification of bases in DNA by the superoxide radical-generating system hypoxanthine/xanthine oxidase. *J Biol Chem* 264: 13024–13028.
20. Halliwell B and Gutteridge (1990) Role of free radicals and catalytic metal ions in human disease: an overview. *Methods Enzymol* 186: 1–85.
21. Inoue S and Kawanishi S (1987) Hydroxyl radical production and human DNA damage induced by ferric nitrilotriacetate and hydrogen peroxide. *Cancer Res* 47: 6522–6527.

22. Ebina Y, Okada S, Hamazaki S, Ogino F, Li JL and Midorikawa O (1986) Nephrotoxicity and renal cell carcinoma after use of iron- and aluminum-nitrilotriacetate complexes in rats. *J Nat Cancer Inst* 76: 107–113.
23. Aruoma OI, Halliwell B, Gajewski E and Dizdaroglu M (1989) Damage to the bases in DNA induced by hydrogen peroxide and ferric ion chelates. *J Biol Chem* 264: 20509–20512.
24. Cheng KC, Cahill DS, Kasai H, Nishimura S and Loeb LA (1992) 8-Hydroxyguanine, an abundant form of oxidative DNA damage, causes G——T and A——C substitutions. *J Biol Chem* 267: 166–172.
25. Sakurai H, Satoh H, Hatanaka A, Sawada T, Kawano K, Hagino T and Nakajima K (1994) Unusual generation of hydroxyl radicals in hepatic copper-metallothionein of LEC (Long-Evans cinnamon) rats in the presence of hydrogen peroxide. *Biochem Biophys Res Commun* 199: 313–318.
26. Suzuki KT, Rui M, Ueda J and Ozawa T (1996) Production of hydroxyl radicals by copper-containing metallothionein: roles as prooxidant. *Toxicol Appl Pharmacol* 141: 231–237.
27. Nakamura M, Nakayama K, Shishido N, Yumino K and Ohyama T (1997) Metal-induced hydroxyl radical generation by Cu+-metallothionein from LEC rat liver. *Biochem Biophys Res Commun* 231: 549–552.
28. Muller T, Schuckelt R and Jaenicke L (1991) Cadmium/zinc-metallothionein induces DNA strand breaks *in vitro*. Arch. Toxicol 65: 20–26.

Metallothionein IV
C. Klaassen (ed.)
© 1999 Birkhäuser Verlag Basel/Switzerland

Effect of metallothionein on cardiac reperfusion injury in rats

Seiki Minamide, Masashi Okamoto, Fumio Naganuma, Katsuyuki Nakajima and Tadashi Suzuki

Department of Laboratory Sciences, Gunma University School of Health Sciences, 3-39-15, Showa machi, Maebashi, Gunma, 371 Japan

Summary. There are some reports that metallothionein, a low molecular metal-binding protein, inhibits cell injury caused by free radicals. It is known that zinc induces metallothionein in the liver and heart. Metallothionein may inhibit myocardial reperfusion injury caused by free radicals.
Eighty two adult Wistar male rats were used in this research. The animals were treated with zinc acetate or metallothionein and the left coronary artery was ligated for 90 min following reperfusion. The myocardial infarct size was measured morphometrically by nitro-blue tetrazolium staining. The amount of metallothionein in the myocardium was measured by radioimmunoassay, and plasma zinc levels were ascertained by atomic absorption spectrophotometry.
Metallothionein and zinc acetate administered before ligation reduced the infarct size in reperfision injury in rat. However, zinc acetate administered just before and after ligation had no effect on reperfusion injury. The levels of metallothionein in animals pretreated with metallothionein or zinc were higher than that of control. These results suggest that metallothionein took effect directly, and zinc indirectly, through induction of metallothionein at the site of myocardial reperfusion injury.

Introduction

Biological roles of metallothionein, a low molecular metal-binding protein, include decreasing the effects of acute heavy metal toxicity and involvement in the metabolism of zinc and copper. Metallothionein has also been reported to inhibit cell injury caused by free radicals [1, 2]. That zinc induces metallothionein in various organs is also well known.

Both myocardial cells and vascular endothelial cells are involved as sources and/or targets of free radical generation, during myocardial ischemia, infarction and subsequent reperfusion. Free radicals played important roles in myocadial tissue injury, particularly during the phase of myocardial reoxygenation [3]. Metallothionein may inhibit myocardial reperfusion injury. This paper examines the effect of metallothionein and zinc on myocardial reperfusion injury in the rat model.

Materials and methods

Animals

Eighty two adult Wistar male rats purchased from Charles River Japan Co ltd., (Kanagawa, Japan) weighing 250–280 g were divided into 4 groups. In the first group (Zn-1), 23 animals were administered zinc acetate (Sigma chemical company, St. Louis, Mo. USA) subcutaneously at a dose of 400 µmol/kg for 3 days before coronary artery ligation. In the second group (Zn-2), 9 animals were administered zinc acetate subcutaneously at the same dose just before liga-tion and at 12 and 24 h thereafter. In the third group (MT), 16 animals were administered met-allothionein-II (Sigma chemical company, St. Louis, Mo. USA) intravenouly at a dose of 5.0 mg/kg just before the coronary artery ligation. In the fourth group (control), 34 animals pre-pared as controls were not treated with zinc or metallothionein either before or after coronary artery ligation.

The animals had the left coronary arteries occluded for 90 min following reperfusion, accord-ing to the method of Selye et al. [4]. During the experiment an electrocardiogram was used to monitor the development of myocardial infarction by coronary artery occlusion. Animals were kept in stainless steel mesh rat cages at $24 \pm 2\,°C$ with $55 \pm 10\%$ relative humidity and venti-lation 10–5 times per hour. Artificial light was provided for 12 h between 8:00 and 20:00. The animals were fed commercial diet (Oriental Yeast Co. ltd., Kanagawa, Japan) and given tap water *ad libitum*.

Drug

Zinc acetate and metallothionein-II were dissolved in distilled water at concentrations of 1.0 m mol/ml and 10.0 mg/ml, respectively.

Methods

All animals were euthanitized 48 h after coronary artery ligation. The hearts were removed for pathological examination and analysis of metallothionein levels in cardiac tissue, and blood was collected for analysis of plasma zinc levels. In morphometric studies, the left ventricles

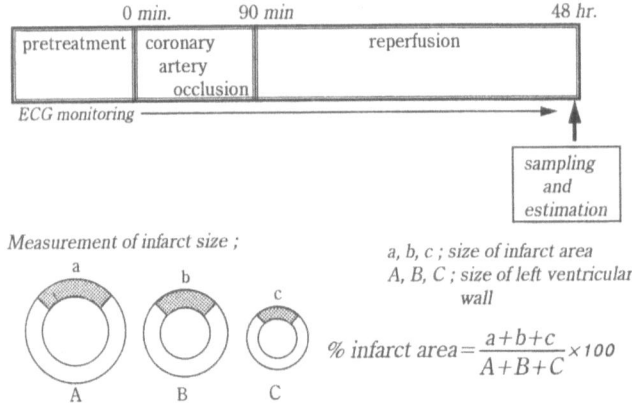

Figure 1. Protocol and formula for calculating total area of infarct size.

were cut transversely into 4 portions and stained with nitro-blue tetrazolium (NBT) and the myocardial infarct area in the left ventricular wall was measured. The total infarct size was calculated by the formula shown in Figure 1. The metallothionein levels in the myocardium were measured by radioimmunoassay, and plasma zinc levels by atomic absorption spectrophotometry [5]. The cardiac tissue was stained with hematoxylin-eosin according to the conventional procedure. Immunohistochemistry for metallothionein in the myocardium was carried out by the streptavidin-biotin peroxidase complex method using rabbit anti-metallothionein antibodies.

Student's t-test was used for statistical analysis of data for the different groups of animals and data expressed as means ± SE

Results

The macroscopic findings showed the infarct area clearly as that area unstained by NBT in the left ventricular wall (Fig. 2). In morphmetric studies, the infarct size in Zn-1 was smaller than that of control (27.2 ± 2.8% vs. 39.7 ± 2.2%, p < 0.01), and also in MT (27.8 ± 3.0% vs. 39.7 ± 2.2%, p < 0.01). On the other hand, the infarct size in Zn-2 was equivalent to that of control (38.5 ± 3.9% vs. 39.7 ± 2.2%, NS) (Fig. 3). The plasma zinc levels of Zn-1 and Zn-2 were significantly higher than that of control (288.0 ± 16.0 vs. 92.3 ± 6.1 µg/dl, p < 0.01. 364.0 ± 23.0 vs. 92.3 ± 6.1 µg/dl, p < 0.01, respectively). The metallothionein levels in the myocardium in Zn-1 and MT were significantly higher than that in control (140.0 ± 17.0 vs. 64.0 ± 10.0 µg/100g, p < 0.01. 227.0 ± 45.0 vs. 64.0 ± 10.0 µg/100g, p < 0.01, respectively). Immunohistochemistry for metallothionein showed deep staining in the myocardial ischemic region in Zn-1 (Fig. 4).

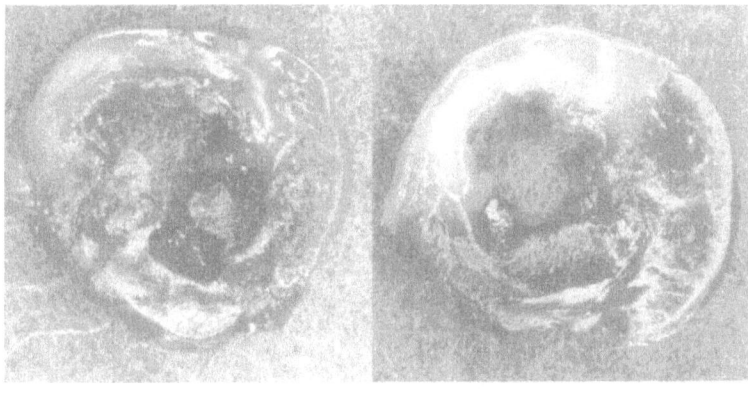

Zn-1   Control

Figure 2. Macroscopic findings in the infarct region in the left ventricle by staining with NBT. The infarct zone is shown as the unstained area.

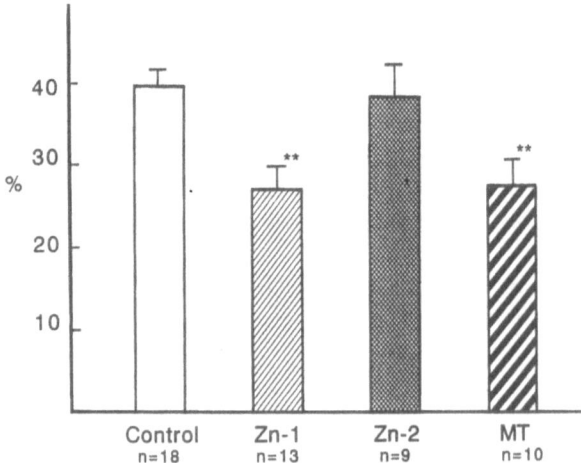

Figure 3.Myocardial infarct size in the left ventricle **Significantly different from control($p < 0.01$). The values are mean ± SE.

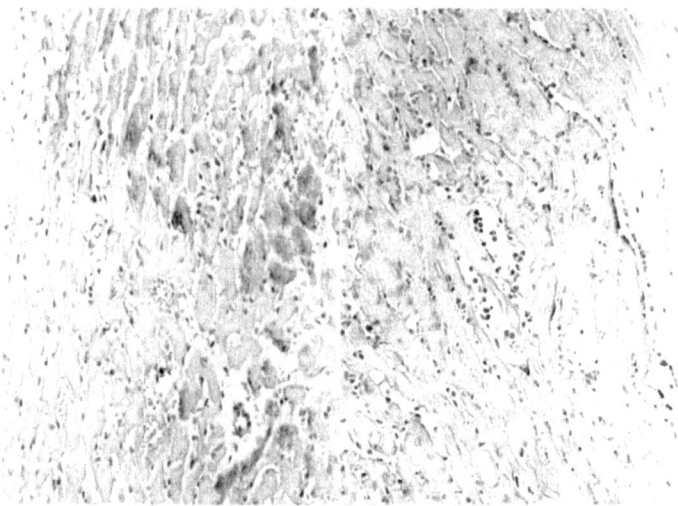

Figure 4. Metallothionein immunohistochemistry of myocardial tissue of rat treated with zinc.

Discussion

Recently, reperfusion therapy in the acute phase of myocardial infarction has been established as a primary therapy [6]. Now. myocardial injury caused by reperfusion itself, so called "reperfusion injury", has become clinically important. The pathogenic mechanism of reperfusion injury is still unclear. Recently there has been a growing body of evidence for the role of free radicals in mediating myocardial tissue injury during myocardial ischemia, particularly during the phase of myocardial reoxygenation [7]. Metallothionein experimentally protects cell injury caused by free radicals, such as adriamycin cardiotoxicity and carbon tetrachloride-induced hepatotoxicity [1, 2, 8]. It may not be unreasonable to assume that the action of metallothionein as a radical scavenger contributes to the protection against cell injury in these pathophysiologic conditions. Biosynthesis of metallothionein can be induced by many compounds, particularly metals such as zinc, copper and cadmium. It may also be induced by stress. In the past, there has been no report studying the protective effect of metallothionein against myocardial reperfusion injury. In this study, we have investigated the protective effect of metallothionein and also of preinduction of metallothionein by zinc, on myocardial infarction in the rat model. The data of this study clearly indicates that pretreatment with zinc or metallothionein reduced the size of the myocardial infarction. A significant increase in the levels of cardiac metallothionein was observed in rats pretreated with zinc and metallothionein. On the other hand, it is known that zinc contributes to the stability of various biomembranes, thus affecting cell and tissue reactivity towards noxious agents. It has also been reported that zinc protects the heart against isoproterenol induced myocardial necrosis [9]. This effect may result from antagonizing various ca 2+ dependant effects. In this study, zinc treatment just before ligation did not reduce myocardial infarction size. It is suggested that the protective effect of zinc pretreatment may result from the induction of metallothionein

References

1. Naganuma, A. Satoh, M. Imura, N. Specific reduction of toxic side effects of Adriamycin by induction of metallothionein in mice, *Jpn J Cancer Res* 79: 406–411, 1988.
2. Thornally, P.J. and Vašák, M. Possible role for metallothionein in protect- ion against radiation-induced oxidative stress. Kinetics and mechanism of its reaction with superoxide and hydrogen radicals. *Biochim Biophys Acta* 827: 36–44,1985.
3. Simpson, P.J. and Lucchesi, B.R. Free radicals and myocardial ischemia and reperfusion injury. *J Lab Clin Med* 110: 13–30, 1987.
4. Selye,H. Bajusz,E. Grasso,S. Mendell,P. Simple techniques for the surgical occlusion of coronary vessels in the rat. *Angiology*, 11: 398–407, 1960.
5. Nakajima, K. Suzuki, K. Otaki, N. and Kimura, M. Epitope mapping of metallothionein antibodies. *Methods Enzymol* 205: 174–189, 1991.
6. Rentrop, K.P. Blanke, H. Karsch, K.R. Wiegand, V. Kostering, H. Oster, H. Leitz, K. Acute myocardial infarction: Intracoronary application of nitroglycerine and streptokinase. *Clin Cardiol* 2: 354–363, 1979.
7. McCord, J.M. Oxygen-derived free radicals in postischemic tissue injury. *N Engl J Med* 312: 159–163, 1985.
8. Cagen,S.Z. and Klaassen,C.D. Protection of carbon tetrachloride-induced hepatotoxicity by zinc: role of metallothionein. *Toxicol Appl Pharmacol* 51: 107–116, 1979.
9. Chvapil,M. and Owen,J.A. Effect of zinc on acute and chronic isoproterenol induced heart injury. *J Mol Cell Cardiol* 9: 151–159, 1977.

Metallothionein IV
C. Klaassen (ed.)
© 1999 Birkhäuser Verlag Basel/Switzerland

Susceptibility to metals and radical-inducing chemicals of metallothionein-null mice

Masahiko Satoh and Chiharu Tohyama

Environmental Health Sciences Division, National Institute for Environmental Studies, Tsukuba, Ibaraki 305-0053, Japan

Introduction

Two research groups produced metallothionein (MT) gene knock-out mice (MT-null mice) with null mutation of MT-I and -II genes (Tab. 1) [1, 2]: MT-null (Aus) mice with a mixed genetic background of 129 Ola and C57BL/6 strains [1] and MT-null (J) mice with a genetic background of 129/Sv strain [2]. Using these mice it has been shown that MT-null mice have an increased sensitivity to cadmium [1–3], zinc [4] and paraquat [5], but not cadmium-MT complex [3]. In *in vitro* studies, primary cultured MT-null cells derived from MT-null mouse embryos were found to be more sensitive to cadmium, oxidative stress, mutagens and anticancer drugs [6, 7]. Hepatocytes from MT-null mice have also been shown to have increased sensitivity to cadmium and oxidative stress, but not to alkylating agents [8].

Since we have been investigating physiological roles of MT using MT-null mice, we present recent results in our laboratory. MT-null (Aus) mice and MT-null (J) mice were kindly provided by Dr. K. H. A. Choo and purchased from Jackson Laboratory (Bar Harbor, ME, USA), respectively. The MT levels, which were measured by radioimmunoassay, in the various tissues such as liver, kidney, lung, heart, pancreas and bone marrow of these MT-null mice were below the limit of detection (< 0.2 µg/g tissue) and could not be induced by cadmium, mercury, zinc or bismuth.

Mercury

Inorganic mercury that causes severe kidney damage after acute and chronic exposure, is well-established as a toxicant to human health. We studied the susceptibility of MT-null (Aus) mice

Table 1. Charactersitics of MT-null mice

Mice	Targetted gene	Strain	Coat color
MT-null (Aus)	MT-I and -II	C67BL/6 and 129 Ola	Nonaguti (black) and Agouti
MT-null (J)	MT-I and -II	129/Sv	Agouti

to the renal toxicity of inorganic mercury [9]. C57BL/6J mice and 129/Sv mice were used as wild-type control mice because MT-null (Aus) mice were of a mixed genetic background of 129 and C57BL/6 strains. Blood urea nitrogen (BUN) and plasma creatinine values, indicators of renal toxicity, were significantly elevated by injection of mercuric chloride in MT-null (Aus) mice compared to those in C57BL/6J mice. In contrast, 129/Sv mice were as susceptible as MT-null (Aus) mice in terms of mercury nephrotoxicity. Thus, sensitivity to inorganic mercury in the MT-null (Aus) mice could not be evaluated because of a remarkable strain difference between C57BL/6J mice and 129/Sv mice.

In the production of MT-null (Aus) mice, embryonic stem cells derived from 129 Ola strain and blastocysts from C57BL/6 strain were used to produce chimeric mice, which were crossed with C57BL/6 strain. In many laboratories, C57BL/6 strain mice have been generally used as control to MT-null (Aus) mice. When MT-null (Aus) mice are to be used, it is necessary to confirm that there is no significant strain difference between the above two strains with regard to the particular toxicity of interest; alternatively, use B6129Sv mice that were produced to have a mixed genetic background of C57BL/6 and 129/Sv as wild-type control mice.

Since B6129Sv mice were not available at the time of this study, we utilized MT-null (J) mice, which have a genetic background of 129/Sv strain only, and 129/Sv mice as wild-type control mice for further study. Mercuric chloride-caused renal toxicity, which was estimated by BUN and plasma creatinine values. The toxicitywas markedly enhanced in MT-null (J) mice compared to 129/Sv mice (Tab. 2). The susceptibility of the kidney of MT-null (J) mice to inorganic mercury was further confirmed by histopathological observations: in MT-null (J) mice, marked morphological changes, such as degeneration and necrosis of proximal tubular cells and dilation of tubular lumen, were produced mercury injection at a dose of 20 μmol/kg. In contrast, almost no damage was observed in the renal tubules of 129/Sv mice after the same mercury dose. Therefore, we conclude that MT-null (J) mice have increased sensitivity to inorganic mercury.

In both MT-null (J) and 129/Sv mice, renal mercury levels increased considerably by 4 h after injection of mercuric chloride, followed by a subsequent decrease. At 4 h, similar amounts of mercury were retained in the kidneys of MT-null (J) mice and 129/Sv mice. At 24 and 72 h after injection, however, mercury accumulation in the kidney was significantly lower in MT-null (J)

Table 2. Renal toxicity of mercuric chloride in MT-null (J) mice and 129/Sv mice

Dose of HgCl$_2$	BUN (mg/100 ml)		Plasma creatinine (mg/100 ml)	
	129/Sv	MT-null(J)	129/Sv	MT-null(J)
-	27.8 ± 2.2	30.3 ± 5.1	0.45 ± 0.06	0.48 ± 0.05
20	27.0 ± 3.8	$133.3 \pm 35.1^{a, b}$	0.45 ± 0.06	$1.00 \pm 0.27^{a, b}$
30	71.3 ± 8.5^{a}	$331.7 \pm 93.9^{a, b}$	0.65 ± 0.10^{a}	$1.57 \pm 0.21^{a, b}$
40	265.0 ± 75.0^{a}	died	1.17 ± 0.25^{a}	died

Mice were given s.c. injections of mercuric chloride (20, 30 and 40 μmol/kg). Renal toxicity was determined 3 days after injection. The values are mean ± SD for four mice.
[a]Significantly different from the corresponding untreated group (P <0.05).
[b]Significantly different from the specified group (P <0.05).

mice than in 129/Sv mice. These results indicate that MT does not affect the initial distribution of mercury to the kidney, but plays a major role in the retention of this metal in the kidney.

Cadmium

Cadmium, an environmentally harmful metal like mercury, has a potential to cause various toxic effects such as liver and kidney damage and lung tumors. Cadmium accumulates rapidly in the liver and kidney after injection in experimental animals. However, the possible role of MT in the uptake and retention of cadmium in these organs, has not been studied. Table 3 shows the distribution of cadmium in the liver of MT-null (Aus) mice. When mice were administered with cadmium chloride at a single s.c. dose of 1.0 mg Cd/kg, cadmium accumulated mainly in the liver by 4 h after injection, with no significant difference between the MT-null (Aus) mice and C57BL/6J mice used as wild-type control mice, although possible strain differences in the cadmium distribution between C57BL/6J mice and 129/Sv mice remain to be studied [10]. There was a marked elimination of cadmium from the liver of MT-null mice by 21 days after administration, compared with relatively slow cadmium elimination in the C57BL/6J mice. Under the condition that no significant liver or kidney damage was observed, MT was considered to play a significant role in the retention of cadmium in the liver but not in the uptake of this metal. Liu et al. [11] also demonstrated results similar to our findings. Moreover, it has been indicated, by using MT-null mice, that MT synthesis is essential for endotoxin-induced hepatic zinc accumulation [12]. The utilization of MT-null mice might help to clarify possible roles of MT in the metabolism of heavy metals.

Table 3. Cadmium concentrations in the liver of MT-null (Aus) mice and C57BL/6J mice at various times after injection of cadmium chloride

Time after injection	Cadmium (μg/g tissue)	
	C57BL/6J	MT-null(Aus)
-	0.12 ± 0.05	0.11 ± 0.03
4 h	8.89 ± 0.71	8.78 ± 0.23
1 day	10.96 ± 0.92	$5.86 \pm 0.74^*$
7 days	11.69 ± 0.64	$4.95 \pm 0.47^*$
21 days	9.40 ± 0.16	$2.92 \pm 0.47^*$

Mice were given s.c. injections of cadmium chloride (1.0 mg Cd/kg). Cadmium concentrations in the liver were determined at specified times after injection. The values are mean ± SD for three mice.
*Significantly different from the corresponding untreated group (P <0.05).

Streptozotocin

Streptozotocin (STZ) causes experimental diabetes by destroying beta cells of the islets of Langerhans. It was demonstrated that pretreatment with zinc, an MT inducer, prevents the development of diabetes caused by STZ in rats [13]. We investigated possible involvement of zinc and/or MT in the pathogenesis of STZ-caused insulin-dependent diabetes mellitus (IDDM) in MT-null mice [14]. It was found that Zn pretreatment has a unique inhibitory effect on IDDM development in MT-null mice in contrast to no marked effect in C57BL/6J mice, suggesting that Zn ions free from MT molecules exerted this protective effect. The highest Zn dose (10 mg/kg) fully suppressed development of hyperglycemia in both types of mice. Pretreatment with Zn partially led to recovery of superoxide dismutase activities in the liver and pancreas in which STZ administration suppressed superoxide dismutase activity in both types of mice. The present study suggests that Zn plays an important protective role in the pathogenesis of IDDM, although the possible involvement of MT in the protection of STZ-caused IDDM cannot be excluded.

Cisplatin

Cisplatin (cis-DDP) is one of the most effective antineoplastic agents, but the clinical use of cis-DDP is limited by its side effects such as severe nephrotoxicity. Because no significant difference in the cis-DDP nephrotoxicity between C57BL/6J mice and 129/Sv mice was observed, C57BL/6J mice were used as wild-type control. We found that MT-null (Aus) mice developed cis-DDP-induced kidney damage (characterized by BUN, plasma creatinine and histopathology) more conspicuously than C57BL/6J mice, and that the same trend was true for lethal toxicity [15].

On the other hand, the cis-DDP nephrotoxicity was prevented by pretreatment with zinc sulfate or bismuth nitrate in the C57BL/6J mice, but not in the MT-null mice (Tab. 4). The renal MT levels in the MT-null (Aus) mice were not increased by either zinc or bismuth treatment

Table 4. Effect of pretreatment with metal compounds on renal toxicity of cisplatin in MT-null (Aus) mice and C57BL/6J mice

Treatments	BUN (mg/100 ml)		Plasma creatinine (mg/100 ml)	
	C57BL/6J	MT-null(Aus)	C57BL/6J	MT-null(Aus)
Contol	25.8 ± 5.3	28.8 ± 5.6	0.43 ± 0.05	0.45 ± 0.06
Cisplatin	98.5 ± 23.3^a	$262.5 \pm 45.7^{a, b}$	0.78 ± 0.17^a	$1.25 \pm 0.19^{a, b}$
Zn + cisplatin	30.8 ± 7.3	$256.3 \pm 46.8^{a, b}$	0.48 ± 0.05	$1.20 \pm 016^{a, b}$
Bi + cisplatin	28.5 ± 8.9	$251.3 \pm 61.8^{a, b}$	0.45 ± 0.06	$1.20 \pm 016^{a, b}$
Zn	31.0 ± 6.9	29.3 ± 6.8	0.43 ± 0.05	0.45 ± 0.06
Bi	29.5 ± 6.9	27.5 ± 6.8	0.48 ± 0.05	0.43 ± 0.05

Mice were given s.c. injections of zinc sulfate (100 μmol/kg) or bismuth nitrate (50 μmol/kg) once a day for 2 days. These metals-injected mice were treated i.p. with cisplatin (40 μmol/kg) 24 h after the last injection of each metal compound. Renal toxicity was determined 4 days after the cisplatin injection. The values are mean ± SD for four mice.
[a]Significantly different from the corresponding untreated group (P <0.05).
[b]Significantly different from the specified group (P <0.05).

although they were increased in the C57BL/6J mice by these treatments. These results indicate that MT plays a defensive role against severe kidney damage caused by cis-DDP, and that the protective effect of metals against cis-DDP nephrotoxicity is due to preinduction of MT synthesis in the target organs.

X-ray radiation

Irradiation by γ, X and UV rays produces active oxygen species in exposed target sites in living cells. MT has been proposed to protect the cells from the oxidative stress by irradiation. We have studied possible involvement of MT in the protection against X-ray irradiation. The number of total leukocytes, an indicator of bone marrow injury, in both MT-null (Aus) mice and their wild-type control mice (B6129Sv mice), a strain with mixed genetic backgrounds of C57BL/6 and 129/Sv, were significantly decreased by X-ray irradiation in a dose dependent manner. X-ray irradiation was found to cause bone marrow injury in MT-null (Aus) mice compared to wild-type control mice. Interestingly, the number of total leukocytes in X-irradiated MT-null (Aus) mice were significantly lower than those of wild-type control mice at relatively low doses between 0.1 and 1.0 Gy but not at and above 3.0 Gy, suggesting that MT may act as an effective antioxidative protein under physiological conditions.

Chemical carcinogen (DMBA)

7,12-dimethylbenz[a]anthracene (DMBA), a polycyclic aromatic hydrocarbon, is a potent skin carcinogen. Oxidative stress has been suggested to play a role in the processes of DMBA tumorigenesis. Since MT can be involved in protection against oxidative stress, we examined the tumorigenicity in the skin of MT-null (Aus) mice after treatment with DMBA [16]. One week after the topical application of DMBA on the shaved dorsal skin, MT-null (Aus) mice had a strong irritant response (erosion and ulceration) in the skin. The DMBA treatment resulted in an increase in papilloma in MT-null (Aus) mice in a dose-dependent manner 14 weeks after the treatment. On the other hand, no change was observed in the skin of wild-type control mice (B6129Sv mice) by the same DMBA treatment. These results indicate that MT-null mice have an increased sensitivity to skin tumorigenesis produced by DMBA.

Conclusion

Our present studies using MT-null mice clearly indicate that MT is an important protective factor against the renal toxicity caused by inorganic mercury and cis-DDP, bone marrow injury induced by X-irradiation and skin tumorigenesis produced by DMBA. Moreover, it was found that MT plays a major role in the retention of mercury and cadmium in the kidney and liver, respectively. On the other hand, it was indicated by using MT-null mice that zinc ions can suppress development of IDDM without major participation of MT. Thus MT-null mice are thought to be a very useful model to study the biological functions and physiological roles of MT.

Acknowledgments

We thank Drs. N. Nishimura (Sydney University, Australia), A. Naganuma (Tohoku University, Japan), K. Shibuya (Kitasato Institute Medical Center Hospital, Japan), M.D. Apostolova, B. Zhang, H. Sone, Y. Aoki (National Institute for Environmental Studies, Japan) and other colleagues for their contribution to the studies compiled in this article.

References

1. Michalska AE, Choo KHA (1993) Targeting and germ-line transmission of a null mutation at the metallothionein I and II loci in mouse. *Proc Natl. Acad Sci USA* 90: 8088–8092.
2. Masters BA, Kelly EJ, Quaife CJ, Brinster RL, Palmiter RD (1994) Targeted disruption of metallothionein I and II genes increases sensitivity to cadmium. *Proc Natl Acad Sci USA* 91: 584–588.
3. Liu J, Liu Y, Michalska AE, Choo KHA, Klaassen CD (1996) Metallothionein plays less of a protective role on cadmium-metallothionein-induced nephrotoxicity than in cadmium chloride-induced hepatotoxicity. *J Pharmacol Exp Ther* 276: 1216–1223.
4. Kelly EJ, Quaife CJ, Froelick GJ, Palmiter RD (1996) Metallothionein I and II protect against zinc deficiency and zinc toxicity in mice. *J Nutr* 126: 1782–1790.
5. Sato M, Apostolova MD, Hamaya M, Yamaki J, Choo KHA, Michalska AE, Kodama N, Tohyama C (1996) Susceptibility of metallothionein-null mice to paraquat. *Environ Toxicol Pharmacol* 1: 221–225.
6. Lazo JS, Kondo Y, Dellapiazza D, Michalska AE, Choo KHA, Pitt BR (1995) Enhanced sensitivity to oxidative stress in cultured embryonic cells from transgenic mice deficient in metallothionein I and II genes. *J Biol Chem* 270: 5506–5510.
7. Kondo Y, Woo ES, Michalska AE, Choo KHA, Lazo JS (1995) Metallothionein null cells have increased sensitivity to anticancer drugs *Cancer Res* 55: 2021–2023.
8. Zheng H, Liu J, Liu Y, Klaassen CD (1996) Hepatocytes from metallothionein I and II knock-out mice are sensitive to cadmium- and tert-butylhydroperoxide-induced cytotoxicity. *Toxicol Lett* 87: 139–145.
9.. Satoh M, Nishimura N, Kanayama Y, Naganuma A, Suzuki T, Tohyama C (1997) Enhanced renal toxicity by inorganic mercury in metallothionein-null mice. *J Pharmacol Exp Ther* 283: 1529–1533.
10. Tohyama C, Satoh M, Kodama N, Nishimura H, Choo A, Michalska A, Kanayama Y, Naganuma A (1996) Reduced retention of cadmium in the liver of metallothionein-null mice. *Environ Toxicol Pharmacol* 1: 213–216
11. Liu J, Liu Y, Michalska AE, Choo KHA, Klaassen CD (1996) Distribution and retention of cadmium in metallothionein I and II null mice. *Toxicol Appl Pharmacol* 136: 260–268
12. Philcox JC, Coyle P, Michalska AE, Choo KHA, Rofe, AM (1995) Endotoxin-induced inflammation does not cause hepatic zinc accumulation in mice lacking metallothionein gene expression. *Biochem J* 308: 543–546
13. Yang J, Cherian MG (1994) Protective effects of metallothionein on streptozotocin-induced diabetes in rats. *Life Sci* 55: 43–51
14. Apostolova MD, Choo KHA, Michalska AE, Tohyama C (1997) Analysis of the possible protective role of metallothionein in streptozotocin-induced diabetes using metallothionein-null mice. *J Trace Elements Med Biol* 11: 1–7
15. Satoh M, Aoki Y, Tohyama C (1997) Protective role of metallothionein in renal toxicity of cisplatinum. *Cancer Chemother Pharmacol* 40: 358–362

Metallothionein IV
C. Klaassen (ed.)
© 1999 Birkhäuser Verlag Basel/Switzerland

Metallothionein-I/II knockout mice are sensitive to acetaminophen-induced hepatotoxicity

Jie Liu[1], Yaping Liu[1], Curtis D. Klaassen[1], Stacey E. Shehin-Johnson[2], Angela Lucas[2] and Steven D. Cohen[2]

[1]*University of Kansas Medical Center, Kansas City, KS 66160, USA*
[2]*University of Connecticut, Storrs, CT 06269, USA*

Introduction

Metallothioneins (MTs) are a group of low-molecular weight, cysteine-rich, metal-binding proteins. MT has an unusually high sulfhydryl content, and thus it has been suggested to react with free radicals and electrophiles [1–2].

Acetaminophen is a widely used analgesic drug. Acetaminophen is safe at therapeutic doses, however, at higher doses it produces hepatic injury in both human and experimental animals [3]. The hepatic injury produced by acetaminophen is generally thought to be initiated by a reactive toxic intermediate, *N*-acetyl-*p*-benzoquinoneimine (NAPQI), formed by cytochrome P450 [3]. Following an overdose of acetaminophen, both glucuronidation and sulfation are saturated, and the formation of NAPQI is increased, which produces liver injury *via* depletion of cellular glutathione, covalent binding to cellular proteins, recruitment and activation of macrophages, initiation of oxidative stress, oxidation of protein thiols, alteration of calcium homeostasis, and damage to nuclear DNA. Thus, both covalent and noncovalent interactions are involved in acetaminophen-induced hepatotoxicity [3–4].

In the present study, we hypothesize that intracellular MT provides cellular protection against acetaminophen toxicity. Binding of the toxic metabolite NAPQI to MT, or reducing NAPQI-induced oxidative damage may represent additional detoxication mechanisms against acetaminophen toxicity. To test this hypothesis, MT-I and II knock-out (MT-null) [5–6] mice were utilized.

MT-null mice are more susceptible to acetaminophen-induced lethality and hepatotoxicity

Acetaminophen administration produced dose-dependent lethality. MT-null mice were more sensitive than control mice to acetaminophen-induced lethality. For example, acetaminophen injection at a dose of 350 mg/kg, caused 75% mortality in MT-null mice, as compared to 30% mortality in control mice.

Liver injury was more severe with the higher doses of acetaminophen. MT-null mice were more susceptible than controls to acetaminophen-induced hepatotoxicity at the doses of

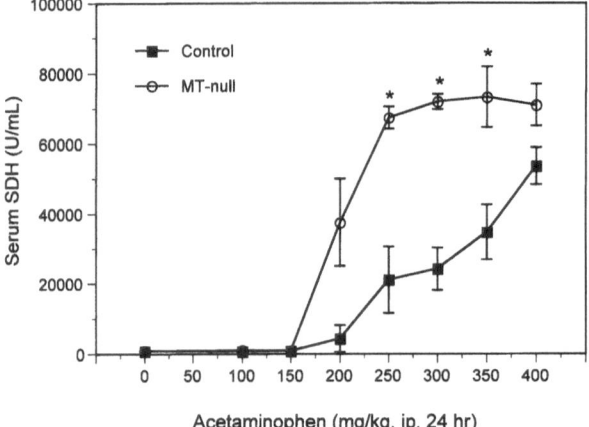

Figure 1. Dose-response of acetaminophen-induced hepatotoxicity in control and MT-null mice. Mice were given an ip injection of acetaminophen (50 to 500 mg/kg, for 24 h). Liver injury in surviving mice was measured by serum sorbitol dehydrogenase (SDH) activity. Values are mean ± SE of 10–6 mice. *Significantly different from control mice.

200–350 mg/kg, as evidenced by marked increases in serum sorbitol dehydrogenase activity (Fig. 1). In parallel with serum enzyme activities, more severe necrosis was observed in MT-null mice than that observed in control mice after acetaminophen administration (data not shown).

Acetaminophen metabolism is not altered in MT-null mice

Hepatic cytochrome P450, b_5 concentration, and cytochrome-c reductase activity were determined. In MT-null mice, total cytochrome P450, b_5 concentrations and cytochrome-c reductase activity were slightly lower than controls (data not shown). Analysis of acetaminophen metabolites in bile and urine were performed by HPLC [7], as shown in Table 1. No difference in biliary and urinary excretion of acetaminophen metabolites was observed between control and MT-null mice after acetaminophen administration (150 mg/kg, ip, 2 h).

Table 1. Biliary and urinary excretion of acetaminophen metabolites two hours after acetaminophen (150 mg/kg, iv) administration. Values are mean ± SE of 4 mice

Group	AA-GSH	AA-cysteine	AA-glucuronide	AA-sulfate
		(µmol/kg)		
Control	310 ± 39	34 ± 5	269 ± 36	16 ± 2
MT-null	310 ± 65	56 ± 10	266 ± 31	17 ± 2

Western blots of liver cytosol indicated that acetaminophen covalent binding at 4 h increased with acetaminophen dose, but there was no consistent difference between control and MT-null mice, suggesting that the net availability of the acetaminophen electrophile was not altered.

Acetaminophen-induced GSH depletion, and lipid peroxidation in control and MT-null mice

Hepatic GSH was determined by enzymatic assay [8]. No difference in hepatic GSH depletion was observed between control and MT-null mice after acetaminophen administration (100 mg/kg, at 1, 2 and 4 h) (Fig. 2, top panel). At a higher dose (300 mg/kg) of acetaminophen, GSH was decreased up to 90% in both control and MT-null mice (data not shown). However, more lipid peroxidation was detected in MT-null mice liver than in controls, as determined by thiobarbituric acid-reactive substances (TBARS) [9] (Fig. 2, bottom panel).

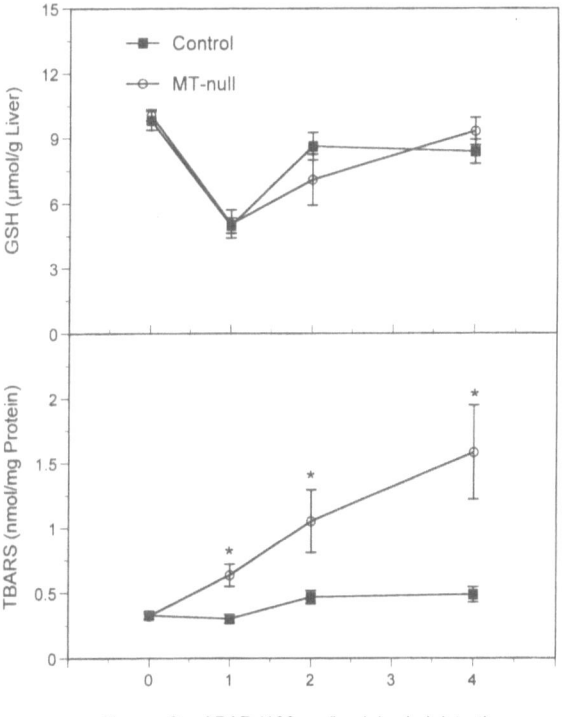

Figure 2. Hepatic concentration of glutathione (top) and thiobarbituric reactive substances (bottom) in control and MT-null mice 1 to 4 h after acetaminophen (100 mg/kg) administration to control and MT-null mice. Values are mean ± SE of 4–6 mice. *Significantly different from controls.

MT-null hepatocytes are more susceptible to NAPQI-induced cytotoxicity and oxidative stress

Mouse hepatocytes were isolated from control and MT-null mice by EGTA-collagenase perfusion method [10]. When hepatocytes were incubated with NAPQI, those from MT-null mice were more sensitive to NAPQI-induced cytotoxicity than those from control mice (Fig. 3, top panel), as evidenced by increased leakage of lactate dehydrogenase (LDH). MT-null hepatocytes were also more sensitive than controls to NAPQI-induced oxidative stress (Fig. 3, bottom panel), as determined by cellular oxidation of 2',7'-dichlorofluorescin diacetate (DCFH-DA) to DCF [11].

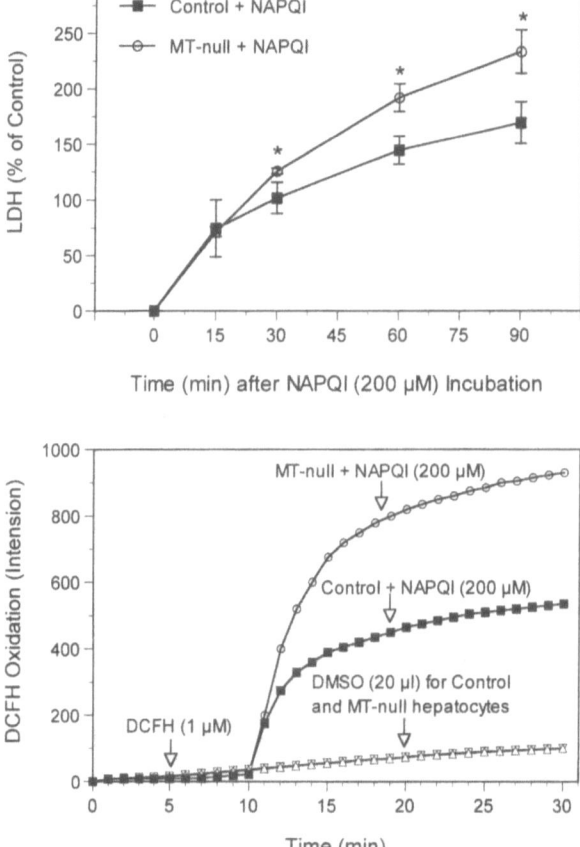

Figure 3. Top panel: Cytotoxicity of *N*-acetyl-*p*-benzoquinoneimine (NAPQI) in cultured hepatocytes isolated from control and MT-null mice. Values are mean ± SE of 4 hepatocyte preparations. *Significantly different from controls. Bottom panel: Representative dichlorofluroscein fluorescence intensity graph following exposure of cultured hepatocytes to NAPQI (200 μM). More oxidative stress was consistently observed in MT-null hepatocytes than controls following NAPQI (50 to 200 μM) exposure.

Discussion

This study demonstrated that disruption of MT I and II genes increases acetaminophen-induced liver injury, suggesting that intracellular MT not only plays a protective role against heavy metal toxicity, but also may protect against organic chemical-induced liver injury.

Acetaminophen-induced hepatotoxicity is thought to be mediated by a cytochrome P450-generated intermediate, NAPQI [3]. Therefore, our initial efforts were directed at determining whether MT-null mice have altered metabolism of acetaminophen. Our results on hepatic cytochrome P450 analysis, as well as quantification of the urinary and biliary metabolites of acetaminophen suggest that MT-null mice metabolize acetaminophen similarly as controls. Further studies on acetaminophen-induced GSH depletion and covalent binding to cytosolic 44KD and 58 KD proteins support the conclusion that the net availability of the acetaminophen intermediates (NAPQI) was not altered. Thus, the greater susceptibility of MT-null mice is not due to the increased production of acetaminophen toxic metabolites, nor the decreased glucuronidation and sulfation of acetaminophen.

Oxidative damage has been proposed as a mechanism of toxicity for a wide range of chemicals. Lipid peroxidation may occur as a result of chemicals being converted to free radicals (i.e. carbon tetrachloride) or via the formation of reactive oxygen species through redox cycling (i.e. quinones and acetaminophen). Oxidative stress may also contribute to acetaminophen hepatotoxicity [3–4]. Following acetaminophen-induced depletion of glutathione, lipid peroxidation occurs [12, 13]. Protection against acetaminophen hepatotoxicity by desferrioxamine, an iron chelator, or by allopurinol, a xanthine oxidase inhibitor, is thought to be mediated by inhibiting lipid peroxidation and oxidative stress, without preventing the depletion of glutathione [13, 14].

MT has been suggested to react with free radicals and electrophiles [1, 2]. Indeed, MT can serve as a sacrificial scavenger for hydroxyl radicals *in vitro* and thus protect against free radical-induced DNA damage [2] and *tert*-butylhydroperoxide-induced lipid peroxidation [15]. In the present study, it was demonstrated that MT-deficient mice are more sensitive to acetaminophen-, and NAPQI-induced lipid peroxidation and oxidative stress, supporting a protective role of intracellular MT in acetaminophen/NAPQI-induced oxidative damage.

Acknowledgements
This work was supported by NIH grant ES-06190, ES-01142 and ES-07163.

References

1. Klaassen CD and Cagen SZ (1981) Metallothionein as a trap for reactive organic intermediates. *Adv Exp Med Biol* 136 Pt A: 633–646.
2. Sato M and Bremner I (1993) Oxygen free radicals and metallothionein. *Free Rad Biol Med* 14: 325–337.
3. Cohen SD, Hoivik DJ and Khairallah EA (1997) Acetaminophen. *In*: Liver and gastrointestinal systems. *In*: *Comprehensive Toxicology*. Elsevier Science, Oxford. pp 329–343.
4. Cohen SD and Khairallah EA (1997) Selective protein arylation and acetaminophen-induced hepatotoxicity. *Drug Metabol Rev* 29: 59–77.

5. Michalska AE and Choo KHA (1993) Targeting and germ-line transmission of a null mutation at the metal-
    lothionein I and II loci in mouse. *Proc Natl Acad Sci USA* 90: 8088–8092.
6. Masters BA, Kelly EJ, Quaife CJ, Brinster RL and Palmiter RD (1994) Targeted disruption of metallothionein
    I and II genes increases sensitivity to cadmium. *Proc Natl Acad Sci USA* 91: 584–588.
7. Howie D, Adriaenssens P and Prescott LF (1977) Paracetamol metabolism following overdosage: Application
    of high performance liquid chromatography. *J Pharm Pharmacol* 29: 235–237.
8. Tietze F (1969) Enzymic method for quantitative determination of nanogram amounts of total and oxidized
    glutathione: Applications to mammalian blood and other tissues. *Anal Biochem* 27: 502–521.
9. Hiroshi O, Nobuko O and Kunio Y (1979) Assay for lipid peroxides in animal tissues by thiobarbituric acid
    reaction. *Anal Biochem* 95: 351–358.
10. Berry MN and Friend DS (1969) High-yield preparation of isolated rat liver parenchymal cell. *J Cell Biol* 43:
    506–520.
11. Zhu H, He M, Bannenberg GL, Moldéus P and Shertzer HG (1996) Effects of glutathione and pH on the oxi-
    dation of biomarkers of cellular oxidative stress. *Arch Toxicol* 70: 628–634.
12. Mason RP and Fisher V (1986) Free radicals of acetaminophen: their subsequent reactions and toxicological
    significance. *Federation Proc* 45: 2493–2499.
13. Jaeschke H (1990) Glutathione disulfide formation and oxidant stress during acetaminophen-induced hepato-
    toxicity in mice *in vivo*: The protective effect of allopurinol. *J Pharmacol Exp Ther* 255: 935–941.
14. Sakaida I, Kayino K, Wasaki S, Nagatomi A, Matsumura Y and Okita K (1995) Protection against
    acetaminophen-induced liver injury *in vivo* by an iron chelator, deferoxamine. *Scand J Gastroenterol* 30:
    61–67.
15. Zheng H, Liu J, Liu YP and Klaassen CD (1996) Hepatocytes from metallothionein knock-out mice are sen-
    sitive to cytotoxicity of Cd and *tert*-butylhydroperoxide. *Toxicol Lett* 87, 139–145.

Metallothionein IV
C. Klaassen (ed.)
© 1999 Birkhäuser Verlag Basel/Switzerland

Effect of oxidative stress and bioavailable zinc on metallothionein in cultured human retinal pigment epithelial cells (RPE)

David J. Tate, Jr.[1], Michael V. Miceli[1, 2] and David A. Newsome[1, 2]

[1]Sensory and Electrophysiology Research Unit, Touro Infirmary, New Orleans, Louisiana 70115, USA
[2]Department of Ophthalmology, Tulane University School of Medicine, New Orleans, LA 70115, USA

Summary. Numerous reports have shown metallothionein protects tissue from reactive oxygen intermediates, considered to be a factor in aging. Because zinc stimulates MT induction and because loss of metal increases proteolysis of the apoprotein, zinc may have a protective role and may indirectly protect against aging. Our laboratory has shown age-related decreases in catalase, zinc and metallothionein in human retinal pigment epithelium. This study was undertaken to investigate the role of bioavailable zinc on MT content in RPE cells subjected to an oxidative stress.

Reactive oxygen intermediates have been implicated in aging and degenerative diseases of the eye, e.g. macular degeneration and cataractogenesis. Catalase, zinc and metallothionein (MT), cellular antioxidants, decrease with age in macular retinal pigment epithelium (RPE) [1–3]. Human RPE cells have high concentrations of MT [3]. This may be due to the rich oxygen environment and the strong metabolic demands of ingestion and digestion of photoreceptor outer segments which are associated with the generation of reactive oxygen intermediates [4]. We have shown that bovine rod outer segment (BROS) phagocytosis is an oxidative stress and increases intracellular H_2O_2 as the reactive oxygen species [4]. Ingestion of nonmetabolizable latex beads increases extracellular H_2O_2 and induces MT [5]. However, ingestion of oxidizable ROS increases H_2O_2 and induces MT to a much greater degree. This suggests a dual source of H_2O_2.

This study investigates the role of bioavailable zinc on MT content in RPE cells subjected to an oxidative stress. We have demonstrated that Chelex treatment of culture medium significantly reduces the zinc available to RPE cells [6]. It has previously been demonstrated in primary cultures of rat hepatocytes that cells cultured in medium containing 1 µM zinc maintain both cellular zinc levels and δ-aminolevulinic acid dehydratase activity, although MT and MT mRNA levels were reduced in comparison to cells cultured in medium containing 16 µM zinc [7]. *In vivo*, MT concentrations in rat hepatocytes are profoundly affected by the level of bioavailable zinc [8–10]. In our study, we also demonstrated a large effect of medium zinc on MT concentration which may be due in part to an accelerated degradation rate of MT apoprotein in the absence of zinc [11]. MT may function by delivering bound zinc to apoenzymes, thereby reconstituting their enzymatic activities [12–14]. Consequently, reduced levels of zinc-MT may reduce the bioavailability and intracellular distribution of zinc. In addition, numerous studies have suggested that MT acts as a free radical scavenger due to its abundance of thiol groups with their high affinity for hydroxyl radicals [12, 15, 16]. Therefore, lowering of MT may directly affect cellular antioxidant functions.

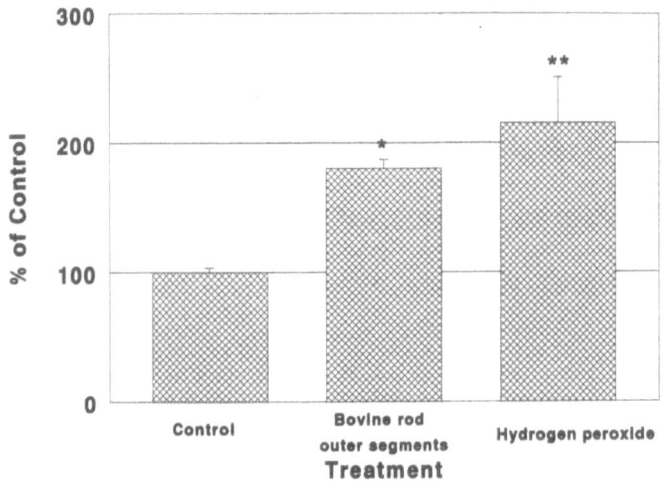

Figure 1. Human RPE cells were cultured (5 days) in Chelex-treated zinc-free DMEM (2.5% FBS) + 10 µM zinc. Phagocytosis of BROS (n = 7, *p = 0.001) or H_2O_2 treatment (0.5 mM, 24 h) (n = 7,** p = 0.002) elevated MT significantly above control levels.

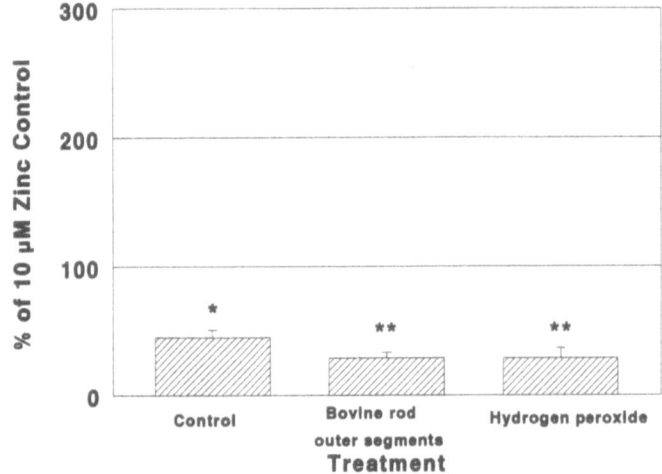

Figure 2. Human RPE cells were cultured (5 days) in Chelex-treated DMEM (2.5% FBS). MT levels in low zinc treated cells were decreased significantly by the oxidative stress of phagocytosis (n = 6; *p = 0.04) and H_2O_2 treatment (n = 8; **p = 0.02). MT levels in low zinc treated cells decreased significantly below those of RPE cells cultured in low zinc + 10 µM zinc.

Recent findings in our laboratory show MT expression is compromised in RPE cells cultured in Chelex-treated Dulbecco's Modified Eagles Medium (DMEM, Sigma Chemical Co., St. Louis, MO) prepared as previously described [6]. Human RPE cells were cultured (5 days) in Chelex-treated zinc-free DMEM (2.5% FBS) or chelex-treated medium + 10 µM zinc. The

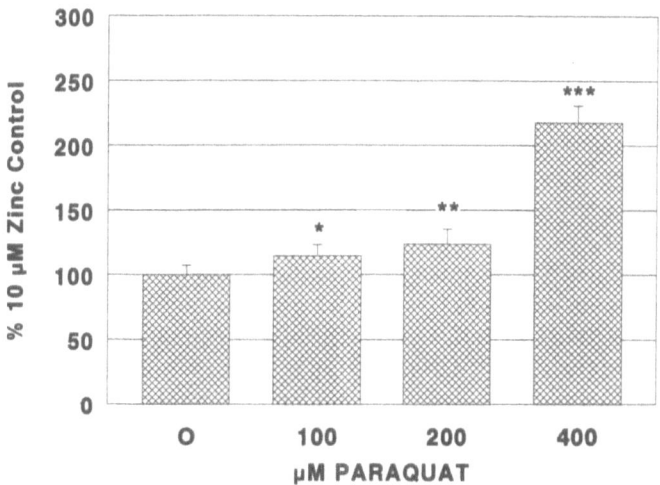

Figure 3. Human RPE cells were cultured (5 days) in Chelex-treated zinc-free DMEM (2.5% FBS) + 10 µM zinc. RPE cells were treated with 0, 100, 200 or 400 µM paraquat for one week. Metallothionein was induced significantly by 100 (n = 6, *p = 0.01), 200 (n = 6, **p = 0.001) and 400 µM paraquat (n = 6, ***p = 0.004).

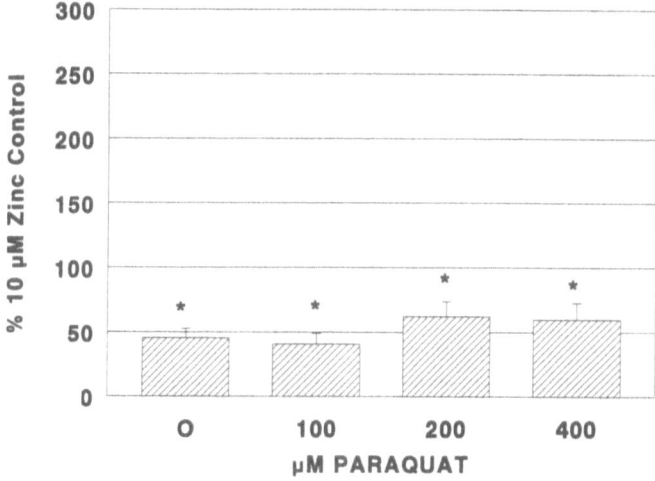

Figure 4. Human RPE cells were cultured (5 days) in Chelex-treated zinc-free DMEM (2.5% FBS). MT levels in low zinc treated cells decreased significantly below MT levels in control RPE cells. The oxidative stress of paraquat did not induce MT levels in the low zinc treated cells (n = 6, *p = 0.001).

cells were treated with 0.1, 0.2 or 0.4 mM paraquat for 7 days or treated with 20 µg/ml bovine rod outer segments (BROS) or 0.5 mM H_2O_2 for 24 h MT content was determined by a Cd\Hemoglobin assay [17].

In cells cultured in 10 µM zinc, MT content was elevated 1.8 and 2.2-fold by phagocytosis of BROS or H_2O_2 treatment, respectively. MT in RPE cells cultured in low zinc medium decreased 55% compared to control. Low zinc cells treated with BROS or H_2O_2 contained 72% less MT compared to control cells.

Paraquat, an herbicide that is toxic to both plants and animals by its generation of O_2^-, increased MT significantly with 0.1, 0.2 and 0.4 mM paraquat in cells cultured in low zinc + 10 µM $ZnCl_2$. RPE cells cultured in low zinc DMEM contained less MT and was not induced by the oxidative stress of paraquat as in control cells. Satoh and colleagues [18] have shown zinc induction of MT can protect mice from paraquat toxicity.

MT gene expression was evaluated using *in situ* hybridization. RPE cells were cultured in low zinc DMEM ± 10 µM $ZnCl_2$ on alkyl-silane coated slides for one week and then treated with paraquat (0–400 µM) for 7 days or H_2O_2 (0.5 mM) for 6 h. Cells were then fixed in Histochoice MB fixative and hybridized for 18 h. Control slides were treated with 40 µg/ml RNAase A for 2 h at 37 °C prior to hybridization. Slides were hybridized with biotinylated MT cDNA probe. RPE cells cultured in low zinc + 10 µM $ZnCl_2$ showed staining for MT mRNA synthesis, which was increased by the addition of 0.1–0.4 mM paraquat and 0.5 mM H_2O_2 (Fig. 5A-5E). MT mRNA in RPE cells cultured in low zinc DMEM demonstrated very little staining with a modest increase when treated with paraquat or H_2O_2 (Fig. 6A–E).

Based on our previous work, we believe it is important to define the role of zinc homeostasis in normal functions of RPE. We have previously reported that human macular RPE shows an age-related decline of both catalase activity [1] and MT content [3] with the greatest decline seen in donors with macular degeneration. More recently, a similar study measuring the zinc content of RPE cells from human donor eyes showed an age-related decline in macular zinc and a negative correlation between RPE cytoplasmic zinc and signs of age-related macular degeneration [2]. The importance of zinc to the efficient functioning of RPE is underscored by a report that zinc supplementation may slow the progression of vision loss due to age-related macular degeneration [19]. Nicolas and colleagues [20] have developed a primate model with signs of macular degeneration. They detected less catalase, zinc, glutathione peroxidase and MT in the retinas of these animals.

The decrease of these important antioxidants with age may produce deleterious effects to the human macula which may not be able to mount a sufficient defense against oxidative stress. The decreased ability of cells to respond to stress during aging could be the result of a decrease in transcriptional regulation by specific nuclear regulatory factors [21–22].

Further studies are in progress to determine how antioxidants and other factors mediate RPE response to reactive oxygen intermediates generated during phagocytosis of photoreceptor outer segments. Learning more about this response may be critical in protecting the RPE from lifelong oxidative stress.

Acknowledgement
Supported by National Eye Institute grants EY-06677 (D.A.N.).

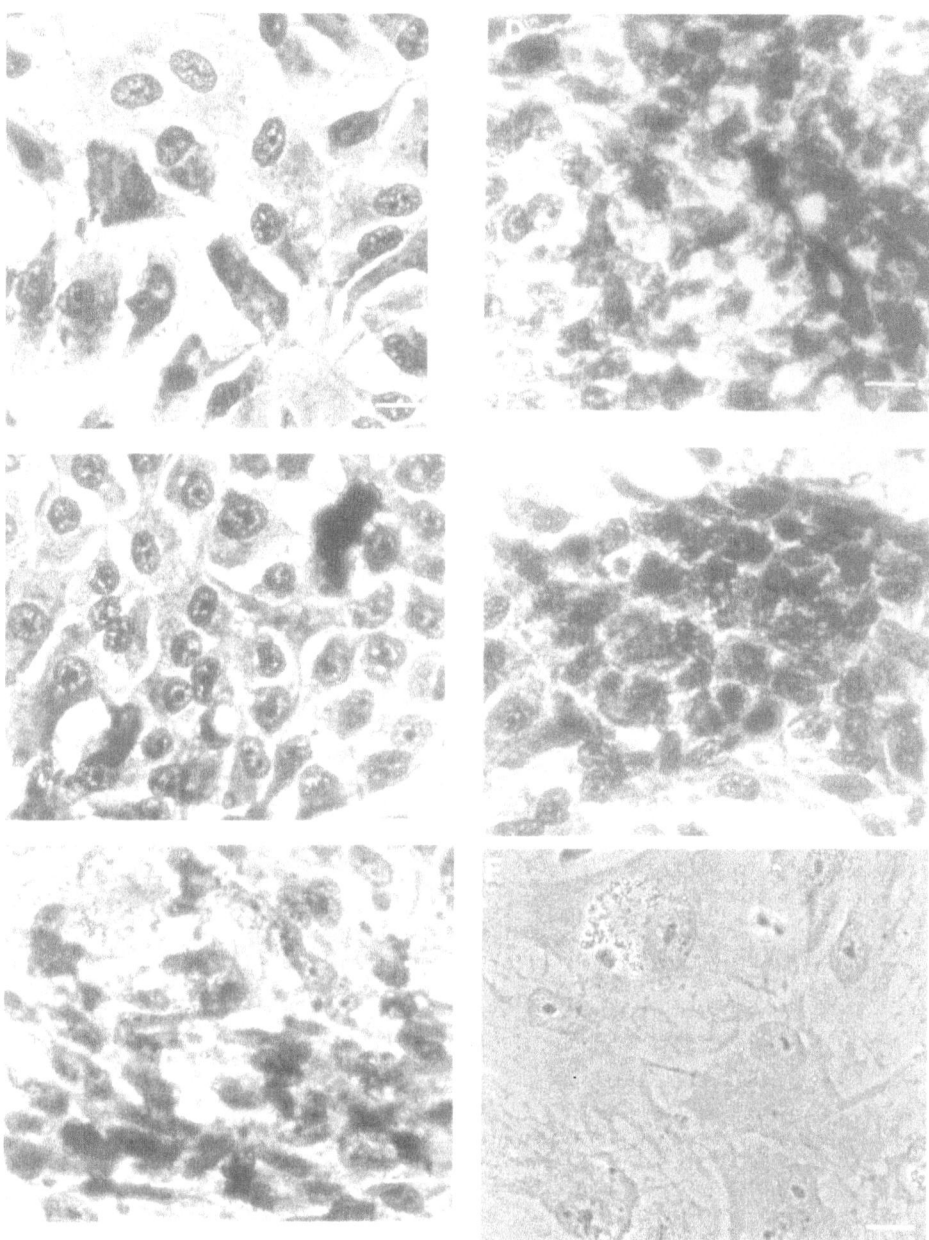

Figure 5. RPE cells were cultured in low zinc DMEM ± 10 µM ZnCl$_2$ on alkyl-silane coated slides for one week and then treated with paraquat (0–400 µM) or H$_2$O$_2$ (0.5 mM) for 6 h Cells were then fixed in Histochoice MB fixative and hybridized for 18 h Control slides were treated with 40 µg/ml RNAase A for 2 h at 37 °C prior to hybridization. Slides were hybridized with biotinylated metallothionein cDNA probe: A): Non-treated RPE cells; B): 100 µM paraquat; C): 200 µM paraquat; D): 400 µM paraquat; E): 0.5 mM H$_2$O$_2$; F): 400 µM paraquat treated plus RNAase A (40 µg/ml). Bar = 20 µm.

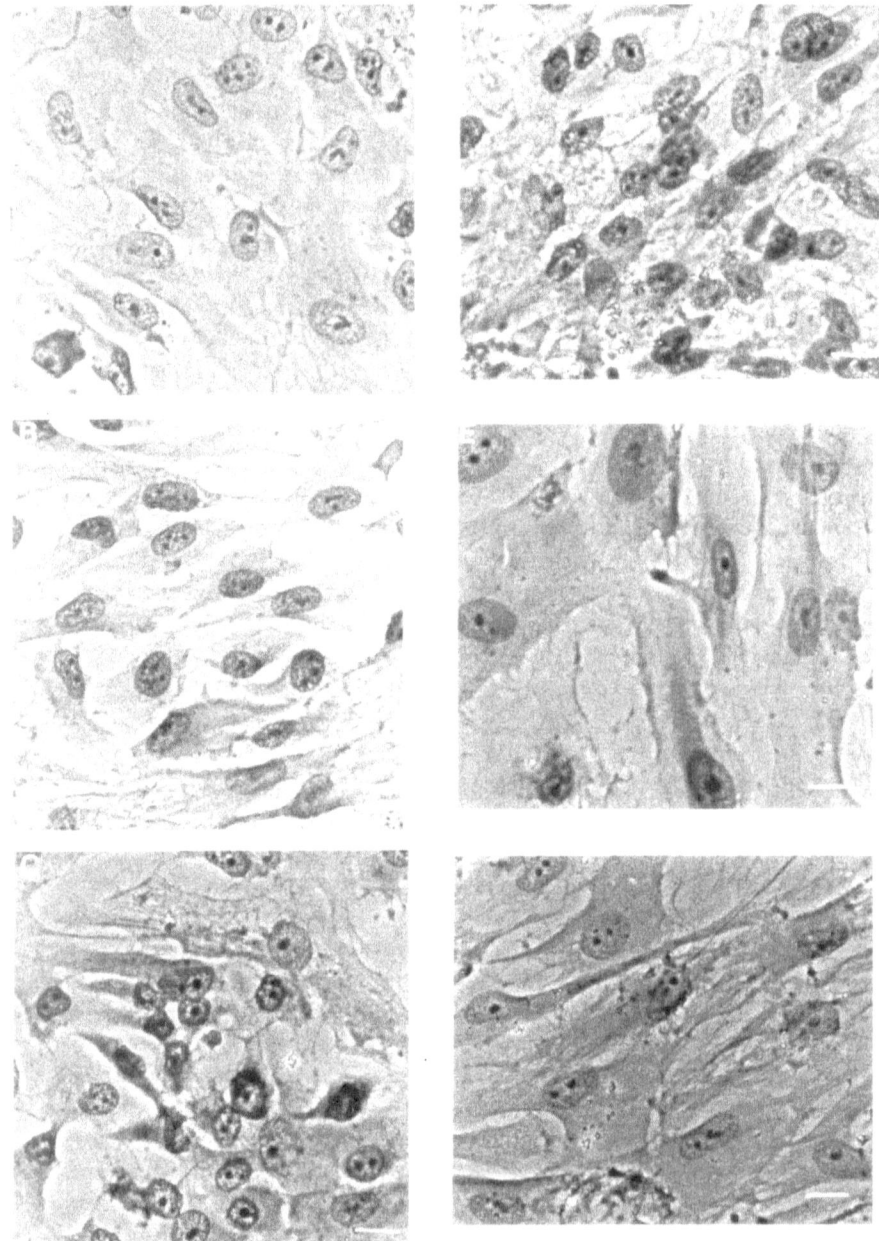

Figure 6. RPE cells were cultured in low zinc DMEM on alkyl-silane coated slides for one week and then treat-ed with paraquat (0–400 µM) or H_2O_2 (0.5 mM) for 6 h Cells were then fixed in Histochoice MB fixative and hybridized for 18 h Control slides were treated with 40 µg/ml RNAase A for 2 h at 37 °C prior to hybridization. Slides were hybridized with biotinylated metallothionein cDNA probe: A): Non-treated RPE cells; B): 100 µM paraquat; C): 200 µM paraquat; D): 400 µM paraquat; E): 0.5 mM H_2O_2; F): 400 µM paraquat treated plus RNAase A (40 µg/ml). Bar = 20 µm.

References

1. Liles MR, Newsome DA, Oliver PD (1991) Antioxidant enzymes in the aging human retinal pigment epithelium. *Arch Ophthalmol* 109: 1285–1288.
2. Newsome DA, Miceli MV, Tate Jr DJ, Alcock NW, Oliver PD (1995) Zinc content of human retinal pigment epithelium decreases with age and macular degeneration, but superoxide dismutase activity increases. *J Trace Elem Exp Med* 8: 193–199.
3. Tate Jr DJ, Newsome DA, Oliver PD (1993) Metallothionein shows an age-related decrease in human macular retinal pigment epithelium. *Invest Ophthalmol Vis Sci* 34: 2348–2351.
4. Miceli MV, Liles MR, Newsome DA (1994) Evaluation of oxidative processes in human pigment epithelial cells associated with retinal outer segments phagocytosis. *Exp Cell Res* 214: 242–249.
5. Tate Jr DJ, Miceli MV, Newsome DA (1995) Phagocytosis and H_2O_2 induce catalase and metallothionein gene expression in human retinal pigment epithelial cells. *Invest Ophthalmol Vis Sci* 36: 1271–1279.
6. Tate Jr DJ, Miceli MV, Newsome DA, Alcock NW, Oliver PD (1995) Influence of zinc on selected cellular functions of cultured retinal pigment epithelium. *Curr Eye Res* 14: 897–903.
7. Schroeder JJ, Cousins RJ (1990) Interleukin 6 regulates metallothionein zinc expression and zinc metabolism in hepatocyte monolayer cultures. *Proc Natl Acad Sci USA* 87: 3137–3141.
8. McCormick CC, Menard MP, Cousins RJ (1981) Induction of hepatic metallothionein by feeding zinc to rats of depleted zinc status. *Amer J Physiol* 240:E414-E421.
9. Bremner I, Davies NT (1975) The induction of metallothionein in rat liver by zinc injection and restriction of food intake. *Biochem J* 149: 733–738.
10. Sato M, Mehra RK, Bremner I (1984) Measurement of plasma metallothionein-I in the assessment of zinc status of zinc deficient and stressed rats. *J Nutr* 114: 1683–1689.
11. Krezoski SK, Villalobos J, Shaw III CF, Petering DH (1988) Kinetic lability of zinc bound to metallothionein in Ehrlich cells. *Biochem J* 255: 483–491.
12. Udom AO, Brady FO (1980) Reactivation *in vitro* of zinc-requiring apo-enzymes by rat liver zinc-thioneine. *Biochem J* 187: 329–335.
13. Li TY, Minkel DT, Shaw III CF, Petering DH (1981) On the reactivity of metallothioneins with 5,5'-dithiobis-(2-nitrobenzoic acid). *Biochem J* 193: 441–446.
14. Hidalgo J, Campmany L, Borras M, Garvey JS, Armario A (1988) Metallothionein response to stress in rats: Role of free radical scavenging. *Amer J Physiol* 254:E71–78.
15. Sato M, Bremner I (1993) Oxygen free radicals and metallothionein. *Free Radicals Biol and Med* 14: 325–337.
16. Thomas JP, Bachowski GJ, Girotti AW (1986) Inhibition of cell membrane lipid peroxidation by cadmium- and zinc-metallothionein. *Biochim Biophys Acta* 884: 448–461.
17. Tate Jr DJ, Miceli MV, Newsome DA (1995) Phagocytosis and H_2O_2 induce catalase and metallothionein gene expression in human retinal pigment epithelial cells. *Invest Ophthalmol Vis Sci* 36: 1271–1279.
18. Satoh M, Naganuma A, Imura N (1992) Effect of preinduction of metallothionein on paraquat toxicity in mice *Arch Toxicol* 66: 145–148.
19. Newsome DA, Swartz M, Leone NC, Elston RC, Miller ED (1988) Oral zinc in macular degeneration. *Arch Ophthalmol* 106: 192–198.
20. Nicolas MG, Fujiki K, Muryama K, Suzuki MT, Shindo N, Hotta Y, Iwata F, Fujimura T, Yasuhiro Y, Cho F et al (1996) Studies on the mechanism of early onset macular degeneration in cynomolgus monkeys. II. Suppression of metallothionein synthesis in the retina in oxidative stress. *Exp Eye Res* 62: 399–408.
21. Luce MC, Schyberg JP, Bunn CL (1993) Metallothionein expression and stress responses in aging human diploid fibroblasts. *Exp Gerontol* 28: 17–38.
22. Sikora E, Kaminska B, Radzisewka E, Kaczmarek L (1992) Loss of transcription factor AP-1 DNA binding activity during lymphocyte aging *in vivo*. *FEBS Lett* 312: 179–182.

Metallothionein IV
C. Klaassen (ed.)
© 1999 Birkhäuser Verlag Basel/Switzerland

Specific binding sites of metallothionein on endothelium of small vessels in myocardium

Shi Cheng, Ge Shujun, Zhong Chongxia, He Qihua and Hou Lin

Department of Biophysics, Beijing Medical University, Beijing, 100083, P.R. of China

MT is primarily an intracellular protein; however, the finding that under conditions of cadmium poisoning, MT occurs in blood plasma and in urine indicates that there must be a pathway out of cells. Of course, some of the extracellular MT could originate from cell destruction. However, addition of MT-3 affects nerve cell growth indicating an extracellular role for MT. Perfusion of MT through the coronary artery quenches free radical signals and protects myocardium from postischemic reperfusion injury [1, 2] suggesting its entry in cells. MT lacks in hydrophobic amino acid residues. It seems resonable to ask: *How does MT get across cell membrane and enter cells?*

Specific binding sites of MT on endothelial cells

To answer the question of how MT enter cells, the interaction between MT and endothelial cells of blood vessels in myocardium was observed by colloidal gold labelled trace technique [3]. MT was conjugated with the non-specific protein bovine serum albumin BSA) to yield MT-BSA complex. Heterobifunctional reagent N-succinimidyl 3-(2-pyridyl dithio)propionate (SPDP) was used as a conjugate reagent [4]. The principle of the reactions is as follows:
1, SPDP reacts with the amino group of MT, yielding MT-PDP;

2, An excess of dithiothreitol is added in order to reduce the disulfide bond of MT-PDP, yielding thiolated MT (MT-SH);

3, Similarly,SPDP reacts with the amino group of BSA, yielding BSA-PDP;

4, MT-SH and BSA-PDP are conjugated.

As a negative control, BSA-BSA complex and a positive control, insulin-BSA complex conjugated by SPDP were prepared. MT-BSA, BSA-BSA and insulin-BSA were labelled with colloidal gold (15 nm in diameter) as a maker for transmission electron microscope (TEM), and MT-BSA*BSA-BSA* and insulin-BSA* were prepared.

Using non-recirculating Langendorff method, isolated rat heart was perfused with Krebs-Henseleit buffer to which MT-BSA* (BSA-BSA* and insulin-BSA* respectively) were added. The concentration of MT was 2×10^{-6} M. After 30 min of perfusion, a strip of left ventricular tissue was excised and processed for TEM. Capillaries and small vessels in myocardium on thin sections were carefully searched for. Gold particles attached to the surface of endothelial cells were observed (Fig. 1). On the majority of vessels the attached gold particles were separate, some of which were engulfed and sequestered in vesicles in endothelial cells. Clusters of particles were distributed on the surface of endothelium in only a few sections. However, no particles were observed after perfusion with BSA-BSA*. Futhermore, perfusion with MT-BSA* following preperfusion with unlabelled MT in high concentration (5×10^{-5} M) could only yield a few particles. However, preperfusion with BSA in high concentration followed by perfusion with MT-BSA* could not block gold particles from attaching to endothelial cells. It is well known that there are receptors of insulin on endothelium of vessels. After insulin-BSA* was perfused gold particles attaching to endothelial surface could be seen (Fig. 2). Preperfusion with unlabelled insulin in high concentration followed by perfusion with insulin-BSA* succeeded in blocking particle from attaching to the cell surface. However, when preperfusion with unlabelled MT in high concentration was followed by perfusion with insulin-BSA*, many gold particles attached to the endothelial surface.

The results of this study including negative (BSA-BSA*) and positive (insulin-BSA*)control experiments, specific (MT vs MT-BSA*) and nonspecific (BSA vs MT-BSA*) blocking

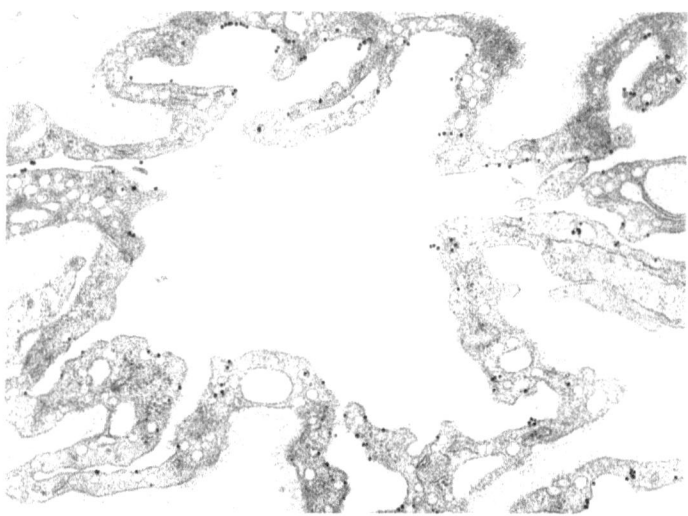

Figure 1. A small vessel of myocardium perfused with MT-BSA*. Many gold particles attached to the endothelial surface can be seen. Some of particles are engulfed and sequestered in vesicles of endothelial cells. 35000 ×.

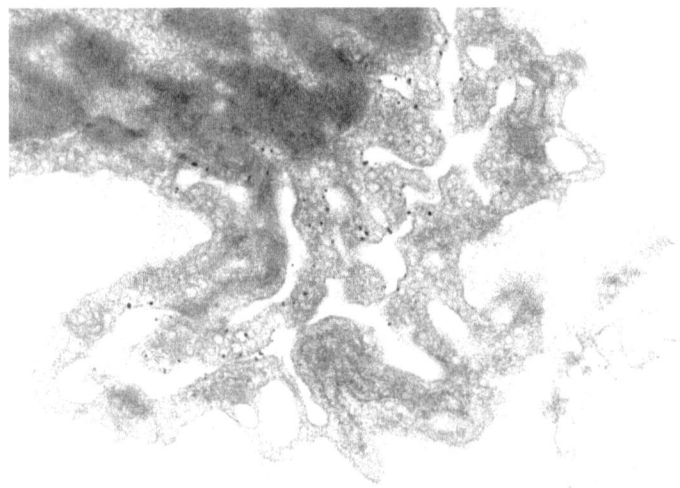

Figure 2. A small vessel in myocardium perfused with insulin-BSA*. The attachment and distribution of gold particles are the same as that of Figure 1. 37000 ×.

experiments, suggest that attachment of gold particles originated from perfusion with MT-BSA*, is specific. Therefore, it is rational to conclude that there are specific binding sites (receptors) for MT on endothelial cells of capillaries and small vessels in myocardium.

Specific binding of metallothionein on endothelial cells in ischemic myocardium

After 30 or 60 min of ischemia, isolated rat hearts were reperfused for 30 min with Krebs-Henseleit buffer containing MT-BSA*. Myocardium was excised for electron microscopic morphometry to compare the number of gold particle attaching to myocardium following ischemia for different periods. Results are shown in Table 1. The frequency with which attached gold particles could be detected, was represented as gold particle density, that is, the particle number per unit area of endothelial surface. Clearly, from Table 1 after 30 min of ischemia, although endothelial cells had already been injured, the particle density did not reduce significantly. However, ischemia for 60 min produced a remarkable reduction of gold particle density.

Physiological implication

Although MT has been on the biochemical scene for the past 40 years, its physiological function is still not well undestood. Its role in metal detoxification is well accepted. Some authors have suggested that MT may contribute or transportat zinc into cells. Being active in

Table 1. Gold particle density (X ± SE)

Groups	Density
Normal control	40.76 ± 4.21
ischemia for 30 min	34.84 ± 4.25
ischemia for 60 min	$11.58 \pm 2 08^{*}$

The freshly excised hearts were perfused in Langendorff model with Krebs- Henseleit buffer continuously gassed with 95% O_2-5% CO_2 (pH 7.4, 37 °C, constant flow rate was about 10 ml/min). The hearts were allowed to stabilize for 10 min and then:
(1) perfused with K-H buffer containing MT-BSA* for 30 min (control); (2) submitted to global ischemia at 37 °C by totally occluding coronary flow. 30 or 60 min later, the hearts were reperfused with K-H buffer containing MT-BSA* for 30 min. A strip of ventricular tissue was excised and processed for morphometric electron microscopy. Using Image Analytic System Q 550 IW, the length (B) of the endothelial boundaries and the number (N) of gold particles were measured. Section thickness (t) was measured by determining the width of small creases of section. Gold particle density was calculated by the formula, $D = (\pi/4t)\cdot(\Sigma N/\Sigma B)$ [5]. n = 48 (the number of photographs, from 4 hearts), compared with normal control * p < 0.001.

metabolism, endothelial cells of blood vessels perform critically important endocrine function that regulate vascular tone, reactivity and permeability. Because endothelial cells are present throughout the vasculature, they are vulnerable to mechanical, chemical and biological insults. Various oxidants including oxidized -LDL, and oxy-cholestereols are also poisonous. However, endothelial cells have a strong ability to engage in self-repair. MT taken up by endothelial cells may contribute zinc to maintain the broad range of zinc-dependent functions. Some authors have proposed that endothelial cells seem to have substantial thio-dependent antioxidant activity [6]. The endothelial cells may simutaneously obtain many thiolate groups from MT to control intracellular redox state [7, 8] and resist a variety of oxidant stresses. Following ischemia, the ability of endothelial cells to take up MT was reduced. However, endothelial cells exposed to ischemia for short periods, still could take up MT. The possibility that mildly injured endothelial cells obtain nutrient zinc and abundant thiolate groups from MT to meet the demand of self-repar, needs further study. It is possible that, on other cells, the specific binding sites for MT also exists. MT as a donor of zinc and thiolate groups entering cells, regulates cellular activity.

Acknowledgement
This work is supported by National Natural Science Funds 39470344.

References

1. Cheng S, Zhang YZ, Wang DY, An RS (1992) Quench effect of metallothionein on free radical signal in postischemic reperfusion of isolated rat hearts. *Natl Med J China*. 72: 749–51.
2. Cheng S, Tang CS, (1990) Metallothionein as a scavenger of free radicals. *In*: Collery P, Poirier LA, Manfait M, Etienne J-Claude (eds): *Metal Ions*. Joholibbey, Paris, pp. 21–26.
3. Roth J (1983) The colloidal gold maker system for light and electron microscopic cytochemistry. *In*: Bullock GR, Petrusz P (eds): *Techniques in Immunocytochemistry*. Academic Press. New York, pp. 217–83.

4. Carlsson J, Drevin H, Axen R (1978) Protein thiolation and reversible protein-protein conjugation. *Biochem J* 173: 723–37.
5. Weibel ER, (ed): (1979) *Stereological Methods Vol 1, Practical Methods for Biological Morphometry.* Academic Press, New York.
6. Elliott SJ, Koliward SK, (1995) Oxidant stress and endothelial membrane transport. *Free Rad Biol Med* 19: 649–58.
7. Vallee BL,(1987) Implication and inferences of metallothionein structure. *In*: Kagi JHR, Kojima Y (eds): Metallothionein II, Birkhäuser Verlag, Basel, pp. 5–16.
8. Vallee BL, Maret W (1993) The functional potential and potential function of metallothionein: A personal perspective. *In*: Suzuki KT, Imura N, Kimura M (eds): *Metallothionein III*, Birkhäuser Verlag, Basel, pp. 1–27.

Metallothionein and carcinogenesis

Further studies on the role of metallothionein in the antitumor effects of cadmium

Michael P. Waalkes[1], Takeaki Nagamine[1], Hideaki Shimada[1], Bhalchandra A. Diwan[2], Timothy P. Coogan[1], Noriyuki Shiraishi[1], M. George Cherian[3] and Robert A. Goyer[4]

[1]*National Cancer Institute at the National Institute of Environmental Health Sciences, Research Triangle Park, NC, USA*
[2]*SAIC Frederick, NCI-FCRDC, Frederick, MD, USA*
[3]*University of Western Ontario, London, ON, Canada*
[4]*National Institute of Environmental Health Sciences, Research Triangle Park, NC, USA*

Metallothionein (MT) is a metal-inducible protein that binds cadmium and is known as a main factor in protection against many of the aspects of cadmium toxicity, including hepatocellular necrosis [1–3]. The basal or initial levels of hepatic MT appear to be very important in dictating the final level of resistance to the hepatotoxic effects of cadmium [1–3]. Furthermore, tolerance to the acute hepatotoxicity of cadmium can be induced by pre-activation of MT gene expression by any number of various stimuli *in vivo* [4]. Similarly, enhanced MT expression clearly reduces the *in vitro* cytotoxic effects of cadmium [5].

Hepatocellular carcinoma or hepatoma is a liver malignancy which is normally very difficult to treat, and will often show relapse following surgery. Moreover, hepatocellular carcinoma is often associated with cirrhosis and many patients are inoperable because of severe liver disease, and non-invasive therapies for hepatocellular carcinoma have been developed and applied clinically [6]. However, patients with hepatocellular carcinoma typically have a very poor prognosis and the development of new therapies is clearly desired.

Previously, we found that oral cadmium treatment either abolished or substantially reduced N-nitrosodiethylamine (NDEA)-induced and spontaneous tumor formation in mouse liver or lung regardless of exposure interval and even when the metal was given well after tumors were formed [7, 8]. Since cadmium salts are powerful emetics, oral exposure in humans would be impractical. Thus, we have recently studied suppression of NDEA-initiated tumors in mice by a single i.v. injection of cadmium [9]. NDEA (776 µmol/kg) was given i.p. at time 0 to male B6C3F1 mice followed by cadmium chloride ($CdCl_2$; 16 µmol/kg) given i.v. 40 weeks after the nitrosamine. This dose of cadmium was essentially non-toxic and had no effect on body weights through the conclusion of the study at 52 weeks. NDEA caused clear increases in hepatic tumor incidence (19 tumor bearing mice/22 mice at risk, 86%) over control (5/24, 21%) that were substantially reduced by exposure to the single, non-toxic dose of cadmium (13/27, 48%, p < 0.05). Multiplicity and size of liver tumors induced by NDEA were also substantially reduced by cadmium exposure. NDEA-induced lung tumor incidence and multiplicity were also reduced by cadmium, although not to as great an extent as for liver tumors. When the tumor-bearing livers were observed soon after the cadmium treatment (~24 h), it was clear that the metal had induced multifocal necrosis specifically in the hepatic neoplasms, indicating a hypersensitivity of the tumors to the metal. These results indicate that a single, non-toxic dose of cadmium dramatically reduces liver tumor burden apparently through tumor cell specific necrosis [9]. This would be consistent with a chemotherapeutic potential for cadmium against liver tumors and perhaps lung tumors [7–9].

Since the expression of MT often dictates the sensitivity of a given cell or tissue to cadmium toxicity [1–5], the levels of MT were also assessed in livers bearing NDEA-induced tumors by immunohistochemical techniques [9]. Metallothionein was not detected immunohistochemically in the NDEA-induced mouse liver tumors, even after cadmium exposure, while surrounding normal cells showed high MT levels [9]. Likewise, in human hepatocellular carcinomas MT was poorly expressed relative to normal liver [9]. Additionally, adenomas induced by NDEA treatment in mice were dissected from the normal liver and MT was measured biochemically by the Cadmium-Hemoglobin method [10]. The basal MT levels in the hepatic adenomas are clearly much lower in the tumor than in the normal tissue (Fig. 1). Furthermore, the induction of MT synthesis after cadmium exposure is markedly reduced in the tumorous tissue when compared to normal liver. These results indicate that MT is poorly expressed in liver tumor cells in comparison to normal liver.

In other studies we have determined the comparative toxicity of cadmium in normal (non-transformed) and tumorous liver cells. For this purpose HepG2 cells, originally derived from a human hepatocellular carcinoma, and TRL 1215 cells, an immortilized, non-tumorigenic cell line derived from newborn rat liver, were compared. Cultures were exposed to cadmium, as $CdCl_2$, for a 3 h period at concentrations ranging from 0 to 100 mM. Cytolysis began at 10 mM cadmium in HepG2 cells and was apparently complete at 25 mM. In TRL 1215 cells these changes were shown only with doses of over 50 mM cadmium. When levels of MT were com-

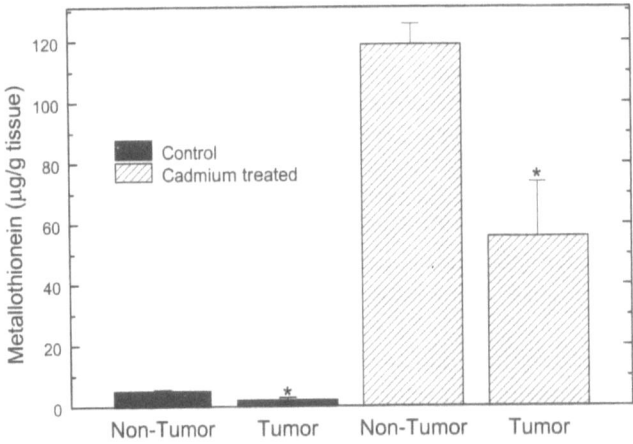

Figure 1. Basal and induced levels of metallothionein in NDEA-induced liver adenomas and surrounding liver tissue in mice. MT levels were assessed with or without cadmium treatment (10 μmol/kg, s.c. 24 h prior to determination) in adenomas or control areas (n = 3). Tumors were confirmed by histopathological analysis. The basal and induced levels of MT were significantly less (as indicated by asterisks; p < 0.05 by Student's t-test) in the tumors than in the surrounding normal tissue.

pared, basal MT concentrations were much higher in TRL 1215 cells than in HepG2 cells. Incubation of HepG2 cells with cadmium caused only minimal elevations of MT protein levels in HepG2 cells while in TRL 1215 cells MT levels were increased dramatically with cadmium exposure. HepG2 cells were also hypersensitive to cadmium toxicity when compared to non-transformed liver cells (TRL 1215). Thus, this cellular hypersensitivity is associated with much lower basal levels of MT and a minimal induction of MT after cadmium exposure in the HepG2 cells. Therefore, the hypersensitivity of liver tumors may be due to a poor expression of MT in transformed cells. In this regard, when the TRL 1215 cells undergo spontaneous transformation to a recognizable malignant phenotype, there is a marked suppression of the basal expression of the MT gene [11]. The transformation of TRL 1215 cells also is accompanied by a enhanced sensitivity to cadmium-induced cytotoxicity, very likely due to diminished MT expression.

In an attempt to define the possible utility of cadmium against human tumors, xenograft studies using human pulmonary tumor cell lines in Nude mice were performed. These studies tested the ability of cadmium to inhibit growth and progression of these transplanted human tumor xenografts. Male athymic nude mice (NCr-*nu*) were inoculated with H460 cells, originally derived from a human non-small cell pulmonary carcinoma. After mice (n = 9) were anesthetized, the left kidney was exposed and 1×10^6 H460 cells in 25 µl were injected under the renal capsule. The contralateral kidney was untreated and served as the control for analysis of tumor growth. After 1 week mice received 0, 125 or 250 ppm cadmium (as $CdCl_2$) in the drinking water. These levels of the metal were essentially non-toxic based on the lack of effect on growth or survival. Mice were observed for an additional 4 weeks after inoculation. At necropsy all mice had tumors in the left kidney but cadmium had caused clear, dose-related reductions in tumor mass (125 ppm, 37%; 250 ppm, 46%). Control mice had a 56% rate of lung metastases which was reduced with cadmium in a dose-related fashion (125 ppm, 33%; 250 ppm, 22%). In a similar study male NCr-*nu* were inoculated with with DMS 114 cells, which were originally derived from a human small cell lung carcinoma. Over a 100 day observation period, tumor weight resulting from DMS 114 inoculation was markedly reduced in a cadmium dose-related fashion (125 ppm, 40%; 250 ppm, 70%) compared to control. Metastases were also reduced by cadmium in this study. An additional experiment gave cadmium as an i.v. loading dose (20 µmol/kg) 4 days after renal capsule inoculation with H460 cells followed by 200 ppm cadmium in the drinking water from 7 days onward. After 4 weeks cadmium had reduced tumor weight by nearly 50% while the lung metastases rate dropped from 52% in controls to 30% in cadmium treated mice. These studies show cadmium can effectively reduce both the growth and metastases of human lung carcinoma xenografts, as evidenced by a suppression of metastatic rate by up to 50%. Other studies indicate that when the normally tumorigenic L6 cells are exposed to cadmium *in vitro* prior to inoculation, there is a marked suppression of tumor progression *in vivo* after inoculation [12]. Together these results indicate that there is a direct effect of cadmium on tumor cells to reduce progression as reflected in reduced growth and metastases. In further study, we find that MT levels in some metastatic tumors are much less than in the primary source tumor, an observation which is true for primary fibrosarcomas and their pulmonary metastases in rats. Thus, the diminished expression of MT in some primary and metastatic tumors may render them sensitive to destruction by cadmium.

In summary, even though cadmium is itself a known carcinogen, it also inhibits formation and growth of chemically-induced or spontaneously occurring liver and lung tumors in mice. In addition, cadmium can dramatically reduce human tumor growth and progression in xenograft model systems. These tumor suppressive effects of cadmium occur with non-toxic doses of cadmium indicating a heightened sensitivity of tumor cells to the metal. In this regard, it appears that MT expression is reduced by malignant transformation in liver cells, a factor that may well contribute to this sensitivity.

Acknowledgments
The authors thank George T. Smith, Robert M. Bare and Larry Claggett for excellent technical assistance. We also thank Dr. Katsuyuki Nakajima and Otsuka America Pharmaceutical, Inc. for their gracious support of Dr. Nagamine.

References

1. Liu Y, Liu J, Iszard MB, Andrews GK, Palmiter RD, Klaassen CD (1995) Transgenic mice that overexpress metallothionein-1 are protected from cadmium lethality and hepatotoxicity. *Toxicol Appl Pharmacol* 135: 222–228.
2. Masters BA, Kelly EJ, Quaife CJ, Brinster RL, Palmiter RD (1994) Targeted distruption of metallothionein I and II genes increases sensitivity to cadmium. *Proc Natl Acad Sci USA* 91: 584–588.
3. Goering PL, Klaassen CD (1984) Resistance to cadmium-induced hepatotoxicity in immature rats. *Toxicol Appl Pharmacol* 74: 321–329.
4. Goering PL, Waalkes MP, Klaassen CD (1994) Toxicology of cadmium. *In*: Goyer, RA, and Cherian, MG (eds): *Handbook of Experimental Pharmacology; Toxicology of Metals, Biochemical Effects*. Vol. 115. New York, Springer-Verlag, pp 189–214.
5. Waalkes MP, Miller MS, Wilson MJ, Bare RM, McDowell AE (1988) Increased metallothionein gene expression in 5-aza-2'-deoxycytidine induced resistance to cadmium cytotoxicity. *Chem -Biol Interact* 66: 189–204.
6.. Okuda K, Kojiro M, Okuda H. Neoplasmas of the liver. *In*: Schiff L, Schiff ER (eds): *Disease of the Liver*. Seventh edition. Philadelphia: Lippincott Company, pp. 1236–1296.
7. Waalkes MP, Diwan BA, Weghorst CM, Bare RM, Ward JM and Rice JM (1991) Anti-carcinogenic effects of cadmium in B6C3F1 mouse liver and lung. *Toxicol Appl Pharmacol* 110: 327–335.
8. Waalkes MP, Diwan BA, Weghorst CM, Ward JM, Rice JM, Cherian MG, Goyer RA (1993) Further evidence of the tumor-suppressive effects of cadmium in the B6C3F1 mouse liver and lung; Late stage vulnerability of tumors to cadmium and the role of metallothionein. *J Pharmacol Exp Ther* 266: 1656–1663.
9. Waalkes MP, Diwan BA, Rehm S, Ward JM, Moussa M, Cherian MG, Goyer RA (1996) Down-regulation of metallothionein expression in human and murine hepatocellular tumors; Association with the tumor-necrotizing and antineoplastic effect of cadmium in mice. *J Pharmacol Exp Ther* 277: 1026–1033.
10. Onosaka S, Tanaka K, Doi M, Okahara K (1978) A simplified procedure for determination of metallothionein in animal tissues. *Eisei Kagaku* 24: 128–131.
11. Zhao CQ, Young MR, Diwan BA, Waalkes MP (1997) Association of arsenic-induced malignant transformation with DNA hypomethylation and aberrant gene expression. *Proc Natl Acad Sci USA* 94: 10907–10912.
12. Abshire MK, Devor DE, Diwan BA, Shaughnessy Jr JD, Waalkes MP (1996) *In vitro* exposure to cadmium in rat L6 myoblasts can result in both enhancement and suppression of malignant progression *in vivo*. *Carcinogenesis* 17: 1349–1356.

Metallothionein IV
C. Klaassen (ed.)
© 1999 Birkhäuser Verlag Basel/Switzerland

Effects of metallothionein expression on development of drug resistance

Toby G. Rossman and Ekaterina I. Goncharova

Nelson Institute of Environmental Medicine and Kaplan Cancer Center, New York University Medical Center, 550 First Avenue, New York, NY 10016, USA

Metallothioneins (MTs) are low molecular weight metal-binding proteins with about 30% cysteine content [1]. MTs have been postulated to function in the protection against heavy metals, scavenging of free radicals, and control of homeostasis of essential metals. We have shown that metallothionein expression protects cells against spontaneous mutagenesis [2], and suggest that low MT expression might be a risk factor for cancer. High expression of metallothionein in certain tumors was associated with resistance to chemotherpeutic agents [3, 4]. The purpose of this study was to determine whether the antimutagenic action of MT might be useful in preventing tumor cells from developing mutation to drug resistance. In order for this to be feasible, the action of the chemotherapeutic agent cannot itself be blocked by high MT expression.

As a model system, we used the G12 cell line, a Chinese hamster V79 cell containing a single copy of the *E. coli gpt* gene as a target for mutagenesis [5]. G12 cells have a very low level of endogenous expression of MT. A number of G12-derived MT-overproducing cell lines were obtained after transfection with the mouse MTI gene [2]. One of these transfectant lines, MT1-2A, is used in these studies.

Compared with the parental G12 cells, MT1-2A cells are more resistant to the mutagenic and cytotoxic effect of cisplatin and some topoisomerase inhibitors (amsacrine, doxorubicin, ellipticine). However, MT expression has little effect on topoisomerase inhibitors which do not generate free radicals (etoposide and novobiocin) or on 5-fluorouracil (FU) (Table I). The antimetabolite FU is one of the more prominent clinical antitumor agents and one of the few drugs that displays significant activity toward colorectal cancer.

In the cell, FU is converted to fluorouridine monophosphate (FUMP). Two mechanisms are thought to contribute to FU's antitumor action: 1) phosphorylation to FUMP to FUTP, which

Table 1. Effects of MT expression on cytotoxicity and mutagenicity of chemotherapeutic agents

Agent	Oxidant generator	Effects of MT on	
		survival	mutagenesis
cisplatin	+	increases	decreases
amsacrine	+	increases	decreases
doxorubicin	+	none	decreases
ellipticine	+	increases	decreases
etoposide	–	none	none
novobiocin	–	none	not mutagenic
5-fluorouracil	–	none	not mutagenic

is then incorporated into RNA with subsequent deleterious effects (e.g. inhibition of rRNA maturation); 2) conversion of FUMP to FdUMP, a potent inhibitor of thymidylate synthetase (TS),
the rate-limiting *de novo* enzyme for synthesis of thymine nucleotides. Inhibition of TS results
in thymidine starvation and inhibition of DNA synthesis [6, 7]. Although small amounts of FU
(via FdUTP) get incorporated into DNA, FU has negligible mutagenicity [8], probably because
any FdUTP incorporated into DNA can be removed by uracil glycosylase [9].

As with other anticancer drugs, tumors initially responsive to FU acquire resistance to it after
several treatments. Resistance has been correlated with defective transport, reductions in the
activities of uridine phosphorylase, uridine kinase, pyrimidine phosporibosyl transferase or
uridylate kinase as well as increases in TS or reduction of its affinity to FdUMP [10]. The latter may be a major resistance mechanism for FU in patients [7, 11].

When grown in the presence of FU, G12 cells develop FU-resistant (large) clones whereas
their MTI-2A transfectants do not (Tab. 2), most likely a result of the antimutagenic effect of
the MT expression in MT1-2A. By examining the background of the stained petri dishes, it is
possible to observe microcolonies of wild type (FU-sensitive) cells. It is clear that FU-resistant
(FU^r) colonies appear only among G12 cell populations that are able to devide a few times in
the presence of FU (i.e. in 15 µM FU and lower), but not among those where growth is completely inhibited (>15 µM). Six FU^r colonies were isolated, grown in the absence of FU, and
tested for their sensitivities to FU. Figure 1 shows 2 examples typical of all 6. These cells are

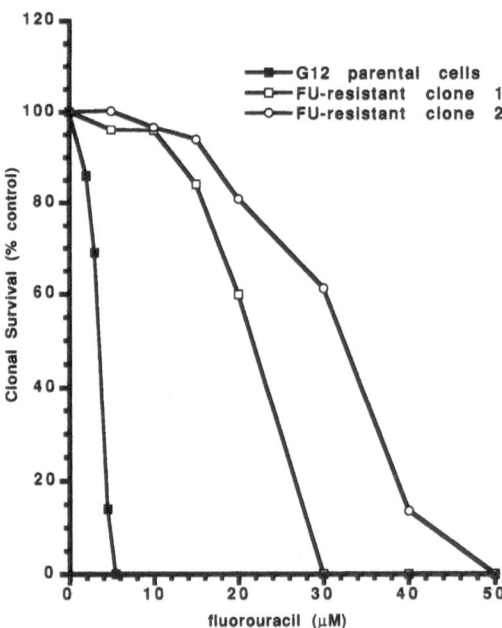

Figure 1. Sensitivity of wild type V79 cells and FU-resistant variant cells to FU. Cells are seeded in the presence
of FU, and colonies are stained after 7 days.

Table 2. Appearance of FU-resistant colonies in plates seeded with 10^4 cells and grown for 10 days in FU

5 fluorouracil (µM)	FU-resistant colonies/plate	
	G12 (parental cells)	MTI-2A (MT transfectant)
7.5	29	0
10.0	3	0
15.0	1	0
17.5	0	0
20.0	0	0

resistant to concentrations of FU which completely block growth of the parental G12 cells, as should be expected, since the colonies were picked from plates containing 10–15 µM FU. The fact that these cells are resistant to 20 µM FU, yet do not arise among cells cultured in above 15 µM FU, supports our interpretation that FU^r clones cannot arise in a wild type population unless there is sufficient growth of the wild type cells (prior to growth cessation) for mutagenesis to take place.

Since FU itself is not mutagenic, the appearance of FU-resistant colonies in G12 cells grown in FU must have arisen as a result of spontaneous (endogenous) mutagenesis during just a few generations of growth. MT blocks spontaneous mutagenesis, which may be primarily caused by endogenous oxidants, since antioxidants also block it [12]. The lack of FU-resistant clones arising in MTI-2A cells is most likely a result of the antimutagenic effect of their MT content.

Since up-regulation of MT in the tumor cells does not limit the toxicity of FU (data not shown), the antimutagenic effects of MT-up-regulation may provide a strategy for preventing FU-resistance in colon tumors. MT's are inducible by a wide variety of compounds, and some combinations are synergistic [13, 14]. It should therefore be possible to up-regulate MT expression in tumors prior to and concommitant with FU chemotherapy. Studies are now underway using human colon cancer cells to prevent mutagenesis to FU-resistance by MT induction using single agents as well as mixtures.

Acknowledgements
This study was supported by NCI grant CA61319 and Kaplan Cancer Center grant CA13343, and is part of New York University Medical Center's NIEHS Center supported by grant ES00260.

References

1. Kägi JHR (1991) Overview of metallothionein. *Meth Enzymol* 205: 613–626.
2. Goncharova EI, Rossman TG (1994) A role for metallothionein and zinc in spontaneous mutagenesis. *Cancer Res* 54: 5318–5323.
3. Satoh M, Cherian MG, Imura N, Shimizu M (1994) Modulation of resistance to anticancer drugs by inhibition of metallothionein synthesis. *Cancer Res* 54: 5255–5257.
4. Ebadi M, Iversen PL (1994) Metallothionein in carcinogenesis and cancer chemotherapy. *General Pharmacol* 25: 1297–1310.
5. Klein CB, Rossman TG (1990) Transgenic Chinese hamster V79 cell lines which exhibit variable levels of *gpt* mutagenesis. *Environ Mol Mutagen* 16: 1–12.

6. Pinedo HM, Peters GJ (1988) 5-Fluorouracil: biochemistry and pharmacology. *J Clin Oncol* 6: 1653–1664.
7. Peters GJ, Van der Wilt CL, Van Groeningen CJ, Meijer S, Smid K, Pinedo HM (1994) Thymidylate synthase inhibition after administration of 5-fluorouracil. *J Clin Oncol* 12: 2035–2042.
8. Landolph JR, Jones PA (1982) Mutagenicity of 5-azacytidine and related nucleosides in C3H/10T 1/2 clone 8 and V79 cells. *Cancer Res* 42(3): 817–823.
9. Warner HR, Rockstroh PA (1980) Incorporation and excision of 5-fluorouracil from deoxyribonucleic acid in Escherichia coli. *J Bacteriol* 14(2): 680–686.
10. Soborero AF, Aschele C, Guglielmi AP, Mori AM, Melioli GG, Rosso R, Bertino JR (1993) Synergism and lack of cross-resistance between short-term and continuous exposure to fluorouracil in human colon adeno-carcinoma cells. *J Nat Cancer Inst* 85(23): 1937–1944.
11. Spears CP, Gustavsson BG, Berne M, Frosing R, Bernstein L, Hayes AA (1994) Mechanisms of innate resistance to thymidylate synthase inhibition after 5-fluorouracil. *Cancer Res* 48: 5894–5900.
12. Goncharova EI, Nádas A, Rossman TG (1996) Serum deprivation, but not inhibition of growth *per se*, induces a hypermutable state in Chinese hamster G12 cells. *Cancer Res* 56: 752–756.
13. Bauman JWE, Liu Y, Liu YP, Klaassen CD (1991) Increase in metallothionein produced by chemicals that induce oxidative stress. *Toxicol Applied Pharmacol* 110: 3 47–354.
14. Coyle P, Philcox JC, Rofe AM (1993) Corticosterone enhances the zinc and interluekin-6-mediated induction of metallothionein in cultured rat hepatocytes. *J Nutr* 123(9): 1464–1470.

Metallothionein IV
C. Klaassen (ed.)
© 1999 Birkhäuser Verlag Basel/Switzerland

A chemical mechanism for inactivation of anticancer drugs by metallothionein

Catherine Fenselau, Miquel D. Antoine, Daniele Fabris, Yetrib Hathout, Tao He and Joseph Zaia

Department of Chemistry and Biochemistry, University of Maryland Baltimore County, 1000 Hilltop Circle, Baltimore, MD 21250, USA

A role for inducible metallothionein in acquired drug resistance has been proposed by several laboratories [1, 2 and others]. We have undertaken to test the hypothesis that deactivation of therapeutic alkylating agents occurs by covalent sequestration by metallothionein. We propose that these reactions make zinc more available to activate the regulatory elements that control transcription and, in turn, the synthesis of more protein. It should be pointed out that other inducible proteins are already considered to act in acquired drug resistance, including glycoprotein-53, and the glutathione-S-transferase family. The latter inactivates therapeutic nitrogen mustards by catalyzing covalent bond formation with the thiol group of the cofactor glutathione.

Earlier studies from this laboratory have confirmed the formation of covalent adducts between rabbit liver metallothionein ñ2A and the anticancer agents melphalan [3] and chlorambucil [4]. Surprisingly, these reactions did not occur randomly across the twenty thiol groups in metallothionein, but rather, both nitrogen mustards reacted selectively with cysteine residues 33 and 48 (Tab. 1). Molecular modeling indicates that a favorable binding site for the reactive aziridinium ion intermediates formed from both chlorambucil and melphalan exists near the zinc ion that is co-chelated by the terminal thiolates in Cys-48 and Cys-33. An apparent first order kinetic rate, measured by monitoring the reaction on-line with electrospray ionization mass spectrometry [5], is consistent with the hypothesis that alkylation takes place within a non-covalent complex preformed between metallothionein and the nitrogen mustard aziridinium ion [3, 4]. It is important to note that one and two drug moieties were found to be covalently bound without demetallation of the protein. This metal retention by the Cys chelating residues was also observed in electrospray mass spectrometry studies of a holo zinc finger polypeptide after methylation [6]. The highly resistant holo-MT becomes more susceptible to proteolysis by trypsin when it is alkylated [3]. Nonetheless, drug-alkylated holo-MT is still far more resistant to proteolysis than denatured MT.

Table 1. Extent of alkylation at cysteine residues in MT-2A

Drug	Cys-33	Cys-48
Melphalan	23%	66%
Chlorambucil	26%	66%
Mechlorethamine	-	> 90%

Recent *in vitro* studies monitored on-line with electrospray mass spectrometry indicate that the small therapeutic mustard mechlorethamine initially forms a single bond with rabbit liver metallothionein, retaining seven metal ions [7]. Tryptic peptides from the denatured methylated product include both MT alkylated selectively at Cys-48 and also MT cross-linked between Cys-48 and a residue near the carboxyl terminus. Again molecular modeling indicates a favorable binding site for the mechlorethamine aziridinium intermediate near Cys-48 [8]. Thus the interaction of rabbit liver MT with all three therapeutic agents is consistent with the hypothesis that induced MT contributes to acquired drug resistance in cancer patients by the mechanism of covalent sequestration.

These coherent observations made *in vitro* have recently been tested in human bladder tumor T 24 cells in culture [9]. These cells were chosen because they have high levels of basal MT. Additional MT synthesis was induced by exposure to cadmium chloride. Isoforms of MT were identified, and relative amounts were quantified before and after induction. Analysis by capillary electrophoresis indicated that the hMT-2A isoform was the most abundant before induction and after induction (> 50%). Consequently, our analytical efforts focussed on drug adducts of hMT-2A. In addition to MT-1A, MT-1F, MT-1G, MT-1X and MT-2A, all with acetylated amino termini, unacetylated forms of MT-1X and MT-2A were detected in cadmium ion-induced cells. When the induced cells were exposed to chlorambucil, MT was isolated and found to be carrying up to four molecules of the drug per molecule of protein. Semiquantitative analysis of peptides produced by tryptic digestion indicates that the majority of the drug is bonded to peptides that contain Cys-33 and Cys-48, consistent with *in vitro* studies.

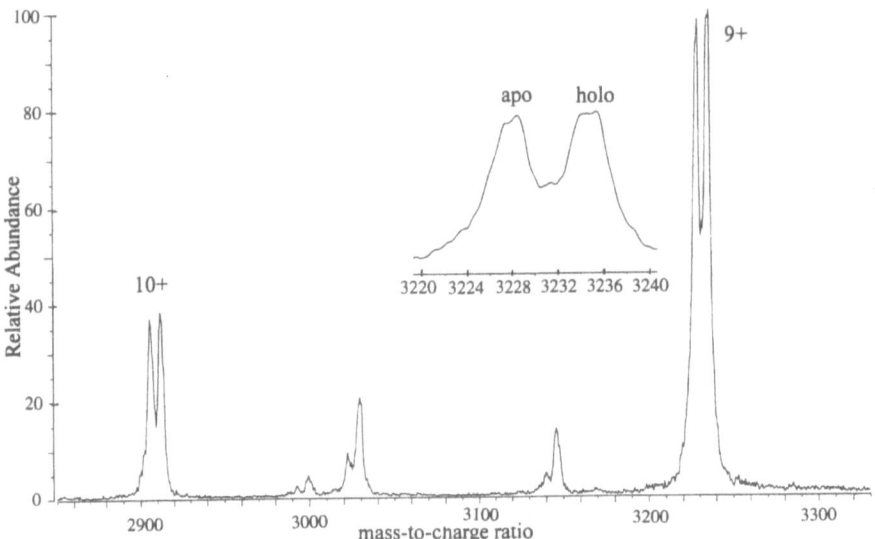

Figure 1. Partial electrospray mass spectrum of carbonic anhydrase (4.5 µM) incubated on line with rabbit Zn7-MT-2A (1.5 µM) for 43.5 min in aqueous solution, pH 7.0.

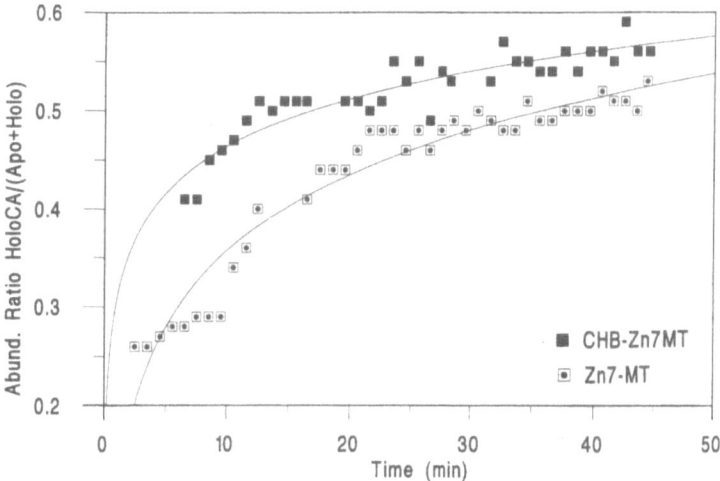

Figure 2. The ratio of holo-carbonic anhydrase to total carbonic anhydrase during incubation with rabbit Zn7-MT-2A (bottom curve) and chlorambucil-alkylated Zn7-MT-2A (top curve). Conditions as in Figure 1.

A role in acquired drug resistance requires that MT be continuously available. Many studies of MT genetics indicate that the release of zinc ions from MT can activate transcription via one or more zinc-binding regulatory proteins [10, 11 and others]. Although alkylation by one or two drug moieties did not directly release metal ions, increased proteolytic susceptibility was observed, which can lead to metal ion release. It is also likely that multiple alkylation eventually denatures the protein. However, the transfer of metal ions from MT to many proteins is thought to occur by direct displacement, not by release to the medium. Consequently, the transfer of zinc ions from MT was compared to that from chlorambucil-alkylated MT, using apo-carbonic anhydrase and the metal-free HIV nucleocapsid RNA binding protein as recipients. The method is illustrated in Figure 1, with electrospray mass spectra of apo-carbonic anhydrase recorded at 43 min as its reaction with Zn7-MT was carried out connected on-line to the electrospray mass spectrometer. The resolution is sufficient to allow the +9 ions of apo-CA and holo-CA to be distinguished. The molecular mass observed for holo-carbonic anhydrase (29105, calculated 29108.1) includes the water molecule that forms the fourth zinc ion ligand. A comparison of the rates of metal ion transfer to carbonic anhydrase is shown in Figure 2, where it can be seen that zinc ions are more readily available from the drug-modified Zn7-MT [12]. A related study with the HIV nucleocapsid protein (Nc p7) is shown in Figure 3, where again zinc ions are more extensively transferred from drug-modified rabbit liver MT-2A [13]. These experiments are consistent with a resistance mechanism involving covalent sequestration and zinc ion activated transcription of new protein.

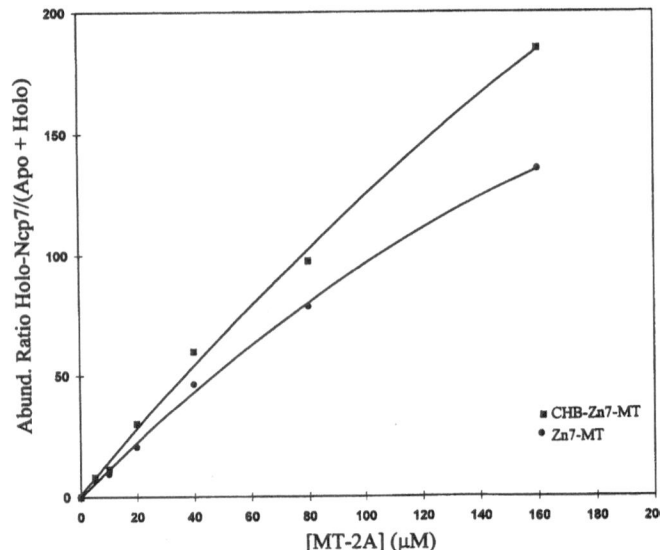

Figure 3. The ratio of holo Nc p7 to total Nc p7 as a function of concentration of rabbit Zn7-MT-2A (bottom curve) and concentration of chlorambucil-alkylated rabbit Zn7-MT-2A (top curve). Aqueous solutions pH 7.0 contained 100 μM apo Nc p7 protein [4-55]. Spectra were measured at 2 min, after which no change was observed.

References

1.   Endresen L, Bakka A, Rugstad H (1983) Increased Resistance to Chlorambucil in Cultured Cells with a High Concentration of Cytoplasmic Metallothionein. *Cancer Res* 43: 2918–2926.
2.   Kelley S, Basu A, Teicher B, Hacker M, Hamer D, Lazo J (1988) Overexpression of Metallothionein Confers Resistance to Anticancer Drugs. *Science* 241: 1813–1815.
3.   Yu X, Wu Z, Fenselau C (1995) Covalent Sequestration of Melphalan by Metallothionein and Selective Alkylation of Cysteines *in vitro. Biochemistry* 34: 3377–3385.
4.   Zaia J, Jiang L, Han M, Tabb J, Wu Z, Fabris D, Fenselau C (1996) A Binding Site for Chlorambucil on Metallothionein. *Biochemistry* 35: 2830–2835.
5.   Yu X, Wojciechowski M, Fenselau C (1993) Assessment of Metals in a Metalloprotein by Electrospray Mass Spectrometry. *Anal Chem* 65: 1355–1359.
6.   Fabris D, Zaia J, Hathout Y, Fenselau C (1996) Retention of Thiol Protons in Two Classes of Protein Zinc Ion Coordination Centers. *J Amer Chem Soc* 118: 12242–12243.
7.   Antoine M (1997) Selective Alkylation of Metallothionein by the Nitrogen Mustard Mechlorethamine. Ph.D. Thesis, University of Maryland Baltimore County, Baltimore, Maryland.
8.   Antoine M, Fabris D, Fenselau C (1998) Covalent sequestration of the nitrogen mustard mechlorethamine by metallothionein. *Drug Metab Dispos* 26: 921–926.
9.   He T (1997) Interaction of Metallothionein and the Anti-Cancer Drug Chlorambucil in Cultured Tumor Cells. Ph.D. Thesis, University of Maryland Baltimore County, Baltimore, Maryland.
10. Maret W (1994) Oxidative metal release from metallothionein via zinc-thiol/disulfide interchange. *Proc Natl Acad Sci USA* 91: 237–241.
11. Palmiter R (1994) Regulation of metallothionein genes by heavy metals appears to be mediated by a zinc-sensitive inhibitor that interacts with a constitutively active transcription factor, MTF-1. *Proc Natl Acad Sci USA* 91: 1219–1223.
12. zaia J, Fabris D, Wei D, Karpel R, Fenselau C (1998) Biomedical Mass Spectrometry: Blood, Sweat and Urine. Proceedings of the 44th ASMS Conference on Mass Spectrometry and Allied Topics, Portland, OR, May 12–16.
13. Hathout Y, Fabris D, Fenselau C (1997) Monitoring metal ion flux in reactions of metallothionein and drug-modified metallothionein by electtrospray mass spectroscopy. *Protein Science* 7 November.

The possible role of metallothionein in cancer therapy by copper acetate

Qing Huai[1], Xingwang Fang[1], Jilan Wu[1], Wenqing Wang[1], Guojian Huang[2] and Shu Zhou[2]

[1]Department of Technical Physics, Peking University, Beijing 100871, P.R. China
[2]Institute of Clinical Medical Sciences, China-Japan Friendship Hospital, Beijing 100029, P.R. China

In traditional Chinese medicine, copper has been used in the treatment of diseases including cancer, for thousands of years. Since 1978, cupric recipes consisting of copper acetate and Chinese herbs have been proven to be clinically effective in the treatment of some human neoplasms without significant adverse side-effect, and it is more efficient when combined with chemotherapy [1]. Experimental results show that the average inhibitory rate for mice liver solid tumor (H_{22}) growth by copper acetate with oral administration in dosage of 150 mg/kg was 55%. Why is it that copper can be used as an agent for eliminating evils and is itself not harmful to the body vitality? The aim of this paper is to study the anticarcinogenic mechanism of copper, and the relationship between copper and metallothionein (MT).

Copper and MT level in tumor bearing mice after oral administration of copper

The contents of copper and metallothionein (MT) in tumor-bearing mice before and after oral administration of copper were examined in comparison with normal mice. After 10 days' oral administration of copper acetate in doses of 150 mg/kg, the level of copper in liver and tumor of the tumor-bearing mice increased significantly, but didn't change in kidney. The concentration of MT in mice liver increased after copper administration and most of the copper in liver was bound to MT, but MT level in tumor decreased.

Why does MT level decrease in tumor after ten days' copper administration and how is the metallothionein-bound copper released? Several events could lead to cleavage of the copper-thiolate center in MT, including the copper-sulphur clusters destroyed by activated leukocytes and/or H_2O_2-generating enzymes. According to the results of Szatrowski et al. [2], we suggested that the reason for the release of MT-bound copper may be that H_2O_2 level in tumor cells is higher than in normal cells.

Naganuma et. al. [3] reported that pre-injection of $CuSO_4$ (40 μmol/kg) could markedly reduce the renal toxicity of CDDP but scarcely affects its anti-cancer activity. Our experiments reveal that copper administration results in marked decrease in MT in tumor but increase in liver, which suggests that MT plays an important role in the detoxification of copper in liver but without reduction in the antitumor activity of copper.

Level of ·OH generating from the Cu^{2+}/H_2O_2 system in the presence of MT

Table 1 shows the amount of ·OH generated from Cu^{2+}/H_2O_2 system in the presence of ZnMT by the 2-hydroterephthalate dosimeter [4]. In Table 1, the content of.OH increases with the increasing incubation time. When small amounts of MT ($[MT]/[Cu^{2+}] = 1/48$, 1/24 and 1/12) was given in the Cu^{2+}/H_2O_2 system, the OH generated was more than in the absence of MT. But the generated ·OH decreases with increasing concentration of MT. When the concentration of MT was one-third that of Cu^{2+}, ·OH could not be detected in the system.

Table 1. Production of.OH radical from Cu^{2+}/H_2O_2 system in the presence of MT[a]

$[MT]/[Cu^{2+}]$		0	1/48	1/24	1/12	1/8	1/6	1/3
	30 min[b]	0.25	0.80	0.78	0.54	0.15	0.07	< 0.06
2-hydroterephthalate	60 min[b]	0.50	1.50	1.49	1.09	0.60	0.15	< 0.08
($\times 10^{-6}$ mol·dm^{-3})	100 min[b]	0.95	2.50	2.35	1.96	1.20	0.39	< 0.08

[a]Initial concentration of terephthalate, 1×10^{-3} mol·dm^{-3}; Cu^{2+}, 1×10^{-4} mol·dm^{-3}; H_2O_2, 2×10^{-4} mol·dm^{-3}.
[b]Incubation time.

Due to the slow reduction of Cu^{2+} by H_2O_2 [5] and the presence of Fenton reaction, the amount of ·OH generated from the Cu^{2+}/H_2O_2 system increases slowly with increasing incubation time.

$$Cu^{2+} + H_2O_2 \rightarrow Cu^+ + HO_2 (H^+ + O_2^{·-}) + H^+ \qquad \text{(slow reaction)} \qquad (1)$$

$$Cu^+ + H_2O_2 \rightarrow Cu^{2+} + ·OH + OH^- \qquad \text{(Fenton reaction)} \qquad (2)$$

Addition of metallothionein can alter the rate of scavenging.OH. When MT and Cu^{2+} were both present together, MT (-SH) was oxidized by Cu^{2+} (cysteine changes to cystine) and Cu^{2+} was reduced to Cu^+ (reaction (3)), which accelerated the.OH formation rate. When the content of cysteine residues in MT is reduced, its ability of scavenge ·OH is negligible at the ratio of $[MT]/[Cu^{2+}] < 1/20$. However, at the ratio of $[MT]/[Cu^{2+}] > 1/6$, the ·OH was obviously inhibited by MT.

$$RS^- + Cu^{2+} \rightarrow RS· + Cu^+ \qquad (3)$$

$$MT (RS^-) + ·OH \rightarrow RS· + OH^- \qquad (4)$$

$$RS· + RS^- = RSS^-·R \qquad (5)$$

$$RS· + RSS^-·R \rightarrow RSSR + RS^- \qquad (6)$$

$$RS· + RS· \rightarrow RSSR \qquad (7)$$

The hydroxyl radical reacts at extremely high rate constants with biological molecules in its vicinity. If the concentration of ·OH is higher than normal level, it has strong toxicity to cells, so it may play an important role in carcinoma development and cancer therapy. When both H_2O_2 and copper ion are available *in vitro* and/or *in vivo*, the ·OH radicals are generated by the Fenton reaction. It means that MT not only scavenges ·OH at its high concentration but also promotes generation of ·OH at its low concentration in Cu^{2+}/H_2O_2 system.

Copper-catalyzed DNA damage in the presence of ZnMT

By comparing Cu^+/H_2O_2 with Cu^{2+}/H_2O_2 system, Cu^+ is found to be more effective than Cu^{2+} in preventing the damage to DNA. In the presence of ZnMT, the damage to DNA increases greatly in the Cu^{2+}/H_2O_2 system; however, the damage decreases in the Cu^+/H_2O_2 system. The relationship between DNA damage and the ratio of $[Cu^+]/[MT]$ or $[Cu^{2+}]/[MT]$ has been established.

The results show that DNA damage decreases with increasing concentration of ZnMT in the Cu^+/H_2O_2 system. It can be explained as follows: MT is both an effective ·OH scavenger and a copper chelator, and thus it inhibits DNA damage in the Cu^+/H_2O_2 system. But in the Cu^{2+}/H_2O_2 system, when the ratio of $[Cu^{2+}]/[MT]$ is equal to 12, the damage to DNA reaches a maximum.

Active oxygen species have been assumed to play an important role in oxidative DNA damage. Stoewe et. al. estimated the efficiency of DNA damage via reaction (9) and (10) [6] and found that the main differences between the two models of.OH-induced damage are that.OH scavengers can efficiently prevent reaction (10) but hardly reaction (9). Reaction (9) may occur with some region-specificity near the sites of Cu(I) fixation.

$$DNA + Cu(I) = DNA\text{-}Cu(I) \tag{8}$$

$$DNA\text{-}Cu(I) + H_2O_2 \rightarrow DNA\text{-}Cu(II)\text{-}\cdot OH + OH^- \rightarrow damage \tag{9}$$

$$DNA + \cdot OH \rightarrow DNA\text{-}\cdot OH \rightarrow damage \tag{10}$$

Since MT bears 2–3 negative charges at physiological pH, the negatively charged thiols are kept away from DNA. In order to improve the protective effect of MT on DNA we demonstrated that MT is partially oxidized by.OH radical, reasulting in the unsaturated coordination of Zn^{2+} (Cu^+) in MT. Accordingly oxygen atoms of phosphate groups in DNA may take part in the coordination of Zn^{2+} (Cu^+), leading to the binding of MT to DNA. The extent of this binding depends on the extent of oxidation of MT [7].

Therefore, in Cu^{2+}/H_2O_2 system, when the concentration of MT is low, MT can reduce Cu^{2+} to Cu^+, and Cu^{2+} can oxidize $MT(RS^-)$ to RSSR, thereby partly destroying the structure of MT. Then Cu^+ can bind to DNA so that site-specific damage to DNA takes place. When the concentration of MT is high, MT not only reduces Cu^{2+} to Cu^+ but also binds with Cu^+. As the binding stability constant of Cu+ with MT is much higher than that with DNA, Cu+ is mainly bound to MT in solution. Besides, MT also can scavenge OH. Thus, the degree of DNA damage decreases with increasing MT concentration [8, 9].

It can be concluded that MT may play an important role in the anticancer effect of copper; not only can it detoxify copper in normal tissue but it also makes copper more toxic to tumor tissue.

Acknowledgement
This work was supported by the National Natural Science Foundation of China.

References

1. Liu TQ (1996) *Discovery of the Mystery of the Cancer.* International Culture Press, Beijing.
2. Szatrowski TP and Nathan CF (1991) Production of Large Amounts of Hydrogen Peroxide by Human Tumor Cell. *Cancer Res* 51: 794–798.
3. Naganuma A, Satoh M, Imura N (1984) Effect of Copper Pretreatment on Toxicity and Antitumor Activity. *Res Commun Chem Pathol Pharmacol* 46(2): 265.
4. Matthews RW (1980) The Radiation Chemistry of the Terephthalate Dosimeter. *Radiat Res* 83: 27–41.
5. Florence TM (1984) The Production of Hydroxyl Radical from Hydrogen Peroxide. *J Inorg Biochem* 22: 221–230.
6. Stoewe R, Prutz WA (1987) Copper-Catalyzed DNA Damage by Ascorbate and Hydrogen peroxide: Kinetics and Yield. *Free Radical Biol Med* 3: 95–105.
7. Prutz WA, Butler J, Land EJ (1990) Interaction of Copper (I) with Nucleic Acids. *Int J.Radiat Biol* 58: 215–234.
8. Fang XW, Wu JL and Wei GS (1997) The association of metallothionein with Phosphate. *Radiat Phys Chem* 46: 111–113.
9. Fang XW, Wu JL and Wei GS (1997) Irradiation-induced binding of metallothionein to DNA. *Radiat Phys Chem* 50: 471–473.

Interaction of metallothionein with the carcinogenic metals Ni(II), Cr(VI) and As(III)

Dean E. Wilcox, Laura L. Bennett, Elizabeth H. Cox, George Haleblian, Brian T. Hill, Eric P. Kowack, Xiaoyan Liu, Jane S. Merkel, Amy E. Palmer, Matthew C. Posewitz, Joann E. Roy and Karen E. Wetterhahn

Department of Chemistry, Dartmouth College, Hanover, NH 03755, USA

Summary. Through a combination of cell culture studies and *in vitro* experiments we have begun to investigate whether MT plays a role in modulating the toxicity and genotoxicity of Ni(II), Cr(VI) and As(III). Our data suggest that MT participates in oxidant-generating redox reactions with Cr(VI) and As(III) that increase the toxicity of these metals and create potentially DNA-damaging species. In the case of Ni(II), MT appears to play a protective role, scavenging radical(s) generated by Ni(II) and H_2O_2, and preventing Ni(II) from replacing Zn(II) in zinc fingers of transcription factors.

Introduction

Since its discovery 40 years ago, metallothionein (MT) has been studied not only for its role in the detoxification of heavy metals but also in a variety of other physiological processes, including metabolism of the essential metals zinc and copper, cellular response to oxidative stress, and interaction with Pt(II) and Au(I) metallotheraputic agents [1]. MT may also be involved in carcinogenesis by metals such as Ni(II), Cr(VI) and As(III).

Certain carcinogenic metals are believed to be genotoxic through redox reactions that lead to DNA lesions [2]. This has been demonstrated most clearly for Cr(VI) and is explained by the uptake-reduction model [3], which postulates that chromate (CrO_4^{-2}) is readily transported into cells where its reduction to Cr(III) by intracellular reductants, such as glutathione (GSH) and ascorbate, generates oxidizing and radical species that damage DNA. Although less is known about the mechanism of Ni(II) carcinogenicity, Ni(II) can produce reactive oxygen species that damage DNA [4]. Finally, the co-carcinogen As(III) has a high affinity for thiols and appears to interact predominantly with DNA repair enzymes [5], but it is also a potent inducer of MT biosynthesis [6].

The high Cys content of MT suggests that it could interact in a number of ways with these carcinogenic metals, serving as a reductant for Cr(VI), scavenging radicals generated by Ni(II) or Cr(VI) redox reactions, and sequestering As(III). We have begun to study the role of MT in Ni(II), Cr(VI) and As(III) carcinogenesis by: 1) cell culture studies that evaluate whether MT modulates metal cytotoxicity and oxidative damage, and 2) *in vitro* studies that examine the direct interaction of MT with these carcinogenic metal ions.

Results and discussion

Cell culture studies

These studies used CHO cells (K1-2 cell lines) transfected with multiple copies of the human MT IIA gene and overexpressing human MT upon metal pressure [7]. High MT levels were maintained in transfected cells by treatment with either 15 µM CdCl$_2$ and 40 µM ZnCl$_2$ (K1-2MT) or just 40 µM ZnCl$_2$ (K1-2MT(Zn)), providing elevated cellular levels of predominantly Cd$_5$Zn$_2$MT and Zn$_7$MT, respectively. Cytotoxicity was measured with the calcein-AM cell viability assay [8]. L-buthionine-[S,R]-sulfoximine (BSO) treatment was used to lower glutathione (GSH) levels, allowing the effects of MT to be distinguished from those of GSH. Cellular oxidants were quantified by the 2',7'-dichlorofluorescin (DCF) assay [9].

Ni(II)

Parental and MT-transfected cells showed nearly identical dose-dependent sensitivity to Ni(II), both with and without GSH depletion. However, when 50 µM H$_2$O$_2$ was administered with the Ni(II), a significant decrease in Ni(II) toxicity was observed in the MT-transfected cells (Fig. 1), both with and without GSH depletion. Addition of the same dose of Zn(II) to the parental cell media provided no protection against Ni(II) toxicity. Cellular oxidant levels were somewhat higher in MT-transfected cells than in parental cells, but oxidant levels in all cells did not change significantly upon treatment with cytotoxic levels of Ni(II). However, treatment with 50 µM H$_2$O$_2$ doubled the level of cellular oxidants in all cells and oxidant levels increased somewhat further with increasing Ni(II) dose. GSH depletion only slightly increased the oxidant levels,

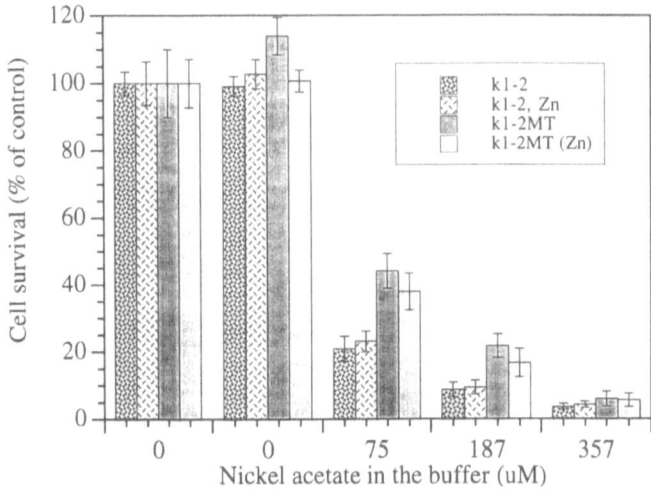

Figure 1. Relative survival of CHO cells after 90 min treatment with 50 µM H$_2$O$_2$ and the indicated dose of nickel acetate; control group (first "0") received no peroxide treatment.

indicating that GSH does not provide significant protection from elevated H_2O_2 levels in these cells. Thus, Zn_7MT or Cd_5Zn_2MT do not effectively scavenge oxidants associated with H_2O_2 treatment but appear to protect CHO cells from damage by specific oxidant(s), possibly OH·, produced by redox reaction(s) of Ni(II) and H_2O_2.

Cr(VI)

The MT-transfected cells are more sensitive to Cr(VI) than are parental cells, and levels of cytosolic oxidants associated with Cr(VI) treatment are higher in MT-transfected cells (Fig. 2), both with and without GSH depletion. In this case, MT appears to be involved in a redox reaction with Cr(VI) that generates oxidants, which are not effectively scavenged by GSH and which appear to correlate with the enhanced sensitivity of MT-transfected cells to Cr(VI).

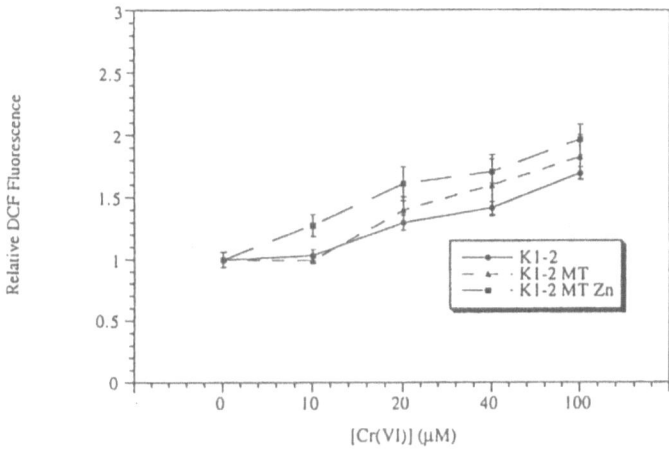

Figure 2. Relative cytosolic oxidant levels in CHO cells as indicated by DCF fluorescence after 30 min of incubation with 5 µM DCF and 90 min treatment with the indicated dose of Cr(VI) as sodium dichromate.

As(III)

The MT-transfected cells are significantly more sensitive to As(III) than are parental cells (Fig. 3), both with and without GSH depletion, and addition of the same Zn(II) dose to the parental cell media did not significantly alter the As(III) toxicity. The level of cellular oxidants in all cells remains fairly constant with increasing As(III) dose, except for a noticeable increase in MT-transfected GSH-depleted cells (Fig. 4). (Kowack, E.P., Liu, X., Lohrer, H., Wetterhahn, K.E., Wilcox, D.E., unpublished results) Thus, MT appears to be involved in a redox reaction with As(III) that generates oxidants, which are observed upon GSH depletion but are effectively scavenged by GSH at its normal cellular level. The increased sensitivity of MT-transfected cells to As(III), however, is not directly due to these oxidants because they are effectively scavenged by GSH in cells not treated with BSO but untreated MT-transfected cells are still more sensi-

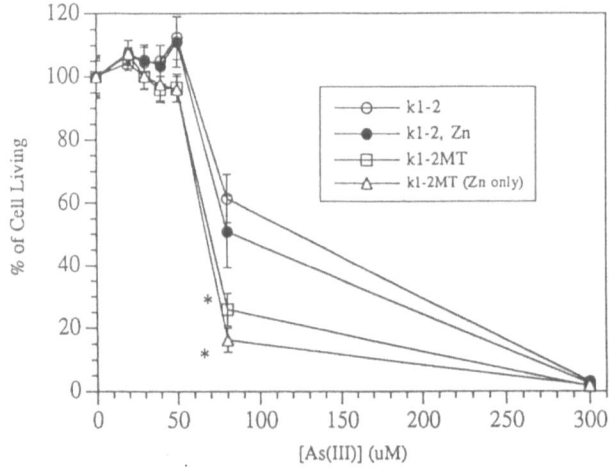

Figure 3. Relative survival of CHO cells after 24 h treatment with the indicated dose of sodium arsenite.

tive to As(III) than are untreated parental cells. Thus, conditions that lower GSH levels (reaction with As(III)-generated oxidants or BSO treatment) appear to supress GSH-dependent As(III) detoxification pathways (e.g. multidrug resistance [10] and As methylation [11]).

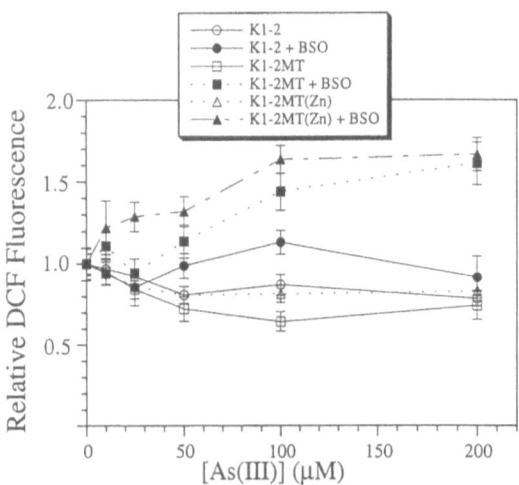

Figure 4. Relative cytosolic oxidant levels in CHO cells as indicated by DCF fluorescence after 30 min of incubation with 10 µM DCF and 60 min treatment with the indicated dose of sodium arsenite.

Mechanistic studies

Ni(II)

Nearly a decade ago, Sunderman and Barber suggested that substitution of carcinogenic metals for Zn(II) in zinc-binding domains of transcription factors may be a mechanism for metal carcinogensis [12]. Since MT binds carcinogenic metal ions, it could play a role in modulating their availability for these transcription factors and thereby affect metal carcinogenicity. Key observations pertinent to this model are: 1) Ni(II), Co(II), Cd(II), As(III) and other carcinogenic metals induce MT biosynthesis [1], and 2) it has been shown that MT is capable of eliminating the ability of the transcription factors Sp1 [13] and TFIIIA [14] to bind to DNA, presumably by sequestering Zn(II) and eliminating the protein structure required for DNA recognition and binding. These results form the basis of the following hypothesis: Ni(II) induces the biosynthesis of MT, which removes Zn(II) from "zinc fingers" in transcription factors, allowing Ni(II) to bind to these proteins and resulting in altered gene expression and/or redox damage to DNA or the transcription factor.

To test this hypothesis, we have studied initially the metal-binding properties of zinc finger peptides for comparison to the metal ion affinity of MT [15]. The zinc finger peptides were prepared by automated solid phase peptide synthesis with Fmoc protected amino acids, purified by reverse phase HPLC, and characterized by analytical HPLC and mass spectrometry. The peptide sequences are: finger #3 from Sp1 (Sp1-3): K F A *C* P E *C* P K R F M R S D H L S K *H* I K T *H* Q N K; finger #2 from myelin transcription factor 1 (MYT1-2): L K *C* P T P G *C* T G Q G H V N S N R N T *H* R S L S G *C* P I; and finger #2 from glucocorticoid receptor (GR-2): Y L *C* A G R N D *C* I I D K I R R K N C P A C R Y R K C L Q A G. For comparison, we note the sequence of the consensus peptide (CP-1) prepared by J. M. Berg and co-workers [16]: P Y K *C* P G *C* G K S F S Q K S D L V K *H* G R T *H* T G.

The UV CD spectrum of Sp1-3 (Fig. 5) shows the appearance of negative intensity in the 220–240 nm range upon binding Zn(II), Ni(II) and Co(II), indicating formation of α helix and β sheet 2° structure expected for this classical zinc finger. A similar result is found for GR-2, but not for MYT1-2, indicating that it does not adopt the expected structure upon binding Zn(II) and other metal ions. UV-visible absorption spectra of the Cd(II)-, Co(II)- and Ni(II)-bound forms of these three peptides show intense thiolate-to-metal charge transfer transitions in the near-UV, and the Co(II)-bound forms also exhibit relatively strong d-d transitions in the visible region that are indicative of tetrahedral coordination. These features, which are components of the $^3T_1(P) \leftarrow {}^3A_2$ transition, shift to lower energy in the series Sp1-3 > MYT1-2 > GR-2 (Fig. 6), consistent with an increasing number of "soft" thiolate ligands.

Spectrophotometric titrations of these absorption bands have been used to determine metal affinities of these peptides and Table 1 compares these values to those of other zinc finger peptides with similar metal coordination. In general, the optimized "consensus peptides" prepared by Berg and co-workers have higher metal affinities than the naturally occuring sequences. Zn(II) is bound with higher affinity than Co(II), Ni(II) or Cd(II), except for the tetra-thiolate peptides GR-2 and CP-1(CCCC) that bind Cd(II) more tightly.

Competition experiments have been used to determine the relative metal ion affinities of zinc finger peptides and MT. ^{1}H NMR data (Fig. 7) show that substoichiometric Zn(II) is distribut-

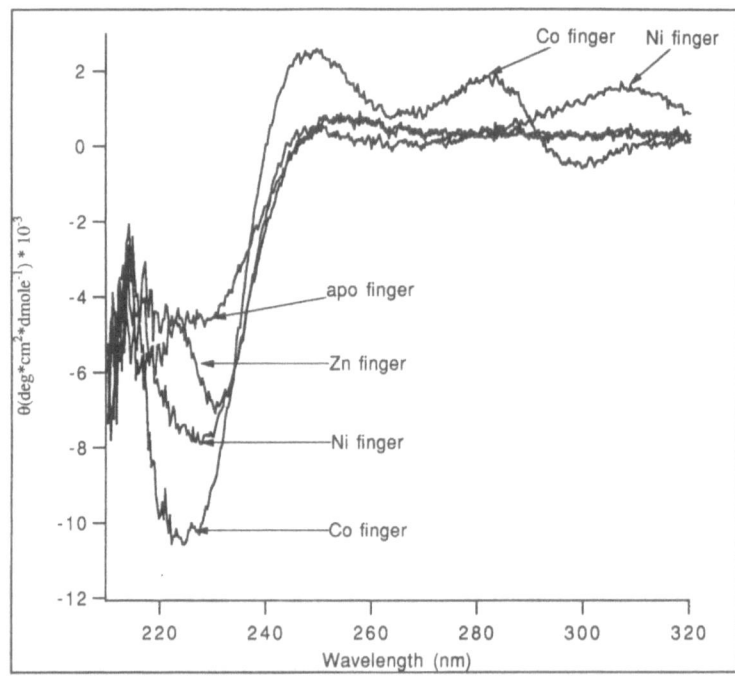

Figure 5. CD spectra of 43 µM Sp1-3 peptide and the peptide after the addition of 65 µM Zn(II), Ni(II) or Co(II) [15].

ed between the peptides in a mixture of Sp1-3 and the α domain of human MT-2 (αhMT2) (i.e. middle spectrum of Figure 7 shows features of both Sp1-3 and Zn-Sp1-3). This indicates that Sp1-3 and αhMT2 have a similar affinity for Zn(II) (Fig. 7) (K ~ 2 × 10^9), which is lower than

Table 1. Equilibrium metal ion binding constants of zinc finger peptides

	Zn(II)	Co(II)	Ni(II)	Cd(II)
CCHH ligands				
Sp1-3[a]	2×10^9	3×10^6	3×10^5	—
TFIIIA[b]	3.6×10^8	2.6×10^5	—	—
CP-1[c,d]	1.8×10^{11}	1.6×10^7	6.3×10^5	5×10^8
CCHC ligands				
MYTI-2[e]	2×10^9	2×10^6	5×10^5	9×10^8
CP-1(CCHC)[c,d]	3.1×10^{11}	1.6×10^7	8.3×10^5	1.6×10^{11}
CCCC ligands				
GR-2[f]	6×10^{10}	8×10^7	2×10^5	8×10^{16}
CP-1(CCCC)[c]	9.1×10^{11}	2.8×10^6	—	2.5×10^{13}

[a][15]; [b][17]; [c][18]; [d][19]; [e]Cox, E.H., Wilcox, D.E., unpublished results; [f]Hill, B.H., Merkel, J.S., Wilcox, D.E., unpublished results.

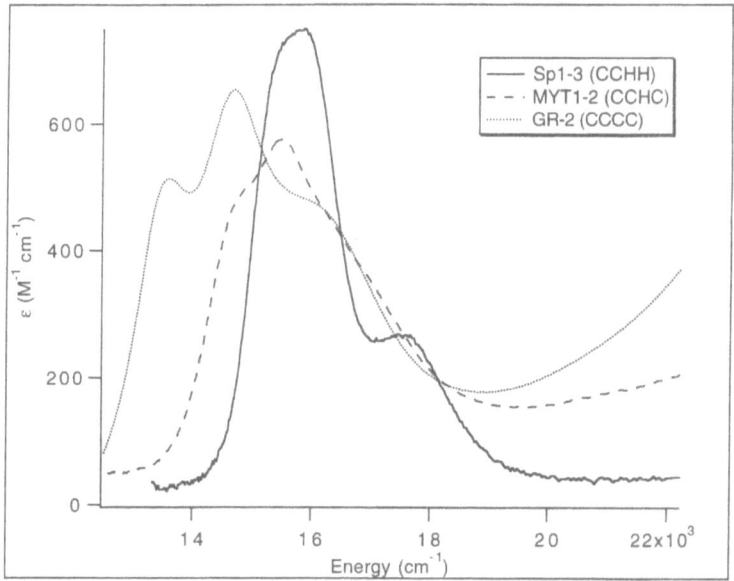

Figure 6. The visible absorption spectra of Co(II) bound to the Sp1-3, MYT1-2 and GR-2 peptides.

previously reported values for the Zn(II) affinity of MT [20, 21]. Both ^{1}H NMR (Fig. 7) and UV absorption (Fig. 8) have been used to monitor competitive binding of Zn(II) and Ni(II) by Sp1-3 and MT. These experiments show that the relative Ni(II) affinity (MT > Sp1-3) results in Ni(II) bound to MT and Zn(II) bound to Sp1-3. A similar result is predicted for MYTI-2 and GR-2 based on their metal ion affinities (Tab. 1). While this suggests that MT plays a protective role in this model for Ni carcinogenesis, relative *in vivo* concentrations of transcription factors, MT, Zn(II) and carcinogenic metals will dictate metal ion distribution among the proteins.

Cr(VI)
In the uptake-reduction model for Cr(VI) carcinogenesis [3], intracellular reductants react with chromate to generate DNA-damaging oxidants and radical species. Since MT is rich in thiols, it can be viewed as a metal-stabilized "electron reservoir" that could serve as a Cr(VI) reductant. Our cell culture results are consistent with this hypothesis, in that MT-transfected cells are more sensitive to Cr(VI) toxicity and have higher levels of oxidants. We have examined the direct redox reaction of sodium dichromate with Zn_7MT, Cd_5Zn_2MT and apo-MT, and find that only apo-MT is able to reduce Cr(VI), which it does at a rate comparable to that of Cys or GSH. These results suggest that apo-MT is available for this redox reaction in the MT-transfected CHO cells.

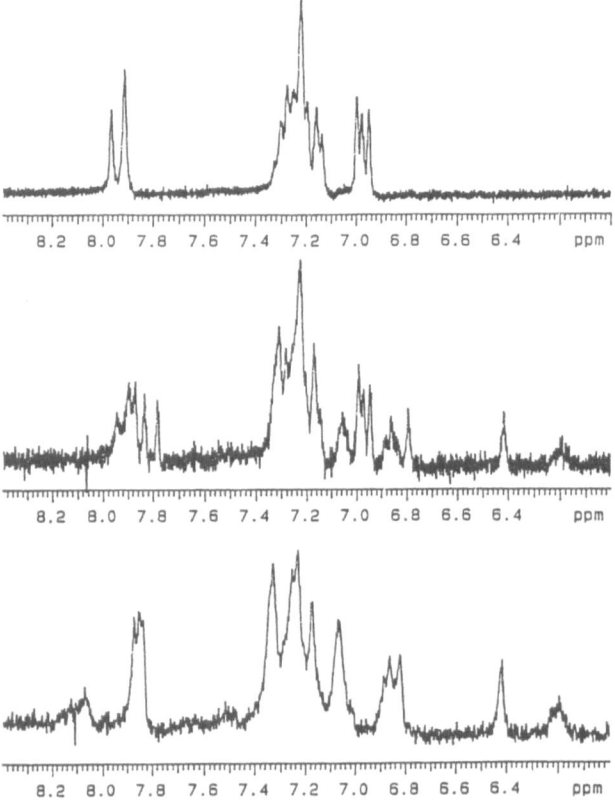

Figure 7. The ^{1}H NMR spectra in the aromatic region of (top) 3.5 mM Sp1-3 peptide, (middle) a mixture consisting of 1.7 mM Sp1-3 peptide and 0.42 mM Zn$_{3.8}$-αhMT2, and (bottom) the middle sample after addition of Ni(II) to a final concentration of 1.9 mM [15]. The bottom spectrum is identical to that of Zn-Sp1-3.

As(III)

Our cell culture results indicated that As(III) and MT generate intracellular oxidants, implicating a direct interaction of As(III) with Zn$_7$MT or Cd$_5$Zn$_2$MT in these cells. To examine this we studied As(III) binding to Zn$_7$MT, using the chromophoric chelate Zincon to quantify displaced Zn(II) [22]. We find that As(III) does not displace Zn(II) from MT even after 24 h of incubation at pH 7. Thus, a specific interaction of As(III) with MT in these cells could involve As(III) binding to metallated forms of MT, apo-MT or possibly to MT under acidic conditions (e.g. lysosomes), since it has been shown that MT has a higher affinity for As(III) than Zn(II) at pH < 5 [6].

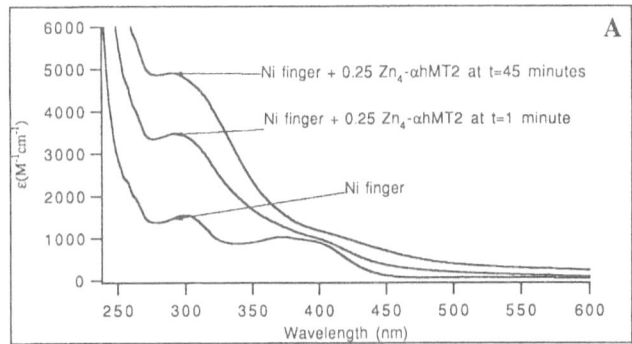

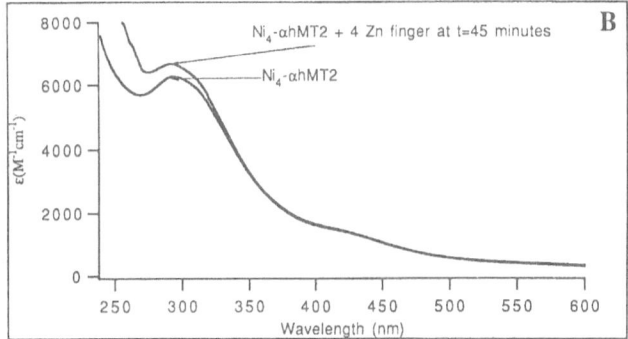

Figure 8. UV absorption spectra of (A) 149 μM Ni-Sp1-3 finger peptide and a mixture of 75 μM Ni-Sp1-3 finger peptide and 17 μM Zn_4-αhMT2 1 min and 45 min after mixing, and (B) 34 μM Ni_4-αhMT2 and a mixture of 17 μM Ni_4-αhMT2 and 75 μM Zn-Sp1-3 finger peptide 45 min after mixing [15].

Acknowledgements
This article is dedicated to our colleague, mentor and friend, Karen E. Wetterhahn (1948–1997). We thank NIH (grant CA61349) for support of this research and H. Lohrer for the K1-2 and K1-2MT cell lines.

References

1. Vallee BL (1991) Introduction to metallothionein. *Methods Enzymol* 205: 3–10.
2. A) Kasprzak KS (1991) The role of oxidative damage in metal carcinogenicity *Chem Res Toxicol* 4: 604–615, B) Klein KB, Frenkel K, Costa M (1991) The role of oxidative processess in metal carcinogenesis. *Chem Res Toxicol* 4: 592–604.
3. Wetterhahn KE, Hamilton JW (1989) Molecular basis of hexavalent chromium carcinogenicity: effect on gene expression. *Sci Total Envir* 86: 113–129.
4. Datta AK, Misra M, North SL, Kasprzak KS (1992) Enhancement by nickel(II) and L-histidine of 2'-deoxyguanosine oxidation with hydrogen peroxide. *Carcinogenesis* 13: 283–287.
5. Hartwig A (1995) Current aspects in metal genotoxicity. *BioMetals* 8: 3–1, and references therein.

6. Albores A, Koropatnick J, Cherian MG, Zelazowski AJ (1992) Arsenic induces and enhances rat hepatic met-
    allothionein production *in vivo*. *Chem-Biol Interact* 85: 127–140.
7. Kaina B, Lohrer H, Karin M, Herrlich P (1990) Overexpressed human metallothionein IIA protects Chinese
    hamster ovary cells from killing by alkylating agents. *Proc Natl Acad Sci USA* 87: 2710–2714.
8. MacCoubrey IC, Moore PL, Haugland RP (1990) Quantitative fluorescence measurements of cell viability
    (cytotoxicity) with a multi-well plate scanner. *J Cell Biol* 111: 58.
9. LeBel C, Ischiropoulos H, Bondy S (1992) Evaluation of the probe 2',7'-dichlorofluorescin as an indicator of
    reactive oxygen species formation and oxidative stress. *Chem Res Toxicol* 5: 227–231.
10. Zaman GJR, Lankelma J, van Tellingen O, Beijnens J, Dekker H, Paulusma C, Oude Elferink RPJ, Baas F,
    Borst P (1995) Role of glutathione in the export of compounds from cells by the multidrug-resistance-asso-
    ciated protein. *Proc Natl Acad Sci USA* 92: 7690–7694.
11. Zakharyan R, Wu Y, Bogdan GM, Aposhian HV (1995) Enzymatic methylation of arsenic compounds: assay,
    partial purification and properties of arsenite methyltransferase and monomethylarsonic acid methyltrans-
    ferase of rabbit liver. *Chem Res Toxicol* 8: 1029–1038.
12. Sunderman FW, Jr, Barber AM (1988) Finger-loops, oncogenes and metals. *Annal Clin Lab Sci* 18: 267–288.
13. Zeng J, Heuchel R, Schaffner W, Kägi JHR (1991) Thionein (apometallothionein) can modulate DNA bind-
    ing and transcription activation by zinc finger containing factor Sp1. *FEBS Lett* 279: 310–312.
14. Zeng J, Vallee BL, Kägi JHR (1991) Zinc transfer from transcription factor IIIA fingers to thionein clusters.
    Proc Natl Acad Sci USA 88: 9984–9988.
15. Posewitz MC, Wilcox DE (1995) Properties of the Sp1 zinc finger #3 peptide: coordination chemistry, redox
    reactions and metal binding competition with metallothionein. *Chem Res Toxicol* 8: 1020–1028.
16. Krizek BA, Amann BT, Kilfoil VJ, Merkle DL, Berg JM (1991) A consensus zinc finger peptide: design, high-
    affinity metal binding, a pH-dependent structure, and a His to Cys sequence variant. *J Amer Chem Soc* 113:
    4518–4523.
17. Berg JM, Merkle DL (1989) On the metal ion specificity of "zinc finger" proteins. *J Amer Chem Soc* 111:
    3759–3761.
18. Krizek BA, Merkle DL, Berg JM (1993) Ligand variation and metal ion binding specificity in zinc finger pep-
    tides. *Inorg Chem* 32: 937–940.
19 Krizek BA, Berg JM (1992) Complexes of zinc finger peptides with Ni(II) and Fe(II). *Inorg Chem* 31:
    2984–2986.
20. Vašák M, Kägi JHR (1983) Spectroscopic properties of metallothionein. *Metal Ions Biol Syst* 15: 213–273.
21. Petering DH, Shaw CF, III (1991) Stability constants and related equilibrium properties of metallothioneins.
    Methods Enzymol 205: 475–484.
22. Shaw CF, III, Laib JE, Savas MM, Petering DH (1990) Biphasic kinetics of aurothionein formation from gold
    sodium thiomalate: a novel metallochromic technique to probe Zn^{2+} and Cd^{2+} displacement from metalloth-
    ionein. *Inorg Chem* 29: 403–408.

Metallothionein IV
C. Klaassen (ed.)
© 1999 Birkhäuser Verlag Basel/Switzerland

A novel metallothionein gene therapy approach for chemo- and radio-protection

Tupur Husain[1], Asim B. Abdel-Mageed[2], Alan M. Miller[1, 3], Laura S. Levy[1, 3] and Krishna C. Agrawal[1, 3, 4]

[1]*Molecular and Cellular Biology Program, [2]Department of Urology, [3]Tulane Cancer Center, and [4]Department of Pharmacology, Tulane University School of Medicine, New Orleans, LA 70112, USA*

Introduction

Bone marrow toxicity induced by specific anticancer agents and/or ionizing radiation in the treatment of cancer remains a limitation in the clinical management of cancer. Hematopoietic growth factors have been employed but only partially alleviate and do not prevent the drug-induced toxicity [1] Gene therapy approaches have been attempted using a retrovirus encoding the MDR1 gene [2]. Other attempts to overcome the specific drug induced myelotoxicity observed in hematopoietic stem cells include utilizing the dihydrofolate reductase gene [3], and the aldehyde dehydrogenase gene [4].

Metallothionein (MT) is known to be a scavenger of free radicals generated upon radiation treatments and can also act as a nucleophilic protein to react with electrophilic anticancer drugs such as, cisplatinum, melphalan and cyclophosphamide. Factors that have been shown to induce metallothionein synthesis include metal ions, growth factors, cytotoxic agents, stress-producing conditions, and X irradiation [5]. Certain cytokines or exposure to radiation cause increased production of oxygen free radicals which have negative effects on DNA and cell membranes. Cytoplasmic MT has been demonstrated to interact with nitric oxide and reduce its cytotoxicity [6]. Physiologic elevations in cytoplasmic MT have an antioxidant role by interfering with membrane peroxidation [7]. Pretreatment of mice with heavy metals led to the hypothesis that intracellular MT may determine the toxicity of anticancer agents to non-malignant tissues [8]. Resistance to anticancer drugs may be modulated by altering MT synthesis [9] which may even be present at elevated levels in breast cancer cells [10]. Studies using tumor cell lines with acquired resistance to cisplatin (CDDP) and carcinoma cell lines chronically exposed to heavy metals demonstrated that overexpression of MT-IIa confers resistance to anticancer drugs [11, 12]. The effects of radiation along with alkylating agents remain elusive, however, and a further understanding of MT overexpression is necessary for the development of therapeutic approaches employing this mechanism of resistance.

Retroviral vectors have been utilized for gene transfer into multiple cell types [13]. A disadvantage of recent gene therapy attempts has been the inactivation of retroviral vectors due to DNA methylation [14]. This however may be overcome using 5-azacytidine. The manipulation of MT levels in both normal and neoplastic tissue has the potential to provide more successful chemotherapeutic intervention [15]. It has been demonstrated that human hematopoietic stem cells and cell lines displaying progenitor cell phenotypes can be models for optimal *in vitro* transduction [16] and more recently *in vivo* use [17]. We have recently initiated a novel approach by retrovirally transfecting the human metallothionein gene (MT-IIa) [18] delete into the U937 myeloid progenitor and K562 erythroid progenitor cell lines. This can lead to the development and implementation of a gene therapy protocol using retroviral vectors for MT-IIa to protect bone marrow stem cells from radiation and/or chemotherapy induced injury.

Methods and results

Retroviral vector and transfection procedure

The retroviral vector designated pLMTSN contains an insert encompassing the coding region of MT-IIa under the control of LXSN LTR promoter and a neomycin resistance gene for selection (Fig. 1). The control construct (pLXSN) contains only an antibiotic (G418) selectable marker. The vectors containing MT-IIa (pLMTSNmp and pLMTSNc) correspond to particles obtained from the mass population of cells and a selected clone, respectively. U937 and K562 cells were transfected with pLXSN or pLMTSN viral particles at a cell to viral particle ratio of 1:2. The cells in contact with the viral particles were incubated in the presence of polybrene (4 μg/ml). Cells were selected using G418 (0.5–0.75 mg/ml). Further treatments of cells included 5-azacytidine (azaC) (4 μM) and zinc acetate (100 μM).

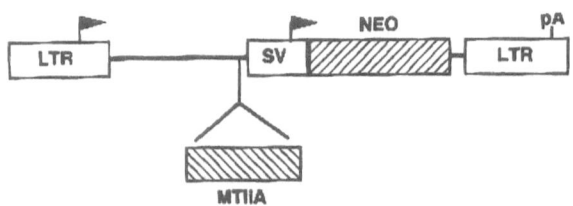

Figure 1. Diagram of the retroviral vector, LMTSN, in which the coding sequence of the human MT-IIa gene has been inserted into the multiple cloning site of the LXSN vector. Arrows indicate transcriptional start sites and direction of transcription. pA indicates the polyadenylation signal. The LTR, based on Moloney murine leukemia virus, directs expression of MT-IIa. SV indicates a fragment from SV40 containing the early promoter. The SV40 promoter directs the expression of the bacterial *neo* gene.

MT RNA and protein detection

The total cellular RNA was isolated and RT-PCR amplification (200 ng of RNA) was performed using methods we described previously [19]. Our initial data using RT-PCR indicate upregulated levels of MT-IIa following transfection of the U937 cells with the MT-IIa gene (Fig. 2). RNA isolated from the cells that were transfected using viral particles obtained from pLMTSN transfected cells had higher levels of MT transcripts when compared to pLXSN transfected cells. Immunocytochemical methods and differential silver staining of cellular lysates (25 μg) after SDS-PAGE was performed to determine protein expression. The former approach indi-

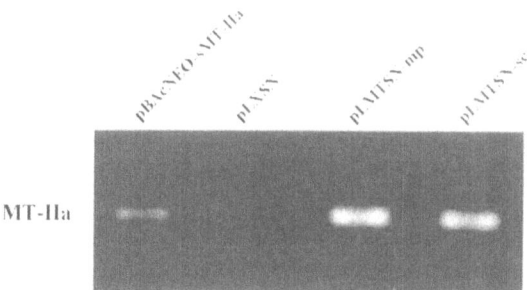

Figure 2. RT-PCR detection of MT-IIa product (171 bp) in U937 cells: pBacNeo-sMT-IIa plasmid (positive control), pLXSN (negative control), pLMTSN (MT-IIa construct – mass population), pLMTSN (MT-IIa construct – selected clone).

cated that intracellular levels of MT protein were higher in U937 cells containing the MT-IIa gene (Fig. 3). The latter approach demonstrated increased levels of MT in U937 and K562 cells. Treatment of azaC increased the expression of MT in K562 but not U937 cells (Fig. 4).

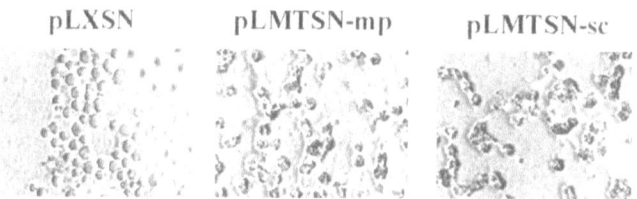

Figure 3. Immunocytochemical staining of U937 cells for the presence of MT: pLXSN (negative control), pLMTSN (MT-IIa construct – mass population), pLMTSN (MT-IIa construct – selected clone).

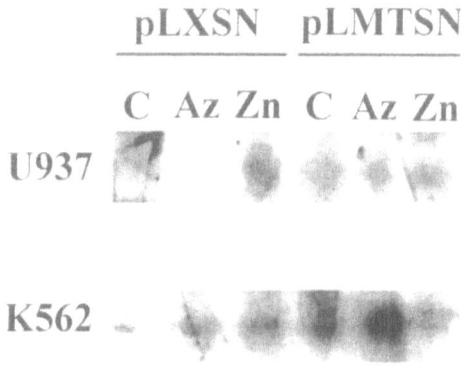

Figure 4. SDS-PAGE and differential silver staining for MT in U937 and K562 cells: pLXSN (negative controls), pLMTSN (MT-IIa); absence or presence of azacytidine (4 µM) or zinc acetate (100 µM).

Colony forming unit survival assays

U937 cell were plated in 0.3% Bacto-agar/RPMI medium, colony forming units (cfu) were quantitated after 7d by light microscopy. The preliminary results indicated a 5.1–7.4 fold increase in survival of the cells transfected with MT-IIa subjected to 100 cGy (Fig. 5). Similar treatments using the drug cisplatinum (10–1000 ng/ml) were conducted. Initial data demonstrate a trend in increased cfu survival potential in MT tansfected cells after exposure to the drugs (Fig. 6).

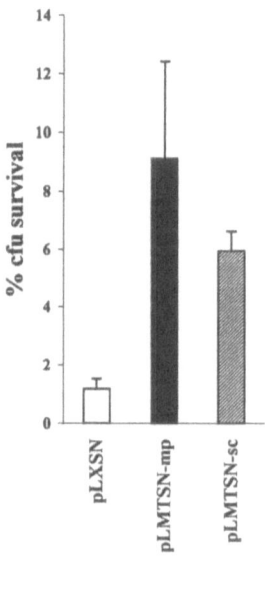

Figure 5. Colony forming unit survival of transfected U937 cells after exposure to 100 cGy: negative control (□), pLMTSN-mp (■), pLMTSN-sc (▨).

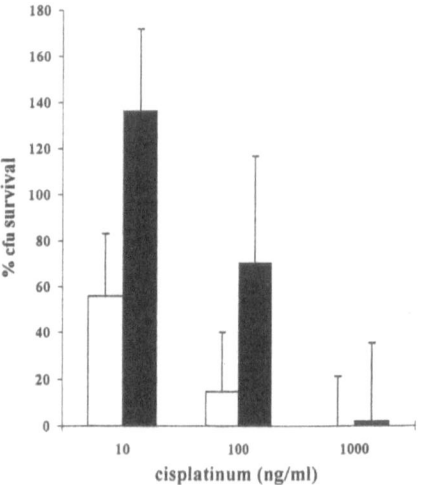

Figure 6. Colony forming unit survival of transfected U937 cells after treatments with cisplatinum (10, 100, 1000 ng/ml).: pLXSN (negative controls) (□), pLMTSN (MT-IIa) (■).

Effects of MT on reactive oxygen species (ROS)

Cell lysates (8 µg) were tested for their ability to scavenge ROS using a chemiluminescence assay by addition of hypoxanthine substrate, a chemiluminescent luminol/MCLA probe, and xanthine oxidase. ROS levels were determined for 20 min. Initial experiments using K562 cells transfected with MT-IIa, demonstrated a 13% increase in ROS scavenging capacity, a 60% and 39% increase with additional treatments of azaC and zinc respectively, when compared to their negative controls (Fig. 7).

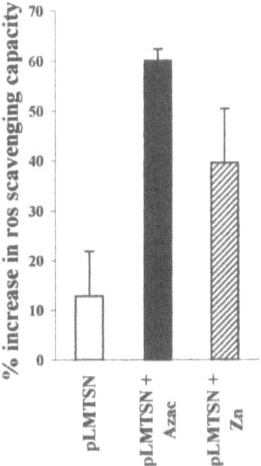

Figure 7. Percent increases in reactive oxygen species scavenging capacity of K562 cells transfected with pLMTSN compared to negative controls at final chemiluminescent assay time point (20 min) in the absence or presence of azacytidine (4 µM) or zinc acetate (100 µM).

Discussion

The preliminary data suggests that retroviral transfection of the human MT-IIa gene into U937 and K562 cells with low ratios of viral particles to host cells in the presence of a protein/peptide carrier polybrene provide a feasible approach for the insertion and expression of this gene into myeloid progenitor cells. The initial data suggest that azacytidine may be able to increase the expression of the transgene in the erythroid progenitor cells, possibly by reversing endogenous cellular DNA methylation. Transfection of the U937 myeloid progenitor cells also appears to confer protective properties to the cells upon exposure to low doses of ionizing radiation and cisplatinum. The presence of the transgene and monitoring of ROS suggests a positive correlation between antioxidant potential of the K562 cells and expression of MT-IIa. Increased cell survival after radio-chemotherapy treatment of MT-IIa-transfected cells suggests that intracellular MT-IIa may be vital for reducing the myelotoxic effects of ionizing radiation and/or chemotherapy. Further studies will be conducted to determine the mechanism(s) by which MT-IIa increases cell survival.

Acknowledgements
This work was supported by a grant from the Tulane Cancer Center.

References

1. Gianni AM, Bregni M, Siena S, Orazi A, Stern AC, Gandola L, Bonadonna G (1990) Recombinant Human Granulocyte-Macrophage Colony-Stimulating Factor Reduces Hematologic Toxicity and Widens Clinical Applicability of High-Dose Cyclophosphamide Treatment in Breast Cancer and Non-Hodgkin's Lymphoma. *J Clin Oncol* 8: 768–778.
2. Pastan I, Gottesman MM, Ueda K, Lovelace E, Rutherford AV, Willingham MC (1988) A retrovirus carrying an MDR1 cDNA confers multidrug resistance and polarized expression of P-glycoprotein in MDCK cells. *Proc Natl Acad Sci USA* 85: 4486–4490.
3. Corey CA, DeSilva AD, Holland CA, Williams DA (1990) Serial Transplantation of Methotrexate-Resistant Bone Marrow: Protection of Murine Recipients from Drug Toxicity by Progeny of Transduced Stem Cells. *Blood* 75: 337–343.
4. Magni M, Shammah S, Schiro R, Mellado W, Dalla-Favera R, Gianni AM (1996) Induction of Cyclophosphamide-Resistance by Aldehyde-Dehydrogenase Gene Transfer. *Blood* 87: 1097–1103.
5. Shibuya K, Satoh M, Muraoka M, Watanabe Y, Oida M, Shimizu H (1995) Induction of metallothionein synthesis in transplanted murine tumors by X irradiation. *Radiat Res* 143 (1: 54–57.
6. Schwarz MA, Lazo JS, Yalowich JC, Allen WP, Whitmore M, Bergonia HA, Tzeng E, Billiar TR, Robbins PD, Lancaster Jr JR, Pitt BR (1995) Metallothionein protects against the cytotoxic and DNA-damaging effects of nitric oxide. *Proc Natl Acad Sci USA* 92: 4452–4456.
7. Schwarz MA, Lazo JS, Yalowich JC, reynolds I, Kagan VE, Tyurin V, Kim YM, Watkins SC, Pitt BR (1994) Cytoplasmic Metallothionein Overexpression Protects NIH 3T3 Cells from *tert*-Butyl Hydroperoxide Toxicity. *J Biol Chem* 269: 15238–15243.
8. Basu A, Lazo JS (1990) A hypothesis regarding the protective role of metallothioneins against the toxicity of DNA interactive anticancer drugs. *Toxicol Lett* 50 (2–3: 121–122.
9. Satoh M, Cherian G, Imura N, Shimizu H (1994) Modulation of Resistance to Anticancer Drugs by Inhibition of Metallothionein Synthesis. *Cancer Res* 54: 5255–5257.
10. Oyama T, Takei H, Hikino T, Iino Y, Nakajima T (1996) Immunohistochemical Expression of Metallothionein in Invasive Breast Cancer in Relation to Proliferative Activity, Histology and Prognosis. *Oncology* 53: 112–117.
11. Kelley SL, Basu A, Teicher BA, Hacker MP, Hamer DH, Lazo JS (1988) Overexpression of Metallothionein Confers Resistance to Anticancer Drugs. *Science* 241: 1813–1815.
12. Yang YY, Woo ES, Reese CE, Bahnson RR, Saijo N, Lazo JS (1994) Human Metallothionein Isoform Gene Expression in Cisplatinum Sensitive and Resistant Cells. *Mol Pharmacol* 45: 453–460.
13. Afione SA, Conrad CK, Flotte TR (1995) Gene Therapy Vectors As Drug Delivery Systems. *Clin Pharmacokinet* 28 (3: 181–189.
14. Hoeben RC, Migchielsen AAJ, van der Jagt RCM, van Ormondt H, van der EBAJ (1991) Inactivation of the Moloney Murine Leukemia Virus Long Terminal Repeat in Murine Fibroblast Cell Lines Is Associated with Methylation and Dependent on Its Chromosomal Position. *J Virol* 65 (2: 904–912.
15. Cherian MG, Howell SB, Imura N, Klaassen CD, Koropatnick J, Lazo JS, Waalkes MP (1994) Role of Metallothionein in Carcinogenesis. *Toxicol Appl Pharmacol* 126: 1–5.
16. Bauer TR, Hickstein DD (1997) Transduction of Human Hematopoietic Cells and Cell Lines Using a Retroviral Vector Containing a Modified Murine CD4 Reporter Gene. *Human Gene Ther* 8: 243–252.
17. Nelson DM, Metzger ME, Donahue RE, Morgan RA (1997) *In Vivo* Retrovirus-Mediated Gene Transfer into Multiple Hematopoietic Lineages in Rabbits Without Preconditioning. *Human Gene Ther* 8: 747–754.
18. Karin M, Richards RI (1982) Human metallothionein genes – primary structure of the metallothionein-II gene and a related processed gene. *Nature* 299: 797–802.
19. Abdel-Mageed A, Agrawal KC (1997) Antisense down-regulation of metallothionein induces growth arrest and apoptosis in human breast carcinoma cells. Cancer *Gene Ther* 4: 199–207.

Metallothionein immunostaining as a prognostic indicator in canine mammary tumours

I. Carmen Fuentealba and Julia E. Mullins

Department of Pathology and Microbiology, Atlantic Veterinary College, University of Prince Edward Island, Charlottetown, Prince Edward Island, C1A 4P3, Canada

Research on metallothionein (MT) has been traditionally focused on its role in metal homeostasis and toxicity [1]. In recent years, the role of MT in carcinogenesis [2] and its potential applications differentiating malignant from benign neoplasms has received increased attention. Metallothionein is elevated in ovarian neoplastic cells [3] and its over expression in some cancer cell lines has been found to confer resistance to chemotherapy [4, 5, 6]. Increased expression of MT has been described in some types of human cancer including thyroid carcinomas [7], malignant melanomas [8] and mammary cancer [9, 10, 11, 12, 13, 14]. In human breast cancer studies, tumours with a better prognosis had lower levels of MT expression than those tumours with a less favourable prognosis [9, 10, 11, 13, 14].

The aim of this study was to determine the immunohistochemical location of MT in canine mammary tumours and its possible correlation with morphological characteristic of these tumours. Surgical specimens from spontaneous malignant (n = 20), and benign mammary tumours (n = 20) were selected. The paraffin blocks of mammary tissue were retrieved for immunohistochemistry and histology. An indirect immunoperoxidase technique, using monoclonal antibody E9 against horse MT (a kind gift from A. Cryer, J. Kay and J.M. Stark, University of Wales College of Medicine, Cardiff, UK) was employed. This antibody is reactive against a conserved epitope shared by the I and II isoforms of human, rat and horse metallothionein [15].

Slides were assessed for staining intensity and MT overexpression. Intensity of the stain was classified as weak (+), moderate (++) and strong (+++) immunoreactivity. Metallothionein overexpression was defined as those cases with more than 10% of positive tumour cells [9, 12].

Results

Results of MT overexpression in specific benign and malignant mammary tumours are presented in Table 1. In normal mammary tissue, most alveolar and ductal epithelial cells were weakly positive for MT but sporadic cells stained intensely. Myoepithelial cells in normal and neoplastic mammary tissue did not stain with MT. Intensity of the immunostain in normal tissue often correlated with that of the adjacent neoplastic tissue.

Papillary adenoma was diagnosed in 3 of 20 cases. MT overexpression (more than 10% of MT positive tumour cells) was detected in all the cases examined. Staining intensity in neo-

Table 1. Classification of canine mammary tumours and immunohistochemical overexpression of Metallothionein (MT)

Type of tumour	MT overexpression
Benign tumours	
Papillary adenoma	3/3
Complex adenoma	2/4
Benign mixed tumour	4/13
Malignant tumours	
Solid adenocarcinoma	3/5
Papillary adenocarcinoma	1/2
Tubular adenocarcinoma	1/7
Complex adenocarcinoma	1/2
Malignant mixed tumour	1/1
Anaplastic carcinoma	0/3

plastic epithelial cells was moderate, but in those cells lining papillary projections the staining intensity was strong (Fig. 1A). Connective tissue cells were negative.

Metallothionein overexpression was observed in 2/4 cases diagnosed as complex adenoma. Epithelial cells arranged in tubules and acini stained mild to moderate for MT, but the papillated parts stained strongly. Mesenchymal cells were negative.

Benign mixed mammary tumour was diagnosed in 13 of the 20 cases. The proportion of immunopositive cells correlated with the proportion of luminal epithelium present in the

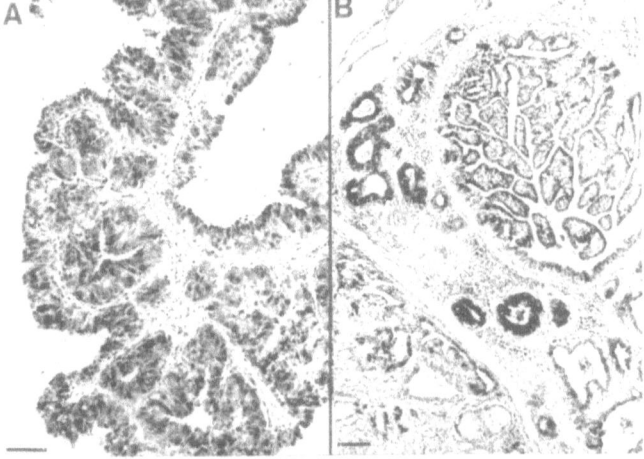

Figure 1. Metallothionein immunostaining in canine mammary tumours using an indirect immunoperoxidase method and E9 monoclonal antibody against horse metallothionein and haematoxylin counterstain. A) Intense MT immunostaining in a papillary adenoma. Bar = 100 µm. B) Variable MT immunostaining in a papillary adenocarcinoma. Bar = 160 µm.

tumour, and varied therefore from 1 to 50%. MT overexpression was detected in 4 out of 13 cases. Intensity of the immunostain in epithelial cells was moderate to strong. Spindle shaped myoepithelial cells stained negative. Star-shaped vacuolated myoepithelial cells had faint MT stain. Myxomatous, fibrous and cartilaginous tissues were negative except for a few chondrocytes which stained positive.

Solid carcinoma was diagnosed in 5 of the 20 cases. Overall staining varied from 0 to 50% of the tumour and MT overexpression was observed in 3 cases. Staining intensity in immunopositive epithelial cells varied from weak to strong.

Metallothionein overexpression was observed in 1 out of 2 cases diagnosed as papillary adenocarcinoma. MT immunostaining intensity was weak. Myoepithelial cells were negative. Metastatic luminal epithelial cells within a regional lymph node stained weakly positive for MT.

Tubular adenocarcinoma was found in 7 of the 20 cases of malignant mammary tumours. The proportion of immunopositive cells in the tumour varied from 0–50% (Fig. 1B) and MT overexpression was found in only 1 of the 7 cases examined. The staining intensity was weak to moderate and heterogeneous within the neoplasm, with large negative central areas and scattered immunopositive epithelial cells at the periphery of the tumour.

Complex carcinoma was diagnosed in 2 of the 20 cases of malignant mammary tumours. MT overexpression was detected in 1 case. A few epithelial cells at the centre of neoplastic lobules stained moderately. Weak immunostaining was seen in the case with less than 10% immunopositive cells.

MT overexpression was detected in the malignant mixed mammary tumour examined. Immunostaining intensity was weak.

Anaplastic carcinoma was diagnosed in 3 of the 20 cases of malignant mammary tumours. Negative staining was seen in one case, and less than 10% of the cells stained in the other 2 cases examined. The immunostain intensity was weak to moderate. Tumour emboli within lymphatics also stained moderately. Fibrous connective tissue was negative for MT.

Discussion

Spontaneous canine mammary tumours have been regarded as a suitable model for the study of human breast cancer [16]. Neoplastic transformation of human and canine mammary epithelium is associated with similar antigenic changes, suggesting that in both species tumorigenesis may be characterized by common pathogenic pathways [17]. However, the frequency of the various histologic types of mammary tumours in dogs is different to that found in human breast tumours [18], and less than 50% of histologically diagnosed carcinomas in dogs result in tumour-associated death [19].

Recent research suggests that MT is associated with cell proliferation and differentiation which are both occurring at an accelerated rate in growing tumours [2]. The presence of MT in the developing liver and the endodermal yolk sac of mice suggests that the developmental profile of MT is similar to other oncofetal gene products such as alpha-fetoprotein and that it has potential use as a marker for aggressive tumour behaviour [9]. However, it is not known whether tumours with MT overexpression originate from cells expressing MT already in their normal state or MT expression should be regarded as a secondary phenomenon caused by one or more factors responsible for MT induction [11]. Synthesis of MT can be induced in tissues by metal ions such as zinc, copper, and cadmium as well as by endogenous factors such as glucocorticoids, interferon, interleukin-1, lipopolysaccharide, ultraviolet light, progesterone, and vitamin D3 [1, 7].

Metallothionein has the ability to bind to metal ions, which emphasize its function as an intracellular reservoir for essential ions. Metallothionein plays an important role in detoxification of toxic metals such as zinc and copper [1, 20]. Histochemically detectable copper, zinc and iron have been found in malignant melanoma [21]. Furthermore, studies qualifying trace metal concentrations in benign and malignant tumours, including breast carcinomas, have demonstrated increased levels of copper or zinc in malignant tumours when compared with corresponding benign tumours [22]. The presence of high levels of copper in tumour tissue is interesting since this metal has been implicated in the generation of hydroxyl free radicals that can damage DNA. These radicals may cause double-stranded DNA breakage, which is not repairable by cellular mechanisms [23].

Metallothionein also has the capacity to scavenge free radicals [24]. Cells under oxidative stress are likely to overexpress MT as a protective response to the DNA-damaging influence of hydroxyl radicals [24]. Decrease in the cytotoxic activity of certain anticancer drugs and increased resistance in MT-rich cells occurs during exposure to ionizing radiation. However, the mechanism by which MT contributes to this protection is still unclear. Cancer cell lines originating from leukemic cells [4], ovarian carcinoma [5, 6], prostatic carcinoma [25] with increased Mt expression are more likely to be resistant to chemotherapy. Ovarian carcinoma cells with acquired resistance to anticancer agents often show an increase in MT mRNA and MT content [6].

In the present study the staining intensity was higher in benign compared to malignant mammary tumours, however, a few malignant tumours stained strongly. We used a similar score and definition of MT overexpression to that described in human breast cancer [9, 12], and detected immunohistochemical MT overexpression in both benign and malignant canine mammary tumours. Conversely, in human breast tumours MT immunostaining has been consistently found to correlate with a poor prognosis in invasive duct carcinoma whereas lack of MT expression has been found in normal lobular epithelium and little or no expression in lobular carcinomas [9–11, 13, 14]. Although breast tumours of favourable prognostic type have significantly lower levels of MT expression than tumours of poor prognostic subtypes, invasive lobular carcinomas are constantly MT-negative irrespective of their clinical course [10]. The cause for the inconsistency between MT immunopositivity results in canine and human mammary tumours is not clear. Most immunohistochemical studies of MT have focused in breast cancer [9, 10,12–14], thus it is possible that, similar to the situation in canine mammary tumours, MT may also be present in benign breast tumours. For example, in addition to ductal carcinoma *in situ* and invasive ductal carcinoma, MT immunohistochemical expression also occurs in benign breast lesions such as adenosis, scleradenosis, papillomas and epitheliosis, which has been interpreted as a possible indication that MT overexpression may occur during cell proliferation and tumour progression [11]. Furthermore, Cherian (1994) reported high levels of MT in both benign and malignant epithelial tumours and suggested that the presence of MT may depend on type of tumour, cellular origin, morphological heterogeneity or stage of growth.

Finally, the significance of the lack of correlation between staining intensity and percentage of stained cells is unclear. Our finding of variations in the intensity and distribution of the stained cells within the same specimen is shared by others. Invasive ductal carcinoma and ductular carcinoma *in situ* have a heterogenous MT immunostaining pattern, named mosaic pattern due to presence of strongly positive malignant cells or ducts adjacent to unstained malignant cells or ducts [9, 13]. Variations in MT immunostaining may represent heterogeneity in the stage of growth of neoplastic cells [2]. Douglas-Jones et al. (1995) suggested that immunostaining heterogeneity is a reflection of a wide variety of expression of MT among malignant cells in the same duct, indicating the possibility of a clonal selection process to occur during progression of the tumour.

In this study, MT staining and overexpression in canine mammary tumours was inconsistent and occurred in both benign and malignant neoplasms. In general, however, benign mammary tumours stained strongly whereas most malignant neoplasms stained weakly. We concluded that, in contrast with reported findings in human breast cancer studies, MT immunostaining appears to have a limited value to predict biological behaviour of canine mammary tumours. Further investigations are needed to clarify the role of MT in neoplastic transformation and to determine whether differences in prognosis of some human breast cancers and canine malignant mammary tumours are due to factors others than MT content.

References

1. Hamer D (1986) Metallothionein. *Annu Rev Biochem* 55: 913–951.
2. Cherian MG (1994) The significance of the nuclear and cytoplasmic localization of metallothionein in human liver and tumor cells. *Environ Health Perspect* 102: 131–135.
3. Murphy D, McGowan AT, Crowther D, Mander A, Fox BW (1991) Metallothionein levels in ovarian tumours before and after chemotherapy. *Brit J Cancer* 63: 711–714.
4. Beck WT, Muller TJ, Tamzer LR (1979) Altered surface membrane glycoproteins in vinca alkaloid-resistance human leukemic cells. *Cancer Res* 39: 2070–2075.
5. Andrews PA, Murphy MP, Howell SB (1987) Metallothionein-mediated resistance in human ovarian carcinoma cells. *Cancer Chemother Pharmacol* 19: 149–154.
6. Kelleey SL, Basu A, Teicher BA, Halker MP, Hamer DH, Lazo JS (1988) Overexpression on MT confers resistance to anticancer drugs. *Science* 241: 1813–1815.
7. Nartey N, Cherian MG, Banerjee D (1987) Immunohistochemical localization of metallothionein in human thyroid tumours. *Amer J Pathol* 129: 177–182.
8. Zelger B, Hittmair A, Schir M, Ofner D, Fritsch PO, Bocker W, Jasani B, Schmid KW (1993) Immunohistochemically demonstrated metallothionein expression in malignant melanoma. *Histopathology* 23: 257–264.
9. Fresno M, Wu W, Rodriguez JM, Nadji M (1993) Localization of metallothionein in breast carcinoma. An immunohistochemical study. *Virchows Arch A: Pathol Anat Histopathol* 423: 215–219.
10. Schmid KW, Ellis IO, Gee JMW, Darke BM, Lees WE, Kay J, Cryer A, Stark JM, Hittmair A, Ofner D et al (1993) Presence and possible significance of immunocytochemically demonstrable metallothionein overexpression in primary invasive ductal carcinoma of the breast. *Virchows Arch A: Pathol Anat Histopathol* 422: 153–159.
11. Bier B, Douglas-Jones AG, Totsch M, Dockhorn-Dworniczak B, Bocker W, Jasani B, Schmid KW (1994) Immunohistochemical demonstration of metallothionein in normal human breast tissue and benign and malignant breast lesions. *Breast Cancer Res Treat* 30: 213–221.
12. Haerslev T, Jacobsen GK, Zedeler K (1995) The prognostic significance of immunohistochemically detectable metallothionein in primary breast carcinomas. *APMIS* 103: 279–285.
13. Douglas-Jones AG, Schmid KW, Bier B, Horgan K, Lyons K, Dallimore ND, Moneypenny IJ, Jasani B (1995)

Metallothionein expression in duct carcinoma *in situ* of the breast. *Hum Pathol* 26; 217–222.

14. Goulding H, Jasani B, Pereira H, Reid A, Galea M, Bell JA, Elston CW, Robertson JF, Blamey RW, Nicholson RA et al (1995) Metallothionein expression in human breast cancer. *Brit J Cancer* 72: 968–972.
15. Jasani B, Elmes ME (1991) Immunohistochemical detection of metallothionein. *Methods Enzymol* 205: 95–107.
16. Hampe JE, Misdorp W (1974) Tumours and dysplasias of the mammary gland. *Bull World Health Organ* 50: 111–133.
17. Mottolese M, Morelli L, Agrimi U, Benevolo M, Sciarretta F, Antonucci Natali PG (1994) Spontaneous canine mammary tumors. A model for monoclonal antibody diagnosis and treatment of human breast cancer. *Lab Invest* 71: 182–187.
18. Nerurkar VR, Chitale AR, Jalnapurkar BV, Naik SN, Lalitha VS (1989) Comparative pathology of canine mammary tumours. *J Comp Pathol* 101: 389–397.
19. Brodey RS, Goldschmidt MH, Roszel JR (1983) Canine mammary gland neoplasms. *J Amer Anim Hosp Assoc* 19: 61–90.
20. Goering PL, Klaassen CD (1984) Zinc-induced tolerance to cadmium hepatotoxicity. *Toxicol Appl Pharmacol* 74: 299–307.
21. Bedrick AE, Ramasamy G, Tchertkoff V (1971) Histochemical determinations of copper, zinc, and iron in pigmented nevi and melanoma. *Amer J Dermatopathol* 13: 575–578.
22. Margalioth EJ, Schenker JG, Chevion M (1983) Copper and zinc levels in normal and malignant tissues. *Cancer* 52: 868–872.
23. Samuni A, Chevion M, Czapski G (1981) Unusual copper-induced sensitization of the biological damage due to superoxide radicals. *J Biol Chem* 256: 12632–12635.
24. Thornalley PJ, Vašák M (1985) Possible role for metallothionein in protection against radiation-induced oxidative stress. Kinetics and mechanism of its reaction with superoxide and hydroxyl radicals. *Biochim Biophys Acta* 827: 36–44.
25. Webber MM, Rehman SMM, James GT (1988) Metallothionein induction and deinduction in human prostatic carcinoma cells: relationship with resistance and sensitivity to adriamycin. *Cancer Res* 48: 4503–4508.

Metallothionein IV
C. Klaassen (ed.)
© 1999 Birkhäuser Verlag Basel/Switzerland

Metallothionein isoform gene expression in four human bladder cancer cell lines

Scott H. Garrett[1], Seema Somji[1], John H. Todd[1], Donald A. Sens[1], Donald L. Lamm[2] and Mary Ann Sens[1]

Departments of Pathology[1] and Urology[2], Robert C. Byrd Health Sciences Center, West Virginia University, Morgantown, West Virginia, USA

Introduction

The expression of metallothionein (MT) has been demonstrated using immunohistochemical methods in a number of human tumors, including: testicular embryonal carcinomas, thyroid tumors, transitional cell carcinomas of the bladder, *in situ* and invasive breast carcinomas, malignant melanomas, pancreatic carcinomas, and salivary gland tumors. In the most widely studied instance, ductal breast carcinoma, the expression of immunoreactive MT has been shown to correlate with a poor disease prognosis [1–4]. In more limited studies, the immuno-histochemical expression of MT in transitional cell carcinomas of the urinary tract has been correlated with tumor resistance to chemotherapeutic regimens employing cisplatin or cisplatin alone [5, 6]. However, the direct correlation of MT expression with a poor prognosis is not universal. In colorectal adenocarcinoma, an inverse correlation of MT expression to tumor stage and lymph node involvement was demonstrated [7]. While these studies demonstrate the potential use of MT as a prognostic indicator, interpretation is limited by the fact that the MT antibody reveals expression of a family of genes, not a specific gene product. In humans, the MTs are encoded by a family of genes located at 16q13 that contains 10 functional and 7 non-functional MT isoforms [8–11]. Based on charge characteristics, the genes are divided into four classes designated 1–4. With the exception of MT-4 which has not been extensively studied, the other MT genes have been shown to exhibit inducer-, tissue-, and developmental-specific patterns of gene regulation. Questions remain as to which MT gene products are being assessed in protocols using MT immunolocalization to define tumor prognosis and if expression patterns vary within and among tumors. As an initial step in addressing whether different MT isoforms are expressed in transitional cell carcinomas of the urinary tract, the present study uses reverse transcription-polymerase chain reaction (RT-PCR) technology as described previously to determine the expression of the human MT isoforms in four bladder cancer cell lines [12].

Results and discussion

The major goal of the study was to determine the expression of each MT isoform in 4 bladder cancer cell lines. It was demonstrated that the MT-2A, MT-1X, MT-1A, and MT-3 genes were

MT Isoform

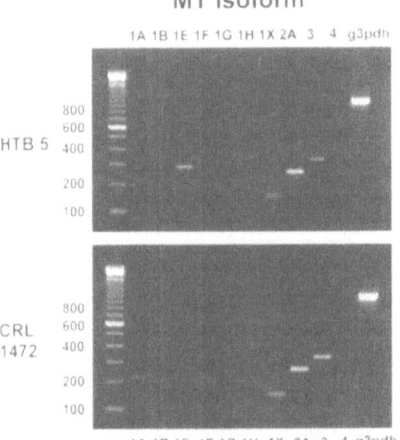

Figure 1. RT-PCR analysis of MT isoforms for two of the four bladder cancer cell lines used in this study. Expression profiles are shown for HTB 5, and CRL 1472 following 40 reaction cycles. The profiles of the other two lines (HTB1 and HTB2) are similar except no MT-4 or MT-1E expression was detected (data not shown).

expressed in all 4 cell lines under basal conditions. The additional expression of MT-1E was demonstrated in one cell line and MT-4 in another cell line (Fig. 1). An additional goal of this study was to determine if the MT isoforms varied in the level of mRNA expression among the isolates. For the MT-3 isoform, no expression of mRNA was found for any of the 4 cell lines at 30 cycles (Fig. 2). At 35 cycles, expression of MT-3 mRNA was demonstrated in the HTB 2, HTB 5, and CRL 1472 cell lines, while no expression was noted in the HTB 1 cell line. At 40 cycles, all 4 cell lines expressed MT-3. From these findings it can be concluded that the HTB 1 cell line had the lowest level of MT-3 mRNA expression. The CRL 1472 cell line appeared to have the highest level of MT-3 mRNA expression, while expression in the HTB 2 and HTB

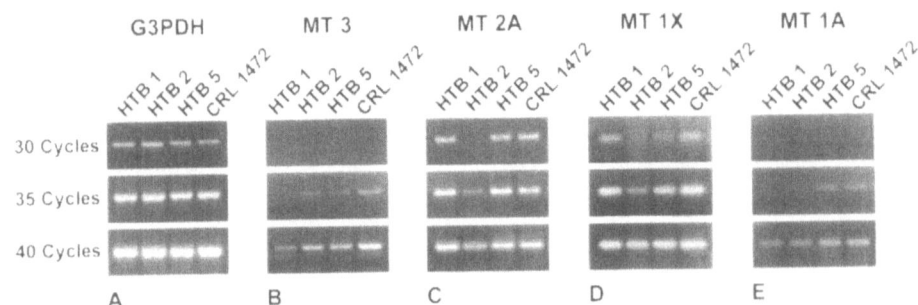

Figure 2. RT-PCR analysis of (A) glyceraldehyde 3-phosphate dehydrogenase (G3PDH), (B) MT 3, (C) MT 2A, (D) MT 1X, and (E) MT 1A in the bladder cancer cell lines HTB 1, HTB 2, HTB 5, and CRL 1472. Results are shown for samples of the reaction products removed at 30, 35, and 40 reaction cycles.

5 lines was roughly equivalent. An identical protocol was utilized to determine expression of the remaining isoforms among the 4 cell lines (Fig. 2).

An examination of the MT isoform-specific reaction products at 35 PCR cycles yields a qualitative indication of the overall level of MT mRNA expression among the 4 cell isolates (Tab. 1). Judged on a qualitative scale of +1 for low intensity and + 3 for maximum intensity, it is apparent that the HTB 2 cell line expresses a lower amount of MT mRNA than the other 3 cell lines. HTB 1, HTB 5, and CRL 1472 cell lines show similar expression of the MT-2A and MT-1X isoforms, although expression of the other MT isoforms reveals distinct patterns. The HTB 5 cell line, due to a high level of MT-1E mRNA expression, exhibits the greatest overall expression of MT mRNA. The CRL 1472 cell line, due to the additional expression of MT-3 and MT-1A mRNA as compared to HTB 1, shows the second greatest MT RNA expression. The order of total expression of MT mRNA is then estimated to be: HTB 2 < HTB 1 < CRL 1472 < HTB 5.

Table 1. MT gene isoform expression for bladder cancer cell lines HTB 1, HTB 2, HTB 5, and CRL 1472. MT1 isoform-specific RT-PCR reaction products were examined at 35 reaction cycles. The intensity of gel bands was judged on a qualitative scale and assigned +1 for low intensity, +2 for moderate intensity, and +3 for high intensity. A minus sign indicates a band was not present

	Metallothionein isoform									
	1A	1B	1E	1F	1G	1H	1X	2A	3	4
HTB 1	–	–	–	–	–	–	++	++	–	–
HTB 2	–	–	–	–	–	–	+	+	+	–
HTB 5	+	–	++	–	–	–	++	++	+	–
CRL 1472	+	–	–	–	–	–	+++	++	+	–

In an effort to determine if MT mRNA abundance translated into increases in MT protein, MT protein was examined in cell homogenates from the 4 cell lines by immunoblotting with MT antibody. Unfortunately, no definitive conclusions could be drawn. The cellular content of MT protein in the 4 cell lines was found to be at the limit of detection of the assay, between 0.25 and 0.50 ng of MT/μg of total protein. Immunolocalization using MT antibody, a more sensitive technique for detection of the MT protein, did yield a qualitative indication of the localization of the MT protein. In all 4 cell lines, MT was diffusely localized throughout the cytoplasm, with no evident nuclear localization (data not shown).

It is difficult to speculate on the significance of these expression patterns in terms of cellular function or tumor biology due to the limited studies assessing human MT gene expression in human cells and tissues. One reason for the lack of studies is the fact that several of the MT genes (MT-1H, MT-1X, MT-3, and MT-4) have only recently been isolated. Another reason is the evolutionary event in the human MT-1 gene resulting in duplications that increased the number of functional MT-1 gene isoforms to 7 active and 6 pseudogenes in the human, as compared to 1 in the mouse. Currently, it is not known if these multiple isoforms provide expanded function or simply represent an example of biological redundancy. The fact that many more MT-1 genes exist in the human than in the mouse, the animal model used for most molecular MT research, renders extrapolation of rodent MT-1 isoform data to humans problematic.

One of the few human studies to date examined the role of human MT isoform gene expression as a possible mediator of cisplatin resistance in 3 non-bladder human carcinoma cell lines [13]. In this study, the pattern of gene expression was determined for the MT-2A, MT-1A, B, E, F, and G genes in 3 pairs of cell lines where one member of each pair was resistant to cisplatin. It was found that the most highly overexpressed MT isoform in all 3 cisplatin resistant human carcinoma cell pairs was MT-2A and that no other isoform was universally overexpressed. This finding was interpreted as being consistent with a role for the MT-2A isoform in cisplatin resistance. Furthermore, since no mRNA was detected for the MT-1A, B, F, or G genes it was concluded that these could be ruled out as potential mediators of cisplatin resistance. These findings are important for the present study because in all 4 bladder cancer cell lines, MT-2A mRNA was shown to have the highest level of expression compared to other members of the MT gene family. The 4 bladder cancer cell lines have an appreciable basal expression of a gene proposed to be a mediator of cisplatin resistance.

The bladder cancer cell lines were shown not to express mRNA for the MT-1B, F, and G genes, in agreement with the findings of Yang and coworkers [13]. In contrast, the 4 bladder cancer cell lines were noted to express mRNA for the MT-1A gene. This finding is interesting for several reasons. First, it has been shown that the MT-1A gene has a different pattern of regulation compared to the MT-2A gene [14]. Whereas, the MT-2A gene was induced by cadmium, zinc, and glucocorticoids, the MT-1A gene was induced only by the heavy metal cadmium. This differential response was demonstrated to be due to functional differences in the respective promoter/regulatory regions of the two genes. Additionally, in cultures of human proximal tubule cells, the expression of the MT-1A gene has been shown to be induced by $CdCl_2$ only at an interval of exposure that occurs at, or immediately precedes, the initiation of cell death [15]. This is in contrast to the other expressed MT genes (MT-2A, 1E, and 1F) whose expression in these cells did not correlate to $CdCl_2$-induced cell death, but did correlate to exposure [16]. The finding that the 4 bladder cancer cell lines express MT-1A mRNA in the absence of metal exposure, could be an indication that these cells are capable of increased expression of MT-1A, and thus resistance, when challenged with a heavy metal agent such as cisplatin. The other 2 MT-1 isoforms, MT-1H and MT-1X, have only recently been isolated [9] and only one other study has assessed the expression of these genes in humans. In this study it was demonstrated that mRNA for the MT-1X, but not the MT-1H gene, was expressed in the developing and adult human kidney [13]. The present finding that MT-1X had a notable level of expression in all 4 bladder cell lines suggests a role similar to that proposed by Yang and coworkers [13] for MT-2A in mediating resistance to chemotherapeutic drugs.

The MT-3 gene was originally reported to be expressed only in neuronal tissues in both man and animals [11]; however, its expression has recently been shown in the human kidney [17] and in the mouse maternal deciduum [18]. The finding of MT-3 mRNA expression in all 4 bladder cancer cell lines is of particular interest. This is due to the fact that MT-3 in the neural system has been shown to have a growth inhibitory activity which is not duplicated by any other human MT gene [19, 20]. There is no literature explaining how this growth inhibitory activity might relate to tumor cell growth or tumor biology in the bladder or other tissues. While the expression of mRNA for the MT-2A, MT-3, MT-1A, and MT-1X genes was demonstrated in all 4 bladder cancer cell lines at varying levels, in 2 cell lines there was expression of mRNA for an additional MT gene. MT-4 was found in the CRL 1472 cell line in marginal quantity, as noted by only a light reaction product band on EtBr-stained gels after 40 PCR cycles. Little is known about the MT-4 gene beyond the information in the original report describing the isolation of the gene and restriction of expression to squamous epithelia [10]. Due to the marginal expression in the CRL 1472 cell line and the sensitivity of RT-PCR in detecting rare transcripts, it is unlikely that this level of expression would be functionally significant in this cell line. However, this is not the case for expression of mRNA for the MT-1E gene found only in the HTB 5 cell line, where MT-1E expression was qualitatively similar to MT-2A expression. The expression of MT-1E mRNA in the HTB-5 cells is interesting for several reasons. First, it has been shown that cells transformed with the c-Ha-*ras* oncogene atypically express high basal levels of the MT-1E transcript, possibly reflecting a role of normal *ras* genes in the control of MT gene expression [21]. Furthermore, in the studies of Yang and coworkers [13], one of the 3 cell pairs matched for sensitivity and resistance to cisplatin overexpressed the MT-1E mRNA transcript. While the fact that all 3 cell lines did not express MT-1E is evidence for non-involvement of this gene in cisplatin resistance, this would not extend to the aggressiveness of the individual tumor cell lines. As introduced in the study by Yang and coworkers [13], the fact that acute cisplatin treatment cannot be shown to repeatably induce MT in cultured cells suggests that the elevated levels of MT may reflect a cell selection process. As such, it may be that the expression of an additional isoform of MT can have an affect on the tumor biology of the cell independent of a role in chemotherapeutic sensitivity.

Acknowledgment
This publication was made possible by grant number ESO7687 from the National Institute of Environmental Health Sciences, NIH.

References

1. Schmid KW, Ellis IO, Gee JMW, Darke BM, Lees WE, Kay J, Cryer A, Stark JM, Hittmair A, Ofner D, Dunser M, Margreiter R, Daxenbichler G, Nicholson RI, Bier B, Bocker W, Jasani B (1993) Presence and possible significance of immunocytochemically demonstratable metallothionein over-expression in primary invasive ductal carcinoma of the breast. *Virchows Arch A Pathol Anat* 422: 153–159.
2. Haerslev T, Jacobsen K, Nedergaard LZedeler K (1994) Immunohistochemical detection of metallothionein in primary breast carcinomas and their axillary lymph node metastases. *Path Res Pract* 190: 675–681.
3. Douglas-Jones AG, Schmid KW, Bier B, Horgan K, Lyons K, Dallimore ND, Moneypenny IJ, Jasani B (1995) Metallothionein expression in duct carcinoma *in situ* of the breast. *Hum Pathol* 26: 217–222.
4. Goulding H, Jasani B, Pereira H, Reid A, Galea M, Bell JA, Elston CW, Robertson JF, Blamey RW, Nicholson RA, Schmid KW, Ellis IO (1995) Metallothionein expression in human breast cancer. *Brit J Cancer* 72: 968–972.
5. Bahnson RR, Becich M, Ernstoff MS, Sandlow J, Cohen MB, Williams RD (1994) Absence of immunohistochemical metallothionein staining in bladder tumor specimens predicts response to neoadjuvant cisplatin, methotrexate and vinblastine chemotherapy. *J Urol* 152: 2272–2275.
6. Kotoh S, Naito S, Sakamoto N, Goto K, Kumazawa J (1994) Metallothionein expression is correlated with cisplatin resistance in transitional cell carcinoma of the urinary tract. *J Urol* 152: 1267–1270.
7. Öfner D, Maier H, Riedmann B, Bammer T, Rumer A, Winde G, Böcker W, Jasani B, Schmid KW (1994) Immunohistochemical metallothionein expression in colorectal adenocarcinoma: correlation with tumor stage and patient survival. *Virchows Arch.* 425: 491–197.

8. West AK, Stallings R, Hildebrand CE, Chiu R, Karin M, Richards R (1990) Human metallothionein genes: structure of the functional locus at 16q13. *Genomics* 8: 513–518.
9. Stennard FA, Holloway AF, Hamilton J, West AK (1994) Characterization of six additional human metallothionein genes. *Biochim Biophys Acta* 1218: 357–365.
10. Quaife CJ, Findley SD, Erickson JC, Froelick GJ, Kelly EJ, Zambrowicz BP, Palmiter RD (1994) Induction of a new metallothionein isoform (MT-IV) occurs during differentiation of stratified squamous epithelia. *Biochemistry* 33: 7250–7259.
11. Palmiter RD, Findley SD, Whitmore TE, Durnam DM (1992) MT-III, a brain-specific member of the metallothionein gene family. *Proc Natl Acad Sci USA* 89: 6333–6337.
12. Mididoddi S, McGuirt JP, Sens MA, Todd JH, Sens DA (1996) Isoform-specific expression of metallothionein mRNA in the developing and adult human kidney. *Toxicol Lett* 85: 17–27.
13. Yang Y-Y, Woo ES, Reese CE, Bahnson RR, Saijo N, Lazo JS (1994) Human metallothionein isoform gene expression in cisplatin-sensitive and resistant cells. *Mol Pharmacol* 45: 453–460.
14. Richards RI, Heguy A, Karin M (1984) Structural and functional analysis of the human metallothionein-I_A gene: differential induction by metal ions and glucocorticoids. *Cell* 37: 263–272.
15. Bylander JE, Li SL, Sens MA, Sens DA (1995) Exposure of human proximal tubule cells to cytotoxic levels of $CdCl_2$ induces the additional expression of metallothionein 1A mRNA. *Toxicol Lett* 76: 209–217.
16. Bylander JE, Li SL, Sens MA, Hazen-Martin DJ, Re GG, Sens DA (1994) Induction of metallothionein mRNA and protein following exposure of cultured human proximal tubule cells to cadmium. *Toxicol Lett* 71: 111–122.
17. Hoey JG, Garrett SH, Sens MA, Todd JH, Sens DA (1997) Expression of MT-3 mRNA in human kidney, proximal tubule cell cultures, and renal cell carcinoma. *Toxicol Lett* 92: 149–160.
18. Liang L, Fu K, Lee DK, Sobieski RJ, Dalton T, Andrews GK (1996) Activation of the complete mouse metallothionein gene locus in the maternal deciduum. *Mol Reprod Develop* 43: 25–37.
19. Erickson JC, Sewell AK, Jensen LT, Winge DR, Palmiter RD (1994) Enhanced neurotrophic activity in Alzheimer's disease cortex is not associated with down-regulation of metallothionein III (GIF). *Brain Res* 649: 297–304.
20. Uchida Y, Takio K, Titani K, Ihara Y, Tomonaga M (1991) The growth inhibitory factor that is deficient in Alzheimer's disease is a 68 amino acid metallothionein-like protein. *Neuron* 7: 337–347.
21. Schmidt CJ, Hamer DH (1986) Cell specificity and an effect of ras on human metallothionein gene expression. *Proc Natl Acad Sci USA* 83: 3346–3350.

The suppression of metallothionein synthesis inhibits the growth of leukemia P388 cells

Atsushi Takeda[1], Harumi Hisada[1], Shoji Okada[1], John E. Mata[2], Patrick L. Iversen[2] and Manuchair Ebadi[2]

[1]*Department of Radiobiochemistry, School of Pharmaceutical Sciences, University of Shizuoka, 52-1 Yada, Shizuoka 422, Japan*
[2]*Department of Pharmacology, University of Nebraska College of Medicine, 600 South 42nd Street, Omaha, NE 68198, USA*

Summary. Metallothionein (MT) isoforms, by regulating the homeostasis of zinc, influence aspects of molecular and cellular biology and regulate events over a wide rage of physiological and/or pathological parameters. For example, tumor cells with acquired resistance to antineoplastic agents over-express MTs. By using MT-I antisense phosphorothioate oligodeoxyribonucleotide (ODN) (18-mer) complementary to the intron 2/exon 3 splice site of the mouse MT-I and human MT-If mRNA, we learned that MT-I antisense ODN caused a suppression of MT synthesis and hence inhibited the growth of Leukemia P388 cells, Ehrlich carcinoma cells and sarcoma 180 cells in a dose- and time- dependent fashion; whereas, MT-II antisense ODN inhibited the growth of Chinese hamster lung V79 cells. Moreover, treatment with MT-I antisense ODN attenuated the growth and progression of tumors in mice. The tumor inhibitory effects of MT-I antisense ODN, evident in both *in vivo* and *in vitro* systems, was specific since ceruloplasmin antisense ODN was devoid of any action. The results of these studies are interpreted to suggest that the growth of neoplastic cells depends on the expression of MT, and in this area, MT-I and the MT-II exhibit distinct actions, which may provide a unique avenue for chemotherapy against certain neoplasms.

The expression of metallothionein in growth and differentiation

Metallothionein (MT), first identified in horse kidneys by Margoshes and Vallee [1], is associated with many physiological processes such as cellular growth and differentiation and/or pathological events such as resistance of tumors to antineoplastic agents [2]. Four isoforms of MT, designated as MT-I—IV, have been identified, but MT-I and MT-II are the most extensively characterized isoforms in mammalian and non-mammalian organisms [3]. MT genes, expressed in tissues of most organisms, are modulated by many factors such as metals, glucocorticoids and cytokines [4]. Moreover, MT induction is closely linked to zinc uptake [5]. MT-III [6], found in high concentrations in the hippocampus, is involved in transport and compartmentation of zinc. MT isoforms, isolated generally from the cytoplasm, are also detected in the nuclei of developing tissues [7, 8] and neoplastic tissues [9]. These tissues require zinc, an essential nutrient in transmitting genetic information and for growth and development [10, 11]. In addition, MT isoforms are involved in the maintenance of zinc homeostasis, in activating zinc metalloenzymes, in detoxification of toxic metals [2, 5], and in synthesizing zinc fingers and other transcription factors in the nucleus. For example, it has been shown that following hepatectomy and hepatic regeneration, MT is translocated from cytoplasm into nucleus [12]. Indeed, the results of studies with human tumor cells have indicated that nuclear retention of MTs is an energy requiring process; and the ability of MTs to accumulate in subcellular compartments against a concentration gradient, may be an important phenomenon in fostering the action of MTs to supply zinc to target sites within subcellular compartments [13].

Utilization of metallothionein antisense oligodeoxyribonucleotide (ODN)

Inhibition of protein synthesis by targeting mRNA transcripts with synthetic ODNs have been demonstrated for ODNs which contain methylphosphonate and phosphorothioate backbones [14–16]. Previous studies have shown that phosphorothioate ODNs which are antisense to the human MT-II transcript, i.e. complementary to exon 1 at a site which is 7 bases downstream from the ATG translational start site or complementary to exon 1/intron 1 splice donor site, produce a dose-dependent cytotoxicity against hamster lung V79 cells [17]. The antisense ODN targeted to the splice donor site showed greater efficacy than the other ODN tested and demonstrated the essentiality of the MT-II gene product. The present study demonstrates the importance of the MT-I gene product using antisense ODNs targeted to the intron 2/exon 3 splice donor site of the human (MT)I-f gene [18]. The sequence (5'-CAGCAGGAGCAGCAGCCT-3'), which is complementary to both the mouse MT-I and human MT-If mRNA, and a random sequence (5'-TCCTAGGTCCATGTCGTACGC-3') were synthesized as phosphothioate ODNs.

Inhibition of leukemia P388 cells with metallothionein-I antisense oligodeoxyribonucleotide

Incubation of leukemia P388 cells with MT-I antisense ODN (1 to 15 µM) for 72 h caused growth inhibition in a dose- and time-dependent fashion (Fig. 1). Similarly, MT-I antisense

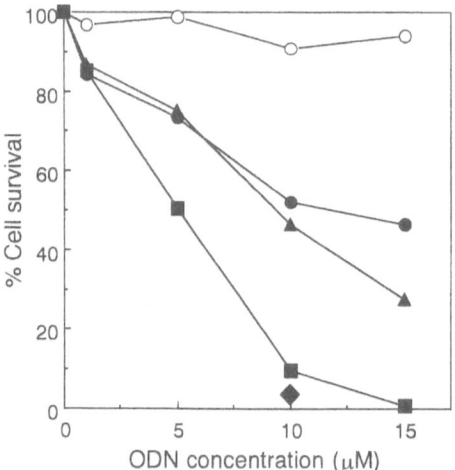

Figure 1. Survival of Leukemia P388 Cells Treated with metallothionein (MT) Antisense oligodeoxyribonucleotide (ODN). Cell survival ratios were determined 24 h (●), 48 h (▲), 72 h (■) and 120 h (◆) after treatment with MT antisense or control ODN at various concentrations. The data determined 24 h after treatment with control ODN was used as control cell survival (○). Each point is the mean value of 6 wells, and was calculated by the formula: MT antisense ODN- or control ODN-treated cell density × 100 untreated cell density.

ODN inhibited the survival of Ehrlich carcinoma and sarcoma 180. In addition, subcutaneous administration of MT-I antisense ODN to mice (1.2 mg/kg five times daily for two days) in areas where tumor cells were transplanted, caused a significant reduction in tumor weight (0.51 ± 0.18 g), compared to control ODN (0.68 ± 0.10 g), which had no reducing effect on tumor weight [19]. The body weights of control group and MT-I antisense ODN group remained identical. The tumor inhibitory effects of MT-I antisense ODN, evident in both *in vivo* and *in vitro* systems, was specific since ceruloplasmin antisense ODN was devoid of any action.

The growth of tumor cells may depend in part on the availability of zinc and the proper expression of MT. On the other hand, the expression of MT in tumor cells may be associated with resistance to antineoplastic agents [20–22], since tumor cells with acquired resistance to antineoplastic agents tend to overexpress MTs. For example, the overexpression of human MT-IIa has been observed in cis-diamminedichloroplatinum resistant cells [9, 22].

Determination of the levels of zinc and metallothionein following treatment with MT-I antisense oligodeoxyribonucleotide

By using SDS-PAGE and western blotting [23], the level of MT was determined in P388 cells treated with MT-I antisense ODN. The staining pattern with coomassie brilliant blue of the whole cytosolic proteins of the MT antisense ODN-treated cells, was similar to that of the control ODN-treated cells. However, the MT level of the former was significantly lower than that of the latter (MT antisense ODN/control ODN = 0.67 ± 0.02; Fig. 2).

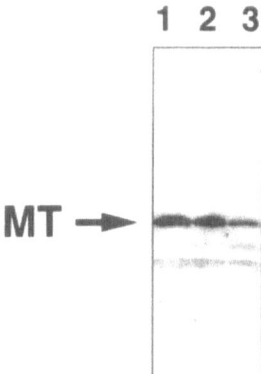

Figure 2. Metallothionein (MT) Levels of control and MT Antisense oligodeoxyribonucleotide (ODN)-treated Cells. P388 cells were treated with 30 μM MT antisense ODN or control ODN for 3 h The cytosol (12.5 μg protein) obtained from the cells of each group was subjected to SDS-PAGE followed by western blotting. After 3 h, the percent survival of MT antisense ODN-treated cells was 50–80%. Quadruplicate experiments were done separately. Lane 1, untreated; Lane 2, control ODN, Lane 3, MT antisense ODN. Relative MT levels (to non-treated cells) of control ODN-treated cells and MT antisense ODN-treated cells, as determined with a densitometer (Shimadzu CS-9000), were 0.82 ± 0.26 and 0.55 ± 0.16, respectively.

The uptake and efflux of ^{65}Zn was measured in the MT-I antisense ODN-treated cells. When the MT-I antisense ODN-treated cells were incubated with ^{65}ZnCl$_2$ for 30 min, ^{65}Zn uptake by the MT-I antisense ODN-treated cells was not significantly decreased compared with control ODN-treated and untreated cells, despite the decrease in MT level. ^{65}Zn efflux from the MT-I antisense ODN-treated cells after incubation with ^{65}ZnCl$_2$ for 24 h, was also nearly to the same extent as those from control ODN-treated and untreated cells. As a function of MTs, the maintenance of intracellular zinc homeostasis has been known for some time. However, in the present study, the change in the uptake and efflux of ^{65}Zn was not observed in the MT-I antisense ODN-treated cells. This is probably due to the fact that the proportion of MT-binding zinc to total cellular zinc is very low [24] and a decrement of MT-I binding zinc might hardly be detected in our assay system. Thus, zinc MT-I and zinc MT-II may differentially play a crucial role for the living processes of these cells through intracellular zinc transfer systems.

Conclusions and speculations

The precise actions of MT-I, MT-II or a combination of MT-I and MT-II in cell growth and in exhibition of resistance to antineoplastic agents, need to be delineated carefully. Although MT may not be necessary for maintaining proper reproductive cycle [25], a full comprehension of association among MT isoforms, cell growth, and tumor suppression, nevertheless, may provide a new avenue for discovery of novel antineoplastic agents.

Acknowledgments
The authors express their heartfelt appreciation to Miss Mai Suzuki for her technical assistance and to Mrs. Lori Clapper for her excellent secretarial skills. This study was completed in part from a grant from the USPHS NS34566-04.

References

1. Margoshes M, Vallee BL (1957) A cadmium protein from equine kidney cortex. *J Amer Chem Soc* 79: 4813–4814.
2. Suzuki KT, Imura N, Kimura M (eds) (1993) *Metallothionein III: Biological Roles and Medical Implications*, Birkhäuser, Basel.
3. Searle PF, Davison BL, Stuart GW, Wilkie TM, Norstedt G, Palmiter RD (1984) Regulation, linkage and sequence of mouse metallothionein I and II genes. *Mol Cell Biol* 4: 1221–1230.
4. Brady FO (1991) Induction of metallothionein in rats. *Methods Enzymol* 205: 559–567.
5. Bremner I, Beattie JH (1990) Metallothionein and the trace metal. *Annu Rev Nutr* 10: 63–83.
6. Palmiter RD, Findley SD, Whitmore TE, Durnam DM (1992) MT-III, a brain-specific member of the metallothionein gene family. *Proc Natl Acad Sci USA* 89: 6333–6337.
7. Nishimura H, Nishimura N, Tohyama C (1989) Immunohistochemical localization of metallothionein in developing rat tissues. *J Histochem Cytochem* 37: 715–722.
8. Elmes ME, Haywood S, Jasani B (1991) A histochemical and immunocytochemical study of hepatic copper and metallothionein in the pre- and post-natal rat. *J Pathol* 164: 83–87.
9. Kondo Y, Kuo S-M, Watkins SC, Lazo JS (1995) Metallothionein localization and cisplatin resistance in human hormone-independent prostatic tumor cell lines. *Cancer Res* 55: 474–477.
10. Ebadi M, Swanson S (1988) The status of zinc, copper and metallothionein in cancer patients. *Prog Clin Biol Res* 259: 161–167.
11. Vallee BL, Falchuk KH (1993) The biochemical basis of zinc physiology. *Physiol Rev* 73: 79–118.

12. Tsujikawa K, Suzuki N, Sagawa K, Itoh M, Sugiyama T, Kohama Y, Otaki N, Kimura M, Kimura T (1994) Induction and subcellular localization of metallothionein in regenerating rat liver. *Eur J Cell Biol* 63: 240–246.
13. Woo ES, Kondo Y, Watkins SC, Hoyt DG, Lazo JS (1996) Nucleophilic distribution of metallothionein in human tumor cells. *Exp Cell Res* 224: 365–371.
14. Iversen PL, Ebadi E (1992) Antisense oligonucleotide-mediated inhibition of metallothionein protein synthesis in Neuroblastoma IMR32 and Chang liver cells in culture. *Biol Signals* 1: 57–64.
15. Takeda A, Norris JS, Iversen PL, Ebadi M (1994) Antisense oligonucleotide of c-myc discriminates between zinc- and dexamethasone-induced synthesis of metallothionein. *Pharmacology* 48: 119–126.
16. DiBaise JK, Ebadi M, Iversen PL Patterns of cellular uptake and effects on cell survival using antimetallothionein oligodeoxyribonucleotide conjugates *in vitro*. *Biol Signals* 3: 140–149.
17. Iversen PL, Mata JE, Ebadi M (1992) Synthetic antisense oligonucleotide probes the essentiality of metallothionein gene. *Biol Signals* 1: 293–299.
18. Takeda A, Hisada H, Okada S, Mata JE, Ebadi M, Iversen PL (1997) Tumor cell growth is inhibited by suppressing metallothionein-I synthesis. *Cancer Lett* 116: 145–149.
19. Hisada H, Takeda A, Suzuki M, Okada S, Mata JE, Iversen PL, Ebadi M (1998) *In vivo* growth inhibition of leukemia P388 by metallothionein-I antisense oligodeoxyribonucleotide. *J Trace Elem Exp Med* 11: 1–4.
20. Satoh M, Kloth DM, Kadhim SA, Chin JL, Naganuma A, Imura N, Cherian MG (1993) Modulation of both cisplatin nephrotoxicity and drug resistance in murine bladder tumor by controlling metallothionein synthesis. *Cancer Res* 53: 1829–1832.
21. Satoh M, Cherian MG, Imura N, Shimizu H (1994) Modulation of resistance to anticancer drugs by inhibition by metallothionein synthesis. *Cancer Res* 54: 5255–5257.
22. Yang Y-Y, Woo ES, Reese CE, Bahnson RR, Saijo N, Lazo JS (1994) Human metallothionein isoform gene expression in cisplatin-sensitive and resistant cells. *Mol Pharmacol* 45: 453–460.
23. Aoki Y, Suzuki KT (1991) Detection of metallothionein by western blotting. *Methods Enzymol* 205: 108–114.
24. Philcox JC, Tilley MH, Coyle P, Rofe AM (1994) Metallothionein and zinc homeostasis during tumor progression. *Biol Trace Elem Res* 40: 295–308.
25. Masters BA, Kelly EJ, Quaife CJ, Brinster RL, Palmiter RD (1994) Targeted disruption of metallothionein I and II genes increases sensitivity to cadmium. *Proc Natl Acad Sci USA* 91: 584–588.

Metallothioneins as biomarkers

Metallothionein IV
C. Klaassen (ed.)
© 1999 Birkhäuser Verlag Basel/Switzerland

Identification and characterisation of metallothioneins from environmental indicator species as potential biomonitors

Ceri A. Morris, Stephen Stürzenbaum, Beate Nicolaus, A. John Morgan, John L. Harwood and Peter Kille

School of Biosciences, University of Wales Cardiff, Cardiff CF1 3US, Wales, UK

Summary. This study has given the first indication that two commonly used environmental indicator species contain members of the metallothionein superfamily. This demonstrates that these species use similar proteins to higher life forms to adapt to heavy metal conditions. Future studies will investigate if the MT proteins are present in the specialised compartments containing high levels of sulphur and copper/cadmium or if other cysteine-rich proteins such as the phytochelatin family are involved. Previously reported functions of the MT proteins in metal homeostasis suggests that MT expression may be up-regulated as a consequence of the high metal concentration of the environment from which these organisms were collected. The use of the MT expression profiles, as sensitive indicators of the metal concentration of the environment, will be investigated.

Introduction

Heavy metals are essential for life. However, a paradox exists in that certain levels of heavy metals become toxic and, therefore, organisms must stringently regulate cellular levels. The aim of the research group in Cardiff is to study the mechanisms biological systems use to handle heavy metals by assessing their impact as pollutants, their involvement in disease and the basic mechanisms by which they are bound within biological systems. This article will focus on the isolation and characterization of metallothionein (MT) from various model bioindicator organisms and the implications for understanding biological response to pollutants and environmental biomonitoring.

Mineral extraction from the earth's crust for utilisation in manufacturing industries inevitably results in the release of a group of naturally ecotoxic substances, the heavy metals. European Union legislation requires that testing of contaminated sites be carried out to assess human health risks. At risk are terrestrial, fresh water and marine ecosystems. Present methods of measuring levels of heavy metal pollution are crude e.g. soil toxicity tests involve measurement of the number of earthworms to survive pollutant exposure or assessing reduction of biodiversity within polluted habitats. Therefore it is necessary to develop clear assay procedures that can provide objective information about heavy metal pollution. Obvious targets to assay are the MT superfamily.

A number of environmental indicator species, commonly used in ecological studies, have been identified to form the centre of these studies. They include the earthworm (*Lumbricus rubellus*), the rainbow trout (*Oncorhynchus mykiss*) and marine macro-algae (*Fucus vesiculosis*) which form model organisms for terrestrial, fresh water and marine ecosystems respectively. *Lumbricus rubellus* is one of the most wide spread invertebrates in Europe and is found in highly polluted environments therefore providing an ideal organism to indicate soil toxicity [1, 2]. Historically, tolerant macro-algae such as *Fucus* spp. have been used as biological indicator organisms due to their resistance to heavy metals and their ability to accumulate metals so that their intracellular composition reflects average pollutant loads in the sea over extended periods [3, 4, 5]. Using macro-algae, therefore, permits evaluation of chronic problems and also makes possible measurements of substances whose seawater concentrations are very low. In contrast *Oncorhynchus mykiss* is very metal sensitive and is mainly used in a laboratory based test with the regulatory levels of waterborne heavy metals being low in comparison to soil levels. Although considerable information is known about the response of *Oncorhynchus mykiss* to heavy metals, little is known concerning the molecular mechanisms underlying *Lumbricus rubellus* and *Fucus* spp. to heavy metals.

Results and discussion

Specialised localisation of metal in indicator species exposed to pollution

The biochemistry/physiology of metal homeostasis is complex due to the differential requirement and toxicity of the different metals. The localisation of metal ions in both the *Lumbricus rubellus* and *Fucus vesiculosis* exposed to polluted sites has been investigated and both species have been found to store metal in specialised compartments.

The localisation of the metal ions in *Lumbricus rubellus* inhabiting a Roman lead/zinc mine has been studied (Fig. 1). In the chloragogenous tissue surrounding the alimentary canal of these earthworms phosphorous, zinc, lead and calcium were co-localised in compartments (chloragosomes) whilst cadmium was located in a distinct sulphur-rich granule, called the "cadmosome". Similarly, *Fucus vesiculosis*, when collected from an estuary (Devoran, U.K.) polluted by the discharge from a disused copper mine, was shown to accumulate high levels of copper and zinc (up to 15 fold greater than the control values) [6]. Previous studies, using microprobe analysis, have shown that copper is localised within the cell in distinct compartments called physodes (Fig. 2, taken from reference 7). These cellular compartments also contain high levels of sulphur. The question that needs to be addressed is whether MT, as one of the most common cysteine-rich proteins involved in metal binding, is involved in metal compartmentalisation in either of these bio-indicator species.

Isolation and characterisation of a cysteine-rich protein from Lumbricus rubellus

The discovery of the "cadmosome", where high levels of cadmium and sulphur were co-localised, sparked off the search for the metallothionein protein from *L. rubellus*. A cysteine-rich protein was extracted by a two-step purification procedure. A protein fraction associated with high cadmium levels was extracted by gel filtration (Sephadex G-75) of the soluble earthworm extract (Fig. 3a). Ion exchange (Q-Sepharose) chromatography was used to purify the protein associated with the cadmium (Fig. 3b). Amino acid hydrolysis showed the purified pro-

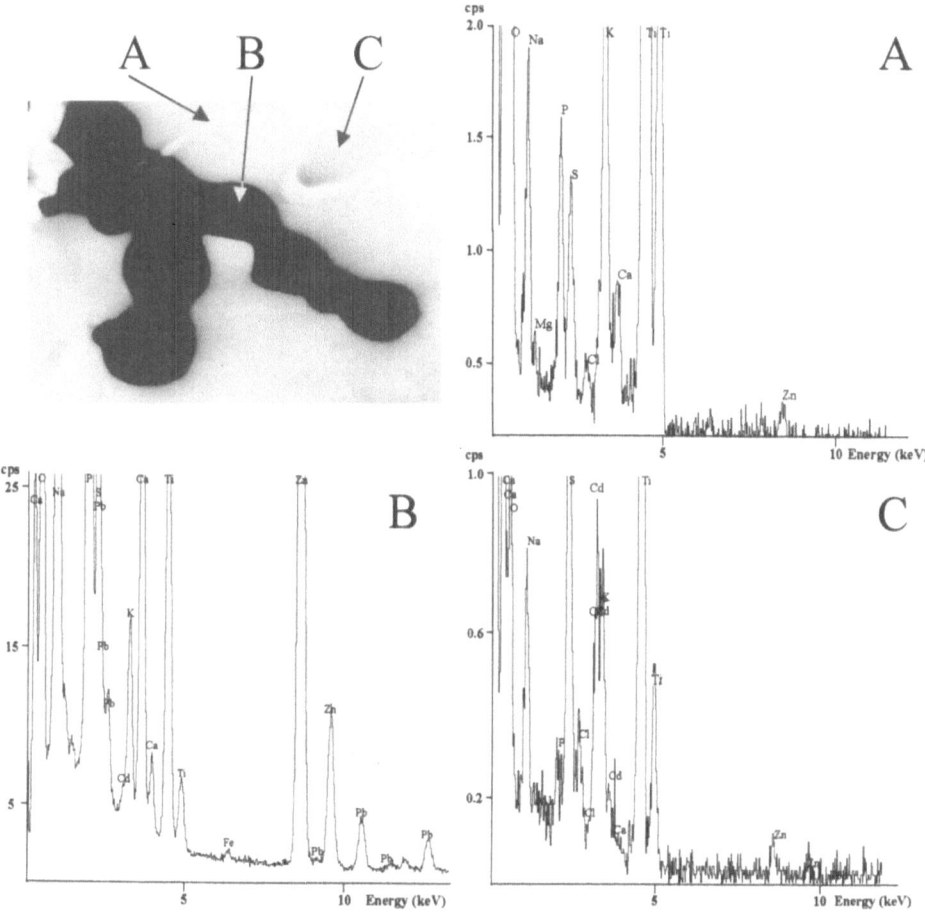

Figure 1. X-ray energy spectrum of the chloragogenous tissue in *L. rubellus* native to a Pb/Zn/Cd mine. Static probe X-ray analysis was performed in the cytosol (A), lead/zinc chloragosomes (B), and cadmium/sulphur-rich "cadmosome" (C). Note the relatively high cadmium and sulphur content in C.

tein contained a high proportion of cysteine residues (25%) with a low number of aromatic residues. After the protein was purified to homogeneity, N-terminal sequence determined that the N-terminus was blocked. The purified protein was treated with proteinase K and the peptide not completely digested was sequenced. Of the nine amino acids sequenced, two were cysteine residues [–N-T-?-**C-C**-G-F-D-A-]. The preliminary data therefore suggests that this protein is a member of the Metallothionein family.

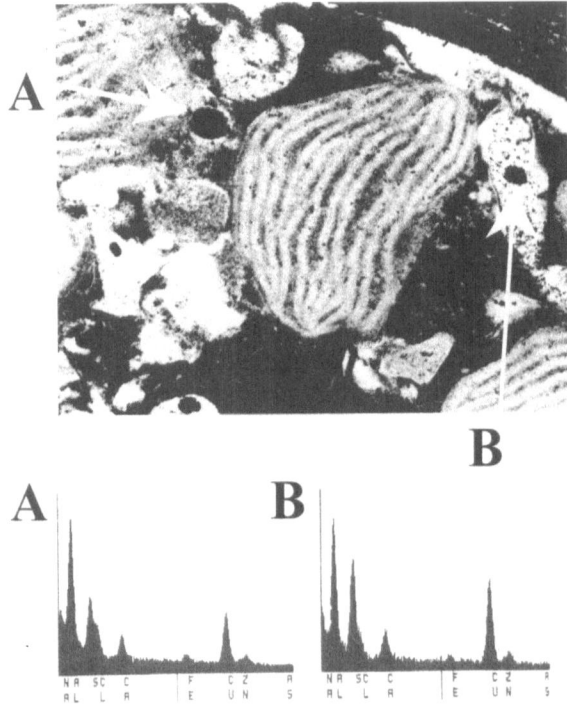

Figure 2. The elemental analysis of metal dense "physodes" (A and B) present in the cytosol of *F. vesiculosis* found in heavy metal contaminated environments show copper to be co-incident with a high level of sulphur.

Identification of the first marine macro-algae metallothionein gene

A cDNA library was constructed from *F. vesiculosis* exposed to high levels of metal for isolation of "marker" genes. To isolate the metallothionein gene, degenerate primers were designed to the conserved cysteine motifs of plant MT and these were used to screen a cDNA library made from metal-exposed *F. vesiculosis* using non-isotopic methods. A full length clone of ~ 1.2 kb was identified and sequenced demonstrating an open reading frame of approximately 200 bp (Fig. 4). The putative protein sequence of 67 amino acid residues was shown to have high identity to plant and mollusc MT (Fig. 5). Comparisons using the FASTA algorithm between the *Fucus* MT sequence and those from *Arabidopsis* (accession number p43392) and *Brassica napus* (accession number p43402) showed sequence identity between 42–38% at the protein level. The *Fucus* MT peptide had 36% identity to the MT from *Helix pomatia* (accession number p33187) and 31% to the MT from *Mytilus edulis* (accession number p80252). A linker region of fourteen amino acids exists between the two domains containing the Cys-X-Cys motifs. These two domains contain three Cys-X-Cys motifs in the N-terminal domain and

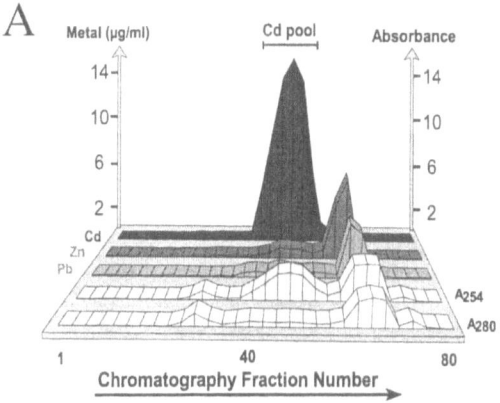

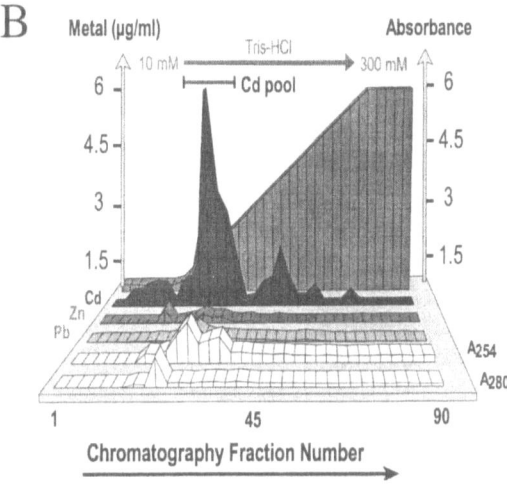

Figure 3. A) Extraction of *L. rubellus* cadmium-binding protein using gel filtration (Sephadex G75) methods from soluble earthworm extract. B) Purification of cadmium-binding protein using ion exchange chromatography.

three Cys-X-Cys with one Cys-X-Cys-Cys motif at the C-terminal. These motifs are presumably involved in the metal binding function.

Characterisation of the metallothionein gene promoter from rainbow trout

Work performed over the last fifteen years has analysed the relationship between waterborne heavy metal contamination and *in vivo* MT expression. The need to study gene expression has

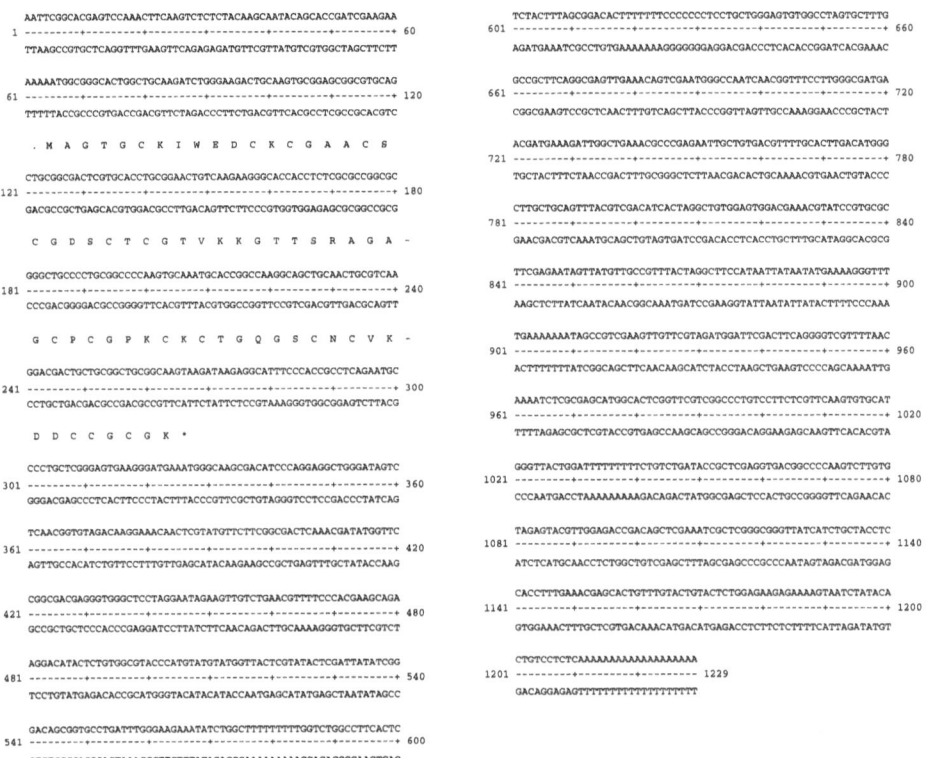

Figure 4. The nucleotide sequence of the *Fucus* MT showing the predicted protein sequence of 67 amino acid residues.

led to the development of routine procedures for the isolation of MT genes from fish species which has given rise to over twenty-two MT sequences in the last three years [8, 9]. In order to determine the molecular mechanisms underlying metal induction of MT in fish, genomic loci have been isolated and characterised [10]. Analysis of the 5' flanking region of fish MT genes

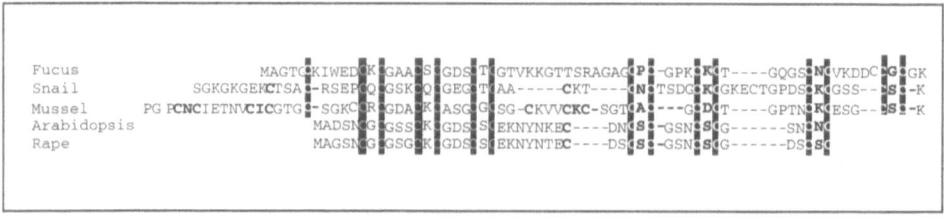

Figure 5. Alignment of the predicted *Fucus* MT protein sequence with MT sequences showing high identity. Black boxes indicate cystine residues, putatively involved in metal binding.

in cell lines demonstrated the presence of functional proximal and distal MRE (metal responsive element) clusters [11, 12]. Future work will focus on two species, firstly as an aquaculture species, the rainbow trout and secondly as a transgenic model species, the zebrafish. By introducing the MT gene control elements upstream from reporter genes and introducing these constructs into fish eggs, it should be possible to study the complex relationships between metal pollution and organic pollutants such as environmental oestrogens.

Acknowledgements
The *Lumbricus rubellus* study was financially supported by the British Natural Environment Research Council (NERC grant GR9/02083) and a Morgan E. Williams Studentship. The *Fucus vesiculosis* study was made possible by the award of an EERO fellowship to Dr. B. Nicolaus with more recent support by the Natural and Environmental Research Council under the ROPA scheme (GR3/R9694). The project supervision was facilitated by the NERC continued support for Dr. Kille through the Advanced Fellowship program (GT5/94/ALS).

References

1. Svendsen C, Meharg AA, Freestone P, Weeks JM (1996) Use of an earthworm lysosomal biomarker for the ecological assessment of pollution from an industrial plastics fire. *Appl Soil Ecol* 3: 99–107.
2. Brennan S, Ataria JA, Booth LH, Eason CT (1997) Cytochrome P450 as a biomarker in earthworms. Second International Workshop on Earthworm Ecotoxicology, Amsterdam, NL. p16.
3. Aderhold D, Williams CJ, Edyvean RGJ (1996) The removal of heavy metal ions by seaweeds and their derivatives. *Bioresource Technology* 58: 1–6.
4. Martin MH (1997) Concentration of cadmuim, copper, lead, nickel and zinc in the alga *Fucus serratus* in the Severn estuary from 1971 to 1995. *Chemosphere* 34: 325–334.
5. Phillips DJH (1994) Macrophytes as biomonitors of trace metals. *In*: KJM Kramer (ed.): *Biomonitoring of Coastal Waters and Estuaries*. CRC Press, Boca Raton, 85–103.
6.. Morris CA, Nicolaus B, Harwood JL, Kille P. Isolation of a member of the 14-3-3 gene family from a marine alga, *Fucus vesiculosis*. *Plant Molecular Biology; submitted.*
7. Smith KL, Hann AC, Harwood JL (1986) The subcellular localisation of absorbed copper in *Fucus*. *Physiol Plant* 66: 692–698.
8. Kille P, Santos CRA, Soltz D, Cosson R, Power D, Olsson P-E (1997) The impact of diverse environments on the primary structure of Teleost Metallothioneins. *Gene; submitted*
9. Kille P, Olsson P-E (1996) The use of Metallothionein genes for determining the phylogenetic and evolutionary relationships between extant Teleosts. *J Mol Evol; in preparation.*
10. Kille P, Kay J, Sweeney GE (1993) Analysis of regulatory elements flanking Metallothionein genes in Cd-tolerant Fish (Pike and Stone Loach). *Biochim Biophys Acta* 1216: 55–64.
11. Olsson P-E, Kling P, Erkell LJ, Kille P (1995) Structural and functional analysis of the rainbow trout (*Oncorhyncus* mykiss) Metallothionein-A gene. *Eur J Biochem* 230; 344–349.
12. Olsson P-E, Kille P (1997) Functional comparison of the metal-regulated transcriptional control regions of Metallothionein genes from cadmium tolerant and sensitive Teleost species. *Biochim Biophys Acta* 1350; 325–334.

Metallothionein IV
C. Klaassen (ed.)
© 1999 Birkhäuser Verlag Basel/Switzerland

Assessment of metal exposure of marine edible mussels by means of a biomarker

Biserka Raspor[1], Jasenka Pavičić[1], Sonja Kozar[1], Željko Kwokal[1], Marina Paić[1], Nikša Odžak[2], Irena Ujević[2] and Zorana Kljaković[2]

[1]Ruđer Bošković Institute, Center for Marine Research Zagreb, HR-10001 Zagreb, P.O. Box 1016, Croatia
[2]Institute of Oceanography and Fisheries, HR-21000 Split, P.O. Box 500, Croatia

Introduction

Preliminary results on the application of the inducible metal binding protein, i.e. metallothionein (MT) as a biomarker for the assessment of metal exposure of indigenous mussels *Mytilus galloprovincialis* Lmk. are presented. This study is an amendment of the long-term monitoring program of the coastal seawater area [1] by which the metal body-burden in edible bivalves has been monitored. An attempt is made to get the additional and relevant information on the biological response of the widespread, sessile and filter-feeding marine organisms, often used to monitor metal pollution of the marine and the estuarine coastal areas [2].

The objective of our study is to apply the reliable analytical and biochemical methods to determine toxic metals (cadmium, mercury) and the MT content in mussel tissues. The gills and the digestive gland are the mussel tissues involved in the process of metal uptake and detoxification [3] and therefore related to the induction of MT. Under laboratory and field conditions, the induction of MT in the digestive gland and gills, as the consequence of bivalves exposure to cadmium and mercury respectively, is well documented [3, 4]. The intention of our study is therefore, to correlate the content of toxic metals with that of MT in gills and the digestive gland in order to assess the exposure of mussels and the biological effects of metal pollution.

Material and methods

Sampling of M. galloprovincialis

Individuals of *M. galloprovincialis* were collected in the coastal area of the Kaštela Bay, which is an industrial and urban center of Dalmatia, Croatia. Sampling took place during winter in 1995, at three distances from the outlet of the ex-chlor-alkali plant, as shown in Table 1 with the Roman numerals. Production in this chlor-alkali plant was terminated in 1990.

At the sampling time the salinity of the coastal seawater was 38×10^{-3}. The mussel condition index determined on 7 specimens was, on average, low (3.0%), reflecting the phase after mussel spawning. Certain biometric data of the *M. galloprovincialis* specimens, collected at the sampling sites I to III, are presented in Table 1.

Table 1. The range of shell length and the total mussel weight of *M. galloprovincialis* specimens collected at three sampling sites in the Kaštela Bay, winter 1995. The weight of the shell and the edible part correspond to the total mussel weight

Sampling sites	Shell length/cm	Total mussel weight/g
I Closest to the ex-chlor-alkali plant	4.5 to 6.8	9.6 to 25*
II At a distance from the ex-chlor-alkali plant	3.7 to 5.0	5.5 to 20*
III Town of Kaštel Sućurac, far away from ex-chlor-alkali plant	3.6 to 5.1	6.6 to 18*

* contribution of the fouling organisms.

Metal content determination

The composite samples of mussel tissue (20 specimens for gills and the digestive gland and 6 specimens for the whole edible part) were analyzed for cadmium and mercury content. Prior to analysis, the tissue samples were digested with Suprapure chemicals (Merck, Germany). Mass fractions of all metals are expressed on wet tissue basis.

Cadmium content has been analyzed in parallel by spectrometric and electrochemical methods (Tab. 2). Tissue samples digested with a mixture of HNO_3 and $HClO_4$ were subsequently analyzed for cadmium with the Perkin Elmer instrument model 1100B, equipped with a graphite atomizer HGA 700 (GFAAS) and the automatic sample changer. After tissue digestion with a mixture of HNO_3 and H_2O_2, cadmium content was determined by an electrochemical method, applying the standard addition procedure. Electrochemical measurements were performed in a differential pulse anodic stripping voltammetric mode (DPASV) with the polarograph PAR model 174[A] at the hanging mercury drop electrode (HMDE). The total mercury content in different mussel tissues was determined with the cold vapor atomic absorption spectrometry (CVAAS) using a Perkin Elmer instrument model 410.

Table 2. The mass fractions of the toxic metals cadmium ($w_{fw}Cd$) and mercury($w_{fw}Hg$) in the composite samples of mussel tissue. Analytical techniques applied for metal content determination are indicated within the brackets. The mass fractions are expressed on wet tissue basis (fw)

Sampling sites	Tissue of *M. galloprovincialis*	$w_{fw}Cd\ 10^6$ (GFAAS)	$w_{fw}Cd\ 10^6$ (DPASV)	$w_{fw}Hg\ 10^6$ (CVAAS)
I	edible part	0.18		2.35
I	gills	0.06	0.04	3.00
I	digestive gland	0.26	0.21	2.80
II	edible part	0.18	0.17	1.50
II	gills	0.05	0.06	2.70
II	digestive gland	0.43	0.42	1.77
III	edible part	0.11		0.074
III	gills	0.10	0.09	0.147
III	digestive gland	0.28	0.22	0.163

MT determination

MT fractions were isolated from cytosolic fractions after tissue homogenization in three volumes of 0.02 M TRIS HCl buffer (pH = 8.6) and subsequent centrifugation at 30000xg for 40 min at 4 °C. In order to remove the interfering high molar mass proteins, the supernatant was thermally treated for 10 min at 70 °C, and in addition centrifuged at 30000xg for 20 min at 4 °C. The MT content was analyzed by an electrochemical method, i.e. the modified Brdička reaction [5, 6] with the polarograph PAR model 174^A in a differential pulse polarographic mode (DPP) on HMDE. Mass fractions of MT presented in Table 3 are expressed on wet tissue basis. The 95% confidence interval of the mean is indicated, too.

Table 3 The mass fraction of MT (w_{fw}MT), expressed on wet tissue basis, determined by modified Brdička reaction, after thermal treatment of the supernatant (30000xg, 4 °C)

Sampling station	Mussel tissue	w_{fw}MT 10^3	
		Mean	95% confidence interval
I	edible part	1.12	1.05 to 1.20
I	gills	0.48	0.45 to 0.51
I	digestive gland	2.85	2.63 to 3.08
II	edible part	0.85	0.80 to 0.90
II	gills	0.45	0.42 to 0.48
II	digestive gland	3.46	3.31 to 3.60
III	edible part	1.25	1.16 to 1.34
III	gills	0.64	0.61 to 0.68
III	digestive gland	4.23	3.79 to 4.68

Results and discussion

Our preliminary results indicate that, in different mussel tissues, the mass fractions of cadmium (w_{fw}Cd, Tab. 2) and MT (w_{fw}MT, Tab. 3) decrease in the same order, i.e.:

DIGESTIVE GLAND > EDIBLE PART OF MUSSEL > GILLS

The distribution of mercury in mussel tissues shows a different pattern (w_{fw}Hg, Tab. 2):

GILLS > DIGESTIVE GLAND > EDIBLE PART OF MUSSEL

The digestive gland of mussel *M. galloprovincialis* has both the highest inducibility and the highest extraction efficiency for MT [7]. Therefore, it is recommended that the assessment of the effects of cadmium on this indicator organisms should be performed by measuring MT and metal content in the digestive gland of *M. galloprovincialis*. In the whole edible part of mussels from different areas of the Adriatic Sea [8] the mass fractions of cadmium amount to 0.12×10^{-6} to 0.19×10^{-6}, similar to the results reported in Table 2 ($(0.11$ to $0.18) \times 10^{-6}$). The sources of cadmium pollution are diverse and unspecific. Therefore, the biological effects cannot be estimated by measuring only the cadmium content. On the contrary, the ex-chlor-alkali plant is still a point source of mercury pollution, even though production was terminated in 1990. It clearly indicates the gradient of mercury content in seawater (unpublished data) as well as in mussels, (Tab. 2). At sampling sites I to III (Tab. 2) in the whole edible part of mussels the mass fractions of mercury decrease from $(2.35, 1.50$ to $0.074) \times 10^{-6}$. In the whole edible part of mussels from the Adriatic areas, remote from the ex-chlor-alkali plant, the mass fractions of mercury amount to $(0.020$ to $0.030) \times 10^{-6}$ [8]. The predominant partition of cadmium in the digestive gland of mussels and of mercury in the gills indicate that the uptake routes of these two toxic metals by *M. galloprovincialis* are different. Long-term studies are needed to assess the basal levels of MT in the digestive gland and the gills of *M. galloprovincialis*. When metals with long biological half-lives are monitored, estimation of the mussel age is still a key problem [4, 8].

Acknowledgment
The authors acknowledge the financial support from the Ministry of Science and Technology, Republic Croatia, for the research project 00981511 "Biomarkers and the biological effects of metals" and from the FAO MAP, Athens for the research project CRO/32 "Metallothionein as indicator of mussel exposure to heavy metals".

References

1. Monitoring programme of the Eastern Adriatic Coastal Area, Report for 1983–1991 (1994) MAP Technical Report Series No. 86. UNEP, Athens.
2. O' Connor T.P, Beliaeff B (1995) Recent trends in coastal environmental quality: Results from the mussel watch project. NOAA, Silver Spring MD.
3. Roesijadi G (1992) Metallothioneins in metal regulation and toxicity in aquatic animals. *Aquat Toxicol* 22: 81–114.
4. Pavičić J, Škreblin M, Raspor B, Branica M, Tušek-Žnidarič M, Kregar I, Stegnar P (1987) Metal pollution assessment of the marine environment by determination of metal-binding proteins in *Mytilus* sp. *Mar Chem* 22: 235–248.
5. Brdička R (1933) Polarographic studies with dropping mercury cathode. Part XXXI- A new test for proteins in the presence of cobalt salts in ammoniacal solutions of ammonium chloride. *Collect Czech Chem Commun* 5: 112–128.
6. Olafson RW, Sim RG (1979) An electrochemical approach to quantitation and characterization of metallothioneins. *Anal Biochem* 100: 343–351.
7. Pavičić J, Raspor B, Martinčić D (1993) Quantitative determination of metallothionein-like proteins in mussels; Methodological approach and field evaluation. *Mar Biol* 115: 435–444.
8. Martinčić D, Kwokal Z, Branica M, Stoeppler M (1987) Trace metals in selected organisms from the Adriatic Sea. *Mar Chem* 22: 207–220.

Metallothionein IV
C. Klaassen (ed.)
© 1999 Birkhäuser Verlag Basel/Switzerland

Metallothionein level estimation in the bioassay method for water quality assessment in freshwater reservoirs polluted by waste waters

Dasha A. Avramova[1], Boris I. Synsynys[1], Valentina A. Romantsova[2] and Gennady M. Rott[2]

[1]*Department of Ecology of the Institute of Nuclear Power Engineering, Studgorodok-1, Obninsk, 249020, Kaluga region, Russia*
[2]*Medical Radiological Research Center of the Russian Academy of Medical Sciences,Obninsk, 249020, Kaluga Region, Russia*

Introdoction

There exist at present about 60 hydrobiological methods for testing the quality of surface waters. Most of these methods make it possible to estimate the degree of water pollution caused by organic substances [1, 2]. However, there appear to be no biological methods for estimating the degree of water contamination caused by heavy metals.

Freshwater and terrestrial molluscs are employed as bioindicators of radiological contamination of various ecosystems [3]. They can also be used for environmental monitoring of heavy metals [4]. We propose a new method of biotesting metal-contaminated freshwater reservoirs without determining the presence of heavy metals in water molluscs. It is based on estimating changes in the content of metallothioneins (MT) in mollusc soft tissues. MT are low- molecular weight metal-binding proteins containing cysteine residues [5]. MTs are easily inducible in cells of living organisms by many heavy metals as well as by ionizing and UV radiation, strong oxidants and other chemicals [6, 7]. New insights into regolation of MT gene expression via a Zn-sensitive inhibitor coupled with the "rescue" hypothesis for metal detoxification by MT provide the basis for interpreting the significance of MT induction in response to toxic metals [8].

The advantages of the method proposed are as follows: (1) the possibility of estimating contamination by a whole range of heavy metals; (2) the possibility of estimating the degree of the biological response, in particular, that of molluscs. Our study presents data on MT content in different mollusc species inhabiting water reservoirs of middle Russia polluted by waste waters.

Approaches

The reservoirs under consideration were the Protva-river and Belkino pond in the vicinity of the town of Obninsk (Kaluga Region). The indicator animals were the following molluscs: *Anodonta* sp., *Shpaerium*, *Planobarins* sp. and *Lymnaea stagnalis*. Molluscs were collected in different locations of the river and pond during 1995-1997. For the monitoring area of the river (water quality class I1) the age population structure of the collected Anodonta sp. molluscs was determined by measuring the maximum shell size. MT in the soft tissues was determined using the radioactive indicator method whereby chelated MT metal ions were substituted for by radioactive ^{109}Cd^{2+} [5]. Before conducting experiments on estimating MT the molluscs were stored at –40°C. The soft tissue of the mollusc was separated from the shell, weighed and homogenized to obtain homogenate in a Potter glass homogenizer. The indicator reagent contained ^{109}CdCl$_2$ ("Cyclotron" Research-and-Production Association, Obninsk, specific activity 7.4 GBq/mg) and CdCl$_2$ in 0.01 M Tris-HC1 buffer, pH 7.4 (the total Cd concentration in the mixture was 2 µg/ml while its radioactivity was 40.0 kBq/ml). 0.2 ml of 0.01 M Tris-HC1 buffer, pH 7.4, and 0.2 ml of the indicator reagent were added to 0.2 ml of the homogenate. The mixture was stirred and incubated for 10 minutes at room temperature. Then 0.1 ml of 2% rabbit haemoglobin was added to the mixture which was stirred, heated for 3 min using boiling water bath, then cooled in ice-bath and centrifuged for 10 min at 7.000 r/min. Adding haemoglobin and the subsequent operations were repeated two more times. After that 0.4 ml of supernatant was placed in a test-tube to count and measure the radioactivity using the Beckman PU 5500 gamma- connter (USA). For each series of measurements a blank sample was analyzed (buffer was used instead of the sample being examined) and the total radioactivity was determined (buffer was used instead of haemoglobin and the sample being examined). The MT concentration was calculated as described in [9].

The water quality in the water reservoirs was determined by using the biotic index of macrozoobenthos organisms [1].

The results were processed by using standard statistical methods, considering the 95% confidence interval of mean values.

Results

Data on the specific MT content in different species of molluscs inhabiting Belkino pond are presented in Figure 1. As can be seen the MT concentration is practically the same for mol-

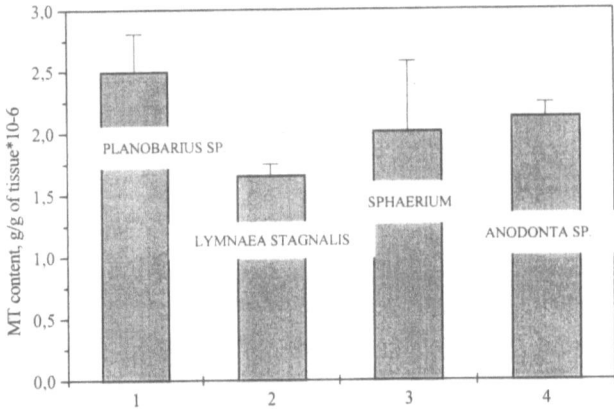

Figure 1. Specific MT content in different mollusc species inhabiting Belkino pond. 1 - Planobarins sp., 2 - Lymnaea stagnalis, 3 - Shpaerium, 4 - Anodonta sp. The ordinate axis shows specific MT content, µg/g of tissue.

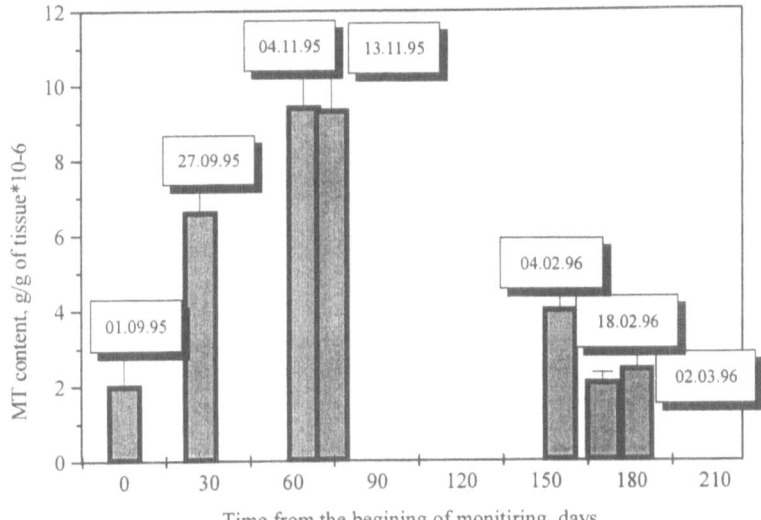

Figure 2. Seasonal changes in specific MT content in Planobarius sp. inhabiting Belkino pond. Collection dates: 01.09.95 (the beginning of monitoring); 27.09.95; 04.11.95; 13.11.95; 04.02.96; 18.02.96; 02.03.96. The abscissa axis shows time from the beginning of monitoring, days. The ordinate axis shows specific MT content, μg/g of tissue.

luscs of different species and amounts to about 2 mg/g of soft tissue. Further examination of the Planobarius sp. mollusc population from September 1995 to March 1996 showed changes in specific MT content (Fig. 2). From the beginning of September to the middle of November 1995 the specific MT content increased by a factor of 5 which was likely to be due to increased rainfall runoff into the pond containing lead and other pollutants. Then the MT concentration in the Planobarius sp. molluscs decreased and by March 1996 it basically amounted to the original value (the same as that at the beginning of the monitoring). At the same time the MT content in the Lymnaea Stagnalis molluscs collected from the same time did not show any changes (Fig. 3). Species variability may be related to both nutrition pattern and inherent features of organisms and to their ability to activate genes regulating MT synthesis. It should be noted that during the examination period (from September to March) the water quality class which is determined in terms of the variety of microzoobenthos changed from class III (moderate contamination) to class VI (heavy contamination).

We also estimated the MT content in the soft tissue of the Anodonta sp. inhabiting the Protva river. The water quality class on the river section determined by the Woodiwiss method [1] changed from class II (clean water) in the autumn to class III (moderately contaminated) in the winter. It should be noted that the MT content in the Anodonta sp. collected from the river was on the average 1.5 times lower than that in the organisms inhabiting the pond which has more contaminated water.

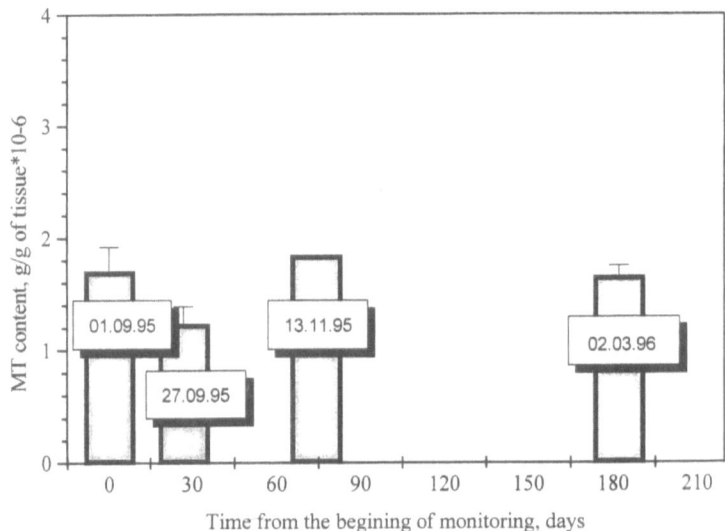

Figure 3. Seasonal changes in specific MT content in Lymnaea stagnalis inhabiting Belkino pond. Collection dates: 01.09.95 (the beginning of monitoring); 27.09.95; 13.11.95; 02.03.96. The abscissa axis shows time from the beginning of monitoring, days; The ordinate axis shows specific MT content, µg/g of tissue.

Apart from seasonal variations in the MT specific content in soft tissue of molluscs we also examined the dependence of the MT content on the habitat location of molluscs in the pond and the river. Fig. 4 presents sampling locations on Belkino pond and Fig. 5 provides the MT specific content in the tissue of the Lyumnaea stagnalis and Planobarins sp. molluscs collected in different locations. As seen from the latter figure the MT concentration in the tissue of the molluscs collected in location 1 is higher than that in locations 2 and 3. This can be due to the proximity of a construction site, of a heavy-traffic highway and a parking lot with a filling station whose runoff waters get into the stream flowing into the pond. Hence the molluscs inhabiting location 1 are exposed to greater anthropogenic stresses while water in location 3 is biologically decontaminated by depositing pollutants in the pond silt, among other factors.

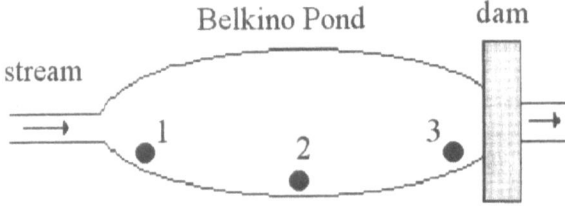

Figure 4. Sampling locations on Belkino pond.

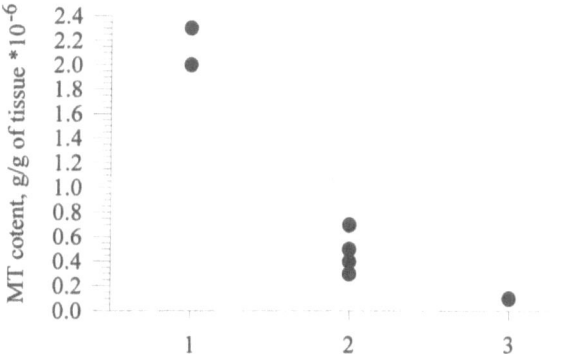

Figure 5. Specific MT content in different mollusc species depending on the sampling location in Belkino pond. The abscissa axis shows the number of a sampling location; The ordinate axis shows specific MT content, µg/g of tissue.

Three sampling locations were selected on the Protva river: 1 – the city beach, 2 – the place of effluents from the Physics and Power Engineering Institute (PPEI) and 3 – a location downstream the place of efffluents from the PPEI (Fig. 6). Data on the MT content in the soft tissues of the Anodonta sp. in these locations are presented for several months in Fig. 7. Comparing the data for the beach location and that for efffluents from the PPEI and considering that no excess activity of radionuclides was found by means of gammaspectrometry in molluscs from the PPEI sampling location one can suggest that metal contaminated effluents were released into the river by the PPEI.

The study revealed that the MT content in the river molluscs of different ages (and hence of different masses) as can be seen from Fig. 8 depends on the soft tissue mass (the correlation coefficient of linear regression is 0.7) whereas the MT specific content per gram of soft tissue is not related to the mass (the correlation coefficient of linear regression is –0.25) and hence, to the Anodonta sp. age (Fig. 9). The data presented in Fig. 8 and Fig. 9 are suggestive of the lack of the cumulative effect which could be expected to occur when heavy metals accumulate in soft tissues of molluscs with increasing age. This suggestion is to be validated in special lab-

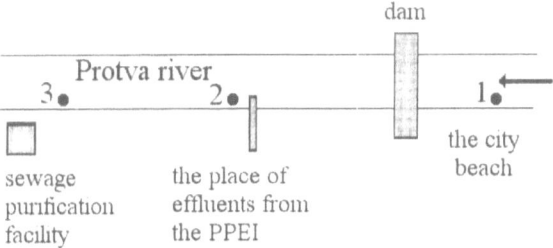

Figure 6. Sampling locations on the Protva river.

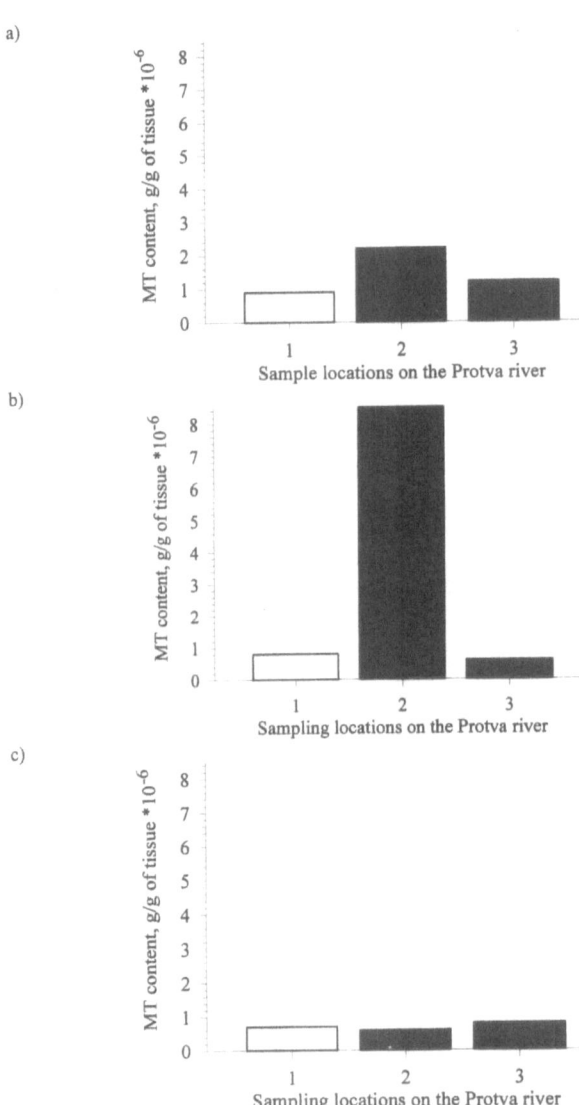

Figure 7. Specific MT content in Anodonta sp. depending on the sampling location in the Protva river for several months: a - September, b - October, c - November. The abscissa axis shows sampling locations on the Protva river; The ordinate axis shows specific MT content, µg/g of tissue.

oratory tests, with the uptake and accumulation of metals in mollusc tissues being controlled simultaneously.

One can suggest that the modeling of the effect of environmental limiting factors on molluscs in a laboratory will make it possible to assess the role of MT synthesis induction in devel-

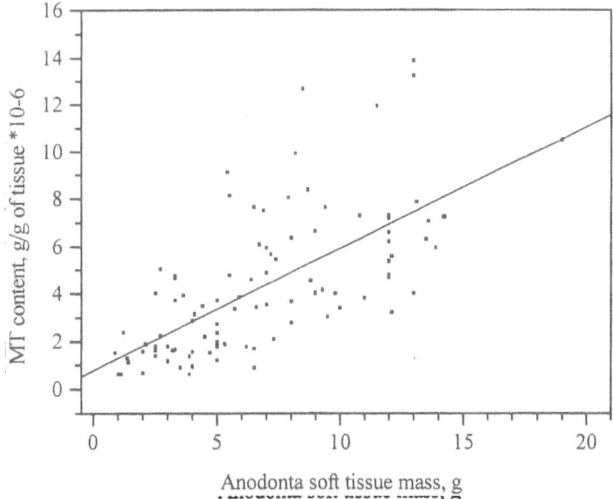

Figure 8. Total MT content in the Anodonta sp. depending on its mass. The abscissa axis shows mass of the mollusc soft tissue; The ordinate axis shows total MT content, µg/g of tissue.

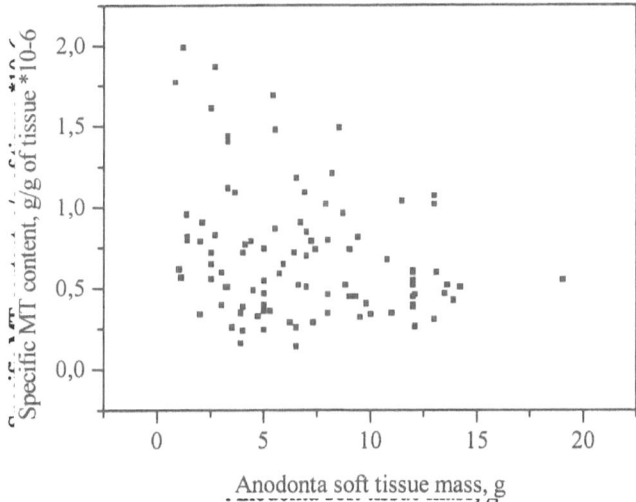

Figure 9. Specific MT content in the Anodonta sp. depending on its mass. The abscissa axis shows mass of the mollusc soft tissue; The ordinate axis shows specific MT content, µg/g of tissue.

oping mechanisms of mollusc adaptation to heavy metal exposure. One can also hope that the established regularities will enable a prompt method for estimating water quality to be created in addition to the already existing hydrobiological methods for testing surface waters.

Acknowledgment
The authors wis
he to thank Natasha V.Shirkina for technical assistance. We thank Zinaida M. Gorobets and Natasha N. Dikonenko
from the municipal water quality laboratory for data on chemical analysis of water from the Protva river.

References

1. Abacumov VA (ed.) (1983) *Manual for hydrobiological methods for analyzing surface waters and sediments.* *Hydrometeoisdat*, Leningrad (in Russian)
2. Mitchel MK, Stapp WB (1994) *Field manual for water quality monitoring.* Dexter, Michigan
3. Frantsevitch II, Pankov IV, Yermakov AA et al. (1995) Molluscs as indicators of environmental contamination caused by radionuclides. *Ecologia* 1: 57–62 (in Russian)
4. Beeby A, Favex SC (1983) Short-term changes in Ca, Pb, Zn and Cd concentrations of the garden snail Helix Aspersa Muller from a central London car park. *Environm Pollut* A30: 233–244
5. Kagi JHR, Schaffer A (1988) Biochemistry of metallothionein. *Biochemistry* 27: 8509–8515
6. Bremner I (1991) Nutritional and physiological significance of metallothionein. *Meth Enzymol* 205: 25–39
7. Klaassen CD, Lehman-McKeeman LD (1989) Induction of metallothionein. *J Amer College Toxicol* 8: 1315–1321
8. Roesijadi G (1996) Metallothionein and its role in taxic metal regulation. *Comp Biochem Phisiol* 113:2:117–123
9. Eaton DL, Cherian MG (1991) Determination metallothioneins in tissues by cadmium hemoglobin affinity assay. *Meth Enzymol* 205: 83–90

Subject index

(The page number refers to the first page of the chapter in which the keyword occurs.)

EXS 85

D-Amino Acids in Sequences of Peptides of Multicellular Organisms

Jollès, P.,
Muséum National d'Histoire Naturelle, Paris, France (Ed.)

Life on earth almost exclusively uses laevorotatory or left-handed amino acids (L-enantiomers), rather than D-enantiomers. Nevertheless, with improved analytical methods, D-amino acids have been detected in a variety of peptides of multicellular organisms during recent years.

This book takes stock of our present knowledge in this rapidly expanding research area. In a series of chapters it discusses the characterization and analysis of D-amino acids, their occurrence and function in animal peptides and proteins, some possible biosynthetic pathways, and their appearance during ageing. Furthermore, one chapter approaches the puzzling question of homochirality and life.

Contents

EXS 85
Jollès, P. (Ed.)
D-Amino Acids in Sequences of Peptides of Multicellular Organisms
1998. 200 pages. Hardcover
ISBN 3-7643-5814-9

BioSciences with Birkhäuser

For orders originating from all over the world except USA and Canada:

For orders originating in the USA and Canada:

(Prices are subject to change without notice. 11/98)

Birkhäuser Verlag AG
P.O. Box 133
CH-4010 Basel / Switzerland
Fax: +41 / 61 / 205 07 92
e-mail: orders@birkhauser.ch

Birkhäuser Boston, Inc.
333 Meadowland Parkway
USA-Secaucus, NJ 07094-2491
Fax: +1 / 201 348 4033
e-mail: orders@birkhauser.com

Birkhäuser

Molecular Aspects of Cancer and its Therapy

Mackiewicz, A.,
University School of Medical Sciences, Poznan, Poland /
Sehgal, P.B.,
New York Medical College, Valhalla, NY (Ed.)

This book highlights recent progress in the molecular, cellular and immunological mechanisms that contribute to the pathophysiology of cancer and the design of therapeutic modalities based upon these molecular insights. Areas of particular emphasis include cancer immunology and the immunotherapy of cancer, the role of cytokines in modulating the social behaviour of cancer cells, the genetic alterations that characterize human cancer and metastasis, and a consideration of the more experimental approaches to cancer therapy, including gene therapy using expression vectors for cytokines and their receptors, antisense RNA therapy, and anti-idiotypic antibody immunization.

This volume serves to introduce the general reader as well as the cancer specialist to personalized perspectives of particular topics in cancer research by leading research groups in the field. The combination of a „reviews"-approach with a more research-oriented approach in discussions of specific research topics provides a stimulating and forward-looking volume which serves to update selected aspects of cancer research today. This combination will be useful to both the beginner as well as the more advanced biomedical scientist.

Tamm, I., Kikuchi, T., Cardinale, I., Murphy, J.S. and Krueger, J.G.:
Cytokines in breast cancer cell dyshesion

Jassem, E. and Jassem, J.
Clinical relevance of genetic alterations in lung cancer

Jasinska, A., Sobczak, K., Koztowski, P., Napierata, M. and Krzyzosiak, W.J.:
Recent advances in understanding funtion and mutations of breast cancer susceptibility

Maraveyas, A., Hrouda, D. and Dalgleish, A.G.:
Tumour immunology

Sehgal, P.B.:
Cytokines in the host-tumor interaction

Wiznerowicz, M., Rose-John, S. and Mackiewicz, A.:
Gene therapy of cancer

Skorski, T., Szczylik, C. and Calabretta, B.:
Antisense strategy for cancer therapy

Ratajczak, M. and Gewirtz, A.M.:
Oligonucleotide therapeutics for human leukemia

Hemstreet, G.P., III:
Genetic instability and tumor cell variation

Subject Index

Detailed information about all our titles also available on the internet:
http://www.birkhauser.ch

Contents

List of contributors

Preface

Backer, J.M. and Hamby, C.V.:
Genetic control of metastasis

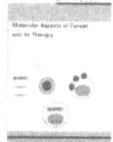

MCBU
Mackiewicz, A. / Sehgal, P.B. (Ed.)
Molecular Aspects of Cancer and its Therapy
1998. 240 pages. Hardcover
ISBN 3-7643-5724-X

BioSciences with Birkhäuser

For orders originating from all over the world except USA and Canada:

For orders originating in the USA and Canada:

(Prices are subject to change without notice. 11/98)

Birkhäuser Verlag AG
P.O. Box 133
CH-4010 Basel / Switzerland
Fax: +41 / 61 / 205 07 92
e-mail: orders@birkhauser.ch

Birkhäuser Boston, Inc.
333 Meadowland Parkway
USA-Secaucus, NJ 07094-2491
Fax: +1 / 201 348 4033
e-mail: orders@birkhauser.com

Birkhäuser

Radioactive Isotopes in Clinical Medicine and Research XXIII
Proceedings of the 23rd International Badgastein Symposium

Bergmann, H.,
Institut für Biomedizische Technik und Physik, Wien, Austria /
Köhn, H.,
Institut für Nuklearmedizin, Wien, Austria /
Sinzinger, H.,
Universitäts-Klinik für Nuklearmedizin, Wien, Austria (Ed.)

Radioactive Isotopes in Clinical Medicine and Research XXIII presents an update in the latest clinical research in nuclear medicine. It provides with in-depth information on the all areas of nuclear medicine. The chapters of this volume have been grouped into the following sections: Neurology/Psychiatry, Therapy, Radiopharmacology, Endocrinology/Thyroid, Oncology/Haematology, Clinical PET, Cardiology, Varia, Physics/Radiation Protection, World Wide Web/WWW demo. Special attention is paid to the virtual media for teaching, training, communication, quality control etc.
Primarily intended for specialists in the nuclear medicine, this volume will also be of considerable interest to clinicians using diagnostic and therapeutic nuclear medicine procedures, including cardiologists, haematologists, neurologists, nephrologists, oncologists, pharmacologists, and psychiatrists.

Detailed information about all our titles also available on the internet:
http://www.birkhauser.ch

Check our Highlights for new and notable titles selected monthly in each field

Bergmann H. / Köhn H. / Sinzinger H. (Ed.)
Radioactive Isotopes in Clinical Medicine and Research XXIII
Proceedings of the 23rd International Badgastein Symposium
1998. Approx. 590 pages. Hardcover
ISBN 3-7643-5967-6
Due in December 1998

BioSciences with Birkhäuser

For orders originating from all over the world except USA and Canada:

For orders originating in the USA and Canada:

(Prices are subject to change without notice. 11/98)

Birkhäuser Verlag AG
P.O. Box 133
CH-4010 Basel / Switzerland
Fax: +41 / 61 / 205 07 92
e-mail: orders@birkhauser.ch

Birkhäuser Boston, Inc.
333 Meadowland Parkway
USA-Secaucus, NJ 07094-2491
Fax: +1 / 201 348 4033
e-mail: orders@birkhauser.com

Birkhäuser